Handbook of Analysis of Oligonucleotides and Related Products

EDITED BY

JOSE V. BONILLA

Girindus America, Cincinnati, Ohio, USA

G. SUSAN SRIVATSA

ElixinPharma, Encinitas, California, USA

CRC Press
Taylor & Francis Group
Boca Raton London New York

CRC Press is an imprint of the
Taylor & Francis Group, an informa business

CRC Press
Taylor & Francis Group
6000 Broken Sound Parkway NW, Suite 300
Boca Raton, FL 33487-2742

First issued in paperback 2016

ISBN 13: 978-1-138-19845-6 (pbk)
ISBN 13: 978-1-4398-1993-7 (hbk)

Library of Congress Cataloging-in-Publication Data

Handbook of analysis of oligonucleotides and related products / edited by Jose V. Bonilla and Susan Srivatsa.
p. ; cm.
Includes bibliographical references and index.
ISBN 978-1-4398-1993-7 (hardcover : alk. paper)
1. Oligonucleotides--Analysis. I. Bonilla, Jose V. II. Srivatsa, Susan.
[DNLM: 1. Oligonucleotides--analysis. QU 57]

QP625.O47H36 2011
572.8'5--dc22 2010033563

Visit the Taylor & Francis Web site at
http://www.taylorandfrancis.com

and the CRC Press Web site at
http://www.crcpress.com

Contents

Preface

The past two decades have seen an explosive growth in the research applications of oligonucleotides. As a direct manifestation of their diverse pharmacology, oligonucleotides represent one of the most significant pharmaceutical breakthroughs in recent years and have the potential to revolutionize biomedical research. Indeed, this unique class of compounds has shown great promise as diagnostic and therapeutic agents for a wide range of human diseases including cancer, cardiovascular disease, diabetes, viral infections, and many other degenerative disorders.

This already popular field has been further energized with the awarding of the 2006 Nobel Prize in Physiology or Medicine to Andrew Fire and Craig C. Mello for their discovery of RNA interference—gene silencing by double-stranded RNA. Much of the current research effort is focused on improving our basic understanding of the chemistry and biology surrounding the various mechanisms of action of oligonucleotides. In spite of the market approval of two oligonucleotide-based drugs, Vitravene™ in 1997 and Macugen™ in 2004, there have been significantly fewer published articles devoted to the practical aspects of the analysis of oligonucleotides in support of pharmaceutical development.

A typical oligonucleotide therapeutic agent is a short chain (*ca.* 7–15 KDa), possibly chemically modified, DNA or RNA sequence manufactured by chemical synthesis utilizing automated synthesizers. Owing to their relatively large sizes as compared with typical small-molecule drugs, there are many technical challenges associated with the analysis of these novel therapeutic products. While there are numerous reports on conventional and innovative techniques that have been applied to oligonucleotide analysis, there is no single publication to date that pulls together the relevant techniques in a single source. With this book, we attempt to fill this void by providing a compilation of state-of-the-art analytical methodologies suitable for the analysis of oligonucleotides in support of both research and development. It is not our intent to present an extensive review of the literature with respect to individual analytical techniques; rather, we would like to provide readers with a practical guide to apply such techniques to oligonucleotides in research, development, and manufacturing settings.

An essential element of establishing the definitive identity of an oligonucleotide is the confirmation of molecular weight and molecular sequence. Strategies for enzymatic or chemical degradation of chemically modified oligonucleotides toward mass spectrometric sequencing are addressed in detail. Purity analysis by chromatographic or electrophoretic methods, another area of great importance in drug development, is detailed in five chapters that cover such key techniques such as RP-HPLC, AX-HPLC, HILIC, SEC, and CGE. Characterization of sequence-related impurities in oligonucleotides is discussed in a section on LC-MS. Structure elucidation, an important part of product characterization, is covered in multiple chapters on base composition analysis, NMR, IR, T_m, and mass spectrometry. Approaches to the accurate determination of molar extinction coefficient, a key parameter used for rapid quantitation of oligonucleotide content, are also presented. Because of the highly hygroscopic and electrostatic nature of oligonucleotides, accurate determination of assay values, a regulatory requirement, can be problematic. A chapter devoted to this topic addresses the unique challenges related to oligonucleotide assays. Specific chapters on determination of endotoxins, heavy metals, and residual solvents address the means of establishing the overall quality of oligonucleotides. An overview of approaches to assessing the chemical stability of oligonucleotides is also discussed.

Analysis of oligonucleotides in biological matrices continues to be a formidable challenge. The use of highly sensitive and specific hybridization techniques for supporting pharmacokinetics and drug metabolism studies in preclinical and clinical development is discussed in detail. Finally, a chapter of this book is devoted to an overview of how the relevant analytical information can be presented in a form that will meet the current regulatory expectations for oligonucleotide therapeutics.

This handbook is a truly unique reference manual on the practical applications of modern and emerging analytical techniques for the analysis of oligonucleotides. It represents the culmination of the collaboration of 30 leading analytical scientists from around the world in this arena. It is our intent to provide the reader with a comprehensive overview of the most commonly used analytical techniques and their strengths and limitations toward assuring the identity, purity, quality, and strength of an oligonucleotide intended for therapeutic use.

G. Susan Srivatsa
Jose V. Bonilla

Editors

Dr. Jose V. Bonilla started his scientific career with the NASA Space Shuttle Project at Argonne National Laboratory. While at Argonne, he also worked in advanced research projects for the Department of Energy and the Department of Defense.

Dr. Bonilla has more than 20 years of industry experience in supporting R&D and manufacturing of a broad variety of products such as specialty plastics for food contact applications and medical devices (GE Advanced Materials), as well as specialty excipients and active pharmaceutical ingredients (APIs) for the pharmaceutical industry (ISP & Girindus).

His career has been dedicated to the introduction and implementation of cutting-edge analytical technologies such as LC-MS, high-speed gas chromatography, high-speed GPC, online GC, online HPLC, and online near-IR. He has extensive experience in the management of industrial analytical laboratories in compliance with regulatory requirements. He is the author and coauthor of several peer-reviewed publications including the *Handbook of Plastics Analysis*. Dr. Bonilla obtained his PhD and MS degrees in Analytical Chemistry from the University of Oklahoma; he started his undergraduate studies at the National University in Colombia and completed his BS degree in Chemistry at Bethel College, Kansas.

Dr. G. Susan Srivatsa is Founder and President of ElixinPharma, a scientific consulting firm dedicated to assisting pharmaceutical companies with the development of oligonucleotide-based therapeutics.

Dr. Srivatsa has more than 20 years of experience (Procter & Gamble, Allergan, Abbott, Telios, and Isis Pharmaceuticals) in the development of small molecules, proteins, peptides, and oligonucleotides. At Isis, Dr. Srivatsa pioneered the regulatory strategy for the quality control of oligonucleotide therapeutics, resulting in the first oligonucleotide drug approval, Vitravene™, in the United States and Europe. Dr. Srivatsa has contributed to the successful development of more than 35 DNA and RNA oligonucleotide drug candidates through various stages of clinical development and has published widely in the area of oligonucleotide analysis in peer-reviewed journals. In 1998, she was elected to the Analytical R&D Steering Committee of PhRMA and served on the Expert Working Group for the ICH Guideline Q6A: Specifications for New Drug Substances and Drug Products.

Dr. Srivatsa received a BS in Chemistry from the California State University, Fullerton and a PhD in Analytical Chemistry from the University of California, Riverside. Under the mentorship of Professor Donald T. Sawyer, her graduate research focused on electrochemical and structural studies of transition metal complexes as models for oxygen binding and electron transfer hemoproteins. She pursued post-doctoral research under Professor Dallas L. Rabenstein on the bioanalytical applications of NMR spectroscopy.

Contributors

Hüseyin Aygün
BioSpring GmbH
Frankfurt, Germany

Jose V. Bonilla
Girindus America, Inc.
Cincinnati, Ohio

Doug Brooks
Regado Biosciences
Durham, North Carolina

Gary Burt
Girindus America, Inc.
Cincinnati, Ohio

Judy Carmody
Avatar Pharmaceutical Services, Inc.
Marlborough, Massachusetts

Sandra Carriero
Immunochemistry, Laboratory Sciences
Charles River Laboratories Preclinical Services
Montreal Inc.
Quebec, Canada

Marvin H. Caruthers
University of Colorado
Boulder, Colorado

Ming Fai Chan
Accugent Laboratories, Inc.
Carlsbad, California

Sky Countryman
Phenomenex
Torrance, California

Hagen Cramer
Girindus America, Inc.
Cincinnati, Ohio

Michele L. DeRider
Catalent Pharma Solutions
Research Triangle Park, North Carolina

Renee N. Easter
University of Cincinnati
Cincinnati, Ohio

Kevin J. Finn
Girindus America, Inc.
Cincinnati, Ohio

Eric Herzberg
Girindus America, Inc.
Cincinnati, Ohio

Helen Legakis
Immunochemistry, Laboratory Sciences
Charles River Laboratories Preclinical Services
Montreal Inc.
Quebec, Canada

Patrick A. Limbach
University of Cincinnati
Cincinnati, Ohio

Barbara J. Markley
Associates of Cape Cod, Inc.
Falmouth, Massachusetts

Sean M. McCarthy
Waters Corporation
Milford, Massachusetts

Dennis P. Michaud
Avecia Biotechnology, Inc.
Milford, Massachusetts

Michael P. Murphy
Intertek Analytical Services
Whitehouse, New Jersey

Veeravagu Murugaiah
Alnylam Pharmaceuticals
Cambridge, Massachusetts

Bernhard Noll
Roche Kulmbach GmbH
Kulmbach, Germany

Soheil Pourshahian
Girindus America, Inc.
Cincinnati, Ohio

Ipsita Roymoulik
Avecia Biotechnology, Inc.
Milford, Massachusetts

G. Susan Srivatsa
ElixinPharma
Encinitas, California

Jim Thayer
Dionex Corporation
Sunnyvale, California

Zoltan Timar
Agilent Technologies, Inc.
Boulder, Colorado

Yansheng Wu
Archemix Corporation
Cambridge, Massachusetts

Huihe Zhu
Girindus America, Inc.
Cincinnati, Ohio

Introduction

Although the first dinucleotide was chemically synthesized 55 years ago,[1] the development of a universally useful chemical methodology for preparing both DNA and RNA had to wait until 1980 when we discovered how to use 2′-deoxynucleoside-3′-phosphoramidites as synthons (Figure 1).[2–4] Briefly, the first step involves condensation of an appropriately protected 5′-dimethoxytrityl-2′-deoxynucleoside-3′-phosphoramidite to a base-protected 2′-deoxynucleoside attached to a controlled pore glass support. This reaction initially used tetrazole as an activator, although a large number of weak acids have since been proposed. Of historical interest and before we published the use of tetrazole, we discovered that several other weakly acidic reagents could be used to activate this reaction, including various chloracetic and sulfonic acids, amine hydrochlorides, and even 2-nitropropane. However, as these reagents were either hygroscopic or highly mutagenic, we never seriously considered them for general use—especially for nonchemists in a machine setting.

The next step of this cycle is to use acetic anhydride in pyridine to acylate any unreactive nucleoside and to remove phosphite adducts from the bases. This step is followed by oxidation with iodine in aqueous lutidine, which converts the phosphite internucleotide linkage to phosphate. I have often been asked why we did not develop a cycle where oxidation was preformed once after completion of oligonucleotide synthesis (thus eliminating one step of the cycle). Unfortunately, this is not possible. Phosphites are very unstable toward acid, which is used in the next step of the cycle to remove the dimethoxytrityl group. By conversion during each cycle to an acid-stable phosphate internucleotide linkage, the instability problem is eliminated. Following removal of the dimethoxytrityl group, the cycle is complete (3–4 minutes), and the resulting dinucleotide is ready for addition of the next synthon.

Once the requisite number of cycles has been completed, the product oligonucleotide is removed from the support and protecting groups eliminated using a mild base such as ammonia. The product can then be purified by reverse phase high-performance liquid chromatography (HPLC) or any number of other approaches. Of interest is that oligomers useful for DNA sequencing or polymerase chain reaction (PCR) can simply be used directly without purification. This is because the repetitive yields are very high, which thus generate the oligonucleotide (even a 20- or 30-mer) as the major reaction mixture product. Generally, this chemistry can be used to synthesize oligomers up to 75 or so nucleotides in length. However, I know of at least one example where the chemistry was repeated 450 cycles, the reaction mixture displayed on a gel, the gel in the vicinity of the product size (a 450-mer, no oligonucleotide bands) cut from the gel, and the correct oligonucleotide isolated after cloning.

The combination of DNA chemistries with automated instruments for the purpose of synthesizing oligonucleotides has been a major focus for many years. Initially, several of these machines used designs similar to the apparatus developed by Bruce Merrifield[5] in the peptide area. These included early nonautomated synthesizers from M. Gait[6] and K. Itakura[7] using the phosphate triester method. A new chemistry based upon the nucleoside chlorophosphite approach[8] was also incorporated in manual[4] and automatic[9] devices. However, none of these machines proved to be commercially viable either because the condensation reactions were slow and incomplete (phosphotriester approach) or the synthons were unstable (chlorophosphite method). It was not until the nucleoside phosphoramidite chemistry was automated that a viable machine could be marketed and used universally by chemists, biologists, and biochemists. The first of these was developed by W. Efcavitch at Applied Biosystems (the 380A Instrument). This machine was revolutionary, as it was not only the first to generate near-quantitative coupling yields but also because it successfully demonstrated the multiple advantages of using nitrogen gas rather than liquids and pumps to manipulate solvents and

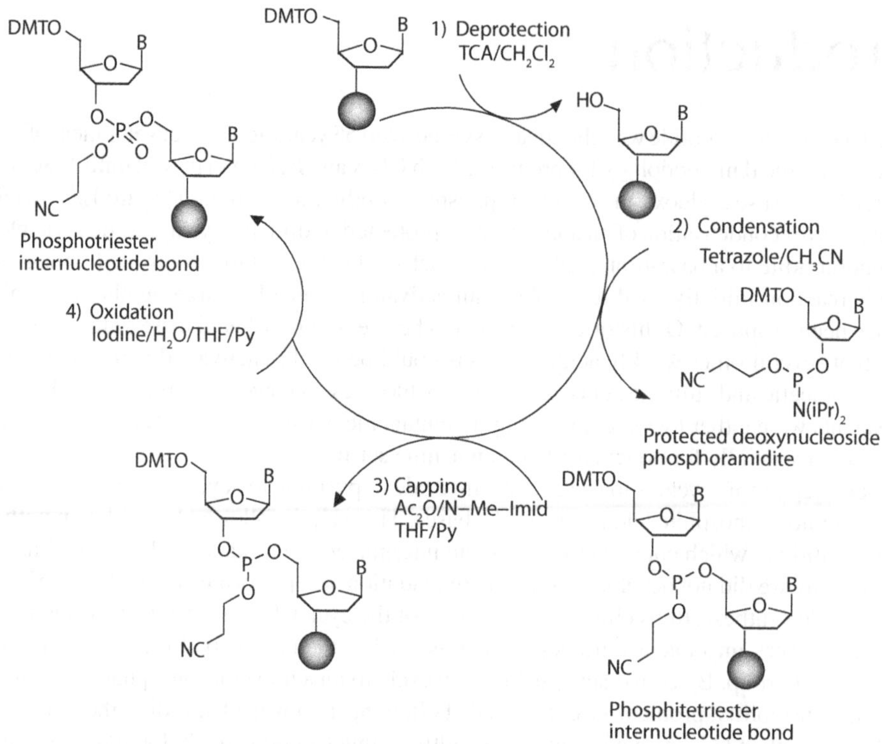

FIGURE 1 The synthesis cycle of preparing oligonucleotides using phosphoramidite chemistry on controlled pore glass supports.

reagents. As the field evolved and ever-larger synthesizers were needed for preparing increasing amounts of oligonucleotides, several instruments were developed for this purpose. Although there are many, perhaps the BioAutomation MerMade series currently enjoys the leadership position. As more and more oligonucleotides enter the clinic in order to be tested for various therapeutic indications (currently in excess of 240 phase 1 and phase 2 trials), the need for ever-increasing quantities will further test our ability to design instruments that can be used successfully for these exciting applications.

Over the years, this methodology has survived virtually without change and proven to be useful, not only for DNA and RNA synthesis but also to prepare base, backbone, and sugar modification of the oligonucleotides as well as many analogues.[10–13] Applications range from the use of these oligomers as diagnostic reagents, including forensics, therapeutic drugs, antisense reagents for studying gene expression and cell differentiation, sequencing, PCR amplification of genes and chromosomes, interfering RNA and/or micro-RNA antagomers, and many other uses. Recently, the phosphoramidite chemistry has been adapted for in situ synthesis of DNA micoarrays. Such synthesis has been achieved by spatial control of one step of the synthesis cycle that results in thousands to hundreds of thousands of unique oligonucleotides distributed on an area of a few square centimeters. The main methods used to achieve spatial control include (1) control of the coupling step by inkjet printing[14] or physical masks,[15] (2) control of the 5′-hydroxyl deblock step by classical[16] and maskless[17] photolithographic deprotection of photolabile monomers, or (3) digital activation of photogenerated acids to carry out standard detritylation.[18] Oligonucleotides made on these commercial microarrays can be prepared at a rate of 15 million unique sequences per week with lengths up to 200 nucleotides and a fidelity approaching 50%.[19] There are many applications for these microarrays. For example, these oligonucleotides can be cleaved from their solid surfaces and pooled to enable new applications such as shRNA libraries[20] that cover all known open reading frames in the human and mouse genome, gene synthesis,[21,22] and large-scale, site-directed mutagenesis.[23]

Thus, in modern pharmaceutical and biotechnology companies, it is clear that synthetic oligonucleotides represent one of the cornerstone technologies used routinely for many applications in applied and basic research. Moreover, they also serve as marketable products in both diagnostic and therapeutic areas. With so many oligonucleotides in clinical studies, the future looks very positive for developing several therapeutic products from these compounds.

Perhaps the major recent advance in nucleic acid synthesis has been the development of precise methods for analyzing and characterizing oligonucleotides. When we first investigated this chemistry in 1980, there were very few analytically useful methods. For example, our work was mainly based upon phosphorus NMR, reverse-phase HPLC, and enzymatic analysis of the product oligonucleotides. Only recently have modern mass spectral techniques proven useful for full characterization of the phosphoramidite synthons.[24] As outlined in this handbook, there have been many new developments for separating and characterizing oligonucleotides. These include new developments in HPLC (RP-HPLC, AEX-HPLC, SEC-HPLC) and in hydrophilic interacting chromatography (HILIC), which are useful for separating oligonucleotides from impurities. Analytical methods as well have advanced considerably in the past few years. Many of these are discussed here and include mass spectral methods (HPLC-MS, MS/MS, MALDI-TOF), NMR, FT-IR, and the analysis of trace metals and solvents. As a result of these developments and others such as the use of various bioanalytical techniques to characterize oligomers, the nucleic acid chemist now has a complete arsenal of methods for separating and fully characterizing oligonucleotides useful for therapeutic and diagnostic applications as well as the synthons needed for development of new, more advanced analogs.

REFERENCES

1. Michelson, A. M., and A. R. Todd. 1955. *J. Chem. Soc.* 2632–2638.
2. Caruthers, M. H. 1985. *Science* 230*:* 281–285.
3. Beaucage, S. L., and M. H. Caruthers. 1981. *Tetrahedron Lett.* 22: 1859–1862.
4. Matteucci, M. D., and M. H. Caruthers. 1981. *J. Am. Chem. Soc.* 103: 3185–3191.
5. Merrifield, R. B., and J. M. Stewart. 1965. *Nature* 207: 522–523.
6. Gait, M. J., and R. C. Sheppard. 1977. *Nucl. Acids Res.* 4: 1135–1158.
7. Ito, H., Y. Ike, S. Ikuta, and K. Itakura. 1982. *Nucleic Acids Res.* 10: 1755–1769.
8. Letsinger, R. L., and W. B. Lunsford. 1976. *J. Am. Chem Soc.* 98: 3655–3661.
9. Alvarado-Urbina, G., G. M. Sathe, W. C. Liu, M. F. Gillen, P. D. Duck, R. Bender, and K. K. Ogilvie. 1981. *Science* 214: 270–274.
10. Leumann, C. J. 2002. *Bioorg. Med. Chem.* 10*:* 841–854.
11. Petersen, M., and J. Wengel. 2003. *Trends Biotechnol. 21:* 74–84.
12. De Mesmaeker, A., K.-H. Altmann, A. Waldner, and S. Wendebon. 1995. *Curr. Opin. Struct. Biol.* 5: 343–355.
13. Beaucage, S. L., and R. P. Iyer. 1992. *Tetrahedron* 48: 2223–2311.
14. Hughes, T. R., M. Mao, A. Jones, et al. 2001. *Nat. Biotechnol.* 19: 342–347.
15. Southern, E. M., U. Maskos, and J. K. Elder. 1992. *Genomics* 13: 1008–1017.
16. Pease, A. C., D. Solas, E. J. Sullivan, M. T. Cronin, C. P. Holmes, and S. P. A. Fodor. 1994. *Proc. Natl. Acad. Sci.* 91: 5022–5026.
17. Singh-Gasson, S., R. D. Green, Y. J. Yue, C. Nelson, F. Blattner, M. R. Sussman, and F. Cerrina. 1999. *Nat. Biotechnol.* 17: 974–978.
18. Gao, X. L., E. Le Proust, H. Zhang, O. Srivannavit, E. Gulari, P. L. Yu, C. Nishiguchi, Q. Xiang, and X. C. Zhou. 2001. *Nucleic Acids Res.* 29: 4744–4750.
19. Leproust, E. M., B. J. Peck, K. Spirin, H. Brummel McCuen, B. Moore, E. Nomsaraev, and M. H. Caruthers. 2010. *Nucl. Acids Res. in press.*
20. Silva, J. M., M. Z. Li, K. Chang, et al. 2005. *Nat. Genet.* 37: 1281–1288.
21. Richmond, K. E., M. H. Li, M. J. Rodesch, et al. 2004. *Nucl. Acids Res.* 32: 5011–5018.
22. Tian, J. D., H. Gong, N. J. Sheng, X. C. Zhou, E. Gulari, X. L. Gao, and G. Church. 2004. *Nature* 432: 1050–1054.

23. Saboulard, D., V. Dugas, M. Jaber, J. Broutin, G. Souteyrand, J. Sylvestre, and M. Delcourt. 2005. *Biotechniques* 39: 363–368.
24. Kupihar, Z., Z. Timar, Z. Darula, D. J. Dellinger, and M. H. Caruthers. 2008. *Rapid Commun. Mass Spectrom.* 22: 533–540.

Marvin H. Caruthers
University of Colorado
Boulder, Colorado

1 Purity Analysis and Impurities Determination by Reversed-Phase High-Performance Liquid Chromatography

Hagen Cramer, Kevin J. Finn, and Eric Herzberg
Girindus America, Inc.

CONTENTS

1.1 INTRODUCTION

The increasing significance of oligonucleotides as therapeutic agents necessitates a high level of quality control. In such protocols, chromatographic analysis of crude and final active pharmaceutical ingredients (API) is necessary to ensure the detection of contaminants at concentration levels down to trace amounts relative to the drug. A combination of chromatographic techniques, in particular reverse-phased high-performance liquid chromatography (RP-HPLC), anion-exchange (AEX) HPLC (Chapter 2), and mass spectrometry (Chapters 4 and 5), is needed for the identification and structural elucidation of by-products and degradation products resulting from the production process. In clinical studies employing oligonucleotides, high sensitivity and low sample requirement for analytical methods are requisite for the identification of metabolites (Chapter 8).

1.2 HISTORICAL ASPECTS

Reversed-phase HPLC (RP-HPLC) is one of the most important techniques for the characterization of oligonucleotides. Over the years, the ability to resolve impurities from the main product peak has increased dramatically by introduction of smaller chromatographic particle sizes. Another advantage of RP-HPLC is that mobile phases that work well for separating impurities and at the same time are compatible with electrospray ionization mass spectrometry (ESI-MS) are available (see Chapter 4).

Just 10–15 years ago, the typical particle size used for analytical reversed-phase columns was 5 μm. To achieve good separations on 5 μm particle-size columns, fairly long columns were needed (up to 250 mm), resulting in long run times. Typically, these columns had a diameter of 4.6 mm. In the late nineties, the first columns with 3.5-μm particle size useful for oligonucleotide analysis came to market. With the smaller particle-size beads, it now was possible to achieve better separations with shorter columns, reducing typical column length to 50 mm (up to 150 mm maximum). This resulted in shorter run times as well. In addition, more accurate HPLC pumps allowed reduction of the column diameter to 2.1 mm, thereby further decreasing buffer consumption. Modern HPLC systems could also tolerate the attendant increased pressure associated with media having smaller particle sizes.

Recently, even smaller particle sizes were introduced with the smallest ones useful for oligonucleotide analysis being 1.7 μm. Particle sizes of below 1.7 μm, while useful for small molecule analytes, result in degradation of the oligonucleotide during analysis owing to the increased shearing forces present at these high back pressures. As it is, 1.7 μm particle-size sorbents generate back pressures of over 400 bar (ca. 6000 psi), making them incompatible with traditional HPLC systems. To achieve similar column plates without increasing back pressures, fused-core (also called core-shell) particle technology was introduced by Joseph J. Kirkland in 2007[1,2] based on his earlier work with poroshell silicas.[3,4] Thereby, the all-porous particle typically used is replaced with a nonporous core surrounded by a porous shell.

While polystyrene-based columns have proven successful on the preparative scale owing to their increased chemical stability, the analytical reversed-phase HPLC market is dominated by silica gel or silica-based resins.

1.3 REVERSE-PHASED HIGH-PERFORMANCE LIQUID CHROMATOGRAPHY COLUMNS

There are multiple companies that offer reversed-phase HPLC columns. Table 1.1 lists commercially available silica-based reversed-phase HPLC columns. Columns most widely used in oligonucleotide separations are available from Agilent (Zorbax), Phenomenex (Clarity), and Waters (X-Terra®, X-Bridge™, Acquity BEH).

Typical particle sizes for columns from all manufacturers used to be 5 μm for many years until about 10–15 years ago. In the mid to late 1990s, smaller particle-size columns were introduced and recently sub-2 μm particles have started gaining in popularity. However, not all column manufacturers made the switch to sub-2 μm particle-size resins. The use of sub-2 μm particle sizes allows the use of much shorter columns without losing separation power because theoretical plates can be maintained.[5,6] A reduction of the particle diameter by 50% results approximately in a doubling of the plate count. Therefore, fast and efficient separations can be achieved because separation time is proportional to column length. A shorter column run at the same velocity as a longer column also uses less solvent. However, the small particle sizes result in high back pressure a traditional HPLC system cannot withstand. New ultra-high pressure systems had to be developed. These new systems are called U-HPLCs or UPLCs and at this time only a handful of companies are offering

TABLE 1.1
HPLC Column Manufacturers

Manufacturer	Brand Name	Sub-2 μm Particles
Advanced Chromatography Technologies	ACE	No (3 μm minimum)
Agilent	Zorbax Eclipse Plus & Extend	Yes
Azko-Nobel	Kromasil	No (2.5 μm minimum)
Bischoff	ProntoPEARL	Yes
Grace/Alltech	Alltime, Vydac, VisionHT	Yes
Beckman-Coulter	Ultrasphere ODS	No (5 μm minimum)
EMD/Merck	LiChrospher	No (5 μm minimum)
GL Sciences	Intersil	No (2 μm minimum)
Interchim	Uptisphere	No (5 μm minimum)
Macherey-Nagel	Nucleosil, Nucleodur	Yes
Phenomenex	Gemini, Luna & Clarity	No (3 μm minimum)
Resek	Allure, Ultra, Pinnacle	Yes
Sepax	GP series	Yes
Shimadzu/Shant	Pathfinder	Yes
Supelco	Ascentis	No (3 μm minimum)
Thermo	Hypersil Gold	Yes
Waters	X-Terra, X-Bridge & Acquity BEH	Yes
YMC	Several	Yes

such systems, which are listed in Table 1.2. Systems are being improved continuously because requirements on components due to the increased pressure are much more rigorous in comparison to traditional HPLC systems.

Silica-based bead chemistry has traditionally been the state of the art because of its good mechanical strength, spherical shape and high chromatographic efficiency, and compatibility with a host of organic solvents. Two chief problems associated with the use of silica-based cores are resolution of basic analytes and stability of the bonded phase toward low- and particularly high-pH mobile phases. Several modifications to the silica-bonded phase have been implemented in order to address these issues,[7] including incorporation of polar functional groups,[8] sterically hindered silanes, bidentate, or hybrid organic–inorganic stationary phases.[9]

The XTerra column, first introduced in 1999 by Waters (Milford, MA), uses patented hybrid particle technology (HPT) to overcome traditional silica's instability to high pH. The core bead is composed of a methylpolyethoxysilane (MPEOS) monomer synthesized by condensation of tetraethoxysilane (TEOS) and methyltriethoxysilane (MTEOS). XTerra's stationary phase demonstrates equivalent efficiency to state of the art silica-based C18 columns while addressing the problem of pH instability. Hybrid particle technology, so named because it combines inorganic (silica) with organic (polymeric) bead chemistry, describes the replacement of one of every three silyl groups with a methyl group.

TABLE 1.2
UPLC Systems

Manufacturer	Name of System	Pressure Limit, Bar
Agilent	1200 Series	600
	1290 Infinity LC	1200
Hitachi	LaChrom ULTRA L-2160U	600
Thermo	Accela	1000
Waters	Acquity	1000

The substitution of the methyl for the more polar silanol dramatically increases the hydrophobicity of the core structure of the particle backbone. The introduction of the hybrid particle offers increased robustness and improved resolution of basic compounds. In 2005, Waters launched a second generation hybrid column called XBridge with bridged ethyl hybrid (BEH) technology. The polyethoxysilane core marketed as BEH technology relies on cross-linking TEOS with bis(triethoxysilyl)ethane (BTEE). The result was a material of much greater mechanical stability owing to the increased level of crosslinking while maintaining the superior pH stability of hybrid columns. Those features make the BEH technology also very attractive for ultra performance liquid chromatography instrument (UPLC) applications, where the increased pressures necessitate increased stability of the beads. BEH columns are available in a variety of particle sizes from 1.7 to 10 μm, which allows the BEH technology to be adapted to both HPLC and UPLC applications. BEH technology based UPLC columns are called Acquity UPLC columns and were introduced shortly after the XBridge HPLC columns. The recommended operating pH range for BEH-based columns is from 1 to 12.

Column lifetime is dramatically impacted by pH, leading to partial hydrolysis of the bonded phase, and resulting in variable retention times and inconsistent performance. The prevailing use of BEH-based columns is mainly due to its robustness toward wide range of pH, stability toward dimethyl sulfoxide (DMSO) (important for analysis of crude RNA), and high mass loading capacity. The separation quality is comparable to capillary gel electrophoresis without compromising yield. The BEH-based columns are advantageous for the analysis of dye or lipidoyl labeled oligos because the added hydrophobicity increases separation efficiency. The BEH-based columns are available in variety of phases (C18, C8, Phenyl, and Shield RP18) and also boast long column lifetime (>1000 injections) at elevated temperatures (60°C).

Zorbax Eclipse Plus columns were introduced by Agilent (Santa Clara, CA) in 2006 and are offered at particle sizes of 1.8, 3, and 5 μm to accommodate a wide range of analytical HPLC applications. Eclipse Plus columns are available at multiple selectivity choices (C18, C8, and Phenyl) for optimized resolution of all sample types and provide high resolution and excellent peak shape of all types of compounds at pH 2–9. Eclipse Plus columns achieve superior performance through extra dense bonding and a precise double-endcapping process. Agilent also offers columns especially developed for low- and high-pH applications. Zorbax SB (StableBond) columns are made using bulky, unique silanes that sterically protect the siloxane bond not including any acid-labile endcapping. The result is vastly improved column life and extraordinary chemical and temperature stability in the pH 1–6 range for a wide variety of phases (SB-C3, SB-CN, SB-Phenyl, SB-C8, and SB-C18). Zorbax Extend-C18 columns incorporate a unique bidentate ring structure, having a propylene bridge in combination with the bulky C18 group, thereby shielding the silica support from dissolution. Such bonded silanes, combined with a double-endcapping process, protect the silica from dissolution at high pH—up to pH 11.5.

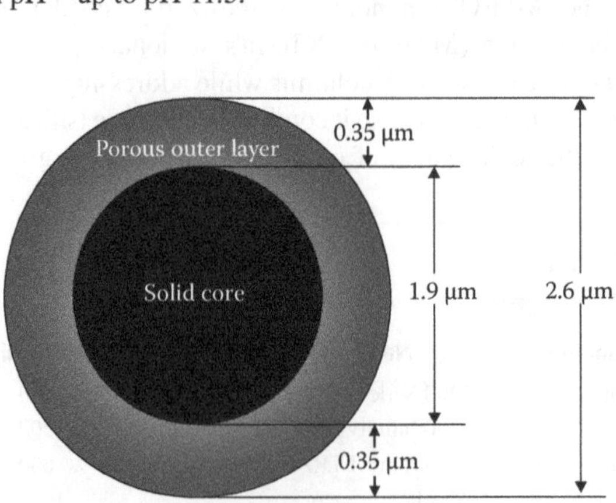

FIGURE 1.1 Fused-core/core-shell particle technology (Kinetex 2.6-μm column dimensions as an example).

TABLE 1.3
Fused-Core/Shell-Core HPLC Columns

Manufacturer	Brand Name	Particle Size (Core/Outer Layer), μm
Agilent	Poroshell 120	2.7 (1.7/2 × 0.5)
MAC-MOD Analytical	Halo	2.7 (1.7/2 × 0.5)
Phenomenex	Kinetex	2.6 (1.9/2 × 0.35)
		1.7 (1.25/2 × 0.23)
Supelco	Ascentis Express	2.7 (1.7/2 × 0.5)

Silica-based sorbents of RP-HPLC column used for oligonucleotide analysis typically have pore sizes of about 100 Å. For the three examples given above, the pore sizes are 120 Å for the XTerra, 135 Å for the XBridge, and 95 Å for the Zorbax Eclipse Plus. However, for larger oligonucleotides, for example aptamers, 300 Å pore size columns tend to yield better results.

Fused-core or shell-core technology provides an elegant way around the pressure limits of traditional HPLC-systems. Particles with a solid core and porous shell behave in regards to pressure like an equivalently sized fully porous particle while theoretical plate numbers are similar to a sub-2 μm particle column (see Figure 1.1). Originally developed by F. F. Kirkland (Advanced Materials Technology, Wilmington, DE),[1,2,10] such columns are now available from several different companies (see Table 1.3) and are comparable in performance to sub-2 μm columns.[11] Kinetix columns are available as C_{18} and pentafluorophenyl (PFP) for a variety of separation applications.

The key feature of the fused-core technology is a spherical porous shell grown on the surface of a solid silica-based bead. It addresses two of the most critical effects of column performance—the eddy diffusion (also known as the multipath effect) and resistance to mass transfer.[12] Shown in Figure 1.2 is a Van Deemter plot[13]—a graphical description of the three parameters that most contribute to band broadening: (1) the eddy diffusion or "A term," governed by the particle size, (2) the longitudinal diffusion or "B term," and finally (3) the "C term," which is related to the kinetics of resistance to mass transfer. By controlling the particle-size distribution, the band broadening effect of eddy diffusion is dramatically diminished. The A term, in accordance with the Van Deempter plot, is independent of mobile phase velocity and is related solely to average diameter of the particle. The C term is also dependent on particle size.

The distribution of analyte molecules in the stationary phase and the mobile phase is governed by the kinetics of diffusion between the two phases for the analyte. By introducing a semi-porous (rather than fully porous shell), the analyte spends less time diffusing in and out of pore on the

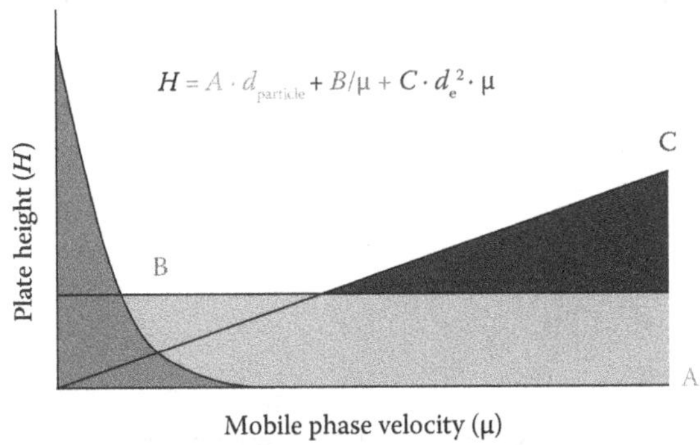

FIGURE 1.2 Van Deempter plot and equation of plate height (H) vs. mobile phase velocity (μ); A: eddy diffusion; B: longitudinal diffusion; C: kinetics of resistance to mass transfer; d_{particle}: particle diameter; d_e: effective particle size. d_e represents the effective particle size and is equal to the particle diameter in the case of fully porous particles.

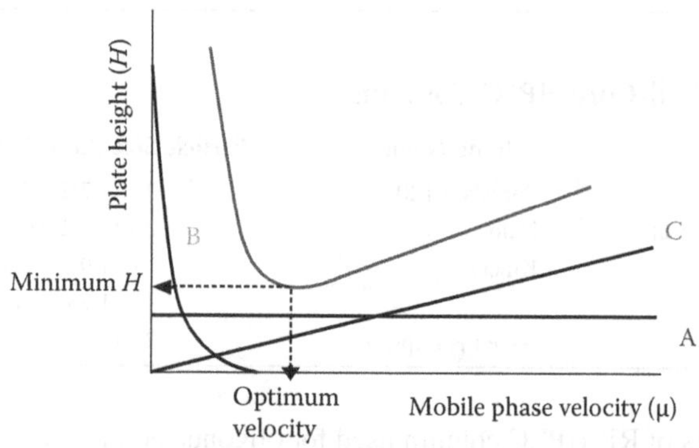

FIGURE 1.3 Typical Van Deempter plot of plate height (H) vs. mobile phase velocity (μ).

stationary phase, thus reducing the dispersive effect known as resistance to mass transfer. Figure 1.3 depicts a typical Van Deempter plot. The resistance to mass transfer, or C term, is carried by the square of the effective particle size and varies sharply at high flow rates.

By minimizing the particle size, the Eddy diffusion A term and mass transfer C terms are minimized and the result is an ability to carry out separation at higher mobile phase flow rates without sacrificing plate height. Figure 1.4 represents a more optimized Van Deempter plot resulting from the use of smaller particle size.

Silica-based resins have dominated the analytical HPLC market of microparticulate sorbents since the inception of HPLC almost 40 years ago. While polystyrene-based particles are widely used in larger-scale purifications, there are only a few reports on their use in analytical HPLC. The PRP-1 column from Hamilton, NV, was introduced in the early 1980s and has since then been employed for preparative,[14–16] as well as analytical oligonucleotide separations.[17,18] The PRP columns are polymeric reversed-phase column and are composed out of a copolymers of styrene and divinylbenzene (PS-DVB) and are available at particle sizes of 5–20 μm and at pore sizes of 100 (PRP-1) and 300 Å (PRP-3). The ruggedness of the PS-DVB particles make such columns an attractive choice for high-pH applications or when the analyte or crude sample is contaminated with other aggressive chemicals not compatible with silica (e.g., RNA purifications).

Over the years others have reported the use of PS-DVB columns for the analysis and purification of oligonucleotides as well. Huber et al. reported good resolution of phosphorylated from dephosphorylated oligonucleotides when using columns filled with PS-DVB acquired from Riedel-de Haën (Seelze, Germany) and adding poly(vinyl alcohol) during polymerization.[19,20] Gelhaus et al. was

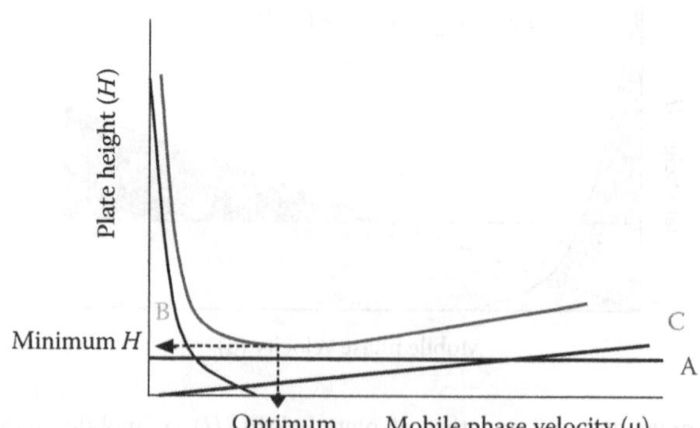

FIGURE 1.4 The effect of smaller particle size on the Van Deempter plot.

able to achieve separation of 18-mers of the same base sequence but with differing alkyl modifications with a OligoSep column (Transgenomics, Omaha, NE) comprised of nonporous, C18 modified polystyrene-divinylbenzene (PS-DVB).[21]

Lloyd et al. reported good resolution for long oligonucleotides and double-stranded DNA ladders using PLRP-S columns (Polymer Laboratories, UK). These columns are based on rigid macroporous reversed-phase poly(styrene–divinylbenzene)–based sorbents and come at many different pore sizes of 100 to 4000 Å.[22]

For more than 40 years, columns packed with microparticulate sorbents have been successfully used in high-performance liquid chromatography (HPLC). Despite many advantages, HPLC columns packed with microparticulate, porous stationary phases have some limitations, such as the relatively large void volume between the packed particles and the slow diffusional mass transfer of solutes into and out of the stagnant mobile phase present in the pores of the separation medium.[23] One approach to diminish the problem of restricted mass transfer and interparticle void volume is the use of monolithic chromatographic beds, in which the separation medium consists of a continuous rod of a rigid, porous polymer that has no interstitial volume but only internal porosity. Because of the absence of interparticle volume, all of the mobile phase is forced to flow through the pores of the separation medium.[24] According to theory, mass transport is enhanced by such convection and enhances chromatographic efficiency.[25] Monolithic chromatographic beds are usually prepared by polymerization of suitable monomers and porogens in a stainless steel or fused silica tube that acts as a mold.[26] The porous structure is achieved as a result of the phase separation that occurs during the polymerization of a monomer or a mixture of both a cross-linking monomer and a porogenic solvent.[27]

Huber and coworkers have demonstrated that the chromatographic separation performance of cross-linked, norbornene-based, monolithic capillary columns prepared via ring-opening metathesis polymerization (ROMP) indicates good separation capabilities for single- and double-stranded nucleic acids.[28] Such monolithic columns were able to separate diastereoisomers of short phosphorothioate oligonucleotides. Longer PS oligomers coalesced into a single peak, where peak widths decreased with increasing length of the oligonucleotides. Four homologous oligodeoxynucleotides, ranging in length from 24 to 27 nucleotides, could be baseline separated within 7 min using a triethylammonium acetate buffer and an acetonitrile gradient. Over the years, Huber and others have published extensively on the analysis of oligonucleotide and nucleic acids using monolithic capillary columns.[29–42]

Further, Huber and coworkers proposed a new model for predicting the retention time of oligonucleotides.[43,44] Their model is based on support vector regression using features derived from base sequence and predicted secondary structure of oligonucleotides. Because of the secondary structure information, their model is applicable even at relatively low temperatures where the secondary structure is not suppressed by thermal denaturing.

1.4 STATIONARY PHASES

The most commonly used stationary phase in reversed-phase HPLC is based on octadecylsilane (ODS) or C18 groups. There are several reasons for this, but one of the strongest is tradition. Early column packings were based on C18 because C18-based silanes were readily available at that time and reasonable in cost. Another reason for the popularity of C18 is the relatively high organic content that can be reacted onto silica supports. In addition the long-chain C18 ligand shows greater stability at both low and higher pH, compared to shorter chain ligands resulting in better separation reproducibility. However, there are some disadvantages to C18 bonded phases packings. Column packings with shorter functional groups can reequilibrate more rapidly after a gradient elution separation. Densely bonded C18 packings can also exhibit *phase collapse* when mobile phases contain a high aqueous content.[45] Often, the starting concentration of organic modifier must be less than 5% for adequate separation of very polar compounds. When exposed to high concentration of aqueous buffer, the densely packed C18 hydrophobic phase tends to minimize its surface area by self-association with attendant dewetting, a phenomenon known as phase collapse (see Figure 1.5). The folding of the stationary phase on itself results in inability of the surface to come into contact with the mobile phase, and the consequence is poor chromatography marked by increased tailing and retention time variability. The phenomenon is much less common when shorter bonded phases are used.

Reversed-phase column packings with aliphatic C18 groups are predominantly used for the analysis of oligonucleotides. While historically ion-pair RP-HPLC was well suited for the separation of phosphodiester oligonucleotides (PO-ONs), initial separation of phosphorothioate oligodeoxynucleotides (PS-ODNs) failed.[46] The difference in retention time between phosphodiester and phosphorothioate oligonucleotides is drastic. The replacement of PO with PS linkages between bases often doubles the retention time owing to the highly lipophilic nature of the phosphorothioate bond. The creation of a chiral center on the phosphorous center upon incorporation of the sulfur atom changes the physical and chemical properties of the modified molecule and leads to the formation of a set of 2^n diastereoisomers, where n is the number of chiral internucleotide linkages. The substitution of PO for PS in the internucleosidic backbone results in substantial peak broadening.[47,48] However, this limitation has changed dramatically over the years, and today the resolving power of RP-HPLC is well suited for PS oligos as well and comparable to the separation efficiency of capillary gel electrophoresis (CGE) (see Figure 1.6). However, HPLC is a much more robust technique than CGE and therefore has replaced CGE in many applications.

The relative differences in the length (hydrophobicity) or charge of N and N–1 oligonucleotides are small. HPLC separation is difficult and becomes more challenging as N increases. In addition, slow mass transfer (diffusion) of high molecular weight analytes within sorbent pores further complicates the separation owing to peak broadening. For that reason, historically, the best resolution of oligonucleotides has been achieved with nonporous[49] or superficially porous chromatographic sorbents (core-shell technology, see Section 1.3).[1,2,10] However, owing to the low mass load capacity of nonporous sorbents, today only porous or superficially porous sorbents are being applied to oligonucleotide analysis. Smaller particle size of the packing material decreases the diffusion path of molecules and provides for high chromatographic performance. In combination with high temperatures, relatively slow flow rates, and shallow gradients, a separation of N from N–1 for up to 60-mer oligonucleotides is achievable.[50]

C18 columns tend to be too lipophilic for the analysis of certain modified oligonucleotides. Therefore, in the case of cholesterol or other lipid conjugates, C8 or sometimes C4 columns are preferred over C18 columns. Recently, however, there has been an increase of stationary phases with other functionalities for use in reversed-phase HPLC. Such stationary phases provide different separation selectivity than traditional C18 or C8 stationary phases and are especially useful when the chromatographer is restricted to using a particular mobile phase such as in LC–MS studies. Varying selectivity by changing the stationary phase is an effective alternative to changing the

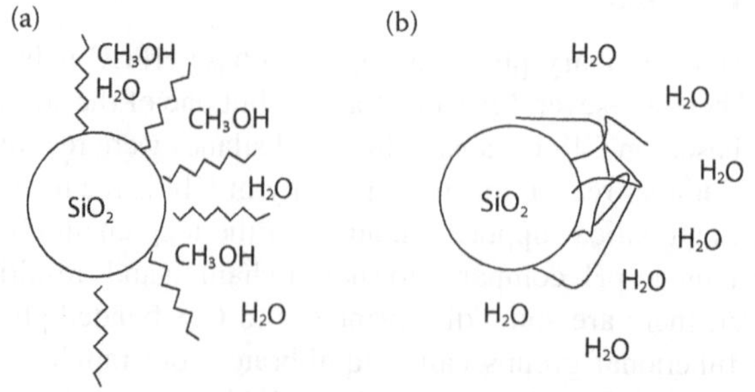

FIGURE 1.5 Depiction of (a) normal interaction of the C_{18} bonded phase with the mobile phase; (b) phase-collapsed situation.

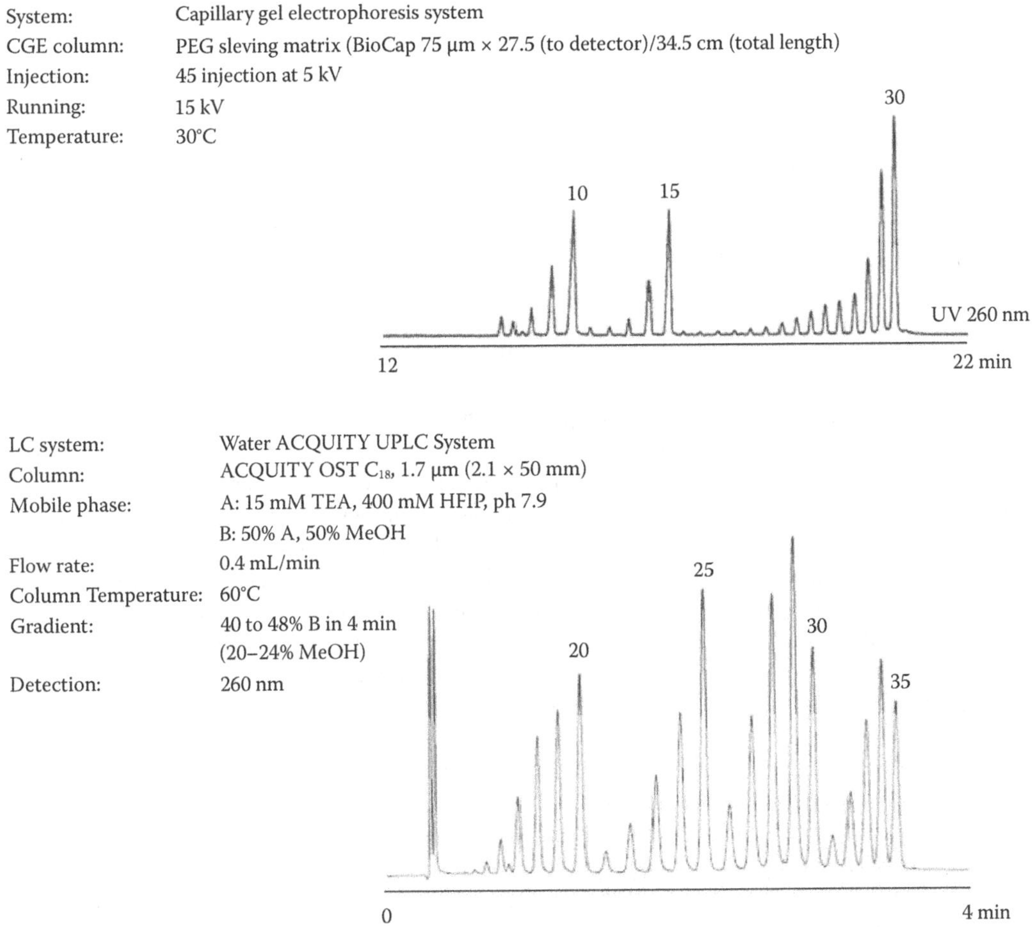

FIGURE 1.6 Comparison of CGE and IP-HPLC separation of deoxythymidine ladders. (Courtesy of Waters Corporation.)

mobile phase, and this approach is being used more frequently. Typically, stationary phases used are C18, C8, phenyl, and fluoro or a combination thereof. Different commercial C18 columns also often show different separation selectivities, usually based on differences in silica supports or stationary phase chemistry.

Typical solute-column interactions are based on hydrophobic, steric resistance, hydrogen bonding, ionic and London dispersion (also called dipole–dipole) forces.[51] Phenyl columns show additional π–π but less hydrophobic interactions. Phenyl columns, such as phenylpropyl or phenylhexyl columns, are less commonly used for RP-LC separation. The selectivity of phenyl columns differs from that for alkyl-silica columns. π-Acids, such as aromatics, are preferentially retained by π–π interactions on phenyl versus alkyl-silica columns. The enhanced retention of π-acids varies with the organic solvent in the mobile phase as: tetrahydrofuran (least) < acetonitrile < methanol (most). Phenyl columns also show stronger dispersion interactions versus (less polarizable) C8 or C18 columns. The reduced hydrophobicity of phenyl versus alkyl groups leads to smaller hydrophobic and steric interactions, possibly because the phenyl groups are more ordered.

Perfluorinated alkyl or phenyl ligands are the basis of so-called fluoro columns. Solutes of lower refractive index (and lower molecular polarizability) are more strongly retained on fluoro-alkyl columns, relative to a C8 or C18 column. Polyaromatics are less retained on the fluoro-alkyl column than substituted benzenes, while aliphatic solutes are more retained. Fluoro-substituted aromatics show even larger retention factor on fluoro columns. This behavior has been attributed to differences in solute–column dispersion interactions for fluoro-alkyl columns, as a result of the much lower polarizability of fluoro-alkyl columns.[51]

1.5 MOBILE PHASES

The term reverse-phased chromatography describes a separation technique utilizing a bonded phase (stationary phase) composed of a polystyrene- or silica-based bead covalently modified with nonpolar groups. The technique differentiates itself from "normal" phase chromatography with alumina or silica in that polar compounds are the first to elute followed by more hydrophobic components. While reverse-phased chromatography relies solely on hydrophobicity as a mechanism of separation, ion-pairing chromatography describes a technique in which a long-chained alkyl amine is added in low concentration to the mobile phase in order to achieve enhanced resolution.[52–54] The exact nature of ion-pairing phenomenon has been the subject of debate for several decades; however, it is now generally accepted that an ion-pairing reagent such as a tri- or tetraalkylammonium salt, when added to the mobile phase, is capable of associating with the nonpolar stationary phase through dynamic hydrophobic interactions as represented in Figure 1.7.

The charged ammonium ion, in turn, acts as an ion exchanger along the surface of the stationary phase and provides a means to separate charged species bearing hydrophobic groups according to charged state. The actual mechanism is certainly more complex, given the both the presence of multiple charged species in the solvent mixture. The retention and order of elution is primarily governed by

1. Charge of the oligonucleotide, whereby retention time increases in proportion to the number of charges in the oligonucleotide.
2. Length of alkyl chain in the ion-pairing reagent; increased hydrophobicity in the ion-pairing reagent leads to extended retention.
3. Proportion of organic solvent in the mobile phase;[55] retention is decreased by higher concentration of organic solvent.

Varying separation selectivity by optimizing the mobile phase is the most powerful approach for optimizing separation resolution.[56] Selection of mobile phase for ion-pair (IP) HPLC is a critical parameter. While triethylammonium acetate (TEAA) is the most commonly used ion-pairing buffer component,[57] a variety of other systems have been employed in IP chromatography of nucleosides and oligonucleotides, including tetrabutylammonium hydrogen sulfate,[58,59] tetrabutylammonium iodide,[60] tetrabutylammonium phosphate,[61] tetrabutylammonium acetate,[16] tetrabutylammonium bromide (TBAB),[62] tributylammonium acetate (TBAA),[63] ethylenediamine acetate,[64] hexylammonium acetate (HAA),[65–67] triethylammonium bicarbonate (TEAB),[42] and triethylamine in combination with hexafluoroisopropanol (HFIP).[12,68–73] The use of HFIP and TEAB as an ion-pairing agent

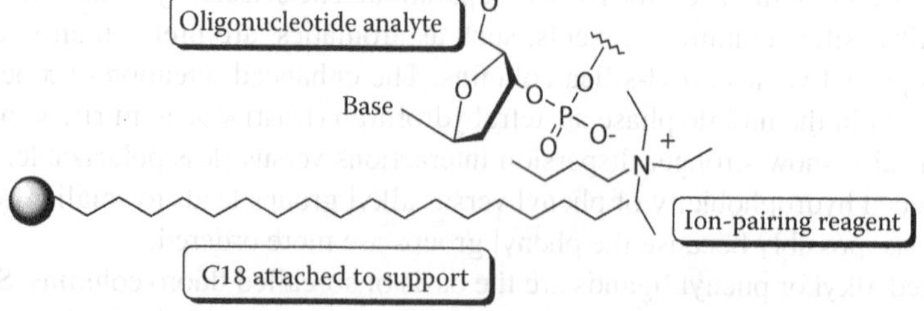

FIGURE 1.7 Association of an oligonucleotide with an ion-pairing reagent at the surface of a C_{18} support.

is particularly attractive because it is compatible with MS coupling to HPLC (refer to Chapter 4) and often provides excellent separation. Once only used for LC/MS separations, HFIP is now used as widely as TEAA because of its unique ion-pairing ability. Gilar et al. found that in hetero-oligonucleotide ladders, oligonucleotides one nucleotide apart could overlap or even reverse retention order when using a TEAA-based buffer system for the separation due to the different hydrophobicity of the different bases (see Figure 1.8) (hydrophobicity increases in the following order: C < G < A < T).[12] Buffer systems based on TEA-HFIP show a less pronounced dependence of the hydrophobicity of the bases and an overlap of two oligonucleotides one nucleotide apart typically does not occur.

The ion suppression that plagued the early analysis of oligonucleotides using LC/MS with high concentrations (normally greater or equal to 100 mM) of TEA can be avoided by simply switching to an HFIP-based ion-pairing mobile phase, where only small concentrations of TEA are being

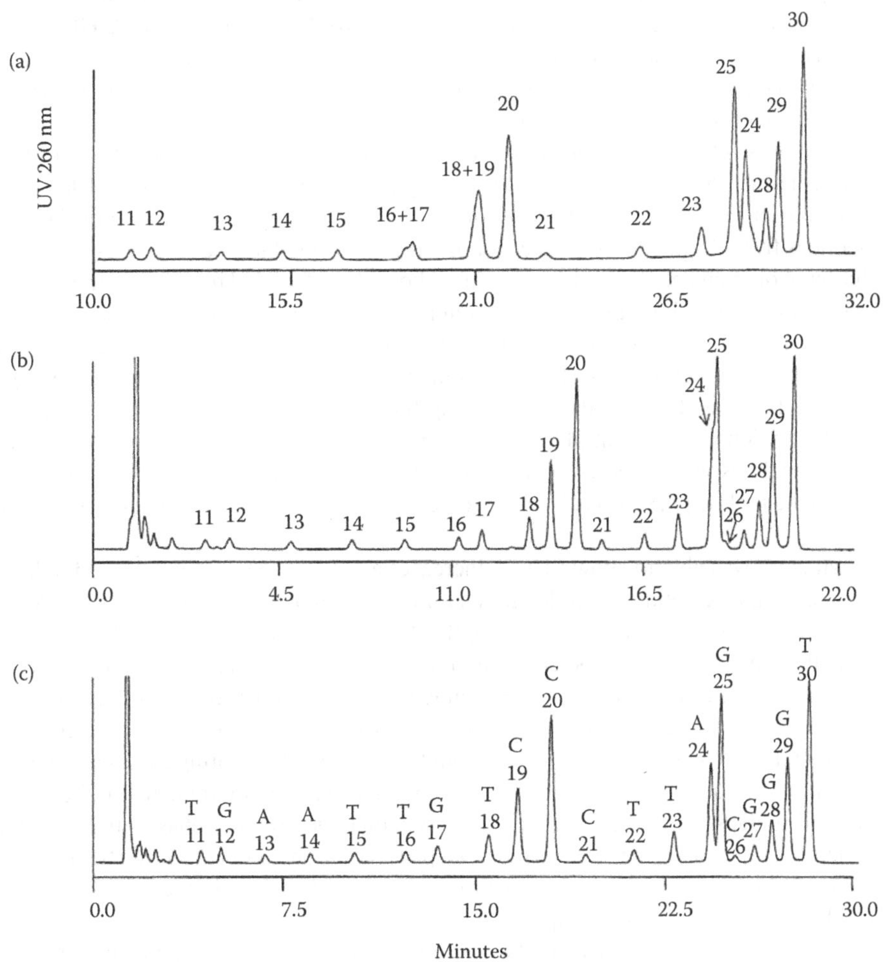

FIGURE 1.8 Separation of a 10-30mer hetero-oligonucleotide ladder using three separation ion-pairing buffer systems. (a) 0.1 M TEAA, pH 7, ion-pairing system. Mobile phase A: 5% acetonitrile in 100 mM TEAA; mobile phase B: 15% acetonitrile in 100 mM TEAA; gradient begins from 5% acetonitrile at a gradient slope of 0.25% acetonitrile/min. (b) 100 mM HFIP ion-pairing buffer, pH 8.2. Mobile phase A: 10% methanol in 4.1 mM TEA/100 mM HFIP; mobile phase B: 40% methanol in 4.1 mM TEA/100 mM HFIP; gradient begins with 10% methanol at gradient slope of 0.25% methanol/min. (c) 16.3 mM TEA/400 mM HFIP pH 7.9 ion-pairing buffer. Mobile A: 10% methanol in 16.3 mM TEA/400 mM HFIP; mobile phase B: 40% methanol in 16.3 mM TEA/400 mM HFIP; gradient begins at 16% methanol at gradient slope 0.23%. All separations utilized an XTerra MS C18, 2.5 μm, 50 mm × 4.6 mm column. (Courtesy of Waters Corporation.)

added. HFIP or other organic polyfluorinated alcohols, also called additives or organic modifiers, can be added to either polar or nonpolar mobile phases. The use of such organic modifiers leads not just to improvements in separation but also to the extension of silica-based column lifetimes.[74]

McCarthy et al. compared several ion-pairing systems for their usefulness in the separation of a homo-oligonucleotide and hetero-oligonucleotide ladder.[75] For their investigation they included TEAA, TEA/HFIP, dimethylbutylammonium acetate (DMBAA), tripropylammonium acetate (TPAA), TBAA, and HAA as ion-pairing reagents. Resolution of the homo-oligonucleotide ladder improved with increasing concentration and hydrophobicity (alkyl chain length) of the ion-pairing reagent. Separation efficiency (or peak capacity) decreased with oligonucleotide length and more hydrophobic ion-pairing reagents such as HAA started to outperform TEA/HFIP system in resolution of longer oligonucleotides (30- to 35-mers). In the separation of hetero-oligonucleotide ladders ion-pairing systems performed better, which separated based predominantly by a charge-based mechanism. Separation improved from TEAA < DMBAA < TPAA < TEA/HFIP ~ HAA.

McKeown et al. looked into the effect of several different parameters on the retention behavior of a series of poly dT oligonucleotides (5- to 18-mer) under isocratic conditions using RP IP-HPLC.[62] They study the effects of temperature, pH, eluent ionic strength, percentage organic modifier, concentration, and alkyl chain length of the ion-pairing reagent using a Kromasil C18, 100 Å, 5 µm particle-size column (250 mm × 4.6 mm). Reversed-phase chromatography using a 100 mM ammonium acetate buffer resulted in broad co-eluting peaks and incomplete resolution of the individual oligonucleotides from the poly dT mixture, confirming the need of ion-pairing reagents for an effective resolution of oligonucleotides, which was also shown by others.[67] The effect of the hydrophobicity of the alkylammonium ion-pair reagent on retention of the oligonucleotides was investigated for five different ion-pairing reagents: tetramethyl- (TMAB), tetraethyl- (TEAB), tetrapropyl- (TPAB), tetrabutyl- (TBAB), and tetrahexyl- (THAB) ammonium bromide. The retention of the oligonucleotides was directly related to the alkyl chain length of the ion-pairing reagent. With the shortest alkyl chain length ion-pair reagent (TMAB) the analytes were all unretained, but complete retention of all the analytes was observed with the longest alkyl chain length (THAB). The percentage acetonitrile in the mobile phase was observed to be of critical importance in the optimization of the separation.[76–78]

McKeown et al. was able to show that an increase in column temperature caused a decrease in the retention of oligonucleotides, with longer chain length oligonucleotides being more affected by changes in temperature than smaller chain lengths. From pH 4.8 to 6.8, he found a decrease in the retention of all oligonucleotides. This retention time shift was unexpected because the charged backbone is comprised of strong acids with a pK_a value of about 1.[79] Over the entire pH range studied, the phosphodiester groups are therefore fully ionized. He therefore concluded that the proportion of ionized silanol groups was reduced at lower mobile phase pH, resulting in a reduction of negative charge on the surface of the silica-based packing material thereby influencing the separations. Such unreacted acidic silanols are known to be present on most reversed-phase silica materials with a wide variety of pK_a values being reported.[80] Depending on the sorbent, only about 25% of total alkylammonium ions perform ion-pair functions with the other 75% interacting and electrostatically neutralizing residual silanol groups.[81]

McKeown et al. found that increasing the concentration of TBAB from 1.5 to 10 mM resulted in an increase in retention for all the oligonucleotides. At higher concentration, sorbent surface might become saturated with the ion-pairing reagent or micelles are being formed in solution, which can lead to a reduced availability of adsorption sites and hence decreased retention. However, because of TBAB's solubility limit of 10 mM in water, McKeown et al. was not able to extend his studies into higher buffer concentrations using this particular ion-pairing buffer.

An efficient oligonucleotide separation is dependent on the concentration of both triethylamine (TEA) and hexafluoroisopropanol (HFIP).[70] The role of the triethylammonium cation, the active ion-pairing agent, is well understood. An increase in TEA concentration improves ion-pairing efficiency and, consequently, the separation selectivity. A more efficient ion-pairing mechanism also results into an increase in retention time. Because the pKa of TEA is 10.7, a side effect of an increased TEA concentration is a rise of mobile phase pH, which may reduce the lifetime of silica-based columns. However, the hybrid organic–inorganic silica of BEH-based columns is highly stable up to a pH of 12.

HFIP's role on the ion-pairing efficiency of the buffer is less clear. An increase in the HFIP concentration from 100 to 400 mM results in better separation efficiency, but because HFIP is not an active ion-pairing agent, its effect could only be indirect. One possible explanation is that the limited solubility of TEA in aqueous HFIP solutions changes the distribution of TEA between the mobile and stationary phases and forces the triethylammonium ion adsorption on the sorbent surface. This, in turn, enhances the ion-pairing retention mechanism and improves the separation performance. In fact, the most successful ion-pairing system represents the maximum concentration of TEA (16.3 mM) that is soluble in 400 mM HFIP aqueous solution at ambient temperature. This buffer provides for more efficient separation than traditional TEAA ion-pairing buffers, in which longer gradients were required to achieve similar column peak capacity. Gilar et al. showed that only marginal separation was achieved using the TEAA buffer system, whereas baseline resolution of all 19- to 25-mer peaks was obtained using a TEA-HFIP mobile phase.[70] A very important feature of HFIP is that it reduces the impact of oligonucleotide hydrophobicity on retention,[12] which appears to be crucial for the separation success of phosphorothioate oligonucleotides (PS-ONs).

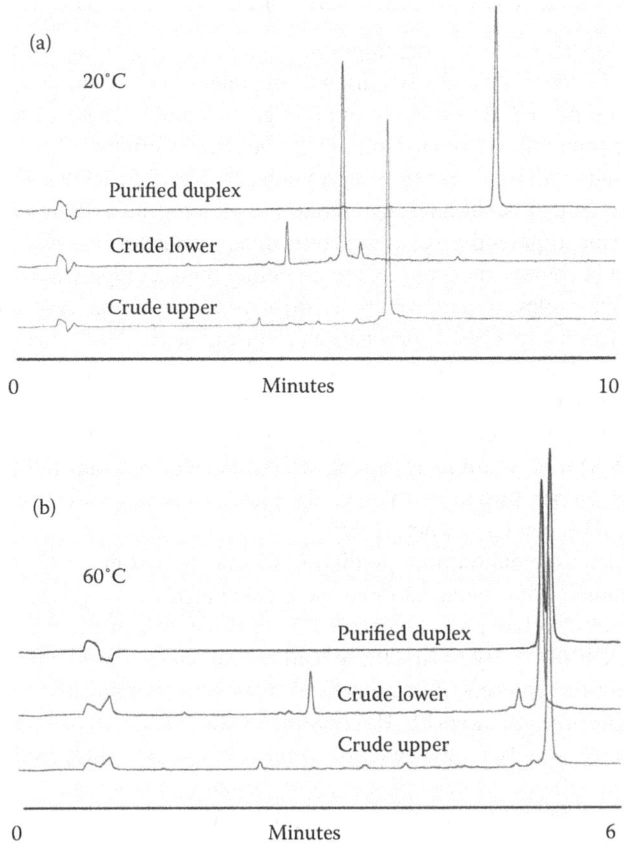

FIGURE 1.9 Comparison of single strand and duplex analysis at 20°C and 60°C. (a) Analysis of single strands and duplex at 20°C; (b) analysis of the single strands and duplex under denaturing (60°C) conditions; Mobile phase A: 25 mM HAA, pH 7.0; mobile phase B: 100% methanol; gradient 30–40% methanol in 10 min; column: Aquity UPLC OST C18. 1.7 µm, 2.1 mm × 50 mm. (Courtesy of Waters Corporation.)

The mobile phase requirements for the analysis of double-stranded oligonucleotides are somewhat different. While there are reports of using the TEA-HFIP ion-pairing system for the analysis of siRNA duplexes,[82,83] because of HFIP denaturing properties, McCarthy et al. relied on TEAA (100 mM) and HAA (25 mM) buffer systems for the analysis and purification of double-stranded RNA and DNA to increase the stability of the duplexes under IP-HPLC conditions.[67] Approaching the melting temperature, dramatic peak broadening occurs, indicating on-column duplex melting. Duplex melting is accompanied by an appearance of complementary oligonucleotides. For this reason, 20°C was selected as a generic separation temperature. While it is well established that retention times of single stranded oligonucleotides are strongly sequence dependent,[12,43,44,50,84] McCarthy et al. found that the retention time of all three double-stranded oligodeoxynucleotides (19-mer dsDNA) used for their investigation were sequence independent. He concluded that IP-HPLC retention of double-stranded oligonucleotides is predominantly driven by charge-to-charge interaction and that dsDNA or siRNA are therefore more retained by the RP-column than their corresponding single strands, making RP IP-HPLC useful for the purification of on-column annealed siRNA. See Figure 1.9 for analytical traces of the crude single strands and the purified duplex at 20°C (non-denaturing conditions; duplex intact) and 60°C (denaturing conditions, duplex elutes as two single strands) using a BEH-based Acquity UPLC OST C18 column (2.1 × 50 mm, 1.7 μm particle size) and a 25 mM HAA buffer system with an acetonitrile gradient.

1.6 RETENTION TIME AND SEPARATION SELECTIVITY PREDICTION MODELS

In HPLC, retention time (RT) is the most important parameter governing the separation of solutes and is often used for the qualitative identification of oligonucleotides. The study of the relationship between the retention time and the sequence of an oligonucleotide can be a useful tool to optimize the conditions for the separation of a particular oligonucleotides mixture.[85]

The commonly used method, linear free energy relationship (LFER) describing the behavior of solute molecules at the liquid–solid interface, models retention time as a sum of individual energy contributions (dispersion, dipole–dipole, π–π, proton donor–acceptor interactions, etc.).[86] However, this prediction model becomes inaccurate when modeling more complex molecules such as oligonucleotides because their relevant parameters are difficult to determine. Alternatively, quantitative structure retention relationship (QSRR) provides a promising method for retention time predictions. Gilar et al. developed models by simple summation of the retention contributions of the individual nucleotides obtained from experimentally determined homo-oligonucleotides.[12] Their model was based only on oligonucleotide length and base composition. Huber and coworkers used support vector regression (SVR) to develop their model, which included oligonucleotides having a length of 15–48 over a wide temperature range. The model took into consideration information of length, sequence, and predicted secondary structure[43,44]

A different approach to retention time prediction was taken by Lei et al.[84] Base sequence autocorrelation (BSA) features for oligonucleotides were calculated by weighting constitutional, topological, geometrical, electrostatic, and quantum-chemical features of the four bases (A, T, C, and G) obtained from CODESSA.[87] By having these features calculated based only on sequence, all the oligonucleotides could be represented in numerical form and optimum models were obtained by employing multiple linear regression (MLR) combined with genetic algorithm (GA) feature selection. The derived linear models showed equally good performance compared to works by Huber and coworkers,[43,44] but without the need of secondary structure prediction. A novel strategy to predict the retention time at any temperature was also proposed.

Gilar et al. developed a separation selectivity (or peak capacity) model for oligonucleotides and compared it to empirical data derived from IP-HPLC analyses of oligonucleotide ladders using a BEH-based Acquity UPLC column (50 mm × 2.1 mm, 1.7 μm) and TEAA or TEA/HFIP buffer systems.[71] They showed that the overall sample peak capacity is nothing but the sum (or the integral) of the resolutions in the HPLC chromatogram. The position of the peak capacity maximum is rather insensitive to the molecular weight of the oligonucleotide. The peak capacity model was developed for homo-oligonucleotides. Separation of hetero-oligonucleotides partially depends on their sequence. Also, peak capacity cannot be reliably calculated for oligonucleotides with strong secondary structure.

The retention factor B is dependent on the molecular weight of the oligonucleotide and the logarithm of this factor B varies linearly with the logarithm oligonucleotide's molecular weight. When plotting the logarithm of oligonucleotide molecular weight as a function of the logarithm of the retention factor, the resulting line was shown to correlate closely with experimental data. The intercept and the slope of the equation may change with temperature, the type of stationary phase, organic modifier (acetonitrile, methanol, isopropanol, etc.), and the nature of the ion-pairing system.

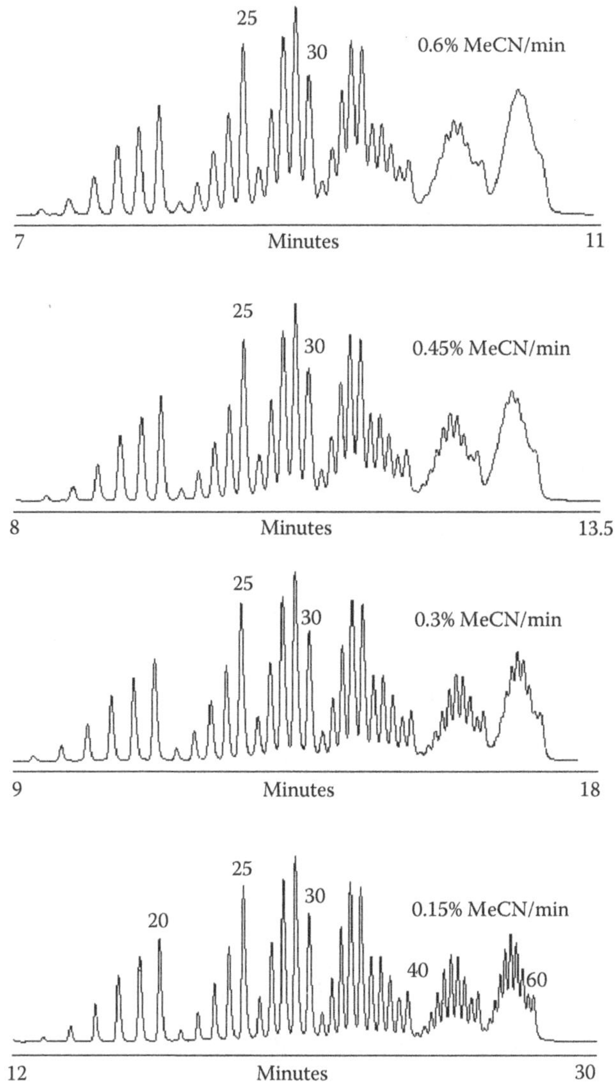

FIGURE 1.10 Decreasing gradient slope increases resolution but negatively impacts analysis run time. Fifteen- to sixty-mer oligodeoxythymidine separation using different gradient slopes; Mobile phase A: 100 mM TEAA; mobile phase B: 20% acetonitrile in 100 mM TEAA; gradient was 40–70% B; Column: Acquity OST C18, 1.7 μm, 2.1 mm × 50 mm at 60°C. (Courtesy of Waters Corporation.)

It was shown theoretically and experimentally that the slope is much steeper for the TEA/HFIP than with the TEAA ion-pairing system, which means that TEA/HFIP is a more efficient system for the separation of oligonucleotides than TEAA.[12,50,71]

Gilar et al. investigated the impact of sorbent particle size, column length, and gradient time on the retention factor and compared theoretical to experimental data.[71] Not surprisingly, the best peak capacity was obtained for the column packed with the smallest particle size sorbent. Intriguingly, gains in resolution for longer columns are not as pronounced as one might expect. At constant gradient run time the gradient slope is proportionally shallower for shorter columns. In other words, the peak capacity of longer columns is reduced by proportionally sharper gradient, which tends to

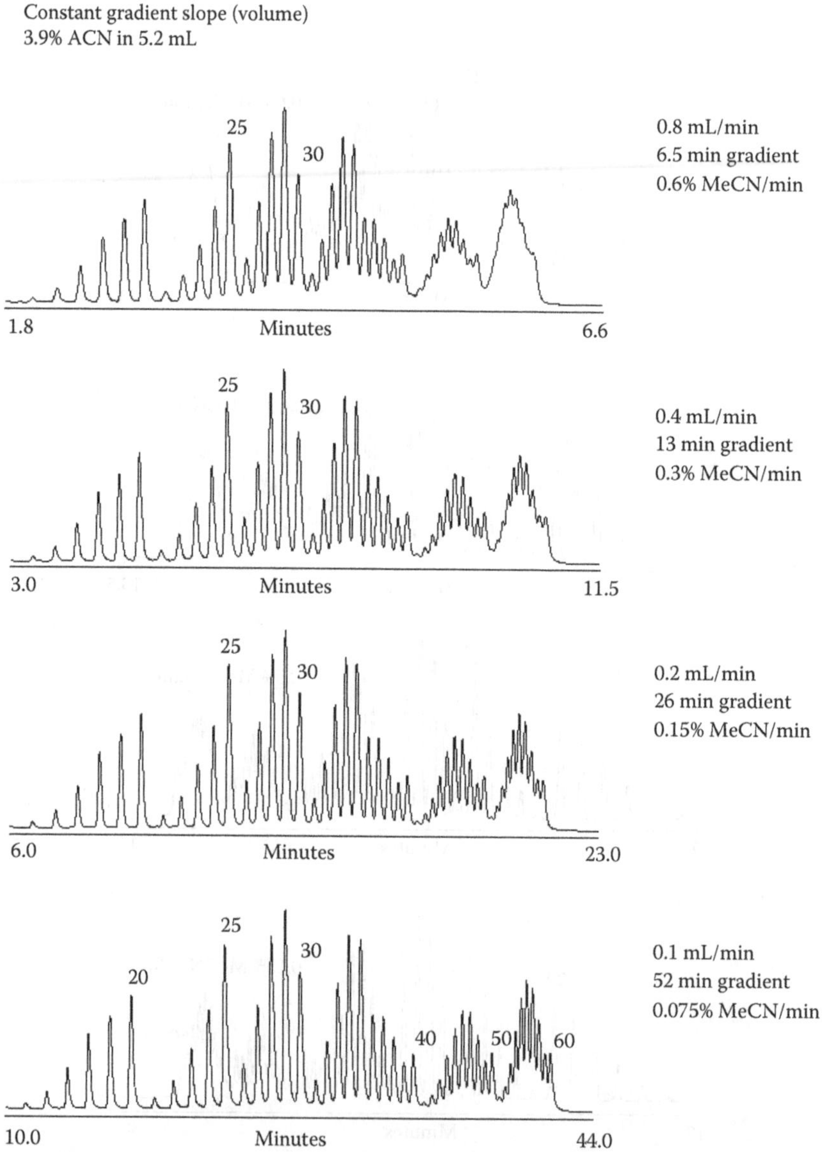

FIGURE 1.11 Maintaining most of the resolution while decreasing analysis time by increasing the flow rate and proportionally reducing the gradient time. Fifteen- to sixty-mer oligodeoxythymidine separation using different gradient slopes; Mobile phase A: 100 mM TEAA; mobile phase B: 20% acetonitrile in 100 mM TEAA; gradient was 40–70% B; Column: Acquity OST C18, 1.7 μm, 2.1 mm × 50 mm at 60°C. (Courtesy of Waters Corporation.)

reduce or eliminate the positive impact of higher column efficiency. The full benefits of longer columns in a gradient separation are only realized when changing the gradient duration in proportion with the column volume (length).

IP RP-HPLC analysis of oligonucleotides is typically performed with shallow gradients. Decreasing gradient slope increases resolution, but negatively impacts analysis throughput by increasing run time (see Figure 1.10).

While an increase in the flow rate decreases the separation efficiency, the resulting loss in peak capacity is less detrimental compared to using sharper gradients. Therefore, for the fast analysis of oligonucleotides it is more practical to maintain a relatively shallow gradient and reduce the analysis time by increasing the flow rate and gradient time proportionally thereby maintaining a constant gradient slope. The number of column volumes remains constant. The separation selectivity remains unchanged with only some loss of resolution (see Figure 1.11).

1.7 SELECTED PRACTICAL EXAMPLES OF IP-HPLC

The choice of column, mobile phase composition, temperature, instrument, and the method of sample preparation are all critical parameters for an effective purity analysis of oligonucleotides. In the following section, the authors have attempted to compile a broad range of spectral data from their own work to demonstrate the use of IP-HPLC as a powerful tool for the analysis of oligonucleotides. These examples aptly illustrate the uniqueness of each oligonucleotide sequence. In our experience, there is no "one size fits all" approach to separation and analysis of oligonucleotides, and this is demonstrated by the necessity of screening a series of columns and mobile phases for use with each sequence of interest.

In the course of our work, we routinely synthesize and characterize sequences belonging to a broad range of oligonucleotide subclasses, such as antisense, immunostimulatory oligonucleotides, aptamers, small interfering RNA (siRNA), microRNA (miRNA), decoys, and splice modulators. The following list gives a brief overview of the kind of modifications that are routinely incorporated into oligonucleotides when making the above mentioned subclasses: phosphodiester and phosphorothioate DNA, duplex and single strand RNA, 2′-modified RNA, LNA, gapmers, chimeric sequences, conjugates, aptamers, PEGylated oligonucleotides, backbone modified sequences, and sequences containing modified or unnatural nucleoside bases. The analysis of modified oligonucleotides, particularly ones bearing lipophilic groups such as cholesterol or long chain fatty acid esters, sometimes require the use of C4 or C8 columns instead of the standard C18 reverse phased column. For the synthesis of duplex RNA, denaturing as well as nondenaturing methods must be available for characterization of the duplex. Large-scale manufacturing of oligonucleotides typically relies on preparative anion exchange purification for several reasons. Preparative anion exchange (AEX) chromatography is generally more efficient than preparative IP RP-HPLC; it converts the oligonucleotide into the sodium form during purification, and it can be performed using low-pressure HPLC equipment. However, the purified oligonucleotide elutes under high salt conditions. Pools or fractions containing high salt levels are sometimes difficult to analyze and require sample preparation prior to IP-HPLC analysis. Some techniques for improved chromatography of high-salt samples are included below. While it is understood that method optimization will be essential for each new sequence, the following sections should be helpful in the selection of parameters that one must consider at the initial stages of IP-HPLC method development.

1.7.1 Column Influence on Chromatography

The delivery of therapeutic oligonucleotides to their desired target is an enormous challenge being addressed in a number of different ways. Increasing the lipophilicity of highly negatively charged oligonucleotides to pass through densely hydrophobic cell membranes, thereby improving their pharmacokinetic properties can be accomplished by attachment of a lipophilic group to either terminus

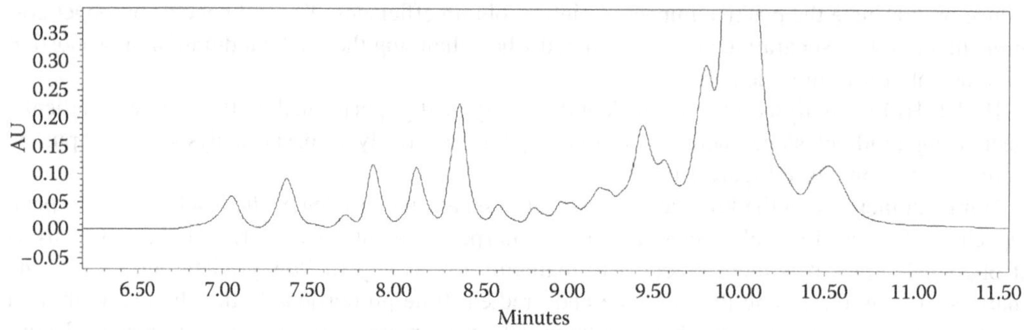

FIGURE 1.12 Separation of a crude RNA 39-mer using XTerra C18, 2.5 μm, 4.6 mm × 50 mm. Mobile phase A: 100 mM HFIP, 7 mM TEA; mobile phase B: methanol.

of the oligonucleotide to form a so-called conjugate. The following example highlights the marked difference in chromatography of a 40-mer RNA/2′-*O*-methyl/2′-fluoro RNA oligonucleotide bearing a cholesterol group at the 5′-terminus compared to the 39-mer prior to conjugation. Increased lipophilicity associated with the cholesterol moiety caused the oligonucleotide to be highly retained on a C18 stationary phase, resulting in extremely poor separation. Lipophilic conjugates particularly in combination with longer sequences often show largely increased retention on RP columns mandating the use of shorter carbon chain stationary phases, such as C4, C8, or C12 columns. C18 columns, however, are better suited to separate a broad range of oligonucleotides prior to the conjugation step. Shown in Figures 1.12 and 1.13 is a comparison of the chromatography of a 39-mer RNA oligonucleotide before and after conjugation with cholesterol.

Note the improved peak shape and resolution of the impurities and full length product (FLP) in the separation using the XTerra C18 column (Figure 1.12) compared to the ACE-3 C4 column (Figure 1.13), indicating the superiority of the C18 column for the separation of the unconjugated 39-mer oligonucleotide. After the conjugation step, however, the C4 column is much better suited for the HPLC analysis of the conjugated oligonucleotide when compared to the C18 column (see Figure 1.14).

Figure 1.14 exemplifies the difficulties encountered when analyzing conjugated oligonucleotides. Failure sequences are not well resolved from the main peak and peak broadening occurs. The chromatography of the conjugate using the C4 column shows a much improved separation. The impurity peaks are well resolved from the main peak (see Figure 1.15).

A C18 column is more effective in separating an unconjugated oligonucleotide in comparison to a C4 or C8 column. When selecting a C18 column, there are a wide variety of columns that can be

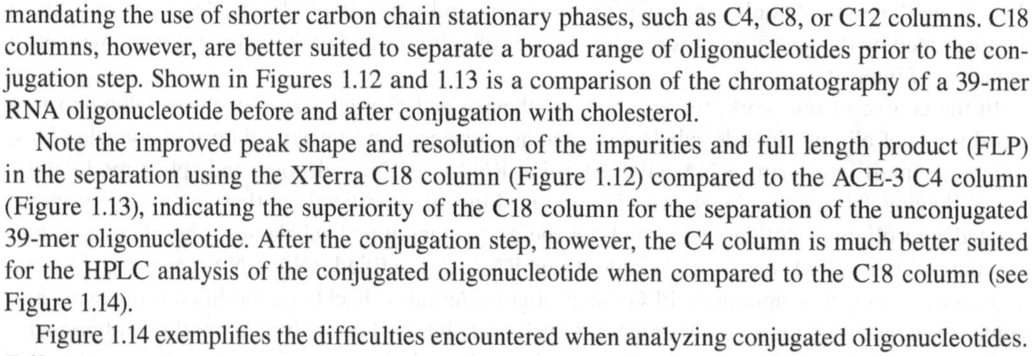

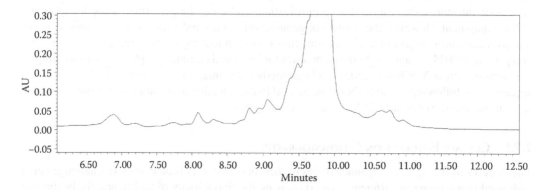

FIGURE 1.13 Separation of a crude RNA 39-mer using ACE-3 C4, 3.5 μm, 2.1 mm × 150 mm. Mobile phase A: 200 mM HFIP, 8 mM TEA, 5% methanol; mobile phase B: 200 mM HFIP, 8 mM TEA, 90% methanol.

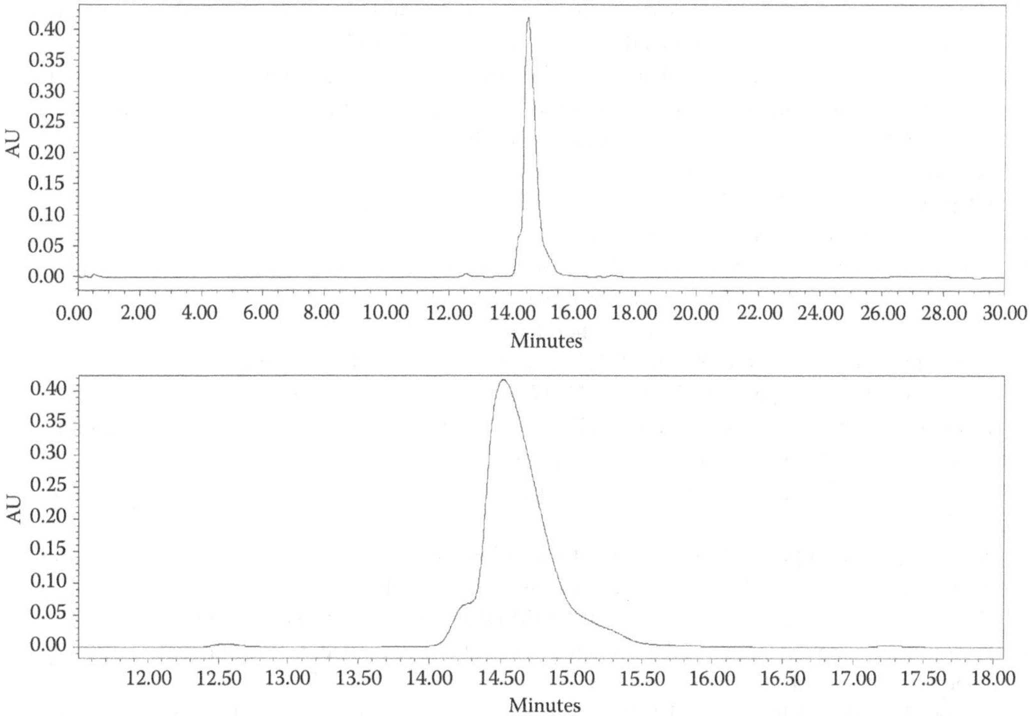

FIGURE 1.14 Separation of a crude RNA 40-mer cholesterol conjugate using XTerra C18, 2.5 μm, 4.6 mm × 50 mm. Mobile phase A: 100 mM HFIP, 7 mM TEA; mobile phase B: methanol.

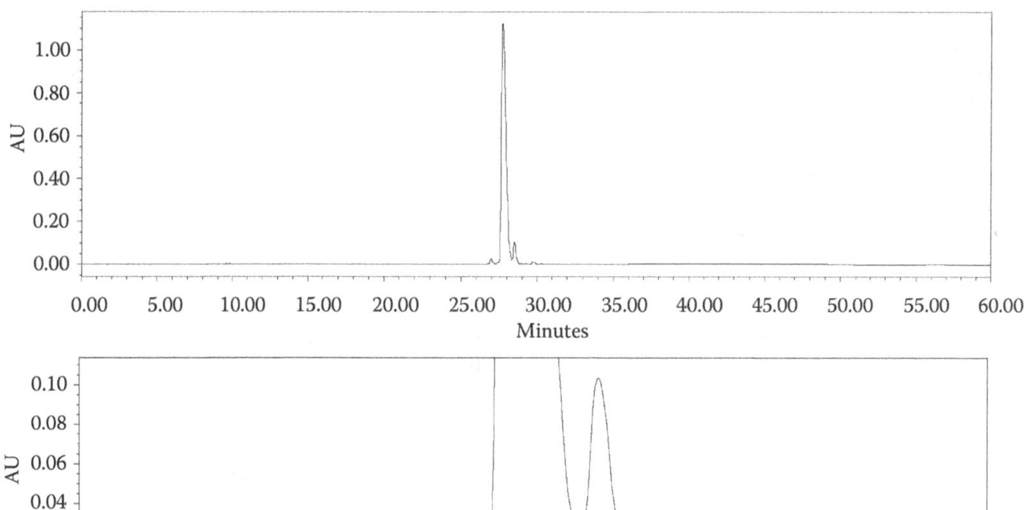

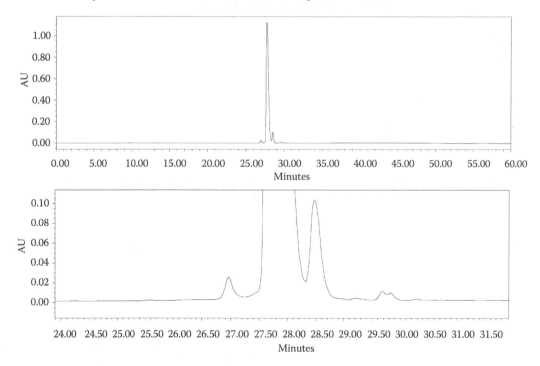

FIGURE 1.15 Separation of a crude RNA 40-mer cholesterol conjugate using ACE-3 C4, 3.5 μm, 2.1 mm × 150 mm. Mobile phase A: 200 mM HFIP, 8 mM TEA, 5% methanol; mobile phase B: 200 mM HFIP, 8 mM TEA, 90% methanol.

chosen from (see Section 1.3). In addition to varying column dimensions and particle sizes, many column manufacturers have introduced modifications to their sorbents that can result in an altered separation of the oligonucleotide of interest. The following examples will demonstrate how using columns of varying dimensions and sorbent modifications for the analysis of a 2′-*O*-methyl phosphorothioate RNA 20-mer can produce drastically different results. In each example, the mobile phase and gradient were adjusted to optimize the separation of the failure sequences from the full length product. (Note: In order to verify impurity resolution, the reference sample used for this investigation was spiked with approximately 3% of the (N–2), (N–1), and (N+1) failure sequences to produce a second reference sample. By overlaying the chromatograms from the reference and spiked sample the quality of the separation was then verified.)

The following chromatograms were generated using a Waters XTerra C18 2.5 µm, 4.6 mm × 50 mm column (100 mM HFIP, 7 mM TEA, ACN gradient). The N–3 impurity is only partially resolved from the main peak, while the N–2, N–1, and N+1 failure sequences fall under the main peak. Owing to this co-elution, the overall purity of the FLP cannot be accurately quantitated. The sorbent particle size and column dimensions of the Waters Xterra column are insufficient to achieve the theoretical plates needed for separating all impurities from the full length product (see Figure 1.16).

In an attempt to improve the resolution of the failure sequences, the Xterra column was replaced with a Waters XBridge OST C18 2.5 µm, 2.1 mm × 50 mm column, which also had a reduced internal diameter, while the buffer conditions (100 mM HFIP, 7mM TEA) were maintained. The change of the column improved the overall chromatography of the sequence, now completely resolving the N–3 and partially resolving the N–2 and N+1 failure sequences (see Figure 1.17).

While the Waters XBridge OST C18 2.5 µm, 2.1 mm × 50 mm improved the overall resolution of this particular sequence when compared to the Waters Xterra C18 2.5 µm, 4.6 mm × 50 mm, the calculated purity was still not completely accurate due to the partial resolution of the N–2 and N+1 peaks and co-elution of the N–1 with the full length product.

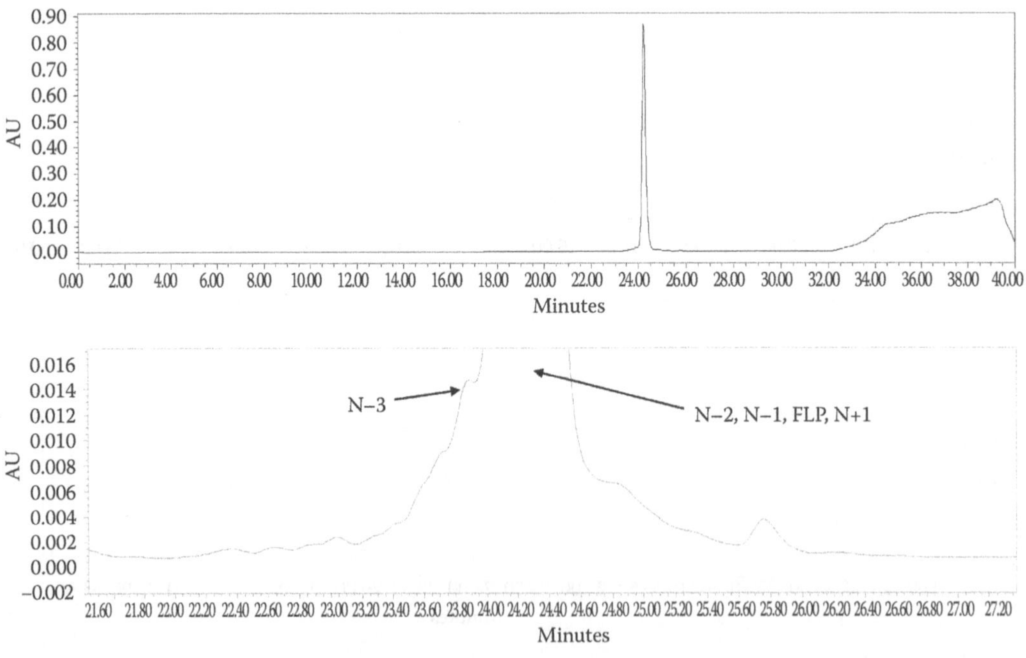

FIGURE 1.16 Chromatography of a 2′-*O*-methyl phosphorothioate RNA 20-mer using a Waters Xterra C18 column (with expansion). Mobile phase A: 400 mM HFIP, 15 mM TEA; mobile phase B: acetonitrile.

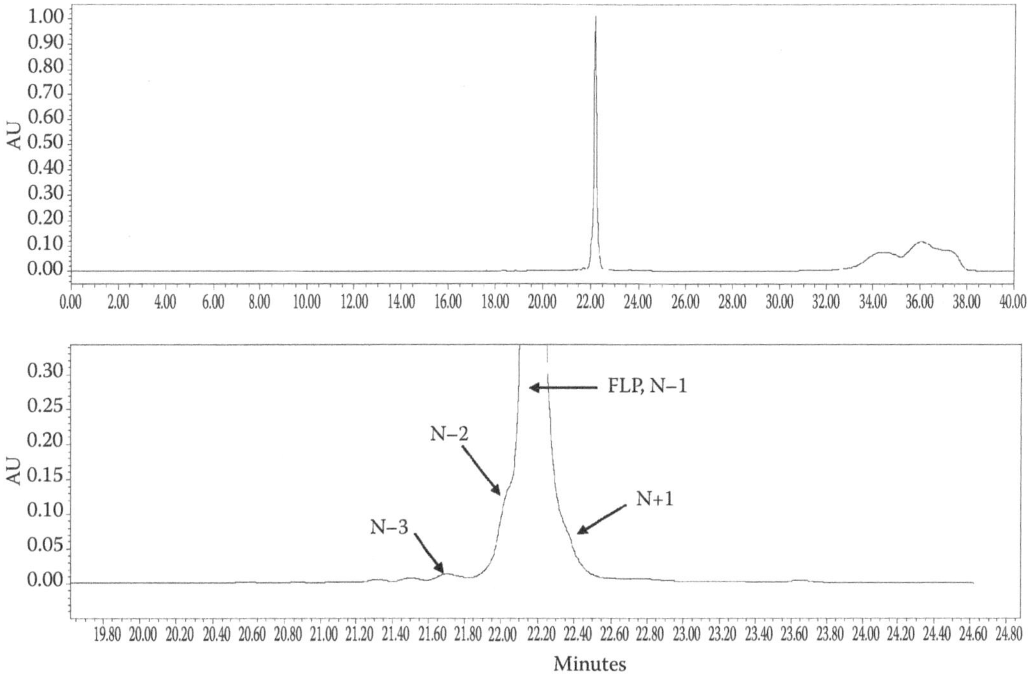

FIGURE 1.17 Chromatography of a 2′-*O*-methyl phosphorothioate RNA 20-mer using a Waters XBridge OST C18 column (with expansion). Mobile phase A: 100 mM HFIP, 7 mM TEA; mobile phase B: acetonitrile.

In an attempt to continue the trend of increasing impurity resolution, the particle size of the sorbent was reduced resulting in an increase of theoretical plates (see Section 1.3). Unfortunately, when using columns of particle sizes less than 2.5 μm, the back pressure on the column tends to exceed the pressure capabilities of a standard HPLC instrument. Typically, when running a flow rate of approximately 0.3 mL/min on a 1.7 μm column, the back pressure is between 4000 and 5500 psi (pressure is dependent on the column temperature and mobile phase being used). A standard HPLC instrument is not capable of handling back pressures of this intensity. Therefore, columns with particle sizes less than 2 μm need to be run on an ultra performance liquid chromatography instrument (UPLC), which is capable of handling back pressures up to 15,000 psi. The following chromatograms were taken from an analysis using a Waters Acquity BEH C18 1.7 μm, 2.1 mm × 50 mm and a UPLC system (100 mM HFIP, 7 mM TEA) (see Figure 1.18).

In this case, the N–1 and N+1 impurities are beginning to resolve from the main peak, thus allowing a more accurate calculation of the main peak purity. While not completely resolved from the main peak, the resolution of N–1 and N+1 in this profile is again superior to the previous example that used a column of increased particle size. These examples demonstrate how changing the column type and reducing the internal diameter and particle size of a column can lead to a more efficient separation of the failure sequences from the full length product of a nonconjugated oligonucleotide.

In addition to comparing the effects of column dimensions and particle sizes on the analysis of a particular compound, it can also be useful to compare the performance of different column manufacturers and their various solid support modifications. The following examples demonstrate that while different column manufacturers can produce columns of identical dimension and particle size, their resultant chromatography can be different enough to justify superiority between them.

The following analyses were performed under optimized conditions, which lead to resolution of N–1 and N+1 failures of a 20-mer 2′-*O*-methyl phosphorothioate RNA sequence. With the

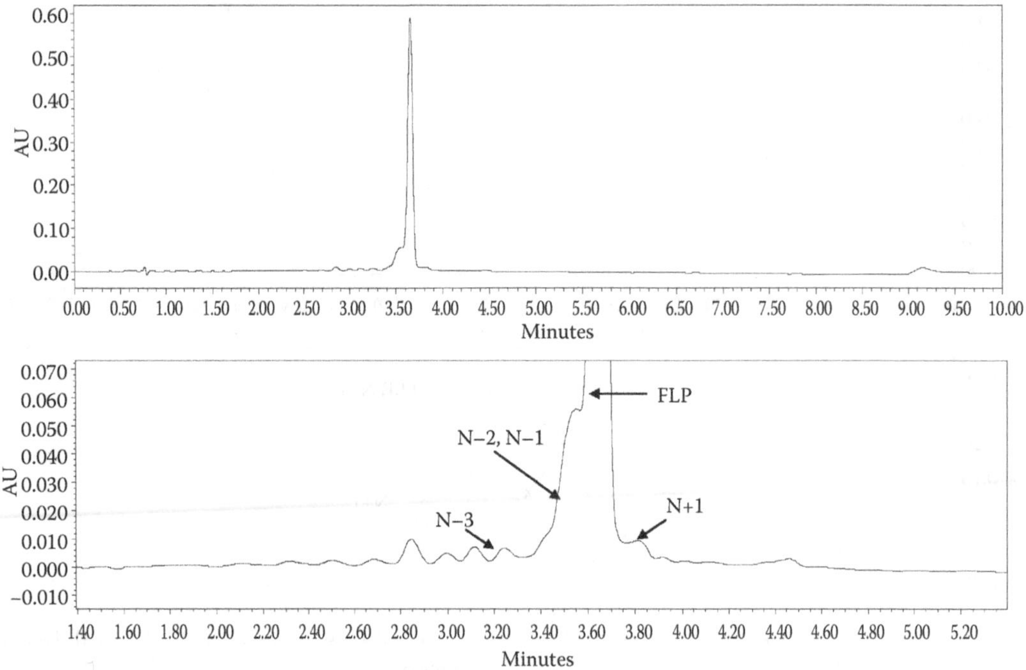

FIGURE 1.18 Chromatography of a 2′-*O*-methyl phosphorothioate RNA 20-mer using a Waters Acquity BEH C18 column (with expansion). Mobile phase A: 400 mM HFIP, 15 mM TEA; mobile phase B: acetonitrile.

separation of this particular sequence optimized, there was an opportunity to directly compare the performance of different column manufacturers and modified solid supports. Under optimized conditions, we achieved sufficient separation (resolution > 1.0) of the N–1 and N+1 failure sequences from the full length product using a Waters Acquity BEH C18 1.7 μm, 2.1 mm × 50 mm column and hexylammonium acetate as the ion-pairing agent. In order to investigate how other column manufacturers compared to the Waters Acquity BEH C18 1.7 μm, 2.1 mm × 50 mm, we selected columns of identical dimensions, and then analyzed the same sample using identical method conditions. The first example compares the performance of a Phenomenex Kinetex C18 1.7 μm, 2.1 mm × 50 mm column to the Waters Acquity BEH C18 1.7 μm, 2.1 mm × 50 mm (see Figure 1.19).

Under these conditions the Waters Acquity column gives a FLP retention time of about 3.50 min, while the Phenomenex Kinetex column shows a FLP retention time of about 2.80 min. In this case, the dimensions and particle sizes of both columns are identical; however, each uses a different particle design that has the potential to lead to different chromatography (see Section 1.3). After performing the analyses, it was concluded that neither column provides a significant advantage over the other in terms of overall separation of the impurity profile. In both cases, N–1 and N+1 resolution is evident, and the overall impurity profile is similar. The only noticeable difference in performance is the resulting retention times of the FLP for each column. The Phenomenex Kinetex produces an FLP retention time of 2.80 min, while the Waters Acquity produces an FLP retention time of 3.43 min. While the difference in retention time is only 0.63 min, the ability of the Waters Acquity to retain the sequence for a slightly longer period under these conditions results in a slightly improved resolution of the N–1 failure sequence when compared to the Phenomenex Kinetex. For the Phenomenex Kinetex, the resolution of N–1 is 1.05. For the Waters Acquity, the resolution of N–1 is 1.29. While this difference in resolution isn't profound, it can prove important when analyzing rather pure oligonucleotides that can produce N–1 levels less than 1.0%.

The previous example compared the chromatography of two columns (Waters Acquity and Phenomenex Kinetex) that had the same dimensions and particle size. These columns also operated

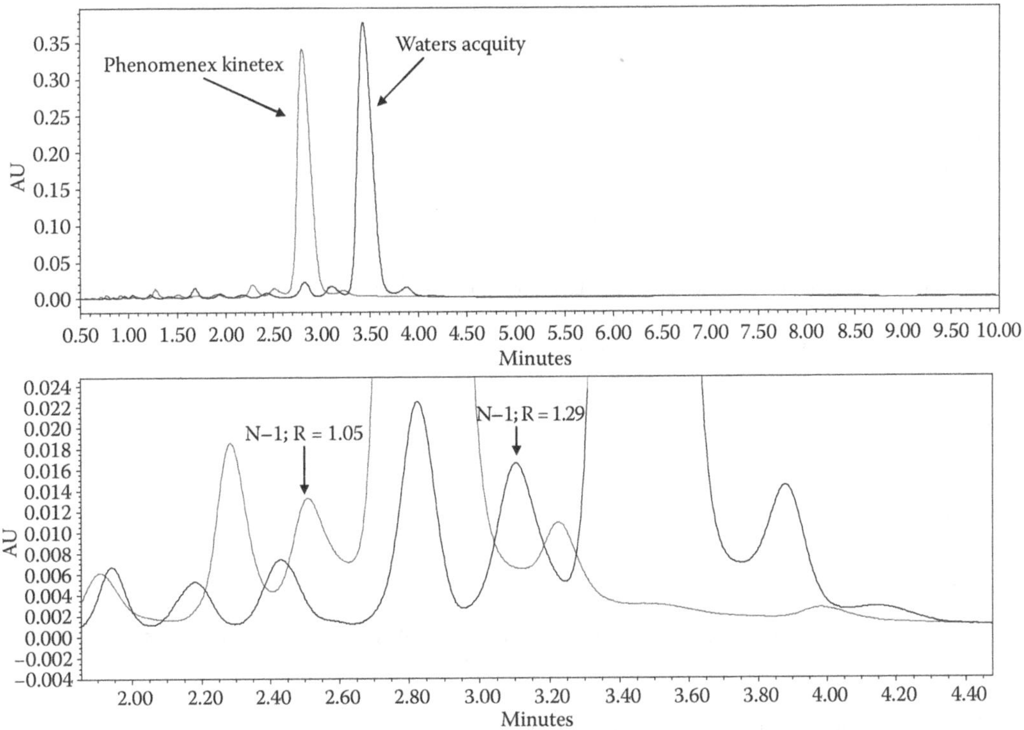

FIGURE 1.19 Chromatography of a 2′-*O*-methyl phosphorothioate RNA 20-mer using Waters Acquity and Phenomnex Kinetex C18 columns. Mobile phase A: 15 mM HAA, 7.0% methanol, 3.0% acetonitrile; mobile phase B: 70% methanol, 30% acetonitrile.

under the same mechanism, that being a hydrophobic interaction between the oligonucleotide of interest and the C18 chain attached to the solid support. The C18 column is currently the preferred phase for chromatographic oligonucleotide analysis; however, various column manufacturers are exploring other phase options that might be the next preferred method in oligonucleotide analysis. One such alternative that is currently being explored is a pentafluorophenyl (PFP) phase, which, similar to the C18 column, interacts via hydrophobic interactions (for details, see Section 1.4). In order to compare how the PFP phase performs in comparison to a C18 phase, the same sequence from the example in Figure 1.19 was run under identical conditions using a Phenomenex Kinetex PFP 1.7 μm, 2.1 mm × 50 mm instead of the Kinetex C18 1.7 μm, 2.1 mm × 50 mm (see Figure 1.20).

On the basis of the overlaid chromatograms of these two analyses (at full scale), the same dimensions and particle sizes of these two columns leads to a nearly identical retention of the oligonucleotide of interest. However, when zooming in on the region of the full length product, it's obvious that there are substantial differences in the interactions of these two phases with this oligonucleotide. Under these conditions, the N–2, N–1, and N+1 failures are well resolved when using the C18 phase. In contrast, when using the column with the PFP phase, the N–2 and N–1 failure sequences co-elute and are not well resolved from the full length product, and the N+1 failure elutes as a shoulder off of the back side of the main peak. In addition to failing to completely resolve the N–1 from the main peak, the failure sequences are not well resolved when using the PFP phase. This can prove detrimental when analyzing samples of low concentrations because the broadening of the impurity profile can cause low level impurities to be lost in the baseline noise, thus leading to an inaccurate depiction of the entire profile of the sample. When analyzing this 2′-*O*-methyl phosphorothioate RNA 20-mer sequence, it is obvious that the PFP column does not resolve the impurity profile as well as the C18 phase, in spite of their similar selection mechanism.

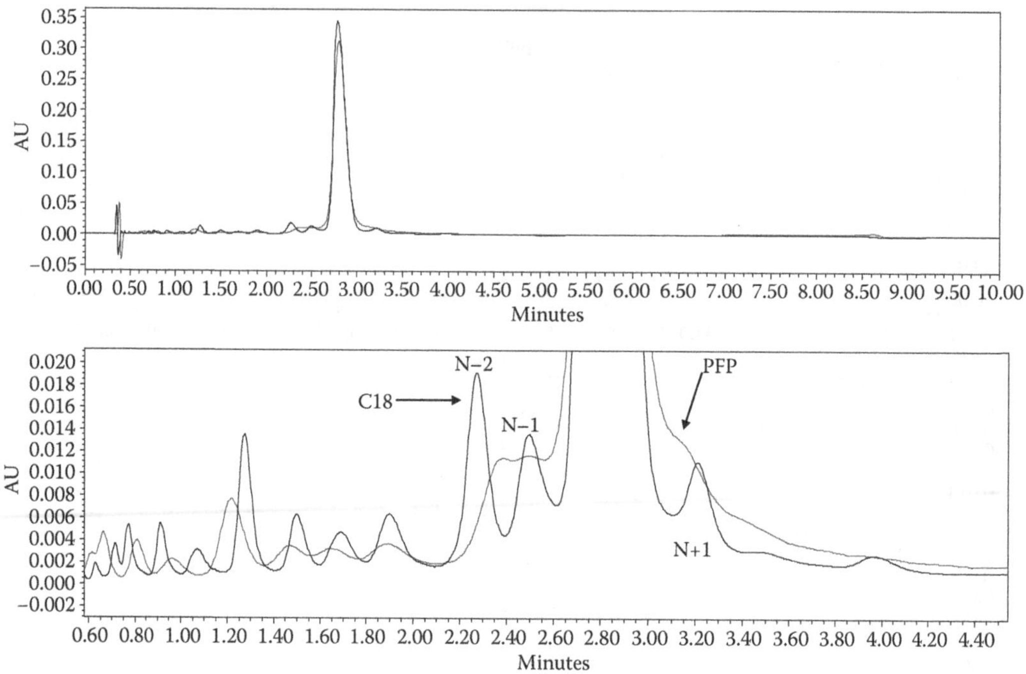

FIGURE 1.20 Chromatography of a 2′-*O*-methyl phosphorothioate RNA 20-mer using a Phenomenex Kinetex C18 and a Phenomenex Kinetex PFP column of identical dimensions (1.7 μm, 2.1 mm × 50 mm). Mobile phase A: 15 mM HAA, 7.0% methanol, 3.0% acetonitrile; mobile phase B: 70% methanol, 30% acetonitrile.

1.7.2 Mobile Phase Influence on Chromatography

As described in the theoretical portion of this chapter, selection of mobile phase, specifically selection of the ion-pairing agent, is just as influential as the selection of the analytical column when analyzing the oligonucleotide of interest. In our experience, the ion-pairing agents of HFIP/TEA and HAA have predominately been the most effective in separating oligonucleotides. In the previous section of this chapter (Section 1.7.1), there was a sequential demonstration of improved impurity resolution on a particular 2′-*O*-methyl phosphorothioate RNA sequence (see Figures 1.16, 1.17, and 1.18). In spite of this increasing trend of impurity profile resolution in these examples, the N–1 and N+1 failure sequences were still not fully resolved from the main peak, leaving an uncertainty in the overall purity calculation. In an attempt to further separate the compound of interest from the failures without increasing the theoretical plates, the ion-pairing agents used in the mobile phases were changed from HFIP/TEA (100 mM HFIP, 7 mM TEA) to hexylammonium acetate (15 mM HAA). The conditions were optimized leading to the following example (see Figure 1.21).

Now the N–1 and N+1 impurities are fully resolved from the main peak, leaving little uncertainty in the purity of the main peak. In this instance, changing the ion-pairing agent in the mobile phase was the key for the complete resolution of the N–1 and N+1 impurities from the full length product. While changing the ion-pairing agent from HFIP/TEA to HAA resulted into a complete resolution of the failure sequences of this particular sequence, it should be noted that HAA is not a better ion-pairing reagent than HFIP/TEA for the analysis of oligonucleotides per se. This example simply demonstrates how ion-pairing agents can have different selectivities with oligonucleotides, thus producing different chromatographic separations.

When selecting ion-pairing agents for chromatographic analyses, one must not only evaluate its ability to separate a sequence. If there is an interest in using MS to delineate structural information, in addition to its chromatographic capabilities, one must also consider the compatibility of the

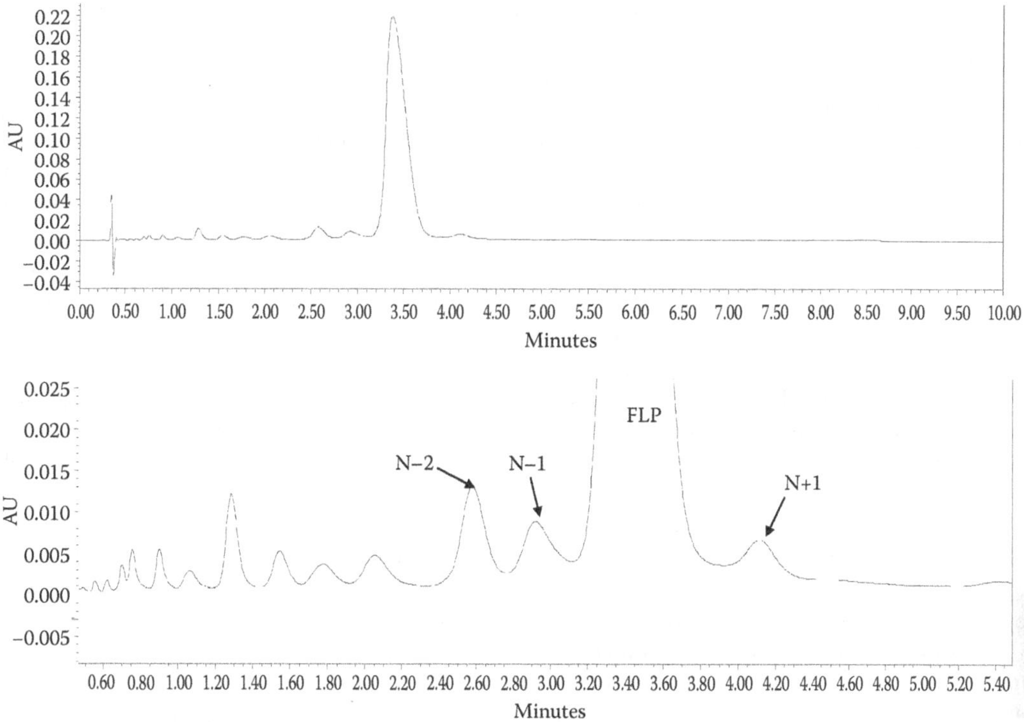

FIGURE 1.21 Chromatography of a 2′-*O*-methyl phosphorothioate RNA 20-mer using a Waters Acquity BEH UPLC C18 column. Mobile phase A: 15 mM HAA, 7.0% methanol, 3.0% acetonitrile; mobile phase B: 70% methanol, 30% acetonitrile.

selected mobile phase with mass spectrometry. Mass spectrometry plays an important role in the process development of an oligonucleotide owing to its ability to identify any impurities or failure sequences that are produced during a synthetic process. When coupled with HPLC (or UPLC), mass spectrometry allows the analyst to directly correlate a chromatographic peak with its atomic mass, thus allowing quantitation and identification of impurities in a sample's profile. Selection of the ion-pairing agent is important when using mass spectrometry. Ion-pairing agents suppress or help ionization of the analyte to different extends resulting into different intensities of the mass signal in the mass spectrometer. In the previous example (see Figure 1.21), when hexylammonium acetate was used, it led to a complete resolution of all impurities for this RNA sequence. However, when this method was transferred to LC/MS, mass identification of low-level impurities was greatly diminished owing to this suppression at the typical column loads of 10 pmol/μL. However, when analyzing the same sequence using an HFIP/TEA buffer system, all low-level impurities were easily detectable. The phenomenon of ion suppression needs to be taken into consideration when selecting an analytical HPLC method and is discussed in greater detail in the mass spectrometry chapter (see Chapter 4). While a particular ion-pairing agent may lead to improved resolution of an oligonucleotide's UV profile, it may decrease the resolution of mass signal in the total ion count (TIC), thus preventing the analyst from learning valuable information about his or her sample and synthetic process.

1.7.3 HPLC versus UPLC

As a result of the continual evolution of therapeutic oligonucleotides in the pharmaceutical industry, demands for the production and impurity characterization of oligonucleotides continues to increase. While the typical HPLC system is capable of adequately analyzing a majority of oligonucleotides, the complexity of highly modified oligonucleotides has stretched the capabilities of high-pressure liquid chromatography, and in some cases, prevented complete impurity characterization. As discussed in Section 1.7.1, decreasing the particle size of columns to sub 2 μm typically leads to an increased resolution of the impurity profiles of oligonucleotides. However, most HPLC instruments are incapable of handling the high back pressures produced when using columns with particles of this size. In response to such needs, instrument manufacturers introduced ultra high-pressure liquid chromatography (UPLC) systems that are capable of handling such back pressures (see Section 1.3 for details).

Ultra high-pressure liquid chromatography systems have a number of advantages when compared to standard HPLC systems. The first, and likely most important, advantage is its ability to increase the resolution of the oligonucleotide profile. An example of such a case can be seen in Figures 1.18 and 1.21 above. As an oligonucleotide successfully travels through clinical trials, the demand for impurity characterization increases. In many cases, the complexity of the sequence and the similarity of the impurities in comparison to the FLP do not permit a standard HPLC from adequately separating the impurity profile. Because of this, the UPLC system appears to be the most effective instrument in chromatographic separation of oligonucleotides.

In addition to improving the overall resolution of an oligonucleotide's impurity profile, the UPLC system is capable of analyzing samples in a fraction of the time that HPLCs require. The following examples depict the analysis of a 2′-*O*-methyl phosphorothioate RNA 20-mer sequence that was optimized using both HPLC and UPLC.

Figures 1.22 and 1.23 depict two analyses of the same 20-mer oligonucleotide sample using optimized HPLC and UPLC methods. As shown above, the HPLC analysis requires a run time of 20 min, while the UPLC analysis requires only 6 min. In spite of the reduced run time of the UPLC

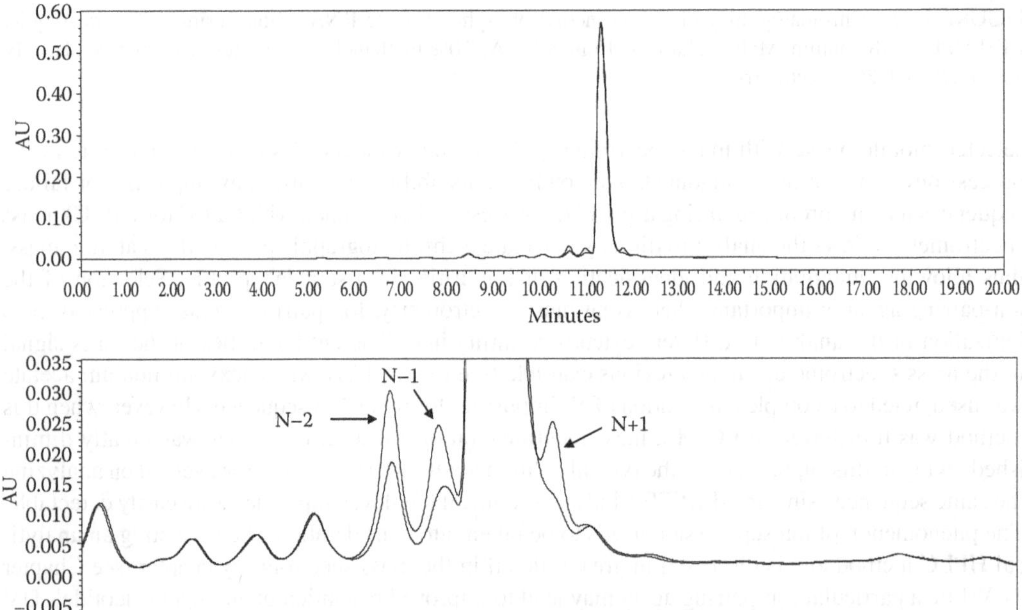

FIGURE 1.22 HPLC analysis of a 2′-*O*-methyl phosphorothioate RNA 20-mer (three percent impurity spike of N−2, N−1, and N + 1 overlayed with reference to demonstrate failure sequence resolution). Mobile phase A: 15 mM HAA, 7.0% methanol, 3.0% acetonitrile; mobile phase B: 70% methanol, 30% acetonitrile; column: Waters Acquity UPLC system with a Waters Acquity BEH UPLC C18 1.7 μm, 2.1 mm × 50 mm (dotted line 3% spike of N−2, N−1, and N+1).

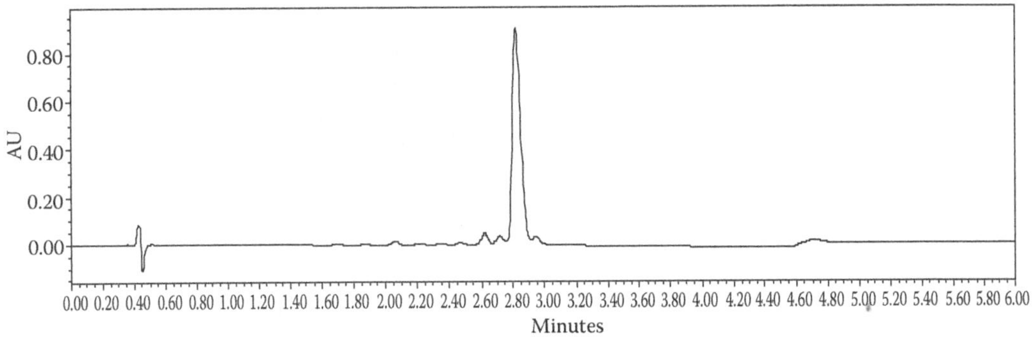

FIGURE 1.23 UPLC analysis of a 2′-*O*-methyl phosphorothioate RNA 20-mer. Mobile phase A: 15 mM HAA, 7.0% methanol, 3.0% acetonitrile; mobile phase B: 70% methanol, 30% acetonitrile; column: Waters Acquity UPLC system with a Waters Acquity BEH UPLC C18 1.7 μm, 2.1 mm × 50 mm.

method, impurity resolution of the N–1 and N+1 failure sequences is maintained. The reduced run times of the UPLC system can prove invaluable when analyzing in process samples that require short turnaround times such as deprotection and stability indicating samples. In addition, at times of high sample volumes, such as during fraction analysis following purification, a fast analytical method is of great importance to increase turnaround time. A UPLC's ability to reduce run times to as little as 10% of a typical HPLC method improves the overall efficiency of both synthesis and analytical laboratories.

In addition to the previous two advantages that UPLC analyses have over HPLC analyses, a third advantage is the flow rate of typical UPLC methods. A majority of the UPLC methods used to analyze oligonucleotides run in the range of 0.2–0.4 mL/min. When compared to reversed-phase HPLC methods that typically run at a range of 0.5–0.75 mL/min, mobile phase consumption can be reduced to approximately 25% when using UPLC. This flow rate reduction not only reduces the amount of ion-pairing agent and solvent that needs to be purchased, but it also reduces the amount of waste that is produced.

1.7.4 Temperature Influences of Chromatography

The retention of oligonucleotides is often greatly affected by temperature. It has been suggested that column temperature will dictate the extent to which the ion-pairing reagent is adsorbed to the stationary phase and thus will impact the electrostatic interaction of the oligonucleotide with its surface.[62] Additionally, increasing the temperature is expected to impact the peak shape. As was discussed in Section 1.3, increasing the temperature will typically improve mass transfer properties. The example shown in Figure 1.24 demonstrates the impact of temperature on the retention time and peak shape of a 16-mer phosphorothioate LNA (locked nucleic acid)/DNA mixed sequence. Buffers were composed of 100 mM HFIP spiked with 7 mM triethylamine in water (mobile phase A) and methanol (mobile phase B). The retention time was shifted by as much as 27% on increasing the column temperature from 30° to 60°C.

Shown in Figure 1.24 is an overlay of four separations carried out on a 16-mer mixed LNA/DNA sequence at four column temperatures using an XTerra MS-C18 column (2.5 μm, 4.6 mm × 50 mm). The temperature also had a pronounced effect on the peak shape, as seen in the expansions below. The sharpness of the peak correlates with an increase in column temperature.

Another phenomenon related to peak shape was uncovered on further expansion of the spectra (see Figure 1.25): splitting of the main peak into leading and tailing shoulders. At low temperature, e.g., 30°C, a shoulder appears at the early retention time relative to the apex. At high temperature (60°C), the shoulder appears at longer retention time compared to the center of the peak. This shouldering effect, when pronounced, presents an obstacle to consistent integration of the full length

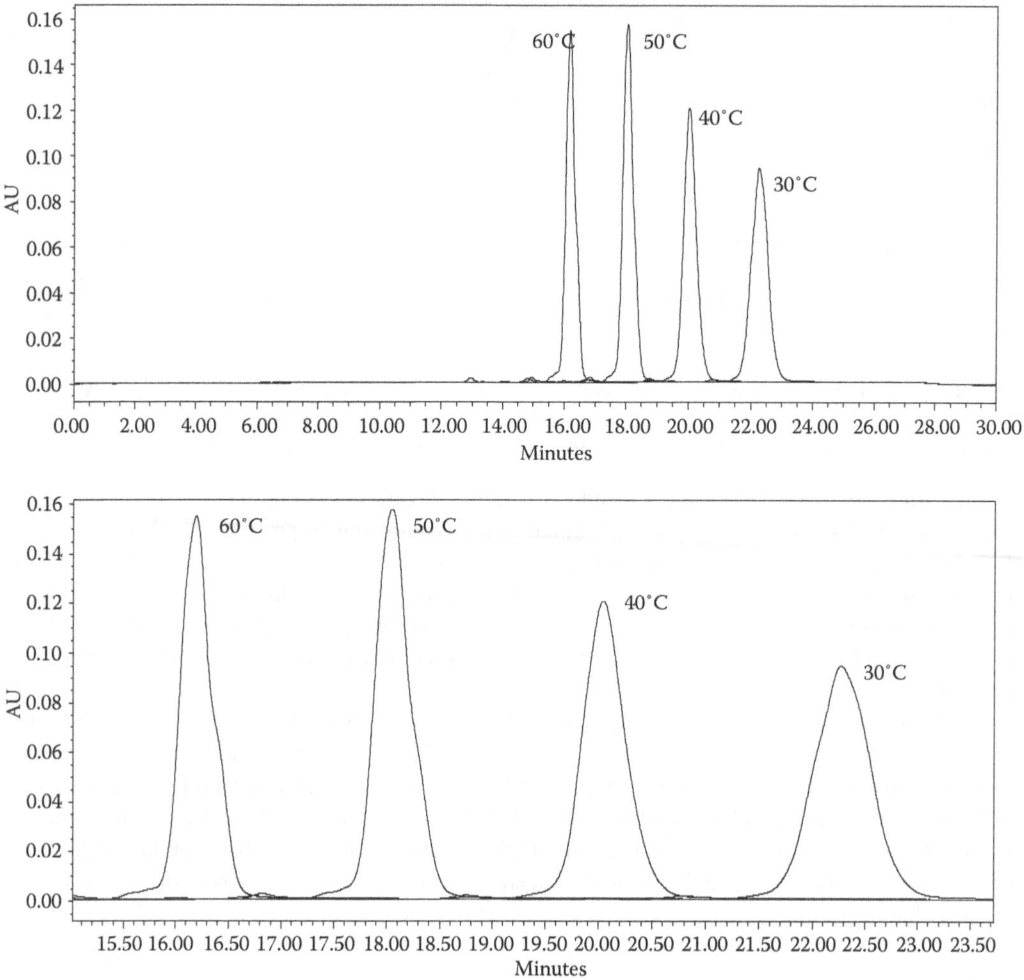

FIGURE 1.24 Overlay of a series of separations of a 16-mer LNA/DNA on an XTerra column at various temperatures. Mobile phase A: 100 mM HFIP spiked with 7 mM triethylamine in water; mobile phase B: 100% methanol; column: Waters XTerra MS-C18, 2.5 μm, 4.6 mm × 50 mm.

product. We chose to carry out separation at 40°C, a temperature that displayed minimal peak splitting.

1.7.5 Sample Preparation

1.7.5.1 Salt Effects

The presence of salts may greatly influence the quality of oligonucleotide separation. Generally, samples containing high salt concentration are encountered when analyzing crude RNA, which typically contains a high amount of fluoride salts and anion-exchange purification fractions containing high concentrations of salt buffer. A different problem represents crude samples of DNA or 2′-modified RNA in concentrated aqueous ammonia. Here the high pH of the sample might influence the analysis. High-salt and high-pH levels interfere with the ability of oligonucleotides to participate in ion pairing and often results in change in retention time, peak splitting, or augmented injection peaks. Several techniques are known to curb these effects, including sample desalting,

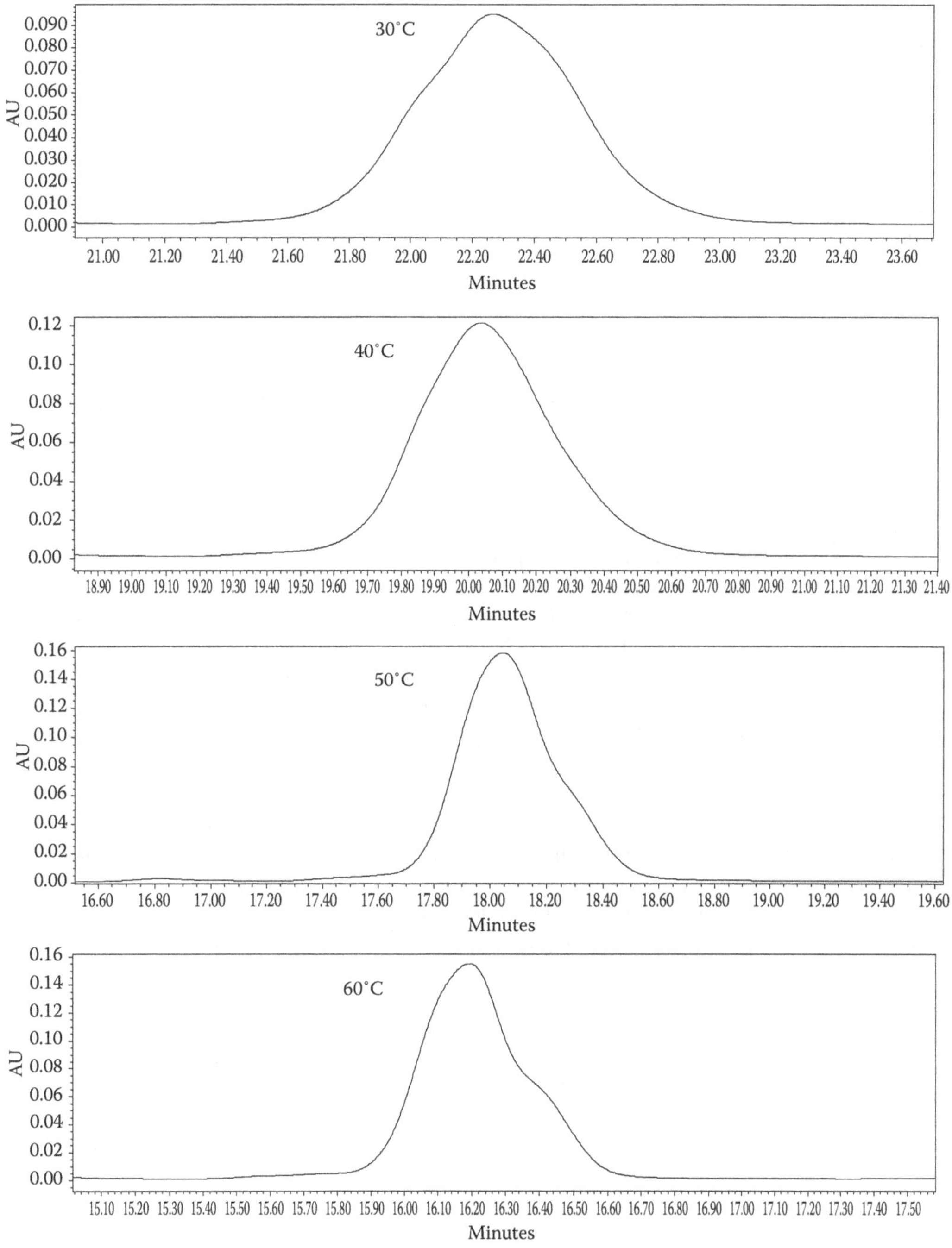

FIGURE 1.25 Peak shape and shouldering at various temperatures of a 16-mer LNA/DNA oligonucleotide. Mobile phase A: 100 mM HFIP spiked with 7 mM triethylamine in water; mobile phase B: 100% methanol; column: Waters XTerra MS-C18, 2.5 μm, 4.6 mm × 50 mm.

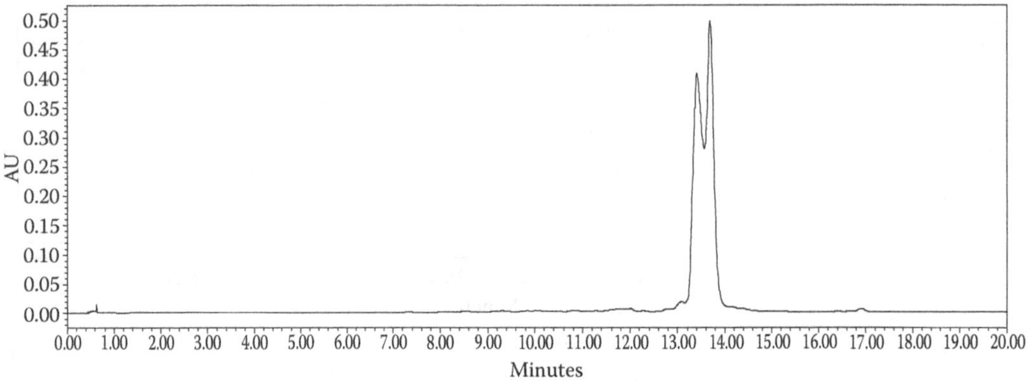

FIGURE 1.26 IP-HPLC analysis of a 21-mer ssRNA sample in a 0.06 M NaBr Purification Buffer. Mobile phase A: 100 mM HFIP/0.1% TEA/1% MeOH in water; mobile phase B: 100 mM HFIP/0.1% TEA/95% MeOH in water; column: XBridge OST C18, 2.5 μm, 2.1 mm × 50 mm at 75°C.

diluting the sample with an ion-pairing agent prior to analysis, or lowering the organic content of the mobile phase used to equilibrate the column prior to injection. The presence of the split peak makes integration of in process samples a challenge and purity analysis inaccurate.

An example of this phenomenon is illustrated in Figure 1.26, which represents an IP-HPLC chromatogram of a 21-mer phosphodiester RNA in an aqueous solution of approximately 0.06 M NaBr derived from anion-exchange purification. The HPLC separation was executed on a Waters Acquity module utilizing a XBridge OST C18 (2.5 μm, 2.1 mm × 50 mm) in conjunction with 100 mM HFIP/0.1% TEA/1% MeOH in water as mobile phase A at 75°C. Mobile phase B consisted of 100 mM HFIP/0.1% TEA/95% MeOH in water.

The same purification fraction was then diluted with TEAA stock solution to 0.1 M concentration of TEAA. Dilution of the sample with ion-pairing reagent, rather than water, allowed for a complete exchange of bound salt for ion-pairing agent, resulting in elution of the FLP as a single peak (see Figure 1.27).

Note that in the expanded chromatogram (Figure 1.28), resolution of the major peak from its N–1 impurity (RT = 13.35 min) is possible only after the sample was diluted with TEAA. This example

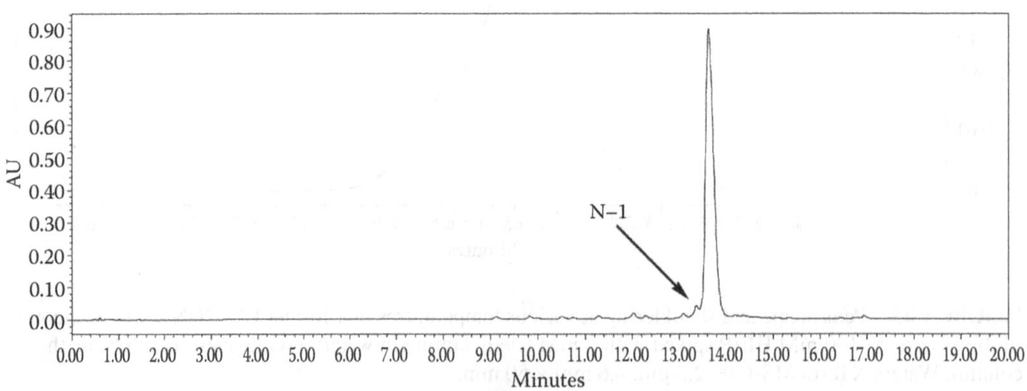

FIGURE 1.27 Preparation of the 21-mer ssRNA sample in 0.06 M NaBr plus 0.1 M TEAA resulted in elution of the FLP as a single peak. Mobile phase A: 100 mM HFIP/0.1% TEA/1% MeOH in water; mobile phase B: 100 mM HFIP/0.1% TEA/95% MeOH in water; column: XBridge OST C18, 2.5 μm, 2.1 mm × 50 mm at 75°C; sample diluted to final concentration of 0.1 M in TEAA.

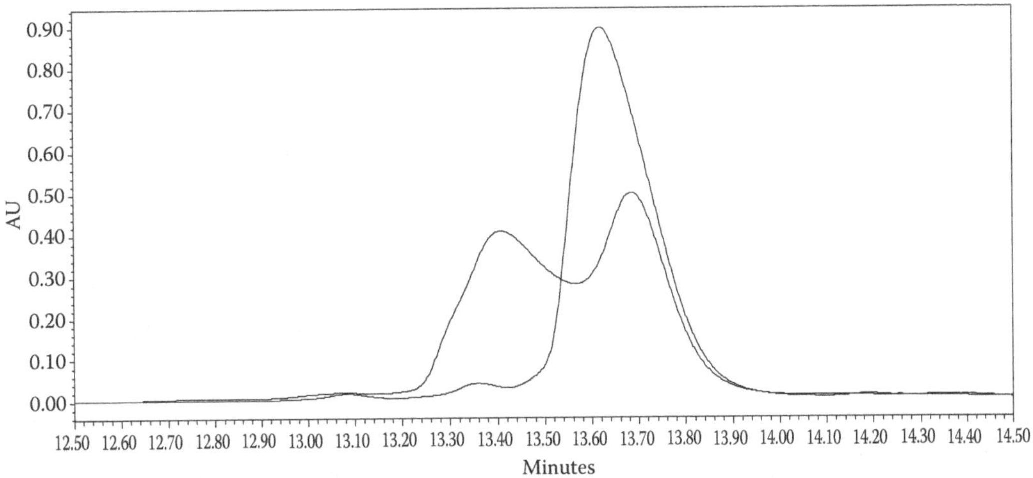

FIGURE 1.28 Expanded overlay of the chromatograms from Figures 1.26 and 1.27 before and after dilution with TEAA. Mobile phase A: 100 mM HFIP/0.1% TEA/1% MeOH in water; mobile phase B: 100 mM HFIP/0.1% TEA/95% MeOH in water; column: XBridge OST C18, 2.5 μm, 2.1 mm × 50 mm at 75°C.

shows that purification fractions containing high salt concentrations could be analyzed effectively by diluting analytical samples with a stock solution of TEAA instead of water.

Another common problem encountered in either the analysis of crude samples in ammonia or the purification fractions in high salt buffer is the presence of a dominant injection peak. Figure 1.29 displays such a chromatogram in which the major oligonucleotide peak is dwarfed by the injection peak at RT = 0.7 min. The sample, a 24-mer phosphorothioate DNA sequence, was obtained from a large-scale manufacturing purification using a buffer composition of 20 mM sodium hydroxide (mobile phase A) and 20 mM sodium hydroxide in 3 M NaCl (mobile phase B). The sample was analyzed on a Waters UPLC system with a Waters OST C18 column (1.7 μm, 2.1 mm × 50 mm) utilizing 100 mM HFIP/0.1% TEA in water as mobile phase A and 100% MeOH as mobile phase B.

A dominant injection peak is particularly undesirable for crude samples because shorter failure sequences often co-elute with the injection peak, resulting in an artificially high percent purity of

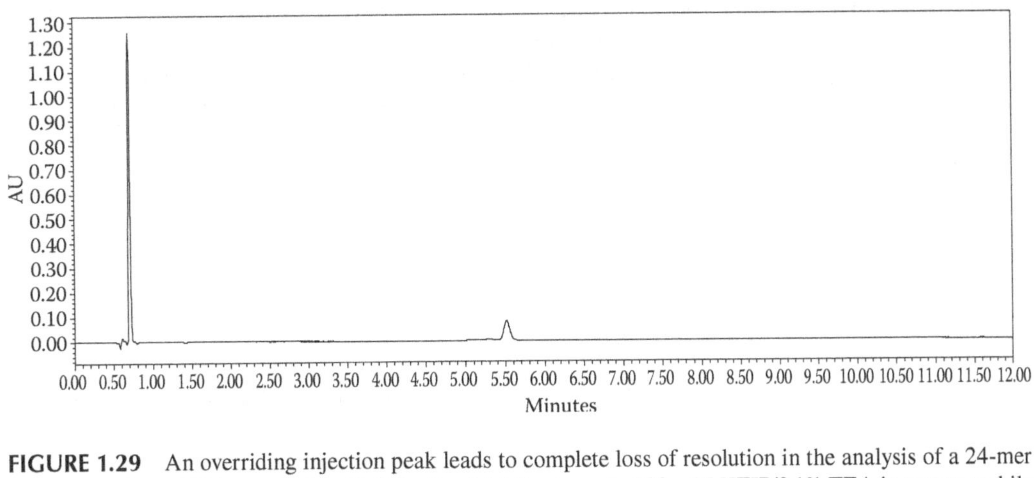

FIGURE 1.29 An overriding injection peak leads to complete loss of resolution in the analysis of a 24-mer phosphorothioate DNA sample in a NaCl buffer. Mobile phase A: 100 mM HFIP/0.1% TEA in water; mobile phase B: 70% MeOH/30% ACN; column: Waters Acquity UPLC system with an Acquity BEH UPLC C18 column 1.7 μm, 2.1 mm × 50 mm 55°C.

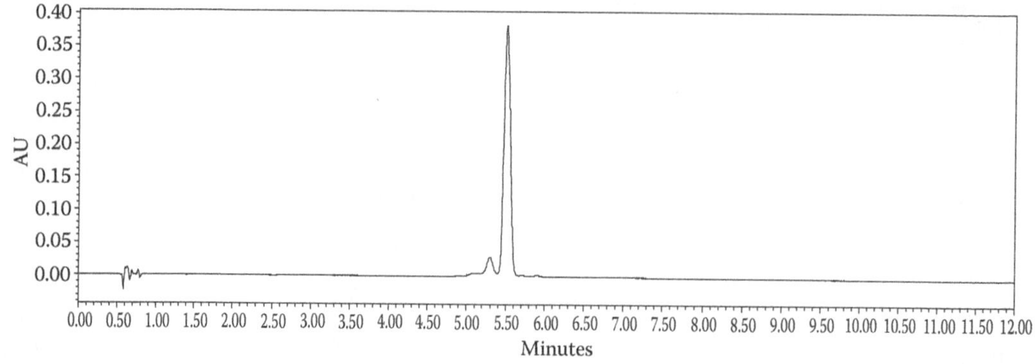

FIGURE 1.30 Dilution of the high salt sample to 0.1 M TEAA permits analysis of the 24-mer phosphorothioate DNA sample. Mobile phase A: 100 mM HFIP/0.1% TEA in water; mobile phase B: 70% MeOH/30% ACN; column: Waters Acquity UPLC system with an Acquity BEH UPLC C18 column 1.7 µm, 2.1 mm × 50 mm 55°C; sample prepared in 0.1 M TEAA.

the major peak by area integration. The 24-mer phosphorothioate DNA product shown in Figure 1.29 was diluted to 0.1 M TEAA and reanalyzed to give the spectrum shown in Figure 1.30. Again, the resolution of the main peak from the major impurity is dramatically improved.

Figure 1.31 depicts the chromatography of a 20-mer 2′-*O*-methyl phosphorothioate RNA whose sample was prepared in water. The sample is representative of a selected pool of purification fractions from AEX purification and therefore containing high salt concentration. The chromatography was carried out on UPLC, Waters Acquity column, 1.7 µm, 2.1 mm × 50 mm (45–55% mobile phase B over 8 min) at 60°C. Mobile phase A was composed of 15 mM HAA in water, and mobile phase B consisted of methanol in acetonitrile.

The typical broadening of the main peak and characteristic injection peak was attributed to high salt content of the pooled purification fractions. The importance of matching the ion-pairing reagent used for dilution with the ion-pairing reagent in the mobile phase was demonstrated by diluting the sample to 0.01 M in TEAA and analyzing the sample with the HAA method described above. The resulting chromatogram is displayed in Figure 1.32.

The spectrum, shown in detail in the lower chromatogram of Figure 1.32, still shows significant peak splitting as well as an attendant injection peak. It is thought that the triethylammonium cation binds strongly to the oligonucleotide and does not allow for effective exchange between the TEAA (used for sample preparation) and HAA used for elution. The split peak can be understood in terms of a sample mixture consisting of TEAA and HAA bound to the oligonucleotide; each having a

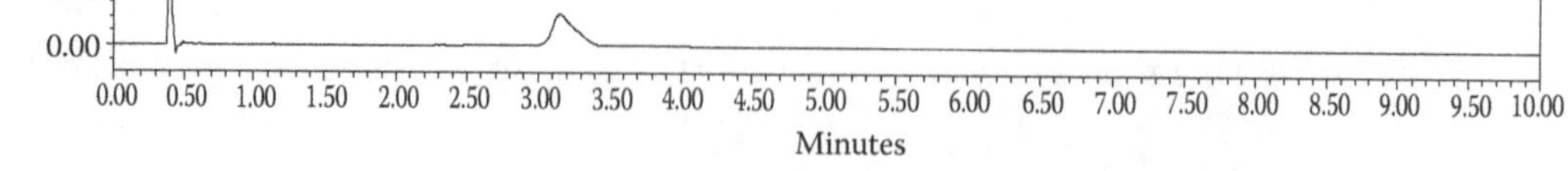

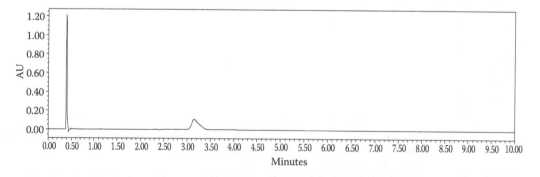

FIGURE 1.31 IP-UPLC chromatography of a 20-mer 2′-*O*-methyl phosphorothioate RNA from AEX purification diluted in water. Mobile phase A: 15 mM HAA in water; mobile phase B: MeOH/ACN; column: Waters Acquity UPLC system with an Acquity BEH UPLC C18 column 1.7 µm, 2.1 mm × 50 mm 60°C.

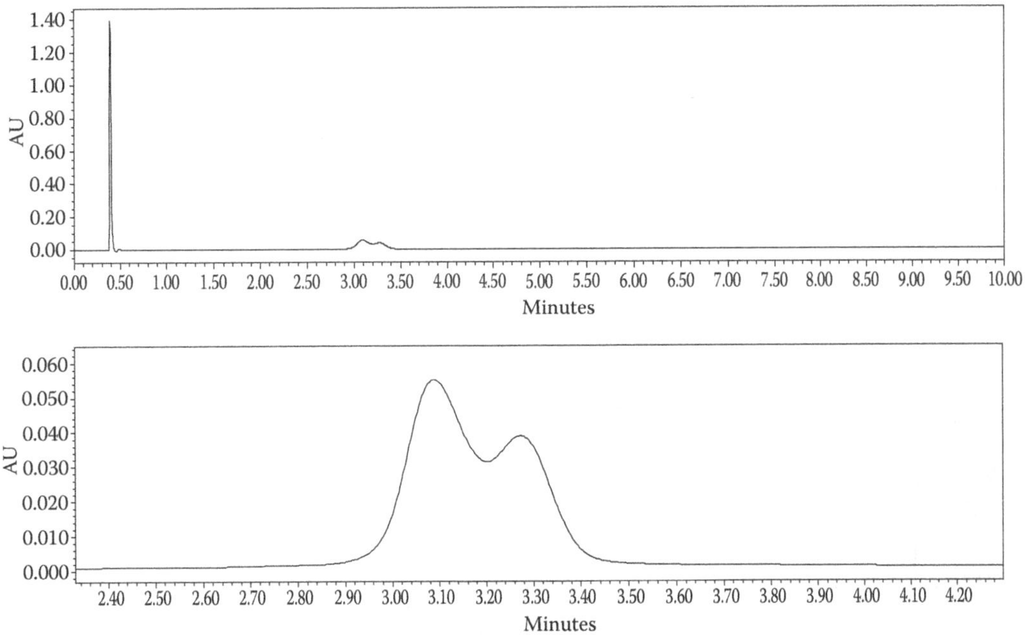

FIGURE 1.32 IP-UPLC chromatography of a 20-mer 2′-*O*-methyl phosphorothioate RNA from AEX purification; the sample was diluted to 0.01 M TEAA (with expansion). Mobile phase A: 15 mM HAA in water; mobile phase B: MeOH/ACN; column: Waters Acquity UPLC system with an Acquity BEH UPLC C18 column 1.7 μm, 2.1 mm × 50 mm 60°C; gradient: 45–55% mobile phase B over 8 min; sample prepared in 0.1 M TEAA; lower panel details expansion of main peak.

unique retention time. The remaining injection peak is due to the concentration of triethylammonium cation in the sample not being sufficient to completely replace the sodium cation, which is in great excess from the purification buffer.

The chromatography of a sample displaying a problematic injection peak can also often be improved by lowering the organic content of the mobile phase used in the column equilibration and at the start of the separation. The example of the 20-mer 2′-*O*-methyl phosphorothioate RNA aptly demonstrates the effect of modified starting organic concentration. Figure 1.33 shows the

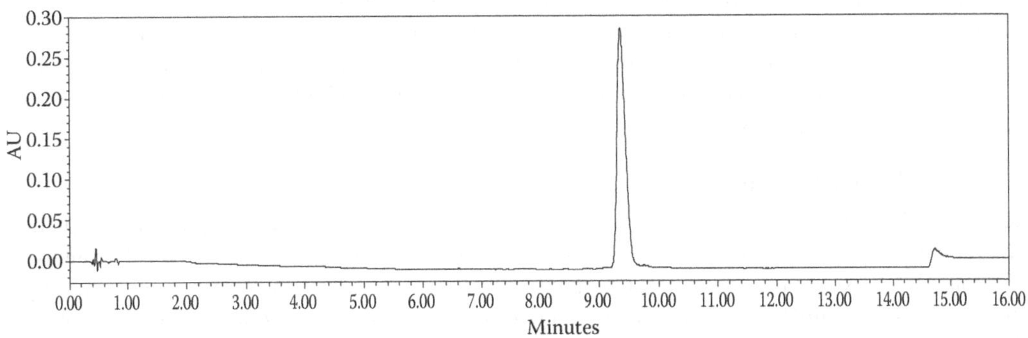

FIGURE 1.33 Improved chromatography of the 20-mer 2′-*O*-methyl phosphorothioate RNA from AEX purification by modification of the gradient and eluent composition. Mobile phase A: 15 mM HAA in water; mobile phase B: MeOH/ACN; column: Waters Acquity UPLC system with an Acquity BEH UPLC C18 column 1.7 μm, 2.1 mm × 50 mm 60°C; gradient: 0–55% mobile phase B over 14 min.

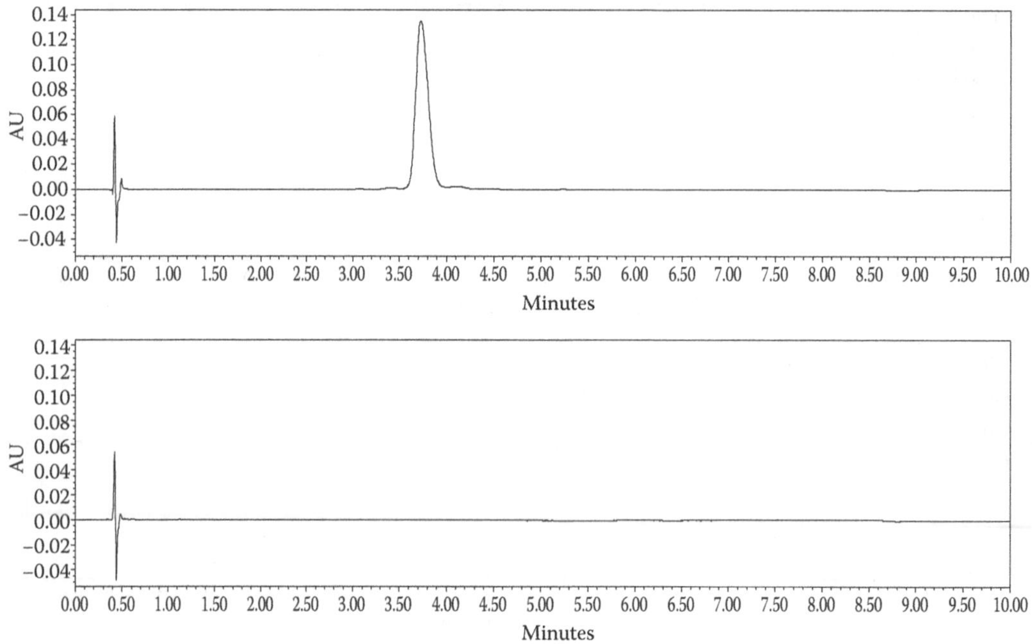

FIGURE 1.34 The preparation of a sample of the 20-mer 2′-*O*-methyl phosphorothioate RNA from AEX purification in 0.1 M HAA eliminates the injection peak and peak splitting. Water blank is shown in the lower panel. Mobile phase A: 15 mM HAA in water; mobile phase B: MeOH/ACN; column: Waters Acquity UPLC system with an Acquity BEH UPLC C18 column 1.7 μm, 2.1 mm × 50 mm 60°C; gradient: 45–55% mobile phase B over 8 min; sample prepared in 0.1 M HAA.

chromatography of the 20-mer 2′-*O*-methyl oligonucleotide on the same system using a gradient of 0–55% B over 14 min instead of 45–55% mobile phase B over 8 min.

Preparation of the sample in 0.1 M HAA allows adequate exchange between bound salt and the ion-pairing buffer. The importance of using the same buffer for dilution as the ion-pairing buffer used for the gradient is highlighted in Figure 1.34.

If dilution of the sample with ion-pairing buffer fails to eliminate the injection peak, the sample may be a candidate for desalting. In this case, a NAP-5™ G-25 Sephadex cartridge (GE Healthcare) was used for rapid sample desalting. The resulting chromatogram is presented in Figure 1.35.

The overall purity of the main component in the desalted sample compared to the sample analyzed after gradient modification is nearly identical: 96.11% FLP in both cases. A comparison of the expanded chromatograms is presented in Figure 1.36.

1.7.5.2 Addition of Buffer

Section 1.7.7 highlights the challenges in the analysis of duplex RNA, specifically the determination of strand excess for the titration of a mixed sequence of RNA and 2′-*O*-methyl RNA. The analysis of duplex RNA is often difficult because many suitable analytical methods for analysis of single strands require high-pH or high-temperature conditions known to cause duplex RNA to denature. Even the use of methods described as “nondenaturing,” generally characterized by neutral pH and low temperature, are sufficiently denaturing to bring about partial denaturation of RNA duplexes owing to the content of organic solvent and ion-pairing reagent of the eluent composition alone. It should be noted that the tendency of the duplex to undergo denaturation is a function of the melting temperature and is highly sequence dependent. The addition of buffering solutions may be

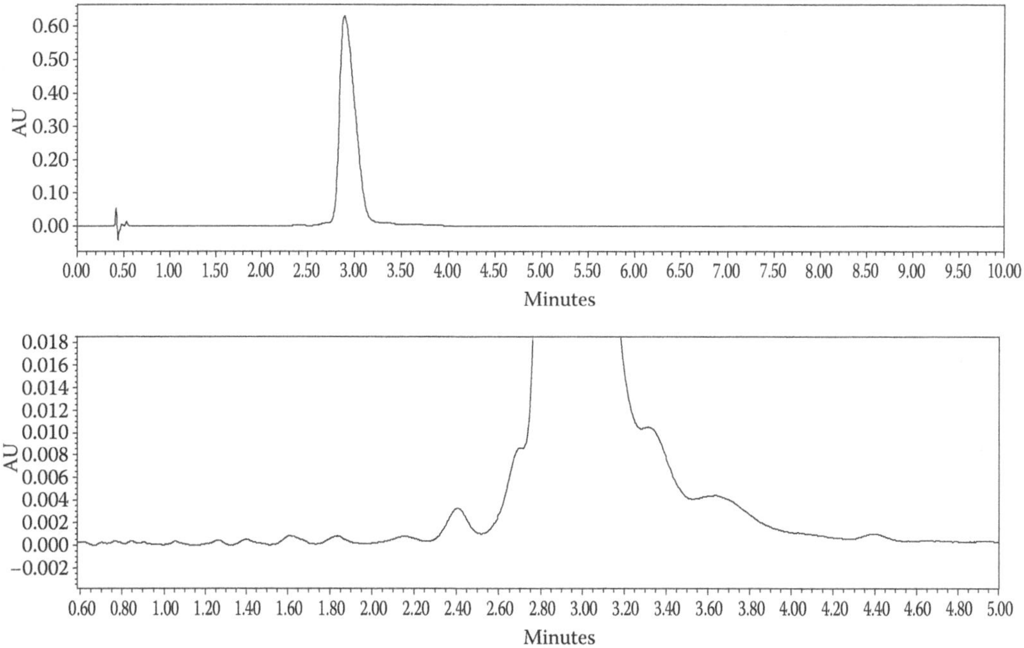

FIGURE 1.35 IP-UPLC chromatography of the 20-mer 2′-*O*-methyl phosphorothioate RNA sample from purification; the sample was desalted prior to analysis. The expanded chromatogram of 20-mer 2′-*O*-methyl phosphorothioate RNA sample is displayed in the lower panel. Mobile phase A: 15 mM HAA in water; mobile phase B: MeOH/ACN; column: Waters Acquity UPLC system with an Acquity BEH UPLC C18 column 1.7 μm, 2.1 mm × 50 mm 60°C; gradient: 45–55% mobile phase B over 8 min; lower panel details expansion of main peak; sample was desalted using Sephadex cartridge before analysis.

necessary to curb the denaturing effects of the IP-HPLC conditions when analyzing duplex RNA. The resulting IP-HPLC chromatograms show mostly duplex with attendant sense and antisense single-strand components separated. The chromatograms shown in Figure 1.37 correspond to a sample prepared in the presence and absence of phosphate-buffered saline (PBS) using a Waters XBridge C18 column and 100 mM HFIP/16 mM TEAA in water/methanol 99:1 as mobile phase A and 100 mM HFIP/16 mM TEAA in water/methanol 5:95 as mobile phase B. The analysis temperature was 20°C for both runs. The first chromatogram (Figure 1.37), representing the sample prepared without PBS buffer, indicates a slight excess of antisense (8.7 min RT) relative to the sense strand (10.9 min RT). However, the presence of both single strands present at the same time indicates that this is not a fully nondenaturing method, which is also called a partially denaturing method. The second chromatogram, showing the analysis of the same sample prepared with PBS, still indicates an excess of antisense strand; however, this time there is only one strand in excess (the antisense strand), indicating that the method is fully nondenaturing, thus all complementary single strand is present in the form of a duplex. Both methods clearly show an excess of antisense strand; however, only when adding buffer to the sample a completely nondenaturing method can be obtained. While a partial denaturing method can be used for determining the excess of a single strand in a duplex, the interpretation of a fully nondenaturing method is easier.

1.7.5.3 Concentration and Injection Volume

The appropriate sample concentration range is necessary for optimum purity analysis. Among the analytical techniques routinely used for oligonucleotide characterization (AEX-HPLC, IP-HPLC,

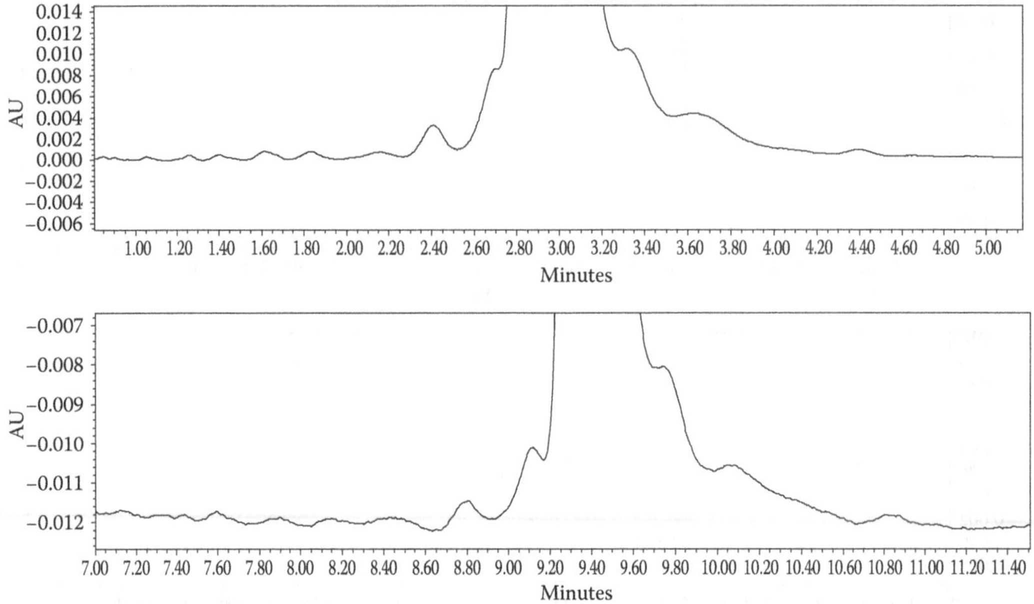

FIGURE 1.36 Comparison of desalting and addition of an ion-pairing reagent for sample preparation; each strategy was effective in improving chromatography of samples with high salt content and/or high pH. Mobile phase A: 15 mM HAA in water; mobile phase B: MeOH/ACN; column: Waters Acquity UPLC system with an Acquity BEH UPLC C18 column 1.7 µm, 2.1 mm × 50 mm 60°C; gradient: 45–55% mobile phase B over 8 min; upper panel details expansion of main peak after desalting using Sephadex cartridge before analysis; lower chromatogram displays the effect of sample preparation in 0.1 M HAA.

capillary gel electrophoresis), IP-HPLC is certainly the least sensitive to concentration effects, which is of practical importance. Despite the robust nature of IP chromatography toward various sample concentrations, the practitioner should be aware of several issues. When too little sample is used, minor impurities often fall below the limit of quantitation and an artificially high purity can be expected for the main peak. Overconcentrated samples, on the other hand, will often result in poor resolution and misshapen peaks. The amount of sample to be injected is related to the column dimensions. Typically, samples having a concentration of approximately 10 OD/mL (or 0.04 mg/mL) generate good results.

1.7.6 Analysis of Oligonucleotides Containing Phosphorothioate versus Phosphate Backbones

Increasing the *in vivo* stability oligonucleotides toward nucleases is a prerequisite for their medicinal use and has been recognized as a significant challenge since the early application of ODNs as antisense therapeutics.[88,89] The substitution of oxygen for its unnatural sulfur congener in the oligonucleotide backbone results in increased nuclease stability; however, the introduction of sulfur at the phosphorous center creates a chiral center. The formation of 2^n diastereomers for n nucleoside bases in phosphorothioate ODNs, coupled with the more lipophilic sulfur atom, results in peak broadening and an increase in retention time relative to their phosphate analogues (refer to Section 1.5 describing mobile phases). The presence of residual phosphate (PO) in a phosphorothioate oligonucleotide can be attributed to either incomplete oxidative thiolation or PO exchange (normally during high pH base deprotection) and is considered a major process-related impurity. As such, its proper characterization is vital. Figure 1.38 shows an overlay of a chromatogram of a 24-mer

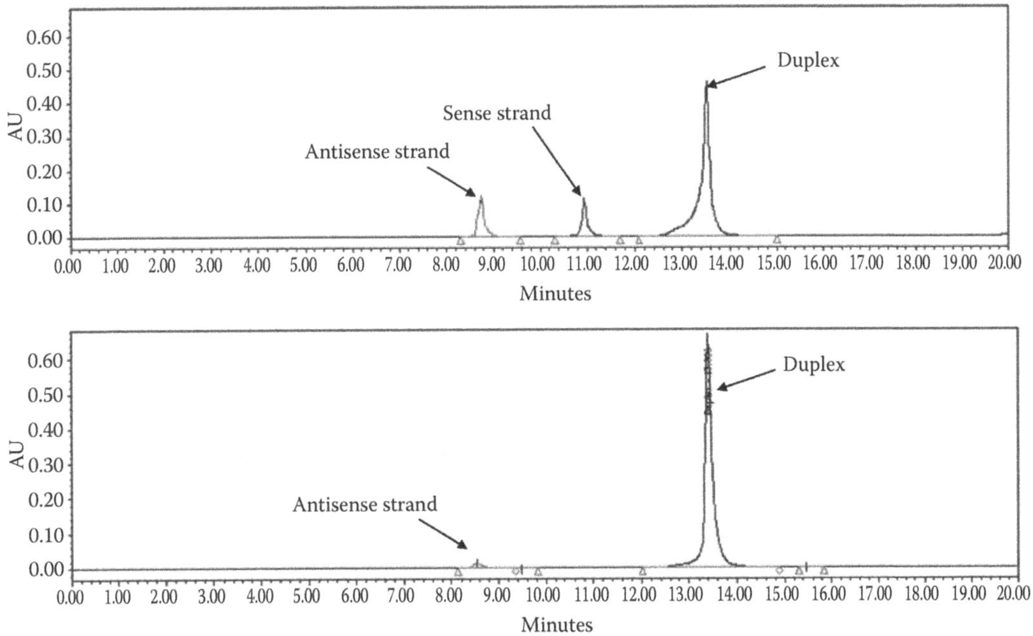

FIGURE 1.37 Samples of duplex siRNA prepared without (top) and with (bottom) PBS buffer. Mobile phase A: 100 mM HFIP/16 mM TEAA in water/methanol 99:1; mobile phase B: 100 mM HFIP/16 mM TEAA in water/methanol 5:95; Waters Acquity UPLC system with a Waters XBridge OST C18, 2.5 μm, 2.1 mm × 50 mm; 23°C; upper panel highlights the denaturing of the duplex in the absence of PBS buffer (antisense, 13.74%; sense, 10.61%; duplex, 75.65); lower panel shows chromatography after preparing the sample with PBS buffer; (antisense, 2.55%; duplex, 97.35).

phosphorothioate DNA oligonucleotide, and its monophosphodiester analogue. Immediately, one notices the shift toward lower retention time for the oligonucleotide containing a single PO linkage (RT = 5.88 min) relative to its phosphorothioate congener (RT = 5.94 min). The method utilized UPLC separation technology as described in the previous section (Waters UPLC, Waters OST C18 column, 1.7 μm, 2.1 mm × 50 mm) HFIP as the ion-pairing agent.

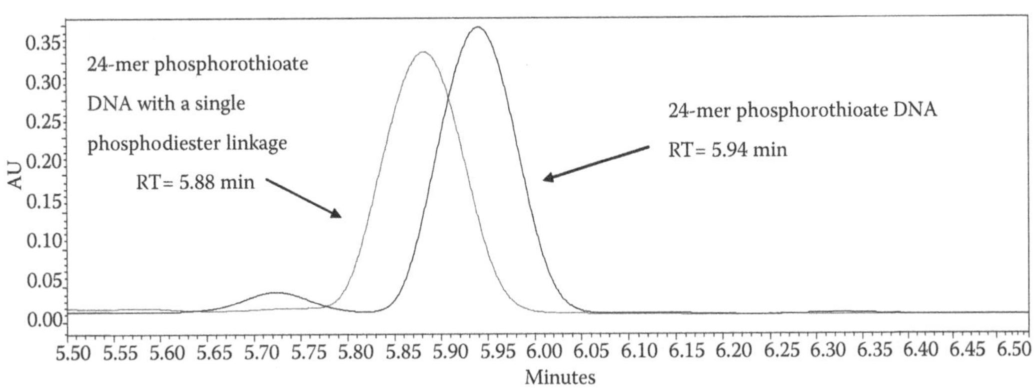

FIGURE 1.38 Phosphorothioate DNA 24-mer and its monophosphodiester analogue. Mobile phase A: 100 mM HFIP/0.1% TEA in water; mobile phase B: 70% MeOH/30% ACN; column: Waters Acquity UPLC system with an Acquity BEH UPLC C18 column 1.7 μm, 2.1 mm × 50 mm 55°C.

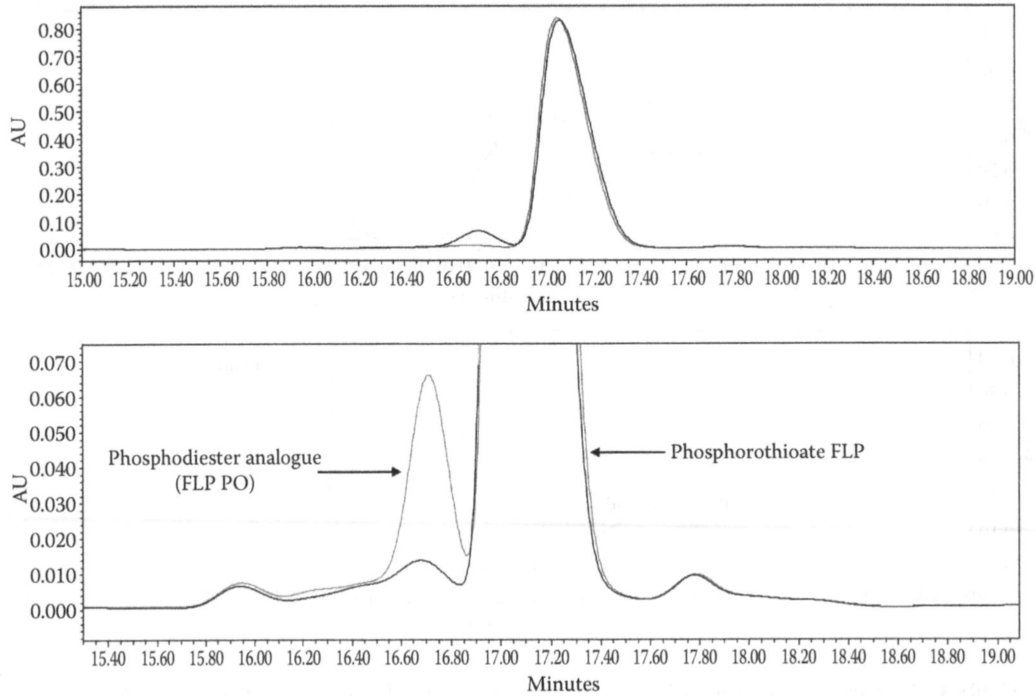

FIGURE 1.39 Overlay of two analyses: a full-length phosphorothioate DNA 18-mer and the same product spiked with 5% PO impurity. Mobile phase A: mM HFIP/0.1% TEA/1.5% ACN in water; mobile phase B: 60% ACN in water; column: XBridge OST C18 2.5 µm, 4.6 mm × 50 mm column at 60°C.

Anion-exchange chromatography, which is based on separation by charge, is normally better suited for the separation of the full length phosphorothioate sequence from its monophosphate impurity. There are, however, examples in which separation of PO impurity from the full length phosphorothioate product are possible. It should be noted that the separation is highly sequence specific and not a general phenomenon. An example of a separation of an 18-mer DNA phosphorothioate oligonucleotide from its major PO containing impurity is highlighted in Figure 1.39. The

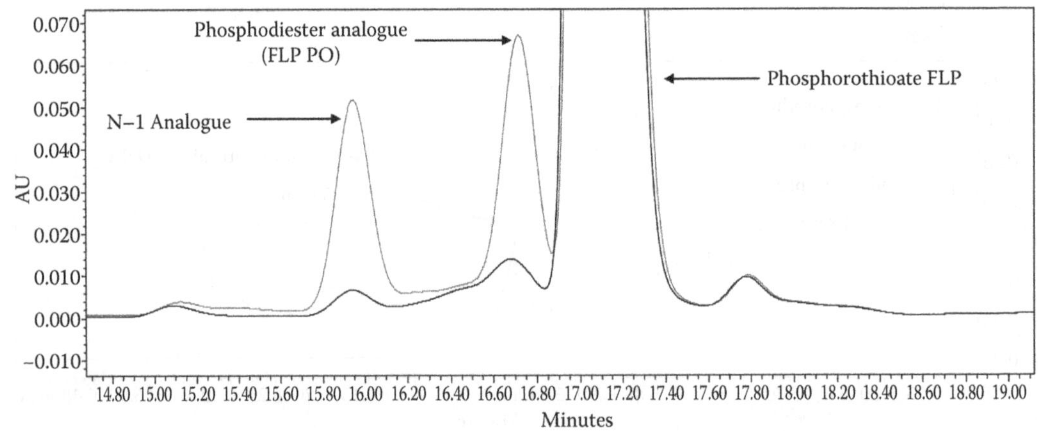

FIGURE 1.40 Resolution of an FLP phosphorothioate DNA 18-mer from its N−1 and PO impurities. Mobile phase A: mM HFIP/0.1% TEA/1.5% ACN in water; mobile phase B: 60% ACN in water; column: XBridge OST C18 2.5 µm, 4.6 mm × 50 mm column at 60°C.

separation was carried out on a Waters HPLC system using 100 mM HFIP/0.1% TEA/1.5% ACN in water (mobile phase A) and 60% ACN in water (mobile phase B) an XBridge OST C18 2.5 μm, 4.6 mm × 50 mm column at 60°C.

A spiking experiment highlighting the ability of an IP-HPLC method to separate the FLP (PO) impurity from N–1 and the full-length phosphorothioate product is shown in Figure 1.40.

1.7.7 Denaturing versus Nondenaturing Ion-Pairing Methods

The advent of siRNA as therapeutics has spurred the development of robust methods for analysis of single strands and the corresponding duplexes. Small interfering RNA is typically manufactured using the following steps: (1) synthesis of both single strands, (2) purification of each single strand separately, (3) desalting of each single strand, and, finally, (4) duplex formation by titration and annealing. Titration describes the process by which the duplex is formed through mixing the sense and antisense single strands in an appropriate ratio, ideally forming the "perfect duplex." The mixture is then heated to break up aggregation of single strands and slowly cooled to allow annealing of the complimentary single strands to yield the duplex. Often the titration ratio is difficult to calculate even with prior knowledge of each component's extinction coefficient. The presence of impurities, particularly shortmers closely related to the full length product (especially N–1), will participate in formation of mismatched duplexes and complicate the process of titration. The necessity of a good analytical method for determination of excess single strand is essential in siRNA manufacturing and the challenges associated with these methods have been recognized.[90] In other words, formation of the perfect duplex is only as perfect as the analytical method. Many HPLC methods, however, require high temperature, high pH, or other conditions that denature, or cause the duplex to lose its secondary and tertiary structure. While these methods may be valuable for purity determination of single-strand components, denaturing of the duplex prohibits accurate measurement of duplex to single strand excess. Instead, a nondenaturing IP-HPLC method needs to be run at low temperature, about neutral pH, and preferably using ion-pairing reagents with less of a tendency to denature. For this reason, McCarthy et al. used HAA buffer in favor of HFIP/TEA for their analyses of siRNA duplex.[67] Sample preparation might also be required for achieving fully nondenaturing conditions. It is typically through the combined use of size exclusion chromatography and nondenaturing anion-exchange and/or nondenaturing IP-HPLC that one may determine single-strand excess (see Chapters 2 and 3).

The following example describes the empirical determination of an endpoint for titration of sense and antisense strands to form the corresponding duplex by using a nondenaturing IP-HPLC method. The sense and antisense strands represent typical 21-mer siRNA single strands, composed of RNA and 2′-*O*-methyl RNA linked by a phosphodiester backbone and having a 2-base phosphorothioate deoxythymidine (dT) overhang. Initially, a broad series of titration ratios were analyzed by varying the relative absorbance-derived OD ratios of sense and antisense strands in 5% increments during a process called microtitration. The samples containing the appropriate ratio of sense to antisense strands were heated to 80°C for a period of 10 min in order to break up any potential self-complimentary secondary structure formed by the individual single strands. After cooling the mixtures, samples were analyzed by denaturing and nondenaturing IP-HPLC (see Figures 1.41 and 1.42 and Table 1.4). Several observations follow from analysis of the data. First of all, the expected OD ratios do not correlate with the percent excess single strands found for sense and antisense using the two parallel methods. It is important to note that the perfect titration ratio is seldom a 1:1 mixture by ODs, even if the extinction coefficient has accurately been determined. This phenomenon is generally due to differences in the impurity profile and tendency to form imperfect duplexes. To further illustrate this point, the OD ratios were adjusted on the basis of calculated extinction coefficients of each strand calculated from the nearest neighbor model. One can see that the experimentally determined titration ratio in this example does not at all correlate to the predicted ratio using extinction coefficients derived from the nearest neighbor model (see Table 1.4).

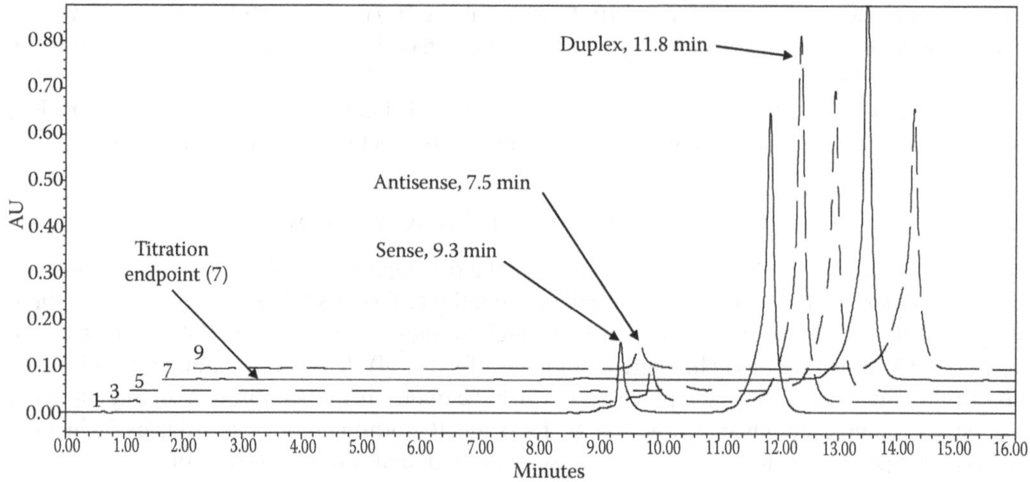

FIGURE 1.41 Overlay of a series of titrations of a 21-mer siRNA showing: (1) excess antisense strand: 7.5 min RT; (2) excess sense strand: 9.3 min RT; (3) duplex: 11.8 min RT. Overlay of six titrations analyzed by non-denaturing IP-HPLC. The endpoint of the titration, represented by Titration 4, indicates a slight (~0.5–1.0%) excess of antisense strand. Mobile phase A: 100 mM HFIP/16 mM TEAA in water/methanol 99:1; mobile phase B: 100 mM HFIP/16 mM TEAA in water/methanol 5:95; Waters Acquity UPLC system with a Waters XBridge OST C18, 2.5 μm, 2.1 mm × 50 mm; 23°C.

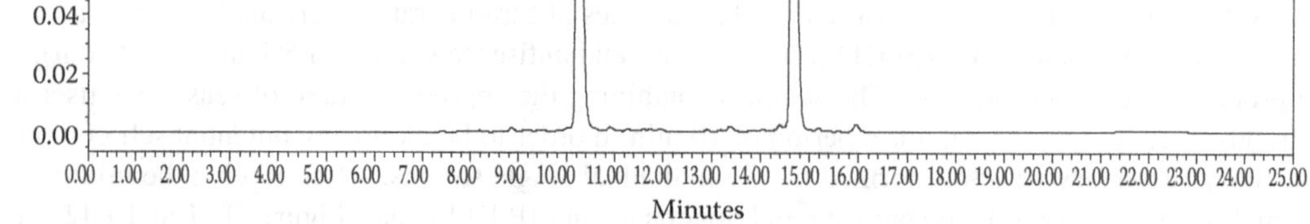

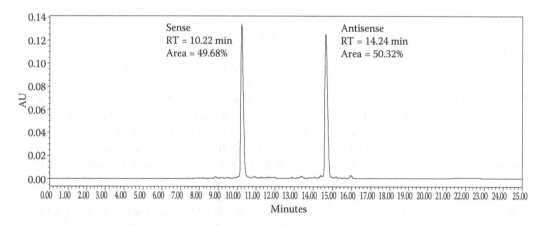

FIGURE 1.42 Denaturing IP-HPLC of the duplex (titration endpoint) from Figure 1.41. A representative separation of the duplex into its complimentary strands by means of a nondenaturing method. In this case, the nondenaturing method is carried out on a Waters XBridge OST C18 column at 75°C using 100 mM HFIP/7 mM TEA in 1% MeOH and 100 mM HFIP/7 mM TEA in 95% MeOH as mobile phase A and B, respectively. The buffer composition is identical to that used for the nondenaturing conditions. The chief difference in the method is the execution of a shallower gradient at significantly higher (75°C) temperature. Area percentages of both main peaks were set to 100% and are not adjusted for extinction coefficients. Results from the denaturing method show antisense strand in slight excess, which is in agreement with results found from the nondenaturing method depicted in Figure 1.41.

TABLE 1.4
Summary of Microtitration Data for a siRNA Duplex

Sample	OD Titration Ratio Sense/ Antisense[a]	Excess in Sense by OD, %	Nondenaturing IP-HPLC		Denaturing IP-HPLC		
			Duplex	% SS excess	% Sense	% AS	% SS Excess
Sense strand	NA	100	NA	94.08	91.26	0	NA
AS strand	NA	0	NA	95.77	0	91.66	NA
Titration 1	1.00/0.85 (*1.00/0.89*)	15 (*12*)	84.53	15.47 (S)	56.97	43.03	13.94 (S)
Titration 2	1.00/0.90 (*1.00/0.94*)	10 (*6*)	87.18	12.82 (S)	55.92	44.08	11.84 (S)
Titration 3	1.00/0.95 (*1.00/0.99*)	5 (*0*)	91.15	8.85 (S)	54.39	45.61	8.78 (S)
Titration 4	1.00/1.00 (*1.00/1.05*)	0 (*–4*)	93.15	6.85 (S)	53.51	46.49	7.02 (S)
Titration 5	1.00/1.05 (*1.00/1.10*)	–5 (*–9*)	97.46	2.54 (S)	51.65	48.35	3.30 (S)
Titration 6	1.00/1.10 (*1.00/1.15*)	–10 (*–13*)	98.63	1.37 (S)	51.12	48.88	2.24 (S)
Titration 7	**1.00/1.15 (1.00/1.20)**	**–15 (–17)**	**98.84**	**1.16 (AS)**	**49.68**	**50.32**	**0.64 (AS)**
Titration 8	1.00/1.20 (*1.00/1.26*)	–20 (*–26*)	95.97	4.03 (AS)	48.45	51.55	3.10 (AS)
Titration 9	1.00/1.25 (*1.00/1.31*)	–25 (*–31*)	92.30	7.70 (AS)	46.92	53.08	6.16 (AS)

Note: Bold-faced data (Titration 7) represents the empirically-derived titration endpoint.

[a] Italicized OD ratios in parentheses reflect calculated differences in extinction coefficient of each strand based on the nearest neighbor model: sense = 201,700 L/mol cm; antisense = 192,700 L/mol cm.

It is also noteworthy to mention that the methods are not sensitive enough to track the 5% incremental increases in amount of antisense strand as prepared during microtitration. The methods do, however, allow one to see the antisense peak begin to grow as the sense peak disappears, and it is at this point that the endpoint of the titration is reached.

The point at which the antisense strand is first seen in excess is generally interpreted as the endpoint of the titration because a slight excess of antisense strand can be tolerated because no off-target effects are expected in contrast to a potential excess of sense strand. In this case, the empirically derived OD ratio of sense to antisense 1.00 to 1.15 (or adjusted for extinction coefficient 1.00 to 1.20) appears to be the ideal titration ratio (see Table 1.4). The denaturing HPLC supports the values derived from nondenaturing HPLC (see Figure 1.42).

1.8 SUMMARY

The increasing significance of oligonucleotides as therapeutics demands a high level of quality control, including accurate purity determination and identification of the related process impurities. The iterative nature of oligonucleotide synthesis inevitably leads to the generation of low-level oligonucleotide by-products such as failure sequences, N+1 impurities, PO impurities (for phosphorothioates), and others. Because of its generality and robustness, IP-HPLC has proven the most important technique for analysis of oligonucleotides, whether they belong to siRNA, phosphorothioate DNA, aptamers, LNA gapmers, or other subclasses.

CGE, AEX, and IP-HPLC are the most widely used analytical tools for the separation of impurities from the full length oligonucleotide. In contrast to AEX-HPLC and CGE we have been witnessing tremendous improvements of resolution in IP-HPLC over the past two decades by moving to smaller particle size sorbents. Also, IP-HPLC tends to be the most robust method of the three in terms of sample concentration, buffer, and instrument requirements. An additional advantage of IP-HPLC is that it can be directly coupled to an ESI-MS instrument, thereby making it feasible to identify each individual peak, which is not easily possible with CGE or AEX-HPLC.

As with any technique, there are many variables affecting the outcome of the separation and the critical parameters and attendant theory are discussed briefly in the introductory section. Drawing from our broad experience in manufacturing and analyzing a host of oligonucleotide subclasses, we present a selection of valuable examples for the practitioner in the experimental section. Though certainly not comprehensive, Section 1.7 provides guidance in column selection, separation of the main peak from its most commonly observed impurities, techniques to curb salt influences, analysis of oligonucleotide small-molecule conjugates, and others.

ACKNOWLEDGMENTS

We thank Ananya Dubey, Sean McCarthy, and Martin Gilar from Waters Corporation for giving us access to RP-HPLC relevant journal articles and application notes including figures.

REFERENCES

1. Cunliffe, J. M., and T. D. Maloney. 2007. Fused-core particle technology as an alternative to sub-2-microm particles to achieve high separation efficiency with low backpressure. *Journal of Separation Science* 30: 3104–3109.
2. Destefano, J. J., T. J. Langlois, and J. J. Kirkland. 2008. Characteristics of superficially-porous silica particles for fast HPLC: some performance comparisons with sub-2-microm particles. *Journal of Chromatographic Science* 46: 254–260.
3. Kirkland, J. J. 2000. Ultrafast reversed-phase high-performance liquid chromatographic separations: an overview. *Journal of Chromatographic Science* 38: 535–544.
4. Kirkland, J. J., F. A. Truszkowski, and R. D. Ricker. 2002. Atypical silica-based column packings for high-performance liquid chromatography. *J. Chromatography A* 965: 25–34.
5. Svec, F. 2008. What is "hot" in column technologies for liquid chromatography. *American Laboratory* 40: 13–17.
6. Majors, R. E. 2006. Fast and ultrafast HPLC on sub-2 um porous particles—where do we go from here? *LC-GC Europe* 19: 352–362.
7. Kirkland, J. J. 2004. Development of some stationary phases for reversed-phase high-performance liquid chromatography. *Journal of Chromatography A* 1060: 9–21.
8. O'Gara, J. E., B. A. Alden, T. H. Walter, J. S. Peterson, C. L. Niederländer, and U. D. Neue. 1995. Simple preparation of a CS HPLC stationary phase with an internal polar functional group. *Analytical Chemistry* 67: 3809–3814.
9. O'Gara, J. E., and K. D. Wyndham. 2006. Porous hybrid organic-inorganic particles in reversed-phase liquid chromatography. *Journal of Liquid Chromatography and Related Technologies* 29: 1025–1045.
10. Kirkland, J. J., F. A. Truszkowski, C. H. Dilks, Jr., and G. S. Engel. 2000. Superficially porous silica microspheres for fast high-performance liquid chromatography of macromolecules. *Journal of Chromatography A* 890: 3–13.
11. Abrahim, A., M. Al-Sayah, P. Skrdla, Y. Bereznitski, Y. Chen, and N. Wu. 2010. Practical comparison of 2.7 microm fused-core silica particles and porous sub-2 microm particles for fast separations in pharmaceutical process development. *Journal of Pharmaceutical and Biomedical Analysis* 51: 131–137.
12. Gilar, M., K. J. Fountain, Y. Budman, et al. 2002. Ion-pair reversed-phase high-performance liquid chromatography analysis of oligonucleotides: retention prediction. *Journal of Chromatography A* 958: 167–182.
13. Van Deemter, J. J., F. J. Zuiderweg, and A. Klinkenberg. 1956. Longitudinal diffusion and resistance to mass transfer as causes of nonideality in chromatography. *Chemical Engineering Science* 5: 271–290.

14. Arghavani, M. B., and L. J. Romano. 1995. A method for the purification of oligonucleotides containing strong intra- or intermolecular interactions by reversed-phase high-performance liquid chromatography. *Analytical Biochemistry* 231: 201–209.
15. Germann, M. W., R. T. Pon, and J. H. van de Sande. 1987. A general method for the purification of synthetic oligodeoxyribonucleotides containing strong secondary structure by reversed-phase high-performance liquid chromatography on PRP-1 resin. *Analytical Biochemistry* 165: 399–405.
16. Swiderski, P. M., E. L. Bertrand, and B. E. Kaplan. 1994. Polystyrene reverse-phase ion-pair chromatography of chimeric ribozymes. *Analytical Biochemistry* 216: 83–88.
17. Johnson, J. L., W. Guo, J. Zang, et al. 2005. Quantification of raf antisense oligonucleotide (rafAON) in biological matrices by LC-MS/MS to support pharmacokinetics of a liposome-entrapped rafAON formulation. *Biomedical Chromatography* 19: 272–278.
18. Lee, D. P. 1988. Chromatographic evaluation of large-pore and non-porous polymeric reversed phases. *Journal of Chromatography* 443: 143–153.
19. Huber, C. G., P. J. Oefner, and G. K. Bonn. 1992. High-performance liquid chromatographic separation of detritylated oligonucleotides on highly cross-linked poly-(styrene-divinylbenzene) particles. *Journal of Chromatography* 599: 113–118.
20. Huber, C. G., P. J. Oefner, and G. K. Bonn. 1993. High-resolution liquid chromatography of oligonucleotides on nonporous alkylated styrene-divinylbenzene copolymers. *Analytical Biochemistry* 212: 351–358.
21. Gelhaus, S. L., and W. R. LaCourse. 2005. Separation of modified 2'-deoxyoligonucleotides using ion-pairing reversed-phase HPLC. *Journal of Chromatography. B, Analytical Technologies in the Biomedical and Life Sciences* 820: 157–163.
22. Lloyd, L. L., M. I. Millichip, and K. J. Mapp. 2003. Rigid polymerics: the future of oligonucleotide analysis and purification. *Journal of Chromatography A* 1009: 223–230.
23. Unger, K. K., ed. 1990. *Packings and Stationary Phases in Chromatographic Techniques*. New York: Marcel Dekker.
24. Petro, M., F. Svec, and J. M. Frechet. 1996. Molded continuous poly(styrene-co-divinylbenzene) rod as a separation medium for the very fast separation of polymers. Comparison of the chromatographic properties of the monolithic rod with columns packed with porous and non-porous beads in high-performance liquid chromatography of polystyrenes. *Journal of Chromatography A* 752: 59–66.
25. Afeyan, N. B., N. F. Gordon, I. Mazsaroff, et al. 1990. Flow-through particles for the high-performance liquid chromatographic separation of biomolecules: perfusion chromatography. *Journal of Chromatography* 519: 1–29.
26. Svec, F., and J. M. J. Frechet. 1992. Continuous rods of macroporous polymer as high-performance liquid chromatography separation media. *Analytical Chemistry* 64: 820–822.
27. Svec, F., and J. M. J. Frechet. 1995. Temperature, a simple and efficient tool for the control of pore size distribution in macroporous polymers. *Macromolecules* 28: 7580–7582.
28. Mayr, B., G. Holzl, K. Eder, M. R. Buchmeiser, and C. G. Huber. 2002. Hydrophobic, pellicular, monolithic capillary columns based on cross-linked polynorbornene for biopolymer separations. *Analytical Chemistry* 74: 6080–6087.
29. Oberacher, H., B. Wellenzohn, and C. G. Huber. 2002. Comparative sequencing of nucleic acids by liquid chromatography-tandem mass spectrometry. *Analytical Chemistry* 74: 211–218.
30. Premstaller, A., H. Oberacher, W. Walcher, et al. 2001. High-performance liquid chromatography-electrospray ionization mass spectrometry using monolithic capillary columns for proteomic studies. *Analytical Chemistry* 73: 2390–2396.
31. Huber, C. G., P. J. Oefner, and G. K. Bonn. 1993. High-resolution liquid chromatography of oligonucleotides on nonporous alkylated styrene-divinylbenzene copolymers. *Analytical Biochemistry* 212: 351–358.
32. Huber, C. G., P. J. Oefner, and G. K. Bonn. 1992. High-performance liquid chromatographic separation of detritylated oligonucleotides on highly cross-linked poly-(styrene-divinylbenzene) particles. *Journal of Chromatography* 599: 113–118.
33. Oberacher, H., A. Premstaller, and C. G. Huber. 2004. Characterization of some physical and chromatographic properties of monolithic poly(styrene-co-divinylbenzene) columns. *Journal of Chromatography A* 1030: 201–208.
34. Premstaller, A., H. Oberacher, and C. G. Huber. 2000. High-performance liquid chromatography-electrospray ionization mass spectrometry of single- and double-stranded nucleic acids using monolithic capillary columns. *Analytical Chemistry* 72: 4386–4393.

35. Bisjak, C. P., L. Trojer, S. H. Lubbad, W. Wieder, and G. K. Bonn. 2007. Influence of different polymerisation parameters on the separation efficiency of monolithic poly(phenyl acrylate-co-1,4-phenylene diacrylate) capillary columns. *Journal of Chromatography A* 1154: 269–276.
36. Wieder, W., C. P. Bisjak, C. W. Huck, R. Bakry, and G. K. Bonn. 2006. Monolithic poly(glycidyl methacrylate-co-divinylbenzene) capillary columns functionalized to strong anion exchangers for nucleotide and oligonucleotide separation. *Journal of Separation Science* 29: 2478–2484.
37. Greiderer, A., S. C. Ligon, Jr., C. W. Huck, and G. K. Bonn. 2009. Monolithic poly(1,2-bis(p-vinylphenyl) ethane) capillary columns for simultaneous separation of low- and high-molecular-weight compounds. *Journal of Separation Science* 32: 2510–2520.
38. Bakry, R., C. W. Huck, and G. K. Bonn. 2009. Recent applications of organic monoliths in capillary liquid chromatographic separation of biomolecules. *Journal of Chromatography Science* 47: 418–431.
39. Wieder, W., S. H. Lubbad, L. Trojer, C. P. Bisjak, and G. K. Bonn. 2008. Novel monolithic poly(p-methylstyrene-co-bis(p-vinylbenzyl)dimethylsilane) capillary columns for biopolymer separation. *Journal of Chromatography A* 1191: 253–262.
40. Trojer, L., S. H. Lubbad, C. P. Bisjak, W. Wieder, and G. K. Bonn. 2007. Comparison between monolithic conventional size, microbore and capillary poly(p-methylstyrene-co-1,2-bis(p-vinylphenyl)ethane) high-performance liquid chromatography columns synthesis, application, long-term stability and reproducibility. *Journal of Chromatography A* 1146: 216–224.
41. Holdsvendova, P., J. Suchankova, M. Buncek, V. Backovska, and P. Coufal. 2007. Hydroxymethyl methacrylate-based monolithic columns designed for separation of oligonucleotides in hydrophilic-interaction capillary liquid chromatography. *Journal of Biochemical and Biophysical Methods* 70: 23–29.
42. Xiong, W., J. Glick, Y. Lin, and P. Vouros. 2007. Separation and sequencing of isomeric oligonucleotide adducts using monolithic columns by ion-pair reversed-phase nano-HPLC coupled to ion trap mass spectrometry. *Analytical Chemistry* 79: 5312–5321.
43. Sturm, M., S. Quinten, C. G. Huber, and O. Kohlbacher. 2007. A statistical learning approach to the modeling of chromatographic retention of oligonucleotides incorporating sequence and secondary structure data. *Nucleic Acids Research* 35: 4195–4202.
44. Kohlbacher, O., S. Quinten, M. Sturm, B. M. Mayr, and C. G. Huber. 2006. Structure-activity relationships in chromatography: retention prediction of oligonucleotides with support vector regression. *Angewwandte Chemie* (International Edition in English) 45: 7009–7012.
45. Przybyciel, M., and R. E. Majors. 2002. Phase collapse in reversed-phase LC. *LC-GC Europe* 15: 652–657.
46. Agrawal, S., J. Y. Tang, and D. M. Brown. 1990. Analytical study of phosphorothioate analogues of oligodeoxynucleotides using high-performance liquid chromatography. *Journal of Chromatography* 509: 396–399.
47. Gilar, M., A. Belenky, and A. S. Cohen. 2000. Polymer solutions as a pseudostationary phase for capillary electrochromatographic separation of DNA diastereomers. *Electrophoresis* 21: 2999–3009.
48. Metelev, V., and S. Agrawal. 1992. Ion-exchange high-performance liquid chromatography analysis of oligodeoxyribonucleotide phosphorothioates. *Analytical Biochemistry* 200: 342–346.
49. Huber, C. G. 1998. Micropellicular stationary phases for high-performance liquid chromatography of double-stranded DNA. *Journal of Chromatography A* 806: 3–30.
50. Gilar, M. 2001. Analysis and purification of synthetic oligonucleotides by reversed-phase high-performance liquid chromatography with photodiode array and mass spectrometry detection. *Analytical Biochemistry* 298: 196–206.
51. Snyder, L. R., J. W. Dolan, and P. W. Carr. 2004. The hydrophobic-subtraction model of reversed-phase column selectivity. *Journal of Chromatography A* 1060: 77–116.
52. Bidlingmeyer, B. A., S. N. Deming, W. P. Price, Jr., B. Sachok, and M. Petrusek. 1979. Retention mechanism for reversed-phase ion-pair liquid chromatography. *Journal of Chromatography* 186: 419–445.
53. Melander, W. R., and C. Horvath. 1980. Mechanistic study on ion-pair reversed-phase chromatography. *Journal of Chromatography* 201: 211–224.
54. Cecchi, T. 2008. Ion pairing chromatography. *Critical Reviews in Analytical Chemistry* 38: 161–214.
55. Jost, W., K. Unger, and G. Schill. 1982. Reverse-phase ion-pair chromatography of polyvalent ions using oligonucleotides as model substances. *Analytical Biochemistry* 119: 214–224.
56. Snyder, L. R., J. J. Kirkland, and J. W. Dolan. 2010. *Introduction to Modern Liquid Chromatography*. 3rd ed. Hoboken, N.J.: John Wiley.
57. Andrus, A., and R. G. Kuimelis. 2001. Analysis and purification of synthetic nucleic acids using HPLC. *Current Protocols in Nucleic Acid Chemistry* Chapter 10: Unit 10.5.

58. Haupt, W. P., and A. Pingoud. 1983. Comparison of several high-performance liquid chromatography techniques for the separation of oligodeoxynucleotides according to their chain lengths. *Journal of Chromatography* 260: 419–428.
59. Hoffman, N. E., and J. C. Liao. 1977. Reversed phase high performance liquid chromatographic separations of nucleotides in the presence of solvophobic ions. *Analytical Chemistry* 49: 2231–2234.
60. Crowther, J. B., R. Jones, and R. A. Hartwick. 1981. High-performance liquid chromatography of the oligonucleotides. *Journal of Chromatography* 217: 479–490.
61. Makino, K., H. Ozaki, T. Matsumoto, H. Imaishia, T. Takeuchi, and T. Fukui. 1987. Reversed-phase ion-pair chromatography of oligodeoxyribonucleotides. *Journal of Chromatography A* 400: 271–277.
62. McKeown, A. P., P. N. Shaw, and D. A. Barrett. 2002. Retention behaviour of an homologous series of oligodeoxythymidilic acids using reversed-phase ion-pair chromatography. *Chromatographia* 55: 271–277.
63. Kurata, C., D. C. Capaldi, Z. Wang, N. Luu, and H. Gaus. 2006. Methods for detection, identification and quantification of impurities. US Patent 2009/0095896, filed on Mar. 31, 2006.
64. Ikuta, S., R. Chattopadhyaya, and R. E. Dickerson. 1984. Reverse-phase polystyrene column for purification and analysis of DNA oligomers. *Analytical Chemistry* 56: 2253–2257.
65. McCarthy, S. M., and M. Gilar. 2010. Hexylammonium acetate as an ion-pairing agent for IP-RP LC analysis of oligonucleotides. Waters Application Note 720003361EN.
66. Gjerde, D. T., L. Hoang, and D. Hornby. 2009. *RNA Purification and Analysis: Sample Preparation, Extraction, Chromatography*. Weinheim: Wiley-VCH.
67. McCarthy, S. M., M. Gilar, and J. Gebler. 2009. Reversed-phase ion-pair liquid chromatography analysis and purification of small interfering RNA. *Analytical Biochemistry* 390: 181–188.
68. Apffel, A., J. A. Chakel, S. Fischer, K. Lichtenwalter, and W. S. Hancock. 1997. New procedure for the use of high-performance liquid chromatography–electrospray ionization mass spectrometry for the analysis of nucleotides and oligonucleotides. *Journal of Chromatography A* 777: 3–21.
69. Apffel, A., J. A. Chakel, S. Fischer, K. Lichtenwalter, and W. S. Hancock. 1997. Analysis of oligonucleotides by HPLC-electrospray ionization mass spectrometry. *Analytical Chemistry* 69: 1320–1325.
70. Gilar, M., K. J. Fountain, Y. Budman, J. L. Holyoke, H. Davoudi, and J. C. Gebler. 2003. Characterization of therapeutic oligonucleotides using liquid chromatography with on-line mass spectrometry detection. *Oligonucleotides* 13: 229–243.
71. Gilar, M., and U. D. Neue. 2007. Peak capacity in gradient reversed-phase liquid chromatography of biopolymers. Theoretical and practical implications for the separation of oligonucleotides. *Journal of Chromatography A* 1169: 139–150.
72. Gaus, H. J., S. R. Owens, M. Winniman, S. Cooper, and L. L. Cummins. 1997. On-line HPLC electrospray mass spectrometry of phosphorothioate oligonucleotide metabolites. *Analytical Chemistry* 69: 313–319.
73. Griffey, R. H., M. J. Greig, H. J. Gaus, et al. 1997. Characterization of oligonucleotide metabolism in vivo via liquid chromatography/electrospray tandem mass spectrometry with a quadrupole ion trap mass spectrometer. *Journal of Mass Spectrometry* 32: 305–313.
74. Bidlingmeyer, B., and Q. Wang. 2006. Additives for reversed-phase HPLC mobile phases. US Patent 7125492, filed July 17, 2003 and issued Jan. 2005.
75. McCarthy, S. M., W. J. Warren, A. Dubey, and M. Gilar. 2008. Ion-pairing systems for reversed-phase chromatography separation of oligonucleotides, paper presented at TIDES Conference in Las Vegas, Nevada.
76. Bartha, Á., and J. Ståhlberg. 1994. Electrostatic retention model of reversed-phase ion-pair chromatography. *Journal of Chromatography A* 668: 255–284.
77. Cruz, E., M. R. Euerby, C. M. Johnson, and C. A. Hackett. 1997. Chromatographic classification of commercially available reverse-phase HPLC columns. *Chromatographia* 44: 151–161.
78. Bidlingmeyer, B. A. 1980. Separation of ionic compounds by reversed-phase liquid chromatography: an update of ion-pairing techniques. *J. Chromatogr. Sci.* 18: 525–539.
79. Cantor, C. R., and P. R. Schimmel. 1980. *The Conformation of Biological Macromolecules*. Part 1. Their Biophysical Chemistry. San Francisco: W. H. Freeman.
80. Nawrocki, J. 1997. The silanol group and its role in liquid chromatography. *Journal of Chromatography A* 779: 29–71.
81. Daignault, L. G., and D. P. Rillema. 1992. Ion-interaction chromatography: a study of the distribution of n-alkylammonium ions on an ODS-2 column. *Journal of Chromatography A* 602: 3–8.

82. Zou, Y., P. Tiller, I. W. Chen, M. Beverly, and J. Hochman. 2008. Metabolite identification of small interfering RNA duplex by high-resolution accurate mass spectrometry. *Rapid Communications in Mass Spectrometry* 22: 1871–1881.
83. Beverly, M., K. Hartsough, and L. Machemer. 2005. Liquid chromatography/electrospray mass spectrometric analysis of metabolites from an inhibitory RNA duplex. *Rapid Communications in Mass Spectrometry* 19: 1675–1682.
84. Lei, B., S. Li, L. Xi, J. Li, H. Liu, and X. Yao. 2009. Novel approaches for retention time prediction of oligonucleotides in ion-pair reversed-phase high-performance liquid chromatography. *Journal of Chromatography A* 1216: 4434–4439.
85. Gelhaus, S. L., W. R. LaCourse, N. A. Hagan, G. K. Amarasinghe, and D. Fabris. 2003. Rapid purification of RNA secondary structures. *Nucleic Acids Research* 31: e135.
86. Tan, L. C., P. W. Carr, and M. H. Abraham. 1996. Study of retention in reversed-phase liquid chromatography using linear solvation energy relationships I. The stationary phase. *Journal of Chromatography A* 752: 1–18.
87. Katritzky, A. R., S. Perumal, R. Petrukhin, and E. Kleinpeter. 2001. Codessa-based theoretical QSPR model for hydantoin HPLC-RT lipophilicities. *Journal of Chemical Information Computer Sciences* 41: 569–574.
88. Uhlmann, E., and A. Peyman. 1990. Antisense oligonucleotides: a new therapeutic principle. *Chemical Reviews* 90: 543–584.
89. Dias, N., and C. A. Stein. 2002. Antisense oligonucleotides: basic concepts and mechanisms. *Molecular Cancer Therapeutics* 1: 347–355.
90. Kreuzian, T. B. 2009. Analysis of duplexes. Paper presented at EuroTides Conference, Dec. 2–3, in Amsterdam, Netherlands.

2 Purity Analysis and Impurities Determination by AEX-HPLC

Jim Thayer
Dionex Corporation

Veeravagu Murugaiah
Alnylam Pharmaceuticals

Yansheng Wu
Archemix Corporation

CONTENTS

2.1 PREFACE

In this chapter on anion-exchange (AE) chromatography (AEC) for oligonucleotide (ON) analysis, we will discuss the utility of ON AEC and its characteristics. We will introduce the topic with a brief background of AEC for synthetic ONs and discuss some rationales for performing analyses of ONs, focusing on those intended for diagnostics and therapeutic applications with special emphasis on siRNA. Because this monograph is intended to provide insight and instruction on development of AEC analyses, we will discuss ON attributes that influence ON–stationary phase interactions, describe the classes of AEC stationary phases, and review the types of studies supported by AEC. We will examine the physical and chemical considerations that impact interactions between ONs and the AEC phases and provide examples of how these are effectively employed. Following that discussion we include a section on methods development, providing a detailed example for a siRNA drug substance. We will also present some advanced example applications including and employing AEC-ESI-MS.

2.2 INTRODUCTION

Anion-exchange chromatography is a mature technology with a long history. However, the development of ON synthesizers is comparatively recent. The first commercially available DNA synthesizers became available in the late 1980s. Among the first uses of AEC for oligodeoxynucleotide (ODN) *purification* was an FPLC method run at high pH to force the ONs into denatured single strands.[1] Shortly thereafter, ON AEC was applied to improving protocols to increase efficiency and yield from DNA synthesizers.[2] Users of the new synthesizers prepared homopolymers to examine instrument performance, evaluating the products by AEC. They encountered the issue of poly-G tetrad ladder formation and observed that high pH separations allowed control of these hydrogen bonds where prior (non-AEC separations) required inclusion of poly C to coax the poly-G tetrad ladders into Watson–Crick H-bonding interactions.[3] In the early 1990s, "antisense" ODNs harboring phosphorothioate (PS) linkages were prepared as therapeutic candidates. The PS linkage results from the replacement of one nonbridging oxygen with a sulfur atom in the phosphodiester (PO) linkage. Separation of the fully PS linked ODN from those harboring one or more PO linkage by AEC was demonstrated,[4] and the high affinity of the PS linkage was employed to perform high-speed separations of PS-ODNs from biological fluids to support studies of ON metabolism.[5] In addition separation of ODNs using both salt *and* pH gradients by AEC was reported.[6] These examples revealed the utility of AEC for ON analysis and purification.

In addition to improving the efficiency and lower the cost of DNA synthesis protocols,[1] automated DNA synthesis, coupled with the rise of the polymerase chain reaction (PCR) process, greatly expanded the numbers and types of chemical and biological questions that could be examined and answered. For PCR-based studies, simple AEC analyses for different length classes were developed.[7] Numerous studies on ODN modifications for antisense ON employed AEC for analysis or purification of PS ODNs[8–18] and phosphoramidate (PN) ODNs.[19–22] The discovery of ribozymes spurred development of RNA synthesis protocols with AEC applications,[23–32] and the widespread dispersion of the HIV virus (AIDS) produced further RNA studies supported by AEC.[33–36] Increased emphasis on RNA synthesis resulted in the realization that phosphoryl migration could introduce aberrant 2′,5′-linkages during synthesis and deprotection of RNA.[37] These linkages were not readily differentiated by other techniques but have been addressable by AEC on *nonporous* AE columns.[38,39] These linkages are found in nature[40] and have been intentionally introduced for certain *RNA interference* (RNAi) applications.[39]

2.3 USE OF AEC IN DEVELOPMENT OF THERAPEUTIC RNAi OLIGONUCLEOTIDES

2.3.1 Overview

Realization that small noncoding RNAs can serve as a natural regulators of [mRNA] promoted intense interest in development of RNAi therapeutics. One attractive aspect of this technology is that it offers promise to aid sufferers of previously "undruggable" illnesses. As with earlier RNA and DNA efforts, AEC has supported many of these developments.[39,41–46] Similarly, AEC has aided development of RNA aptamers as therapeutics.[34,47–50] The siRNAs are short, double stranded (typically, 21–23 nucleotides) molecules. We will present in detail some AEC applications designed to identify and control impurities during siRNA drug substance production and application in the methods development section. Here we detail several rationales for development of the methods.

2.3.1.1 Manufacturing Controls to Minimize Impurities

Every attempt must be taken to control the impurities at the production stage itself. The chemical synthesis of ORN is performed on solid phase using standard β-cyanoethyl phosporamidite chemistry with tert-butyldimethylsilyl (TBDMS) protection of the ribose 2′-hydroxyl group (e.g., see Ref. 23). The synthesis involves a cycle of deprotection, coupling, capping, and oxidation. The chain extension reactions have a repetitive yield of 97–99%. Assuming that in the best-case scenario of 99%, coupling at each and every coupling, a sequence of 21-mer, can have a theoretical purity of only 81%. The impurities remaining must be examined for possible toxicological properties, and these reported in IND filings[51–53] (see also Chapter 19 of this Handbook). However, high-purity ORN may be readily made in small quantities for crystallographic[54–56] and thermodynamic[42,57] studies. With progress through research and toxicology of siRNA candidates (and in preparation for clinical trials), drug substance is made in kilogram scale, which poses a challenge to the manufacturers to meet impurity specifications and to produce high-purity ORNs. When the full cycle of synthesis is completed, the crude product will have a ladder of deletion sequences, as well as N+mers as unavoidable impurities in the drug substance. A well-designed experiment to optimize each step will provide a high-quality product. The ORN differs from ODN synthesis, which needs attention in one or more of the following stages:

1. 2′-protecting groups may slow or reduce coupling reaction rates.[23]
2. 2′-protecting groups may slow cleavage from supports.
3. 2′-protecting groups reduce aqueous solubility.
4. Hydrolysis of 2′-protecting groups may cause chain cleavage and when removed may allow phosphoryl-migration producing 2′,5′-linkages.[37]
5. Secondary structures can hinder deprotection *and* separations.
6. Single stranded RNA is sensitive to enzymatic degradation.
7. Incomplete 2′-*O*-deprotection reduces overall product yield and increases the level of N+ impurities.

The above factors need to be considered for optimization of base deprotection, selection of suitable phosphoramidite activators, and selection of conditions for removal of 2′-*O*-protecting groups.

2.3.1.2 ORN Purification

RNA is almost always purified in a denatured state. Some simple, general precautions must be taken when handling and storing RNA solutions. RNA molecules are much more prone to hydrolysis than DNA molecules. At high pH, cleavage of the backbone occurs through a mechanism involving the phosphoryl migration and strand scission between the 2′ and 3′ hydroxyls of the ribose ring. RNA nucleases are also much more prevalent than DNA nucleases, and steps must be taken to prevent their presence. A sterile technique with fresh clean gloves should be used at all times. All water and buffers to be used with RNA should be treated with RNAse inhibitors and autoclaved where possible. More extensive protocols for eliminating nucleases from solutions and lab-ware can be found in standard laboratory manuals.[58]

Several types of impurities can exist in a preparation of solid phase chemically synthesized RNA. The most prevalent impurities are "truncates": chains that do not couple during one round of synthesis. The truncates are "capped" before the next round of synthesis to minimize unexpected extensions in subsequent coupling cycles. Other impurities result from incomplete deprotection of the bases or the 2′ hydroxyl.

Purification of a long ORN sequence may be accomplished by either a two-step or a single-step chromatographic procedure.[23,59] The two-step procedure requires an initial purification of the molecule on a reversed-phase column with the trityl group at the 5'-position conserved. The trityl group is removed by the addition of an acid followed by neutralization. The final purification is carried out on an anion-exchange column. For the single-step purification, the molecule is first detritylated then purified on an anion-exchange column. A final desalting step completes both purification procedures. So, well-designed synthesis and purification will provide single strands of required purity. Further understanding of synthesis and purification will reveal information on the type of process-related impurities that are expected to end up in the drug substance.

2.3.1.3 Purity of siRNA Drug Substance

The siRNA drug substance is a duplex obtained by annealing essentially equimolar mixtures of passenger (sense) strand and complementary guide (antisense) strand in water or buffer of choice according to standard protocols. There are no covalent bond-forming reactions involved in the annealing of the single strands to form the drug substance. As a result, all the impurities in the drug substance are carried forward from impurities in the single strands. Purity of siRNA duplex in native form is determined by size exclusion chromatography (SEC). The buffer used in the SEC method has a physiological pH that does not denature the duplex. Resolution in SEC is limited and gives rise to a major peak for the duplex and a minor peak arising from excess single strand impurity, when present. However, discrimination of the excess single strand between sense and antisense strand is not obtained. Refer to Chapter 3 for more on SEC. Because of the limitations on the evaluation of purity/impurities of siRNA in its native state by SEC, specifications are set on the limits of SEC impurities and on the individual single strands. The sense and antisense strands are extensively evaluated for their impurity profiles by denaturing AE HPLC methods (vide infra).

The details of RNAi analysis methods we use will be discussed in Section 2.7.3.

2.4 USE OF AEC IN DEVELOPMENT OF THERAPEUTIC APTAMERS

Aptamers are single-stranded structured ONs that interact with targets (proteins, etc.) with high affinity and specificity. Therapeutic aptamers comprise an emerging class of ON based drugs. In contrast with most other classes of ON therapeutics, aptamers form specific recognition configurations through stable and specific higher-order structures to effectively interact with their targets. Therapeutic aptamers have been prepared using sequences from 16 to over 60 nucleotides in length. Many of the drug targets addressed by therapeutic aptamers are extracellular, so PEG (polyethylene glycol)-conjugation is commonly employed to prolong the aptamer's circulation half-life.[60] In 2004, the first aptamer drug, Macugen®, was approved within the United States. This therapeutic is a 40 KDa PEGylated anti–vascular endothelial growth factor (VEGF) aptamer that interacts with and inhibits the activity of VEGF. Several additional PEGylated aptamers are in clinical trials across a variety of therapeutic indications.

Attachment of PEG is typically accomplished using amine linkers. Primary amine linkers can be incorporated into either the 5′ or 3′ ON end or into both ends, via solid phase synthesis. ONs with such linkers are then conjugated with PEG NHS ester (or other PEG reagents) to form PEGylated ON. PEG systems with MW ranges from ~20 to 40 K, both linear and branched, have been used for conjugation of therapeutic aptamers.

To be functional, an aptamer adopts a specific higher-order structure to interact at high affinity with the target. In doing so, the aptamer blocks specific target interactions and inhibits the target's involvement in the expression of the disease. Intramolecular hydrogen bonding, including duplex, triplex, and G-tetrad forms, as well as base stacking, confer stability to the higher-order structures. PEGylated aptamers have thermal melting behaviors similar to their native forms. Upon PEGylation, aptamers intended for therapeutic applications must maintain affinity and specificity toward their targets. We will include further discussion of AEC of aptamers in Section 2.7.4.

2.5 CONSIDERATIONS FOR ANION-EXCHANGE CHROMATOGRAPHY OF ONS

2.5.1 Unwanted or Unpredictable Interactions

AEC relies primarily on electrostatic interactions between the ON and the AE stationary phase. However, all stationary phases harbor potential interactions other than their primary mode (e.g., normal phase, reversed phase, anion exchange, etc.). Modulating those "unwanted" interactions assists with control of selectivity of the components in ON samples. In the case of AEC, the two most common interactions are due to molecular charge (electrostatic) and van der Waals (hydrophobic) interactions. To illustrate the influence of hydrophobic interactions, the example of *weak* anion exchangers is considered: Here, raising the pH to near the stationary phase pK will reduce the positive charge on the phase, will limit the ion exchange capacity, and will increase hydrophobic interactions due to the loss of charge on the tertiary amine. The amines employed for AEC all employ alkyl groups, so raising the pH and reducing the amine charge render the phase increasingly *suitable* for hydrophobic interactions. This can occasionally assist in resolution of ON components but generally not in a predictable manner. In this chapter we will consider primarily *strong* anion-exchange chromatography, which employs quaternary amines because they present positively charged anion-exchange sites under essentially all aqueous chromatography conditions. Because control of hydrogen bonding is often required for AEC of ONs, strong anion exchangers are preferred.

2.5.2 Selection of Anion-Exchange Chromatography Phases (Application Dependent)

2.5.2.1 Porous Phases

AEC was first developed using low cross-linked, fully porous resin beads. These phases were engineered to offer very high capacity, and the ability to retain compounds based primarily on their net negative charge. For this function they work admirably, and because of their high capacity, are often used to *purify* ONs. However, fully porous resins demand that their eluents and chromatographic targets diffuse into and out of the resin beads. Because diffusion rates are inversely proportional to molecular weight (MW), and because oligonucleotides can have relatively high MW, the porous bead AE phases do not provide the best resolution, or sensitivity, owing to diffusion-induced band broadening. In these phases, resolution can be increased by using very low flow rates (e.g., ~0.15 mL/min for 5 mm ID columns), but sensitivity still suffers compared to nonporous phases and throughput becomes very limited. One example of porous bead phases is the GE/Healthcare Mono-Q. This and its derivatives (Source-Q, Resource-Q) are commonly employed for ON *purification* because these resins are available in bulk and hence can be used for *profound* increases in chromatographic scale. However, these phases are used less frequently for ON *analysis*. Examples of ON *analysis* on porous AEC columns include: (1) crude ON analyses,[61] (2) separation of incompletely sulfurized PS ONs,[4] (3) separation of ODNs from ODN-treated human tumors in nude mice as part of a multistep metabolite analysis,[15] (4) use of porous AE beads in pipette tips for retention and desalting of PCR products for ESI-MS,[62] and (5) analysis of RPLC-purified ODNs containing amino- or hydroxyl-modified phosphoramidate linkages.[20]

Examples of ON *purification* on porous AEC columns include: (1) a high pH ON purification,[1,2,14] (2) bulk tRNA[50] and snRNP purifications,[63] (3) purification of 55- and 62-base RNA transcripts,[64] (4) purification of N3′-P5′ phosphoramidates,[25] (5) plasmid purification,[65] ON phosphorothioate and phosphorodithioate purification,[9,50,66] (6) 2–5A purification,[40] (7) pH-gradient purification,[6] (8) purification of ONs with a cleavable disaccharide linkage,[67] and (9) positional and sequence isomer purification.[66] Scale up of purification on porous AE resins are exemplified by GE Healthcare.[11]

2.5.2.2 Nonporous Phases

In order to overcome the mass transfer limitation of porous beads, researchers developed nonporous, or surface functionalized, AE phases. These were designed to place the ion exchange sites only on the bead surface, where mass transfer is dominated by convection, rather than diffusion. Such phases were capable of very high chromatographic efficiency and were quite popular for analytical applications. However, these phases have two disadvantages: (1) they exhibit very low capacity, and (2) they may harbor relatively high nonspecific interactions, especially if they are functionalized with tertiary amines and operated at pH values where the amine is not fully charged (e.g., DEAE-based phases). An example of a nonporous, surface-functionalized resin is the DEAE-NPR (Tosoh). An example of an AE surface-functionalized resin for ON analysis includes the assay of ONs harboring a photoaffinity labeled cross-linking probe.[68]

One path to increased capacity that minimizes nonspecific interactions is to prepare a resin with a strong cation-exchange (anionic) surface and coat that surface with aminated spheres (i.e., cationic nanobeads, or latexes). Because the anionic substrate repels anionic analytes, especially polyanions such as ONs, this approach limits the AE interaction to the volume of the nanobeads on the surface. Where the nanobead functionality is a quaternary amine (strong anion exchanger), this approach also minimizes nonspecific interactions. While these phases may offer slightly better capacity than phases with direct surface functionalization, they still offer very low capacity compared to the fully porous bead phases. Examples of this nonporous approach include the Dionex DNAPac columns (PA100, PA200). There are numerous examples of ON analyses on AE nanobead-coated resins. Typically, these are used for purity analysis of synthetic ON after deprotection and release from synthesizers.[26,69] These phases were also used to purify[27,28] and assay[27,29,55] ribozyme activity. In other assays they were used to resolve *different* "n – 1" impurities from desulfurized ONs,[16] to verify the purity of 5-halogenated uracil-containing ONs[21] and N3′-P5′ phosphoramidates,[59] a dT-EDTA containing probe (as a spin label[70]), a Terbium (chelated through tetraisophthalamide)-linked probe,[71] a caged-morpholino probe,[72] and TNA (L-α-threofuranosyl)-based ONs.[56] They were also employed to characterize ONs intended for generation of reliable microarrays.[73] The DNAPac phase was used to resolve several intermediates in the synthesis of ONs harboring an internal 3′-phosphoglycolate, 5′-phosphate gapped lesion,[74] ONs containing ribodifluorotoluyl nucleotides,[45] and ONs containing 2′-deoxyxanthosine,[22,75] and were used to monitor the conversion of spiroiminodihydantoin to guanidinohydantoin,[77] and to resolve spiroiminodihydantoin and oxaluric acid adducted ONs prior to MALDI-TOF MS.[77] This phase proved capable of resolving 17nt self-complementary ONs containing 1-methyladenine and N6-methylated adenine (converted from 1-methyladenine during deprotection) using an ammonium acetate eluent[57] and resolved diphosphophosphonates, intended as anti-HIV reverse-transcriptase inhibitors, using a very low ionic strength triethylammonium bicarbonate eluent.[35,78]

While these columns may have very low capacity, they are also used for purification, especially in cases where the highest purity is required. DNAPac columns have been employed for purification in support of studies on RNA/DNA crystal structure;[79] Hairpin Ribozyme modeling;[30] stabilization of the HIV TAR hairpin complex;[36] duplex hairpin conversion;[31] cleavage stereospecificity of a *Serratia marcescens* endonuclease;[17] mischarging of class I[80] and recognition of[81] aminoacyl t-RNA synthetases; electrostatic interactions in a peptide-RNA complex;[82] chemical syntheses of substrates inhibiting RNA-RNA ligation;[32] topoisomerase substrate specificity[83] and light-dependent RNAi function using nucleobase-caged RNA.[46] Other important purifications include those of protein- or peptide-associated DNA, such as antisense delivery systems[84] and peptide-DNA footprinting.[85] The above examples involved purification of synthetic DNA after synthesis and/or deprotection. The DNAPac phases have also been used to purify ONs from plasma[5] and tissues[18] in metabolic studies. In a most unusual application, the DNAPac phase was used to purify gold nanocrystal-DNA conjugates away from the gold nanocrystals and nanocrystals coupled to multiple single-strand (ss) DNA.[86]

2.5.2.3 Monolithic and Hybrid Monolithic Phases (Surface-Latexed Monoliths)

Polymer monoliths comprise a single polymer rod with interconnecting pores and channels.[87] The pore structure and surface area can be independently controlled during polymer synthesis.[88,89] Thus, polymer monoliths can be prepared with quite large pores and still produce significantly greater surface area than pellicular phases. This combination results in reasonable capacity with relatively low pressure and can support fast flow rates compared to both porous and pellicular anion exchangers. While there are relatively few polymer monoliths available for nucleic acid separations, two forms of monolith AE columns are commercially available: these are directly aminated and nanobead-coated monoliths. An example of a directly aminated polymer monolith is the CIM disk monolith from BIA (Ljubljana, Slovenia). This was used to purify a hypoviral dsRNA.[90] A hybrid polymer monolith is also available. The hybrid monolith is prepared by functionalizing the monolith to present a cation-exchange (anionic) surface (like the DNAPac), followed by coating the phase with aminated anion-exchange nanobeads. Because the porosity of the monolith can be tailored to accommodate relatively large AE nanobeads, and the nanobeads can be engineered so that their entire volume is accessible to even quite large ONs, increasing the nanobead diameter increases the nucleic acid capacity. Thus a capacity approaching that of the fully porous beads can be prepared, but because the hybrid monolith also benefits from the (faster) convective mass transfer of nonporous phases, the monolith also exhibits the improved resolution, approaching that of nonporous phases. One example of this hybrid approach is the DNASwift SAX-1S monolith (Dionex). This is a new stationary phase but has been used to purify ONs, resolve ONs derived with several different fluorophores, separate phosphorothioate diastereoisomers, as well as 2′,5′ RNA-linkage isomers, and coupled to automated desalting for ESI-MS.[91]

2.6 OLIGONUCLEOTIDE PROPERTIES OF INTEREST FOR ANION-EXCHANGE CHROMATOGRAPHY

2.6.1 Oligonucleotide Anion-Charge Sources

ONs include three basic chemical structure families: phosphate moieties, ribose (or deoxyribose) sugars in furan form, and nitrogenous bases as purines or pyrimidines. In living organisms, the purines exist as guanine (G, dG) or adenine (A, dA) and the pyrimidines are cytosine (C, dC) and thymidine (dT) (or uracil [U] in place of thymidine in RNA). The phosphate moieties are ester-linked to the 5′ and 3′ hydroxyls of ribose or deoxyribose, forming the "backbone" of these nucleic acids. This backbone is highly charged and hydrophilic. Conversely, the nitrogenous bases attached to C-1 of the deoxyribose (or ribose in RNA) are relatively hydrophobic and can participate in pi-pi bond interactions (base stacking). Therefore, in single-stranded form, ONs present both hydrophilic and hydrophobic surfaces. As single strands (ss), the hydrophilic and ionic sugar-phosphate backbone linkages may be free to rotate, allowing the hydrophobic bases to conform to minimum energy configurations, or may be somewhat constrained under low temperature conditions, that can influence their interactions with AE stationary phase surfaces. This rotation is inhibited in ds forms, where the bases are directed inward in the double helix. Thus ds nucleic acids behave as *less* hydrophobic analytes than ssRNA or ssDNA.

2.6.1.1 Sugar-Backbone Linkages

The phosphodiester linkages to ribose contribute a consistent net negative charge at the pH values normally used for ON AEC (pH 6–12; see Section 2.2). At elevated pH values, the presence of a 2′-hydroxyl on ribose (absent in deoxyribose) confers a partial additional charge due to release of the proton, creating a formal oxyanion, but only at *very* high pH values (>12).

2.6.1.2 Tautomeric Oxygens

One of the nitrogenous purine bases (G) and one of the pyrimidines in DNA (T) or RNA (U) harbor a tautomeric oxygen that converts to an oxyanion form as the pH increases from ~7 to ~11.5. Therefore, ONs with identical lengths but different base compositions may be differentially retained at increasing pH values.

2.6.1.3 Chemical Modifications

In order to confer resistance to nucleases found in mammalian tissues, developers of ONs for *therapeutic* applications have prepared numerous chemical modifications to their nucleic acid candidates. These may *introduce* new charges, *increase* anionic affinity, or *mask* the native anionic character of the molecules. Some modifications also introduce chirality that produces diastereoisomers. Under some circumstances, these can be resolved even from one another by anion-exchange chromatography.[9,91] This offers an opportunity to test possible therapeutic efficacy differences between the different diastereoisomers when the ONs are intended for therapeutic applications.

2.6.2 Parameters Influencing Oligonucleotide Retention via Anion Exchange

2.6.2.1 Effects of pH (Influences Net Charge and Hydrogen Bonding)

Net ON charge at pH values of 6–8 derive primarily from the sugar-phosphate backbone. Hence, at these "low" pH values, net charge is proportional to length, and oligonucleotides may elute from AE columns in approximate order of length. As with all chromatographic separations, there are other interactions that influence retention so elution order will not *always* be determined only by the number of phosphodiester linkages, but charge will be the primary mechanism of retention. If one buffers the eluents at higher pH values, retention will increase as the pH increases, but retention will not necessarily increase directly with oligo length. Instead, the retention will increase proportional to the combination of length plus the fractional G + T content (G + U in RNA). This is due to ionization

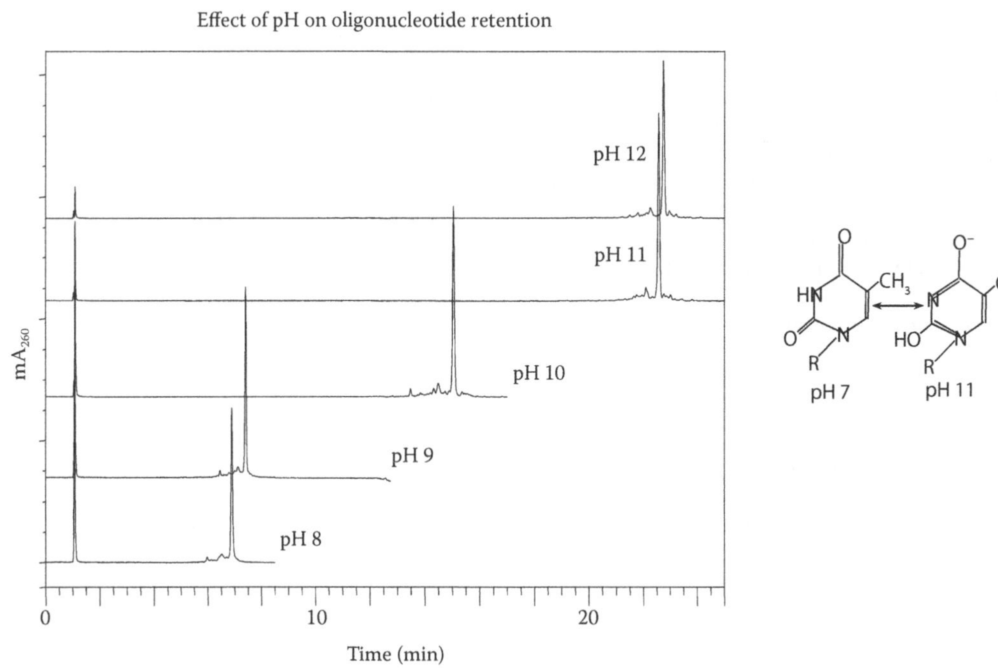

FIGURE 2.1 The effect of pH on retention of a 25-base oligodeoxynucleotide using NaCl eluents without solvent, and at pH values from 8 to 12. System: Dionex DX600 inert quaternary gradient LC system. Column: DNAPac PA200. Gradient: 330–900 mM NaCl in 30 min, 25°C at 1.2 mL/min. The pH values are as indicated. Oligonucleotide sequence: 5′ CTG AAT GTA GGT TCT CTA ACG CTG A 3′. Inset shows ionization of thymidine. Detection here and in most other chromatographic assays is absorbance at 260 nm. (Modified from Thayer, J. R., et al., *Analytical Biochemistry*, 338, 39–47. Copyright 2005, with permission from Elsevier.)

of the tautomeric oxygen on each G and T (or U). Depending on the ON length and adjacent bases in the sequence, pKs for oxyanion formation may be between 9.5 and 10.5. Hence, ONs retention will increase in proportion to the percentage of these bases in the sequence, as well as with length. This is illustrated in Figure 2.1, where a 25-base ON is chromatographed with a common salt gradient at pH values from 8 to 12. The retention of this 25-mer differs only by ~0.5 min between pH 8 and 9, and even less between pH 11 and 12. However, at pH values between 9 and 10, and between 10 and 11, elution of this oligo increases by 6–7 min. This is consistent with ionization of the tautomeric oxygens on T and G, where the pKs for these oxygens are in the range of 9.5–10.5.

The presence of the tautomeric oxygens on the bases at the 3′ and 5′ ends will have a disproportionate influence on the retention because they will have greater opportunity for interactions with the stationary phase than those constrained in the middle of the sequence. Similarly, several consecutive G and/or T bases, even in the middle of the sequence, may tend to align together with the stationary phase to disproportionally increase retention.

In contrast to the effect of pH on charge is its effect on hydrogen bonding. Aptamers are ONs selected for their ability to interact with high specificity and affinity to *target* molecules, such as protein receptors.[48] The capability is produced by the aptamer's secondary structure that may be composed of both Watson–Crick and other hydrogen bonds (e.g., G-tetrad ladder segments, Hoogstein base pairing, etc.). Hence aptamers are usually chromatographed under *denaturing* conditions, as

TABLE 2.1
"Worst-Case" Secondary Structure Oligonucleotides

Poly G: dG_{18}

All are potential self-annealing sites in a tetrad ladder

```
5′GGG GGG GGG GGG GGG GGG  3′       5′  GGG GGG GGG GGG  GGG GGG3′
          5′  GG GGG GGG  GGG GGG GGG G  3′        5′   GGG GGG GGG GGG GGG GGG 3′  etc. →
5′GGG GGG GGG GGG GGG GGG  3′       5′  GGG GGG GGG GGG  GGG GGG 3′
       5′       GG GGG GGG GGG GGG  GGG G′  3′       5′   GGG GGG GGG GGG GGG GGG 3′
```

Mixed base PO_{30} sequence
ACG TAC GTA CGT ACG ACG TAC GTA CGT TCG

Potential hairpin formation

```
5′ ACGTACGTACGTACG\
   ||  |||||||||
 3′ GCTTGCATGCATGC/
```

Potential self-annealing sites

```
5′ACGTACGTACGTACGACGTACGTACGTTCG  3′
    ||||||||||||   ||||||||||||

 3′ GCTTGCATGCATGCAGCATGCATGCATGCA 5′

5′ ACGTACGTACGTACGACGTACGTACGTTCG 3′
                  ||   ||   ||||   ||
             3′   GCTTGCATGCATGCAGCATGCATGCATGCA 5′
                  ||||   ||  ||      |||||||||
5′ACGTACGTACGTACGACGTACGTACGTTCG 3′  5′ ACGTACGTACGTACGACGTACGTACGTTCG 3′
                ||||||||||||       ||||      ||       |||||||
       3′ GCTTGCATGCATGCAGCATGCATGCATGCA5′  3′GCTTGCATGCATGCAGCATGCATGCATGCA
       5′ etc. →
```

ONs with different possible intra- or interstrand complementarity may resolve as multiple forms during chromatography. Example ONs (but not aptamers) that participate in these multiple interactions are illustrated in Table 2.1 and Figures 2.2 and 2.3. Table 2.1 depicts how G-tetrad ladders may form in poly-G sequences and also indicates the multiple Watson–Crick interactions possible in a "worst-case" ON sequence not harboring poly-G sequences. The relative propensities for G-tetrad ladder formation appear to differ in perchlorate eluents having different countercations (e.g., Li^+ versus Na^+).

Figure 2.2 shows that $NaClO_4$ eluents at pH 8 and at 37°–52°C fail to disrupt the G-tetrad ladders, so they chromatograph as very large assemblies and do not typically elute during these "normal" chromatographic conditions. When the same sample is chromatographed at pH 11, the G-tetrad ladders are fully disrupted, and the sample components (here as a set of deoxy-G oligos 12–18 bases long) are readily separated and identified. On the other hand, when the eluent cation is lithium, the formation of G-tetrad ladders is (at least partially) inhibited. With $LiClO_4$ eluent, and under conditions otherwise essentially identical to the $NaClO_4$ eluent system, at least a considerable fraction of the G-tetrad ladders are eluted and identified as the $Pd[G]_{12-18}$ set.

A study using the worst-case mixed-base PO_{30} ODN depicted in Table 2.1 is shown in Figure 2.3. This ON, prone to multiple Watson–Crick interactions, is chromatographed using $LiClO_4$ eluent at pH values of 8, 9, 10, and 11 at both 30° and 60°C. At 30°C (top panel), at least two sets of eluting components are observed, indicating different interactions, appearing at ~8.5 and 11.5 min at pH values of 8, 9, and 10. That these peaks appear at about the same position at pH 8–10 indicates that the tautomeric oxygens on the bases are *not* participating in AE interactions, suggesting that they are embedded in double-helical form. At pH 11 where these hydrogen bonds are broken, peaks eluting at both 8.5 and 11.5 disappear, and a new single major peak elutes at ~ 10.5 min. Because

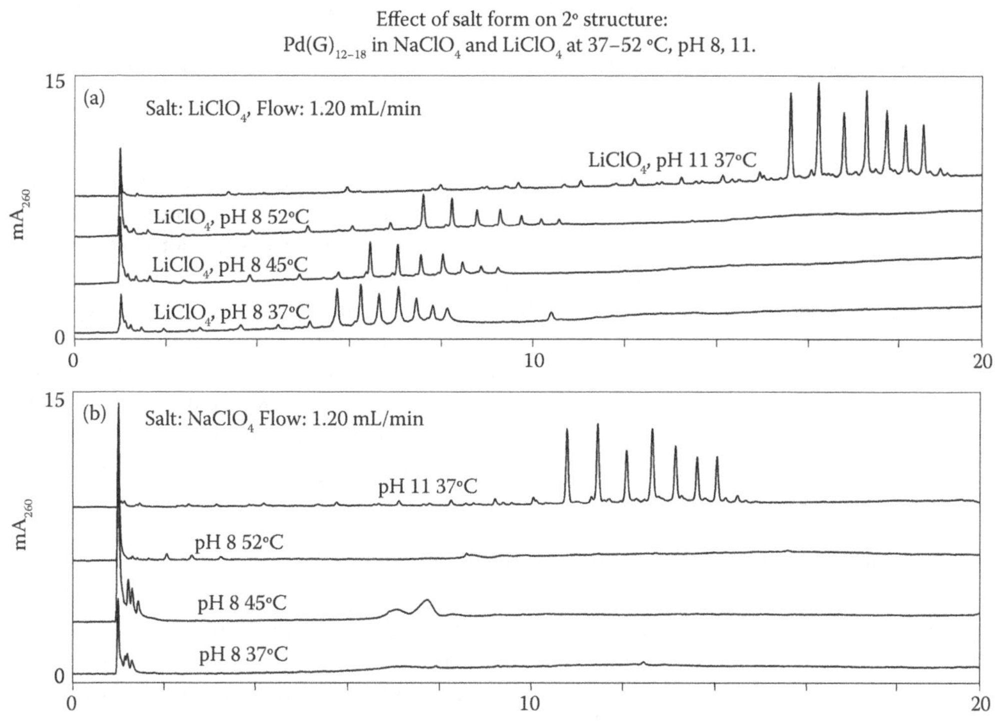

FIGURE 2.2 Effect of pH and salt form on the stability of G-tetrad hydrogen bonds during anion-exchange chromatography. System: Dionex DX600 inert quaternary gradient LC system. Column: DNAPac PA200. Gradient: (a) 62–194 mM $LiClO_4$ in 22 min and 1.2 mL/min; pH and temperature as indicated. (b) 110–218 mM $NaClO_4$ in 18 min and 1.2 mL/min; pH and temperature as indicated.

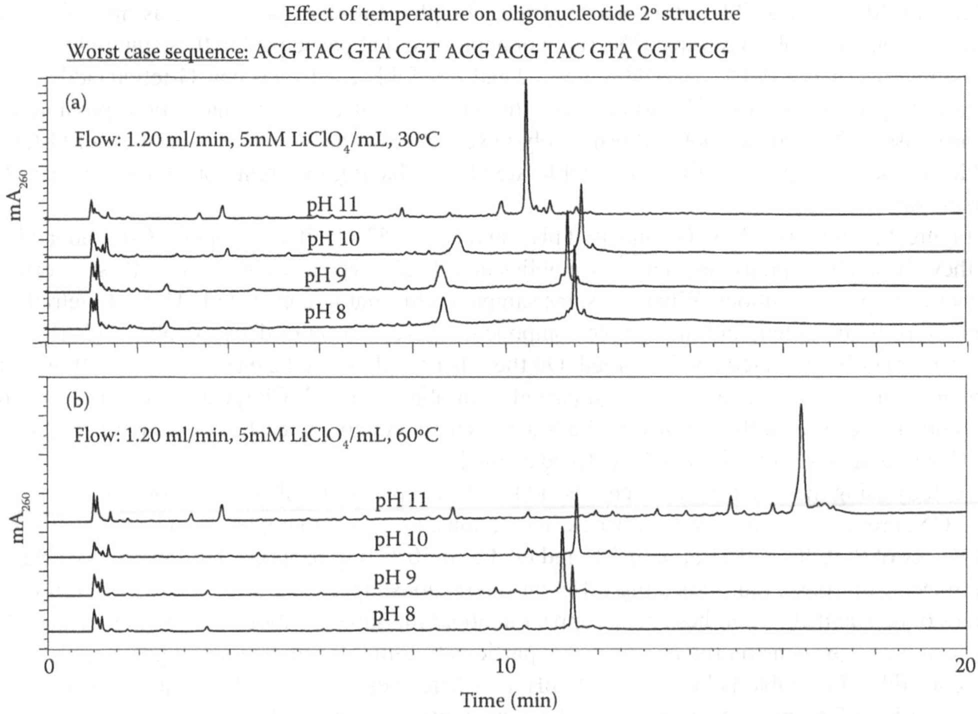

FIGURE 2.3 Effect of pH and temperature on the stability of Watson–Crick hydrogen bonds during anion-exchange chromatography. System: Dionex DX600 inert quaternary gradient LC system. Column: DNAPac PA200 4 mm × 250 mm. Gradient: 80–194 mM $LiClO_4$ in 19 min at 1.2 mL/min; pH and temperature as indicated.

each T and G (15 in the 30-mer) will increase the net charge on the ON to the greatest level at pH 11, the later eluting components (at ~11.5 min) at the lower pH values likely represent *interstrand* Watson–Crick bonds, as they are much more likely to form *longer* ON associations. Those eluting at ~8.5 min likely represent elution of ONs with *intrastrand* Watson–Crick bonds. However, at 60°C (bottom panel) and pH values of 8, 9, and 10, only one set of components appear, and they elute at ~11.5 min. Because retention of these components does not appear to increase consistently with pH they are (again) most probably not linear single strands. This appears to be confirmed at pH 11 as the components eluting at ~11.5 min at the lower pH values are absent, and the primary peak, probably the ss form, elutes later, as expected for a linear ON at elevated temperature, at ~16.5 min.

These examples illustrate the utility of high-pH chromatography for controlling hydrogen bond interactions.

2.6.2.2 Considerations on RNA Chromatography at High pH

We learn through introductory biochemistry classes that RNA degrades upon exposure to pH values as low as 8. Hence pH values above 8 have been considered a poor choice for chromatography of *RNA* ONs. However, pH-induced RNA degradation is a slow process, and does not usually occur in the time frame for ON analysis on column. To evaluate the RNA degradation by pH during chromatography, an RNA sample was run at pH 11 using different flow rates to change the residence time of the RNA on the column (Figure 2.4). The elution profiles (number of peaks and relative peak area for each) remained essentially unchanged for residence times from 7 to 21 min, indicating a lack of on-column degradation for that time period, even at this very high pH. Control of *extensive* hydrogen bonds, such as those formed in G-tetrad ladders of considerable length, may require pH values

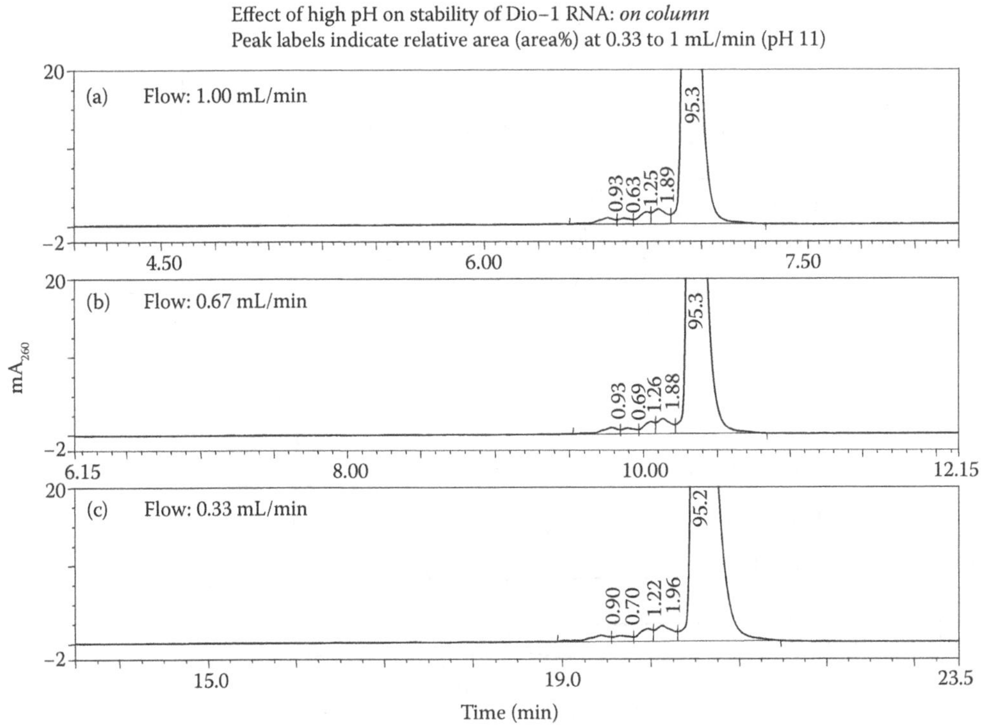

FIGURE 2.4 Chromatography of RNA at high pH on DNAPac PA200. System: Dionex DX600 inert quaternary gradient LC system. Gradient: 99–231 mM $NaClO_4$ in 3.2 column volumes and 30°C. Flow rates: (a) 1.0, (b) 0.67, (c) 0.33 mL/min. (Modified from Thayer, J. R., et al., *Analytical Biochemistry*, 361, 132–139. Copyright 2007, with permission from Elsevier.)

up to pH 12.4, or temperatures above 95°C. When such pH values are used for *purification*, collection of ONs must include immediate buffering to pH values below 8. This is easily accomplished by prefilling the collection tubes with effective concentrations of an appropriate buffer. On the lower pH end, depurination becomes more likely as the pH falls to 6 and below, and because depurination will lead to strand scission, pH values of 6.5 or lower are not recommended for ON AEC. Regarding this lower pH limit for ON chromatography, we do not normally recommend pH values below 6.5.

2.6.2.3 Salt Form: Influence of Hydration

Hatefi and Hanstein described a "chaotropic series" of salts that influences the structure of water during solubilization of biopolymers.[92] They describe how NaCl tends to promote or extend biopolymer hydration shell formation, while use of increasingly "chaotropic salts" such as NaBr, $NaNO_3$, $NaClO_4$, and NaSCN tend to minimize their hydration. Bourque and Cohen,[5] Bergot and Egan,[4] and we[7] employed these salts as eluents to improve resolution of PS ONs (using the higher order chaotropic salts NaBr, $NaClO_4$, and NaSCN). These chaotropic salts have the effect of rendering ON hydration waters more disordered and "Lipophilic." These properties tend to improve ON peak shape (especially on the DNAPac PA100) and also tend to minimize the influence of the nitrogenous bases on retention by AEC when compared to NaCl eluents.[93]

Similarly, the more lipophilic $NaClO_4$ tends to improve the peak shape of fluorophore-derivatized ONs when compared to NaCl under otherwise similar conditions (Figure 2.5).

Use of NaCl and $NaClO_4$ eluent can also aid in resolving "n−1" and "n+1" failure sequences. Figure 2.6 provides an example of such a case (here using very steep, or survey gradients; reducing the gradient slope will improve resolution). Here, the retention of a 24-base ON (Dx86) is compared

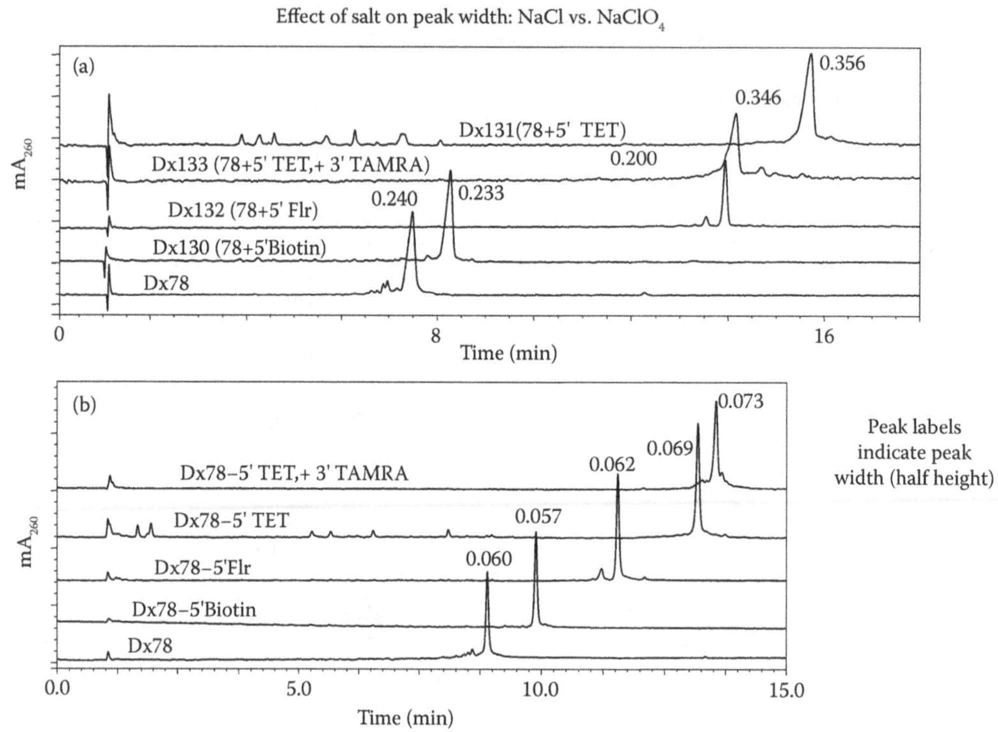

FIGURE 2.5 Effect of salt form on peak width (PW½) of a 25-base variously conjugated oligonucleotide. System: Dionex DX600 inert quaternary gradient LC system. Column DNAPac PA200 4 mm × 250 mm. Gradients: (a) 18 mM NaCl per mL at pH 8 and 35°C; (b) 10 mM $NaClO_4$ per mL at pH 8 and 35°C, both at 1.2 mL/min.

with that of a 23-base "n–1" ON (Dx89). Because the 3′ base in Dx89 is a G, it tends to be retained to a greater degree in NaCl eluent than the longer Dx86 that harbors A at its 3′ terminus, producing insufficient resolution at pH 8. Increasing the pH enhances the influence of the terminal G on Dx89, increasing its retention to elute significantly later than the longer Dx86. Use of $NaClO_4$ as eluent minimized this effect, improving resolution of these two ONs with the shorter Dx89 eluting earlier than the longer Dx86 at pH 8.

One attribute of AEC that impacts its practice is the well understood but often overlooked issue of the sensitivity of AE stationary phases to metal contamination. The demonstrated utility of NaCl and $NaClO_4$ eluents for managing ON resolution leads ON developers naturally to their use. However, NaCl is quite corrosive, and $NaClO_4$ is a moderately strong oxidizer, particularly when in contact with stainless steel (SST), and at extremes of pH. Because many HPLC systems used for AEC employ SST components, care must be taken to limit the exposure and minimize the corrosion that releases metals that foul AE columns. We observe that many, but not all, users of SST systems for AEC of ONs report relatively short column life. In virtually all of the reports we have examined, those users employ NaCl eluents but do not routinely wash and passivate their systems or employ chelating agents in their eluents to minimize fouling of their columns. As an example of this process, one user photographed the active inlet valve of their SST pump after 1 year of use with NaCl eluents. They observed significant corrosion and provided the photograph (see Photograph 1). Likely, the most common metal contamination of AE columns arises from use of SST eluent reservoir filters and *in-line* eluent or sample filters. These components harbor very high surface area and are not often rinsed or replaced. Hence they afford a ready surface for progressive corrosion and leach oxidized metals into NaCl and $NaClO_4$-based eluents. It is likely that use of high pH will accelerate oxidation and exacerbate column fouling. Diligent system rinsing and regular preventive

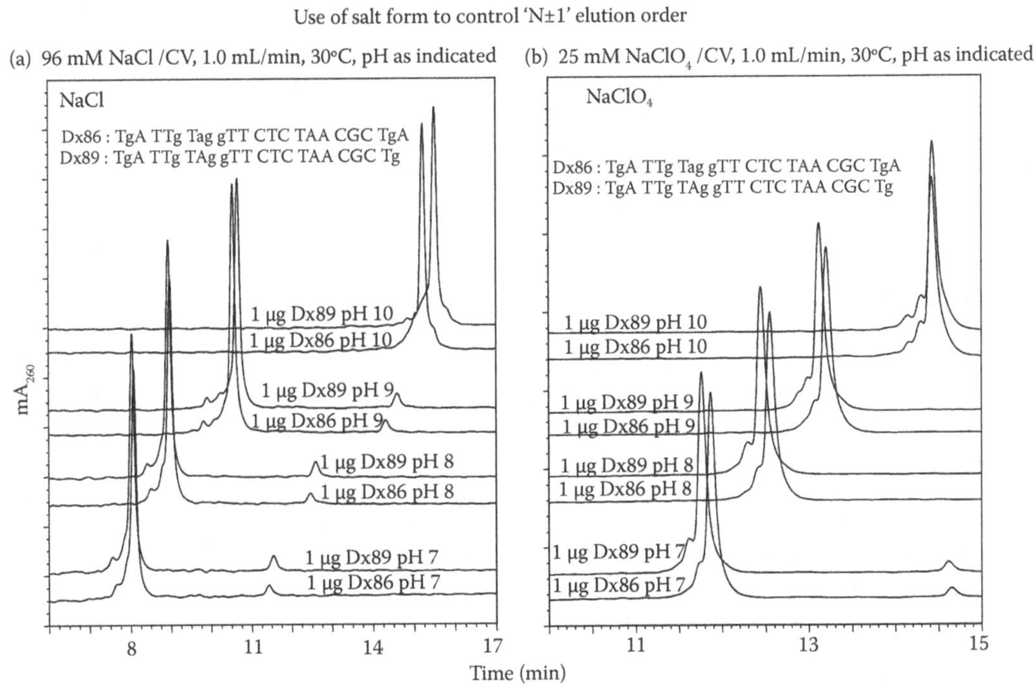

FIGURE 2.6 Use of pH and salt form to control ON *selectivity* on a DNAPac PA200. System: Dionex DX600 inert quaternary gradient LC system. Gradients: (a) 96 mM NaCl/column volume, at 30°C and at the indicated pH; (b) 25 mM $NaClO_4$/column volume, at 30°C and at the indicated pH.

PHOTOGRAPH 1 Example of salt-induced corrosion of stainless steel pump parts by exposure to NaCl eluents. Photo from active inlet valve of a popular SST HPLC pump employed for 1 year with NaCl eluents. Corrosion is evident on all metallic parts.

maintenance, including system passivation and replacement of SST frits and reservoir filters, will help prolong system and column life. Our experience with these issues has led us to consider and use inert HPLC systems with wetted parts being of PEEK (poly-ether-ether-ketone) or titanium, as these are not corroded by chloride or perchlorate salts.

2.6.2.4 Solvent: Influence of Hydrophobic Interactions

Like use of $NaClO_4$, addition of solvent to NaCl eluents suppresses hydrophobic interactions between the ON and the AE phase. In the case of ONs derivatized to very hydrophobic compounds (e.g., fluorophores and/or quenchers used for reporter probes in RT-PCR and microarray assays, or with cholesterol for delivery of RNAi therapeutics), hydrophobic interactions can severely alter chromatographic peak shape, introducing significant peak tailing. In order to control peak tailing, addition of solvents such as MeCN (acetonitrile) is helpful. For reducing the hydrophobicity-induced peak tailing on ONs, 2–30% MeCN may be used. This is illustrated in Figure 2.7, showing how solvent addition affects ON retention and peak width when using NaCl eluents (as in the top panel of Figure 2.5).

Many, but not all, AEC phases are stable to such solvent levels so adherence to manufacturer's recommendations may be noteworthy for improving column longevity. The solvent concentration necessary to minimize peak tailing can be conveniently evaluated with ternary or quaternary pumping systems by proportioning the solvent into the normally aqueous eluents. In most cases, 5–20% MeCN will greatly improve peak shape induced by covalent modification with fluorophores. Very hydrophobic conjugates (e.g., cholesterol) may require higher solvent concentrations, depending on

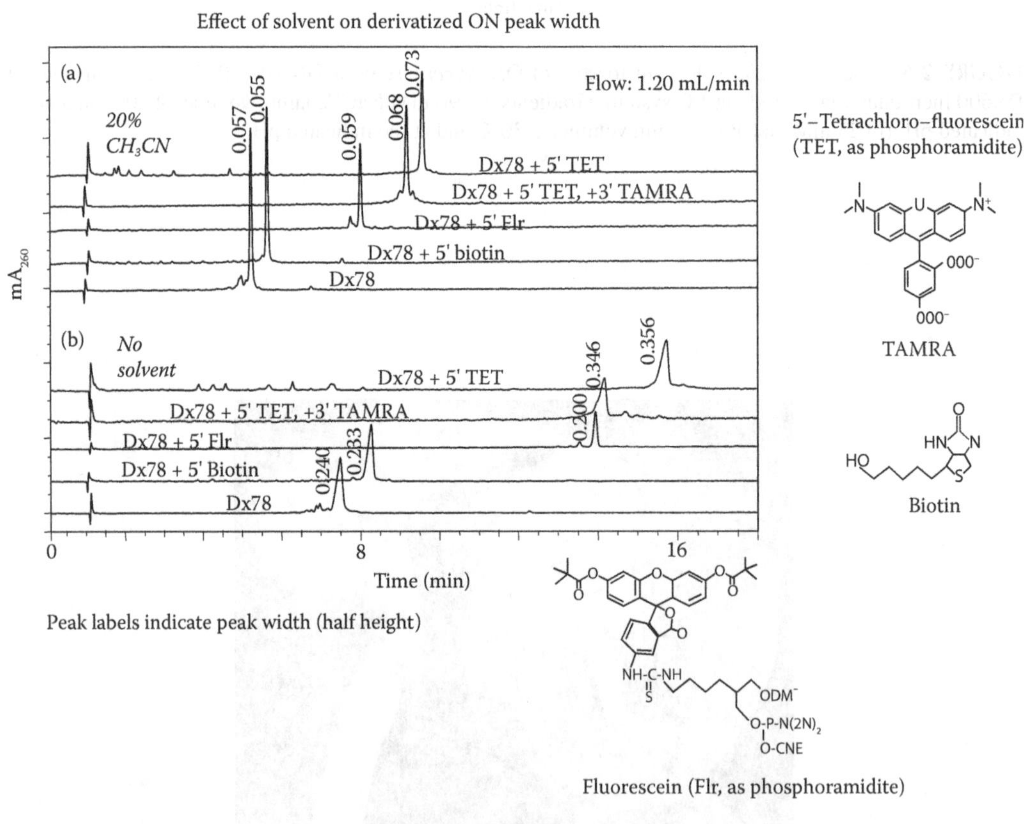

FIGURE 2.7 Effect of solvent on peak width (PW½) of a 25-base variously conjugated oligonucleotide. System: Dionex DX600 inert quaternary gradient LC system. Column: DNAPac PA200 4 mm × 250 mm. Gradient: 18 mM NaCl per mL at pH 8 and 35°C, 1.2 mL/min (a) with 20% acetonitrile in eluent and (b) without solvent in eluent.

the AEC phase. Note that a first use of such solvents will tend to result in significant baseline drift. The drift may be due to release of tightly bound, non-ionic components from previous samples, or even from stationary phases not rinsed with solvent during production. This drift can be eliminated by washing the column with high [solvent] (60–100%) for 1–2 hours.

Hydrophobic interactions between nitrogenous bases and the AEC phase can influence retention without materially altering peak shape. Depending on the sequence, conjugation, and impurities in the synthetic starting materials, ON impurities may elute under the primary synthetic product peak. One method to "tease" impurities away from the main product is to proportion in smaller concentrations of solvent. Typically, we proportion 2–10% MeCN into the eluent to accomplish this effect, but we are aware of some cases where as little as 0.5% MeCN was used to better resolve an impurity from a synthetic primary product. It should be noted that adding solvents to AEC eluents will also cause the compound to elute somewhat earlier than with the same eluents lacking solvents.

2.6.2.5 Temperature: Effect on Dissociation Kinetics and Hydrogen Bonding

Control of temperature during AEC is recommended as this will improve analytical precision and run-to-run reproducibility. However, temperature control may also be helpful for controlling resolution, selectivity, and where necessary, hydrogen bonding.

Resolution of ONs in AEC tends to improve with increasing temperature owing to increases in *retention*, combined with increases in the rates of binding and release from the phase. This combination is often employed to resolve longer ONs from their "n–1" failure sequences. In cases where resolution of multiple components is necessary, increasing or even *de*creasing temperature may improve selectivity (ON analyte spacing) to separate critical ON impurities. Examples of these cases include resolution of multiple primers during QC of ON-based diagnostic kits (having mixtures of ON primers for detecting multiple target sequences) and in some cases to *avoid* separation of diastereoisomers of single sequence oligos (e.g., those introduced by phosphorothioation at specific sites in an ON sequence). Use of elevated temperature to control hydrogen bonding within, or between, ONs is a widely used technique. AEC of dsRNA and DNA/RNA hybrids can also benefit from careful selection of temperature during chromatography. In these analyses, use of temperatures *below* the T_m of the dsRNA or hybrid can result in good resolution of excess single-stranded RNA or DNA from one another. The duplex examples of these approaches will be discussed in Section 2.7.3.

Analyses at multiple temperatures may also reveal the best conditions for resolving duplexes formed between full length and "n±x" oligomers and may distinguish between perfectly complementary duplexes and those harboring mismatches. Finally, AEC at temperatures exceeding the T_m of a duplex will often produce resolution of the sense and antisense (single stranded) components as well as their impurities and/or failure sequences.

2.6.3 Common Oligonucleotide Modifications and Their Influence on AEC

2.6.3.1 Backbone/Linkage Modifications

2.6.3.1.1 Phosphorothioates

Phosphorothioate PS linkages replace a nonbridging oxygen atom in the phosphodiester linkage with a sulfur atom. Unless both nonbridging oxygens are replaced (i.e., phosphoro*di*thioates), this introduces chirality to the linkage and produces a pair of diastereoisomers at each converted linkage. In early antisense ONs, developers usually employed *full* phosphorothioation, where every linkage was a PS. In such cases, the number of possible diastereoisomers was 2^{n-1}, where n indicated the number of linked bases. Thus a 15-mer might harbor 16384 (2^{14}) different diastereoisomers. This is far too many for chromatographic resolution, so their presence results in broad peaks.[7] The presence of the sulfur atom also results in higher binding *affinity* of the PS ONs (versus phosphodiesters) for AEC phases. Because of this profound difference in affinity, fully 'thioated ONs are retained to a much greater degree, allowing resolution of fully PS-linked ONs from incompletely 'thioated

ONs that elute earlier (Figure 2.8). In this case, resolution of the all PS and incompletely 'thioated sequence is dramatically improved at pH 12 compared with pH 8 and this is quite common.

Recently, developers of aptamers and RNAi therapeutics report sparing use of PS linkages in their sequences. Where two PS linkages are inserted, the number of possible diastereoisomers is limited to *four*, and such products have been reported.[41] While RP and IP-RPLC methods were used in reports of the diastereoisomer separations in ON having 2–10 bases,[62,94,96] only AEC has shown the ability to resolve the PS diastereoisomers in longer ONs. Because the R_p and S_p isomers are isobaric, even RPLC coupled to single-stage MS does not allow discrimination of these PS forms. Separation of one 14-base ON with a single PS linkage (contaminating a phosphoro*di*thioate linkage) was demonstrated on a porous AE column,[9] while that of 11- and 12-base ONs with single PS linkages was accomplished on a pellicular AE column.[10] Recently, we demonstrated resolution of four diastereoisomers in a 37-base aptamer harboring two PS linkages using both a pellicular column and a hybrid monolith.[97] We discuss similar work using a 21-base ON in Section 2.8.4.

2.6.3.1.2 Phosphoramidates

Phosphoramidate (PN) linkages result from replacement of a *bridging* oxygen in the phosphodiester with a nitrogen atom. ONs containing these linkages exhibit significantly more stable duplexes with complementary RNA strands, and with complementary strands also composed of these PN linkages. Purity analyses of ONs harboring these linkages have employed AEC, but also CE and RPLC.[19]

2.6.3.1.3 Linkage Isomers (2′,5′-Linkages)

With the potential for RNA therapeutics offered by RNAi, miRNA, and *anti*-miRNA approaches, synthetic RNAs were produced in support for these studies. There is potential for phosphoryl

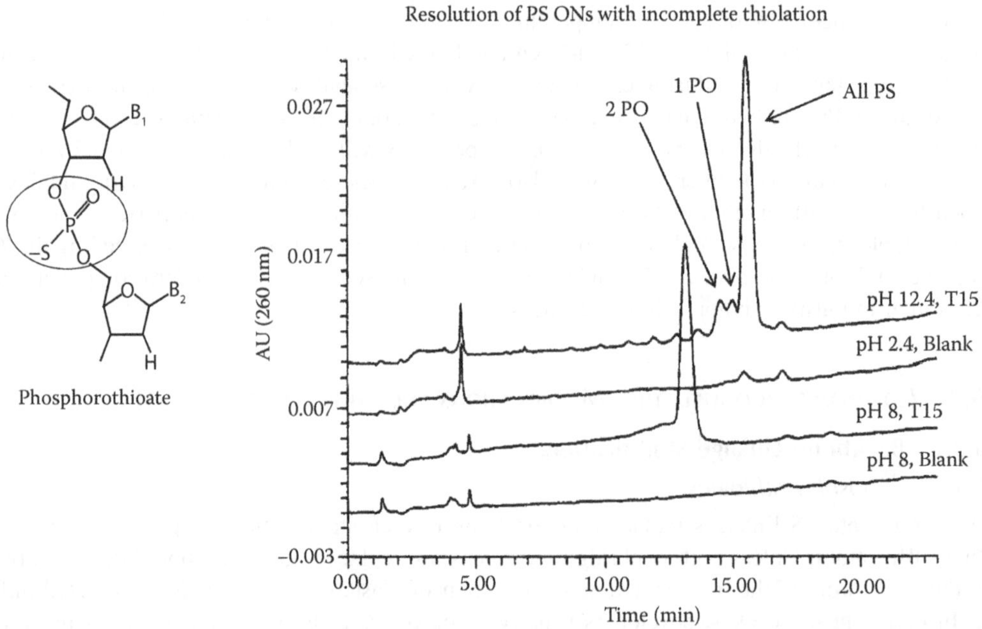

FIGURE 2.8 Anion-exchange chromatography of phosphorothioated ONs. Resolution of fully substituted (all PS) from the incompletely sulfurized components (1PO, 2PO). System: Dionex DX500 inert quaternary gradient LC system. Column: DNAPac PA100. Gradient: 56–323 mM $NaClO_4$ in 20 min at 1.5 mL/min, 30°C and pH 8 (bottom two traces) and 56–330 mM $NaClO_4$ in 20 min and 30°C at pH 12.4 (top two traces). For all traces a curved gradient (4 see Ref. 93 for gradient equation) was used. (Modified from Thayer, J. R., et al., *Methods in Enzymology*, 271, 147–174. Copyright 1996, with permission from Elsevier.)

migration during synthesis, deprotection, and release from the solid phase supports.[37] This can cause formation of aberrant 2′,5′-linkages. Because these present significant probabilities for "off-target" effects (see Ref. 38), methods to identify their presence in therapeutic RNA candidates were sought. These will be discussed further in Section 2.8.2.

2.6.3.2 Sugar Modifications

2.6.3.2.1 RNA versus DNA

As shown above for phosphorothioate linkages, replacement of a single atom in one or two DNA linkages can produce significant ON retention differences. This also applies to the addition (or removal) of oxygen at the 2′ position on the ribose in nucleic acids. Examples of 21-base sense (or guide) sequence synthesized as RNA and DNA were purified on a DNASwift hybrid monolith (Figure 2.9). Under these conditions, the DNA eluted significantly earlier.

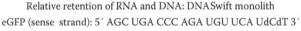

2.6.3.2.2 The 2′ Modifications

The 2′ hydroxyl group is not required for siRNA activity,[44] and AEC has been employed for analyses of oligoribonucleotides harboring 2′ modifications that exhibit improved nuclease resistance[42] and that function in therapeutic RNAi[41,43] but also as Antisense therapeutics[33] and as therapeutic aptamers.[34,47] In one example, PEG is applied to an aptamer containing both 2′-*O*-methyl, and 2′-fluoro modified ONs.[48] AEC purity analyses have also been used in a variety of other common[23,24] and uncommon[54,98] 2′-substituted RNA protecting groups, as well as for DNA/LNA (locked nucleic acid) molecular beacons.[99]

2.6.4 Base Modifications

2.6.4.1 Trityl-on and Fluorophore-Linked ONs

The "Trityl" protecting group employed to prevent multiple coupling reactions during the synthetic cycle represents a hydrophobic "handle" for preliminary ON purification. While this is most

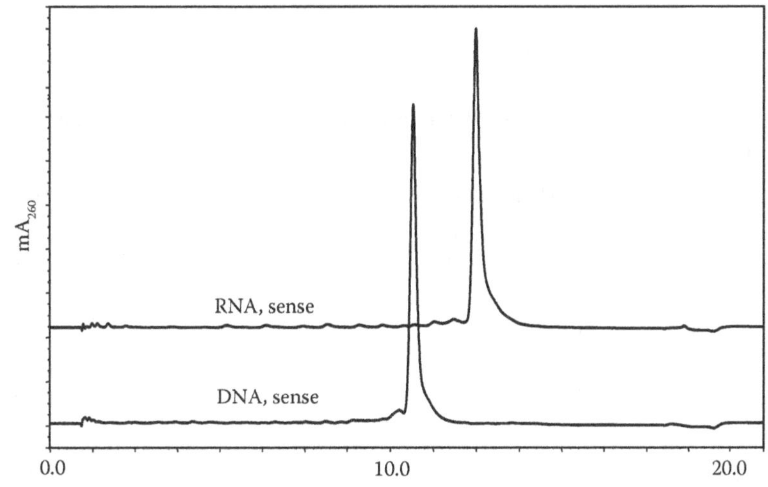

FIGURE 2.9 The relative selectivity of RNA versus DNA on the DNASwift SAX-1S. System: Dionex DX600 inert quaternary gradient LC system. Conditions: 300–600 mM NaCl in 16.7 min, 1.5 mL/min, and 30°C. Sample sequence as indicated.

familiar as "Trityl-On / Trityl-Off" purification on reversed-phase cartridges, this technique has also been used for "on-column" purification using AEC.[11] Similarly, ORNs and ODNs conjugated to various fluorescent dyes are used as probes and molecular beacons in diagnostic applications and in RT-PCR assays. AEC analyses of this ON class are productively influenced by eluent salt, temperature, and solvent, as discussed in Sections 2.6.2.3 and 2.6.2.4.

2.6.4.2 Alternate Bases

One interesting example includes use of alternate bases, such as 2′-deoxyisocytidine, 2′-deoxy-5-methylpseudocytidine, and 2′-deoxyisoguanosine. These were applied to artificially expanded "genetic codes," and this study employed AEC for purification and analysis.[100]

2.7 METHOD DEVELOPMENT

Users of AEC benefit from multiple approaches for the control of ON selectivity. Hence almost any impurity *based on target ON sequence* can be fully or partially resolved. The most effective control one can apply for ON selectivity is the use of pH to control retention. We examined here the retention of *homopolymers* of the various DNA bases (Pd[N]$_{12\text{-}18}$, where N is rA, dA, dT, dG, or dC) using linear gradients of 0.3–1 M NaCl at pH values from 7 to 12. Similarly, we examined these homopolymers for selectivity using 10–195 mM $NaClO_4$. For these salts, RNA (rA) elutes significantly later than DNA (dA, see also Figure 2.9), and the relative contribution of the DNA bases to retention at each pH is provided in Table 2.2. Because these are homopolymers, the effect of neighboring bases is not addressed.

2.7.1 Tailoring Selectivity

2.7.1.1 Retention by Length

Conditions to *minimize* the influence of base interactions with the stationary phases include use of $NaClO_4$ salt (to render the eluent more "lipophilic" than with NaCl), solvent (e.g., 20% MeCN, to minimize hydrophobic and pi bonding interactions between the bases and the phase), and low pH (~6.5 to effectively eliminate partial ionization of the tautomeric oxygens on the G and T/U bases). These conditions were observed to result in elution order governed primarily by length

TABLE 2.2
Relative Retention of the DNA Bases at Various pH Values on the DNASwift SAX-1S Hybrid Monolith[a]

pH	NaCl Elution Order	pH	$NaClO_4$ Elution Order
7	dA < dT ≈ dG < dC	7	dA < dT ≤ dG ≤ dC
8	dA < dT ≈ dG < dC	8	dA < dT ≈ dG ≤ dC
9	dA < dC ≤ dT < dG	9	dA < dC < dT < dG
10	dA < dC « dT « dG	10	dA < dC < dT < dG
11	dA < dC « dT « dG	11	dA < dC « dT < dG
12	dA < dC « dt « dG	12	dA < dC « dT < dG

[a] High pH may be necessary to control ON secondary structure where large segments of an ON sequence are self-complementary or where significant sections of consecutive guanine bases are present. In these cases, high pH can effectively denature Watson–Crick, Hoogstein, and G-tetrad hydrogen bonding.[3] Different AEC columns will harbor different selectivities owing to the different tertiary and quaternary amines, and the substrate polymer chemistry. Hence not all AEC phases will adhere to these elution orders.[91] To the extent that hydrophobic interactions influence this retention order, they can be controlled by addition of varying amounts of solvent (e.g., MeCN) to the eluents.

on the DNAPac PA200.[93] This is demonstrated by retention of ONs with closely related sequence but having different terminal bases in specific order of ON length (Figure 2.10). This experiment involved chromatography of 25 different ONs based on a single sequence and having lengths from 21 to 25 bases. Shown on the slide are only the earliest and latest eluting ONs in each size category. This order of elution does not hold at higher pH values or with NaCl eluents lacking solvent.

2.7.1.2 Control of Elution Order

In cases where ON metabolites are under analysis, ON components of identical lengths may result from differential degradation at both 3′ and 5′ ends. This will result in identical central sequences with differing 3′ and 5′ bases. In some cases one of these ONs will be present in greater abundance than the other, so if the minor abundance ON elutes after the major abundance product, it will likely be more difficult to quantify owing to the major abundance component's peak "tail." In such cases, *reversing* the elution order would be beneficial for resolving the minor failure from the high abundance target sequence. This can sometimes be accomplished by control of pH as shown in Figure 2.11. This figure shows an elution order reversal for two 23-base ONs between pH 8 and 11 on a DNAPac PA200 column and demonstrates sufficient resolution for quantification at both pH 8 and 10 or above when NaCl is the eluent salt. However, when $NaClO_4$ is the eluent salt, this pair of ONs are only effectively resolved at pH 8 and 9 where Dx89 elutes earlier that Dx88. Depending on the column type, an alternate salt (e.g., NaCl instead of $NaClO_4$) can help resolve such metabolites.

Figure 2.12 shows the same pair of ONs as in Figure 2.11 but applied to a DNASwift SAX-1S hybrid monolith. On this column, resolution of the two ONs with Dx88 eluting first is acceptable at pH 10 using NaCl eluents, but at the pH where the elution order is reversed, resolution is not sufficient for either analysis or purification. However, use of $NaClO_4$ as eluent *allows* resolution of the two ONs with Dx89 eluting first, at pH 9.5.

2.7.1.3 Effect of 5′ and 3′ Terminal Bases

As mentioned in Section 2.6.2.1, the 3′ and 5′ terminal bases may exert a disproportionate influence over ON retention. To provide some guidance on this effect, we obtained seven ONs with an

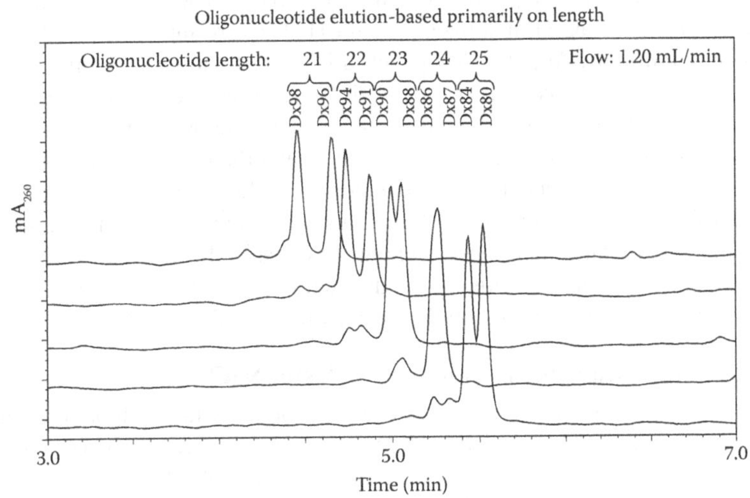

FIGURE 2.10 Control of selectivity for ON elution in order of length. System: Dionex DX600 inert quaternary gradient LC system. Conditions: 69–142 mM $NaClO_4$ in 12 min at 1.2 mL/min and 30°C. Oligonucleotides in each pair represent the earliest- and latest-eluting components (in each size class) of 25 different samples with a common central 21-base sequence. (Modified from Thayer, J. R., et al., *Analytical Biochemistry*, 338, 39–47. Copyright 2005, with permission from Elsevier.)

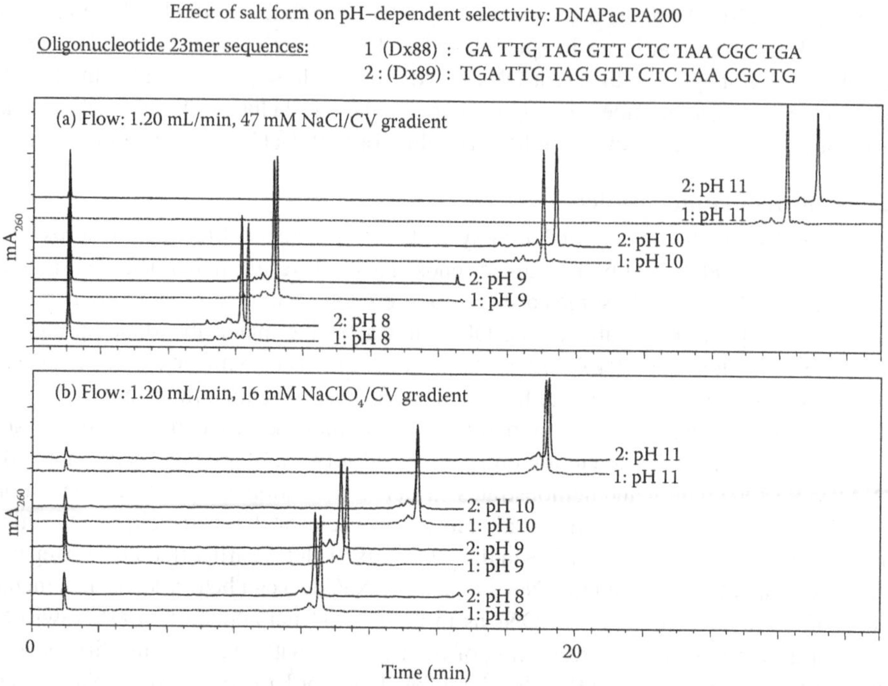

FIGURE 2.11 Use of pH and salt form to adjust oligonucleotide elution order on the DNAPac PA200 (4 mm × 250 mm) column. System: Dionex DX600 inert quaternary gradient LC system. (a) Effect of pH with NaCl eluents; 330–900 mM NaCl in 31.7 min at 30°C and the indicated pH values. (b) Effect of pH with $NaClO_4$ eluents; 70–180 mM $NaClO_4$ in 18 min at 1.2 mL/min at 30°C and the indicated pH values. Both eluents employed *linear* gradients.

identical internal sequence, differing only in their 5′ or 3′ bases. These were chromatographed first using a gradient of 18 mM NaCl/min at pH 9, 10, and 11. These results are presented in Figure 2.13 using a 3 min elution display. At pH 9 (left panel) all 7 eluted within a 0.9 min window, between 7 and 7.9 min. At this pH the ON with "C" on both 5′ and 3′ ends eluted first, and the ON with a 5′G and 3′A eluted last. At pH 10 (middle panel), the elution window for this set was 1.4 min wide, spanning 14.9–16.3 min. ONs containing a T or G eluted later than the same sequence lacking either base at pH 10. At pH 11 (right panel), the elution window was 1.6 minutes, between 22.4 and 24 min. Again, ONs with T or G eluted later. That the elution window increases with pH and that ONs with terminal G and/or T elute later than those without these bases illustrates the effect of tautomeric oxygen ionization on ON retention. This figure also details the differential effect of the four bases on the 5′ end where the 3′ end is A, and on the 3′ end where the 5′ end is C.

2.7.1.4 Controlling Nonspecific Interactions (Solvent, $NaClO_4$)

In Section 2.7.1.2, we described how use of solvent combined with perchlorate eluent minimizes base interactions to promote elution in order of ON length. The separate influences of pH with NaCl and $NaClO_4$ eluent (with and without solvent) on the resolution of ONs differing only in their terminal bases are illustrated in Figures 2.14 and 2.15. In Figure 2.14, the same ONs chromatographed in Figure 2.13, there eluted with NaCl, are chromatographed in NaCl with *20% MeCN*. When this solvent is present, the pH 9 elution window is reduced from 0.9 to 0.6 min, and the window is centered at ~6 min instead of 7.5 min. At pH 10 and 20% MeCN, the elution window is compressed from 1.4 to 0.8 min and centered ~10 min instead of 15 min. At pH 11 and with 20% MeCN, the elution window is compressed from 1.6 to 0.9 min and centered at ~14.5 min instead of ~23 min.

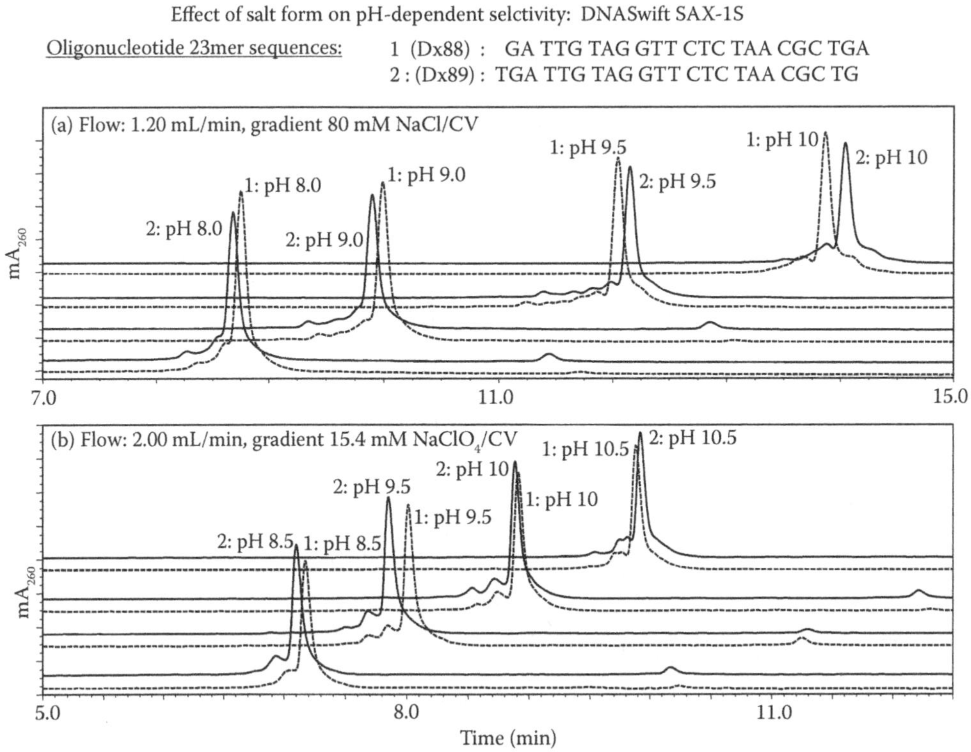

FIGURE 2.12 Use of pH and salt form to adjust oligonucleotide elution order on the DNASwift SAX-1S (5 mm × 150 mm) column. System: Dionex DX600 inert quaternary gradient LC system. (a) 200–1000 mM NaCl in 16.7 min at 1.5 mL/min, at 30°C and at the indicated pH values using a linear gradient. (b) 10–195 mM $NaClO_4$ in 15 min at 2.0 mL/min, at 30°C and at the indicated pH values using a curved gradient (4, see Ref. 93). (Modified from Thayer, J. R., et al., *Journal of Chromatography B*, 338, 39–47. Copyright 2010, with permission from Elsevier.)

This confirms the earlier observation that the bases participate in hydrophobic interactions with the stationary phase and that these can be exploited to control ON selectivity on these AE phases.

In a similar example, Figure 2.15 shows the changes in retention for these same ONs when NaCl is replaced with $NaClO_4$ in the *absence* of solvent. Here, the pH 9 elution window is reduced from 0.9 to 0.5 min; that at pH 10 is reduced from 1.4 to 0.8 min; and the window at pH 11 is compressed from 1.6 to 0.8 min. This is accomplished with a lower gradient slope because $NaClO_4$ is a substantially stronger eluent than NaCl. This figure confirms the importance of ON hydration during chromatography as described in Section 2.6.2.3.

2.7.2 Optimizing Impurity Resolution

ONs may harbor a wide variety of different chemical modifications that may independently interact with a chromatographic stationary phase. Because this can make ON impurity assessments complicated, we have developed approaches to effectively simplify method assessment. These include (1) use of a *quaternary* eluent system that permits direct control of chromatographic pH to automatically scout for optimized ON selectivity, and (2) a different *ternary* eluent system to evaluate the effect of solvent concentration, and HPLC systems designed to program curved gradients that facilitate separation of similar sequence ONs.[93]

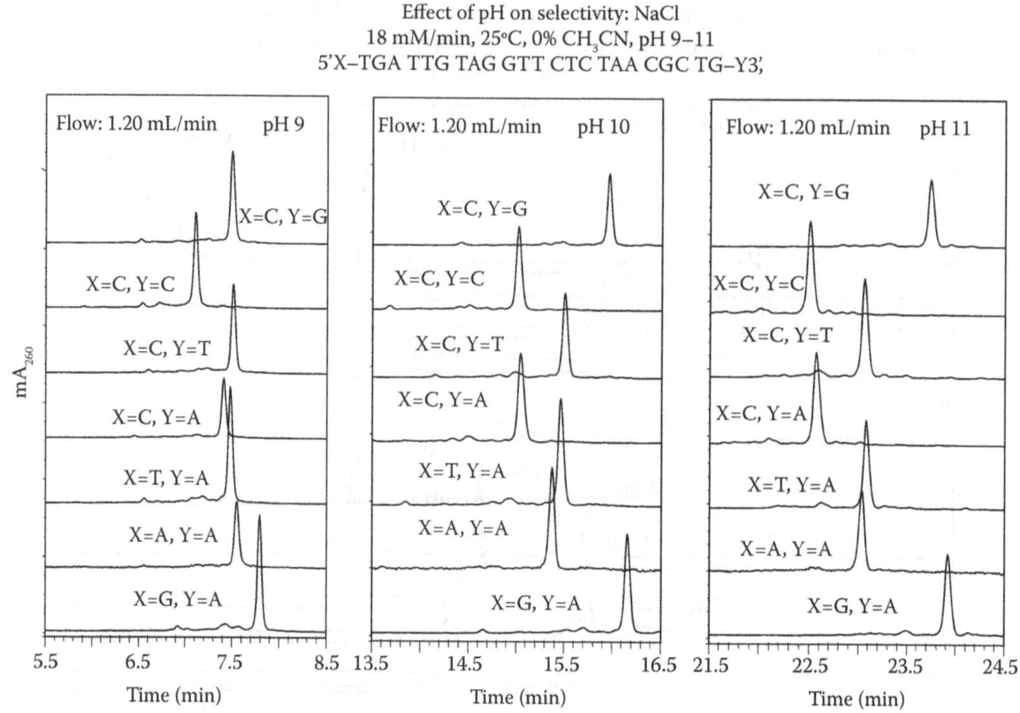

FIGURE 2.13 Analysis of the effect of 5′ and 3′ terminal bases on oligonucleotide retention using the DNAPac PA200 with NaCl eluents. Each of seven ONs differing only in their 3′ and 5′ terminal bases were chromatographed under the conditions shown in Figure 2.11a, here represented at pH 9, 10, and 11.

2.7.2.1 The pH Adjustment (Dial-a-pH)

Because ON retention is profoundly influenced by pH between 6.5 and 12.4, we employ a method to quickly and reliably deliver eluents at pH values in that range. While it is possible to employ eluent selection valves to deliver premade eluents at various pH values, these are cumbersome to execute as each eluent must be separately prepared and multiple valve systems must be employed. Our approach was to develop a programmable eluent system that employs quaternary eluent pumps. This allows use of one pair of eluents for pH control, and the other pair to provide a salt gradient. Typically, the salt gradient eluents are deionized water, and a salt solution (e.g., 1.25 M NaCl). The other two eluents are the pH-forming set (e.g., eluent 2 is 0.2 M NaOH, and eluent 3 contains 0.2 M each of 3–4 buffers with pK values spanning the target pH range). Proportioning between eluents 2 and 3 allows delivery of a target pH. In our method we use a fixed fraction of the total delivered eluent (e.g., 20% to generate the desired pH). An advantage of this system is that the salt form can be quickly changed by preparing a single new eluent (e.g., switching eluent 4 to 0.33 M $LiClO_4$).

To determine the proportion of each of the *pH-buffering* eluents, we prepare a standard curve from the pH delivered by each proportion (0–20%) of each of these two eluents. We monitor the pH with a calibrated pH detector placed in the eluent line after the column and UV detector. Because there are 3–4 buffers used, the relationship of eluent 2 and 3 proportion to pH is *not* linear. Each buffer contributes a distinct pH curve, and these are designed to overlap. Our pH versus relative proportion data are converted into the standard curve using a sixth-order polynomial regression using Microsoft Excel, and the data with regression line is printed to use for day-to-day experiments. In most of our work we use sodium salts, and these are known to influence pH measurements; and we use salt concentrations that are relatively high, usually between 50 mM and 1 M. To examine

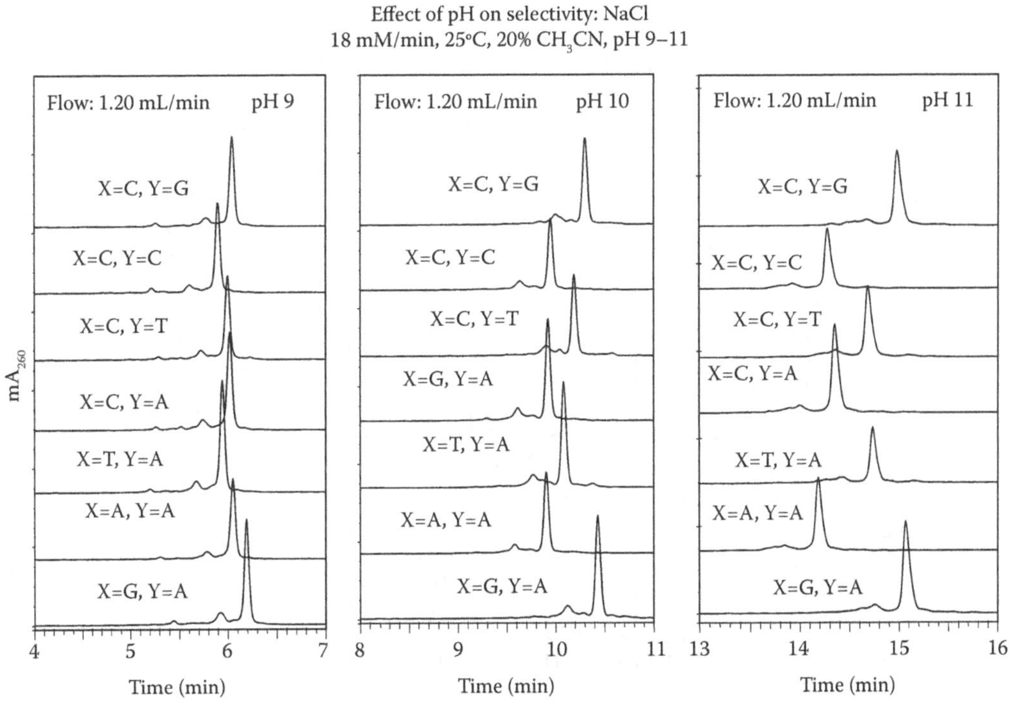

FIGURE 2.14 Analysis of the effect of 5′ and 3′ terminal bases on oligonucleotide retention using the DNAPac PA200 with NaCl eluents *including acetonitrile*. System: Dionex DX600 inert quaternary gradient LC system. Each of seven ON differing only in their 3′ and 5′ terminal bases were chromatographed under the conditions shown in Figure 2.11a, except that 20% acetonitrile was included in the eluents.

the effect of sodium ion on the pH measurement, we repeat the pH measurements using salt concentrations of 100, 400, and 800 mM and find that the pH versus eluent 3 to eluent 4 proportion is not dramatically different under each condition. Figure 2.16 illustrates an example standard curve prepared with NaCl where eluent 2 is 0.2 M NaOH, and eluent 3 is "TAD" buffer, a combination of Tris-base, 2-amino-2-methyl-1-propanol, and di-isopropylamine, each at 0.2 M, and adjusted with methane sulfonic acid (MSA) to pH 7.2. While HCl is typically used to adjust pH for eluents, we find that HCl adjusTment of this buffer set tends to *promote,* while pH adjusTment with MSA tends to *inhibit,* microbial growth.

Figure 2.16 also provides the Excel-derived sixth-order polynomial equation. That equation is then employed to calculate the proportions of eluents 2 and 3 to deliver a desired eluent pH. That the desired pH is correctly delivered, we employ a calibrated, in-line pH electrode to verify the chromatographic pH. Using this system one can program a desired pH with eluents 2 and 3, and use eluents 1 and 4 to prepare simple or complex multistep gradients. The chromatograms in Figures 2.1 through 2.3 and 2.11 through 2.15 were prepared using this approach. Note that we have observed that quaternary gradient calibrations obtained on systems from different manufacturers do not produce identical calibration curves, so users are advised to prepare calibrations on the systems they use.

2.7.2.2 Solvent Options

In some cases addition of solvent is helpful to control ON peak width during chromatography. The work in Figures 2.7, 2.10 and 2.14 employ 20% acetonitrile as solvent. These were prepared using

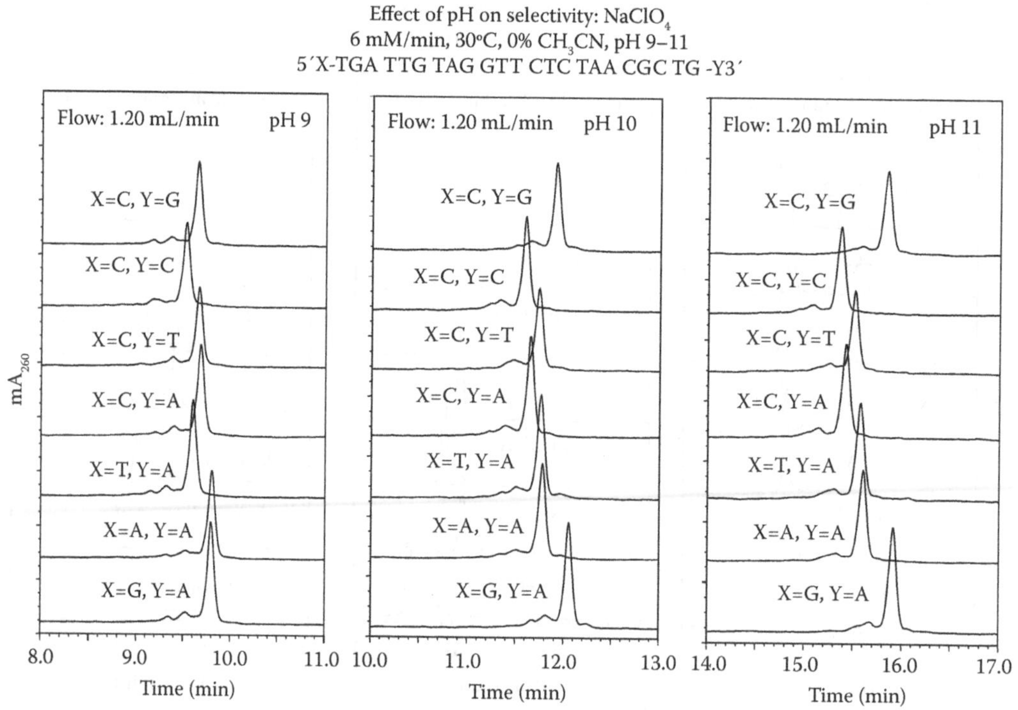

FIGURE 2.15 Analysis of the effect of 5′ and 3′ terminal bases on oligonucleotide retention using the DNAPac PA200 *with $NaClO_4$ eluents*. Seven ONs differing only in their 3′ and 5′ terminal bases were chromatographed. System: Dionex DX600 inert quaternary gradient LC system. Conditions: 70–178 mM $NaClO_4$ in 18 min at 1.2 mL/min and 30°C using the pH values indicated.

the TAD buffer system with eluents 1 and 4 containing 23.5% MeCN. Because 80% of the eluent was delivered from reservoirs 1 and 4, the resulting MeCN concentration was 20%. Where modulation of ON selectivity by solvent concentration is considered, an eluent may be first optimized with respect to pH using the "TAD" buffer system, then further optimized by preparing a binary eluent system with an appropriate buffer in both low, and high, salt eluents and proportioning in solvent from a third reservoir. As with the pH control system, the impact of solvent concentration can be optimized by this simple approach.

2.7.2.3 Temperature Effects

While AEC of mono- and divalent anions tends to exhibit *decreasing* retention with increases in temperature, AEC of polyanions exhibits *increasing* retention with rising temperatures. Because elevated temperatures also increase mass transfer rates, the combination of increased retention and increased mass transfer rates results in minor, but predictable (and hence useful) improvements to ON resolution. This is illustrated in Figure 2.17, where a 20-base ON is chromatographed on a DNASwift monolithic hybrid phase at temperatures from 30° to 70°C. Examination of the baseline peaks eluting just before the target 20-mer, reveals useful improvements in resolution of these components. At 30°, several "n – x" components co-elute, and many of these are at least partially resolved at 70°C.

The pellicular and hybrid monolith AEC columns are compatible with temperature of at least 85°C, and some have been used at temperatures up to 95°C to extract maximal resolution, or to control hydrogen bonding.

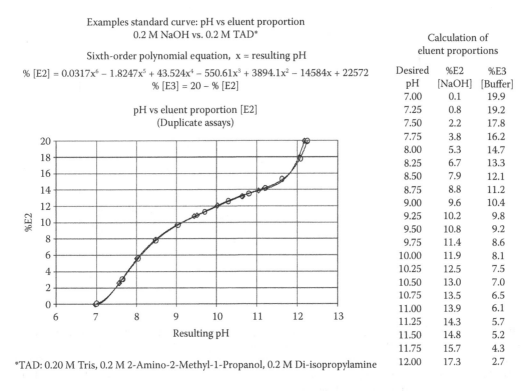

Examples standard curve: pH vs eluent proportion
0.2 M NaOH vs. 0.2 M TAD*

Sixth-order polynomial equation, x = resulting pH

$$\% [E2] = 0.0317x^6 - 1.8247x^5 + 43.524x^4 - 550.61x^3 + 3894.1x^2 - 14584x + 22572$$

$$\% [E3] = 20 - \% [E2]$$

pH vs eluent proportion [E2]
(Duplicate assays)

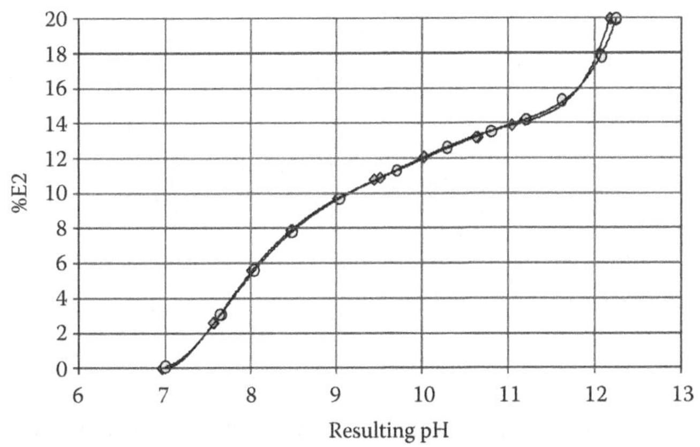

*TAD: 0.20 M Tris, 0.2 M 2-Amino-2-Methyl-1-Propanol, 0.2 M Di-isopropylamine

Calculation of eluent proportions

Desired pH	%E2 [NaOH]	%E3 [Buffer]
7.00	0.1	19.9
7.25	0.8	19.2
7.50	2.2	17.8
7.75	3.8	16.2
8.00	5.3	14.7
8.25	6.7	13.3
8.50	7.9	12.1
8.75	8.8	11.2
9.00	9.6	10.4
9.25	10.2	9.8
9.50	10.8	9.2
9.75	11.4	8.6
10.00	11.9	8.1
10.25	12.5	7.5
10.50	13.0	7.0
10.75	13.5	6.5
11.00	13.9	6.1
11.25	14.3	5.7
11.50	14.8	5.2
11.75	15.7	4.3
12.00	17.3	2.7

FIGURE 2.16 Calibration curve for eluent pH versus proportioned amount of eluent 2. Here eluent 2 (0.2 M NaOH) and eluent 3 (0.2 M Tris, 0.2 M AMP, and 0.2 M DIPA at pH 7.2) were combined in a total eluent fraction of 20%. Eluents 1 (DI H_2O) and 4 (1.25 M NaCl or 0.33 M $NaClO_4$) were used for salt gradient formation. Curve fitting was with a sixth-order polynomial best fit, calculated using Microsoft Excel. On the basis of this calibration, a table of proportions necessary for delivery of the indicated pH values is given to the right of the calibration plot. System: Dionex ICS3000 inert quaternary gradient LC system.

2.7.2.3.1 Use of Temperature for Duplex RNAi Assays

The retention of different sequences may change by different amounts as the temperature increases. Therefore, use of temperature to control the relative retention of guide and passenger strands of duplex RNAs can provide conditions where titration of excess ssRNA can be readily assessed. Figure 2.18 shows an example of this process, where the sense (s) and antisense (as) strands, and the duplex (d) of an eGFP siRNA are chromatographed separately at temperatures from 30° to 80°C (sequences in Figures 2.32 and 2.37, Table 2.4). At 30°C, the antisense strand elutes first, followed by the duplex with the sense strand eluting last. At 40°C and above, the duplex strand elutes first. Also at 40°C, the antisense and sense strands are well resolved, and elute in the same order as at 30°C. Between 50° and 60°C, the elution order of the sense and antisense strands reverse, and at 70°C, both the sense and antisense strands elute slightly later than at 60°C. At 80°C the duplex is melting, so the duplex sample trace shows *both* the sense and antisense strands, co-eluting with their counterparts injected separately. For the purpose of titrating this duplex sample, 40°C offers the best overall resolution of all three forms, allowing facile identification of which single strand is in excess. In this case the antisense strand is in excess. Because duplex RNA developers will have the molar extinction coefficients of the component strands of their samples (see Chapter 12 for more on this topic), analysis of the peak area of the excess component allows calculation of how much alternate strand (in this case, sense strand) to add to meet formulation specifications.

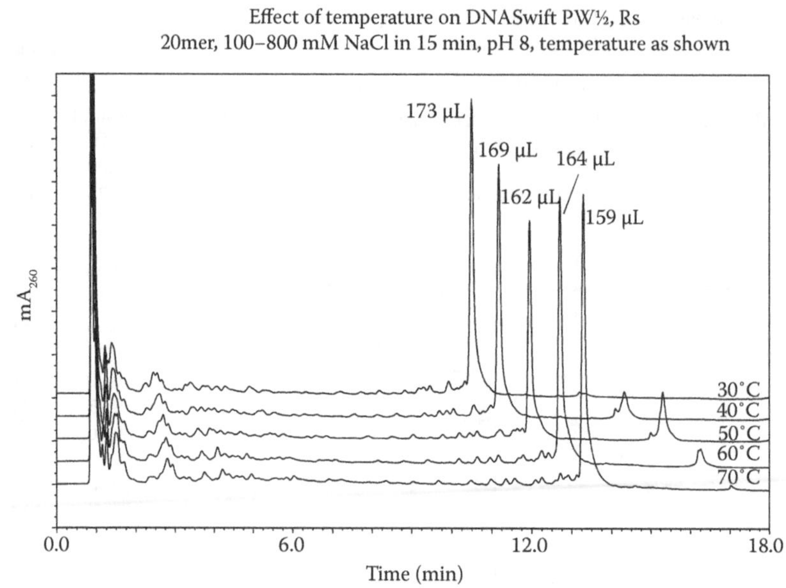

FIGURE 2.17 Effect of temperature on retention and peak width of a 20-base ODN. Column: DNASwift SAX-1S. System: Dionex DX600 inert quaternary gradient LC system. Gradient conditions: 100–800 mM NaCl in 15 min at 1.77 mL/min, pH 8, and at the indicated temperatures. Sample sequence: 5′ GGG ATG CAG ATC ACT TTC CG 3′. (Reprinted from Thayer, J. R., et al., *Journal of Chromatography B*, 878, 933–941. Copyright 2010, with permission from Elsevier.)

2.7.2.4 Linear versus Curved Gradients

AEC of ONs allows retention based on ON length because the charge on these biopolymers arises from the phosphodiester linkages (see Section 2.6.1.1). However, the *difference* in net charge between ONs becomes less as the ON length increases. Hence linear salt gradients tend to produce separations of decreasing resolution as the ON length increases. Users may program multistep salt gradients to serially decrease the salt gradient slope in order to improve resolution of longer ONs. However, some HPLC vendors support *curved* gradients (e.g., Ref. 101) via programming to simplify this task. While this approach is usually employed to resolve impurities in preparations of longer ONs, we present an example of curved gradient elution to improve resolution in Section 2.8.2, where this technique is used to elute ON fragments generated from a nuclease digest.

2.7.3 Development of Methods for siRNA Drug Substance

2.7.3.1 General Considerations

RNA sequences are not simply long strands of nucleotides. Rather, intrastrand base pairing will produce secondary structures such as the hairpin structure as shown in Figure 2.19.

In RNA, guanine and cytosine (GC) pair by forming a triple hydrogen bond, and adenine and uracil (AU) pair by a double hydrogen bond; additionally, guanine and uracil can form a single hydrogen bond base pair. The stability of a particular secondary structure is a function of several constraints:

1. The number of GC versus AU and GU base pairs. Higher-energy bonds form more stable structures.

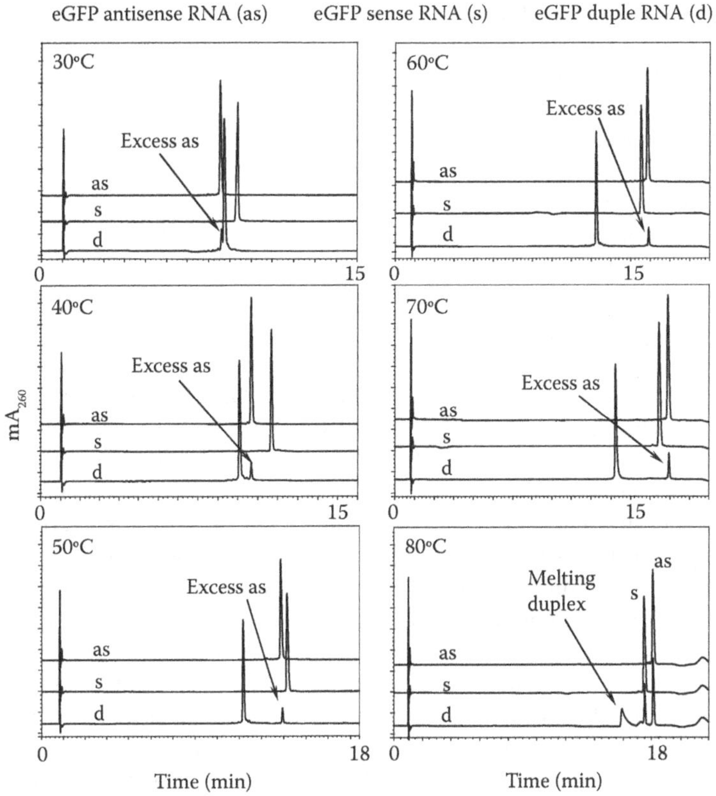

FIGURE 2.18 Effect of temperature on resolution of sense, antisense and duplex RNA. The sense sequence is 5′-AGCUGACCCUGAAGUUCAUdCdT-3′, and the antisense sequence is that shown in Table 2.4. Column: DNAPac PA-200. System: Dionex UltiMate 3000 Titanium inert quaternary gradient LC system. Gradient conditions: 325–750 mM NaCl in 17.2 min at 300 μL/min, pH 7, and at the indicated temperatures.

2. The number of base pairs in a stem region. Longer stems result in more bonds.
3. The number of base pairs in a hairpin loop region. Formation of loops with more than 10 or less than 5 bases requires more energy.
4. The number of unpaired bases, whether interior loops or bulges. Unpaired bases decrease the stability of the structure.

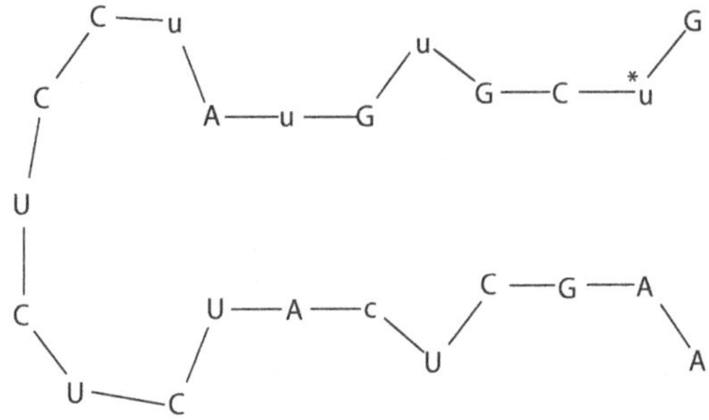

FIGURE 2.19 Hairpin structure of the sequence: 5′-AAGCUcAUCUCUCCuAuGuGCu*G-3′. The lowercase letters represent 2′-*O*-methyl-modified nucleotides; the asterisk near 3′ end represents a phosphorothioate linkage.

The stability of a secondary structure is quantified as the amount of free energy released or used by forming base pairs. The T_m value indicates the stability of the secondary structure.

The 23-mer antisense strand, 5′-AAGCUcAUCUCUCCuAuGuGCu*G-3′, exhibits a hairpin secondary structure as shown in Figure 2.19.

In the above sequence, lowercase letters represent 2′-*O*-methyl-modified nucleotides; asterisk represents a phosphorothioate linkage. This stem loop ON has a T_m of 48.1°C in 100 mM NaCl and introduces chirality to the strand potentially complicating purification. During single-strand purification, selection of the appropriate strand, or selection of conditions *not* resolving the diastereoisomers is necessary. Refer to Chapter 6 for more on T_m.

As discussed earlier, AEX-HPLC is an effective analytical tool if the secondary structure of the molecules can be effectively disrupted. The elution position of the molecules depends on their secondary and tertiary structure. A mixture of different secondary structures can lead to very complicated chromatograms. Denaturing the RNA produces a relatively simple chromatogram typically with sharper peaks. Therefore, AEX-HPLC is usually preferred over reversed-phase HPLC because the anion-exchange columns can be heated to higher temperatures, which more effectively disrupts secondary structures. Denaturing conditions can be achieved with appropriately buffered mobile phase at high pH, elevated temperature, a combination of pH and temperature, or inclusion of chaotropic agents (e.g., urea or formamide) in the eluents. Note that for some anion exchangers, the combination of elevated temperatures and high-pH values, or inclusion of chaotropic agents, may limit column longevity. In a quality-controlled (QC) laboratory, one would like to have a method that can produce narrow peaks and reproducible chromatography. Denaturing methods (for single-strand analyses) are employed for two conditions: A high pH method at moderate column temperature and a low pH method at high-temperature conditions. This dual-pH approach can be tested in a variety of columns that are available in the market. The single strand with the secondary structure, as shown in Figure 2.19, was tested with a Dionex DNAPac PA-200 column with pH 8 buffer at temperatures ranging from 35° to 75°C. Figure 2.20 illustrates that at low temperatures the isomers are separated into two distinct peaks and the separation gradually decreases to become a sharp single peak at 75°C, where the diastereoisomers formed from the phosphorothioate linkage co-elute. For a molecule that exhibits such a behavior, the chromatographic conditions that give a single sharp peak should be selected for QC analysis. Elution of all isomers under one single peak makes quantification simple.

2.7.3.2 Single-Strand Intermediates Method

We demonstrate method development for impurity control in a drug substance with a siRNA example. The ON *ALN-RSV01* is being developed for the treaTment of RSV infection.[102,103] This synthetic double-stranded RNA ON is formed by the hybridization of two partially complementary single-strand RNAs. Each of the single-strand RNAs is composed of 19 ribonucleotides with two thymidine units at the 3′ end. The 19 ribonucleotides of one of the strands hybridize with the complementary 19 ribonucleotides of the other strand, thus forming 19 ribonucleotide pairs and a bis-thymidine overhang at each end of the duplex.

The composition of the sense and antisense strands is given below:

Passenger (sense): 5′-GGC UCU UAG CAA AGU CAA GTT-3′
Guide (antisense): 5′- CUU GAC UUU GCU AAG AGC CTT -3′

Criteria for the selection of a good method are as follows:

1. Sharp baseline resolved peaks
2. Resolution between critical pairs of impurities
3. Ideally, use the same method for both single strands and duplex
4. Reduce method induced degradation

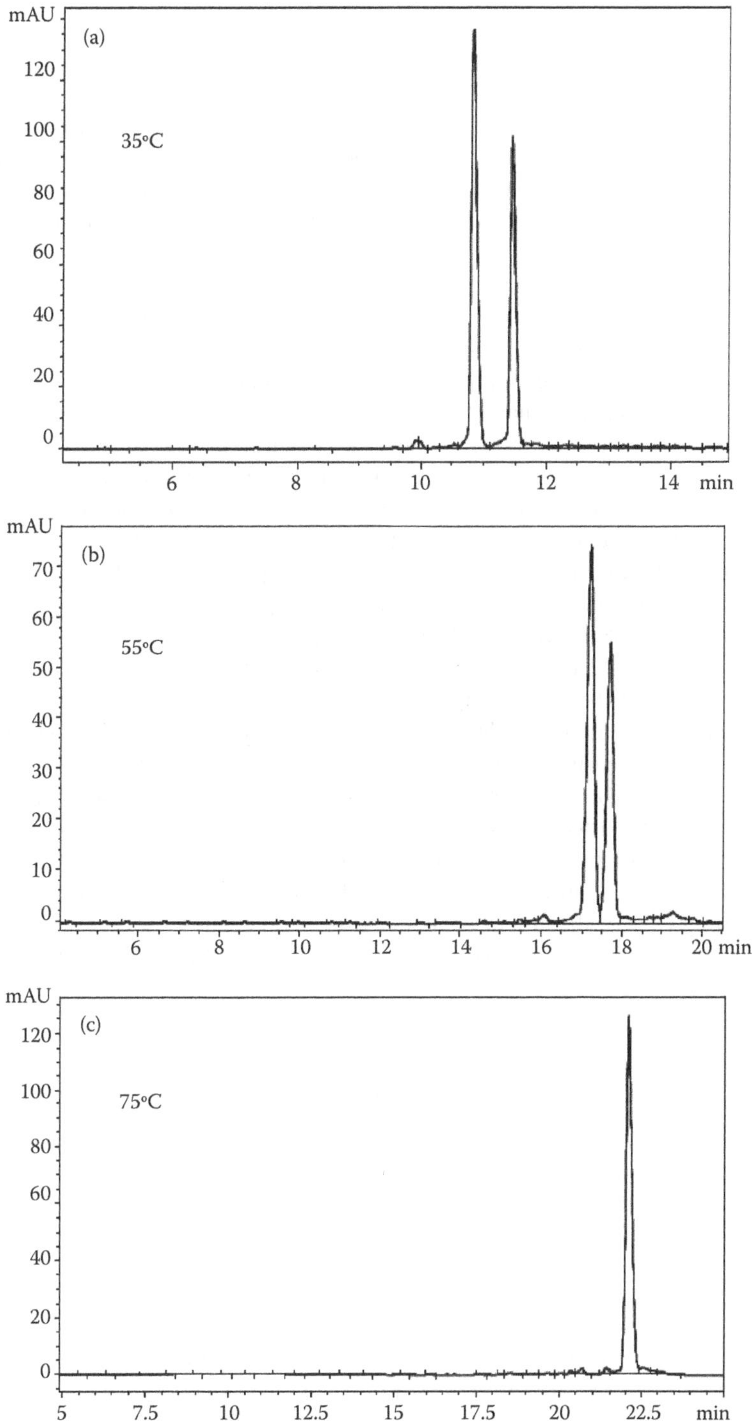

FIGURE 2.20 Chromatograms of the sample shown in Figure 2.19. System: Agilent 1100 HPLC, Column: Dionex DNAPac PA-200, Buffer A: 10% CH_3CN, 25 mM Tris-HCl buffer pH at 8, Buffer B: 10% CH_3CN, 25 mM Tris-HCl buffer pH at 8 and 1 M NaBr, Gradient: 25–55% B in 35 min. (a) At column temperature 35°C, diastereoisomers are resolved, (b) at column temperature 55°C, diastereoisomers are partially resolved and, (c) at column temperature 75°C diastereoisomers elute as single peak.

5. Reliable enough to be transferable to different laboratories
6. Suitability for validation
7. Possess stability indicating characteristics (refer to Chapter 15 for more on stability)

Success of AEX-HPLC mainly relies on the columns and mobile phases and requires tuning the separation with an optimized gradient. The type of columns was discussed in Section 2.5.2. Table 2.3 provides a set of mobile phase combinations available to begin a method scouting and, based on the results, one could modify composition, gradient, and temperature and also other mobile phase combinations.

A typical starting point for a denaturing method is to use pH 11 buffer with salt gradient as discussed in Section 2.6.2.2. It should be noted that during the buffer preparation, pH should be adjusted in aqueous medium before adding organic solvent. To accurately assess the success or failure of new synthetic or deprotection protocol approaches taken for a pH 11 method, optimization of the sense and antisense strands of ALN-RSV01 is provided. Analysis of the crude synthesis by AEX-HPLC combined with spectrophotometric quantification at 260 nm provides the most accurate measurement of full length product of guide (antisense) strand with the above sequence. A Dionex DNAPac PA-100 column at 40°C, used with sodium phosphate and 10% acetonitrile (MeCN), gave optimal resolution. Figure 2.21 illustrates typical chromatograms of a pH 11 method for (a) crude synthesis followed with deprotection, (b) mock pool of a combination of several fractions during a purification process, and (c) final purified sample. Note that Dionex DNAPac PA-200 column provided comparable results as in Dionex DNAPac PA-100 column.

For reference, Figure 2.22 illustrates typical chromatograms for a pH 8 method on a Dionex PA-200 column at 75°C of the same sample set given above. Close examination of both methods indicates that the pH 11 method provides a narrower peak width than the pH 8 method. The pH 8 method results in a peak nearly 4.4 times wider than the pH 11 method for injection of the same sample size. This is due to both the effect of high pH and that of column efficiency, the PA200 being significantly better than the PA100 in this regard.

Further, close examination of several analyses revealed that the "N+x-mers" were well resolved from the full length (N-mer) and each other at pH 11 but eluted as a group of partially resolved peaks on the trailing edge of the N-mer peak under the pH 8 conditions. As a result, pH 8 method overestimates the purity of the final purified product as shown in Figure 2.22c compared to pH 11

TABLE 2.3
Sample Mobile Phase Composition for AEX-HPLC

Mobile Phase A "Weak Solvent"		Mobile Phase B "Strong Solvent"
20 mM Na_3PO_4, 1 mM EDTA pH 11 +	0% Organic	20 mM Na_3PO_4, 1 mM EDTA, 1M NaBr, pH 11 + 0% organic
20 mM Na_3PO_4, 1 mM EDTA pH 11 +	0% Organic	20 mM Na_3PO_4, 1 mM EDTA, 0.5 M $NaClO_4$, pH 11 + 0% organic
20 mM Na_3PO_4, 1 mM EDTA pH 11 +	0% MeCN	20 mM Na_3PO_4, 1 mM EDTA, 1M NaBr, pH 11 + 10% MeCN
20 mM sodium phosphate, pH 8 (mix 0.1282 % w/v monosodium phosphate +0.354 % w/v trisodium phosphate dodecahydrate) + 10% MeCN		20 mM sodium phosphate, 1 M NaCl, pH 8 + 10% MeCN
20 mM sodium phosphate, pH 8 + 10% MeCN		20 mM sodium phosphate, 1 M NaBr, pH 8 + 10% MeCN
25 mM Tris.HCl, 1 mM EDTA, pH 8 + 10% MeCN		25 mM Tris.HCl, 1 mM EDTA, 1 M NaBr, pH 8 + 10% MeCN
25 mM Tris.HCl, 1 mM EDTA, pH 8 + 10% MeCN		25 mM Tris.HCl, 1 mM EDTA, 0.5 M $NaClO_4$, pH 8 + 10% MeCN

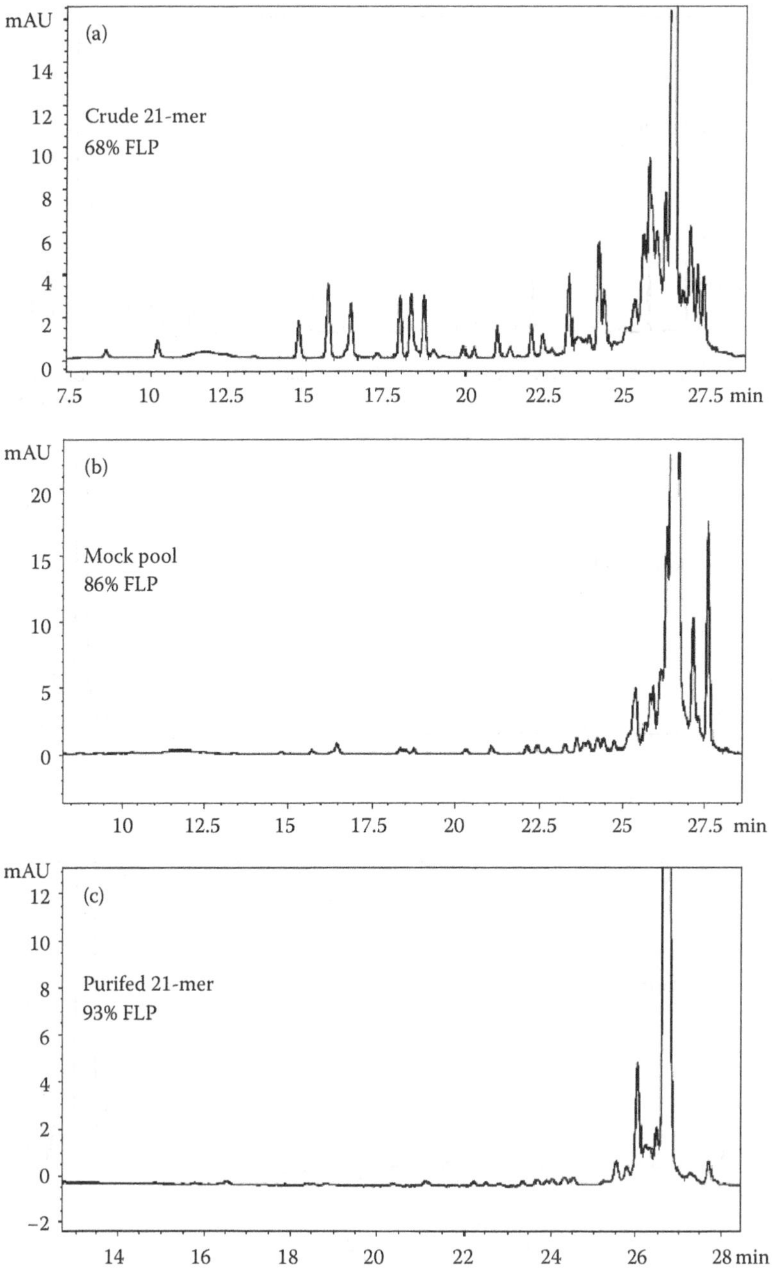

FIGURE 2.21 (a) Chromatogram of a crude synthesis, (b) chromatogram of a mock pool, and (c) purified sample. System: Agilent 1100 HPLC; Column: Dionex DNAPac PA-100; Column temperature: 40°C; Buffer A: 20 mM sodium phosphate, 10% CH_3CN at pH 11; Buffer B: 20 mM sodium phosphate, 1 M NaBr, 10% CH_3CN at pH 11; Gradient: 15–57% B in 30 min. 5′- CUU GAC UUU GCU AAG AGC CTT -3′.

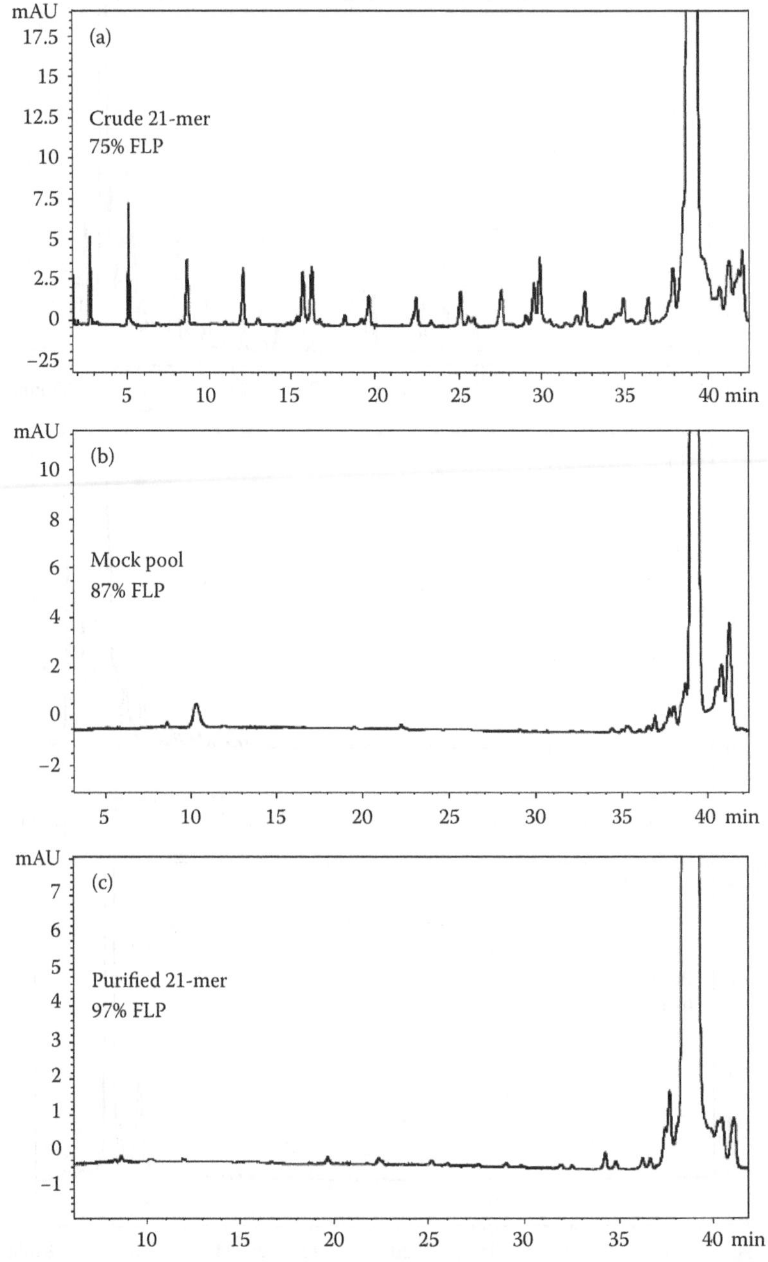

FIGURE 2.22 (a) Chromatogram of a crude synthesis, (b) chromatogram of a mock pool, and (c) purified sample. System: Agilent 1100 HPLC; Column: Dionex DNAPac PA-200; Column temperature: 75°C; Buffer A: 10% CH_3CN, 25 mM Tris-HCl buffer at pH 8; Buffer B: 10% CH_3CN, 25 mM Tris-HCl buffer at pH 8 and 1 M NaBr; Gradient: 20–42% B in 45 min. Sample sequence: 5′- CUU GAC UUU GCU AAG AGC CTT -3′.

method shown in Figure 2.21c. Hence pH 11 method is a method of choice in routine analysis by QC laboratory. Further, resolution characteristics of both methods are detailed below.

All purification protocols considered would normally separate much lower deletion products. Because the separation on AEX-HPLC method is based on the charge, the base composition, the sequence, and the size of the ORN, it becomes increasingly difficult to resolve impurities closer to full length product, such as N-1, N-2, and N+x (where x is a combination of overcoupling products such as N+G, and unprotected oligomers). Critical impurities representative of N-1, N-2, and N+G forms were synthesized for the sense and antisense sequences. Injection of a spiked mixture of N-1, N-2, and N+G with full length product revealed that the pH 11 method did not resolve N-1 peak from full length product for antisense strand as shown in Figure 2.23.

It should be noted that N-1, N-2, and N+G impurities of *sense* strand *were* resolved from the full length product (not shown). However, the pH 8 method resolved N-1, N-2, and N+G from full length product (see Figure 2.23). Until sufficient manufacturing experience reveals that impurity levels are controlled and can be fully documented by one method, we recommend use of both pH 8 and 11 as

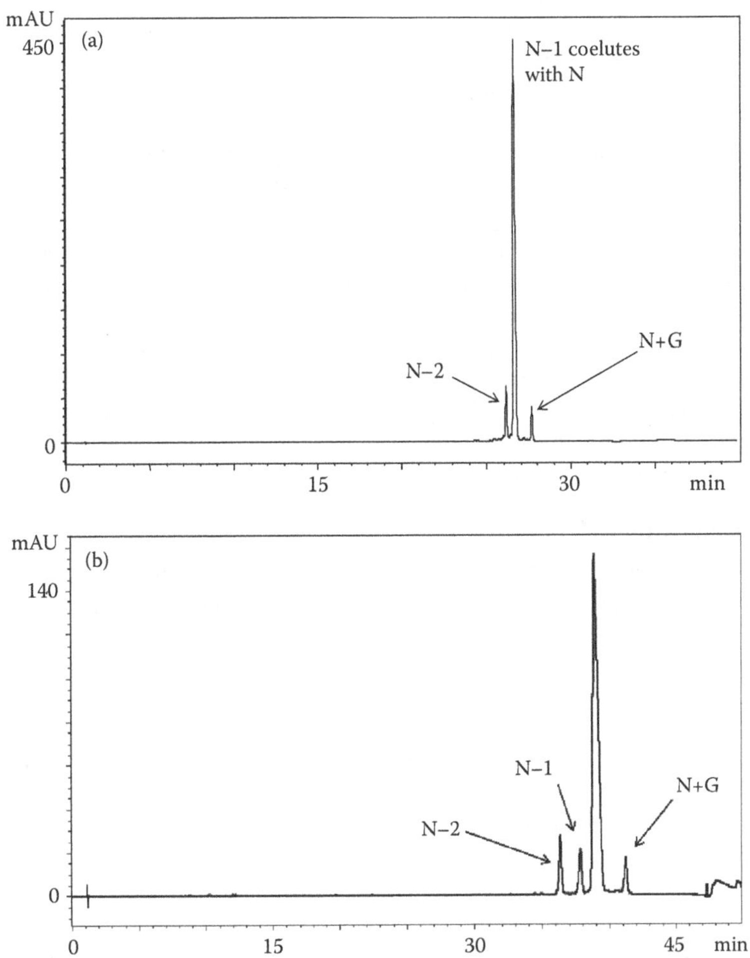

FIGURE 2.23 (a) Chromatogram of N-mer spiked with 5′N-1-mer, 5′N-2-mer, and N+G under pH 11 conditions: System: Agilent 1100 HPLC; Column: Dionex DNAPac PA-100; Column temperature: 40°C; Buffer A: 20 mM sodium phosphate, 10% CH_3CN pH at 11; Buffer B: 20 mM sodium phosphate, 1 M NaBr, 10% CH_3CN pH at 11; Gradient: 15% B to 57% B in 30 min. (b) Chromatogram of N-mer spiked with 5′N-1-mer, 5′N-2-mer and N+G under pH 8 conditions: Column temperature: 75°C; Buffer A: 10% CH_3CN, 25 mM Tris buffer pH at 8; Buffer B: 10% CH_3CN, 25 mM Tris buffer pH at 8 and 1 M NaBr; Gradient: 20% to 42% B in 45 min. N-mer: 5′-CUU GAC UUU GCU AAG AGC CTT -3′, 5′N-1: 5′-UU GAC UUU GCU AAG AGC CTT-3′, 5′N-2: 5′-U GAC UUU GCU AAG AGC CTT -3′ and N+G: 5′-CUU GGAC UUU GCU AAG AGC CTT-3′

a combined dual-pH method. Further, an examination of area-% of full length products and post full length products (N+x peaks) need to be analyzed case by case in selecting a final method.

2.7.3.3 Drug Substance Duplex Method

Because duplexes are large molecules, characterization of the drug substance is generally performed on single strands. Single-strand impurity profiles carry forward to the duplex drug substance. Therefore, a stability indicating denaturing method should be suitable for the analysis of drug substance. Refer to Chapter 15 for more on the stability indicating methods. As far as possible, single-strand methods should be used for the duplex analysis. However, a major difference in the objective of the method is that the full length sense and antisense strands should be resolved enough to monitor any degradation products as contrasted with single nucleotide resolution for single strands. Figure 2.24 is an illustration of the effect of temperature under pH 8 conditions. The nearly equimolar mixture of sense and antisense strands solution is chromatographed on a Dionex DNAPac PA-200 column with the pH 8 method. In repetitive samplings the column temperature is increased in steps between 30° and 75°C to optimize resolution of excess sense and antisense strands and to observe partial denaturing of the duplex.

At temperatures as low as 30°, 40°, and 50°C, a mixture of sense and antisense strand remains as duplex and duplex variants. These variants (or failure duplexes) are formed by the truncate impurities present in both the sense and antisense single strands. As the temperature of the column is increased, the duplex variants and aggregates melt to form a cluster of impurities as illustrated in Figure 2.24a. As the chromatographic temperature increases the duplex will eventually denature into the corresponding single strands as shown in Figure 2.24b. This event is sequence dependent. If the duplex has a very high Tm, aggressive chromatographic conditions such as the addition of formamide or urea may be needed to fully denature the duplex. The method conditions either under pH 8 or 11, which provide full denaturation of duplex, is suitable as a stability indicating method as is elaborated in Chapter 15. The partial denaturation of duplex in the high-temperature pH 8 method finds an important application in making the duplex drug substance, providing the ability to reveal the presence of excess guide or passenger single strands.

2.7.3.4 Annealing Method

The ultra-filtered solutions of the sense and antisense strand are combined in the desired proportions in order to form an equimolar mixture of the two intermediates. The required amounts of each single-strand ON are calculated based on a UV assay using theoretical or experimentally determined molar extinction coefficients and their molecular weights. Refer to Chapter 12 for more on derivation and use of extinction coefficients. To assure better control, the calculated amount of the first strand is mixed with less than the calculated amount of the second strand.

The conditions for AEX-HPLC analysis should be such that a sample of the equimolar mixture of sense and antisense strand shows a well-resolved peak for the excess of the first strand together with a peak for the duplex. An example of such a resolution with slight excess of antisense strand is shown in Figure 2.25a.

An additional amount of the second strand is added and a sample is analyzed again. The mixture that has an excess of sense strand is shown in Figure 2.25b. This "titration" process is repeated until equimolar ratio is achieved and no measurable amounts of excess single strand are detected, or the process specification set to for optimal levels of antisense or sense strands is achieved, as shown in Figure 2.25c. The resulting ratio of single strands is then utilized by manufacturing to perform annealing of the duplex.

2.7.3.5 Impurity Profile of Drug Products

Development is in progress to formulate siRNA in the most deliverable format. The siRNA drug substance is, in general, formulated in aqueous buffers such as phosphate buffered saline (PBS) or encapsulated in cationic liposomes.[104,105] One would not expect process-related impurities during

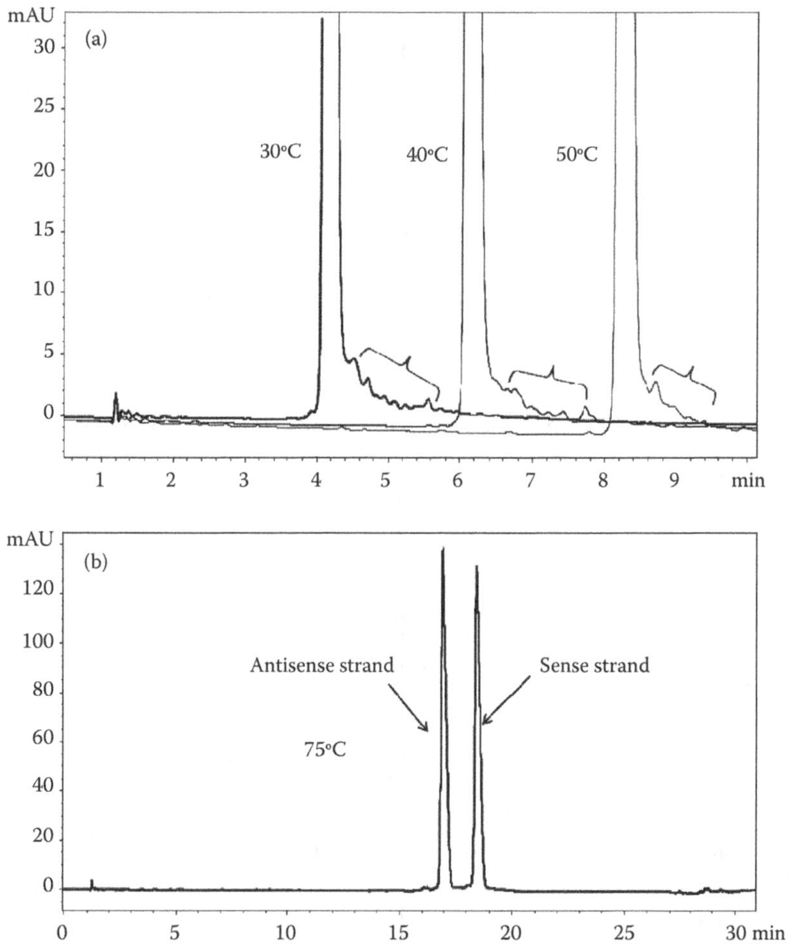

FIGURE 2.24 (a) Chromatogram of a mixture of sense strand and antisense strand that forms the duplex. System: Agilent 1100 HPLC; Column: Dionex DNAPac PA-200; 40°C; Buffer A: 10% CH_3CN, 25 mM Tris-HCl buffer at pH 8; Buffer B: 10% CH_3CN, 25 mM Tris-HCl buffer at pH 8 and 1 M NaBr; Gradient: 30–50% B in 25 min. As the column temperature is increased, "failure duplexes" in the brackets melt. (b) At a high enough temperature, 75°C duplex is fully denatured into antisense strand, sense strand, and deletion impurities.

formulation in PBS buffers. However, potential new impurities introduced during *liposome* formulation need to be evaluated on a case-by-case basis.

As the drug discovery program advances, it becomes a necessary task to identify and characterize the impurities in the drug substance. Conventional AEX-HPLC, or any other single-stage method, is not suitable to identify impurities, as the method either fails to provide molecular identity, or fails to resolve isobaric components, such as linkage isomers. One approach to extend the utility of AEC is to collect impurities as fractions, then desalt and analyze by ESI-MS, as described below in Sections 2.8.

2.7.4 Analysis of PEGylated Aptamers

As discussed earlier in Section 2.4, aptamers form specific recognition configurations through stable and specific higher-order structures to effectively interact with their targets.

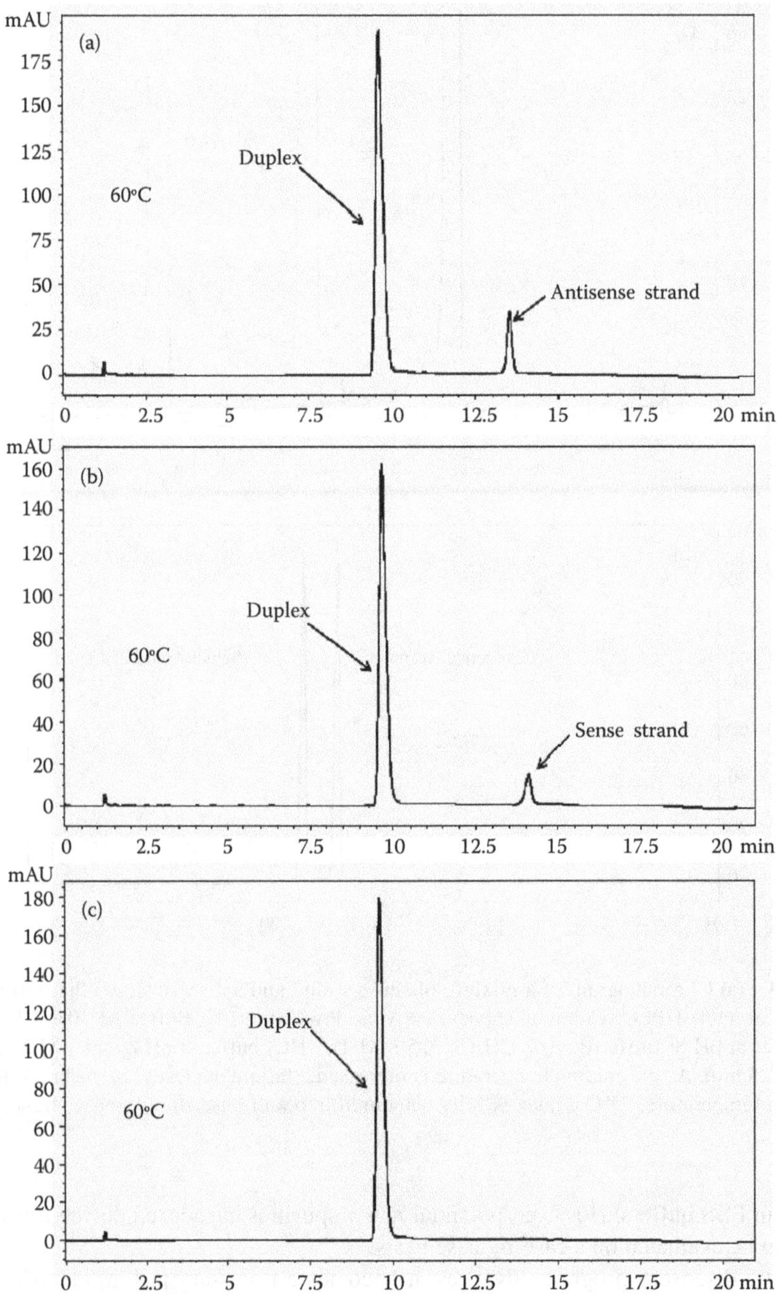

FIGURE 2.25 (a) Chromatogram of a mixture of sense strand and antisense strand that forms the duplex. System: Agilent 1100 HPLC; Column: Dionex DNAPac PA-200; 60°C; Buffer A: 10% CH_3CN, 25 mM Tris-HCl buffer at pH 8; Buffer B: 10% CH_3CN, 25 mM Tris-HCl buffer at pH 8 and 1 M NaBr; Gradient: 30–41% B in 16 min. Antisense strand is slightly in excess. (b) Sense strand is slightly in excess. (c) Sense strand and antisense strand in almost equal proportion. Sequences of sense strand and antisense strands are given in the text.

2.7.4.1 Denaturation of Aptamers

Because of their adoption of higher-order structures, denaturing analytical methods are necessary for aptamer analysis, especially analyses of chemical purity and impurity monitoring.

When using AEC, this is accomplished by means of high-temperature operation or use of high-pH eluents. Thermal and chemical (high pH) options are useful for both native and PEGylated aptamers. High temperature provides energy to break the hydrogen bonds between the nucleobases, causes the aptamers to adopt essentially linear structures, and promotes the nucleic acid–stationary phase interactions described in Section 2.6. The higher temperatures also improve resolution as discussed in the methods development (Section 2.7.2.3). Figure 2.26 illustrates the effect of temperature on AEC of Aptamers without (A), and with (B), PEG conjugation. Typically, when column temperature is above the melting temperature of the aptamers, the resolution increases.

2.7.4.2 PEG Hydration

In aqueous buffered solutions like PBS, PEGylated aptamers eluted significantly earlier during size-exclusion chromatography (SEC) than native aptamers, indicating the hydrodynamic volume (because of increased molecular mass and hydration) of PEGylated aptamers to be significantly larger than that of the native aptamers. Because of the larger mass and hydration, diffusion of PEGylated aptamers will be much slower than the native ON.

2.7.4.3 Polydispersity of PEG

PEG preparations harbor polydispersity with respect to the distribution of PEG length, so the PEG-ON conjugate represents a controlled distribution of molecules with different numbers of

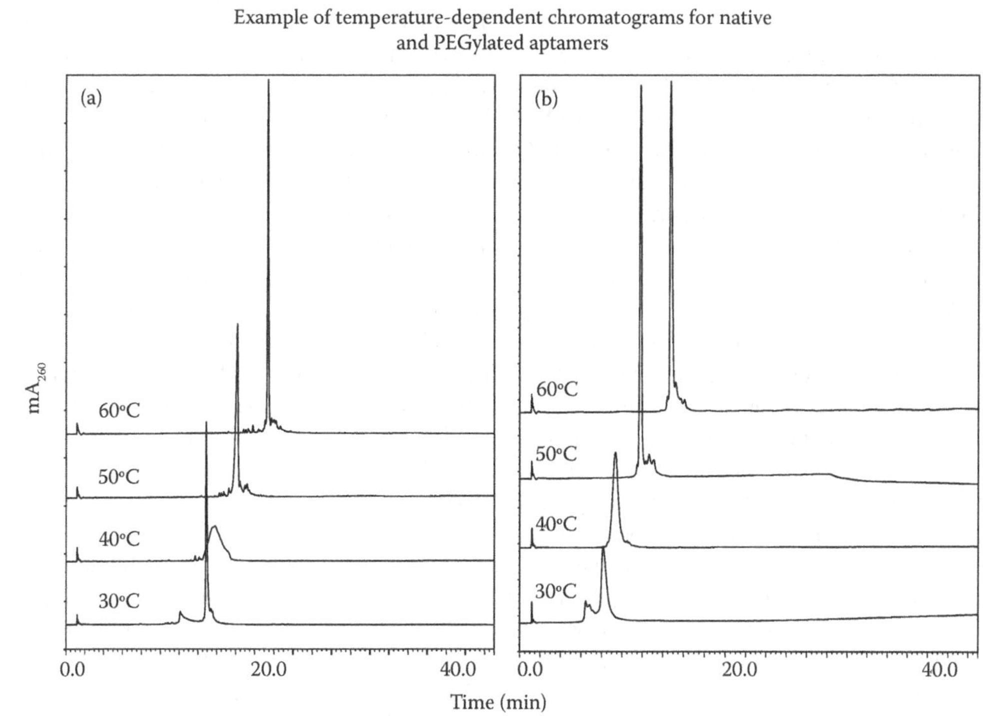

FIGURE 2.26 Chromatography of aptamer ONs (a) without and (b) with PEG conjugation, showing effect of temperature on elution patterns. Column: DNAPac PA200.

polyethylene-glycol units attached to the single aptamer. This will also contribute to chromatographic band-broadening for a number of HPLC approaches. For more on SEC analysis of PEGylated oligonucleotides, see Chapter 3.

2.7.4.4 Steric Hindrance

PEG does not harbor a chromophore, so it does not absorb UV light. Therefore, standard HPLC UV detection only measures the ON part of PEGylated aptamer. While attachment of PEG to the aptamer does not contribute to UV absorption, it does impact adsorption to HPLC columns, likely owing to steric hindrance of electrostatic or hydrophobic interactions. Hence, when similar amounts of an ON are injected to generate similar UV response, caution is necessary to avoid overloading of the column. The peak broadening due to reduced mass transfer and steric hindrance may result in decreased resolution and poorer detection sensitivity for PEGylated aptamers by HPLC methods.

Reversed-phase ion-pair HPLC and anion-exchange HPLC are widely used methods for ON therapeutics. The reversed-phase methods are compatible with on-line MS detectors, a powerful tool to identify ON impurities. However, LC/MS of PEGylated aptamer samples becomes extremely difficult owing to the fact that PEGylated aptamer samples no longer represent a single molecular species but a polydisperse set of molecules with the mass difference of the ethylene glycol unit (based on the PEG polydispersity). PEG also shows greater hydrophobicity under reverse-phase ion-pair HPLC conditions than the native ON. Because reverse-phase columns separate analytes based on differences in hydrophobicity of the analytes, hydrophobicity of PEG becomes the dominant factor influencing retention and resolution of PEGylated aptamers. Hence, under reversed-phase conditions, N-1 and n+1 impurities are no longer readily separated from the full length product.

On the other hand, AEC ON separations are based primarily on the number of charges on the ONs. The PEG chain harbors no charge under most anion-exchange conditions. Therefore, the impact of PEG on AE retention is less significant than under RP IP HPLC conditions. Nonporous strong anion-exchange columns offer convection (rather than diffusion) dominated mass transfer, so are preferred for analysis of PEGylated aptamers. For example, resolution of aptamers on Dionex DNAPac100 and 200 and Tosoh TSKGel DNA-NPR (nonporous) is excellent and suitable for exploring n-1 versus n separation, even for PEGylated sequences. However, during AEC the separation of n-1 from full length PEGylated aptamer is not as great as for the native aptamers, but n-1, and n+1 PEGylated impurities *can* be separated from full length products (Figure 2.27).

PEGylated aptamers elute earlier than their native counterparts on anion-exchange columns, likely because the PEG adducts introduce steric hindrance to the aptamer-stationary phase interaction, thus limiting both electrostatic and hydrophobic interactions with the column. The impact of PEGylation will depend on the PEG molecular weight, the PEG architecture, and how the PEG is conjugated to the aptamer. The influence of PEG length, PEG branching, and the number of attachments to an aptamer is shown in Figure 2.28. From that figure we can extract several observations: (1) Linear PEGs reduce retention less than branched PEGs, (2) larger PEGs reduce retention more than shorter PEGs (e.g., the 20K, 30K and 40K linear PEGylated aptamers elute with decreasing retention time in inverse order of length); and (3) aptamers with PEG on both 3′ and 5′ end cause the shortest retention and greatest band broadening.

Resolution by strong anion-exchange chromatography does decrease when aptamers are PEGylated. A specific challenge for separation of PEGylated aptamers involves resolution of phosphorothioate diastereoisomers. While resolution of PS diastereoisomers of native ONs is demonstrated (see Section 2.8.4), this becomes more difficult after PEGylation. Because reduced but partial resolution is observed for the Aptamer after PEGylation (Figure 2.29), one approach would be to serially connect two or more columns in the attempt to improve this separation. Other avenues include altering the pH, salt form, or temperature to enhance resolution of these components.

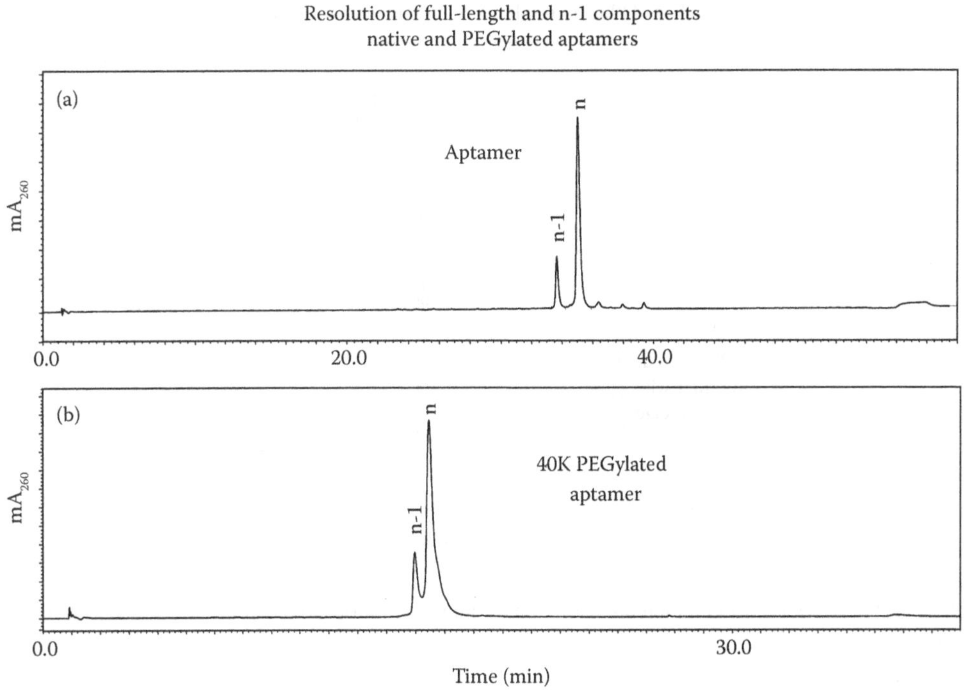

FIGURE 2.27 Example chromatograms of aptamers (a) without and (b) with PEG conjugation, showing resolution of full length from "n–1" truncates. Column: DNAPac PA200. Different optimized gradients were used. NaCl/ACN mobile phases, buffered by sodium phosphate at pH 7.

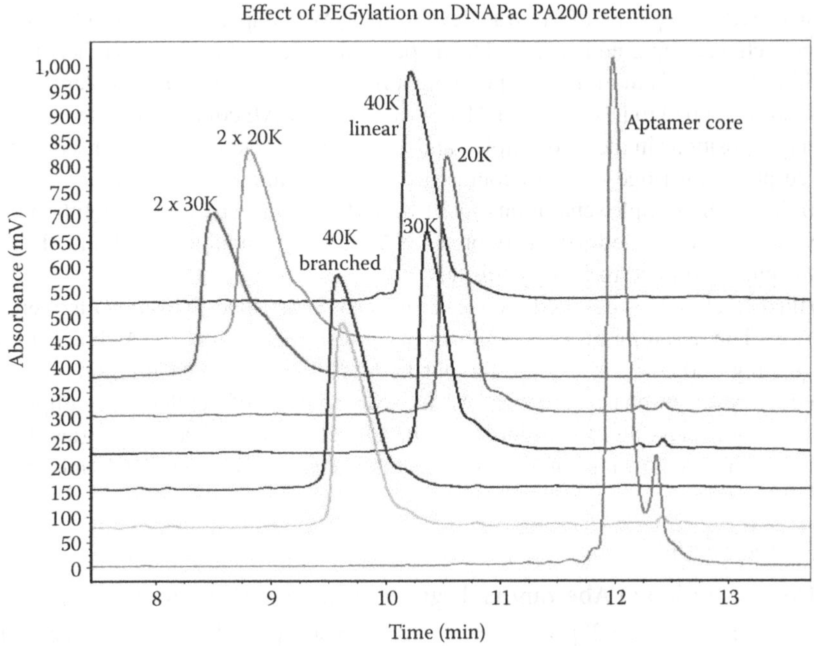

FIGURE 2.28 Effect of PEGylation on retention on the DNAPac PA200 column.

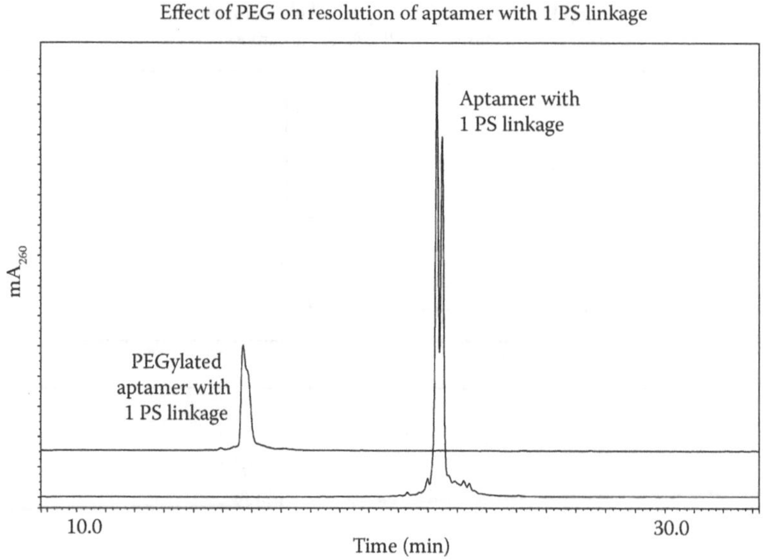

FIGURE 2.29 Effect of PEGylation on resolution of phosphorothioate diastereoisomers. Column: DNAPac PA200.

2.8 ADVANCED APPLICATIONS

2.8.1 Oligonucleotide Desalting: AXLC-MS

The development of new mass spectrometry capabilities were applied to both AE and reversed-phase separated ONs after manual desalting on C18 cartridges by ESI[12,47] and MALDI-TOF approaches.[106] These reports demonstrated the value of MS techniques and indicated a method for identifying impurities. For this purpose, ESI was shown to provide superior results, and a method for automated desalting of IP-RPLC-separated ONs was developed.[107] We adapted that method to automatically desalt anion-exchange separated ONs.[108] The automated system employed a bioinert HPLC system (Dionex Ultimate-3000 Titanium) equipped with a fraction-collecting autosampler (Schematic 1).

The ONs were applied to DNAPac or DNASwift pellicular AE column, detected peaks were collected to unique positions in the autosampler, and the collected fractions were subsequently desalted on a reversed-phase cartridge or guard column using ammonium formate and methanol as eluents. Figure 2.30 shows an example separation of a 21-base RNA on a pellicular monolith, where the collected fractions were evaluated for purity on a DNAPac PA200 analytical column. The separation increased the purity from 78 to 97% by this test.

The purified fraction was desalted on a reversed-phase cartridge (Novatia OligoSep, or Dionex Acclaim PA-2). The desalting process is shown in Figure 2.31. Figure 2.31A shows the separation of the salt (eluting at 0.2–0.4 min) from the purified ON (eluting at ~1.1 min) on an Acclaim PA2 polar-embedded reversed-phase cartridge. An ESI-MS analysis of the desalted sample reveals the expected full length mass and two minor sodium adducts, but no other impurities. We used this approach to identify the positions of aberrant linkage isomers in Sections 2.8.2 and 2.8.3.

2.8.2 Alternate Linkages

2.8.2.1 Demonstration of Aberrant Linkages (Phosphoryl Migration)

Phosphodiester ORNs lacking 2′ protection can undergo phosphoryl-migration causing strand scission and formation of aberrant 2′,5′-linkages.[37] Such linkages can also be *intentionally* introduced to both RNA and DNA for interfering RNA approaches.[39] These *aberrantly linked* ONs do not

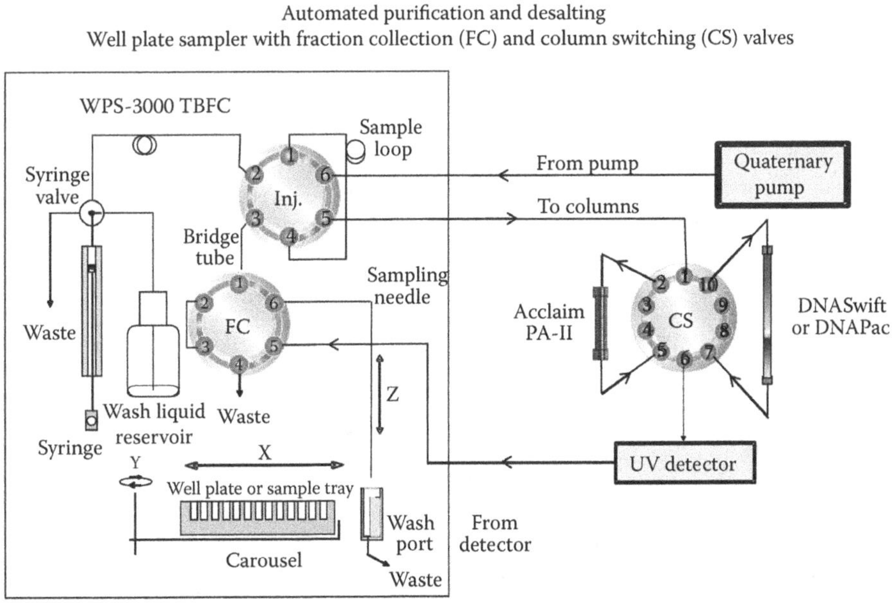

SCHEMATIC 1 HPLC configuration for automated purification and desalting of ONs for salt-sensitive downstream applications (e.g., ESI-MS). This configuration is based on the Dionex UltiMate 3000 Titanium system with WPS BTFC fraction collecting autosampler. (Reprinted from Thayer, J. R., et al., *Analytical Biochemistry*, 399, 110–117. Copyright 2009, with permission from Elsevier.)

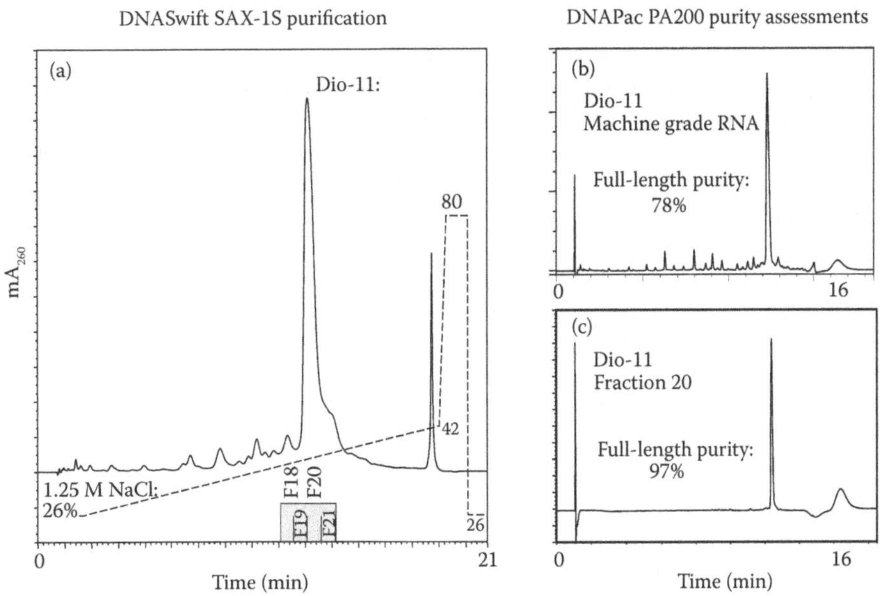

FIGURE 2.30 Purification of RNA using the DNASwift SAX-1S (5 mm × 150 mm) hybrid monolith. (a) A gradient of 235–525 mM NaCl, resolves the full length RNA from failure sequences. System: Dionex UltiMate 3000 Titanium inert quaternary gradient LC system. DNAPac PA200 assays of the (b) commercially obtained and (c) purified RNA show sample enrichment from ~78–97% purity. (Reprinted from Thayer, J. R., et al., *Analytical Biochemistry*, 399, 110–117. Copyright 2009, with permission from Elsevier.)

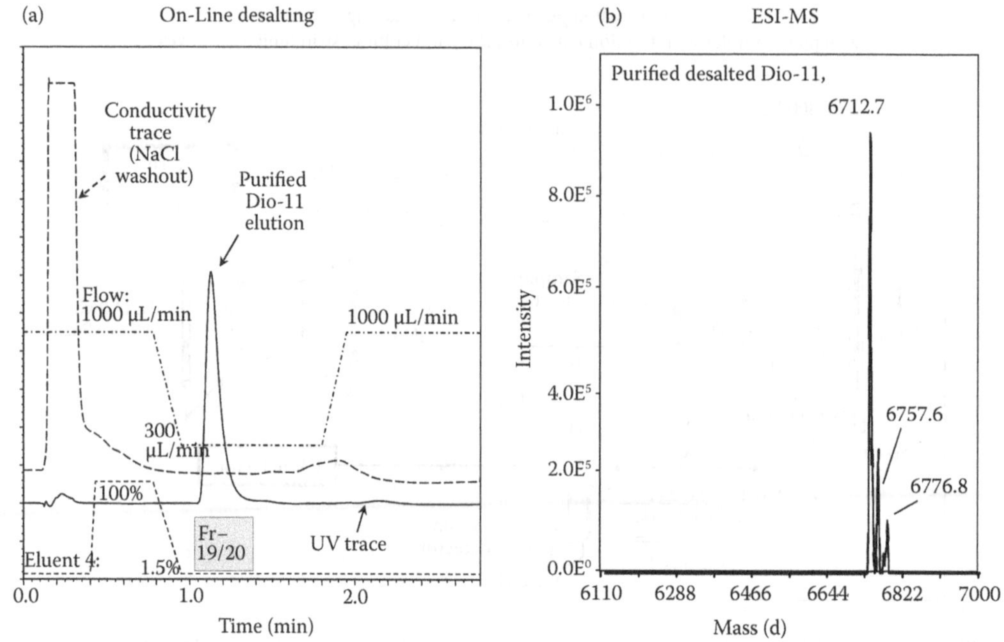

FIGURE 2.31 Desalting and ESI-MS analysis of DNASwift-purified RNA from Figure 2.30. (a) Separation of co-collected salt from the ON using an Acclaim PA-2 column; (b) Deconvoluted ESI-MS spectrum of the desalted RNA. System: Dionex UltiMate 3000 Titanium inert quaternary gradient LC system. (Reprinted from Thayer, J. R., et al., *Analytical Biochemistry*, 399, 110–117. Copyright 2009, with permission from Elsevier.)

differ from their normally linked counterparts by length, net charge, or even mass, so they were not anticipated to be separable by anion exchange, reversed-phase chromatography, or capillary electrophoresis. However, ONs harboring 2′,5′-linkages do assume different solution conformations than ONs harboring only 3′,5′ linkages.[109] Hence, to the degree that they may interact differentially with stationary phases, they can be resolved (see Ref. 38). Our observation is that resolution of ORNs harboring one or more 2′,5′-linkages requires either high resolution (fast mass transfer) phases or long "shallow" (≥40 min) gradients on porous bead AE stationary phases. Even with high-resolution pellicular AEC phases, elution of 2′,5′-linked ONs depends on the *position* of the linkage in the sequence. An example of such a separation is provided in Figure 2.32, where 11 ON samples having the same sequence were chromatographed under identical conditions.

Except for Dio-1, which has no 2′,5′-linkages, these samples each harbor aberrant linkage(s) at independent positions in the sequence (indicated as linkage from the 5′ end in parentheses (see also Table 2.4). The elution positions of these ONs indicate that (1) samples with one or more aberrant linkages in the last few positions from the 5′ end (Dio-10, -11, -12) are not resolved from, or elute only *slightly* earlier than the same sequence without any aberrant linkages; (2) samples with aberrant linkages within 5 bases of the 5′ end (Dio-2, -3, -4, -8) elute *earlier*; (3) samples with the aberrant linkage(s) at positions in the middle of the sequence (Dio-5, -6, -7) elute *significantly* earlier; and (4) the sample with an aberrant linkage at position 15 in this sequence elutes *significantly later* than the same RNA sequence without aberrant linkages. Each of these samples was confirmed to have aberrant linkages by treaTment with phosphodiesterase-II (PDase-II, Calf Spleen exonuclease, IUB code 3.1.16.1). This di-esterase requires a 5′-hydroxyl, proceeds in the 5′ to 3′ direction, produces nucleoside 3′ monophosphates (NMPs), and is incapable of cleaving 2′,5′-linkages. Digestion of Dio-1 produces only nucleotide monophosphates (Figure 2.33). However, digestion of each of

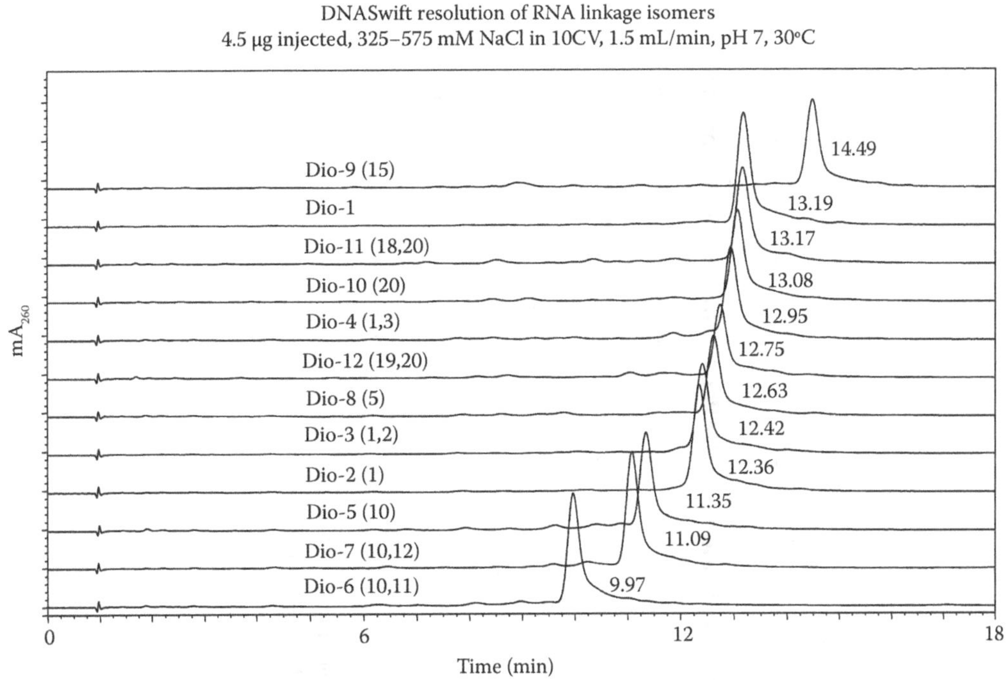

FIGURE 2.32 Chromatography of 12 RNA samples having a single common sequence on the DNASwift SAX-1S. Each sample harbors 0, 1, or 2 aberrant 2′,5′-linkages at unique position(s) in the sequence, resulting in differential retention. System: Dionex DX600 inert quaternary gradient LC system. (Reprinted from Thayer, J. R., et al., *Journal of Chromatography B*, 878, 933–941. Copyright 2010, with permission from Elsevier.)

the other samples produces a fragment eluting in a position distinct from those of all the other samples[38,110] except for Dio-10. Because Dio-10 harbors a single aberrant linkage at position 20, failure to cleave that linkage produces a *di* nucleotide monophosphate, which, at pH 8 is expected to elute as an NMP. In this case the dinucleoside monophosphate contains both a U and a G so chromatography at elevated pH will induce two additional charges that will cause it to elute later than all of the NMPs, so it was demonstrated to harbor an aberrant linkage at that pH.

2.8.2.2 Identification of Aberrant Linkage Position

After digestion of RNA ONs with PDase-II, the fragments can be isolated away from the NMPs using either a DNAPac or DNASwift phase, desalted and examined by ESI-MS. Example purifications are shown in Figure 2.34.

Desalting these fragment followed by ESI-MS (Figure 2.35) reveals their base composition: Assessment of the composition with the sequence allows identification of the linkage position, provided the sequence is not too repetitive.

The example raw mass spectrum of the Dio-3 fragment 1 (top panel, Figure 2.35) reveals a molecular mass of 1327.2 amu. This is consistent with a base composition of $rA_2rU_1rG_1$. Given the RNA sequence, the only position in which these bases are contiguous comprises the four bases at the 5′ end of the sequence. This indicates the aberrant linkage to be at position 1 or at 1 and at other positions in the first four bases. Similarly, fragment-2 from Dio-5 reveals a molecular mass of 691.3, consistent with a single rA and one rG. This fragment could arise from aberrant linkages at positions 3, 10, or 17 in this RNA sequence, so for very short fragments (dimers) not all linkage isomers can be confidently assigned.

TABLE 2.4
Identification of Phosphodiesterase-II RNA Fragments Based on Original Sequence, Elution on DNAPac PA200, and Resulting Mass[a]

Name	Fr'n	Sequence and Position of Digest Fragment								Fragment ID	Mass
Dio1	none	5′ –AUG	AAC	UUC	AGG	GUC	AGC	UUG	–3′	None (only NMPs)	—
Dio3	Fr 1	5′ –A*U*G	AAC	UUC	AGG	GUC	AGC	UUG	–3′	ApUpGpAp,	1327.2
Dio5	Fr 2	5′ –AUG	AAC	UUC	A*GG	GUC	AGC	UUG	–3′	ApGp	691.2
"	Fr 3	5′ –AUG	AAC	UUC	A*GG	GUC	AGC	UUG	–3′	ApGpGp	1037.2
Dio6	Fr 4	5′ –AUG	AAC	UUC	A*G*G	GUC	AGC	UUG	–3′	ApGpGp	1037.2
Dio7	Fr 5	5′ –AUG	AAC	UUC	A*GG*	GUC	AGC	UUG	–3′	ApGpGpGp	1382.2
Dio8	Fr 6	5′ –AUG	AA*C	UUC	AGG	GUC	AGC	UUG	–3′	ApCpUp	958.1
Dio9	Fr 7	5′ –AUG	AAC	UUC	AGG	GUC*	AGC	UUG	–3′	CpAp	652.1
Dio11	Fr 8	5′ –AUG	AAC	UUC	AGG	GUC	AGC*	UU*G	–3′	CpUpUpG–OH	1200.2

[a] eGFP antisense strand: 5′ –AUG AAC UUC AGG GUC AGC UUG–3′. Modifications: Underlined residues have 2′–5′ linkages.

Source: Reprinted from Thayer, J. R., et al., *Analytical Biochemistry*, 399, 110–117. Copyright 2009, with permission from Elsevier.

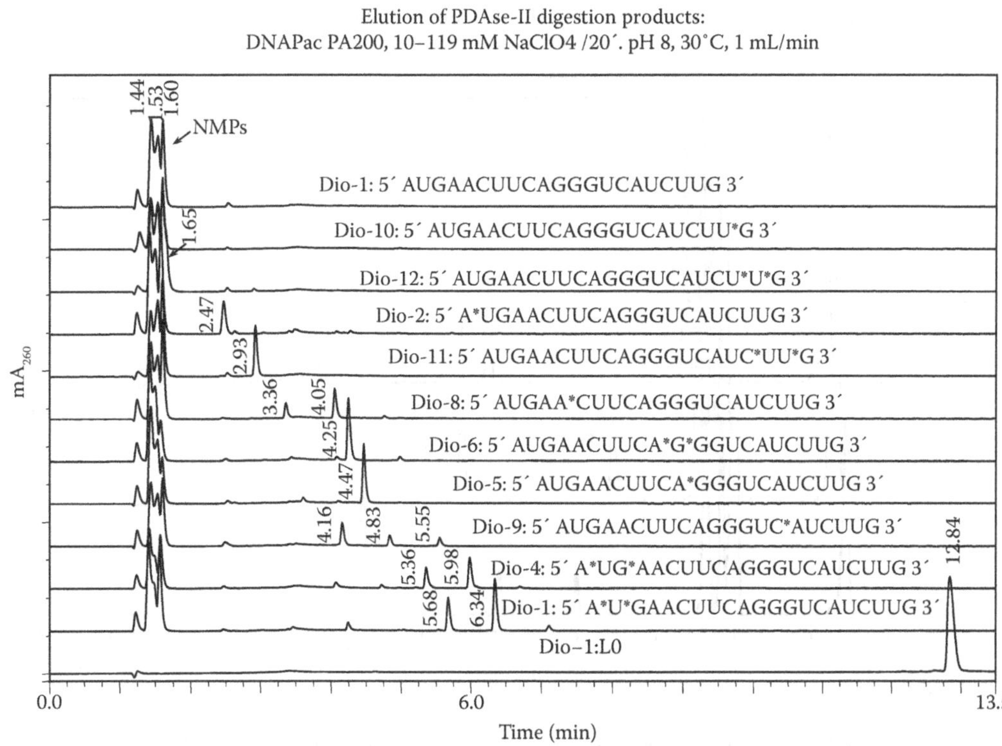

FIGURE 2.33 Anion-exchange chromatography of phosphodiesterase II digestion products of the 12 RNA samples purified as in Figure 2.30. The digestion products from these samples, each with specifically positioned 2′-5′-linkages, are retained to unique positions. This demonstrates resolvable, and hence sequence-dependent, digestion products. System: Dionex DX600 inert quaternary gradient LC system. (Reprinted from Thayer, J. R., et al., *Analytical Biochemistry*, 361, 132–139. Copyright 2007, with permission from Elsevier.)

RNA samples with the aberrant linkage at different positions produce fragments with different base compositions, as summarized in Table 2.4.

For samples Dio-2, Dio-5, Dio-8, Dio-9, and Dio-10 (each harboring only one 2′,5′-linkage), the major digestion products are dinucleotide diphosphates, indicating that the enzyme ratchets over the aberrant linkage and continues downstream cleavage one base to the 3′ side of the aberrant linkage. Samples Dio-3, -4, -6, -7, -11, and -12, produce major digestion products that are tri- or tetra-nucleotide phosphates, indicating that the enzyme skips the aberrant linkage, fails to cleave a phosphodiester sandwiched between two aberrant linkages, and again continues digestion one base to the 3′ side of the 2′,5′-linkage. That this process correctly identifies the positions of the aberrant linkages in these oligoribonucleotides (ORNs) with intentionally introduced 2′,5′-linkages indicates that the process will usually allow identification of the position(s) of the aberrant linkages in unknown samples.[38,108]

2.8.3 Target and Impurity Identification: AXLC-MS

One of the most powerful techniques for both confirmation of the target ON as well as for determination of impurities is LC-MS. Impurity analysis by LC-MS supports identification of minor impurities in both resolved components and those "hiding" under the target ON peak.[13] In that study, ON

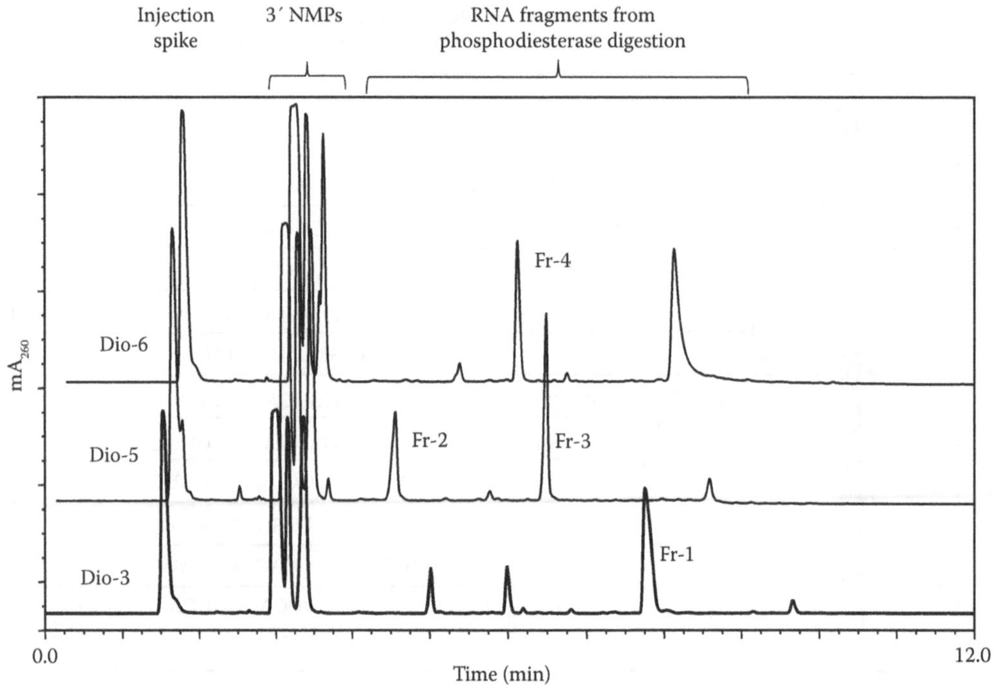

FIGURE 2.34 Purification of phosphodiesterase II digestion products by AEC on the DNAPac PA200 column. Digestion products from the indicated RNA samples, labeled as fragments 1 through 4 were collected for desalting and ESI-MS analysis. System: Dionex UltiMate 3000 Titanium inert quaternary gradient LC system. Gradient: 0–125 mM $NaClO_4$ in 25 min with a curved gradient (4, see Ref. 93). (Reprinted from Thayer, J. R., et al., *Analytical Biochemistry*, 399, 110–117. Copyright 2009, with permission from Elsevier.)

metabolites extracted from mammalian tissue were separated by AEC on a porous AE phase, and fractions from that separation subsequently analyzed by CE and ion-pair LC with ESI/MS detection. The CE separation revealed two or three major, and at least three minor, components in each of the porous AE-separated fractions. ESI/MS of *each* HPLC-separated component revealed *8–10* partially resolved components exhibiting up to five charge states each. Hence these initial forays into IP-RPLC-ESI/MS required a very high level of sophistication to analyze and interpret. This state of affairs has improved, and the reader is referred to Chapters 1 and 4 for further information on this approach.

Automated desalting coupled to ESI/MS as described above (Section 2.8.1) is also useful for identifying impurities in components isolated by AEC. As an example, a 21-base ODN was separated from several impurities on a 2 mm ID DNAPac PA200 as shown in Figure 2.36.

The primary peak was collected as shown and desalted using the automated method of Schematic 1 and Figures 2.30 through 2.37. The resulting ESI-MS (deconvolution report shown in Figure 2.37) reveals the presence of, and distinguishes between, two "n–1" contaminants (an "n-C" and an "n-A"), as well as minor sodium adduction in the full length product.

2.8.4 Resolution of R_P and S_P Phosphorothioate Isomers

As described earlier in Section 2.2.1 PS linkages are commonly inserted in therapeutic ONs. Developers of aptamers and RNAi therapeutics may employ only one or a few PS linkages, and each introduces a chiral center. In one case[41] two PS linkages are inserted, so the number of possible

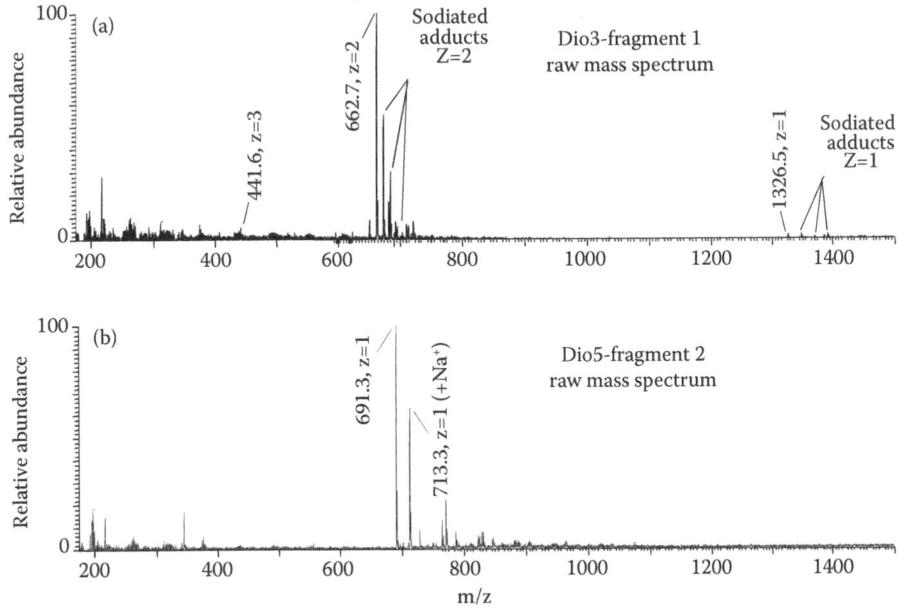

FIGURE 2.35 ESI-MS spectra of collected desalted fragments from samples purified as shown in Figure 2.33. Fragments 1 (from Dio-2) and 2 (from Dio-5) show molecular mass values of 1326.5 and 691.3, respectively. (Reprinted from Thayer, J. R., et al., *Analytical Biochemistry*, 399, 110–117. Copyright 2009, with permission from Elsevier.)

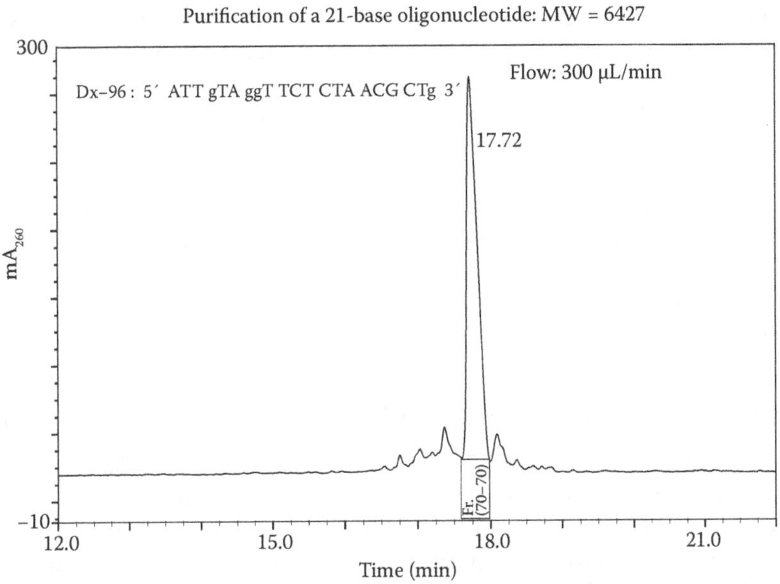

FIGURE 2.36 Purification of a 21-base oligodeoxynucleotide on a 2 mm × 250 mm DNAPac PA200. System: Dionex UltiMate 3000 Titanium inert quaternary gradient LC system. Gradient: 80–195 mM $NaClO_4$ in 15 min at pH 12, 300 μL/min, 30°C.

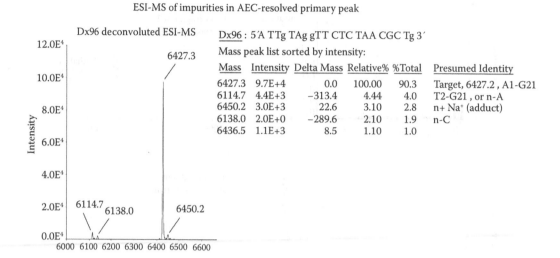

ESI-MS of impurities in AEC-resolved primary peak

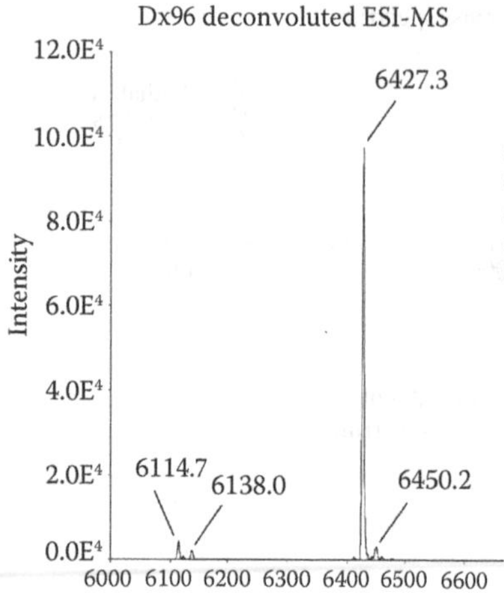

Dx96 : 5′A TTg TAg gTT CTC TAA CGC Tg 3′

Mass peak list sorted by intensity:

Mass	Intensity	Delta Mass	Relative%	%Total	Presumed Identity
6427.3	9.7E+4	0.0	100.00	90.3	Target, 6427.2 , A1-G21
6114.7	4.4E+3	−313.4	4.44	4.0	T2-G21 , or n-A
6450.2	3.0E+3	22.6	3.10	2.8	n+ Na^+ (adduct)
6138.0	2.0E+0	−289.6	2.10	1.9	n-C
6436.5	1.1E+3	8.5	1.10	1.0	

FIGURE 2.37 Deconvoluted ESI-MS analysis of the ODN purified in Figure 36. This purified sample co-eluted with low levels (<5%) of one "n-A" and one "n-C" contaminant that are identified and differentiated by mass spectrometry. ESI MS employed a Thermo Quantum Access instrument operated in negative ion mode. Sample injection and carrier solution was that employed in Ref. 105.

diastereoisomers is limited to *four*. Chromatographic techniques other than AEC do not reliably resolve these diastereoisomers unless the ONs are less than ~10 bases. Because the R_p and S_p isomers are isobaric, even coupling to single-stage MS does not allow discrimination of these PS forms. However, AEC appears able to resolve these in RNA, and partially so in DNA.[9,10] Recently, we demonstrated resolution of four diastereoisomers, in a 37-base RNA aptamer harboring two PS linkages using both a pellicular column and a hybrid monolith. That report is in preparation. To verify that this is a general characteristic, we acquired 21-base ONs as both RNA and DNA with either two or no PS linkages. AEC of the PO samples is shown in Figure 2.9, and that of the PS-containing samples (both on a hybrid monolith) is shown in Figure 2.38.

The samples *without* the PS linkages each produce a single peak for the full length ONs (Figure 2.9). Conversely, the DNA ON with PS linkages produces a triplet peak, and the RNA with PS linkages produces four separable components. After purification, each component, upon de- and renaturation, still produces only a single peak by AEC, so these are not alternate conformations. Each peak was desalted and produced the identical (and expected) full length mass upon ESI-MS, so these are not incompletely, 'thioated isomers. Because we also observe this pattern using longer ORN aptamer, we conclude that resolution of PS diastereoisomers is a general feature of *pellicular* (nonporous, nanobead-coated, and hybrid monolith AEC phases) AEC.

2.8.5 Resolution of Mono-, Di-, and Tri-Nucleoside Phosphates at High pH

While not strictly an *oligo*nucleotide separation, the resolution of these different *NTPs* and *dNTPs* further illustrates the use of high pH to confer different charge states on nucleic acid bases and the use of temperature for further optimization. In Figure 2.39, each of the mono-, di-, and tri-nucleoside phosphates was chromatographed at pH 12 using a gradient of 30–150 mM NaCl at 1.4 mL/min. At this pH, each U, T, and G base harbors an additional negative charge due to tautomeric oxygens on these bases. However, the A and C nucleoside phosphates do not harbor these additional charges so their charges arise primarily from the (mono-, di-, or tri-) phosphates. Because chloride is the eluent, hydrophobic differences between the C and A bases are not suppressed with this eluent

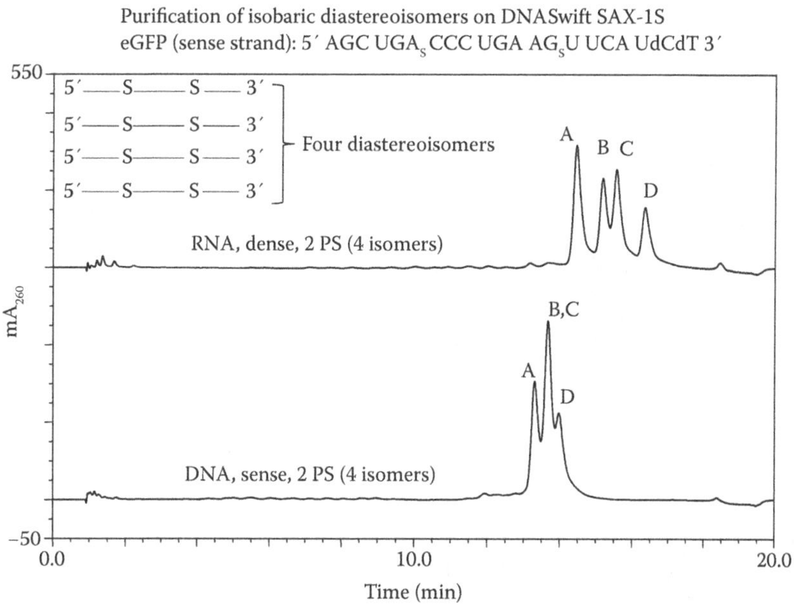

FIGURE 2.38 Separation of *isobaric* phosphorothioate diastereoisomers on the DNASwift SAX-1X. Chromatography of a commercial sample of the indicated sequence with phosphorothioate linkages only at positions 6 and 14 resolves three peaks in DNA form and resolves all four in the RNA sample. System: Dionex DX600 inert quaternary gradient LC system. Gradient: 300–600 mM NaCl in 16.7 min at 1.5 mL/min, 30°C, and pH 7. (Modified from Thayer, J. R., et al., *Journal of Chromatography B*, 878, 933–941. Copyright 2010, with permission from Elsevier.)

salt, so allow their separation at each charge state. Similarly, the differences between the G and U bases in this salt allow their resolution at each charge state.

In the bottom trace, the temperature was 23°C, where all the deoxynucleoside phosphates are essentially resolved, but some nucleoside diphosphates (ADP and UDP) are not. In the top trace, the temperature was 27°C, where all the nucleoside phosphates are essentially resolved, but some nucleoside monophosphates (dTmP and dAMP) are not. Hence, use of the same gradient at slightly different temperatures allows for analysis of the two different sets (RNA versus DNA) nucleoside phosphates.

2.9 CONCLUSIONS

In this chapter we reviewed the development of ON chromatography by anion-exchange phases to the current date, and cite numerous examples of AEC as applied to *analysis* and to a limited degree *purification*. We specifically discuss issues pertinent to development of analyses for *therapeutic* ONs, and describe specific eluent systems and approaches for both single- and double-stranded RNA. These methods and approaches are, of course, applicable to ssDNA, dsDNA and dsDNA/RNA hybrids. We discuss different types of AEC stationary phases, recommend applications for each, and suggest that there is room for overlap between the different phase types and the recommended applications, as that is evident in the literature.

In order to "arm" the reader with information useful for development of new AEC ON analyses, we discuss in some detail those ON attributes that contribute to retention on AEC phases and the means to emphasize or suppress the effect of those attributes in order to optimize *resolution* of target ON sequences and their impurities. Salt form, pH, solvent concentration, and even counter

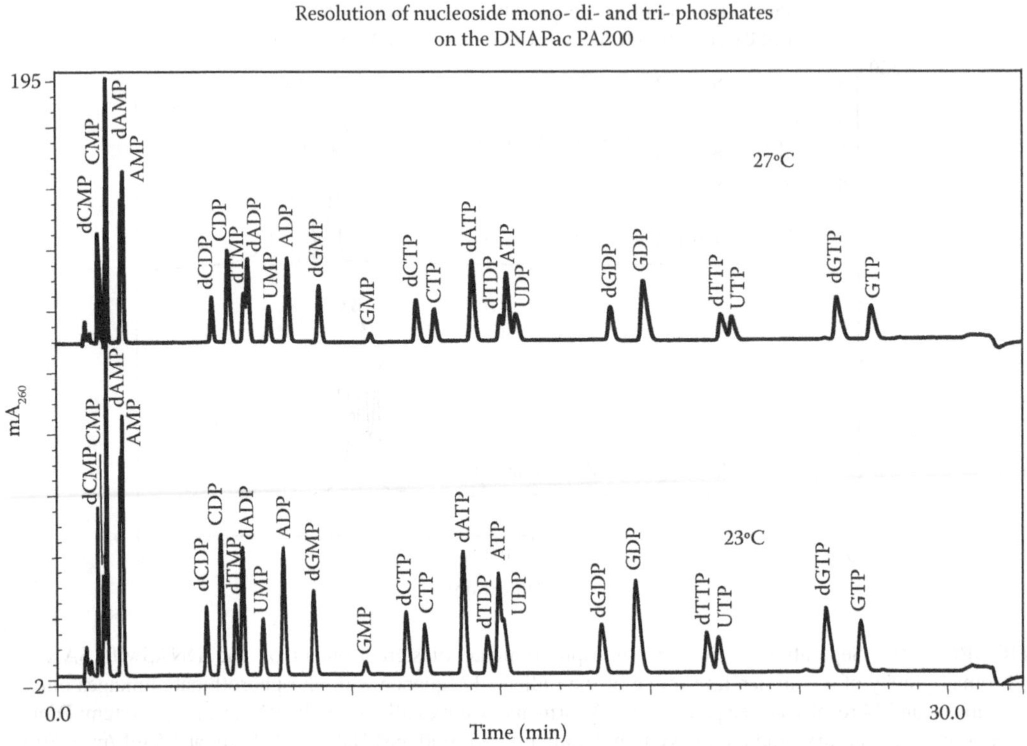

FIGURE 2.39 Analysis of the nucleoside mono-, di-, and tri-phosphates of both RNA and DNA on the DNAPac PA200. System: Dionex Dx600 inert quaternary gradient LC system. Gradient: 30–150 mM NaCl in 30 min at pH 12 and 1.4 mL/min. All DNA components are resolved at 23°C (bottom trace) and all RNA components (top trace) are resolved at 27°C. Note: Minor changes to tubing length, mixer design, and column oven may to different results for different systems.

cation in AEC eluents each provide *independent* means to extract very high resolution separation of target and critical impurities by AEC, especially in pellicular format.

Because ONs intended for therapeutic use typically harbor one or more chemical modifications to increase their stability and biological half-life, and because these modifications may influence the ON interaction with different stationary phases, we discuss the impact of several modifications on this process and the means to manage their influence during AEC. In this effort we employed specific examples to illustrate the use of eluent composition and temperature control for a number of ONs. For those interested primarily in development of AEC methods for therapeutic ONs, we present a set of example applications to optimize resolution and analysis of both ss- and dsRNA in a manufacturing environment, as applied to development of siRNA drug substance. Included in these applications are recommendations for use of high-resolution analyses as such *and* for use as stability-indicating methods (see also Chapter 15).

Finally, we included a section on "advanced" applications that include methods for *automated* collection and desalting of AE-separated ONs and their impurities, prior to downstream, salt-sensitive analyses. As one such method, we demonstrate off-line coupling of AEC to ESI-MS analyses. This approach revealed that some AE-purified ONs may have no ON-derived impurities (see Figures 2.30 and 2.31), while others may harbor "n-1" impurities that are readily distinguished by ESI-MS (Figures 2.36 and 2.37). We couple the ability of *pellicular* AEC to resolve isobaric linkage isomers, to AE purification of PDAse-II digests to identify, then purify, aberrant 2′,5′-linkage isomers, and to determine their base-composition by ESI-MS. This process will typically allow identification of the *position* of the aberrant linkage in the sequence. Also in this section we demonstrate the ability of AEC to resolve isobaric diastereoisomers of ONs harboring 1 or 2 PS linkages, and these are employed in siRNA and aptamer therapeutics.

We believe this information will provide effective instruction for developers of ON analyses for a wide variety of medical, biochemical, and other research applications.

REFERENCES

1. Rahman, K., U. Voss, P. J. Nicholls, and A. D. B. Malcolm. 1988. Nucleic acid purification by Fast Protein liquid chromatography. *Biochemical Society Transactions* 16: 368.
2. Yeung, A. T., and C. G. Miller. 1990. A general method of optimizing automated DNA synthesis to decrease chemical consumption to less than half. *Analytical Biochemistry* 187: 66–75.
3. Stribling, R. 1991. High-performance liquid chromatography of oligoguanylates at high pH. *Journal of Chromatography* 538: 474–479.
4. Bergot, B. J., and W. Egan. 1992. Separation of synthetic phosphorothioate oligodeoxynucleotides from their oxygenated (phosphodiester) defect species by strong-anion-exchange high-performance liquid chromatography. *Journal of Chromatography* 599: 35–42.
5. Bourque, A. J., and A. S. Cohen. 1994. Quantitative analysis of phosphorothioate oligonucleotides in biological fluids using direct injection fast anion-exchange chromatography and capillary gel electrophoresis. *Journal of Chromatography B* 662: 343–349.
6. Lu, T., and H. B. Gray, Jr. 1994. Combined pH gradient and anion-exchange high performance liquid chromatographic separation of oligodeoxyribonucleotides. *Journal of Chromatography A* 686: 339–343.
7. Thayer, J. R., R. M. McCormick, and N. Avdalovic. 1996. High resolution nucleic acid separations by high performance liquid chromatography. In *Methods in Enzymol,* ed. B. Karger and B. Hancock, 271, 147–174. New York: Academic Press.
8. Kanehara, H., M. Mizuguchi, and K. Makino. 1996. Isolation of oligodeoxynucleoside phosphorothioate diastereomers by the combination of DEAE ion-exchange and reversed-phase chromatography. *Nucleosides and Nucleotides* 15: 399–406.
9. Yang, X., R. P. Hodge, B. A. Luxon, R. Shope, and D. G. Gorenstein. 2002. Separation of synthetic oligonucleotide dithioates from monothiophosphate impurities by anion-exchange chromatography on a mono-q column. *Analytical Biochemistry* 306: 92–99.
10. Grant, G. P. G., A. Popova, and P. Z. Qin. 2008. Diastereomer characterizations of nitroxide-labeled nucleic acids. *Biochemical and Biophysical Research Communications* 371: 451–455.
11. GE Healthcare Application Note 18-1116-00, Strategies for large scale purification of synthetic oligonucleotides: http://www1.gelifesciences.com/aptrix/upp00919.nsf/Content/D14A5305E5FA4C74C125 7628001CC150/$file/18111600AA.pdf, and GE Healthcare Application Note 18-1111-25, Automated pilot scale purification of synthetic phosphorothioate oligonucleotides http://www1.gelifesciences.com/aptrix/upp00919.nsf/Content/7B387A892A39BACCC1257628001CCB09/$file/18111125AC.pdf.
12. Xu, Q., K. Musier-Forsyth, R. P. Hammer, and G. Barany. 1996. Use of 1,2,4-dithiazolidine-3,5-dione (DtsNH) and 3-ethoxy-1,2,4-dithiazoline-5-one (EDITH) for synthesis of phosphorothioate-containing oligodeoxyribonucleotides. *Nucleic Acids Research* 24: 1602–1607.
13. Cummins, L. L., M. Winniman, and H. J. Gaus. 1997. Phosphorothioate oligonucleotide metabolism: characterization of the "N+"-mer by CE and HPLC-ESI/MS. *Bioorganic and Medicinal Chemistry Letters* 7: 1225–1230.
14. Wang, X. D., and P. R. Gou. 2009. Polymerase-endonuclease amplification reaction for large-scale enzymatic production on antisense oligonucleotides. *Nature Proceedings* (hdl:10101/npre.2009.3711.1, posted 2 Sept. 2009).
15. Dean, N., R. McKay, L. Miraglia, et al. 1996. Inhibition of growth of human tumor cell lines in nude mice by an antisense oligonucleotide inhibitor of protein kinase C-α expression. *Cancer Research* 56: 3499–3507.
16. Fearon, K. L., J. T. Stults, B. J. Bergot, L. M. Christensen, and A. M. Raible. 1995. Investigation of the 'n–1' impurity in phosphorothioate oligodeoxynucleotides synthesized by the solid-phase β-cyanoethyl phosphoramidite method using stepwise sulfurization. *Nucleic Acids Research* 23: 2754–2761.
17. Koziolkiewicz, M., A. Owczarek, K. Domański, M. Nowak, P. Guga, and W. J. Stec. 2001. Stereochemistry of cleavage of internucleotide bonds by *Serratia marcescens* endonuclease. *Bioorganic and Medicinal Chemistry* 9: 2403–2409.

18. Chen, S.-H., M. Qian, J. M. Brennan, and J. M. Gallo. 1997. Determination of antisense phosphorothioate oligonucleotides and catabolites in biological fluids and tissue extracts using anion-exchange high-performance liquid chromatography and capillary gel electrophoresis. *Journal of Chromatography B* 692: 43–51.
19. Chen, J.-K., R. G. Schultz, D. H. Lloyd, and S. Gryaznov. 1995. Synthesis of oligodeoxyribonucleotide N3′-P5′ phosphoramidates. *Nucleic Acids Research* 23: 2661–2668.
20. Asseline, U., M. Chassignol, J. Draus, M. Durand, and J-C. Maurizot. 2003. Synthesis and properties of oligo-2′-deoxyribonucleotides containing internucleotidic phosphoramidate linkages modified with pendant groups ending with either two amino or two hydroxyl functions. *Bioorganic and Medicinal Chemistry* 11: 3499–3511.
21. Morgan, M. T., M. T. Bennet, and A. C. Drohat. 2007. Excision of 5-halogenated uracils by human thymine DNA glycosylase. *Journal of Biological Chemistry* 282: 27578–27586.
22. Jurczyk, S. C., J. Horlacher, K. G. Devined, S. A. Benner, and T. R. Battersby. 2000. Synthesis and characterization of oligonucleotides containing 2′-deoxyxanthosine using phosphoramidite chemistry. *Helvetica Chimica Acta* 83: 1517–1524.
23. Wincott, F., A. DiRenzo, C. Shaffer, S. Grimm, D. Tracz, C. Workman, D. Sweedler, C. Gonzalez, S. Scaringe, and N. Usman. 1995. Synthesis, deprotection, analysis and purification of RNA and ribozymes. *Nucleic Acids Research* 23: 2677–2684.
24. Grünewald, C., T. Kwon, N. Piton, U. Förster, J. Wachtveitl, and J. W. Engels. 2008. RNA as scaffold for pyrene excited complexes. *Bioorganics and Medicinal Chemistry* 16: 19–26.
25. Matray, T. J., and S. M. Gryaznov. 1999. Synthesis and properties of RNA analogs oligoribonucleotide N3′-P5′ phosphoramidates. *Nucleic Acids Research* 27: 3976–3985.
26. Tsou, D., A. Hampel, A. Andrus, and R. Vinayak. 1995. Large scale synthesis of oligoribonucleotides on a high-loaded polystyrene (HLP) support. *Nucleosides and Nucleotides* 14: 1481–1492.
27. Vinayak, R., A. Andrus, N. D. Sinha, and A. Hampel. 1995. Assay of ribozyme-substrate cleavage by anion-exchange high-performance, liquid chromatography. *Analytical Biochemistry* 232: 204–209.
28. Sproat, B. S., T. Rupp, N. Menhardt, D. Keane, and B. Beijer. 1999. Fast and simple purification of chemically modified hammerhead ribozymes using a lipophilic capture tag. *Nucleic Acids Research* 27: 1950–1955.
29. Murray, J. B., C. M. Dunham, and W. G. Scott. 2002. A pH-dependent conformational change, rather than the chemical step, appears to be rate-limiting in the hammerhead ribozyme cleavage reaction. *Journal of Molecular Biology* 315: 121–130.
30. Earnshaw, D. J., B. Masquida, S. Müller, S. Th. Sigurdsson, F. Eckstein, E. Westhof, and M. J. Gait. 1997. Inter-domain cross-linking and molecular modeling of the hairpin ribozyme. *Journal of Molecular Biology* 274: 197–212.
31. Micura, R., W. Pils, C. Höbartner, K. Grubmayr, M-O. Ebert, and B. Jaun. 2001. Methylation of the nucleobases in RNA oligonucleotides mediates duplex-hairpin conversion. *Nucleic Acids Research.* 29: 3997–4005.
32. Massey, A. P., and S. Th. Sigurdsson. 2004. Chemical syntheses of inhibitory substrates of the RNA-RNA ligation reaction catalyzed by the hairpin ribozyme. *Nucleic Acids Research* 32: 2017–2022.
33. Prater, C. E., A. D. Saleh, M. P. Wear, and P. S. Miller. 2007. Chimeric RNAse H-competent oligonucleotides directed to the HIV-1 Rev response element. *Bioorganic and Medicinal Chemistry* 15: 5386–5395.
34. Cohen, C., M. Forzan, B. Sproat, R. Pantophlet, I. McGowan, D. Burton, and W. James. 2008. An aptamer that neutralizes R5 strains of HIV-1 binds to core residues of gp120 in the CCR5 binding site. *Virology* 381: 46–54.
35. Mackman, R. L., L. Zhang, V. Prasad, et al. 2007. Synthesis, anti-HIV activity, and resistance profile of thymidine phosphonomethoxy nucleosides and their bis-isopropyloxymethylcarbonyl (bisPOC) prodrugs. *Bioorganic and Medicinal Chemistry* 15: 5519–5528.
36. Nair, T. M., D. G. Myszka, and D. R. Davis. 2000. Surface plasmon resonance kinetic studies of the HIV TAR RNA kissing hairpin complex and its stabilization by 2-thiouridine modification. *Nucleic Acids Research* 28: 1935–1940.
37. M. A. Morgan, S. A. Kazakov, and S. M. Hecht. 1995. Phosphoryl migration during the chemical synthesis of RNA, *Nucleic Acids Research* 23: 3949–3953.
38. Thayer, J. R., S. Rao, N. Puri, C. A. Burnett, and M. Young. 2007. Identification of aberrant 2′-5′ RNA link age isomers by pellicular anion-exchange chromatography. *Analytical Biochemistry* 361: 132–139.
39. T. P. Prakash, B. Kraynack, B. F. Baker, E. E. Swayze, and B. Bhat. 2006. RNA interference by 2′,5′-linked nucleic acid duplexes in mammalian cells. *Bioorganic and Medicinal Chemistry Letters* 16: 3238–3240.

40. Hovanessian, A. G., and J. Justesen. 2007. The human 2′-5′ oligoadenylate synthetase family: unique interferon-inducible enzymes catalyzing 2′-5′ instead of 3′-5′ phosphodiester bond formation. *Biochemie* 89: 779–788.
41. Soutschek, J., A. Akinc, B. Bramlage, et al. 2004. Therapeutic silencing of an endogenous gene by systemic administration of modified siRNAs. *Nature* 432: 173–178.
42. Pham, J. W., I. Radhakrishnan, and E. J. Sontheimer. 2004. Thermodynamic and structural characterization of 2′-nitrogen-modified RNA duplexes. *Nucleic Acids Research* 32: 3446–3455.
43. Frank-Kamentetsy, M., A. Grefhorst, N. N. Anderson, et al. 2008. Therapeutic RNAi targeting PCSK9 acutely lowers plasma cholesterol in rodents and LDL cholesterol in nonhuman primates. *Proceedings of the National Academy of Sciences United States of America* 105, 11915–11920.
44. Chiu, Y.-L., and T. M. Rana. 2003. siRNA function in RNAi: a chemical modification analysis. *RNA* 9, 1034–1048.
45. Li, F., P. S. Pallan, M. A. Maier, et al. 2007. Crystal structure, stability and *in vitro* RNAi activity of oligoribonucleotides containing the ribo-difluorotoluyl nucleotide: Insights into substrate requirements by the human RISC Ago2 enzyme. *Nucleic Acids Research.* 35: 6424–6438.
46. Mikat, V., and A. Heckel. 2007. Light-dependent RNA interference with nucleobase-caged siRNAs. *RNA* 13: 2341–2347.
47. Rhodes, A., A. Deakin, J. Spaull, B. Coomber, A. Aitken, P. Life, and S. Rees. 2000. The generation and characterization of antagonist RNA Aptamers to human oncostatin M. *Journal of Biological Chemistry* 275, 28555–28561.
48. Floege, J., T. Ostendorf, U. Janssen, M. Burg, H. H. Radeke, C. Vargeese, S. C. Gill, L. S. Green, and N. Janjic. 1999. Novel approach to specific growth factor inhibition in vivo: antagonism of platelet-derived growth factor in glomerulonephritis by aptamers. *American Journal of Pathology* 154: 169–179.
49. Gilbert, J. C., T. DeFeo-Fraulini, R. M. Hutabarat, C. J. Horvath, P. G. Merlino, H. N. Marsh, J. M. Healy, S. BouFakhreddine, T.V. Holohan, and R. G. Schaub. 2009. First-in-human evaluation of anti von Willebrand Factor therapeutic aptamer ARC1779 in healthy volunteers. *Circulation* 116: 2678–2686.
50. Fennewald, S. M., E. P. Scott, L. Zhang, et al. 2007. Thioaptamer decoy targeting of AP-1 proteins influences cytokine expression and the outcome of arenavirus infections. *Journal of General Virology* 88: 981–990.
51. Jackson-Kram, D., and T. McGovern. 2007. Toxicological overview of impurities in pharmaceutical products. *Advanced Drug Delivery Reviews*, 59: 38–42.
52. McGovern, T., and D. Jackson-Kram. 2006. Regulation of genotoxic and carcinogenic impurities in drug substances and products. *Trends in Analytical Chemistry* 25: 790–795.
53. U.S. DeparTment of Health and Human Services, FDA (CDER). 2006. Guidance for Industry, Investigators and Reviewers: Exploratory IND Studies, Pharmacology and Toxicology. http://www.fda.gov/downloads/Drugs/GuidanceComplianceRegulatoryInformation/Guidances/UCM078933.pdf.
54. Puffer, B., H. Moroder, M. Aigner, and R. Micura. 2008. 2′-Methylseleno-modified oligoribonucleotides for X-ray crystallography synthesized by the ACE RNA solid-phase approach. *Nucleic Acids Research* 36: 970–983.
55. Murray, J. B., H. Szöke, A. Szöke, and W. G. Scott. 2000. Capture and visualization of a catalytic RNA enzyme-product complex using crystal lattice trapping and X-ray holographic reconstruction. *Molecular Cell* 5: 279–287.
56. Pallan, P. S., C. H. Wilds, Z. Wawrzak, R. Krishnamurthy, A. Eschenmoser, and M. Egli. 2003. Why does TNA cross-pair more strongly with RNA than with DNA? An answer from X-ray analysis. *Angewandte Chemie (International Edition in English)* 42: 5893–5895.
57. Yang, H., and S. L. Lam. 2009. Effect of 1-methyladenine on the thermodynamic stabilities of double-helical DNA structures. *FEBS Letters* 583: 1548–1553.
58. Sambrook, J., E. F. Fritsch, and T. Maniatis. 1989. *Molecular Cloning: A Laboratory Manual*, 2nd ed., Cold Spring Harbor, New York: Cold Spring Harbor Laboratory Press.
59. Fearon, K. L., B. L. Hirschbein, J. S. Nelson, et al. 1998. An improved synthesis of oligodeoxynucleotide N3′-P5′ phosphoramidates and their chimera using hindered phosphoramidite monomers and a novel handle for reverse phase purification. *Nucleic Acids Research* 26: 3813–3824.
60. Tucker, C. E., L.-S. Chen, M. B. Judkins, J. A. Farmer, S. C. Gill, and D. W. Drolet. 1999. Detection and plasma pharmacokinetics of an anti-vascular endothelial growth factor oligonucleotide-aptamer (NX1838) in rhesus monkeys. *Journal of Chromatography B* 732: 203–212. Also Bouchard, P. R., R. M. Hutabarat, and K. M. Thompson. 2010. Discovery and development of therapeutic aptamers. *Annual Review of Pharmacology and Toxicology* 50: 237–257.

61. Van Aerschot, A., and J. Rozenski. 2006. Characterization and sequence verification of thiolated deoxyoligonucleotides used for microarray construction. *Journal of the American Society for Mass Spectrometry* 17: 1396–1400.
62. Jiang, Y., and S. A. Hofstadler. 2003. A highly efficient and automated method of purifying and desalting PCR products for analysis by electrospray ionization mass spectrometry. *Analytical Biochemistry* 316: 50–57.
63. Urlaub, H., K. HarTmuth, S. Kostka, G. Grelle, and R. Lührmann. 2000. A general approach for identification of RNA-protein cross-linking sites within native human spliceosomal small nuclear ribonucleoproteins (snRNPs). *Journal of Biological Chemistry* 52: 41458–41468.
64. Lee, B. M., J. Xu, B. K. Clarkson, M. A. Martinez-Yamout, H. J. Dyson, D. A. Case, J. M. Gottesfeld, and P. E. Wright. 2006. Induced fit and "lock and key" recognition of 5S RNA by zinc fingers of transcription factor IIIA. *Journal of Molecular Biology* 357: 275–291.
65. Eon-Duval, A., and G. Burke. 2004. Purification of pharmaceutical-grade plasmid DNA by anion-exchange chromatography in and RNAse-free process. *Journal of Chromatography B* 804: 327–335.
66. Crary, S. M., J. C. Kurz, and C. A. Fierke. 2002. Specific phosphorothioate substitutions probe the active site of *Bacillus subtilis* Ribonuclease P. *RNA* 8: 933–947.
67. Nauwelaerts, K., K.VasTmans, M. Froeyen, et al. 2003. Cleavage of DNA without loss of genetic information by incorporation on a disaccharide nucleoside. *Nucleic Acids Research* 31: 6758–6769.
68. Yamaguchi, T., K. Suyama, K. Narita, S. Kohgo, A. Tomikawa, and M. Saneyoshi. 1997. Synthesis and evaluation of oligodeoxyribonucleotides containing an aryl(trifluoromethyl) diazirine moiety as the cross-linking probe: Photoaffinity labeling of mammalian DNA polymerase β. *Nucleic Acids Research* 25: 2352–2358.
69. Harsch, A., L. A. Marzill, R. C. Bunt, J. Stubbe, and P. Vouros. 2000. Accurate and rapid modeling of iron-bleomycin-induced DNA damage using tethered duplex oligonucleotides and electrospray ionization ion trap mass spectrometric analysis. *Nucleic Acids Research* 28: 1978–1995.
70. Biczo, R., and D. J. Hirsh. 2009. Structure and dynamics of a DNA-based model system for the study of electron spin-spin interactions. *Journal of Inorganic Biochemistry* 103: 362–372.
71. Johansson, M. K., R. M. Cook, J. Xu, and K. N. Raymond. 2004. Time gating improves sensitivity in energy transfer assays with terbium chelate/dark quencher oligonucleotide probes. *Journal of the American Chemical Society* 126: 16451–16455.
72. Ouyang, X., I. A. Shestopalov, S. Sinha, G. Zheng, C. L. W. Pitt, W-H. Li, A. J. Olson, and J. K. Chen. 2009. Versatile synthesis and rational design of caged morpholinos. *Journal of the American Chemical Society* 131: 13255–13269.
73. Semenyuk, A., M. Ahnfelt, C. EsTmer, X. Yong-Hao, A. Földesi, Y.-S. Kao, H.-H. Chen, W.-C. Kao, K. Peck, and M. Kwiatkowski. 2006. Cartridge-based high-throughput purification of oligonucleotides for reliable oligonucleotide arrays. *Analytical Biochemistry* 356: 132–141.
74. Junker, H-D., S. T. Hoehn, R. C. Bunt, V. Marathius, J. Chen, C. J. Turner, and J. Stubbe. 2002. Synthesis, characterization and solution structure of tethered oligonucleotides containing an internal 3′-phosphoglycolate, 5′-phosphate gapped lesion. *Nucleic Acids Research* 30: 5497–5508.
75. Gill, S., R. O'Neill, R. J. Lewis, and B. A. Connolly. 2007. Interaction of the Family-B DNA polymerase from the archaeon *Pyrococcus furiosus* with deaminated bases. *Journal of Molecular Biology* 372: 855–863.
76. Ye, Y., B. H. Munk, J. G. Muller, A. Cogbill, C. J. Burrows, and B. Schlegel. 2009. Mechanistic aspects of the formation of guanidinohydantoin from spiroiminodihydantoin under acidic conditions. *Chemical Research Toxicology* 22: 526–535.
77. Misiaszek, R., Y. Uvaydov, C. Crean, N. E. Geacintov, and V. Shafirovich. 2005. Combination reactions of superoxide with 8-oxo-7,8-dihydroguanine radicals in DNA. *Journal of Biological Chemistry* 280: 6293–6300.
78. Boojamra, C. G., J. P. Parrish, D. Sperandio, et al. 2009. Design synthesis and anti-HIC activity of 4′-modified carbocyclic nucleoside phosphonate reverse transcriptase inhibitors. *Bioorganic and Medicinal Chemistry* 17: 1739–1746.
79. Conn, G. L., T. Brown, and G. A. Leonard. 1999. The crystal structure of the RNA/DNA hybrid r(GAAGAGAAGC·d(GCTTCTCTTC) shows significant differences to that found in solution. *Nucleic Acids Research* 27: 555–561.
80. Nordin, B. E., and P. Schimmel. 2002. Plasticity of recognition of the 3′-end of mischarged tRNA by class-I aminoacyl-tRNA synthetases. *Journal of Biological Chemistry* 277: 20510–20517.
81. Fonvielle, M., M. Chemama, R. Villet, M. Lecerf, A. Bouhss, J-M. Valéry, M. Ethève-Quelquejeu, and M. Arthur. 2009. Aminoacyl-tRNA recognition by the FemX$_{Wv}$ transferase for bacterial cell wall synthesis. *Nucleic Acids Research* 37: 1589–1601.

82. Garcia-Garcia, C., and D. E. Draper. 2003. Electrostatic interactions in a peptide-RNA complex. *Journal of Molecular Biology* 331: 75–88.
83. Perry, K., Y. Hwang, F. D. Bushman, and G. D. Van Duyne. 2006. Structural basis for specificity in the Poxvirus topoisomerase. *Molecular Cell* 213: 343–354.
84. Lochmann, D., J. Weyermann, C. Georgens, R. Prassl, and A. Zimmer. 2005. Albumin-protamine-oligonucleotide nanoparticles as a new antisense delivery system. Part 1: Physicochemical characterization. *European Journal of Pharmaceuticals and Biopharmaceuticals* 59: 419–429.
85. Allen T. D., K. L. Wick, and K. S. Matthews. 1991. Identification of amino acids in *lac* repressor protein cross-linked to operator DNA specifically substituted with bromodeoxyuridine. *Journal of Biological Chemistry* 266: 6113–6119.
86. Matthew-Fenn, R. S., R. Das, J. A. Silverman, P. A. Walker, and P. A. B. Harbury. 2008. A molecular ruler for measuring quantitative distance distributions. *PLoS ONE* 3(10): e3229. doi:10.1371/journal.pone.0003229.
87. Svec, F., and J. M. J. Frechet. 1992. Continuous rods of macroporous polymer as high-performance liquid chromatography separation media. *Analytical Chemistry* 64: 820–822.
88. Peters, E. C., M. Petro, F. Svec, and J. M. J. Frechet. 1998. Molded rigid polymer monoliths as separation media for capillary electrochromatography. 2. Effect of chromatographic conditions on the separation. *Analytical Chemistry* 70: 2296–2302.
89. Svec, F., and C. G. Huber. 2006. Monolithic materials: promises, challenges, achievements. *Analytical Chemistry* 78: 2100–2107.
90. Perica, M. C., I. Sola, L. Urbas, F. Smrekar, and M. Krajacic. 2009. Separation of hypoviral double-stranded RNA on monolithic chromatographic supports. *Journal of Chromatography A* 1216: 2712–2716.
91. Thayer, J. R., K. J. Flook, A. Woodruff, S. Rao, and C. A. Pohl. 2010. New monolith technology for automated anion-exchange purification of nucleic acids. *Journal of Chromatography B*, 878: 933–941, doi:10.1016/j.jchromb.2010.01.030.
92. Hatefi, Y., and W. G. Hanstein. 1969. Solubilization of particulate proteins and nonelectrolytes by chaotropic agents. *Proceedings of the National Academy of Sciences of the United States of America* 62:1129–1136.
93. Thayer, J. R., V. Barreto, S. Rao, and C. Pohl. 2005. Control of oligonucleotide retention on a pH-stabilized strong anion exchange column. *Analytical Biochemistry* 338: 39–47.
94. Murakami, A., Y. Tamura, H. Wada, and K. Makino. 1994. Separation and characterization of diastereoisomers of antisense oligodeoxyribonucleoside phosphorothioates. *Analytical Biochemistry* 223: 285–290.
95. Patil, S. V., B. Mane, and M. M. Salunkhe. 1994. Synthesis and properties of oligothymidylate analogs containing stereoregulated phosphorothioate and phosphodiester linkages in an alternating manner. *Bioorganic and Medicinal Chemistry Letters* 4: 2663–2666.
96. Iyer, R. P., M.-J. Guo, D. Yu, and S. Agrawal. 1998. Solid phase stereoselective synthesis of oligonucleoside phosphorothioates: The nucleoside bicyclic oxazaphospholidines as novel synthons. *Tetrahedron Letters* 39: 2491–2494.
97. J. R. Thayer, Y. Wu, E. Hansen, M. D. Angelino, and S. Rao. 2010. Separation of oligonucleotide phosphorothioate diastereoisomers by anion-exchange chromatography. *Analytical Chemistry*. In preparation.
98. Li, H., P. S. Miller, and M. M. Seidman. 2008. Selectivity and affinity of DNA triplex forming oligonucleotides containing the nucleoside analogues 2′-*O*-methyl-5-(3-amino-1-propynyl)uridine and 2′-*O*-methyl-5-propynyluridine. *Organic and Biomolecular Chemistry* 6: 4212–4217.
99. Yang, C. J., L. Wang, Y. Wu, Y. Kim, C. D. Medley, H. Lin, and W. Tan. 2007. Synthesis and investigation of deoxyribonucleic acid/locked nucleic acid chimeric molecular beacons. *Nucleic Acids Research* 35: 4030–4041.
100. Kim, H.-J., N. A. Leal, and S. A. Benner. 2009. 2′-Deoxy-1-methylpseudocytidine, a stable analog of 2′-deoxy-5-methyl*iso*cytidine. *Bioorganic and Medicinal Chemistry* 17: 3728–3732.
101. Baba, Y., M. Fukuda, and N. Yoza. 1988. Computer-assisted retention prediction system for oligonucleotides in gradient anion-exchange chromatography. *Journal of Chromatography* 458: 385–394.
102. Alvarez, R., S. Elbashir, T. Borland, et al. 2009. RNA interference-mediated silencing of the respiratory syncytial virus nucleocapsid defines a potent antiviral strategy. *Antimicrobial Agents and Chemotherapy* 53: 3952–3962.
103. De Fougerolles, A., and T. Novobrantseva. 2008. siRNA and the lung: research tool or therapeutic drug. *Current Opinion in Pharmacology* 8: 280–285.
104. Akinc, A., M. Goldberg, J. Qin, et al. 2009. Development of lipidoid–siRNA formulations for systemic delivery to the liver. *Molecular Therapy* 17: 872–879.

105. Nguyen, T., E. M. Menocal, J. Harborth, and J. H. Fruehauf. 2008. RNAi therapeutics: an update on delivery. *Current Opinions in Molecular Therapeutics* 10: 158–167.
106. Siegel, R. W., L. Bellon, L. Beigelman, and C. C. Kao. 1998. Moieties in an RNA promoter specifically recognized by a viral RNA-dependent RNA polymerase. *Proceedings of the National Academy of Sciences of the United States of America* 95: 11613–11618.
107. Hail, M., B. Elliot, and K. Anderson. 2004. High-throughput analysis of oligonucleotides using automated electrospray ionization mass spectrometry. *American Biotechnology Laboratory* 22:12–14.
108. Thayer, J. R., N. Puri, C. Burnett, M. E. Hail, and S. Rao. 2010. Identification of RNA linkage isomers by anion-exchange purification with ESI-MS of automatically desalted phosphodiesterase-II digests. *Analytical Biochemistry* 399: 110–117.
109. Schleich, T., B. P. Cross, and I. C. P. Smith. 1976. A conformational study of adenylyl-(3′,5′)-adenosine and adenylyl-(2′,5′)-adenosine in aqueous solution by carbon-13 magnetic resonance spectroscopy. *Nucleic Acids Research* 3: 355–370.
110. Thayer, J. R., S. Rao, and N. Puri. 2008. Detection of aberrant 2′-5′ linkages in RNA by anion exchange. In *Current Protocols in Nucleic Acid Chemistry*, ed. S. Beaucage, 10.13.1–11. New York: Wiley Interscience.

3 Purity Analysis and Molecular Weight Determination by Size Exclusion HPLC

Ming Fai Chan
Accugent Laboratories, Inc.
Ipsita Roymoulik
Avecia Biotechnology, Inc.

CONTENTS

3.1 INTRODUCTION

Size exclusion chromatography (SEC), also known as gel permeation chromatography (GPC) or gel filtration chromatography (GFC), is a powerful technique for the purification, separation, and characterization of natural and synthetic polymers. Over the years, the method has gained wide acceptance as one of the major purification methods for proteins, polysaccharides, and nucleic acids. Although the technique was first developed more than 50 years ago,[1] the original gel matrix, Sephadex (a cross-linked dextran), is still widely used for protein purification today. Given the popularity of SEC in protein purification, it is not surprising that the technique has also been widely used in the purification of DNA,[2–8] RNA,[9–14] oligonucleotides,[15,16] oligonucleotide complexes,[17] and oligonucleotide conjugates.[18,19]

Another area in which SEC finds application is the determination of molecular weight of polymers.[20] This is usually accomplished by comparing the unknown sample against a calibration curve (log M_p versus retention time/volume) constructed with reference standards of known molecular weight. The approach is sometimes referred to as conventional SEC and is usually carried out with a concentration detector such as refractive index (RI) or ultraviolet (UV). With the introduction of light scattering (LS) and viscosity detectors, the molecular weight of a polymer can now be determined without a calibration curve. This mode of SEC is sometimes referred to as multidetector SEC because a mass detector is used in conjunction with a concentration detector for molecular weight determination.

With recent interest in the development of complex oligonucleotide therapeutics such as siRNA duplexes[21] and oligonucleotide conjugates such as PEGylated aptamers,[22] SEC has proved to be an invaluable analytical tool. The analysis of intact oligonucleotide duplexes (and other multiplexes) has been a challenge owing to their susceptibility to dissociation in conventional separation systems that employ organic modifiers such as ion-pairing agents and denaturing conditions such as high pH or high temperatures. SEC is particularly suited for the analysis of duplexes because it can be run in most aqueous salt solutions as mobile phase under very mild conditions. Furthermore, good separation is usually achieved owing to the large molecular weight differences between the single strand, duplex, and multiplexes. For these reasons, SEC is often the method of choice for simultaneously providing identity, quality, and quantitative information of siRNA, and other oligonucleotide complexes under simulated physiological conditions. In the case of PEGylated aptamers, the presence of a polydispersed synthetic polymer, polyethyleneglycol (PEG), greatly complicates the analysis by mass spectrometry. SEC is one of only a handful of techniques that can readily characterize these compounds.

This chapter is focused on the practical aspects of contemporary oligonucleotide analysis by SEC. Because of space limitations, detailed theoretical discussions, particularly those concerning molar mass determination using light scattering and viscometry detection, will not be presented. Instead, a qualitative description will be used wherever appropriate. Readers who are interested in the theoretical background of SEC are encouraged to consult recent review articles[23–25] and excellent treatises[26–30] in this area.

3.2 SEPARATION MECHANISM

Traditional chromatographic separation is usually understood as a phenomenon arising out of the preferential retention of molecules on the column stationary phase. As the molecules move along with the mobile phase, they interact with the surface or molecules present on the surface of the stationary phase. The molecules are subsequently released from the column as the interaction between the mobile phase and the molecules overcomes the interaction between the stationary phase and the molecules, resulting in separation.[31]

In SEC, separation is thought to arise from a geometry-dependent differential diffusion of molecules of different sizes (or more precisely, *hydrodynamic volume*) across the pores present in the stationary phase particles. Molecules too large to fit in the pores pass straight through the interstitial space defined by the *total exclusion or void* volume (V_0) of the SEC column. On the other hand, small molecules with complete access to the pores spend more time traversing both the pores of the stationary phase (with a *pore* volume of V_i) and the interstitial space and elute at the *total permeation* volume V_t. Molecules of intermediate sizes spend varying amounts of time proportional to their size traversing the pores. As a result, separation is achieved in which large molecules are eluted before smaller ones. This mode of separation is opposite to most other chromatographic techniques and is the hallmark of SEC.

Chromatographically, V_0 can be defined as the retention volume at which all molecules above a certain size elute as a single peak owing to complete inaccessibility to the pores of the stationary phase particles. It represents the shortest residence time on the column. Likewise, V_t can also be viewed as the retention volume at which all molecules smaller than a particular size elute as a single peak due to complete access to the pores of the stationary phase particles and hence have the most residence time on the column. Thus the permeation volume can be expressed as

$$V_t = V_0 + V_i \tag{3.1}$$

For an analyte separable by SEC, the retention volume can be described by the general equation

$$V_R = V_0 + K_{SEC}V_i \tag{3.2}$$

where K_{SEC} is the SEC distribution coefficient and can be defined as

$$K_{SEC} = \frac{[S]_i}{[S]_0} \tag{3.3}$$

where $[S]_i$ is the average concentration of solute in the pore volume V_i and $[S]_0$ is the average concentration of solute in the exclusion volume V_0. K_{SEC} values can range from 0 to 1. When $K_{SEC} = 0$, the sample elutes in the exclusion volume, V_0 and when $K_{SEC} = 1$, the sample elutes in the permeation volume V_t. No chromatographic separation will be observed in either case. When $K_{SEC} > 1$, i.e., when the sample elutes later than the *total permeation* volume V_t, other mechanisms (reversed phase, ion exchange, etc.) are in play and the separation is not necessarily determined by size exclusion.

The *hydrodynamic volume* is defined as the volume of a molecule in solution, with its associated molecules of solvation. It is a measure of the size of a molecule in terms of its shape or volume rather than just its mass. Water soluble polymers such as proteins, oligonucleotides, and polyethylene glycol (PEG) are usually hydrated in solution and contain a large number of solvating water molecules. *The hydrodynamic volume* can be expressed in terms of the Stokes radius (R_H), i.e., radius of a hypothetical sphere that diffuses at the same rate as the analyte of interest,

$$R_H = \frac{\kappa_B T}{6\pi\eta D} \tag{3.4}$$

where κ_B is the Boltzmann constant, T is temperature in Kelvin, η is the viscosity, and D is the diffusion coefficient.

The *fractionation range* for a particular column is defined as the molar mass or size range of molecules that can be separated within the exclusion (V_0) and permeation volume (V_t). The pore size, and to a lesser extent, the size of the stationary phase particles, can be varied to achieve separation of molecules within the desired fractionation range. To further illustrate the physicochemical events in SEC, a diagrammatic representation of the phenomenon is provided in Figure 3.1.

3.3 DEFINITIONS

Because SEC separates molecules according to their size, and by extrapolation, their molar mass, it is a useful technique for characterization of polymers and generating information such as *molar*

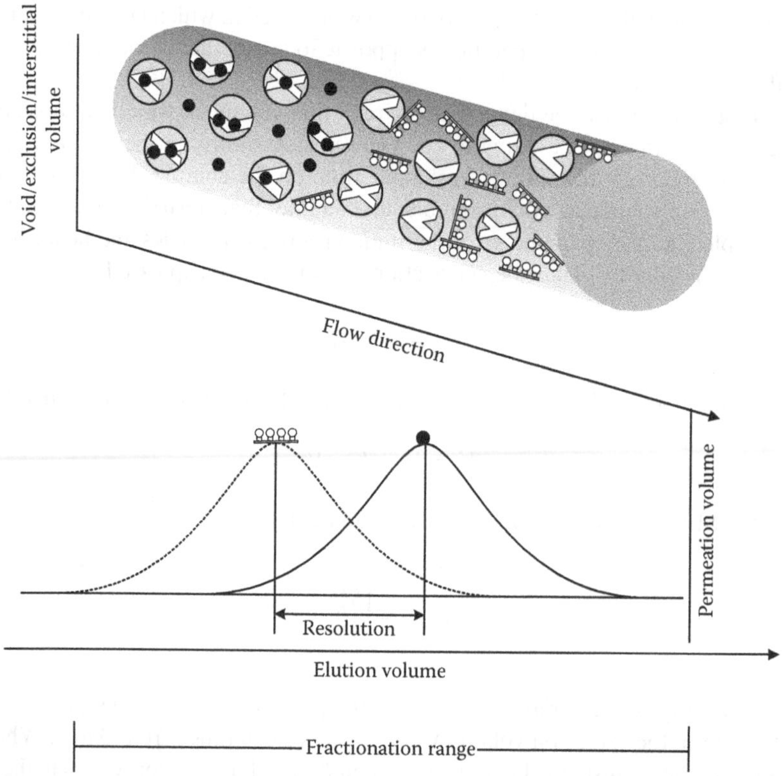

FIGURE 3.1 Schematic representation of SEC separation. Larger molecules navigate around the stationary phase particles and elute earlier, whereas the smaller molecules diffuse through the pores in the stationary phase particles in addition to flowing through the interstitial space and thereby elute later.

mass averages (M_n, M_w, M_z, M_v), *molar mass distribution* and *polydispersity*.[26,27] Theoretically, these values are obtained by analyzing the distribution of molecules with a particular weight within a large collection of molecules of different sizes. In practice, it is more convenient to obtain these values by analyzing the SEC chromatogram. Conceptually, the chromatogram can be divided into thin vertical "slices" (i.e., small increment in the retention time/volume), each of which can be regarded as a collection of molecules with a very narrow molar mass distribution. If the chromatogram is recorded with a concentration detector (such as RI or UV), the height of each of these slices will correspond to the number of molecules in each slice. By taking advantage of this approximation, the properties of a polymer can be readily determined. Some of the values useful in polymer characterization are summarized below:

Number-average molar mass (M_n) is defined as the sum of the mass of samples (N_iM_i) in grams divided by the total number (N_i) of molecules present in the sample:

$$M_n = \frac{\sum N_i M_i}{\sum N_i} = \frac{\sum W_i}{\sum \frac{W_i}{M_i}} \tag{3.5}$$

and from SEC

$$M_n = \frac{\sum_{i=1}^{N} h_i}{\sum_{i=1}^{N} \frac{h_i}{M_i}} \tag{3.6}$$

where h_i is the height of the ith "slice" (retention volume increment) and M_i is the molar mass of the species eluted at the ith "slice." Thus species at the lower end of the molar mass distribution exert a bigger influence on M_n than those at the higher end. Experimentally, M_n values can be conveniently measured from the colligative properties of the polymer such as freezing point depression or osmometry.

Weight-average molar mass (M_w) is defined as

$$M_w = \frac{\sum N_i M_i^2}{\sum N_i M_i} = \frac{\sum W_i M_i}{\sum W_i} \tag{3.7}$$

and from SEC

$$M_w = \frac{\sum_{i=1}^{N} h_i M_i}{\sum_{i=1}^{N} h_i} \tag{3.8}$$

Species at the higher end of the molar mass distribution have a higher impact on the value of M_w. M_w is determined in the laboratory from static light scattering and ultracentrifugation measurements.

Peak molar mass (M_p) is simply the molar mass at the peak apex of an SEC chromatogram. It is used in the construction of calibration curves using polymer standards with narrow molar mass distribution.

Viscosity-average molar mass (M_ν also known as M_η) is defined as

$$M_v = M_\eta = \left[\frac{\sum_{i=1}^{N} h_i (M_i)^a}{\sum h_i} \right]^{1/a} \tag{3.9}$$

where the term a corresponds to the exponent in the Mark–Houwink equation. As the name implies, M_ν can be determined by viscosity measurements.

The *polydispersity* of a polymer is a measure of the breadth of molar mass distribution of a polymer and is expressed as

$$\text{Polydispersity} = \frac{M_w}{M_n} \tag{3.10}$$

The polydispersity of a polymer is usually greater than unity because M_w has always been experimentally observed to be greater than M_n in polydispersed systems. *Polydispersity* is unity for monodispersed systems and close to unity for polymers with narrow molar mass distribution. Cross-linked polymers have unusually large polydispersity.

The *molar mass distribution* (or molecular weight distribution) of a polymer describes the relationship between the number of each polymer species (N_i) and the molar mass (M_i) of that species. A representative molar mass distribution plot of a PEGylated oligonucleotide can be found in Figure 3.20. Together, the chromatographic profile, molar mass averages, and polydispersity provide comprehensive information on the identity and quality of a polymer.

3.4 INSTRUMENTATION AND OPERATION

Because of their sensitive nature, SEC of oligonucleotides is usually performed at or near neutral pH. Isocratic elution is employed at a temperature close to ambient. The basic SEC setup is relatively simple, consisting of a solvent delivery system, a suitable detector, and a recorder or a data system. For improved performance and automation, a degasser, an autosampler, and a column thermostat are usually added to the system. A typical instrument setup is shown in Figure 3.2.

3.4.1 Degasser

Removal of dissolved gases in the mobile phase improves the performance of a SEC system by reducing pulsation and preventing bubble formation in the detector. Because solvated oxygen complexes absorb significantly in the UV range, removal of oxygen reduces baseline drift and improves the sensitivity of detectors that operate in this wavelength range. Installation of a vacuum membrane degasser between the mobile phase reservoir and the solvent pump is a very convenient and effective way to provide a constant supply of degassed solvent.

3.4.2 Solvent Delivery System

The performance of the pump is critical to the success of SEC. The ability to deliver accurate, reproducible, and pulse-free flow of mobile phase is perhaps the single most important requirement for molecular weight determination. Because the calibration curve is a semi-log plot, slight variations

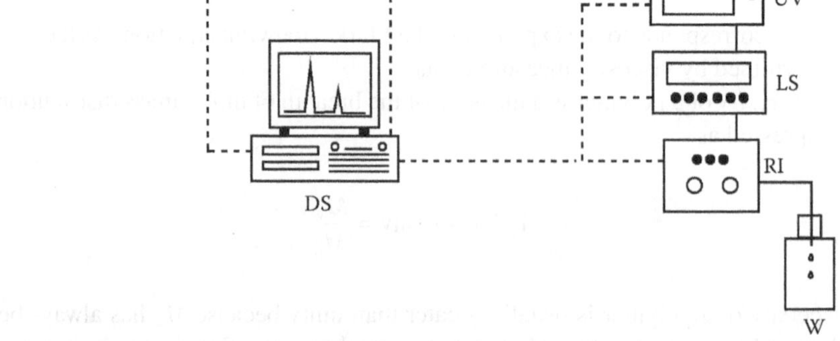

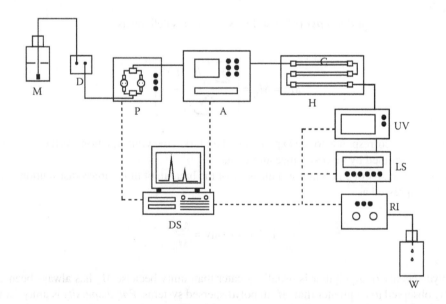

FIGURE 3.2 Typical instrument setup for SEC: A, autosampler; C, SEC Column(s); D, degasser; H, column heater; M, mobile phase reservoir; P, HPLC pump; W, waste reservoir; DS, chromatographic data system; LS, light scattering detector; RI, refractive index detector; UV, UV detector.

in the retention time, sometimes by as little as a few seconds, can translate into significant error in the estimated MW. Most modern HPLC pumps with reciprocating piston design would be suitable for SEC. Single piston pumps could also be employed, provided that a pulse damper is installed to mitigate the pulsation, which may otherwise cause excessive baseline noise in some detectors, particularly the RI detector.

3.4.3 Detectors

Depending on the application, either a concentration detector or a combination of a concentration detector and a mass detector can be used in SEC. For most applications, a concentration detector (UV, RI, ELS) provides adequate qualitative and quantitative information that is useful in the identification and characterization of oligonucleotides. A mass detector (LS, viscosity), when used with a concentration detector, provides additional information on the size (MW and root mean square radius), thus allowing the "absolute" molecular weight of an oligonucleotide to be determined. This is particularly useful as an independent confirmation of the calculated MW.

3.4.3.1 UV Detector

One of the most widely used modes of chromatographic detection is by measuring the UV absorption of the analyte. The relationship between the UV absorbance and concentration is governed by Beer's law, which can be rearranged to give the following equation:

$$c = I_{\mathrm{UV}} \cdot \frac{\mathrm{UV_{RF}}}{\varepsilon \cdot l} \tag{3.11}$$

where c is the concentration of the analyte, I_{UV} is the intensity of the UV signal, $\mathrm{UV_{RF}}$ is the response factor of the UV detector, ε is the molar extinct coefficient, and l is the path length of the detector cell.

Because of the presence of purine and pyrimidine bases, oligonucleotides have strong UV absorption ($\varepsilon \approx 10^5\ \mathrm{M^{-1}cm^{-1}}$) between 255 and 265 nm. It is therefore not surprising that UV is one of the most popular detection methods for this class of compounds. Ideally, the detector is set at the absorption maximum to achieve greatest sensitivity. In practice, however, a default wavelength of 260 nm is usually adopted because the absorption maximum is not always measured beforehand.

3.4.3.2 Refractive Index Detector

A refractive index (RI) detector, or more precisely differential refractometer, measures the amount of solute coming out of the column by comparing the refractive index of the mobile phase with that of the column effluent. This difference in RI detector response is proportional to the amount of solute present. For most dilute solutions encountered in SEC, the amount of solute is very small (≤1% by weight), and hence the RI has to be very sensitive. Most modern RI detectors are capable of measuring RI differences between 5×10^{-3} and 5×10^{-8} RIU. Because the refractive index of a liquid is extremely sensitive to temperature changes, the RI detector must be maintained at a very steady temperature.

The relationship between the concentration of the analyte (c) and the intensity of the RI signal (I_{RI}) can be expressed as

$$c = I_{\mathrm{RI}} \cdot \frac{\mathrm{RI_{CC}}}{dn/dc} \tag{3.12}$$

where c is the concentration of the analyte, $\mathrm{RI_{CC}}$ is the calibration constant for the RI detector, and dn/dc is the specific refractive index increment of the analyte in the mobile phase.

In spite of the strong UV absorbance of oligonucleotides, RI detectors are often used in their analysis by SEC. This is primarily due to the lack of UV absorbance in the most commonly used calibration standards, polyethylene glycol (PEG), and polyethylene oxide (PEO). As a result, RI detectors are used whenever the construction of calibration curves is required. RI detectors are also used for determining the specific refractive index increment (*dn/dc*) of oligonucleotides used in the determination of MW by light scattering detectors.

A baseline drift in RI detectors is usually caused by a fluctuation in temperature. Even though most modern RI detectors are equipped with thermostated flow cells, precautions must still be exercised to avoid exposing the detector to sudden temperature changes such as placement under direct sunlight or directly under a vent. Another problem commonly encountered in RI detectors is wavy baseline. There are several possible causes, but the most common are a trapped bubble or pump pulsation. Pump pulsation can be minimized by using a pulse damper (see Section 3.4.2) while the use of a degasser can minimize trapped bubbles.

3.4.3.3 Evaporative Light Scattering Detector

Evaporative light scattering (ELS) detectors measure the amount of nonvolatile analyte by measuring the intensity of scattered light after removal of volatile components from the column effluent by atomization (nebulization) and evaporation under a stream of nitrogen at elevated temperatures. The ELS detector is less useful when nonvolatile components such as inorganic buffer salts are present in the mobile phase because they tend to leave a residue, leading to high background signal. The advantages of the ELS detector include the ability to accommodate gradient elution, the ability to handle highly UV absorbing mobile phases, and the ability to detect analytes with similar or identical refractive indices to that of the mobile phase. The major drawback of the ELS detector is its nonlinear response to analyte concentration, which necessitates a calibration curve for quantitative analysis.

3.4.3.4 Light Scattering Detector

When light passes through a transparent medium, a portion of it is scattered by particles present in that medium. The intensity of the scattered light is dependent on the scattering angle (θ), and varies with the molecular size, shape, and the concentration of the particles present. For *dilute* polymer solutions such as those typically encountered in SEC, the relationship between the weight-average molar mass, M_w, and excess scattered light intensity $R(\theta)$ at a scattering angle θ, is given by the following equation derived from the *Rayleigh–Gans–Debye approximation*:

$$\frac{K^* c}{R(\theta)} \approx \frac{1}{P(\theta)} \frac{1}{M_w} \tag{3.13}$$

where c is the concentration of polymer solution and

$$\frac{1}{P(\theta)} = 1 + \frac{16\pi^2}{3\lambda^2} \left\langle r^2 \right\rangle_z \sin^2 \frac{\theta}{2} + \cdots \tag{3.14}$$

$$K^* = \frac{4\pi^2 n_0^2 \left(dn/dc \right)^2}{\lambda_0^4 N_A} \tag{3.15}$$

in which N_A is Avogadro's number, n_0 the refractive index of the solvent, λ_0 the wavelength of the incident wavelength in vacuum, *dn/dc* the specific refractive index increment, and $\left\langle r^2 \right\rangle_z^{1/2}$ is the root-mean-square radius of the biopolymer.

Light scattering detectors deduce molecular weight information of an analyte in SEC by measuring *excess* light scattering from the column effluent. In early designs, light scattering is measured at a single, specially chosen angle to facilitate the solution of M_w.[32,34] For example, in low angle LS detectors (LALLS),[32,33] the scattered light is measured at a very small angle ($\theta \leq 7°$), such that $\sin^2\frac{\theta}{2} \approx 0$ and the $\frac{1}{P(\theta)}$ term is reduced to unity, and Equation (3.13) is simplified for dilute solutions to

$$\frac{K^*c}{R(\theta)} = \frac{1}{M_w} \tag{3.16}$$

In another design, light scattering is measured at right angles to the incident radiation (and hence the name RALLS).[34] The single angle approach, while useful, does not usually give information on the root-mean-square radius $\langle r^2 \rangle_z^{1/2}$.

In multiangle light laser scattering (MALLS) detectors, light scattering is measured at several different angles. The light scattering data are then extrapolated to zero angle ($\theta = 0$) and zero concentration (Zimm plot) to determine M_w as well as the root mean square radius $\langle r^2 \rangle_z^{1/2}$. A detailed discussion on LS detectors is beyond the scope of this chapter. Readers who are interested in these subjects are encouraged to consult many excellent review articles[35] and specialty monographs, some of which are listed in the reference section.

When coupled with a RI detector (for *dn/dc* determination), a LS detector can give information on the molecular weight (M_w) of the analyte without the need to construct a calibration curve. This technique, sometimes referred to as "absolute" molecular weight determination, has found use in the analysis of proteins and other biopolymers. When a UV detector is used in conjunction with the LS and RI detector in the three-detector approach,[36–38] the molecular weight can be determined according to the following formula:

$$M_W = \frac{k(\text{Output})_{LS}(\text{Output})_{UV}}{A(\text{Output})_{RI}^2} \tag{3.17}$$

where k is a constant specific for the instrument setup, A is the extinction coefficient, and $(\text{Output})_x$ is the instrument output of the respective detector x. The three-detector approach enables the molecular weight determination of a biopolymer without prior knowledge of the specific refractive index (*dn/dc*). This approach is particularly valuable in cases where *dn/dc* is not known or not easily determined, e.g., oligonucleotides conjugated to PEG or proteins. The molecular weight determination of a PEGylated oligonucleotide by the three-detector approach is shown in Figure 3.3. In this case, the PEGylated oligonucleotide and the PEGylating agent were not readily separated by SEC owing to the small difference in their molecular weight. However, they were readily distinguished from each other by the molecular weights determined.

Because the intensity of light scattering signal is proportional to M_W of the analyte (Equation 3.16), LS detectors are less useful for most single-strand therapeutic oligonucleotides currently under development owing to their relatively small sizes (6–10 KDa). However, LS detectors have unusually strong signals for large molecules and are thus particularly well suited for detecting oligonucleotide multiplexes and aggregates that are easily overlooked because of their relatively weak signals from concentration detectors. The dependence of LS signal on molecular weight is shown in Figure 3.4. Whereas the RI response from each of the three PEG/PEO reference standards (approximately equal weight injected) is proportional to the sample weight, the LS signal decreases significantly with decreasing molecular weight.

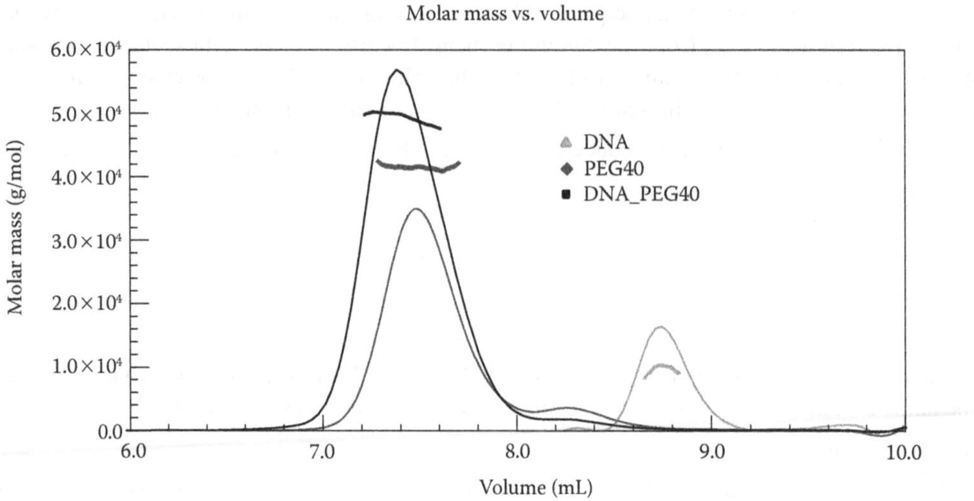

FIGURE 3.3 SEC-MALLS of an oligonucleotide before (8.75 mL, MW 9.7 kDa) and after PEGylation (7.35 mL, MW 49.3 kDa). The PEGylating agent (7.45 mL, MW 41.5 kDa), which is not easily distinguished from the PEGylated oligonucleotides by retention volume alone, is readily identified by its MW (41.5 kDa versus 49.3 kDa for the conjugate). SEC conditions: Tosoh TSKgel 4000 (7.8 mm × 300 mm); elunet: 20 mM HEPES, pH 7.5, 150 mM NaCl, 1 mM EDTA; flow rate 0.5 mL/min; UV, RI, and MALLS detection. (Courtesy of Dr. Ewa Folta-Stogniew, Yale University.)

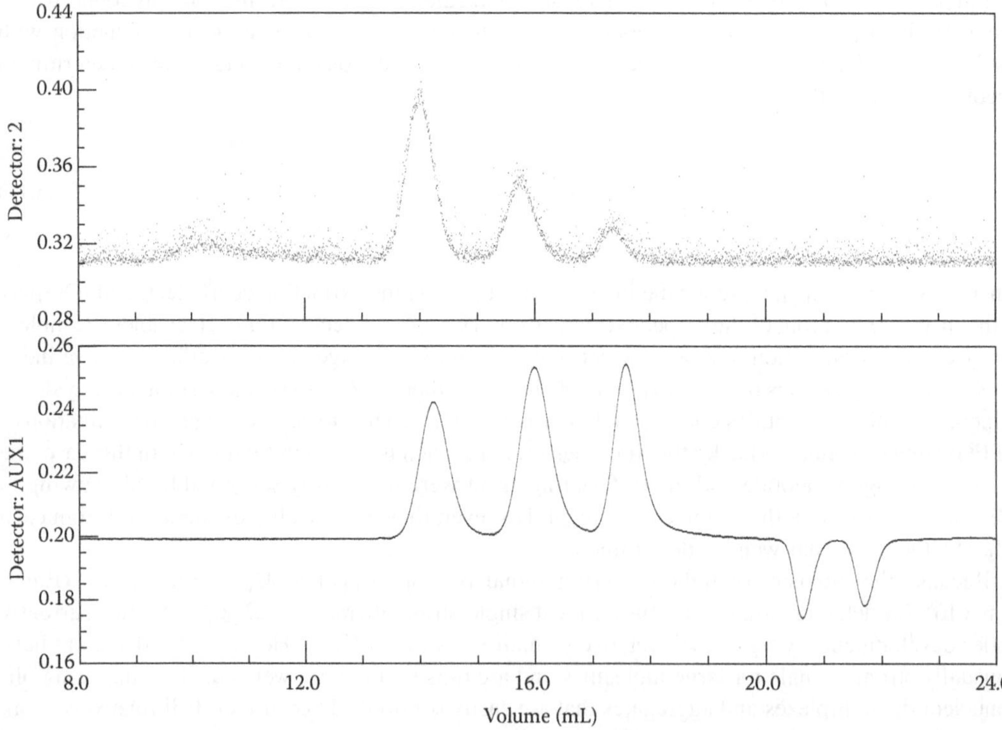

FIGURE 3.4 Comparison of LS and RI signals from PEG/PEO narrow reference standards, M_p 81.4, 32.4, and 12.1 KDa (from left to right). Response from the LS detector (measured at 90°, upper trace) is much stronger for the high M_p PEG, whereas the RI response (lower trace) is about the same for all three.

3.4.3.5 Other Detectors

Several other chromatographic detectors that have been found to be useful in SEC but have not yet found extensive use in oligonucleotide analysis should also be mentioned. These include the *viscosity detector*,[39] a mass detector that yields information on the molar mass average M_v, also known as M_η. Recently, the use of the mass spectrometer (MS) as a detector in SEC has been reported.[40] SEC-MS could potentially yield very useful information about oligonucleotide aggregation states in a liquid dosage form if the formulation buffer does not have a deleterious effect on the ionization.

3.4.4 Column

The columns that have been most commonly employed in SEC of oligonucleotides fall into two categories: chemically bonded porous silica gel columns and hydrophilic polymeric resin columns. Each type of column offers distinct advantages, and the choice is usually dictated by the application.

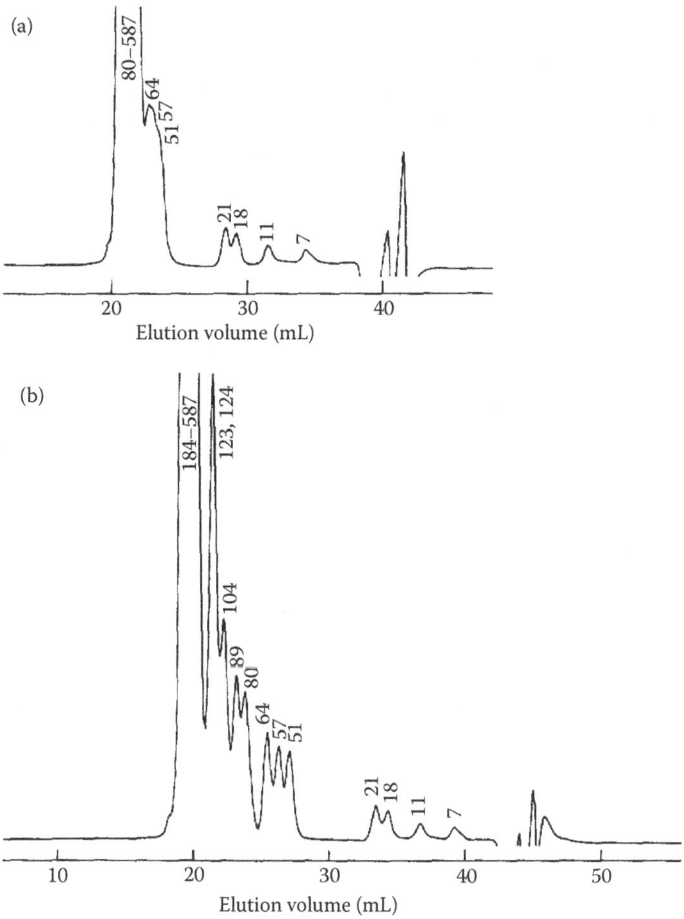

FIGURE 3.5 Effect of pore size on the SEC separation of DNA fragments. 1. SEC of HaeIII-cleaved pBR322 DNA obtained on a (a) G2000SW and (b) G3000SW two-column system. Note the substantial better separation of 51–124 base pair DNA fragments. Chromatographic conditions: 7.5 mm × 600 mm columns eluted with 0.1 M phosphate buffer (pH 7.0) containing 0.1M NaCl and 1 mM EDTA at 0.33 mL/min at 25°C. Detection: UV at 260 nm. (Reprinted from Kato, Y., et al., *Journal of Chromatography A*, 266, 341–349. Copyright 1983, with permission from Elsevier.)

Silica-based columns offer smaller pore sizes (125 Å), can withstand higher pressure (up to 1000 psi), and are available in small particle sizes (≥4 μm). However, they are sensitive to base and high-pH mobile phases should be avoided. Polymeric resins, on the other hand, offer a wider range of pore sizes (120–2000 Å) and can be operated under a wider pH range (2–12). Because of the fragility of resin beads, these columns can only withstand moderate pressure, which limits the flow rate and consequently results in relatively long analysis time.

The selection of a column for SEC separation is primarily determined by the fractionation range, which, in turn, is determined by the pore sizes of the stationary phase. For most of the synthetic therapeutic oligonucleotides currently being developed (MW 7–15 KDa), silica-based columns are a good choice because they typically have a fractionation range (5–100 KDa) that encompasses the MW range of single-strand (*ca.* 7 KDa), double strand (*ca.* 15 KDa), and multiplex (>15 KDa).

For PEGylated oligonucleotides, oligonucleotide conjugates, and multiplexes with high molecular weight, polymeric resin columns are often a better choice. These columns come in a wide range of pore sizes, allowing for separation across a wide range of molecular sizes. Polymer columns can accommodate mobile phases with high ionic strengths and are less susceptible to basic pH.

The effect of pore size on the separation of oligonucleotides has been studied by Kato et al.[41] Using double-stranded DNA fragments (32 fragments ranging from 7 to 587 base pairs) obtained from restriction endonuclease cleavage of plasmid DNA pBR322, these authors showed that the separation efficiency depended greatly on the pore size (Figure 3.5). While both the G2000SW and G3000SW columns (silica-based SEC columns with exclusion limits of 50 and 100 KDa, respectively) were able to separate small fragments, separation of intermediate fragments (51–124 base pairs) was much better achieved with the latter. Likewise, larger DNA fragments (13–1857 base pairs) were better resolved with the polymer-based G5000PW column with an exclusion limit of 1000 KDa (Figure 3.6).

The presence of residual silanol groups in silica-based columns causes the highly charged oligonucleotides to interact with the stationary phase. Thus, it is recommended to condition these columns prior to routine use. Overloading the column with the oligonucleotide of interest for the first few injections can significantly reduce some of these nonspecific interactions, thereby reducing

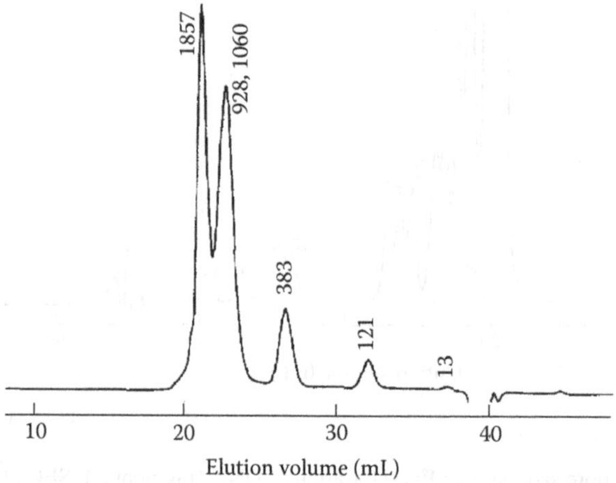

FIGURE 3.6 Effect of pore size on the SEC separation of DNA fragments. 2. Size exclusion chromatogram of BstNI-cleaved pBR322 DNA on a G5000PW (7.5 mm × 600 mm) two column system. Chromatographic conditions are the same as those in Figure 5. (Reprinted from Kato, Y., et al., *Journal of Chromatography A*, 266, 341–349. Copyright 1983, with permission from Elsevier.)

peak tailing. In addition, increasing the ionic strength of the mobile phase by adding millimolar concentrations of inorganic salts also helps to mitigate this problem.

SEC columns have a relatively short lifetime. As a result it is important to store them appropriately. Column fouling is a common problem because SEC is predominantly performed in aqueous media. Microbial growth can be inhibited by storing the columns in the presence of a biocide solution such as 0.01% sodium azide or 20% ethanol. The latter storage solution is usually preferred because it is safer to handle and easier to dispose.

For applications in a development setting, a consistent system suitability routine should be adopted for all columns using an appropriate standard and well-established United States Pharmacopeia (USP) criteria for the various chromatographic parameters such as precision, resolution, and tailing. It is also very important to monitor the plate counts of the column over time. As a general rule, SEC columns that have lost half of their initial plate counts should be retired from use.

When multiple SEC columns are used in series to improve resolution, the order of connection is critical. The columns must be connected in decreasing pore size. This order is necessitated by the sieving mechanism of SEC and allows larger size molecules to have a longer residence time inside the columns. If the columns are connected in reversed order, i.e., smallest pore size first, the resi-

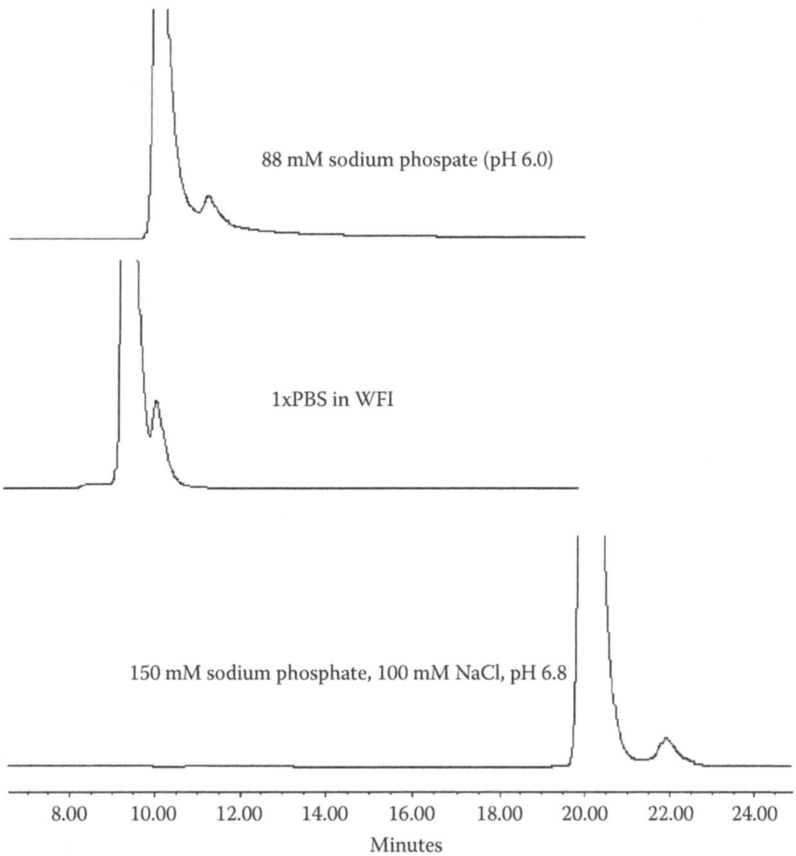

FIGURE 3.7 Effect of inorganic salt on duplex equilibrium and tailing in the SEC of duplex RNAs. Top trace: Single TSKGel SW2000 column separates duplex RNA (main peak) from single strand (smaller peak) RNA but exhibits tailing. Middle trace: Single TSKGel SW2000 column separates duplex RNA (main peak) from single strand (smaller peak) RNA with minimal tailing in the presence of salt in the buffer. Bottom trace: Tandem TSKGel SW2000 columns separate duplex RNA (main peak) from single strand (smaller peak) RNA with minimal tailing and improved resolution between the peaks.

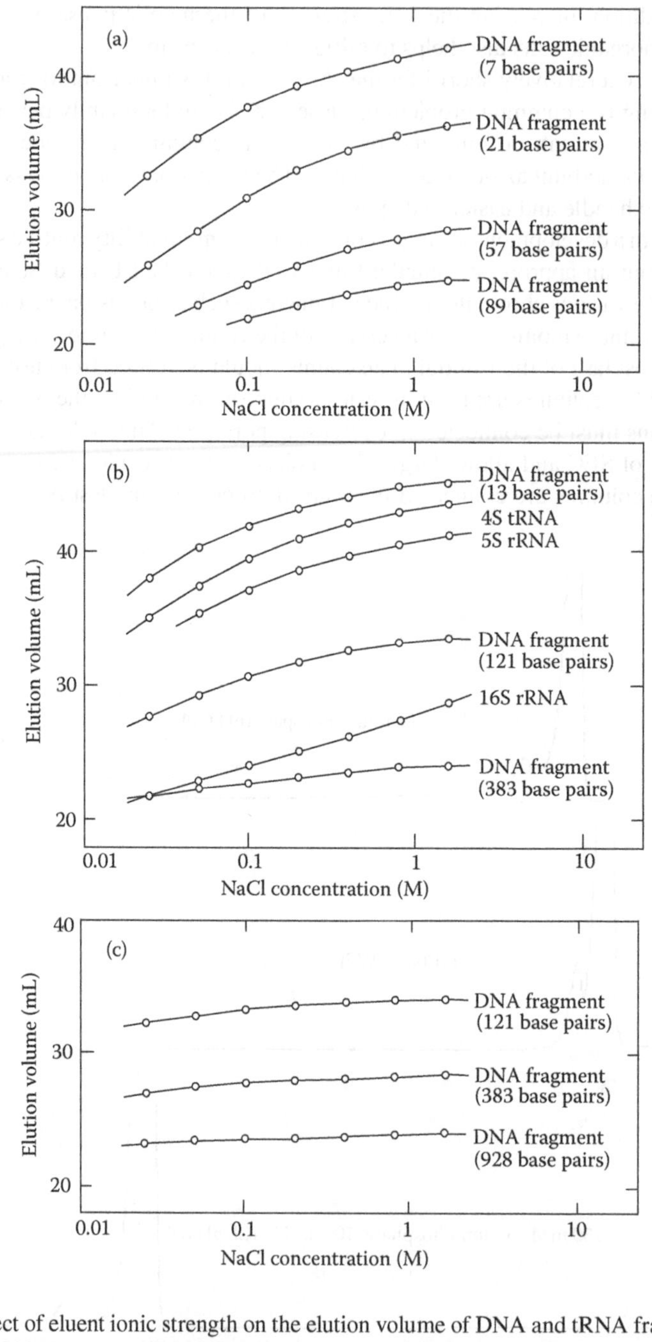

FIGURE 3.8 Effect of eluent ionic strength on the elution volume of DNA and tRNA fragments obtained on (a) G2000SW, (b) G3000SW, and (c) G5000PW two-column system (7.5 mm × 600 mm each). Chromatographic conditions: 0.01 M Tris-HCl buffer (pH 7.5) containing 0.025–1.6 M NaCl and 1 mM EDTA at a flow rate of 1 mL/min. (Reprinted from Kato, Y., et al., *Journal of Chromatography A*, 266, 341–349. Copyright 1983, with permission from Elsevier.)

dence time of larger molecules will be made inordinately short and the effect of the larger pore size column will be largely lost.

3.4.5 Mobile Phase

The choice of mobile phase depends on the nature of the oligonucleotide being analyzed (i.e., single strand, duplex, oligonucleotide conjugates, or aggregates) and on the solvent compatibility of the column (i.e., silica-based columns versus polymer-based columns). Water, buffer solutions such as sodium phosphate and Tris HCl, and inorganic salt solutions such as sodium chloride and potassium chloride have all been used as mobile phases for SEC of oligonucleotides. When silica-based columns are employed, care must be taken to keep the mobile phase between pH 2 and 8 to avoid degradation of the packing material. A dilute buffer at or near-neutral pH is generally preferred to pure water.

Being highly charged molecules, oligonucleotides are also affected by the ionic strength of the mobile phase, which should be carefully controlled. Phosphate-buffered saline (PBS), a commonly employed medium in parenteral and ophthalmic formulations, can also be used as a mobile phase in SEC. Thus SEC is perhaps the only chromatographic technique that allows oligonucleotide drug product to be analyzed in their native formulation. The propensity of oligonucleotide to form duplex in solution depends on the ionic strength. It is generally observed that addition of inorganic salts to the mobile phase not only maintains duplex equilibrium but also reduces tailing arising from unwanted interaction with silica based stationary phase (Figure 3.7).

A systematic study of the effect of ionic strength of the mobile phase on the elution volume of oligonucleotides in SEC has been reported.[41] The elution volume was found to increase with increasing ionic strength. This was attributed to the repulsion between the negatively charged oligonucleotides and the residual silanol or carboxyl groups in the stationary phase. At higher ionic strength this repulsion was disrupted, allowing the oligonucleotides better access to the pores, and hence longer residence time. In general, the effect was more pronounced with silica based SEC columns (Figure 3.8a and b) and larger oligonucleotides seemed to be less affected (Figure 3.8c).

The dependence of height equivalent of theoretical plate (HETP) on mobile phase flow rate is shown in Figure 3.9. In general, HETP decreased (i.e., resolution increases) with decreasing flow rate throughout the range studied. This was found to be especially true in the case of high molecular weight samples where the HETP increased almost sixfold from 0.1 to 1 mL/min. While the best

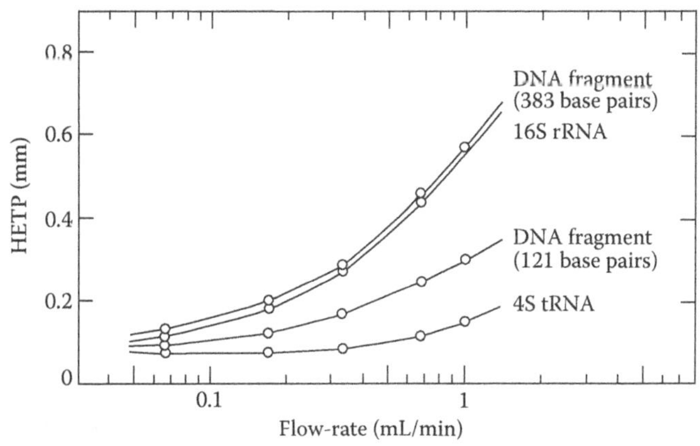

FIGURE 3.9 Effect of flow rate on HETP for DNA and RNA fragments. For RNA fragments, two G5000PW columns (7.5 mm × 600 mm) are used while those used for RNAs are two G4000SW columns (7.5 mm × 600 mm). (Reprinted from Kato, Y., et al., *Journal of Chromatography A*, 266, 341–349. Copyright 1983, with permission from Elsevier.)

resolution was achieved at a flow rate below 0.1 mL/min, flow rates of 0.3–0.5 mL/min seemed to be a good compromise between separation time and resolution.

3.4.6 Data Acquisition and Processing

The instrument control, data acquisition, and data processing functions of modern HPLC instruments are usually handled by computer software supplied by the manufacturer. This software package is adequate for qualitative SEC analysis such as monitoring siRNA duplex formation and simple MW determination by means of a calibration curve. When more sophisticated analyses such as molar mass averages, molar mass distribution, and polydispersity index are required, a specialized SEC/GPC software package should be used because these calculations involve complicated statistical manipulations that are not easily carried out manually. Most HPLC manufacturers offer SEC software modules as an add-on to their basic software. SEC analysis software packages from third-party vendors are also available that work alongside HPLC control software. Usually, the analogue signals from the concentration detectors (RI, UV, and/or ELS) are fed into an A/D converter that comes with third-party SEC software. The A/D converter acts a bridge between the HPLC and the SEC analysis software by digitizing the detector signal before it is sent to the computer for analysis.

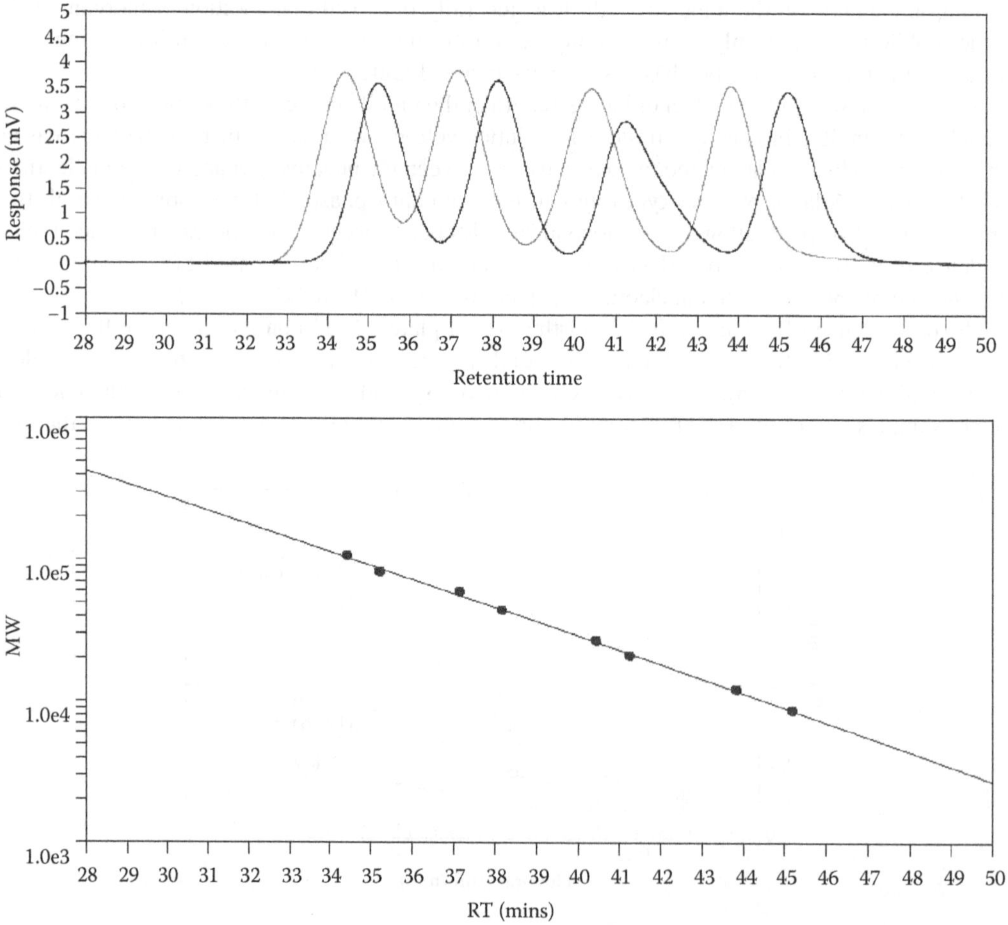

FIGURE 3.10 Typical calibration curve for MW determination of PEGylated oligonucleotides. The PEG/PEO reference standards were analyzed in two groups for better separation (top panel). The combined results were used to construct a calibration curve (lower panel). Second degree curve fitting was used ($r^2 = 0.9996$).

For LS detectors, particularly those involving multiple-angle detection (MALLS), the software supplied by the instrument vendor is usually the best choice.

3.4.7 Calibration Curve

As mentioned before, one of the greatest advantages of SEC is its ability to determine MW information with a relatively simple instrument setup. However, unless a mass detector such as a multiangle laser light scattering (MALLS) detector is coupled with the instrument, SEC cannot be used for absolute MW determination. Instead, a calibration curve constructed from polymer standards of known MW must be used. This is usually achieved by injecting a series of calibration standards (polymers of well-defined M_p with narrow molar mass distribution). Typically, a group of calibration standards with MW range spanning that of the unknown is chosen. The reference standards are injected in groups of 3–5 each to achieve good resolution. Once the standard injections are complete, the results are combined for the calibration curve. The log M_p of the calibration standards is then plotted against the retention time (or volume) to give a calibration curve. In most cases, good correlation is achieved with a first or second degree fitting protocol. The MW of the unknown sample can be readily determined from its retention time or retention volume. The unknown can be further characterized by determining the molar mass averages, polydispersity, and molar mass distribution. A typical calibration curve is shown in Figure 3.10.

SEC determines the MW by measuring the *hydrodynamic volume* of the analyte. The choice of calibration standards is thus crucial to the accuracy of the MW determined. Oligonucleotides, being highly charged, usually attract a large number of water molecules in solution. The calibration standards should have similar physicochemical properties to minimize the impact of differences in hydrodynamic volume and the potential for non-SEC interaction with the column. Ideally, oligonucleotides of well-defined MW would be used as calibration standards for other oligonucleotides. However, such compounds are difficult to come by and surrogates are usually used. For example, narrow polyethylene glycol (PEG) and polyethylene oxide (PEO) calibration standards have been used successfully with PEGylated-oligonucleotides.

3.5 APPLICATIONS OF SEC IN OLIGONUCLEOTIDE THERAPEUTICS

As the field of oligonucleotide therapeutics matures with more complex oligonucleotide multiplexes or polymer conjugates in development, the demand for SEC as an analytical tool is expected to increase. For such compounds, SEC is a critical tool in support of manufacturing, quality control, and stability of drug substances and formulated drug products. The scope of the method may vary to meet the regulatory requirements for *identity*, *purity*, and *assay*. Oligonucleotide identity determination can be achieved by either comparing the retention time/volume of a sample against that of a well-characterized reference standard or through the use of a calibration curve with appropriate molecular weight standards. Alternatively, identity information can be obtained by measuring absolute molecular weights through the use of light scattering detectors.

By their very nature, SEC identity methods can be readily adapted for purity determination as long as the species of interest fall within the fractionating range of the column. This approach is particularly useful for siRNA and other oligonucleotide complexes where the species of interest (monomers, dimers, and multiplexes) have very different molecular weights. The purity can be determined by the relative area under the corresponding peaks. The purity obtained by SEC can be correlated with other weight-by-weight assay methods such as UV to provide a measure of the desired oligonucleotide content in the active pharmaceutical ingredient (API) or drug product. For more detailed discussion of assay of oligonucleotides, refer to Chapter 9.

A sampling of the SEC analysis of oligonucleotides reported in the literature is given in Table 3.1. To further illustrate the usefulness of SEC in oligonucleotide analysis, selective examples drawn from the published literature and the authors' own research are discussed in detail below.

TABLE 3.1
Application of SEC in Oligonucleotide Analysis

Application	Column	Condition	Detection	Reference
Purification of plasmid DNA and DNA fragments	TSK-G5000 PW gel (3.2 mm × 200 mm)	0.05 M Tris, pH 7.4 at 0.5 mL/min	UV at 260 nm	2
Investigation of separation ranges of SEC columns for ds DNA and tRNA	Eight 7.5 mm × 600 mm columns (2×TSKgel G2000SW, 2×G3000SW, 2×G4000SW and 2×G5000PW)	0.1 M phosphate buffer (pH 7.0), 0.1 M NaCl, 1 mM EDTA at 0.33–1.0 mL/min	UV at 260 nm	41
Separation of large DNA restriction fragments	Four TSKgel DNA-PW (7.8 mm × 300 mm)	0.1 M Tris-HCl (pH 7.5), 0.3 M NaCl, 1 mM EDTA at 0.15–0.5 mL/min	UV at 260 nm	4
Molar mass characterization of DNA fragments	Toyo Soda GWPH G4000, G5000 and G6000	0.1 M $NaNO_3$	RI, LALLS	50
Analysis of ds DNA restriction fragments	Superose 6 (10 mm × 106 mm and 10 mm × 300 mm)	0.02 M Tris-HCl, pH 7.6, 0.15 M NaCl	UV at 254 nm	51
Hybridization of DNA with ^{32}P labeled probes	Superose 12 (1000–300,000 range)	0.78 M NaCl, 0.075 M sodium citrate at 0.5 mL/min	Radioactivity ^{32}P	52
Hybridization determination of fluorescein labeled and biotinylated oligos	Bio-Rad Bio-Sil SEC125 column	0.1 M sodium phosphate, pH 7.0 at 1.0 mL/min	UV at 260 nm	53
Monitor binding stoichiometry of gene V protein and 16-mer oligonucleotide	Sephacryl S-200	0.1 M NaCl, 0.02 M Tris HCl, pH 7.0, 8°C at 0.5 mL/min	UV at 280 nm	45
Purification of AS ODN-avidin-cobalamine complex	Superose 12 (10 mm × 250 mm)	20 mM TRIS, 150mM NaCl, pH 7.4 at 0.5 mL/min	UV at 260 nm	54
Analysis of oligo-polymer conjugate	Ultrahydrogel 500 and Ultrahyogel 2000	0.1 M phosphate buffer, pH 6.8; 0.1 M phosphate buffer/20 mM NaCl, pH 6.6	UV at 260 nm with inline RI, MALLS and VISC detectors	18

Determination of radio-labeled PS ODN in association with plasma components	Superose 6 precision column (3.2 mm × 300 mm)	PBS at 0.05 mL/min	Radioactivity (^{3}H and ^{125}I)	55
Estimation of association of ODN-glycopolymer with half sliding complementary	JASCO SB803-HQ and SB804-HQ columns	PBS, pH 7.4 at 0.5 mL/min	RI - 930	56
Labeling efficiency of S-acetyl NHS-MAG3-conjugated morpholino oligomers	Superdex peptide column (separation 1 × 10^2 to 7 × 10^3 Da)	0.10 M pH 7.2 phosphate buffer at 0.60 mL/min	Inline UV and radioactivity (^{99m}Tc)	57
Study transition from antiparallel to parallel G-quadruplex	TSK Gel 2000	50 mM MES or 100 mM NaCl and 50 mM MES or 50 mM CaCl2, 100 mM NaCl and 50 mM MES at 1.0 mL/min	UV at 260 nm	58
Large-scale purification of polyacrylamide free RNA oligo (from plasmid DNA)	Bio-Rad Biogel A 50m	150 mM NaCl, 50 mM sodium phosphate, pH 6.5, 0.1 mM EDTA at 0.2 mL/min	UV at 260 nm	9
Determination of supramolecular assembly and self-assembly of oligo-DNA	Sephacryl S-400	50 mM Tris-HCl, 500 mM NaCl, pH 7.5	UV at 260 nm	59
Rapid purification of RNA from plasmids and globular proteins	Superdex 200 (26 mm × 60 mm)	10 mM phosphate, 100 mM NaCl, pH 6.5 at 3 mL/min	UV at 260 nm	14
Sample pretreatment of RNA-protein UV cross-linked complex prior to MALDI	Superose 75HR 3.2 mm × 300 mm, Superose 74HR, 10 mm × 300 mm	50 mM Tris HCl, pH 7.5, 150 mM NaCl, 5 mM EDTA at 0.04 mL/min	UV at 260 nm/280 nm	60
Analysis of ODN-schizophyllan (a β-1,3-glucan) complex	Showa Denko OHPak SB-806 and SB-802.5 columns	50 mM phosphate buffer at 1.0 mL/min	UV at 260 nm	47

3.5.1 Oligonucleotide Duplexes

3.5.1.1 Monitoring of Annealing during Manufacturing

During the manufacture of a duplex oligonucleotide, the single-strand oligonucleotides are mixed together and allowed to form a duplex in a process called annealing. Prior to annealing, the individual single strands are denatured by exposure to high temperatures, but annealing is usually performed below the melting temperature (T_m). Depending on the process, the end point of annealing may be reached when the equilibrium mixture contains the maximum amount of duplex and the minimum amount of single strand/multiplex species as determined by a given analytical technique. In theory, it should be possible to bring together a 1:1 molar mixture of the sense and antisense strands to form the perfect duplex. In practice, this is not always the case because the single-strand oligonucleotides invariably contain significant amounts of failure sequences such as n-1 and higher order deletion sequences due to the sequential nature of their synthetic process. Many of these deletion sequences are capable of forming stable duplexes with the complementary strand although they may not be as stable as the perfect full length duplexes. Thus the molar equivalents needed for optimum duplex formation depends not only on the absolute full length purity of the two strands but also on the nature of the impurities present, whether they are hybridizable to the complementary sequence or not. For a detailed discussion of hybridization, refer to Chapter 6.

For well-established manufacturing processes with consistent single-strand impurity profiles, it should be possible to predict the molar equivalents required to reach duplexation endpoint without the need of extensive in-process analytical controls. However, in early stage development when the manufacturing process is still evolving, SEC is an excellent tool to follow the mixing stoichiometries of the sense and antisense strands during the annealing process. This can be achieved in real time by following the SEC peak areas of the single-strand oligonucleotides being mixed at process scale. The ratio of the single-strand oligonucleotides can then be adjusted in an iterative mixing and analysis procedure to reach the duplexation end point. An alternative technique is to analyze varying molar ratios of the two single strands at analytical scale by SEC to determine the duplexation end point (lowest single strand peak area), which can then be translated to process scale. Figure 3.11 provides an example of the utility of SEC in a siRNA analytical scale duplexation model study. In this case, the end point of duplexation was reached when a 5% molar excess of the sense strand was used.

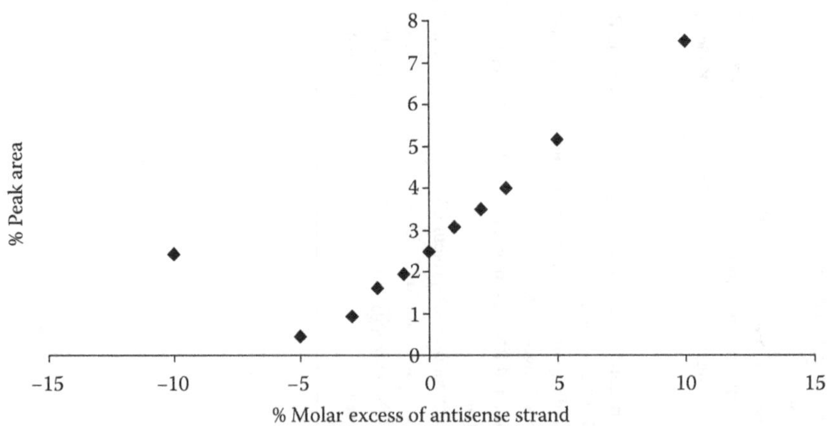

FIGURE 3.11 Process development in duplex formation monitored by SEC. Duplex quality was determined by the amount of unhybridized single strands remaining after varying amounts of sense and antisense single strand oligonucleotides were mixed. The amount of single strand remaining was analyzed by SEC.

3.5.1.2 Analysis of Liposome Encapsulated siRNA

One of the biggest challenges in developing oligonucleotide-based therapeutic agents is their delivery to the appropriate tissues within the human body. Oligonucleotides have poor bioavailability owing to their polyanionic nature. They also have a very short half-life in the body owing to their susceptibility to nuclease enzymes. Among the strategies that have been advanced to enhance the pharmacokinetic profile of oligonucleotides, the formation of liposomes and nanoparticles have attracted the most attention.[42–44]

The release of siRNA from liposome particles has been studied by SEC. Free and liposome-encapsulated siRNA were readily separated by SEC because of significant size difference. Upon disruption of the liposomes by the addition of organic solvents, the encapsulated siRNA was released as shown in Figure 3.12. Interestingly, the siRNA retained its strong UV absorption even in the encapsulated state, which greatly facilitated its detection.

3.5.1.3 Stability Monitoring of DNA and RNA Duplexes

Typically, high-resolution denaturing methods such as strong anion exchange (SAX) or reversed phase ion-pair (RP-IP) chromatography are used to monitor the stability of duplex oligonucleotide API or drug products. The primary objective of this type of analysis is to determine the chemical degradation of the individual single strands in the duplex oligonucleotide. In addition, it is also important to analyze the duplex oligonucleotides under nondenaturing conditions to obtain an understanding of the stability of the duplex form upon storage. SEC has been shown to be stability indicating for monitoring the dissociation of the duplex into the corresponding single strands over time, an indication of the stability of the duplex in the selected formulation. The scope of this analysis is to follow the content of intact duplex oligonucleotide with respect to the presence of single-strand molecules (and their degradation products if possible) and multiplexes. SEC is probably the most valuable nondenaturing stability indicating method for this purpose as it can be performed in the formulation buffer, thereby providing an assay of duplex content and an estimate of the viability of the duplex in the formulation.

SEC methods can also be employed to detect degradation in duplex oligonucleotides. Figure 3.13 shows the SEC chromatograms of a DNA duplex upon exposure to acid and base. As is evident, the sample is acid labile, giving rise to both smaller (hydrolyzed) and larger (cross-linked) degradation

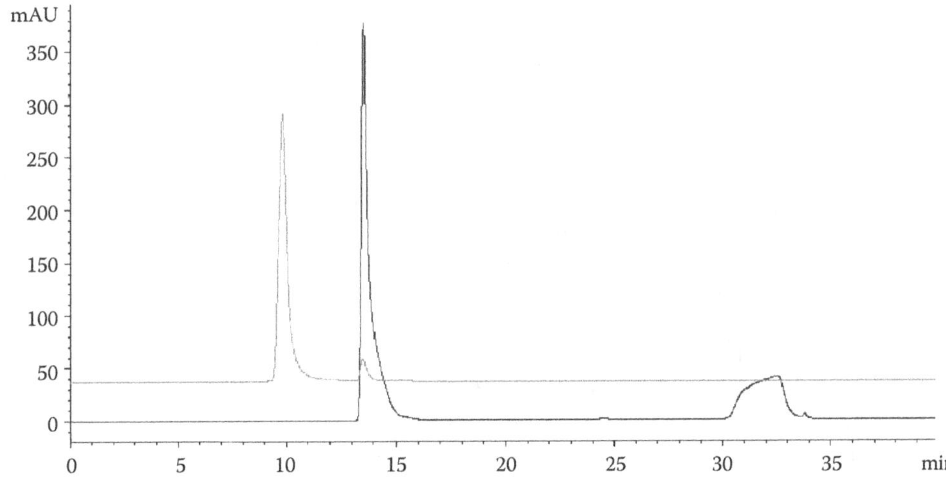

FIGURE 3.12 SEC analysis of liposome encapsulated siRNA. Upper trace: formulation containing liposome encapsulated siRNA (10 min) and a small amount of free siRNA (13.5 min). Upon disruption of the liposomes, all the siRNA was released as free siRNA (lower trace). Chromatographic conditions: 2xTSKgel Super SW2000 (4.6 mm × 300 mm), PBS (pH 7.4), 0.4 mL/min, 25°C, UV detection. (Courtesy of Alnylam Pharmaceuticals, Inc.)

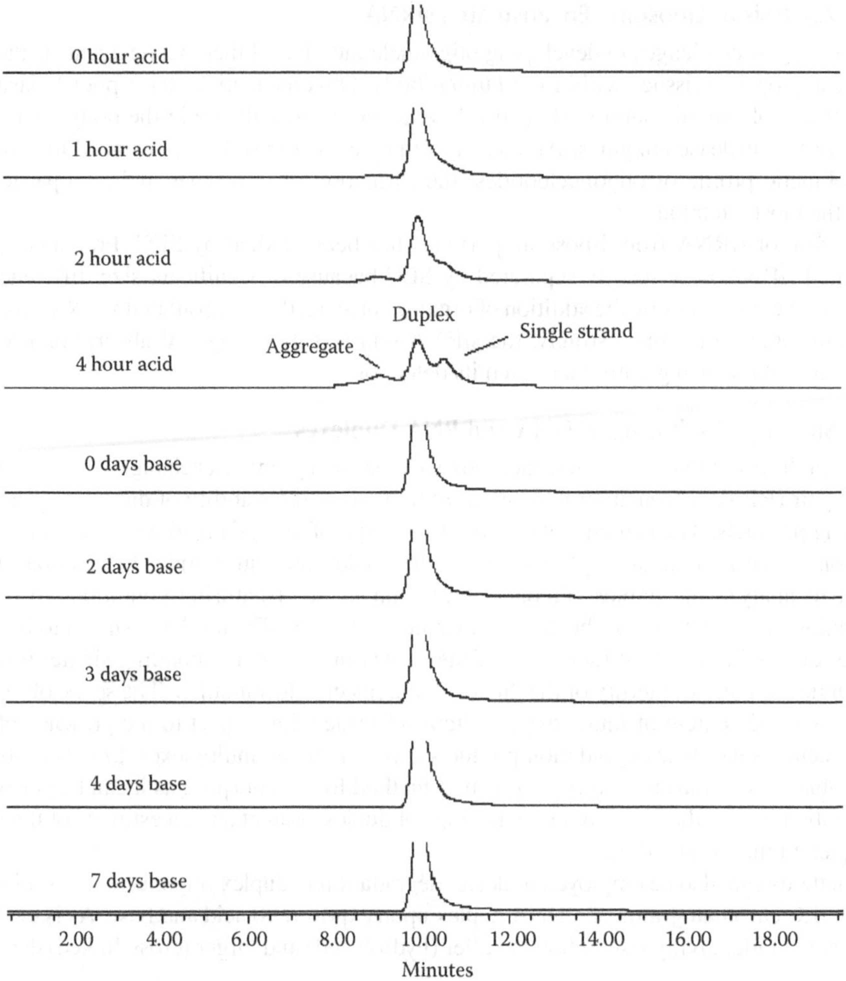

FIGURE 3.13 Degradation of DNA duplex. DNA duplexes rapidly degraded into cross-linked products, unhybridizable single strands and hydrolyzed products within hours of exposure to acidic conditions. However, they were stable to basic conditions even after a week of exposure.

products within hours of exposure. As expected, the same DNA sample is stable under basic conditions after days of exposure. An example of a RNA duplex being stressed in the presence of base is shown in Figure 3.14. As is evident, RNA molecules are base labile due to the presence of the 2′-OH group. Smaller (hydrolyzed) degradation products are observed within hours with complete degradation by 24 hours. These examples serve to highlight the usefulness of SEC in monitoring the stability of duplex oligonucleotides.

Nondenaturing SAX-HPLC and RP-IP HPLC techniques can also be used in monitoring the stability of duplex oligonucleotides (see Chapter 15). However, owing to the composition of mobile phases used, an assay of duplex content performed under these conditions might not be reflective of the duplex stability in the formulation of interest.

3.5.2 Conjugated Oligonucleotides and Oligonucleotide Complexes

The binding stoichiometry of gene V protein (9.7 KDa) from bacteriophage f1 to a 16-mer oligonucleotide (4.9 KDa) was studied by SEC as well as ESI-MS. A Sephacryl S-200 column was

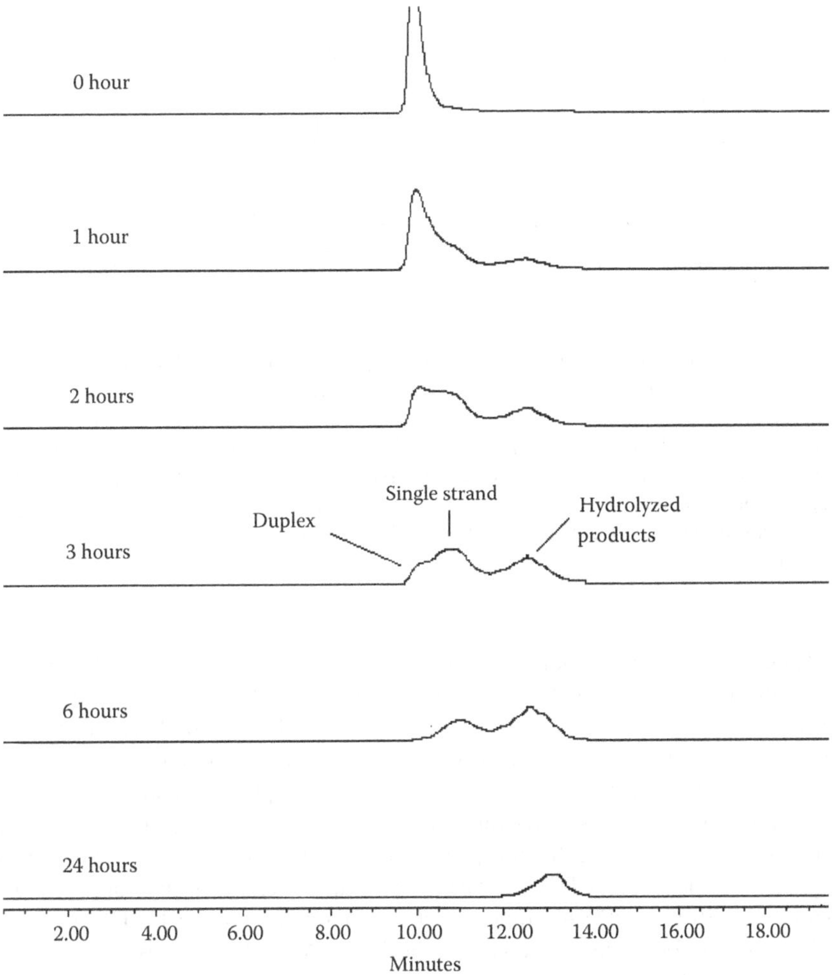

FIGURE 3.14 Base-induced degradation of RNA duplex. RNA duplex rapidly degraded into unhybridizable single strands, and hydrolyzed products within hours of exposure to basic conditions.

employed for SEC (0.02M Tris HCl, 0.1M NaCl at pH 7.0), with the proteins RNase A (13.7 KDa), chymotrypsinogen A (25 KDa), carbonic anhydrase (29 KDa), ovalbumin (43 KDa), and albumin (67 KDa) as MW markers. The protein-oligonucleotide complex was found to elute at nearly the same retention volume as ovalbumin (Figure 3.15). The MW thus determined (43 KDa) corresponded to a 4:1 complex (protein monomer:oligonucleotide) and agreed very well with that determined by ESI-MS (43.6 KDa).[45]

Schizophyllan, a fungal polysaccharide consisting of a 1,3-β-D-linked backbone of glucose residues with 1,6-β-D-glucosyl side groups, forms a complex with oligonucleotides through hydrogen bonding between two main chain glucoses and one nucleotide base.[46] The complex can be used as carriers for therapeutic oligonucleotides such as antisense DNA, siRNA, and CpG ODN (oligodeoxynucleotides). The formation of schizophyllan-oligonucleotide complexes has been demonstrated by SEC in 50 mM phosphate buffer mobile phase using UV detection at 260 nm. Upon treatment with schizophyllan, the well-defined peak of an AS-ODN was replaced by early eluting peaks of higher MW from the conjugate (Figure 3.16). SEC was the only method that could demonstrate the formation of high MW complexes between the two species.[47]

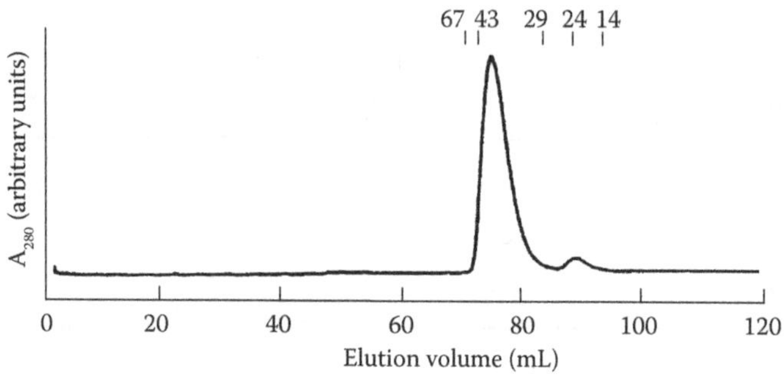

FIGURE 3.15 SEC of gene V protein-oligonucleotide complex (Sephacryl S-200 column, 20 mM Tris HCl, 100 mM NaCl, pH 7.0, 0.5 mL/min at 8°C). The elution positions of molecular weight standards applied separately to the column are indicated above the chromatogram. The standards were RNase A (13.7 kDa), chymotrypsinogen A (25 kDa), carbonic anhydrase (29 kDa), ovalbumin (43 kDa), and albumin (67 kDa). Gene V protein alone had an elution volume of 90 mL; the oligonucleotide I alone had an elution volume of 89 mL. (Reproduced with permission from Chen, X., et al., *Proceedings of the National Academy of Sciences of the United States of America*, 93, 7022–7027. Copyright 1996, National Academy of Sciences of the United States of America.)

The usefulness of SEC as an analytical and preparative technique was clearly demonstrated by the length sorting and purification of DNA-wrapped carbon nanotubes (CNT).[48] Single-strand DNA (ssDNA) was found to wrap around CNT to form a DNA-CNT hybrid. This effectively dispersed CNT into aqueous solution and facilitated their sorting by diameters and chirality. The structurally sorted DNA-CNT species is invaluable in elucidating the physical and chemical properties of CNT. Commercial SEC columns were found to be unsuitable for analyzing DNA-CRT owing to irreversible adsorption. However, the nonspecific adsorption was largely eliminated by using specially designed columns (Sepax CNT-SEC) bearing negatively charged functional groups. The SEC chromatogram of a 30-mer ssDNA-wrapped CNT is shown in Figure 3.17. This method has been successfully scaled up for the length separation of the DNA-CNT. Analysis of randomly collected fractions showed that the length sorting was very effective, resulting in a size distribution of less

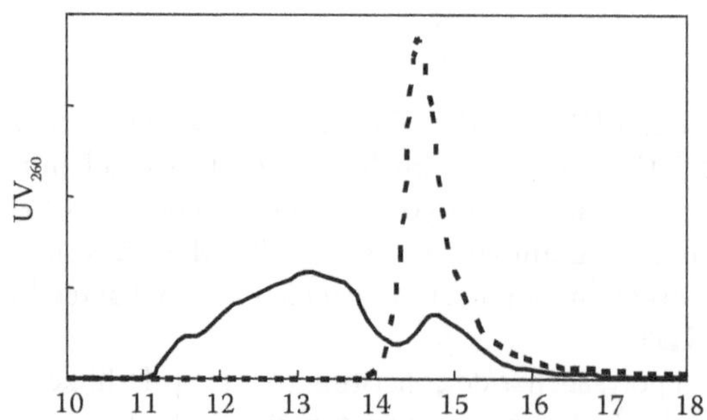

FIGURE 3.16 Size exclusion chromatogram of anti-sense oligodexoynucleotides (AS-ODN, dotted line) and the reaction mixture of AS-ODN and schizophyllan (solid line), UV detection at 260 nm. (From Mochizuki, S., et al., *Advances in Material Design for Regenerative Medicine, Drug Delivery and Targeting/Imaging*, MRS Proceedings 1140, HH05–17. With permission, Materials Research Society, Warrendale, PA, 2009.)

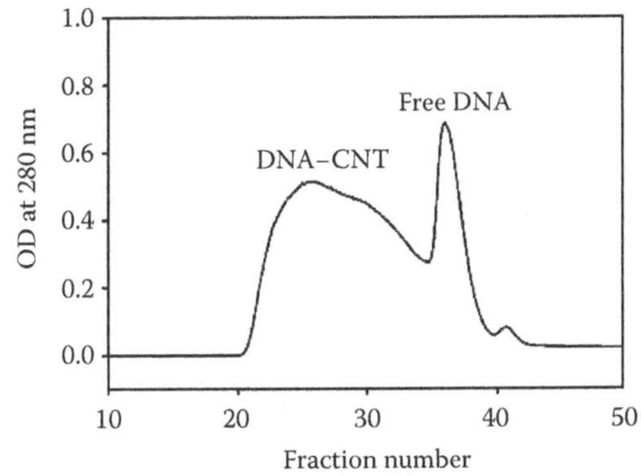

FIGURE 3.17 SEC separation of single-strand oligonucleotide-carbon nanotube (ssDNA-CNT) (3 Sepax CNT-SEC columns in series eluted with 40 mM Tris pH 7, 0.5 mM EDTA, and 0. 2 M NaCl at a flow rate of 0.25 mL/min., UV detection at 280 nm. (Reprinted with permission from Huang, X., et al., *Analytical Chemistry*, 77, 6225–6228. Copyright 2005, American Chemical Society.)

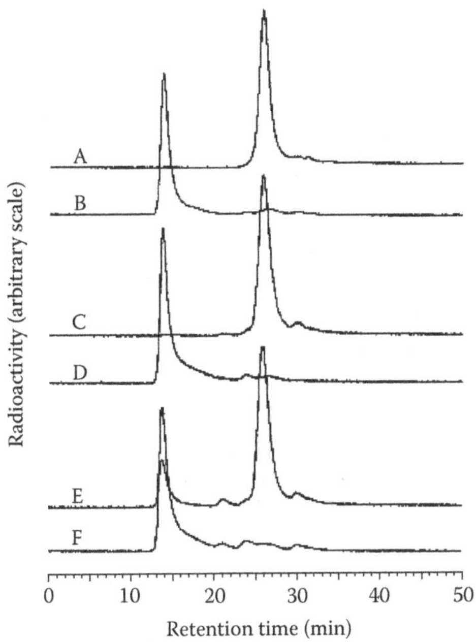

FIGURE 3.18 Solution and *in vitro* serum stability of ^{111}In radio-labeled MORF monitored by SEC. ^{111}In-MORF shows (A) a single peak after purification, (C) incubation for 48 hours in saline at room temperature and (E) for 48 hours at 37°C in serum. (B, D, F) In each case, the radioactivity shifts to higher molecular weight following addition of cMORF-PA polymer. (Reprinted from Liu, C. B., et al., *Nuclear Medicine and Biology*, 30, 207–214. Copyright 2003, with permission from Elsevier.)

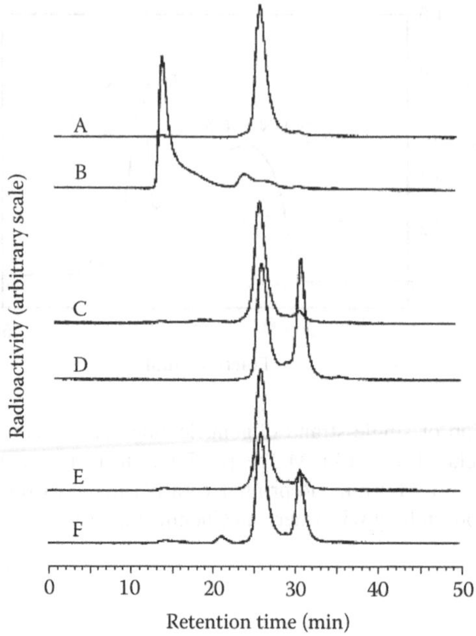

FIGURE 3.19 Solution and *in vitro* serum stability of ^{90}Y radio-labeled MORF monitored by SEC. ^{90}Y-MORF showed a single peak (A) after purification and (B) an almost complete shift to higher molecular weight following addition of cMORF-PA polymer. A low molecular weight peak started to appear after incubation in saline at room temperature for (C) 2 hours and (D) 48 hours and (E) after incubation at 37°C in serum for 12 hours and (F) 48 hours. (Reprinted from Liu, C. B., et al., *Nuclear Medicine and Biology*, 30, 207–214. Copyright 2003, with permission from Elsevier.)

than 10% as determined by atomic force microscopy. This separation was considerably better than 30–80% variation reported with other methods.

Morpholinos (MORFs) are novel synthetic oligonucleotide analogues in which the ribose is replaced by a morpholine. They are potential therapeutic agents because of their resistance to enzyme degradation, low protein binding, and good water solubility. Similar to their oligonucleotide counterparts, MORF segments can be designed to bind tightly to a target sequence by hybridization through base pairing and have been used as delivery vehicles in radiotherapy. The solution and *in vitro* serum stability of a 25-mer MORF labeled with ^{111}In and ^{90}Y (^{111}In-MORF and ^{90}Y-MORF) was studied by SEC. The radiochemical purity in each case was confirmed by a retention time shift upon addition of a complementary MORF conjugated to a high molecular weight polymer (cMORF-PA). As shown in Figure 3.18, purified ^{111}In-MORF (A) hybridized with cMORF-PA to give a single early eluting (high MW) peak (B). After incubation for 48 hours in saline at room temperature, several small peaks were formed but the ^{111}In-MORF peak remained largely intact (C, D). ^{111}In-MORF was less stable in serum at 37°C as evidenced by the formation of an early-eluting, high molecular weight peak, apparently due to serum protein binding (E). After addition of cMORF-PA, all the radioactivity shifted to the early eluting peak (F). However, this early-eluting peak was somewhat broader than those found in (B) and (D), indicating that it was probably made up of more than one component. A similar study was carried out with ^{99}Y-MORF (Figure 3.19). As observed before, purified ^{99}Y-MORF was almost completely shifted to a higher molecular weight peak after addition of cMORF-PA (A and B). However, a low molecular weight peak (31 min) was formed upon incubation in saline at room temperature for (C) 2 hours and (D) 48 hours and in serum at 37°C for (E) 12 hours and (F) 48 hours. The low molecular weight peak was attributed to radiolysis as it could not hybridize with cMORF-PA. This study clearly demonstrates the power of SEC as an analytical tool for monitoring the stability of oligonucleotides and their analogues.[49]

Because of the presence of a polydispersed polymer, MS analysis of PEGylated oligonucleotides can be challenging. SEC is one of the few simple techniques that can give useful information on this class of compounds. Because well-characterized oligonucleotides with known MW are not easily available, narrow PEG or PEO reference standards are usually used for the calibration curve. A typical SEC calibration curve with second degree fitting can be found in Figure 3.10. The chromatographic profile of a representative PEGylated oligonucleotide along with its molar mass distributions are shown in Figure 3.20.

SEC can also be used to monitor the degradation of PEGylated oligonucleotides. The molar mass distribution and the polydispersity are particularly useful parameters for assessing quality. As the PEGylated oligonucleotide degrades, the molar mass distribution shifts to lower values owing to break down of the PEG chain and the oligonucleotide linkage, resulting in an increase in polydispersity. The SEC profile of a PEGylated oligonucleotide undergoing forced degradation at 40°C is shown in Figure 3.21. Degradation of the PEGylated oligonucleotide is clearly shown by broadening

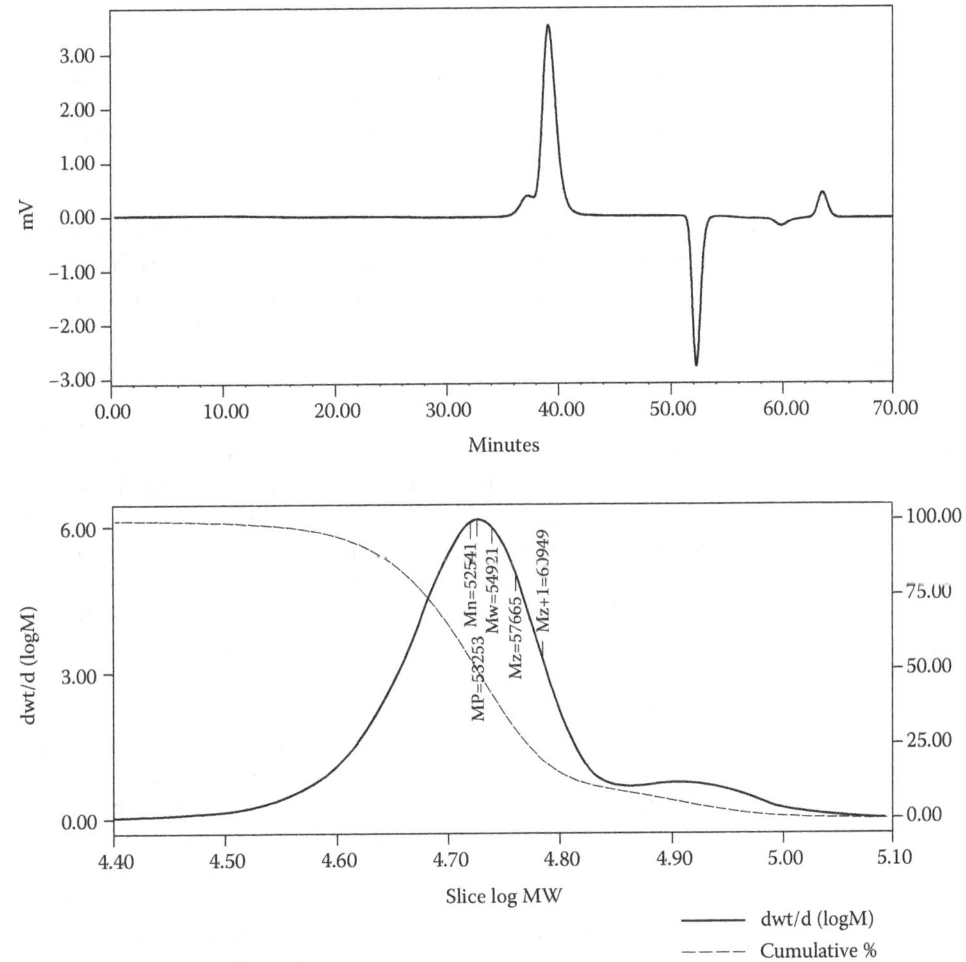

FIGURE 3.20 SEC analysis of a (upper panel) PEGylated oligonucleotide and the (lower panel) corresponding molar mass distribution plot.

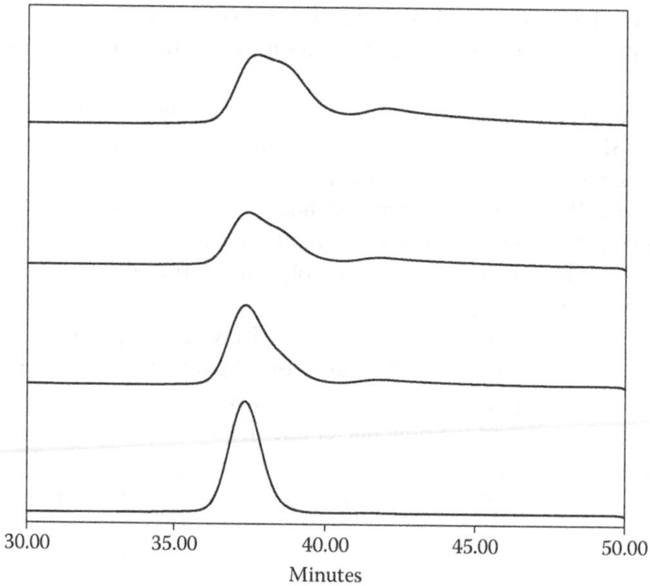

FIGURE 3.21 SEC of PEGylated oligonucleotides undergoing forced degradation. After prolonged thermal exposure, the PEGylated oligonucleotide gradually degrades to lower molecular weight species. From bottom to top: $t = 0$, 3, 5, and 18 weeks of exposure at 40°C. Chromatographic conditions: 2 × Waters Ultrahydrogel columns, phosphate-buffered saline, 0.4 mL/min, 30°C.

of the main peak, extensive tailing, and the concomitant formation of a lower MW peak at higher retention time. The MW difference between the major and minor peaks (ca. 20 KDa) indicates that loss of one of the 20 KDa PEG chains in the branched PEG is a possible degradation pathway.

3.6 SUMMARY

To fully appreciate the potential of SEC, one has to understand its characteristics and limitations. Because of the unique separation mechanism, SEC may not offer the high resolution of some other chromatographic techniques such as SAX or RP-IP HPLC to detect process related impurities in oligonucleotides. However, SEC is an outstanding technique for the analysis of some of the most important classes of oligonucleotides currently in clinical development, such as DNA duplex decoys, siRNA, aptamers, and oligonucleotide conjugates. The usefulness of SEC in oligonucleotide analysis stems from its mild experimental conditions. It is not uncommon for SEC to be performed at ambient temperature, at neutral pH, in the presence of inorganic salts, and when necessary, even in the presence of added organic solvents. This versatility makes possible the analysis of intact siRNA duplexes and other oligonucleotide drug products in their native formulation. SEC is also the method of choice for analyzing oligonucleotide aggregation. Through the use of a calibration curve or a LS detector, oligonucleotides in different states of aggregation can be readily detected.

SEC is a useful technique for determining the molecular weight of biopolymers such as oligonucleotides. Recent advances in ionization techniques such as electrospray ionization (ESI) or matrix-assisted laser desorption ionization (MALDI) have also enabled the determination of molecular weight of oligonucleotides by mass spectrometry (MS). The results obtained by these techniques are quite different. While MS offers highly accurate molecular weight information that is useful for identification, SEC yields broader information that is useful for identification as well as quality evaluation. Furthermore, for PEGylated oligonucleotides or other oligonucleotides conjugated to a synthetic polymer, MS is not particularly useful because of the presence of a polydispersed macromolecule. SEC, on the other hand, is still capable of providing an array of data (molecular weight, molar mass averages, polydispersity, and molar mass distribution) for comprehensive characterization for these types of compounds.

As more and more complex oligonucleotide therapeutics such as siRNA, aptamer, and conjugates enter development, the demand for SEC as one of the primary analytical methods is expected to increase. With the proper software and good laboratory practices, SEC methods can be easily executed on standard laboratory HPLC equipment and are fairly easy to operate. For this reason, it is well suited for research as well as development laboratories.

REFERENCES

1. Porath, J., and P. Flodin. 1959. Gel filtration: A method for desalting and group separation. *Nature* 183: 1657–1659
2. Wehr, C. T., and S. R. Abbott. 1979. High-speed steric exclusion chromatography of biopolymers. *Journal of Chromatography A* 185: 453–462.
3. Himmel, M. E., P. J. Perna, and M. W. McDonell. 1982. Rapid method for purification of plasmid DNA and DNA fragments from DNA linkers using high-performance liquid chromatography on TSK-PW gel. *Journal of Chromatography A* 240: 155–163.
4. Kato, Y., Y. Yamasaki, T. Hashimoto, T. Murotsu, S. Fukushige, and K. Matsubara. 1985. Separation of large DNA restriction fragments by high-performance gel filtration on TSKgel DNA-pw. *Journal of Chromatography A* 320: 440–444.
5. Edwardson, P. A. D., T. Atkinson, C. R. Lowe, and D. A. P. Small. 1986. A new rapid procedure for the preparation of plasmid DNA. *Analytical Biochemistry* 152: 215–220.
6. Moreau, N., X. Tabary, and F. Le Goffic. 1987. Purification and separation of various plasmid forms by exclusion chromatography. *Analytical Biochemistry* 166: 188–193.
7. Raymond, G. J., P. K. Bryant, A. Nelson, and J. D. Johnson. 1988. Large-scale isolation of covalently closed circular DNA using gel filtration chromatography. *Analytical Biochemistry* 173: 125–133.
8. McClung, J. K., and R. A. Gonzales. 1989. Purification of plasmid DNA by fast protein liquid chromatography on superose 6 preparative grade. *Analytical Biochemistry* 177: 378–382.
9. Uchiyama, S., T. Imamura, S.-I. Nagai, and K. Konishi. 1981. Separation of low molecular weight RNA species by high-speed gel filtration. *Journal of Biochemistry* 90: 643–648.
10. Ogishima, T., T. Okada, and T. Omura. 1984. Fractionation of mammalian tissue mRNAs by high-performance gel filtration chromatography. *Analytical Biochemistry* 138: 309–313.
11. Boyes, B. E., D. G. Walker, and P. L. McGeer. 1988. Separation of large DNA restriction fragments on a size-exclusion column by a nonideal mechanism. *Analytical Biochemistry* 170: 127–134.
12. Pager, J. 1993. A liquid chromatographic preparation of retroviral RNA. *Analytical Biochemistry* 215: 231–235.
13. Lukavsky, P. J., and J. D. Puglisi. 2004. Large-scale preparation and purification of polyacrylamide-free RNA oligonucleotides. *RNA* 10: 889–893.
14. Kim, I., S. A. McKenna, E. Viani Puglisi, and J. D. Puglisi. 2007. Rapid purification of RNAs using fast performance liquid chromatography (FPLC). *RNA* 13: 289–294.
15. Molko, D., R. Derbyshire, A. Guy, A. Roget, R. Teoule, and A. Boucherle. 1981. Exclusion column for high-performance liquid chromatography of oligonucleotides. *Journal of Chromatography A* 206: 493–500.
16. Sawadogo, M., and M. Van Dyke. 1991. A rapid method for the purification of deprotected oligodeoxynucleotides. *Nucleic Acids Research* 19: 674.
17. Huang, X., R. S. McLean, and M. Zheng. 2005. High-resolution length sorting and purification of DNA-wrapped carbon nanotubes by size-exclusion chromatography. *Analytical Chemistry* 77: 6225–6228.
18. Minard-Basquin, C., C. Chaix, C. Pichot, and B. Mandrand. 2000. Oligonucleotide-polymer conjugates: Effect of the method of synthesis on their structure and performance in diagnostic assays. *Bioconjugate Chemistry* 11: 795–804.
19. Wang, C. C., T. S. Seo, Z. Li, H. Ruparel, and J. Ju. 2003. Site-specific fluorescent labeling of DNA using Staudinger ligation. *Bioconjugate Chemistry* 14: 697–701.
20. Moore, J. C. 1964. Gel permeation chromatography. I. A new method for molecular weight distribution of high polymers. *Journal of Polymer Science Part A* 2: 835–843.
21. For a review article, see Rana, T. M. 2007. Illuminating the silence: Understanding the structure and function of small RNAs. *Nature Reviews. Molecular Cell Biology* 8: 23–36.

22. For a review, see Dausse, E., S. Da Rocha Gomes, and J.-J. Toulmé. 2009. Aptamers: A new class of oligonucleotides in the drug discovery pipeline? *Current Opinion in Pharmacology Anti-infectives/New Technologies* 9: 602–607.
23. Striegel, A. M. 2005. Multiple detection in size-exclusion chromatography of macromolecules. *Analytical Chemistry* 77: 104 A–113 A.
24. Winzor, D. J. 2003. Analytical exclusion chromatography. *Journal of Biochemical and Biophysical Methods* 56: 15–52.
25. Barth, H. G., B. E. Boyes, and C. Jackson. 1998. Size exclusion chromatography and related separation techniques. *Analytical Chemistry* 70: 251–278.
26. Striegel, A., W. Yau, J. Kirkland, and D. Bly. 2009. *Modern Size-exclusion Liquid Chromatography: Practice of Gel Permeation and Gel Filtration Chromatography*. New York: John Wiley.
27. Wu, C.-S., ed. 2004. *Handbook of Size Exclusion Chromatography and Related Techniques*, 2nd ed. New York: Marcel Dekker.
28. Mori, S., and H. G. Barth. 1999. *Size Exclusion Chromatography*. Berlin: Springer.
29. Styring, M. G., and A. E. Hamielec. 1989. Determination of molecular weight distribution by gel permeation chromatography. In *Determination of Molecular Weight*, A. Cooper, ed. New York: Wiley-Interscience.
30. Janča, J. 1984. *Steric Exclusion Liquid Chromatography of Polymers*. New York: Marcel Dekker.
31. Snyder, L. R., J. J. Kirkland, and J. L. Glajch. 1997. *Practical HPLC Method Development*, 2nd ed. New York: John Wiley.
32. Kaye, W., A. J. Havlik, and J. B. McDaniel. 1971. Light scattering measurements on liquids at small angles. *Polymer Letters* 9: 695–699.
33. Takagi, T. 1990. Application of low-angle laser light scattering detection in the field of biochemistry: review of recent progress. *Journal of Chromatography A* 506: 409–416.
34. Dollinger, G., B. Cunico, M. Kunitani, D. Johnson, and R. Jones. 1992. Practical on-line determination of biopolymer molecular weights by high-performance liquid chromatography with classical light-scattering detection. *Journal of Chromatography* 592: 215–228.
35. Wyatt, P. J. 1993. Light scattering and the absolute characterization of macromolecules. *Analytica Chimica Acta* 272: 1–40.
36. Hayashi, Y., H. Matsui, and T. Takagi. 1989. Membrane protein molecular weight determined by low-angle laser light-scattering photometry coupled with high-performance gel chromatography. *Methods in Enzymology* 172: 514–528.
37. Wen, J., T. Arakawa, and J. S. Philo. 1996. Size-exclusion chromatography with on-line light-scattering, absorbance, and refractive index detectors for studying proteins and their interactions. *Analytical Biochemistry* 240: 155–166.
38. Folta-Stogniew, E. 2006. Oligomeric states of proteins determined by size-exclusion chromatography coupled with light scattering, absorbance, and refractive index detectors. *Methods in Molecular Biology* 328: 97–112.
39. Haney, M. A. 2003. The differential viscometer. II. On-line viscosity detector for size-exclusion chromatography. *Journal of Applied Polymer Science* 30: 3037–3049.
40. Desmazières, B., W. Buchmann, P. Terrier, and J. Tortajada. 2008. APCI interface for LC- and SEC-MS analysis of synthetic polymers: advantages and limits. *Analytical Chemistry* 80: 783–792.
41. Kato, Y., M. Sasaki, T. Hashimoto, T. Murotsu, S. Fukushige, and K. Matsubara. 1983. Operational variables in high-performance gel filtration of DNA fragments and RNAs. *Journal of Chromatography A* 266: 341–349.
42. Li, W., and F. C. Szoka, Jr. 2007. Lipid-based nanoparticles for nucleic acid delivery. *Pharmaceutical Research* 24: 438–449.
43. Fattal, E., and G. Barratt. 2009. Nanotechnologies and controlled release systems for the delivery of antisense oligonucleotides and small interfering RNA. *British Journal of Pharmacology* 157: 179–194.
44. De Rosa, G., and M. La Rotonda. 2009. Nano and microtechnologies for the delivery of oligonucleotides with gene silencing properties. *Molecules* 14: 2801–2823.
45. Cheng, X., A. C. Harms, P. N. Goudreau, T. C. Terwilliger, and R. D. Smith. 1996. Direct measurement of oligonucleotide binding stoichiometry of gene v protein by mass spectrometry. *Proceedings of the National Academy of Sciences of the United States of America* 93: 7022–7027.
46. Sakurai, K., and S. Shinkai. 2000. Molecular recognition of adenine, cytosine, and uracil in a single-stranded RNA by a natural polysaccharide: Schizophyllan. *Journal of the American Chemical Society* 122: 4520–4521.

47. Mochizuki, S., J. Minari, and K. Sakurai. 2009. Antisense oligonucleotides delivery to antigen presenting cells by using schizophyllan. In *Materials Research Society Symposium Proceedings: Advances in Material Design for Regenerative Medicine, Drug Delivery and Targeting/Imaging*, 1140:HH05-17. V. P. Shastri et al., eds. Warrendale, PA: Materials Research Society.
48. Huang, X., R. S. McLean, and M. Zheng. 2005. High-resolution length sorting and purification of DNA-wrapped carbon nanotubes by size-exclusion chromatography. *Analytical Chemistry* 77: 6225–6228.
49. Liu, C. B., G. Z. Liu, N. Liu, Y. M. Zhang, J. He, M. Rusckowski, and D. J. Hnatowich. 2003. Radiolabeling morpholinos with ^{90}Y, ^{111}In, ^{188}Re and ^{99}Tc. *Nuclear Medicine and Biology* 30: 207–214.
50. Nicolai, T., L. Van Dijk, J. A. P. P. Van Dijk, and J. A. M. Smit. 1987. Molar mass characterization of DNA fragments by gel permeation chromatography using low-angle laser light scattering detector. *Journal of Chromatography* 389: 286–292.
51. Ellegren, H., and T. Låås. 1989. Size-exclusion chromatography of DNA restriction fragments: Fragment length determinations and a comparison with the behaviour of proteins in size-exclusion chromatography. *Journal of Chromatography A* 467: 217–226.
52. Podell, S., W. Maske, E. Ibanez, and E. Jablonski. 1991. Comparison of solution hybridization efficiencies using alkaline phosphatase-labelled and 32P-labelled oligodeoxynucleotide probes. *Molecular Cell Probes* 5: 117–124.
53. Morgan, R., and J. Celebuski. 1991. Large-scale purification of haptenated oligonucleotides using high-performance liquid chromatography. *Journal of Chromatography* 536: 85–93.
54. Guy, M., A. Olszewski, N. Monhoven, F. Namour, J. Gueant, and F. Plénat. 1998. Evaluation of coupling of cobalamin to antisense oligonucleotides by thin-layer and reversed-phase liquid chromatography. *Journal of Chromatography B: Biomedical Sciences and Applications*, 706: 149–156.
55. Bijsterbosch, M. K., E. T. Rump, R. L. De Vrueh, R. Dorland, R. Van Veghel, K. L. Tivel, E. A. Biessen, T. J. Van Berkel, and M. Manoharan. 2000. Modulation of plasma protein binding and in vivo liver cell uptake of phosphorothioate oligodeoxynucleotides by cholesterol conjugation. *Nucleic Acids Research* 28: 2717–2725.
56. Akasaka, T., K. Matsuura, and K. Kobayashi. 2001. Transformation from block-type to graft-type oligonucleotide-glycopolymer conjugates by self-organization with half-sliding complementary oligonucleotides and their lectin recognition. *Bioconjugate Chemistry* 12: 776–785.
57. Liu, G., S. Zhang, J. He, Z. Zhu, M. Rusckowski, and D. J. Hnatowich. 2002. Improving the labeling of S-acetyl NHS-MAG_3-conjugated morpholino oligomers. *Bioconjugate Chemistry* 13: 893–897.
58. Miyoshi, D., A. Nakao, and N. Sugimoto. 2003. Structural transition from antiparallel to parallel G-quadruplex of $d(G_4T_4G_4)$ induced by Ca^{2+}. *Nucleic Acids Research* 31: 1156–1163.
59. Ohya, Y., T. Nishi, T. Nohori, S. Jo, K. Ohta, K. Jozuka, and T. Ouchi. 2007. Construction of supramolecular assemblies and self-organized structures using oligo-DNAs. *Nucleic Acids Symposium Series* 51: 37–38.
60. Urlaub, H., E. Kühn-Hölsken, and R. Lührmann. 2008. Analyzing RNA–protein crosslinking sites in unlabeled ribonucleoprotein complexes by mass spectrometry. *Methods in Molecular Biology* 488: 221–245.

4 Analysis of Oligonucleotides by Liquid Chromatography and Mass Spectrometry

Soheil Pourshahian
Girindus America, Inc.
Sean M. McCarthy
Waters Corporation

CONTENTS

4.1 INTRODUCTION

Solid phase synthesis of oligonucleotides requires stepwise addition of nucleotides to a growing chain. Although this process is highly efficient, failure sequences and process related modifications and impurities are often present in the final product. Depending on the application, particularly when these sequences are intended for use in therapeutic applications, identification of impurities is important for the optimization of the synthetic process. With respect to therapeutic applications, impurities can have toxicological implications; therefore, their identification and minimization in therapeutic oligonucleotides is important.[1]

Matrix assisted laser desorption ionization (MALDI) and electrospray ionization (ESI) mass spectrometry have been used to generate gas phase ions from oligonucleotides and nucleic acids and have made molecular weight determination of nucleic acids routine.[2,3] Hyphenation of liquid chromatography (LC) and mass spectrometry (MS) greatly enhances the power of mass spectrometry for the analysis of simple and complex mixtures. Liquid chromatography has been used with both MALDI[4] and ESI; however, ESI produces gas phase ions from a flowing solution and is more compatible for the direct interfacing with LC without the need for fraction collection and spotting on a MALDI target plate.

Commonly used chromatographic modes of separation for the analysis of oligonucleotides are size exclusion, anion-exchange, and ion-pair reversed phase. Among these, ion-pair reversed phase liquid chromatography (IP-RPLC) and electrospray mass spectrometry (ESI-MS) have been shown to be the most suitable for the direct interfacing of liquid chromatography and mass spectrometry.

The purpose of this chapter is to give the reader an understanding of how mass spectrometry, coupled with liquid chromatography, can allow the practitioner to determine the identity and quantity of oligonucleotides. We will discuss the application of MS to determine the identity of impurities and target molecules, and some of the added complexity introduced by the ionization process itself. This chapter will focus on the use of IP-RPLC–ESI-MS for oligonucleotide characterization. The experimental issues involved in impurity analysis will be discussed, and applications highlighting the use of this technique will be presented in this chapter.

4.2 BACKGROUND

ESI is a commonly used ionization technique for the analysis of oligonucleotides.[5–8] This is largely due to the ability to interface a liquid chromatograph to the ESI source and its soft ionization, which limits unwanted fragmentation of oligonucleotides. Other MS platforms have been utilized

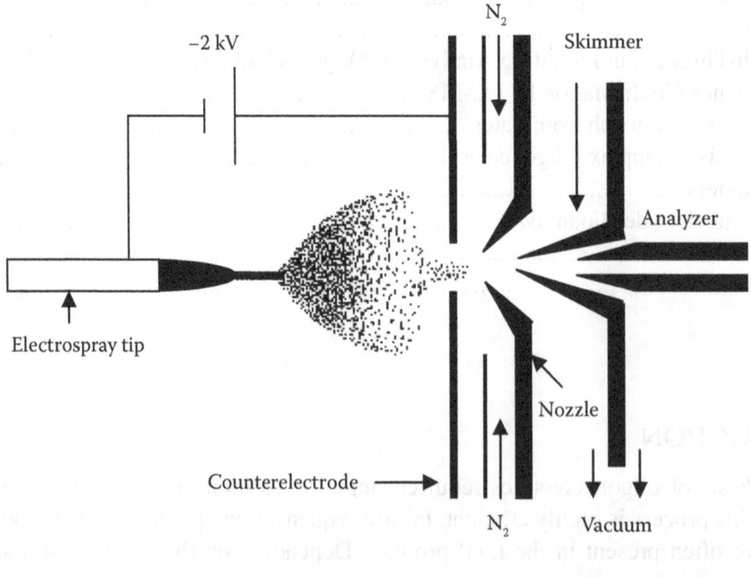

FIGURE 4.1 Schematic of an ESI source.

for oligonucleotide analysis, particularly matrix assisted laser desorption ionization MS (MALDI-MS).[2,4,9,10] While MALDI-MS can process samples in a high throughput manner, sample preparation is needed prior to analysis. Although this process can be automated, the ability of LC-MS methods to analyze crude synthetic samples without the need for any sample preparation has dramatically increased its popularity.

ESI-MS analysis is accomplished by the generation of gas phase ions in atmospheric pressure by applying a high voltage to a capillary tip and a nearby counter electrode (Figure 4.1). While ESI sources can operate in positive and negative ion modes, oligonucleotides are generally analyzed in negative-ion mode owing to their negatively charged phosphate backbone. The required voltage for efficient ESI depends on the solvent properties such as surface tension, viscosity, and volatility. The electrospray process begins with the liquid at the end of the capillary elongating under the pressure of accumulated charges and forming a conical shape called a "Taylor cone." Gas phase ions are generated through production of charged droplets from this cone. These droplets shrink owing to solvent evaporation, which increases coulombic repulsion within the droplet. The droplets then undergo a cascade of ruptures yielding smaller and smaller droplets, which, ultimately leads ions entering the gas phase.[11]

Gas phase ions from large molecules, such as oligonucleotides, can carry multiple charges. This is due to the presence of several ionizable groups. In the case of oligonucleotides, the main ionizable groups are the hydroxyl in the case of phosphodiester, or thiol groups in the case of phosphorothioates of the phosphate linkages.

As a result of multiple ionizable groups, many charge states are present for the same molecule. Masses are displayed in the spectrum as a ratio of the mass to charge (m/z), where m is the mass of the species and z is the charge. The presence of multiply charged ions allows for the detection of high molecular weight compounds within a relatively narrow mass range, which is much lower than the actual mass of the analyte. Generally, more than one charge state is present for a single compound in the gas phase, although the number of charge states can be controlled experimentally and will be discussed later in this chapter. The presence of multiple charge states adds an additional level of complexity to the MS data, but software has been developed to deconvolute MS spectra of multiply charged species to their zero charge mass.

Analysis of nucleic acids by mass spectrometry has been the subject of several reviews.[12–20] There are many factors that can affect the ionization, and therefore the intensity, of oligonucleotides signals. These factors include viscosity, surface tension, volatility, pH, and ionic strength of the electrosprayed solution. Some of these factors can also affect the chromatographic separation, and often times the optimal conditions for mass spectrometric detection do not provide the best chromatographic separation and vice versa. We will review the main considerations needed to develop a robust method for oligonucleotide analysis and then give specific examples targeted toward the main application areas for oligonucleotide analysis.

4.3 CHROMATOGRAPHIC CONSIDERATIONS FOR MS ANALYSIS

Retention of oligonucleotides in RP separations is largely governed by solvatophobic interactions between the hydrophobic bases of the nucleic acids and the stationary phase, generally C18. Elution is accomplished by using a gradient of increased organic solvent, typically acetonitrile or methanol. For RP separations, low ionic strength solutions are used, generally composed of ammonium acetate and alkylammonium salts,[18,21] although the use of unbuffered amines have also been reported.[22] For this reason, RP-LC is fully compatible with MS. The difficulty in using this method is not related to its MS compatibility but rather results from its limited chromatographic resolution of oligonucleotides. For unmodified oligonucleotides, hydrophobicity does not correlate with length, which can lead to co-elutions and retention reversal with shorter sequences eluting after longer ones. Reverse phase separations can yield good separations for single-stranded species, which are modified with hydrophobic groups such as trityl and large aliphatic groups. In these separations, generally, the elution order is reversed compared to separation of the same unmodified oligonucleotides using an IP-RP method.[21] Reversed elution in RP methods with modified oligonucleotides is due to the increased relative hydrophobic character of the molecule as the hydrophilic oligonucleotide decreases in length.

4.3.1 Effect of the Solution pH

Ionization efficiency and charge distribution are affected by the pH of the electrosprayed solution, which can also affect the chromatographic separation and oligonucleotide stability. In general, higher charge states and greater signal intensity are observed when oligonucleotides are electrosprayed from solutions of high pH.[23,24] The effect of solution pH on charge state distribution can be explained by acid-base equilibrium in solution and gas phase. At low pHs it is more likely for oligonucleotides to be protonated and therefore be detected at lower charge states (higher *m*/*z*).

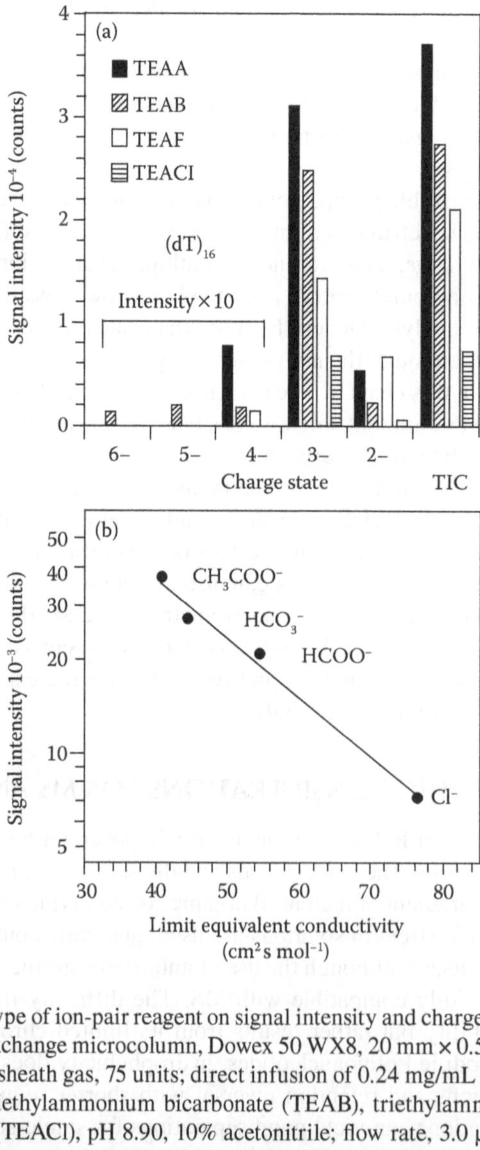

FIGURE 4.2 Influence of type of ion-pair reagent on signal intensity and charge state distribution in ESI-MS of oligonucleotides: cation-exchange microcolumn, Dowex 50 WX8, 20 mm × 0.50 mm; scan, 200–2500 amu; electrospray voltage, 4.0 kV; sheath gas, 75 units; direct infusion of 0.24 mg/mL (dT)16 in 50 mM triethylammonium acetate (TEAA), triethylammonium bicarbonate (TEAB), triethylammonium formate (TEAF), or triethylammonium chloride (TEACl), pH 8.90, 10% acetonitrile; flow rate, 3.0 μL/min. (From Huber, C. G., and A. Krajete, *Analytical Chemistry*, 71, 3730–3739, 1999. With permission.)

The effect of pH of a triethylammonium acetate (TEAA) solution on oligonucleotide mass spectral signal intensity and charge state distribution has been investigated by direct infusion of a polyT oligonucleotide in 50 mM TEAA and 10% acetonitrile.[25] Variation of the pH of TEAA from 6.8 to 8.4 through the addition of acetic acid exhibits no significant effect on signal intensity or charge state distribution of the oligonucleotide. Direct infusion of the oligonucleotide in a solution at pH values higher than 9.00 shows a significant improvement in signal intensities. This change in signal intensity can be explained through the amount of acetic acid needed to titrate triethylamine to a lower pH. At higher pHs the amount of acetate ions present in the solution is less. The acetate anions compete with oligonucleotides for ionization and reduce the signal intensity. Titration of triethylamine with acetic acid exhibits the largest shift between pH 6.00 and 8.50, which is close to the equivalence point, and therefore a small amount of acetic acid is needed to change the pH in this range. This would also explain the insignificant effect of the pH change from 6.8 to 8.4 on mass spectral signal intensity.

4.3.2 Ion-Pairing Reagent

The ion-pairing system used for oligonucleotides affects both the separation and ionization of oligonucleotides. Generally, alkyl amines are used due to their affinity for the stationary phase. Alkyl amines are buffered with an acid and a variety of acids have been used for this purpose including acetic acid, formic acid, carbonic acid, and 1,1,1,3,3,3-hexafluroisopropanol (HFIP) among others.[18] For most applications the appropriate pH is between 6 and 8. The reasons for this are that this pH range results in positively charged ammonium ions that adsorb to the stationary phase of a RP column, generally C18, and interact with the charged phosphate backbone of oligonucleotides.

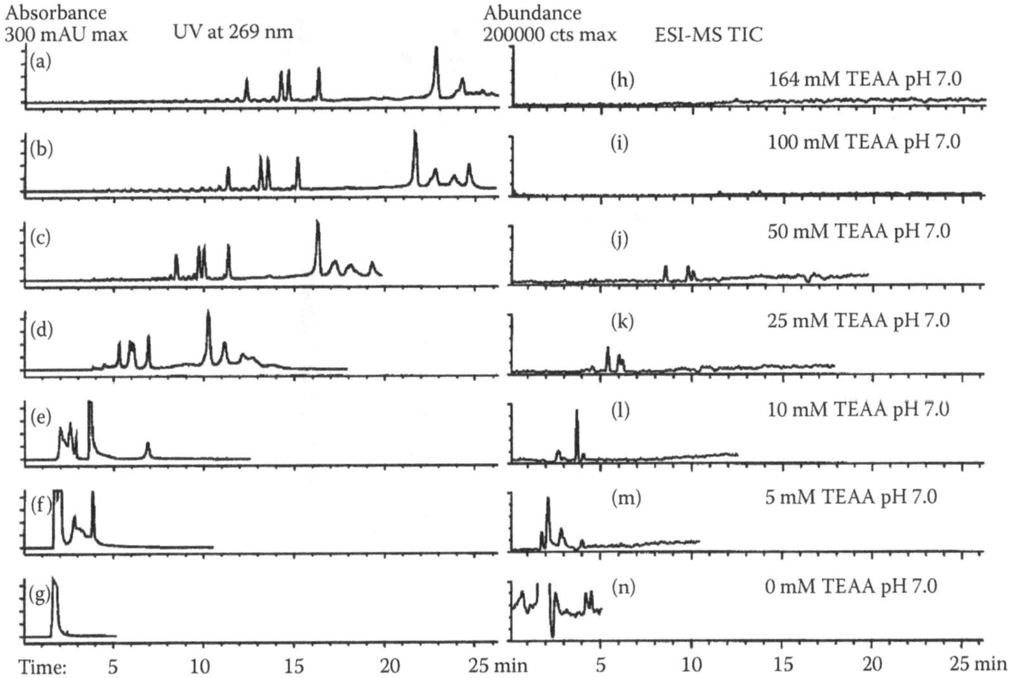

FIGURE 4.3 Effect of TEAA at pH 7 on electrospray signal intensity. Poly T mix (15, 19, 20, 25, 74, 75) 100 pmol/component. LC gradient 10–20% acetonitrile/30 min at 0.2 mL/min. 35°C. (a–g) UV detection at 269 nm. (h–n) ESI total ion current. (a,h) 164 mM TEAA; (b,i) 100 mM TEAA; (c,j) 50 mM TEAA; (d,k) 25 mM TEAA; (e,l) 10 mM TEAA; (f,m) 5 mM TEAA; (g,n) 0 mM TEAA. (Reprinted with permission from Apffel, A., et al., *Analytical Chemistry*, 69, 1320–1325. Copyright 1997, American Chemical Society.)

The IP-RP strategy is not general because modifications, such as large hydrophobic groups, and the nucleobases themselves, in the case of single-stranded species, interact with the stationary phase, which will alter their chromatographic behavior as described above. Details of these effects are described in Chapter 2, and the reader is advised to consider both the chromatographic resolution and MS compatibility when developing a method for oligonucleotide analysis. Here we discuss selected mobile phase compositions as they relate to MS sensitivity and provide some guidance about their applicability to oligonucleotide separations. Details about specific applications will be discussed in the next section.

A commonly used IP system, TEAA, yields low signal in ESI-MS.[23,26,27] The main reason for low signal lies with the use of acetate rather than as a result of TEA.[21,23,28] During the electrospray process there is preferential evaporation of triethylamine from the electrospray microdroplets leaving an excess of acetic acid in the droplets due to acetic acids lower volatility.[25] The use of other acids does not lead to significantly increased signal in the MS. It has been shown that MS signal suffers from a variety of buffering agents including acetic acid (AA), formic acid (FA), HCl, carbonic acid

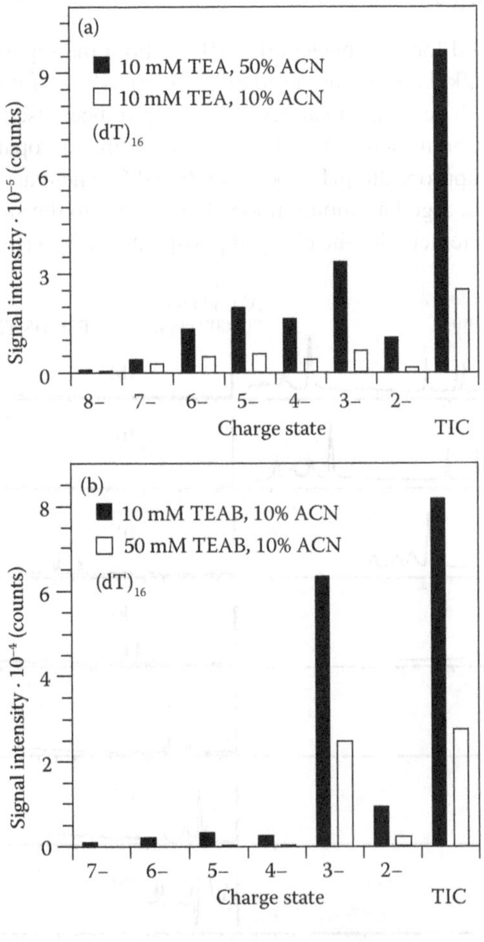

FIGURE 4.4 Influence of (a) acetonitrile concentration and (b) ion-pair reagent concentration on signal intensity and charge state distribution in ESI-MS of oligonucleotides: cation-exchange microcolumn, Dowex 50 WX8, 20 mm × 0.50 mm; scan, 200–2500 amu; electrospray voltage, 4.0 kV; sheath gas, 75 units; direct infusion of 0.24 mg/mL (dT)16 in (a) 10 mM triethylamine, pH 11.30, 10 and 50% acetonitrile, respectively, and (b) 10 and 50 mM TEAB, pH 8.90, respectively, 10% acetonitrile; flow rate, 3.0 μL/min. (Reprinted with permission from Huber, C. G., and A. Krajete, *Analytical Chemistry*, 71, 3730–3739. Copyright 1999, American Chemical Society.)

(B), and phosphoric acid (PA).[24,25] As shown in Figure 4.2, AA provides the greatest signal intensity of the various acids, and the amount of signal suppression correlates well with limit equivalent conductivity of the counterion in the ion-paring reagent. This finding suggests that the volatility of the ion-pairing reagent plays some role on the signal intensity but that other factors also contribute to signal intensity, such as conductivity. A main factor limiting the use of PA is the formation of clusters of these ions in the gas phase, which complicates the MS spectrum. For these reasons the most widely used acid for oligonucleotide MS analysis has been AA.

It has been suggested that the suppression of oligonucleotide signal upon addition of any acid is due to the competition of anions for ionization.[24,26] This effect can be minimized by reducing the concentration of the buffering acid. As shown in Figure 4.3, MS signal intensity increases with decreased TEAA concentration, but this also has a deleterious effect on the chromatographic resolution that is often undesired as chromatographic resolution of impurities is often necessary. Also shown in Figure 4.3, MS signal for shorter oligonucleotides, with a correspondingly lower mass, becomes evident more rapidly at higher concentrations of TEAA than longer sequences, but that the UV intensity remains unaffected. This is largely due to their ionization efficiency, where shorter molecules with correspondingly lower mass are more easily ionized. For the analysis of short (less than 15-mer) oligonucleotides the use of TEAA may be appropriate. Other trialkylammonium salts such as diisopropyl-, tributyl-, dimethylbutyl, and hexyl-ammonium acetate IP systems also have a negative effect on the mass spectral signal[29,30] but show a similar trend when the concentration of acetate is decreased, further confirming that signal suppression is largely due to the buffering acid rather than the amine. A benefit that these more hydrophobic amines offer compared to TEA is that they retain their chromatographic resolution at much lower concentrations than TEA, which is less hydrophobic and therefore a less efficient IP agent.

Another interesting phenomenon occurs when using acetate buffered amines for oligonucleotide analysis. A dramatic reduction in the number of charge states results from the addition of acid compared to the basic conditions in which an amine without pH adjustment is used.[24,25,31] Figure 4.4 shows the influence of pH adjustment of TEA with carbon dioxide gas to form TEAB on the charge state distribution. The bulk of the population of ions resides in the 3- charge state in Figure 4.4b compared to a more uniform distribution in the absence of acid in Figure 4.4a. A reduction in the number of charge states (charge state collapse), coupled with sufficient amount of sample loaded on the column to overcome the signal suppression from acetate, can expedite the identification of impurities present in an oligonucleotide sample. This technique is generally used if sufficient chromatographic resolution of impurities in the sample cannot be obtained. It is also imperative that the electrospray conditions, such as cone voltage, capillary voltage, and temperature settings among others, be adjusted to eliminate the contribution of modifications, such as depurination, introduced by the electrospray process itself, which can complicate the data further. The appropriate conditions are variable and largely depend on the sequence and modifications on the oligonucleotide.

Electrospray ionization is influenced by the physical property of the solvent. The onset of electrospray depends on the surface tension of the solvent. Viscosity of the solvent, on the other hand, affects electrospray droplet size and solvent evaporation from droplets. The concentration of organic solvent in the electrosprayed solution affects MS signal intensity as a consequence of changing the surface tension and evaporation rate of solvent from the electrosprayed droplets. The concentration of organic solvent in the electrosprayed solution also has a dramatic effect on the total ion current, and to a lesser extent, charge state reduction.[25] An increased concentration of organic solvent leads to an increase in the total ion current (Figure 4.4a). While beneficial for the electrospray process and MS signal, increased organic concentration is largely incompatible with IP RPLC separation of oligonucleotides as the concentration needed to elute an oligonucleotide is predetermined by the choice of IP agent and column choice. A solution for this problem is post-column addition of a sheath liquid to the LC eluent before entering the mass spectrometer.[32] A variety of different sheath liquids including methanol, 2-propanol, acetonitrile, hexafluoro-2-propanol, triethylamine, 10 mM triethylamine in acetonitrile, and 400 mM hexafluoro-2-propanol in methanol have been tested,

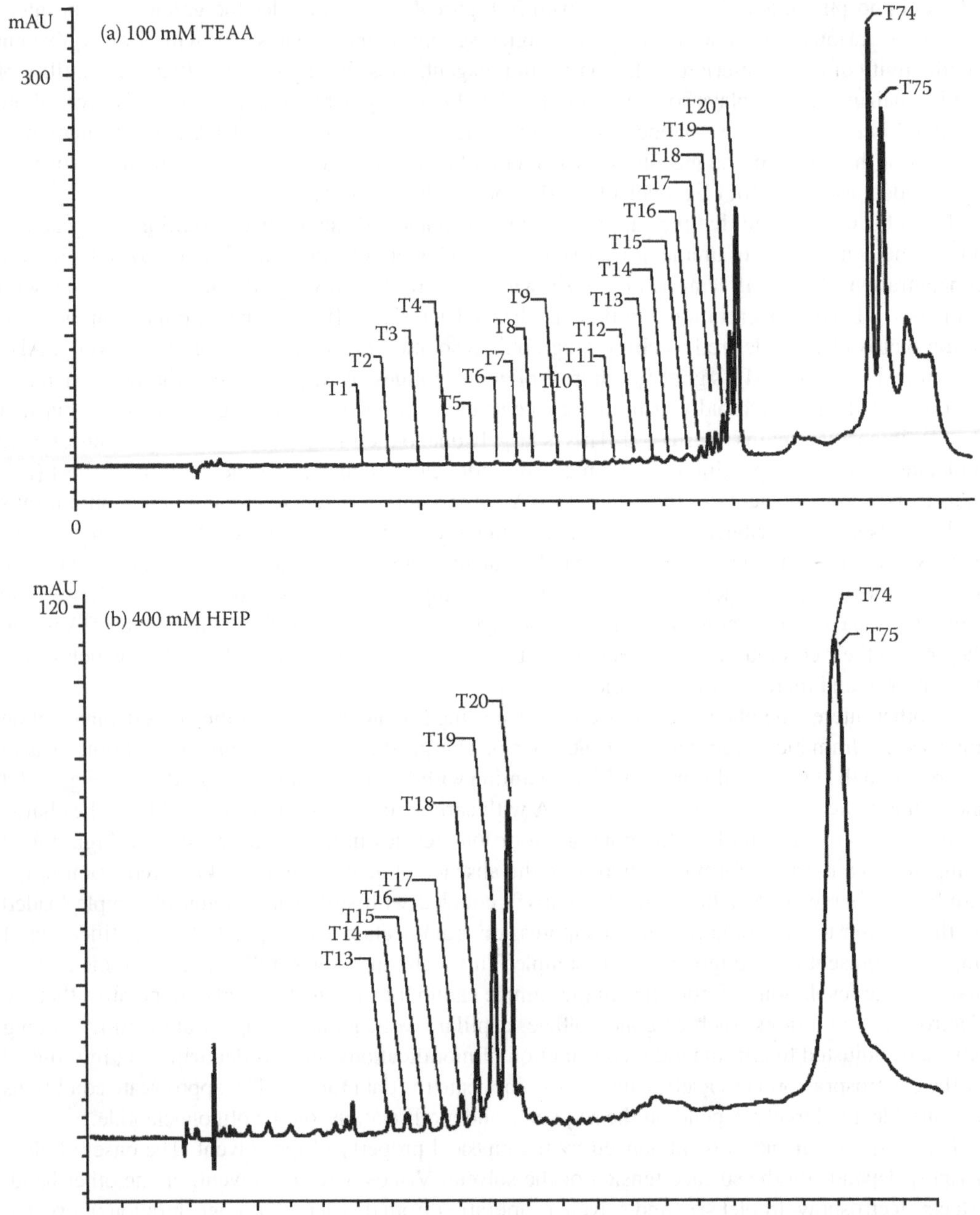

FIGURE 4.5 Typical separation of oligonucleotides using TEAA- and HFIP-based mobile phases. Synthetic mixture of PolyT oligonucleotides (19, 20, 74, 75) at 50 pmol/μL each component. UV detection at 269 nm. (a) 200 mM TEAA, pH 7.0, LC gradient 12–18% acetonitrile/30 min at 200 μL/min, 35°C. (b) 400 mM HFIP adjusted to pH 7.0 with TEA, LC gradient 25–40% methanol/30 min at 200 μL/min, 50°C. (Reprinted with permission from Apffel, A. et al., *Analytical Chemistry*, 69, 1320–1325. Copyright 1997, American Chemical Society.)

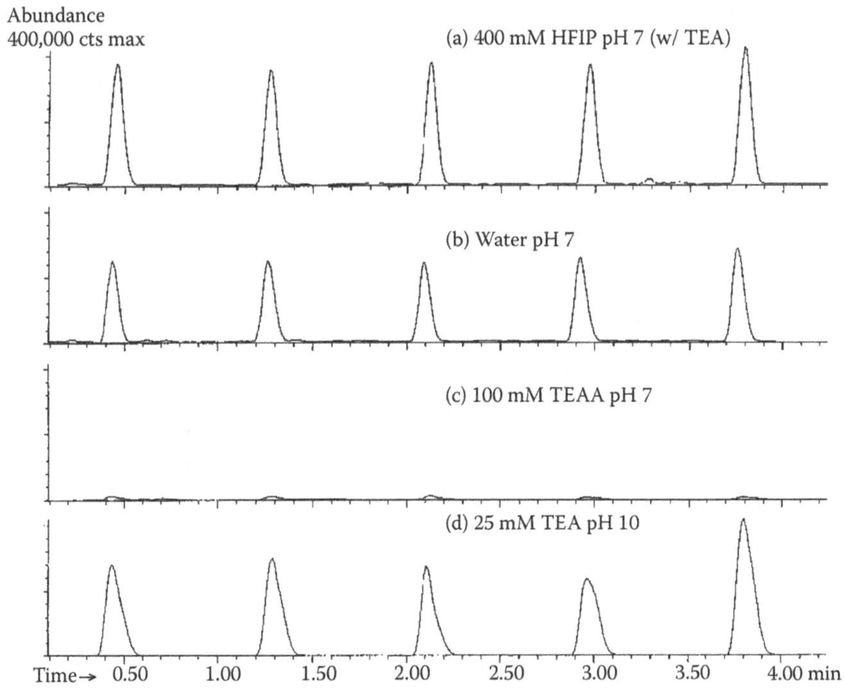

FIGURE 4.6 Flow injection comparison of ESI solvent additives: 50 pmol flow injections of dT20 at 200 μL/min. ESI total ion current. (a) 400 mM HFIP, pH 7.0 (adjusted with TEA); (b) water, pH 7.0; (c) 100 mM TEAA, pH 7.0; (d) 25 mM TEA, pH 10. (Reprinted with permission from Apffel, A., et al., *Analytical Chemistry*, 69, 1320–1325. Copyright 1997, American Chemical Society.)

acetonitrile being the most efficient in improving the signal intensity.[25,33] LC eluent that is mixed with a sheath solution of acetonitrile containing 0.1 M imidazole improves the ESI-MS performance without compromising the LC separation. This method allows the use of a 0.1 M TEAA buffer that otherwise would suppress the ionization to the point that no analyte signal would be detected.[32]

Perhaps the most widely used IP system for MS analysis of oligonucleotides is composed of TEA-HFIP. This IP system has been demonstrated to provide good sensitivity in the mass spectrometer and very good chromatographic resolution of a variety of oligonucleotide sequences.[26,28] The improvement in chromatographic resolution has been attributed to the inherent hydrophobicity imparted by HFIP. This hydrophobic character more efficiently saturates the stationary phase of the column with TEA, leading to higher adsorbed concentration. A higher concentration of TEA on the stationary phase, in turn, leads to greater IP efficiency and better chromatographic resolution for short to moderate length oligonucleotides (Figure 4.5).[18,26,28] Increased chromatographic resolution is only found for oligonucleotide sequences up to 40-mer when compared to acetate buffered mobile phases. As shown in Figure 4.6, MS sensitivity is also dramatically improved compared to TEAA and is better than TEA or water alone, regardless of oligonucleotide length. This improvement in MS sensitivity is attributed to several factors. First HFIP is more volatile than acetate and therefore readily evaporates from the droplets produced during the electrospray process. Second, the pKa of HFIP is much higher than acetic acid, so that in pH 7 buffered solutions a greater portion of HFIP can evaporate because it is incompletely dissociated.

TEA-HFIP provides better resolution and improved MS sensitivity compared to many acetate buffered systems, particularly TEAA, owing to its hydrophobicity and volatility respectively. There are reports on acetate buffered IP systems, such as tetrabutylammonium acetate (TBuAA) and hexylammonium acetate (HAA) at low concentration, which provide good chromatographic resolution, but MS sensitivity still remains lower than with HFIP systems as discussed previously. The HFIP system has been utilized for quantification and characterization of oligonucleotides to identify metabolites, failure sequences, and remaining protecting groups not removed following synthesis.[34] This system is particularly suited to this application as the abundance of contaminant species can be much lower than the target sequence and the added MS sensitivity is necessary in some instances.

4.3.3 Desalting

A major difficulty for the analysis of oligonucleotides comes from their propensity to form adducted species in the gas phase, largely with sodium and potassium, although other adducts are possible.[18] When using IP systems, it is also possible to find adducts resulting from the IP system itself. Adduct formation results from the substitution of protons with cationic species on the negatively charged sugar-phosphate backbone of the oligonucleotide.

It has been reported that precipitation of oligonucleotides from buffered solutions, such as ammonium acetate either with or without the addition of additional chelation agents, can decrease adduct

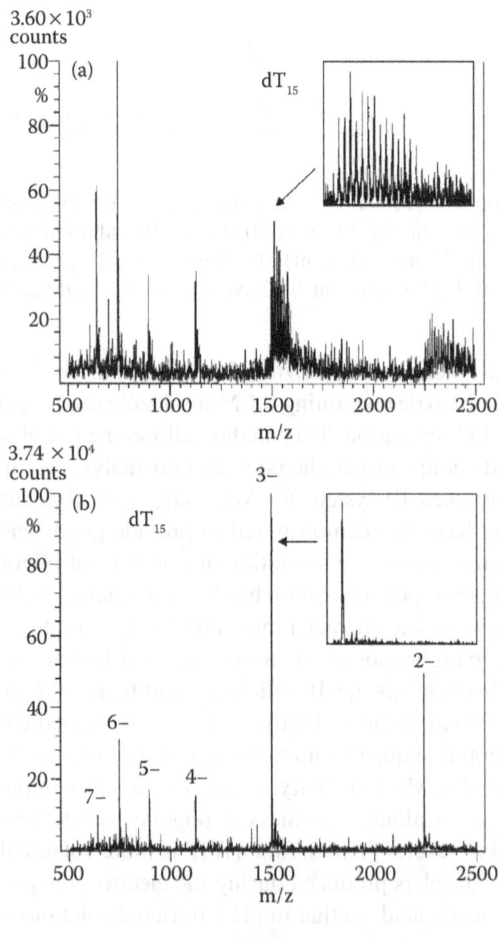

FIGURE 4.7 Mass spectra of dT_{15} (a) without and (b) with on-line cation-exchange sample preparation. Inset, expanded view of the 3- charge state. Cation-exchange microcolumn, ROMP-(COOH)2, 45 mm × 0.8 mm; scan, 500–2500 amu in 5 s; flow injection into 10 mM TEA in 50% water-50% acetonitrile (v/v), 3 μL/min; injection volume, 5 μL; sample, 400 pmol of $dT_{15}Na_{15}$. (Reprinted with permission from Huber, C. G., and M. R. Buchmeiser, *Analytical Chemistry*, 70, 5288–5295. Copyright 1998, American Chemical Society.)

formation from alkali metals significantly.[7,35,36] Cold ethanol precipitation of oligonucleotides in the presence of ammonium acetate is a simple and efficient way of sample desalting. This method is time consuming, largely molecular weight dependent, and can produce variable yields that make quantitation difficult to impossible. Microdialysis has also been used for desalting.[37–43] Microdialysis provides a high-efficiency cleanup of the sample by exchanging nonvolatile cations with ammonium ions. While this approach works well for qualitative work, it is not quantitative and is time consuming because it requires several steps prior to analysis.

Solid phase extraction has also been reported for sample desalting routinely.[44,45] These methods are often used for high-throughput applications where the practitioner must analyze a large number of samples per day. Rapid analysis of oligonucleotides is often required for high-throughput applications such as synthetic oligonucleotide quality control analysis and genotyping. The large volume of samples, often greater than 2000 per day, makes a rapid and robust method necessary. LC-MS is particularly useful for this application because it requires minimal to no sample preparation and is amenable to automation to limit sample handling.

Either in conjunction with, or instead of, physical removal of excess cationic contaminants, adduct formation can be limited by addition of basic modifiers to the chromatographic eluent, or by passing the eluent through a cation-exchange resin, either on-line or off-line, which serves to physically remove contaminant cations.[5,6,21,46–48] Post-column addition of additives have been favored largely owing to the fact that the use of ion-exchange resins deteriorates the chromatographic resolution and they can quickly become saturated if the concentration of metal ions is high.[44] Additives introduced into the chromatographic eluent are robust and allow for more quantitative data because the analyte is used without pre-purification steps and is not detrimental to the chromatographic performance.

A variety of additives have been demonstrated to decrease adduct formation including triethylamine (TEA), imidizole, and ammonium acetate.[5,6] It has been proposed that the mechanism of action for additives is that they preferentially populate the negatively charged sites on the oligonucleotide backbone and efficiently displace metal ions. During the ionization process, the added base is efficiently removed from the analyte because of its volatility, resulting in less adducted species in the mass spectrum. Regardless of the method used for cation removal, the result is increased signal to noise in the obtained spectrum (Figure 4.7).

Removal of excess salt is imperative to obtain sufficient MS signal. Hyphenation of LC with MS is effective at removing low levels of salt because the oligonucleotide is retained on the column while the salt elutes readily. In many cases it is advantageous to desalt samples online to prevent loss of low molecular weight species, which may be relevant to determine the impurity profile of the sample. Samples with high salt concentration require sample desalting prior to analysis,[47,49] but it is advantageous to divert the eluent to waste to prevent contamination of the MS source. It is also imperative that the solutions used for mobile phase preparation are also free from salt. It is highly recommended that the practitioner use deionized water instead of bottled HPLC grade water as the latter often contains a significant amount of salt.

4.4 MOLECULAR WEIGHT DETERMINATION

Determination of the molecular weight of an oligonucleotide is a common application of mass spectrometry. A typical mass spectrum in negative ion mode shows a series of peaks, each representing a multiply charged ion of the intact molecule that has lost protons from the phosphodiester groups (Figure 4.8). The molecular weight of the molecule can be determined from the two adjacent multiply charged signals.

Consider an oligonucleotide with the molecular weight M (Da) that is analyzed in negative ion mode. For an ion with z_1 charges whose determined mass to charge is m_1 (Da) we have

$$z_1 \times m_1 = M - z_1$$

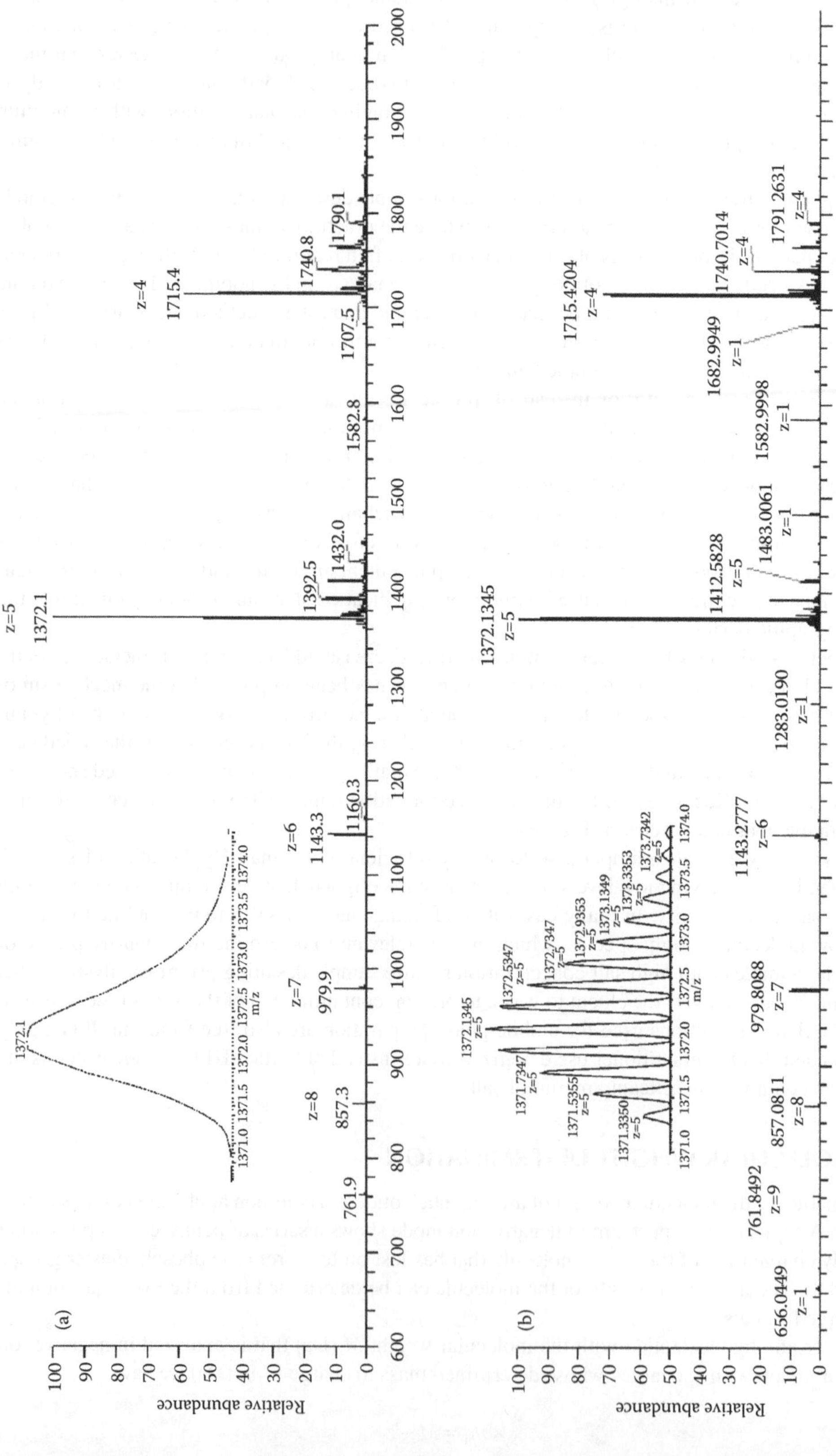

FIGURE 4.8 Mass spectrum of an oligonucleotide obtained in (a) low and (b) high resolution. (z: charge state)

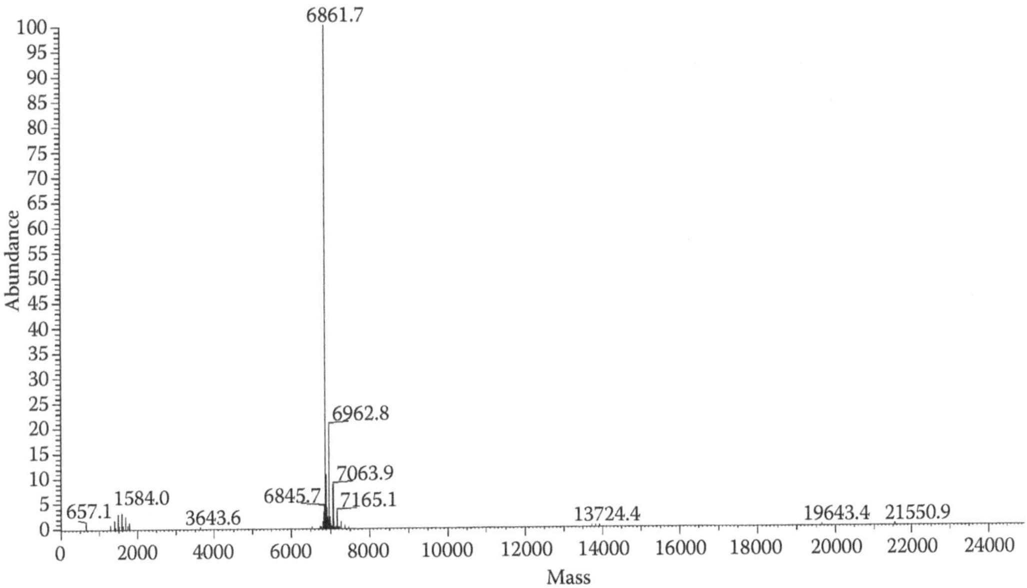

FIGURE 4.9 Deconvoluted mass spectrum in Figure 4.8b.

For an adjacent charge state with a higher *m/z* we have

$$(z_1 - 1) \times m_2 = M - (z_1 - 1)$$

From these two equations we can calculate z_1:

$$z_1 = \frac{(m_2 + 1)}{(m_2 - m_1)} \text{ and } M = (z_1 \times m_1) + z_1$$

A variety of algorithms such as MaxEnt and ZNova have been developed to deconvolute mass spectra of multiply charged species to their zero charge mass (Figure 4.9). There is also software, such as ProMass, that automates the deconvolution procedure and annotates data to minimize the need for manual data interpretation.

Mass accuracy of the determined molecular weight depends on the mass analyzer and can be in the low parts per million range for high-resolution instruments. Depending on the resolution of the MS, one obtains an average or a monoisotopic mass. Average mass is the weighted average of the natural isotopes for atomic mass of each element in the molecule. The molecular weights detected using low-resolution mass analyzers is the average mass (since the determined molecular weight is the apex of the peak rather than weighted average, it is not the average mass but very close to the average mass). Monoisotopic mass is calculated using the exact mass of the predominant isotope of each element. High-resolution mass analyzers are able to detect isotopic distributions hidden under the average mass signal. Figure 4.8 shows the mass spectrum of an oligonucleotide obtained using a low-resolution (Figure 4.8a) and high-resolution (Figure 4.8b) instrument. The -5 charge state at 1372.1 Da in the inset (Figure 4.8a) does not show the isotopic distribution.

4.5 QUANTITATIVE LC-MS

LC-MS is a powerful tool for quantification of analytes in complex mixtures. Detected molecular weight adds another layer of specificity to the analysis because co-eluting analytes can be further

TABLE 4.1
Average and Monoisotopic Mass Differences Relative to the Full Length Product for Common Impurities[a]

Mass Difference (Average)	Mass Difference (Monoisotopic)	Problem	Affected Locations
+96.05	+ 95.943	+PS	Backbone
+ 79.98	+ 79.966	+PO	Backbone
−16.07	−15.977	PS-PO conversion	Backbone
−117.12	−117.044	Depurination	A base
−135.13	−135.054	Depurination	A base
−133.11	−133.039	Depurination	G base
−151.13	−151.049	Depurination	G base
−0.98	−0.984	Deamination	C, A, G bases
+14.03	+14.016	Reaction with methylamine	C base
+42.04	+42.011	+Acetyl	C base
+53.06	+53.027	+Cyanoethyl	Backbone
+70.08	+70.042	+isobutyryl	G base
+147.39	+145.909	+Chloral	Backbone
+114.26	+114.086	+tert-butyl dimethylsilyl	rA, rG, rC, U
+104.11	+104.026	+benzoyl	A base
+80.08	+80.026	Modified C	C base
+302.37	+302.131	+DMT	5′O in all nucleotides
−52.03	−51.995	Depyrimidation	U base
−94.07	−94.016	Depyrimidation	U base
−313.21	−313.058	-dA (PO)	A
−329.21	−329.053	-dG (PO)	G
−289.18	−289.046	-dC (PO)	C
−304.19	−304.046	-dT (PO)	T
−329.21	−329.053	-rA (PO)	rA
−345.21	−345.047	-rG (PO)	rG
−305.18	−305.041	-rC (PO)	rC
−306.17	−306.025	-U (PO)	U
−343.23	−343.068	-2′O-Me-rA (PO)	2′O-Me-rA
−359.23	−359.063	-2′O-Me-rG (PO)	2′O-Me-rG
−319.21	−319.057	-2′O-Me-rC (PO)	2′O-Me-rC
−320.19	−320.041	-2′O-Me-U (PO)	2′O-Me-rU
−331.20	−331.048	-2′F-dA (PO)	2′O-Me-dA
−347.20	−347.043	-2′F-dG (PO)	2′O-Me-dG
−307.17	−307.037	-2′F-dC (PO)	2′O-Me-dC
−308.16	−308.021	-2′F-dU (PO)	2′O-Me-dU
−341.22	−341.053	-LNA-A (PO)	LNA-A
−357.22	−357.047	-LNA-G (PO)	LNA-G
−317.19	−317.041	-LNA-C (PO)	LNA-C
−318.18	−318.025	-LNA-U (PO)	LNA-U

[a] PO, phosphate; PS, phosphothioate; dA, deoxyadenosine; rA, adenosine; 2′F-dA (PO), 2′fluoro-deoxyadenosine phosphate; DMTr, dimethoxytrityl; Me, methyl; LNA, locked nucleic acid.

differentiated by their mass spectral signal. In quantitative analysis the intensity of MS signal is correlated with the quantity of the compound in the sample. Integration of ion chromatograms after normalization gives the percentage of each compound present in the sample in a similar way that one can get area percent of each compound based on a UV chromatogram. This type of relative quantification is done under the assumption that the ionization efficiency of all of the compounds being quantitatively analyzed is the same. On the basis of this assumption, equal concentration of these compounds will generate the same ion intensity as the mass spectrum, but this assumption is limited. Ionization efficiency depends heavily on the hydrophobicity and presence of ionizable groups on the compound. Quantification by MS is often less preferred than UV based quantification but may be necessary if sufficient chromatographic resolution is not possible. Absolute quantification requires a standard sample of the analyte and is done through calibration curve or isotope dilution.

4.6 COMMON IMPURITIES

Several different processes related impurities are generated during the synthesis of oligonucleotides. Most synthetic methods used currently achieve yields of 98–99+% per synthetic cycle, so impurity levels increase as oligonucleotide length increases. These impurities include shortmers, longmers, incomplete removal of protecting groups, and impurities generated as a result of acid or base treatment during synthesis. Degradation of the oligonucleotides after synthesis can generate some of the impurities produced during synthesis as well as new impurities, but these effects can be reduced significantly by proper storage.

4.6.1 SHORTMERS

As mentioned, the yield of stepwise addition of nucleotides to the growing chain of oligonucleotide during solid phase synthesis is not one hundred percent, and yields can vary dramatically depending on sequence and modifications, and as a result failure sequences and shortmers are present in the final product. Shortmers occur because of failure to couple the next nucleotide or incomplete detrytilation and capping.[50,51] The (n-1) failure sequence with "n" representing the number of nucleotides in the desired full length product, results from failure to add a nucleotide at a single location in the sequence. These (n-1) mer impurities can potentially comprise all possible internal and terminal single base deletions.[52,53] Most oligonucleotides are synthesized from 3′ to 5′ end with the first nucleotide attached to a solid support. In the case of 3′ to 5′ synthesis, resulting failure sequences are commonly species that have nucleotides missing from the 5′ end of the full length product. The (n-1) sequence can also be generated as a result of the incoming phosphoramidite reacting with functional groups on the support instead of the 5′-hydroxyl of the first nucleoside on the solid support. In this instance, the oligonucleotide chain would continue growing during subsequent elongation cycles and eventually, after cleavage and deprotection, will produce a 3′ (n-1) sequence with a phosphate group on the 3′ end. The terminal phosphate can be removed during the synthesis, producing a 3′ (n-1) without a terminal phosphate.

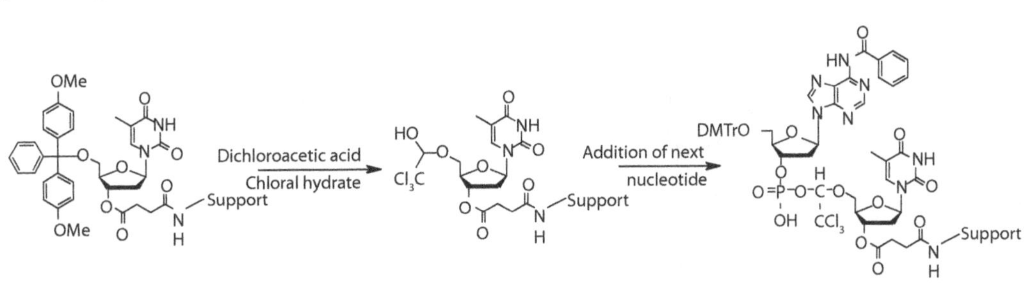

FIGURE 4.10 Incorporation of chloral in the oligonucleotide.

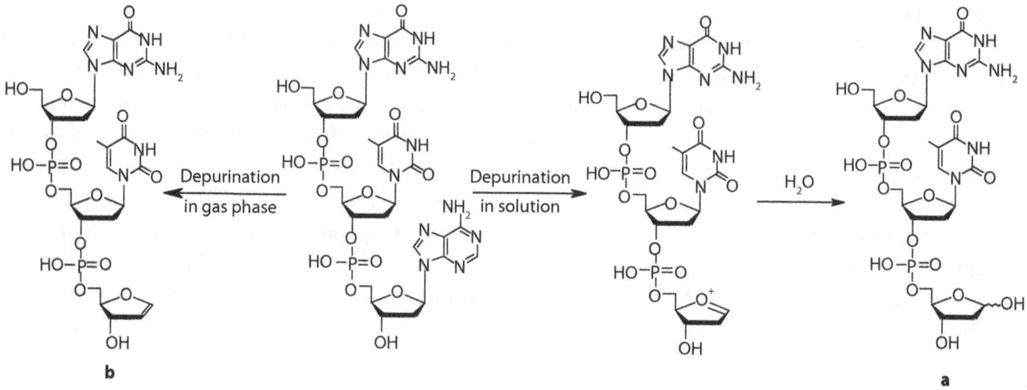

FIGURE 4.11 Depurination of oligonucleotides generated (a) in solution and (b) in the gas phase.

4.6.2 Longmers

Longmers are impurities with greater number of nucleotides and a correspondingly larger molecular weight than the full length product. The formation of longmers ((n+1), (n+2), etc.) may occur in several ways.[54] Protected phosphoramidites are activated with a weak organic acid and then mixed with the solid support for a short time (contact time) to be added to the growing oligonucleotide chain. Phosphoramidites, in the presence of an activator, are known to generate low concentrations of dimers, trimers, etc. Coupling of these oligomeric amidites, instead of a monomeric species, results in the addition of more than one nucleotide to the growing chain during a single coupling step. Longmer formation can also result from increased contact time between the phosphoramidite solution and the solid support. In this case more than one monomer phosphoramidite can be added to the chain in a single coupling step.

Growth of the oligonucleotide chain from multiple locations, such as unprotected groups on the oligonucleotide, leads to branched oligonucleotides with molecular weights much higher than the desired full length product.[55] These impurities are much less common using current amidite chemistries and are rarely seen in using current synthetic procedures.

4.6.3 Incomplete Removal of Protecting Groups

Nucleophilic centers on the phosphoramidite are chemically protected by a variety of orthogonal protecting groups, which are removed during or at the end of synthesis. Incomplete removal of the protecting groups will generate impurities that have a higher molecular weight than the full length product. A list of common protecting groups and their molecular weight are listed in Table 4.1.

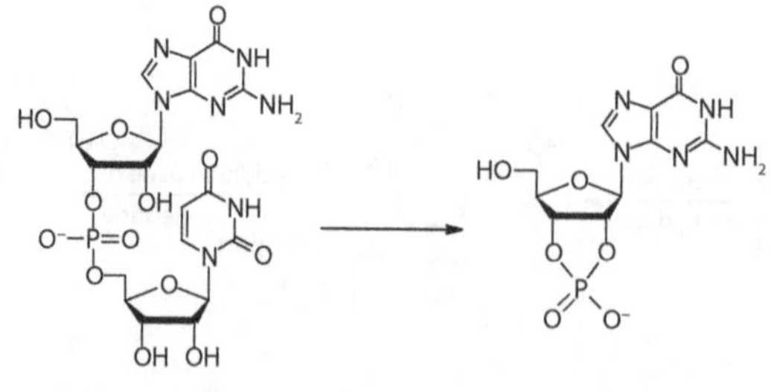

FIGURE 4.12 Degradation of RNA during cleavage and deprotection and formation of cyclic phosphate.

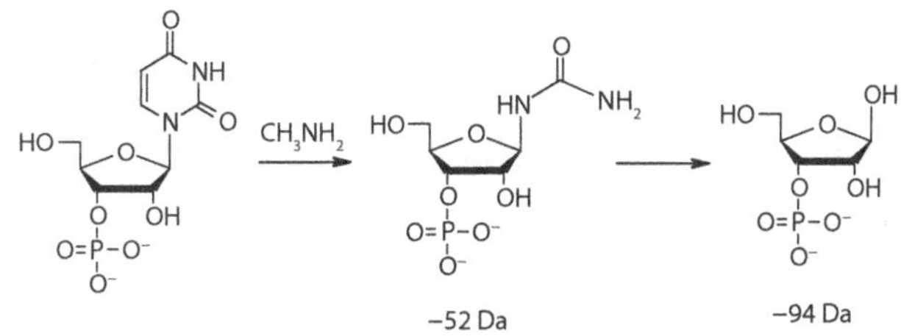

FIGURE 4.13 Depyrimidation impurities generated as a result of reaction with methylamine during cleavage and deprotection.

4.6.4 Acid Treatment Related Impurities

The dimethoxytrityl (DMTr) protecting groups on the 5′ oxygen of the last nucleotide are removed by dichloroacetic acid treatment during synthesis. In this process, called detritylation, the last nucleotide will become available for the next coupling step. Commercial dichloroacetic acids often contains small amounts of chloral (trichloroacetaldehyde), which can be incorporated into the oligonucleotide backbone during synthesis.[56] The resulted impurity will have an addition of Cl_3CCHO to the full length product (Figure 4.10).

Acid treatment during detritylation can lead to depurination of the purines in the oligonucleotide.[9,57] In this process, rapid protonation of the N-7 position in guanine and N-1 position in adenine is followed by the cleavage of the N-glycosidic bond. Following cleavage, water is added to the resulting oxocarbenium ion to yield a 1′-OH ribose.

Depurinated species can also be generated during the electrospray ionization. The voltage difference between the nozzle and the skimmer (Figure 4.1) determines the kinetic energy of the ions entering the mass analyzer. If collisions between an ion and the background gas is strong enough to generate sufficient internal energy in the ion, fragmentation could occur.

This type of fragmentation, called nozzle-skimmer dissociation, has been used for oligonucleotide sequencing.[58] Depurinated species generated during ionization are not solvated, and water is not added to the molecule after depurination. Depurinated species generated in the gas phase will have a molecular weight that is 18 Da less than their solution-based counterparts (Figure 4.11b). It has been reported that guanine depurination in solution yields a mixture of elimination and substitution (addition of water) products.[9]

4.6.5 Base Treatment Related Impurities

Harsh cleavage and deprotection with a base could degrade RNA sequences and as a result, impurities similar to failure sequences but with a cyclic phosphate on the 3′ end can form (Figure 4.12). Phosphorothioate (PS) to phosphate (PO) conversion, depyrimidation (Figure 4.13), and deamination during cleavage and deprotection are possible.

4.6.6 Oligonucleotide Degradation Products

The stability of oligonucleotides under different conditions is an important aspect of drug development. Degradation products generated under acidic, basic, and oxidizing conditions are largely similar to the impurities that have been discussed. Depurination at low pH, deamination at high pH, and sulfur loss (PS to PO conversion) under oxidative stress are common products. Under thermal stress an impurity is generated that has a molecular weight of 80 Da higher than the full length product.[59] This impurity is due to the modification of cytosines as a result of the reaction of cytosine containing sequences with depurinated oligonucleotides (Figure 4.14). A list of common impurities with their corresponding masses is listed in Table 4.1.

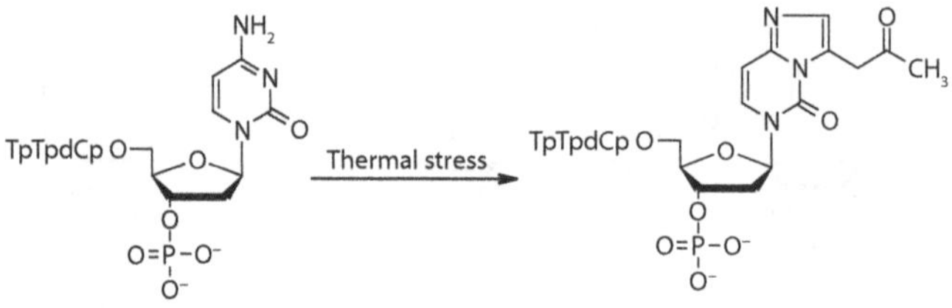

FIGURE 4.14 Modification of cytosine under thermal stress.

4.7 APPLICATIONS

The analysis of oligonucleotides often necessitates MS characterization of species to either confirm or determine their sequence. The extent of MS characterization often depends on the end use of the sequence. More detailed analysis, such as MS/MS, is often reserved for sequences intended for use in therapeutic and diagnostic applications and is discussed elsewhere in this book. Regardless of the final application, often LC-MS analysis is required to confirm that the expected mass is observed or to confirm a sequence based on accurate mass. As discussed previously, oligonucleotides are prepared synthetically, and therefore the sequence is generally known. An LC-MS method can be used to quickly verify that the observed mass matches what is expected for a particular sequence. This assumes that the connectivity is correct as two different oligonucleotide sequences with different connectivity are isobaric and cannot be distinguished from each other by mass alone. Despite this limitation, accurate mass measurement is very useful for sequence confirmation by assignment of failure sequences based on their molecular weight. The latter application necessitates chromatographic conditions that elute species in order of their length, which directly correlates to their charge. Here we will describe methods for both high-throughput analysis of oligonucleotides and more detailed sequence confirmation. These topics will be discussed as they apply to many oligonucleotide species including single-stranded and duplex species.

4.7.1 High-Throughput Desalting Analysis of Oligonucleotides

High-throughput methods are largely used by oligonucleotide manufacturing environments where large numbers of oligonucleotides are synthesized. The number of sequences prepared daily can range from hundreds to thousands. Each of the sequences is different depending on customer requirements and prepared at the low milligram to milligram scale. Syntheses are largely done in parallel, and there is a need for quality control analysis of sequences before they are shipped to customers. For these applications, generally, the practitioner is looking for a target mass and is willing to sacrifice chromatographic resolution for speed of analysis. For this reason, rapid LC-MS methods have been developed for this application where analyses can be completed in as little as 30 seconds without the need for additional desalting steps. MALDI analysis is also used for this application, and analysis time per sample can be much faster than an LC-MS method; however, high-throughput MALDI analysis requires additional equipment, such as a robot, for sample preparation.

For high-throughput desalting LC-MS applications, the analyte is injected on to a short column, generally a reverse phase C18 column. To further increase sample throughput, a switching valve configured with two columns. Using this configuration, one column is being analyzed while the other is being loaded and desalted, which limits downtime. This also eliminates elution of salt to the mass spectrometer, minimizing MS downtime. The length of column is an important consideration when developing a method for high-throughput analysis. Fountain et al. reported that it is necessary to pass 10 column volumes through the column for efficient desalting and elution of oligonucleotides.[60] For this reason the length of the column directly correlates with analysis time.

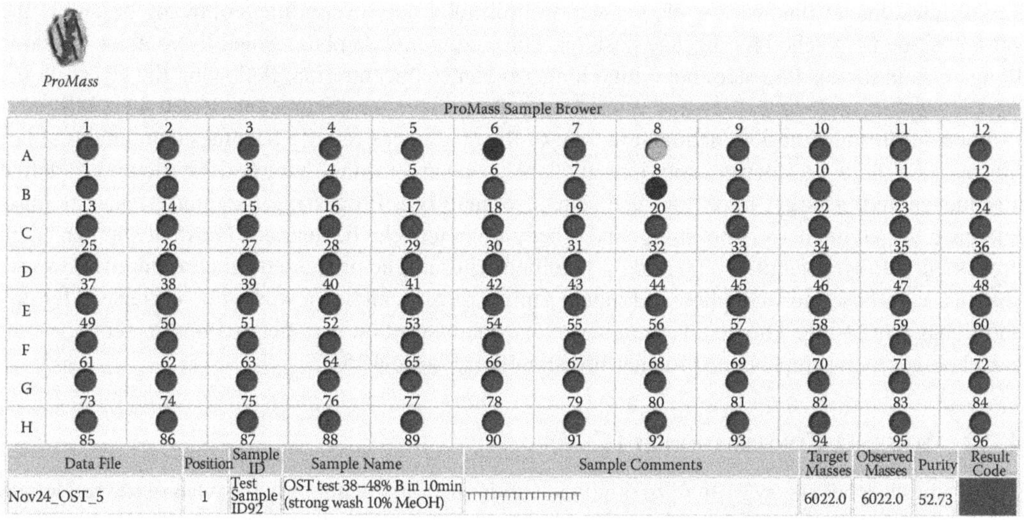

FIGURE 4.15 Screen shot of high throughput LC MS analysis of oligonucleotides using ProMass for MassLynx.

Using a two-valve, two-column configuration requires a multistep protocol. The first step entails loading and desalting of the analyte. Following loading, the analyte is desalted with a high aqueous mobile phase. The mobile phase contains a low concentration of an IP agent, which is largely responsible for retention. The amount of IP agent should be sufficient to retain the oligonucleotide but not excessive to limit ion suppression in the MS. The flow rate for the desalting step is generally

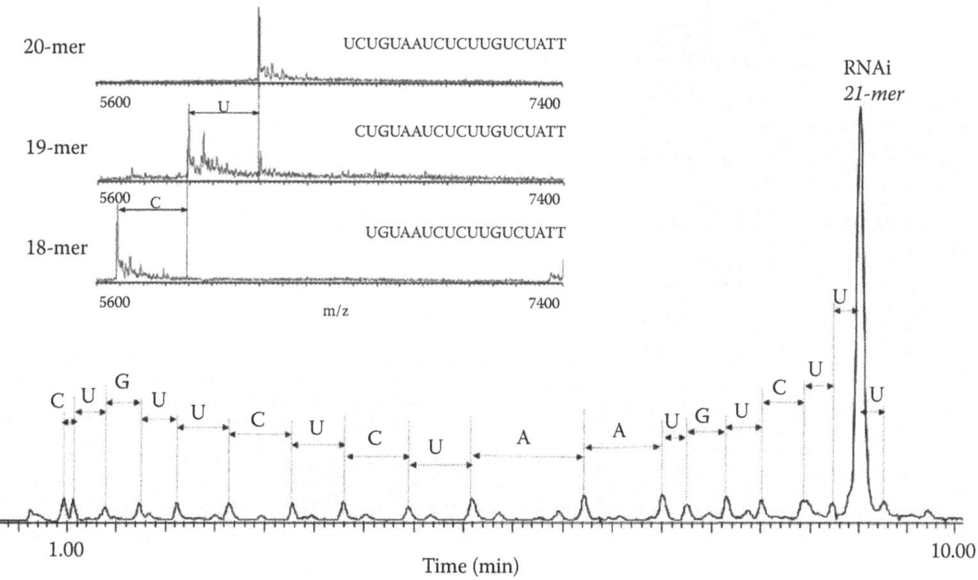

FIGURE 4.16 Assignment of oligonucleotide sequence based on deconvoluted accurate mass data for a high resolution separation of an oligonucleotide. Sequence is assigned by mass difference between adjacent peaks where the target sequence is known.

3 to 4 times greater than for the elution step to limit total duty cycle time. Following desalting, the valve position is switched to the elute position. The eluting mobile phase generally contains the same IP agent as in the loading step, but with a higher organic concentration. Following the switch to the elution position, oligonucleotides are immediately eluted from the column and detected via MS.

For high-throughput data acquisition, one of the challenges rapidly becomes data analysis. For this purpose there are software solutions available, such as ProMass, which allow the user to define a sequence and/or target mass for each sequence analyzed. Processing parameters such as mass tolerance, signal intensity, smoothing, and other parameters can be defined. The data are then batch processed following acquisition based on specified criteria. The processed data are stored and viewable in a Web-based format where individual sample results can be viewed without the need for any additional processing. The results are also color coded based on user defined parameters; the user can choose which samples need further investigation (Figure 4.15).

4.7.2 Sequence Confirmation by LC-MS

Many times it is necessary to analyze oligonucleotides to confirm not only the mass of the full length sequence but also the masses and relative abundance of truncated sequences and other synthetic impurities. As mentioned previously, elution of oligonucleotide sequences in order of their length is desired for applications where confirmation of sequence is needed. High MS sensitivity, mass accuracy, and good chromatographic separation are imperative for the detection and assignment of low-level impurities and their relative abundance in the sample. It is common that a method will perform well chromatographically for one oligonucleotide but poorly for another. The difference in performance is generally related to modifications on the sequences themselves, such as hydrophobic tags, phosphorothioated sequences, and duplex species, but can also be related to the sequence itself. It is common when using less efficient IP systems, such as TEAA, for oligonucleotide sequences to elute according to their base composition rather than their sequence.[61]

Retention of oligonucleotides in order of their sequence allows for rapid confirmation of the sequence based on accurate mass (Figure 4.16). Confirmation of the sequences is accomplished by determining the deconvoluted mass of each peak in the chromatogram. If the sequence is known, peaks can be assigned by comparison of the calculated mass to that experimentally determined. If the sequence is not known, it is possible to infer the sequences by comparing the mass difference between adjacent peaks. In either case, the practitioner should be aware that this method only allows one to assign a preliminary sequence, particularly if there are shorter sequences. This is primarily due to the fact that the data do not give definitive assignment of connectivity but rather allow one to infer connectivity. There is software available to automate this process, such as ProMass, which will interrogate the data, determine the molecular weight of each peak, and assign sequences based on the defined sequence. For complete assignment of the sequence, each of the shorter sequences must be present. In some instances, the synthetic process may be very efficient, and there may be little to no presence of certain failure sequences. In these cases, limited exonuclease digestion can be used to artificially generate a full ladder sequence.

4.7.3 Analysis of Phosphorothioates

Phosphorothioates are a class of oligonucleotides that have been largely used for antisense therapeutics. In fact, both therapeutic oligonucleotides that are available as of this printing are phosphorothioated products. These molecules are favored owing to their extended stability in organisms because of their resistance to nucleases. The degree of phosphorothiation is variable from only a few residues, generally at the termini of the sequences, to fully phosphorothioated sequences where each phosphate linkage is thioated.

The separation of phosphorothioate oligonucleotides can be very challenging owing to very broad peaks.[62–64] Each phosphorothioate (PS) linkage generates two isomers; so for molecules with

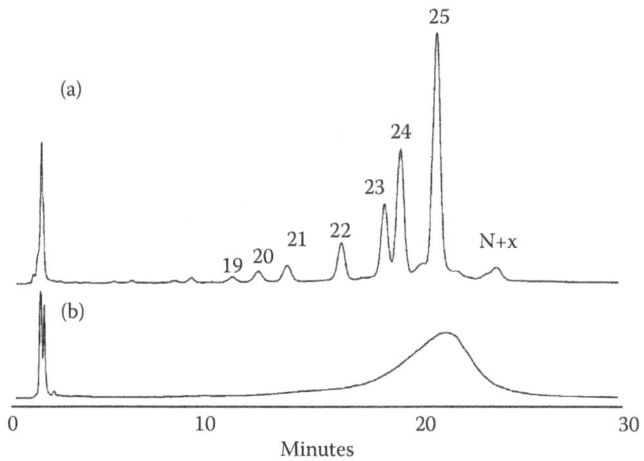

FIGURE 4.17 Impact of ion-pairing buffer on the separation of PS-oligonucleotides. (a) TEA-HFIP buffer. (b) Triethylammonium acetate. The 18–25-mer ladder was prepared by digesting the 25-mer with 39-end exonuclease. LC conditions: 50 mm × 4.6 mm column; 60°C, 0.5 mL/min. Conditions (a): Mobile phase A: 8.6 mM TEA, 100 mM HFIP buffer, pH 8.6; B: methanol, linear gradient from 16% B, slope 0.15% per minute. Conditions (b): Mobile phase A: 0.1 M TEAA, pH 7. Mobile phase B: acetonitrile, gradient starts at 12% B, slope 0.125% B per minute. (From Gilar, M., et al., *Oligonucleotides*, 13, 229–243, 2003. With permission.)

a large number of PS linkages, the number of possible conformations rapidly eclipses the ability of any chromatographic method to resolve them. As a result, broad peaks are generally obtained for chromatographic separation of PS oligonucleotides. Each of the conformation is expected to exhibit the same efficiency in the intended application; therefore, there is no need to chromatographically resolve them. It has been demonstrated that HFIP buffered mobile phases offer a particular advantage for the analysis of phosphorothioates and the compatibility of HFIP with MS make this system a logical choice for analysis. HFIP dramatically reduces isomeric resolution of phosphorothioates by more effectively eliminating hydrophobic contribution from slight variations in PS conformation to a much greater extent than acetate buffered systems (Figure 4.17).[34]

4.7.4 Analysis of Duplex Oligonucleotides

Duplex oligonucleotides have become increasingly popular since the discovery of the RNAi mechanism by Fire and Mello.[65] The RNAi mechanism is a biologically conserved mechanism that regulates gene expression by reducing RNA translation into protein. The molecules used to exploit this mechanism are largely short interfering RNA (siRNA), which are short 21-mer sequences with two base overhangs on each strand, although other strategies exist. Sequences are chosen to target a particular gene and limit unwanted gene silencing by off-target effects. Off-target effects can result if internal base deletions are present rather than simple 3′ or 5′ bases are missing. The reason for this is that a 3′ or 5′ deletion still exhibits the same bulk desired sequence, but its affinity for the target may be diminished; however, an internal deletion results in an entirely different sequence with an entirely different target.

To date, the largest successful application of the RNAi mechanism has been in gene silencing in cell and small animal studies, although they are being heavily investigated for therapeutic use as well with great promise and many clinical trials are currently ongoing. The challenge for therapeutic applications is in confirming the sequence of each single strand and also characterization of the duplex itself to minimize the risk of off target effects as described above. Their analysis by LC-MS has been particularly challenging owing to the fact that duplexes are formed by hydrogen

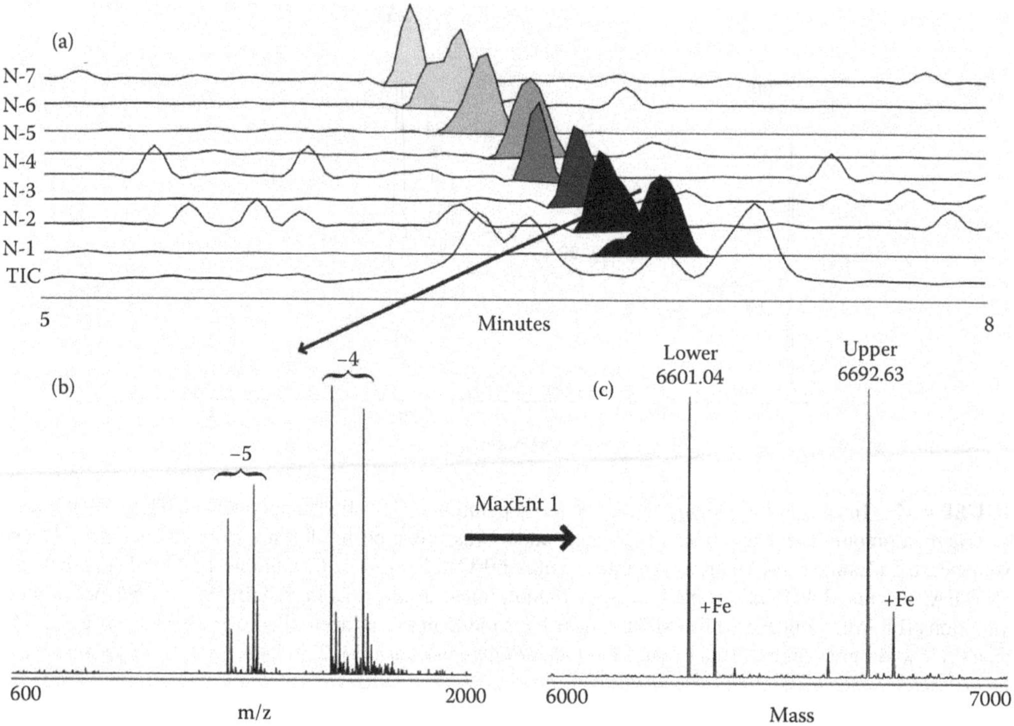

FIGURE 4.18 Identification of siRNA impurities by LC-MS. (a) Total ion chromatogram and extracted ion chromatograms for selected impurities. Selected ion species typically corresponded to $[M-3H]^{-3}$ charge state for truncated RNA species. (b) MS spectrum for U21/L20 truncated duplex. (c) Deconvoluted MS data. Two dominant mass signals correspond to the expected mass of 21 nt upper strand and 20 nt lower strand. The MS was operated at ESI negative mode, capillary voltage was 3 kV, cone voltage was 28 V, extractor voltage was 3 V, source temperature was 150°C, desolvation temperature was 350°C, cone gas flow was 31 L/h, and desolvation gas flow was 700 L/h. (Reprinted from McCarthy, S. M., et al., *Analytical Biochemistry*, 390, 181–188. Copyright 2009, with permission from Elsevier.)

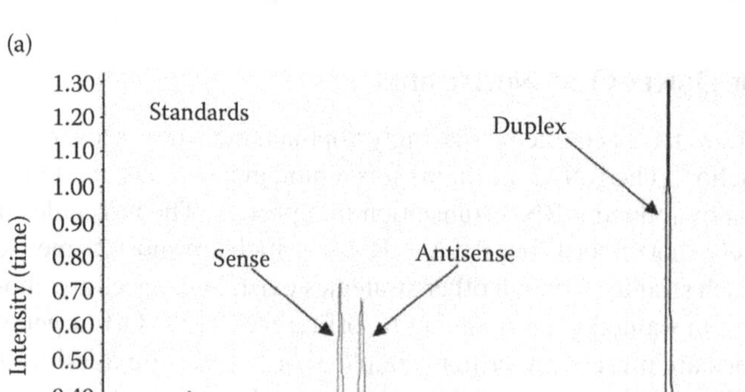

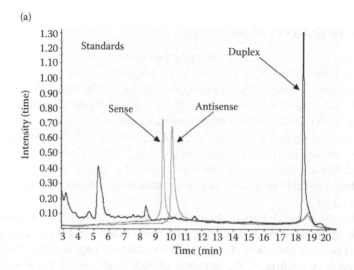

FIGURE 4.19 (a) Overlaid UV 260 nm chromatograms from separate injections of the sense, antisense, and duplex oligonucleotides; 10 mL injections of 5 mM (duplex) and 2.5 mM (single strands).

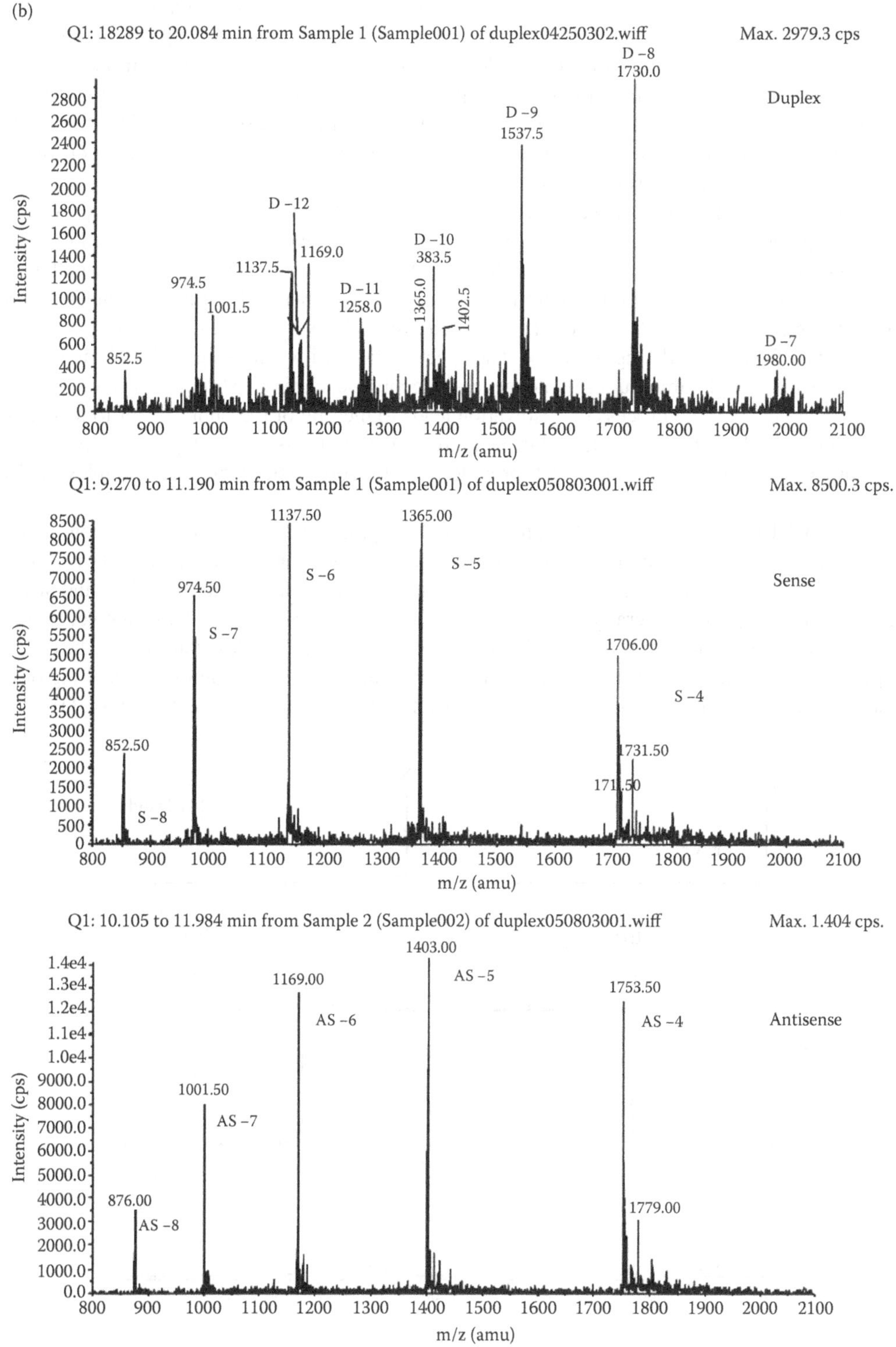

FIGURE 4.19 (Continued) (b) The corresponding ESI mass spectra for each of the chromatograms shown in (a). ESI mass spectra with their corresponding charge states are labeled D for the duplex, AS for the antisense, and A for the sense strand. (From Beverly, M., et al., *Rapid Communications in Mass Spectrometry*, 19, 1675–1682, 2005. Copyright Wiley-VCH Verlag GmbH & Co. KGaA. Reproduced with permission.)

bonding between complementary strands. The melting point, defined as the temperature at which only half of the possible base combinations are interacting, are variable and depend largely on the sequence, although modifications on the nucleotides, such as 2′ modifications, can increase their melting point. In general, the melting point of a particular duplex will range from 50° to 90°C. For this reason it is often possible to chromatographically separate duplexes in their intact state, but MS spectra of the duplex species reveal single-stranded components because they melt during the ionization process itself or during desolvation.

Acetate buffered IP systems have been utilized for the analysis of oligonucleotide duplexes.[66–68] Efficient separation of the intact duplex from single-stranded species is easily achieved owing to the reduction in hydrophobic interactions with the stationary phase by duplexes and the better ion-pairing because the phosphate groups are more exposed. As shown in Figure 4.18, duplex species can be successfully analyzed via LC-MS and truncated duplex species can be chromatographically resolved from the full length duplex. The spectra obtained when using high boiling point IP reagents, such as hexylammonium acetate, generally contain each single strand instead of the intact duplex. It is likely that the process of desolvation leads to melting of the duplex prior to MS analysis. Although the duplex species melts during analysis, the duplex is chromatographically resolved from single-stranded components, which provide unambiguous assignment of the duplex, and its components.

The TEA-HFIP system is very efficient at providing good chromatographic resolution of short to moderate length oligonucleotides, but there are several applications in which its utility is limited. One such application is the analysis of duplex species such as double-stranded DNA and siRNA molecules. These species generally have a melting point between 50° and 90°C, which necessitates a lower column temperature if they are to be separated in their intact duplex state. While the column temperature can be decreased, HFIP itself is denaturing and can lead to significant melting of the duplex species on column. It has also been suggested that a decreased concentration of TEA, and

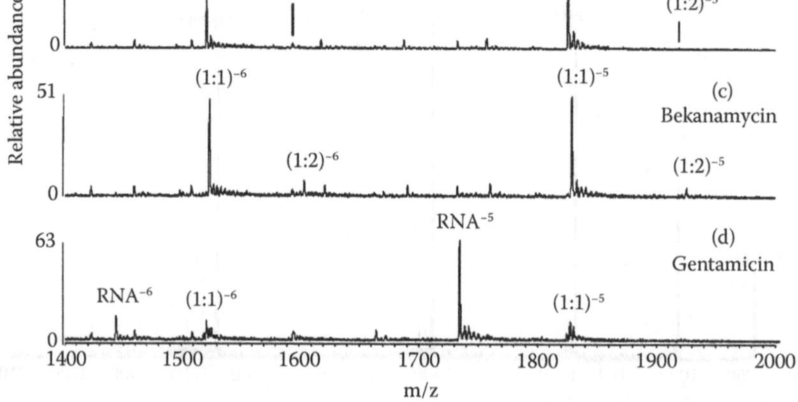

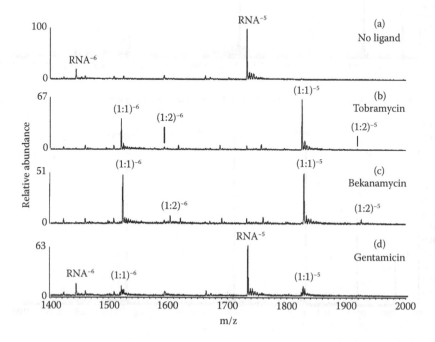

FIGURE 4.20 ESI spectra for samples containing the tobramycin aptamer (RNA). Samples were directly infused into the ESI source in 50 mM ammonium acetate, 30% isopropanol/water. (From Keller, K. M., et al., *Journal of Mass Spectrometry*, 40, 1327–1337, 2005. Copyright Wiley-VCH Verlag GmbH & Co. KGaA. Reproduced with permission.)

therefore decreased ionic strength, can lead to increased melting.[18] Despite these effects, TEA-HFIP has been used successfully for duplex characterization (Figure 4.19a). Beverly et al. have reported the analysis of inhibitory RNA duplexes with no evidence for on column melting.[69,70] In fact, ESI MS of the inhibitory RNA duplex resulted in intact duplex MS signal, although the signal intensity was dramatically reduced and favored lower charge states (Figure 4.19b).

4.7.5 Aptamers

Mass spectrometric analysis of aptamers can be challenging, particularly when analyzing their complexes with molecules.[71–73] In their native form, aptamers range from 20 to 50 nucleotides in length and are designed to adopt specific tertiary structures that interact with molecules through noncovalent interactions. The challenge for MS analysis of aptamers and their complexes comes from the desire to preserve these structures during the ionization process and in the gas phase following ionization. ESI-MS is particularly well suited to this application because it is a relatively gentle ionization technique and leaves the complex largely intact, although weak interacting species may be disrupted. In this way the complexes of aptamers, and their noncovalent interaction with molecules, can be analyzed by mass spectrometry to access stoichiometry and the affinity of host/guest interaction as shown in Figure 4.20. The interaction of this aptamer with tobramycin and bekanamycin is quite strong and not disrupted during the ionization process, while its interaction

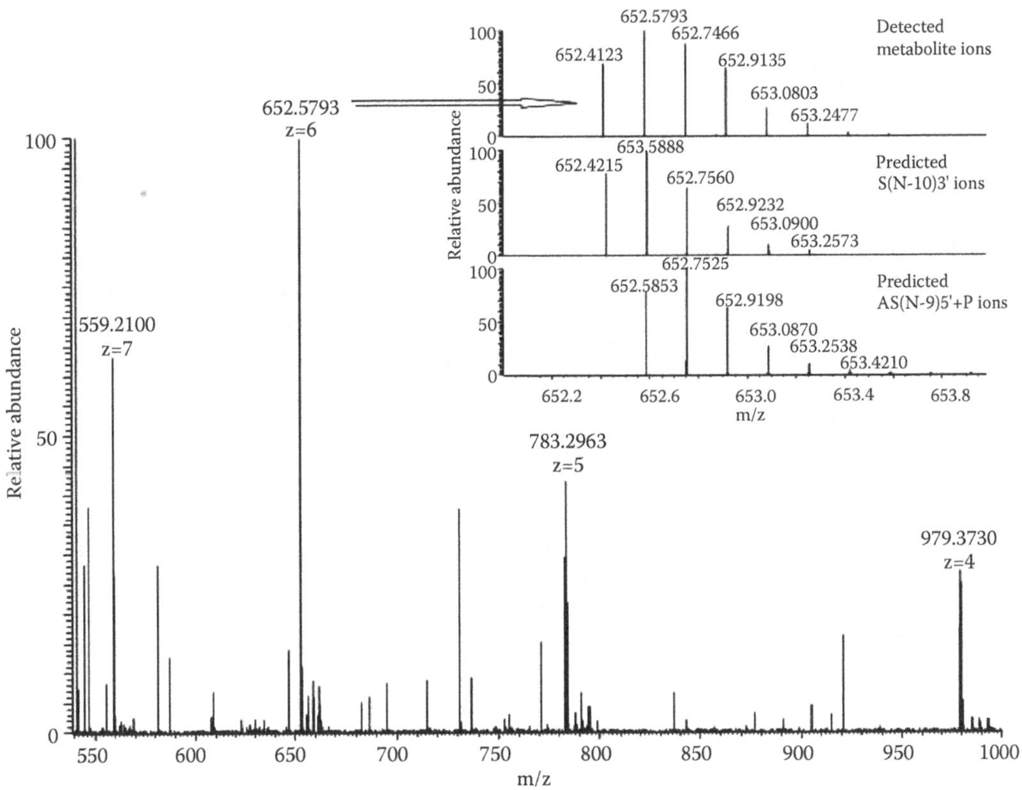

FIGURE 4.21 An example of the metabolite identification using accurate mass measurements. The accurate mass spectrum of a metabolite derived from the test sequence (TS) suggests that this metabolite is identified as S(N-10)3′, but not AS(N-9)5′+P. (From Zou, Y., et al., *Rapid Communications in Mass Spectrometry*, 22, 1871–1881, 2008. Copyright Wiley-VCH Verlag GmbH & Co. KGaA. Reproduced with permission.)

with gantamycin is quite low. These results are consistent with what is known about the binding seen in solution.

4.7.6 Oligonucleotide Bioanalysis

For therapeutic oligonucleotides there is a need to fully characterize oligonucleotides both prior to and after administration. We have discussed various techniques for analyzing oligonucleotides prior to administration to an organism, but these same techniques are not directly compatible with samples obtained from biological sources. One of the main challenges for bioanalysis of oligonucleotides is in their isolation from biological matricies.[70,74–77]

There are considerable matrix effects that need to be considered when attempting to isolate oligonucleotides from biological matricies such as salt, interaction with small organic and inorganic molecules, and interaction with proteins.[77] There have been a number of sample preparation strategies reported, and recovery of the analyte from the matrix depends on the matrix itself but also on the oligonucleotide sequence and modifications.[77]

Many recent reports have utilized HFIP-based mobile phases for chromatographic separation of oligonucleotides following isolation from biological fluids,[74,75,77] although other mobile phases have been reported.[76] Most reports have focused on the use of triple quadrupole MS instruments for identification of metabolites; however, quadrupole ion trap instruments have also been used. Regardless of the MS instrument used, sensitivity and mass accuracy are imperative for unambiguous identification of species. An example of the metabolite identification using accurate mass measurements is shown in Figure 4.21. High MS resolution is imperative for identification of oligonucleotide metabolites. In this example the mass difference between two possible metabolites is less than 1 Da; however, high-quality MS spectra allow preferential assignment of one possibility. The assignment is based on the presence of an isotope present in the spectrum that only appears in one of the possible explanations of the data. The need for high-resolution spectra is imperative for assignment as the mass difference between the two possible explanations is less than 0.2 Da.

4.8 CONCLUSION

LC-MS analysis of oligonucleotides can be challenging because the practitioner needs to carefully balance the chromatographic separation with MS sensitivity. In the chapter we outlined many considerations and their impact on MS analysis and chromatographic resolution. The development of a robust method for oligonucleotide analysis depends, to some extent, on the sequence, modifications, and type of oligonucleotide being analyzed. We outlined many common impurities and the corresponding mass difference observed in mass spectra and the potential source of these impurities.

The level of characterization for an oligonucleotide varies greatly, depending on the intended application. In particular, oligonucleotides intended for use as therapeutic applications will require a much greater degree of characterization than those used as polymerase chain reaction (PCR) primers or exploratory studies. The reader should be aware that the level of characterization required may change as this application area develops. For this reason it is highly suggested that the reader consult recent literature before developing LC-MS methods for oligonucleotides that will be used in a regulated environment.

REFERENCES

1. Levin, A. A., S. P. Henry, D. Monteith, and M. V. Templin. 2001. *Antisense Drug Technology: Principles, Strategies and Applications*. Boca Raton, FL: Taylor & Francis.
2. Pieles, U., W. Zurcher, M. Scharl, and H. E. Moser. 1993. Matrix-assisted laser desorption ionization time-of-flight mass spectrometry: a powerful tool for the mass and sequence analysis of natural and modified oligonucleotides. *Nucleic Acids Research* 21: 3191–3196.

3. Potier, N., A. Van Dorsselaer, Y. Cordier, O. Roch, and R. Bischoff. 1994. Negative electrospray ionization mass spectrometry of synthetic and chemically modified oligonucleotides. *Nucleic Acids Research* 22: 3895–3903.
4. Zhan, O., A. Gusev, and D. M. Hercules. 1999. A novel interface for on-line coupling of liquid capillary chromatography with matrix-assisted laser desorption/ionization detection. *Rapid Communications in Mass Spectrometry* 13: 2273–2278.
5. Greig, M., and R. H. Griffey. 1995. Utility of organic bases for improved electrospray mass spectrometry of oligonucleotides. *Rapid Communications in Mass Spectrometry* 9: 97–102.
6. Greig, M. J., H.-J. Gaus, and R. H. Griffey. 1996. Negative ionization micro electrospray mass spectrometry of oligonucleotides and their complexes. *Rapid Communications in Mass Spectrometry* 10: 47–50.
7. Limbach, P. A., P. F. Crain, and J. A. McCloskey. 1995. Characterization of oligonucleotides and nucleic acids by mass spectrometry. *Current Opinion in Biotechnology* 6(1): 96–102.
8. Stults, J. T., J. C. Marsters, and S. A. Carr. 1991. Improved electrospray ionization of synthetic oligodeoxynucleotides. *Rapid Communications in Mass Spectrometry* 5: 359–363.
9. Gut, I. G., 1997. Depurination of DNA and matrix assisted laser desorption/ionization mass spectrometry. *International Journal of Mass Spectrometry and Ion Processes* 169/170: 313–322.
10. Brüchert, W., R. Krüger, A. Tholey, M. Montes-Bayón, and J. Bettmer. 2008. A novel approach for analysis of oligonucleotide-cisplatin interactions by continuous elution gel electrophoresis coupled to isotope dilution inductively coupled plasma mass spectrometry and matrix-assisted laser desorption/ionization mass spectrometry. *Electrophoresis* 29: 1451–1459.
11. Kebarle, P. 2000. A brief overview of the present status of the mechanisms involved in electrospray mass spectrometry. *Journal of Mass Spectrometry* 35: 804–817.
12. Banoub, J. H., R. P. Newton, E. Esmans, D. F. Ewing, and G. Mackenzie. 2005. Recent developments in mass spectrometry for the characterization of nucleosides, nucleotides, oligonucleotides, and nucleic acids. *Chemical Reviews* 105: 1869–1915.
13. Crain, P. F. 1990. Mass spectrometric techniques in nucleic acids research. *Mass Spectrometry Reviews* 9: 505–554.
14. Nordhoff, E., F. Kirpekar, and P. Roepstorff. 1996. Mass spectrometry of nucleic acids. *Mass Spectrometry Reviews* 15: 67–138.
15. Tost, J., and I. G. Gut. 2002. Genotyping single nucleotide polymorphism by mass spectrometry. *Mass Spectrometry Reviews* 21: 388–411.
16. Hofstadler, A. S., K. A. Sannes-Lowery, and J. C. Hannis. 2005. Analysis of nucleic acids by FTICR MS. *Mass Spectrometry Reviews* 24: 265–285.
17. Thomas, B., and A. V. Akoulitchev. 2006. Mass spectrometry of RNA. *Trends Biomedical Sciences* 31 173–181.
18. Huber, C. G., and H. Oberacher. 2001. Analysis of nucleic acids by on-line liquid chromatography-mass spectrometry. *Mass Spectrometry Reviews* 20: 310–343.
19. Limbach, P. A. 1996. Indirect mass spectrometric methods for characterizing and sequencing oligonucleotides. *Mass Spectrometry Reviews* 15: 297–336.
20. Meng, Z., and P. A. Limbach. 2002. *Genomic Technologies: Present and Future*. 197–233. Norwich, U.K.: Caister Academic Press.
21. Bleicher, K., and E. Bayer. 1994. Analysis of oligonucleotides using coupled high performance liquid chromatography-electrospray mass spectrometry. *Chromatographia* 39(7): 405–408.
22. Gaus, H. J., S. R. Owens, M. Winniman, S. Cooper, and L. L. Cummins. 1997. On-line HPLC electrospray mass spectrometry of phosphorothioate oligonucleotide metabolites. *Analytical Chemistry* 69: 313–319.
23. Bleicher, K., and E. Bayer. 1994. Various factors influencing the signal intensity of oligonucleotides in electrospray mass spectrometry. *Biological Mass Spectrometry* 23: 320–322.
24. Cheng, X., D. C. Gale, H. R. Udseth, and R. D. Smith. 1995. Charge state reduction of oligonucleotide negative ions from electrospray ionization. *Analytical Chemistry* 67: 586–593.
25. Huber, C. G., and A. Krajete. 1999. Analysis of nucleic acids by capillary ion-pair reversed-phase HPLC coupled to negative-ion electrospray ionization mass spectrometry. *Analytical Chemistry* 71: 3730–3739.
26. Apffel, A., J. A. Chakel, S. Fischer, K. Lichtenwalter, and W. S. Hancock. 1997. Analysis of oligonucleotides by HPLC-electrospray ionization mass spectrometry. *Analytical Chemistry* 69: 1320–1325.
27. Bleicher, K., and E. Bayer. 1994. Analysis of oligonucleotides using coupled high performance liquid chromatography-electrospray mass spectrometry. *Chromatographia* 39: 405–408.

28. Apffel, A., J. A. Chakel, S. Fischer, K. Lichtenwalter, and W. S. Hancock. 1997. New procedure for the use of high-performance liquid chromatography-electrospray ionization mass spectrometry for the analysis of nucleotides and oligonucleotides. *Journal of Chromatography A* 777: 3–21.
29. Oberacher, H., W. Parson, R. Muhlmann, and G. G. Huber. 2001. Analysis of polymerase chain reaction products by on-line liquid chromatography—mass spectrometry for genotyping of polymorphic short tandem repeat loci. *Analytical Chemistry* 73: 5109–5115.
30. Bothner, B., K. Chatmann, and G. Siuzdak. 1995. Liquid chromatography mass spectrometry of antisense oligonucleotides. *Bioorganic and Medicinal Chemistry Letters* 5: 2863–2868.
31. Muddiman, D. C., X. Cheng, H. R. Udseth, and R. D. Smith. 1996. Charge-state reduction with improved signal intensity of oligonucleotides in electrospray ionization mass spectrometry. *Journal of the American Society of Mass Spectrometry* 7: 697–706.
32. Deguchi, K., M. Ishikawa, T. Yokokura, I. Ogata, S. Ito, T. Mimura, and C. Ostrander. 2002. Enhanced mass detection of oligonucleotides using reverse-phase high-performance chromatography/electrospray ionization ion-trap mass spectrometry. *Rapid Communications in Mass Spectrometry* 16: 2133–2141.
33. Huber, C. G., and A. Krajete. 2000. Sheath liquid effects in capillary high-performance liquid chromatography–electrospray mass spectrometry of oligonucleotides. *Journal of Chromatography A* 870: 413–424.
34. Gilar, M., K. J. Fountain, Y. Budman, J. L. Holyoke, H. Davoudi, and J. C. Gebler. 2003. Characterization of therapeutic oligonucleotides using liquid chromatography with on-line mass spectrometry detection. *Oligonucleotides* 13: 229–243.
35. Limbach, P. A., P. F. Crain, and J. A. McCloskey. 1995. Molecular mass measurement of intact ribonucleic acids via electrospray ionization quadrupole mass spectrometry. *Journal of the American Society for Mass Spectrometry* 6(1): 27–39.
36. Pomerantz, S. C., and J. A. McCloskey. 1990. Analysis of RNA hydrolyzates by liquid chromatography-mass spectrometry. In *Methods in Enzymology*, 796–824. San Diego, CA: Academic Press.
37. Liu, C., D. C. Muddiman, K. Tang, and R. D. Smith. 1997. Improving the microdialysis procedure for electrospray ionization of biological samples. *Journal of Mass Spectrometry* 32: 425–431.
38. Muddiman, D. C., D. S. Wunschel, C. L. Liu, L. Pasa-Tolic, K. F. Fox, A. Fox, G. A. Anderson, and R. D. Smith. 1996. Characterization of PCR products from bacilli using electrospray ionization FTICR mass spectrometry. *Analytical Chemistry* 68: 3705–3712.
39. Muddiman, D. C., G. A. Anderson, S. A. Hofstadler, and R. D. Smith. 1997. Length and base composition of PCR-amplified nucleic acids using mass measurements from electrospray ionization mass spectrometry. *Analytical Chemistry* 69: 1543–1549.
40. Muddiman, D. C., A. P. Null, and J. C. Hannis. 1999. Precise mass measurement of a double-stranded 500 base-pair (309 KDa) polymerase chain reaction product by negative ion electrospray ionization Fourier transform ion cyclotron resonance mass spectrometry. *Rapid Communications in Mass Spectrometry* 13: 1201–1204.
41. Null, A. P., J. C. Hannis, and D. C. Muddiman. 2000. Preparation of single stranded PCR products for electrospray ionization mass spectrometry using the DNA repair enzyme lambda exonuclease. *Analyst* 125: 619–625.
42. Hannis, J. C., and D. C. Muddiman. 1999. Accurate characterization of the tyrosine hydroxylase forensic allele 9.3 through development of electrospray ionization Fourier transform ion cyclotron resonance mass spectrometry. *Rapid Communications in Mass Spectrometry* 13: 954–962.
43. Hannis, J. C., and D. C. Muddiman. 1999. Characterization of a microdialysis approach to prepare polymerase chain reaction products for electrospray ionization mass spectrometry using on-line ultraviolet absorbance measurements and inductively coupled plasma—atomic emission spectroscopy. *Rapid Communications in Mass Spectrometry* 13: 323–330.
44. Gilar, M., A. Belenky, and B. H. Wang. 2001. High-throughput biopolymer desalting by solid-phase extraction prior to mass spectrometric analysis. *Journal of Chromatography A* 921: 3–13.
45. Gilar, M., Bouvier, E.S.P. Purification of crude DNA oligonucleotides by solid-phase extraction and reversed phase high-performance liquid chromatography. *Journal of Chromatography A* 890: 167–177.
46. Deroussent, A., J.-P. L. Caer, J. Rossier, and A. Gouyette. 1995. Electrospray mass spectrometry for the characterization of the purity of natural and modified oligodeoxynucleotides. *Rapid Communications in Mass Spectrometry* 9: 1–4.
47. Huber, C. G., and M. R. Buchmeiser. 1998. On-line cation exchange for suppression of adduct formation in negative-ion electrospray mass spectrometry of nucleic acids. *Analytical Chemistry* 70: 5288–5295.
48. Liu, C., Q. Wu, A. C. Harms, and R. D. Smith. 1996. On-line microdialysis sample cleanup for electrospray ionization mass spectrometry of nucleic acid samples. *Analytical Chemistry* 68: 3295–3299.

49. Thayer, J. R., N. Puri, C. Burnett, M. Hail, and S. Rao. 2009. Identification of RNA linkage isomers by anion exchange purification with electrospray ionization mass spectrometry of automatically desalted phosphodiesterase-II digests. *Analytical Biochemistry* 399: 110–117.
50. Smith, L. M. 1988. Automated synthesis and sequence analysis of biological macromolecules. *Analytical Chemistry* 60: 381A–390A.
51. Krotz, A. H., P. Klopchin, D. L. Cole, and V. T. Ravikumar. 1997. Phosphorothioate oligonucleotides: Largely reduced (N-1)-mer and phosphodiester content through the use of dimeric phosphoramidite synthons. *Bioorganics and Medicinal Chemistry Letters* 7: 73–78.
52. Chen, D., Z. Yan, D. L. Cole, and G. S. Srivatsa. 1999. Analysis of internal (n-1)mer deletion sequences in synthetic oligodeoxyribonucleotides by hybridization to an immobilized probe array. *Nucleic Acids Research* 27: 389–395.
53. Fearon, K. L., J. T. Stults, B. J. Bergot, L. M. Christensen, and A. M. Raible. Investigation of the 'n–1' impurity in phosphorothioate oligodeoxynucleotides synthesized by the solid-phase β-cyanoethyl phosphoramidite method using stepwise sulfurization. *Nucleic Acids Research* 23: 2754–2761.
54. Krotz, A. H., P. G. Klopchin, K. L. Walker, G. S. Srivasta, D. L. Cole, and V. T. Ravikumar. 1997. On the formation of longmers in phosphorothioate oligodeoxyribonucleotide synthesis. *Tetrahedron Letters* 38: 3875–3878.
55. Kurata, C., K. Bradley, H. Gaus, N. Luu, I. Cedillo, V. T. Ravikumar, K. Van Sooy, J. V. McArdle, and D. C. Capaldi. 2006. Characterization of high molecular weight impurities in synthetic phosphorothioate oligonucleotides. *Bioorganics and Medicinal Chemistry Letters* 16: 607–614.
56. Gaus, H., P. Olsen, K. Van Sooy, C. Rentel, B. Turney, K. L. Walker, J. V. McArdle, and D. C. Capaldi. 2005. Trichloroacetaldehyde modified oligonucleotides. *Bioorganics and Medicinal Chemistry Letters* 15: 4118–4124.
57. Suzuki, T., S. Ohsumi, and K. Makino. 1994. Mechanistic studies on depurination and apurinic site chain breakage in oligodeoxyribonucleotides. *Nucleic Acids Research* 22: 4997–5003.
58. Meng, Z., and P. A. Limbach. 2005. Shotgun sequencing small oligonucleotides by nozzle-skimmer dissociation and electrospray ionization mass spectrometry. *European Journal of Mass Spectrometry* 11: 221–229.
59. Rentel, C., X. Wang, M. Batt, C. Kurata, J. Oliver, H. Gaus, A. H. Krotz, J. V. McArdle, and D. C. Capaldi. 2005. Formation of modified cytosine residues in the presence of depurinated DNA. *Journal of Organic Chemistry* 70: 7841–7845.
60. Fountain, K. J., M. Gilar, and J. C. Gebler. 2004. Electrospray ionization mass spectrometric analysis of nucleic acids using high-throughput on-line desalting. *Rapid Communications in Mass Spectrometry* 18: 1295–1302.
61. Gilar, M., K. J. Fountain, Y. Budman, U. D. Neue, K. R. Yardley, P. D. Rainville, R. J. Russell, and J. C. Gebler. 2002. Ion-pair reversed-phase high-performance liquid chromatography analysis of oligonucleotides: Retention prediction. *Journal of Chromatography A* 958: 167–182.
62. Metelev, V., and S. Agrawal. 1992. Ion-exchange high-performance liquid chromatography analysis of oligodeoxyribonucleotide phosphorothioates. *Analytical Biochemistry* 200: 342–346.
63. Agrawal, S., J. Y. Tang, and D. M. Brown. 1990. Analytical study of phosphorothioate analogues of oligodeoxynucleotides using high-performance liquid chromatography. *Journal of Chromatography A* 509: 396–399.
64. Gilar, M., A. Belenky, and A. S. Cohen. 2000. Polymer solutions as a pseudostationary phase for capillary electrochromatographic separation of DNA diastereomers. *Electrophoresis* 21: 2999–3009.
65. Fire, A., S. Xu, M. K. Montgomery, S. A. Kostas, S. E. Driver, and C. C. Mello. 1998. Potent and specific genetic interference by double-stranded RNA in Caenorhabditis elegans. *Nature* 391: 806–811.
66. McCarthy, S. M., M. Gilar, and J. Gebler. 2009. Reversed-phase ion-pair liquid chromatography analysis and purification of small interfering RNA. *Analytical Biochemistry* 390: 181–188.
67. Xiao, W., and P. J. Oefner. 2001. Denaturing high-performance liquid chromatography: A review. *Human Mutation* 17: 439–474.
68. Oberacher, H., et al. 2001. On-line liquid chromatography-mass spectrometry: a useful tool for the detection of DNA sequence variation. *Angewandte Chemie (International Edition)* 40: 3828–3830.
69. Beverly, M., K. Hartsough, and L. Machemer. 2005. Liquid chromatography/electrospray mass spectrometric analysis of metabolites from an inhibitory RNA duplex. *Rapid Communications in Mass Spectrometry* 19: 1675–1682.
70. Beverly, M., K. Hartsough, L. Machemer, P. Pavco, and J. Lockridge. 2006. Liquid chromatography electrospray ionization mass spectrometry analysis of the ocular metabolites from a short interfering RNA duplex. *Journal of Chromatography B* 835: 62–70.

71. Sperry, J. B., J. M. Wilcox, and M. L. Gross. 2008. Strong anion exchange for studying protein-DNA interactions by H/D exchange mass spectrometry. *Journal of the American Society of Mass Spectrometry* 19: 887–890.
72. Keller, K. M., M. M. Breeden, J. Zhang, A. D. Ellington, and J. S. Brodbelt. 2005. Electrospray ionization of nucleic acid aptamer/small molecule complexes for screening aptamer selectivity. *Journal of Mass Spectrometry* 40: 1327–1337.
73. Cavanagh, J., L. M. Benson, R. Thompson, and S. Naylor. 2003. In-line desalting mass spectrometry for the study of noncovalent biological complexes. *Analytical Chemistry* 75: 3281–3286.
74. Zou, Y., P. Tiller, I.-W. Chen, M. Beverly, and J. Hochman. 2008. Metabolite identification of small interfering RNA duplex by high-resolution accurate mass spectrometry. *Rapid Communications in Mass Spectrometry* 22: 1871–1881.
75. Zhang, G., J. Lin, K. Srinivasan, O. Kavetskaia, and J. N. Duncan. 2007. Strategies for bioanalysis of an oligonucleotide class macromolecule from rat plasma using liquid Chromatography—Tandem Mass Spectrometry. *Analytical Chemistry* 79: 3416–3424.
76. Bigelow, J. C., L. R. Chrin, L. A. Mathews, and J. J. McCormack. 1990. High-performance liquid chromatographic analysis of phosphorothioate analogues of oligodeoxynucleotides in biological fluids. *Journal of Chromatography B: Biomedical Sciences and Applications* 533: 133–140.
77. Lin, Z. J., W. Li, and G. Dai. 2007. Application of LC-MS for quantitative analysis and metabolite identification of therapeutic oligonucleotides. *Journal of Pharmaceutical and Biomedical Analysis* 44: 330–341.

5 Sequence Determination and Confirmation by MS/MS and MALDI-TOF

Zoltan Timar
Agilent Technologies, Inc.

CONTENTS

5.1 INTRODUCTION

The application of mass spectrometry (MS) to protein and peptide characterization and sequencing has revolutionized the field and is essentially responsible for the overwhelming growth of the discipline of proteomics in the past decade. While the application of MS to characterization to nucleic acids is only beginning to be widely accepted, many of the techniques and approaches developed in proteomics have led to significant advances for the analogous approaches in the DNA world.

Nucleic acid primary structure analysis (sequencing) is the process of determining the nucleobase order in a given oligonucleotide fragment. The sequencing of oligonucleotides of biological origin has application in the fields of forensics and genotyping, elucidating structural features and function, while sequencing of oligonucleotide-based active pharmaceutical ingredients (APIs) is regulatory requirement in quality control of manufacturing.

Two major categories of sequencing processes are (1) sequence confirmation (verification) and (2) *de novo* sequencing. Approaches based on sequence verification are appropriate when the expected nucleobase order and modifications are known *prior* to analysis, and *de novo* approaches are required if this information is unavailable or just partially available.

The majority of techniques for oligonucleotide primary structure determination by mass spectrometry rely on the creation of sequence ladders. The sequence ladders consist of shorter oligonucleotides derived from the target oligonucleotide via tandem mass spectrometry (Section 5.2), enzymatic methods such as enzymatic degradation (Section 5.3.3.1) or chain termination synthesis (Section 5.3.4) and chemical cleavage (Section 5.3.3.2). The sequence ladders are identical (degradation, cleavage) or complimentary (terminating method) to a fraction of the original, target sequence. Sorting and identification of the ladders by mass spectrometry provide the information from which the target oligonucleotide sequence can be inferred (Figure 5.1).

Knowledge of the nucleic acid type, modifications, and sequence or the aim of the sequencing is used to determine the approach for generating the sequence ladders and analytical technique by which the fragments are sorted. Often, a combination of different methodologies is necessary for a high level of reliability in the *de novo* sequencing of short, modified oligos.

Mass spectrometric approaches to sequencing rely on similar principles as the electrophoresis methods and can be considered both complimentary and a competitive technology, albeit somewhat lower in performance and currently requiring expertise for data interpretation (base read-out). Gel electrophoresis (or pyrosequencing) methods often fail or give unreliable results in certain areas, of nucleic acid *de novo* sequencing. In cases where gel electrophoresis is ineffective, mass spectrometry based sequencing is preferred. Those areas are determined by limitations of the termination and electrophoresis based sequencing, such as:

- Incorrectly terminated products frequently cannot be distinguished from correctly terminated ones.[1]
- Separation techniques such as electrophoresis are occasionally misleading due to aberrant mobility of certain fragments, leading to erroneous sequence determination.[2]
- No direct applications are possible for determination of nucleic acid modifications.
- Terminating methods are generally unsuitable for the sequencing of short oligonucleotides (<20-mers).
- There is a high error rate for RNA sequence determination.[3–5]

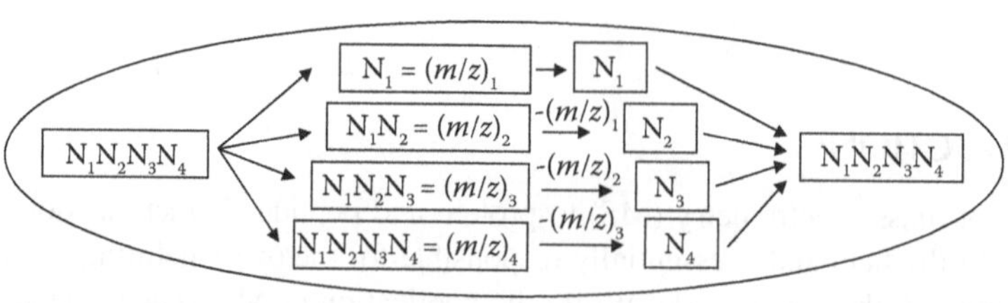

FIGURE 5.1 The sequence ladders method. Oligonucleotide $N_1N_2N_3N_4$ (target) is broken into three fragments: N_1, N_1N_2 and $N_1N_2N_3$. The mass (m/z) differences between ladders correspond to the masses of the terminal nucleotides: N_2, N_3, and N_4 and determine their position, from which the original sequence can be inferred.

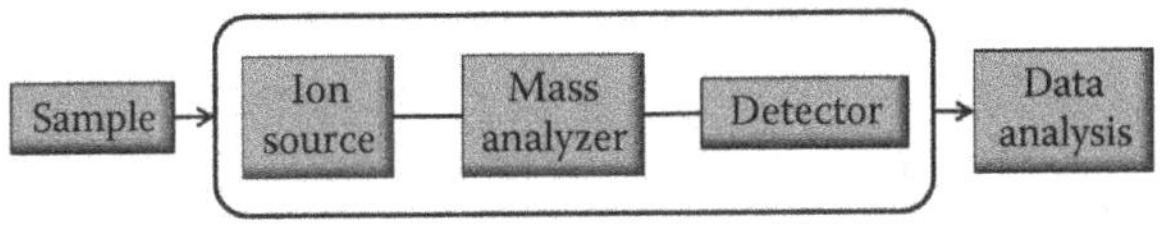

FIGURE 5.2 Schematic diagram of a mass spectrometer.

Advantages of mass spectrometry over electrophoresis separation based sequencing are high speed of fragment sorting (approximately milliseconds), fragment generation in less than a second (MS/MS), and mass determination for each oligonucleotide fragment. Because of factors cited above, examples in which MS sequencing is a better choice than the termination method include heterozygous polymorphism analyses or frame shift mutation,[6] highly structured sequences (RNA),[5] biologically modified nucleic acids,[7] and chemically modified oligonucleotides, such as APIs.[8,9] Transcriptional sequencing in these specific cases is frequently impossible, ambiguous, or cannot provide the required information, and consequently mass spectrometric-based sequencing methodologies may be more attractive.

Theory

The instrumental performance requirements for oligonucleotide sequencing are highly sensitive, accurate mass determination of well-resolved precursor and fragment ions. The ideal analytical instrument for mass determination of molecules is the mass spectrometer (Figure 5.2). A range of mass spectrometer instrumental configurations exist, each with different characteristics of sensitivity, mass accuracy, and resolution.

The analytical capabilities required for this process are separation, desorption, ionization, fragmentation, mass analysis, and detection. In order for chemical compounds to be manipulated in the gas phase in a mass spectrometer, two processes must occur: the oligonucleotide molecules must be ionized usually by a gain or loss of a proton, and they must be transferred to the gas phase (desorption). The ionization of compounds occurs in the ion source of the mass spectrometer. The use of two mild ionization modes are widespread for oligonucleotide mass analyses: matrix assisted laser desorption/ionization (MALDI)[10–12] and electrospray ionization (ESI).[13–15] In these processes, the oligonucleotides are simultaneously ionized and desorbed. MALDI is preferred for complex sample mixtures because it generates mainly singly charged molecular ions, which simplifies data interpretation. By contrast, ESI gives rise to multiply charged ions, resulting in more complex mass spectra than MALDI.

High-resolution mass determination needed for this application can be obtained using a range of commercially available mass spectrometer configurations including time of flight (TOF), Fourier transform ion cyclotron resonance (FTICR),[16] or Orbitrap mass spectrometers.[17] Early research in this area was conducted on quadrupole ion trap (QIT) instruments, which is still an attractive alternative for MS^n-type tandem MS applications. Tandem QIT-MS is a mild method of fragmentation; hence, it results in fewer and more specific sequence ions than collision-induced dissociation (CID), although the resolution and accuracy of QIT-MS is limited. A popular and illustrative configuration for a high-resolution mass analyzer for MALDI and for ESI-TOF is based on the time-of-flight measurement. For multistage MS, the TOF analyzer is frequently coupled with a preliminary lower-resolution mass filter such as a quadrupole, resulting in a quadrupole-time-of-flight (Q-TOF) configuration (Figure 5.3).

In MALDI, the analyte is co-crystallized with a UV absorbing aromatic compound called a matrix (usually an organic acid or amine) onto the solid surface of the sample target plate. A laser beam, tuned to the matrix excitation energy, is then applied to the crystals, vaporizing and ionizing the matrix along with the compound to be analyzed. Usually, a single charge is transferred from the

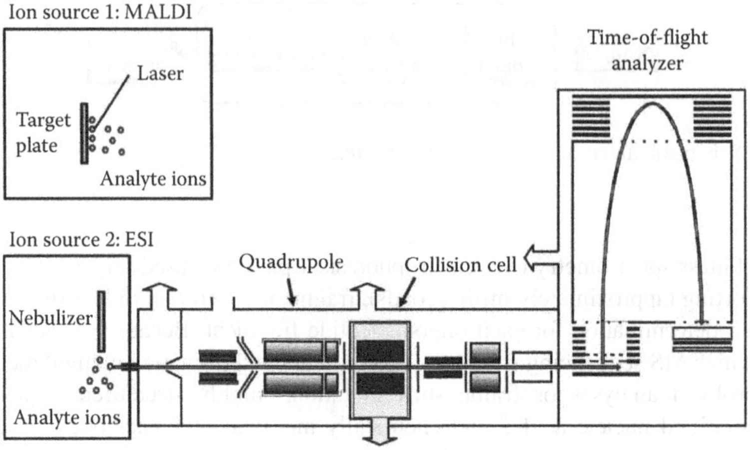

FIGURE 5.3 Schematic construction of a Q-TOF hybrid mass spectrometer set up with electrospray ionization source (ESI) and collision-induced dissociation (CID) cell.

aromatic matrix substance to the analyte in the gas phase, then the charged particles are accelerated by an electrostatic field. Mass determination of ions in MALDI-TOF MS is based on the time it takes the ion to fly through the mass spectrometer to the detector. For MALDI-generated ions, the process is discrete; the TOF start signal is given when the laser pulse is fired, and the stop is when the ions are detected. The distance that all ions need to travel and the acceleration voltage is the same for each individual ion. However, the mass of every species is different; thus their velocity and time-of-flight necessary to reach the detector will be different. In other words, the time necessary for each individual ion to travel the same distance in the flight tube is different and proportional to the square root of their mass/charge value.

$$t = \frac{d}{\sqrt{2 * U}} * \sqrt{\frac{m}{z}}$$

where t is time of flight, d is flight tube length, U is accelerating voltage, m is mass, and z is charge.

The formation of ions in ESI involves a different process. The liquid flow (usually HPLC or infusion) containing the analyte and buffer molecules is dispersed to fine aerosol droplets. The ion formation includes extensive solvent evaporation. The aerosol is sampled into the first vacuum stage of the mass spectrometer through a heated capillary to further evaporate solvent from droplets. Charge accumulation deforms the shrinking droplets (Rayleigh limit), and charged jets are emitted (Rayleigh fission). During the fission, the droplet loses mass and charge. There are two major theories that explain the final production of gas-phase ions:

1. The Ion Evaporation Model (IEM) suggests that as the droplet reaches a certain radius the field strength at the surface of the droplet becomes large enough to assist the field desorption of solvated ions.
2. The Charged Residue Model (CRM) suggests that electrospray droplets undergo evaporation and fission cycles, eventually leading progeny droplets that contain on average one analyte ion or less. The gas phase ions form after the remaining solvent molecules evaporate, leaving the analyte with the charges that the droplet carried.

While there is no definite scientific proof, a large body of indirect evidence suggests that small ions are liberated into the gas phase through the ion evaporation mechanism, while larger ions form by charged residue mechanism.[18]

For ESI-TOF MS, TOF (*t*) is measured in a slightly different manner than in the case of MALDI-TOF MS. The ion generation in the case of ESI is continuous as opposed to discrete ion generation. Thus, following the ionization, desorption, and desolvation process described above, a continuous stream of analyte ions are available for mass analysis and detection. In order to measure specific flight times, the stream of ions is sampled in narrow discrete packets, which are electrostatically pulsed into the TOF flight tube. The TOF start signal is given when the packet is generated and the stop signal is given when the ions are detected.

Electrospray ionization offers excellent sensitivity and wide mass range capabilities. It is possible to analyze biomolecules with molecular weights far in excess of the *m/z* range of a given mass analyzer owing to the ESI phenomenon generating multiple charged forms of a particular ion species. Large biomolecules tend to be represented as a distribution of multiple charge states.

The ions observed by mass spectrometry may be quasi-molecular ions created by the addition of a proton (a hydrogen ion) and denoted $(M+H)^+$, or of another cation such as sodium ion, $(M+Na)^+$, or the removal of a proton, $(M-H)^-$. Multiply charged ions such as $(M+nH)^{n+}$ and $(M\text{-}nH)^{n-}$ are often observed. Oligonucleotides are typically analyzed by ESI using the negative charges formed during ionization in "negative polarity" or "negative ion mode." This results in molecules with the molecular formula $(M\text{-}nH)^{n-}$ and many times as alkali metal adducts abbreviated as $(M\text{-}nH+mMe)^{(n-m)-}$-type ions (n > m; Me: Na^+ or K^+). For large macromolecules, there can be many charge states, resulting in a characteristic charge state envelope.

Because the mass analyzer is used to determine mass on the basis of mass to charge ratio (*m/z*), this multiply charged ion spectrum can be "deconvoluted" to determine an accurate intact molecular weight for the uncharged species. For example, a 21-mer DNA oligonucleotide with a mass of 6483.116 might be observed with 6^- charges at *m/z* 1079.512 $((6483.116-6*1.00728)/6)^{6-}$ and with 5^- charges at *m/z* 1295.616 $[(6483.116-5*1.00728)/5]^{5-}$.

5.2 MS/MS SEQUENCING

5.2.1 Basic Considerations

The ability of tandem mass spectrometry (MS/MS) to generate detailed structural information is based on the capability to perform gas phase chemical reactions in the ion population of analyte molecules. In a typical MS/MS experiment, a mixture of analytes is initially separated in the first stage of MS mass analysis based on their *m/z*. Particular species of interest are isolated and transferred to a collision cell as "parent ions" in which they are fragmented using one of several available fragmentation techniques, such as collision-induced dissociation (CID), infrared multiphoton dissociation (IRMPD), blackbody infrared radiative dissociation (BIRD), sustained off-resonance irradiation (SORI), surface-induced dissociation (SID), ultraviolet photodissociation, electron capture dissociation, and postsource decay (PSD).[19] It has been exploited for oligonucleotide nozzle skimmer (NS) or in-source fragmentation in addition to other fragmentation methods[20–24] that fragmentation in the source of an electrospray system increases with the number of nucleotides in the analyte ions.[25] In the final stage, the resulting "daughter ion" fragments are mass analyzed with high resolution and accuracy in the mass analyzer. This process can be repeated for multiple parent ions for a given sample. Fragmentation is accomplished through collision with neutral or charged molecules or gas atoms, electron beam, or laser light. The energy used to form the ions and for activation for dissociation varies depending on method and specific instrumentation used. The dissociation channels for the given ion are a function of the ion type (positive, negative, charge state, etc.), molecular structure, internal energy distribution of the ion, and the duration of fragmentation reaction. The instrumental time frames associated with studies of oligonucleotide decompositions range from seconds (ion trapping) to some hundreds of microseconds in beam-type instrumentation.

5.2.1.1 Fragmentation Patterns and Nomenclature

The potential for sequence analysis of linear oligonucleotides by tandem mass spectrometry is dependent on the fundamental understanding of their unimolecular dissociation processes. Early studies aimed at understanding these processes were conducted in simple synthetic oligonucleotides, which were used as model compounds. Even though these studies were based on quadrupole ion trap instrumentation, the mechanistic understanding that resulted serves as the basis for understanding fragmentation in other systems. Oligonucleotide fragmentation has been thoroughly investigated by McLuckey and comprehensive nomenclature for fragment (sequence) ions was proposed.[15]

The weakest bond in oligodeoxynucleotides is the N-glycosidic bond, which is usually cleaved first under gas phase fragmentation. The nucleobase loss makes the 3′ carbon-oxygen bond sensitive to a consecutive elimination reaction, giving rise to a-B-type sequence ions. All covalent bonds around the phosphate group have the potential to be broken, resulting in 5′ (a, b, c, d) and 3′ (w, x, y, z) sequence ions. If the phosphate terminated species (d and w) lose a water molecule, the d-H_2O and w-H_2O fragments are observed in the MS/MS spectrum. These fragments have the same masses as the c and x fragments, respectively. Both nomenclatures (c versus d-H_2O and x versus w-H_2O) have widespread use in the scientific literature and refer to the 3′- or 5′-metaphosphoric-acid ester structures.[26] The RNA c (d-H_2O) fragment is likely stabilized as 2′-3′-cyclic phosphate (Figure 5.4).

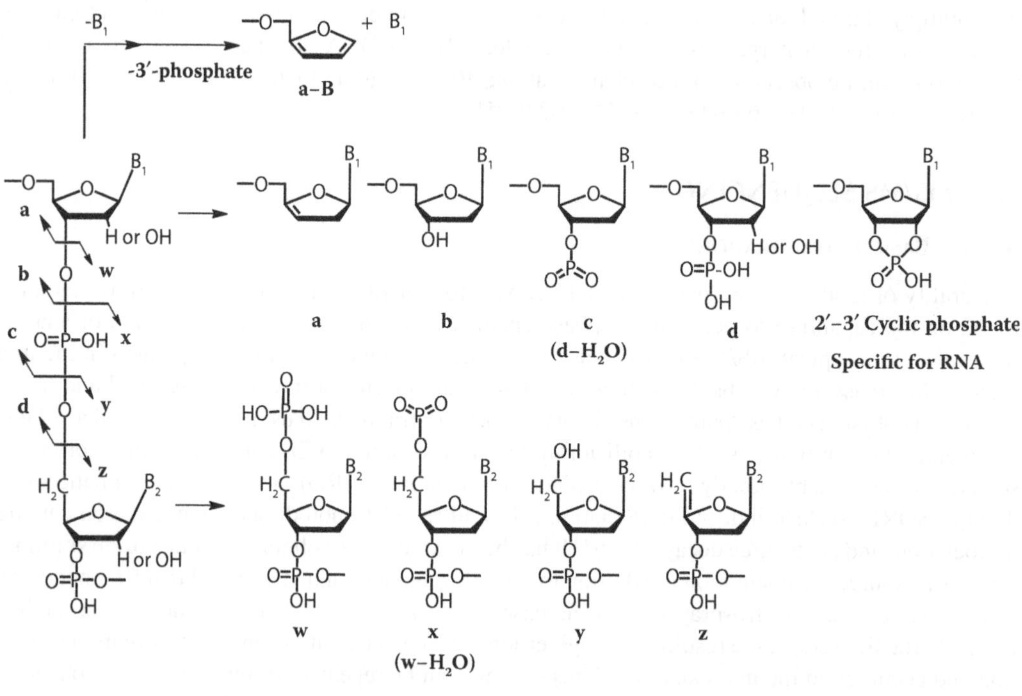

FIGURE 5.4 Nomenclature of oligonucleotide fragments by McLuckey et al.[15,27] Ion species are negatively charged in the $(M\text{-}nH)^{n-}$ type series. B_n: nucleobase at position n. The original naming of a-B ions included the position as subscript of "a" such as a_7-B, where B is lost from position 7 as well; thus, we use $(a\text{-}B)_7$ for this type hereafter. B_1, lost in forming the a-B ion fragment, denotes either charged (positively by protonation or negatively by deprotonation) or neutral nucleobase species. 'c'-fragments are also named as "d-H_2O" and "x" as "w-H_2O" by different research groups. "c"-type RNA fragments are likely in their cyclic phosphate form.

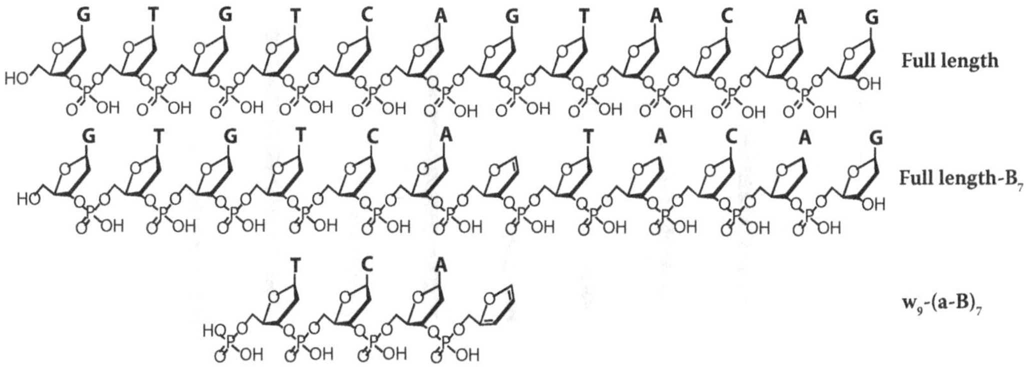

FIGURE 5.5 Nomenclature for nucleobase loss and internal fragmentation.

Base loss from any specific fragment is denoted by the "-B_n" nomenclature. Internal fragments are identified by naming both fragmented ends (e.g., w_9-(a-B)$_7$) (Figure 5.5).

5.2.1.2 Principle of Mass and Charge Conservation

Gas phase fragmentation includes ion dissociation for which process the mass and charge conservation principle applies. Conservation of mass and charge requires that $M^{n-} = m_a^{x-} + m_b^{y-}$ and $n^- = x^- + y^-$, where M is the parent ion mass, n^- is its charge, while m_a and m_b, x^- and y^- are the complementary masses of fragments and charges, respectively. McLuckey[27] demonstrated the mass and charge conservation on highly charged short DNA strands as depicted on Figure 5.6. Consideration of this principle is useful both for fundamental understanding and in specific spectral interpretation.

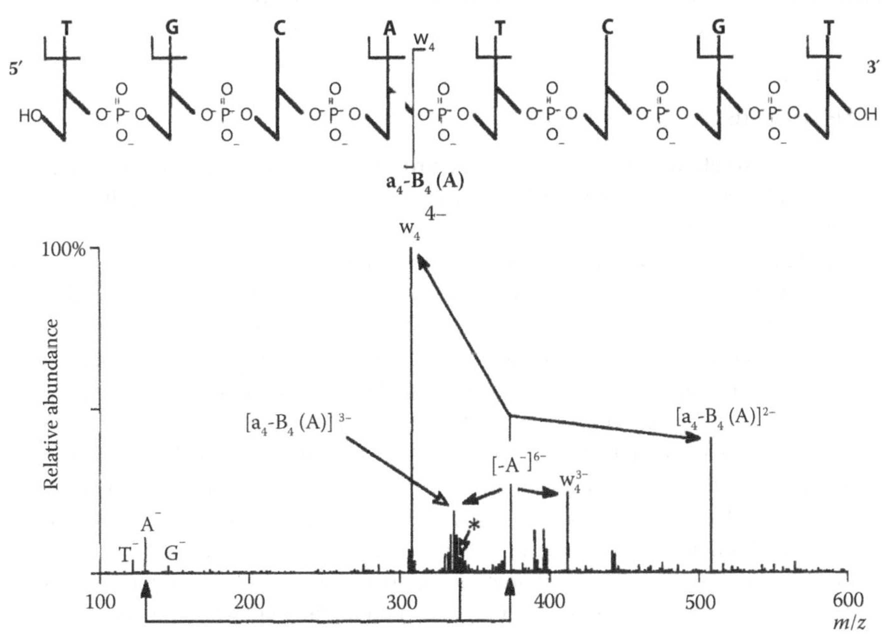

FIGURE 5.6 MS/MS spectrum of the $(M\text{-}7H)^{7-}$-ion, where M = 5′-d(TGCATCGT)-3′. The asterisk indicates the mass/charge location of the parent ion. Open arrowheads with a label at the tail are used to identify some peaks. Closed arrowheads are used to indicate the genealogy of the fragments. (Reprinted with permission from McLuckey, S. A., and S. Habibi-Goudarzi, *Journal of the American Chemical Society*, 115(25), 12085–12095, 1993. Copyright 1993, American Chemical Society.)

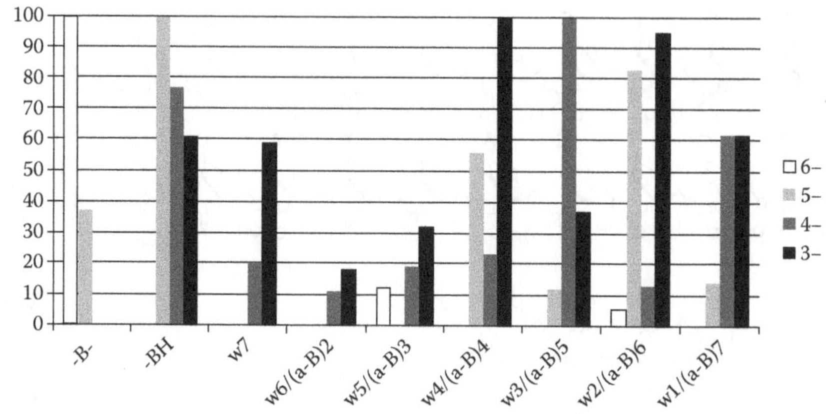

FIGURE 5.7 Ion trap tandem mass spectrometry data reduced to show the relative contributions from the competitive consecutive reactions for 6^- (white), 5^- (pale grey), 4^- (dark grey), and 3^- (black) parent anions of $d(A)_8$. (Reprinted from McLuckey, S. A., and G. Vaidyanathan, *International Journal of Mass Spectrometry and Ion Processes*, 162, 1–16. Copyright 1997, with permission from Elsevier.)

5.2.1.3 Charge Effects

Parent ion charge state plays a major role in fragmentation. In general, the high charge states (5^- and 6^- for dA_8) result primarily in nucleobase anion and neutral nucleobase loss. Lower charge states (3^-) yield in a more even distribution of every kind of a-B→w complementary sequence ions which is more useful for sequence determination. The charge effect on fragmentation by ion trap mass spectrometry has been investigated by McLuckey et al.[28] (Figure 5.7).

5.2.2 Methods

5.2.2.1 Separations

In many cases the resolution provided by mass spectrometry alone is insufficient for identification of all components of interest in a complex mixture. In these cases, it is common to utilize an initial separation technique such as capillary electrophoresis (CE) or high-performance liquid chromatography (HPLC), directly combined with electrospray ionization mass spectrometry. However, for the analysis of oligonucleotides, this presents particular challenges. Ideal conditions for reversed phase liquid chromatography separations require suppression of ionization to enhance hydrophobicity differences. By contrast, ideal conditions for electrospray ionization require ionized species in solution. The need for balancing ionization and separation conditions has led to a number of separation protocols for oligonucleotides. In one approach for negative polarity, oligonucleotide analysis utilizes 50–400 mM 1,1,1,3,3,3-hexafluoro isopropanol (HFIP), 4–12 mM triethylamine (TEA) pH 6–8 in 0–50% MeOH resulting in 1 negative charge for three to six nucleotides on the appropriate C18 reverse phase HPLC column.[29] Best peak shapes are reached at elevated column temperatures (50°–70°C). The method provides excellent chromatographic separation, while the hyphenated electrospray ionization results in a wide charge state range of multiply charged molecular ions at mild ion source conditions. Alkali metal salt formation is typical with this technique and could be somewhat mitigated by application of EDTA (5–10 μM) in the analyte solution and by elevated fragmentor (capillary) voltage. TEA also helps in alkali metal adduct reduction. A comparable separation technique is developed by Huber and Oberacher utilizing butyldimethylammonium bicarbonate (5–100 mM, BDMAB or DMBA) with acetonitrile (ACN) gradient.[30] The BDMAB method results in few charge states only, hence excellent signal intensity in the mass spectrum, while the chromatographic separation is similar to that of the HFIP method. Typical HPLC conditions used by Oberacher et al. is triethylammonium or butyldimethylammonium bicarbonate (25 mM); monolithic capillary column (60 mm × 0.2 mm ID) was prepared according to the published protocol;[31] flow rate 2 μL/min to which ACN (3 μL/min) was added through a T joint before the ion source. Sensitivity: ~300 fmol. CID energy in the IT: 13–20%.[32] Alkali metal salts are usually replaced by the butyldimethylamine during the separation (ion exchange) that is experimentally observed as peak serial with 101 Da difference.

5.2.2.2 Electrospray Ionization—Negative Mode

Electrospray ionization is most commonly used in the negative ion mode owing to the inherent acidity of the oligonucleotide backbone. Various ion fragmentation techniques and mass analyzers have been used to study the fragmentation pathways of oligonucleotide anions produced by ESI, some of which are listed in Table 5.1.

Early studies of fragmentation patterns were conducted using collisionally induced dissociation with quadrupole ion trap mass spectrometry. Fragmentation channels of small oligonucleotides were studied in order to understand gas phase dissociation mechanisms. Mono-,[44] di-,[43] and small oligonucleotide fragmentation[35,45,46] investigations serve as tractable model systems and give insights that are helpful in characterizing the dissociation of larger multiply charged oligonucleotide ions. Habibi-Goudarzi et al. investigated the fragmentation of deoxymono- and dinucleotides subjected to ion trap collisional activation. The major gas phase fragmentation channel of monomers is the neutral nucleobase loss along with the PO_3^- formation from the 5′ terminus in case of 5′-monophosphates. Water elimination from the 3′-4′ has been observed following the neutral base loss in case of 5′-monophosphates. In the ion trap collisionally induced dissociation, the order of loss of the nucleobase from nucleotides (monomers) is A, T followed by C and G. However, G base loss is nonfavored if a phosphate is present at 5′ position, explained by the stabilizing effect of an interaction of the N^2 amino group of guanine with the close 5′ phosphate. Base anion loss, however, follows a different order: $A^- > G^- > T^- > C^-$ and is very dependent on the 3′ adjacent nucleobase.[43] Guanine, again, interacting with the 5′-phosphate disfavors the 5′-base anion dissociation.

McLuckey et al. found that multiply charged oligodeoxynucleotides (4- to 8-mers) also followed the fragmentation pathway of mono- and dinucleotides, the tendency to dissociate an adenine anion, followed by 3′ C-O bond cleavage on the sugar from which the base has been lost,[15] producing an

TABLE 5.1
Activation Method and Mass Analyzer Combinations for Oligonucleotide Sequencing[a]

Activation	Analyzer	Notes	Reference
CID	Q-IT	Multiply charged ions; low energy activation	15
CID	QQQ	Multiply charged ions; low-energy activation	14, 22, 33, 34
NS	EBqQ		22
CID	QhQ, EB-TOF		35
In-source CID, NS, ICR, IRMPD	FT-ICR	8- and 108-mers; In-source fragmentation is very efficient but needs very high resolving power mass analyzer.	20, 23, 24
BIRD	FT-ICR	50- and 100-mer ssDNA versus 64-mer dsDNA	36, 37
CID	IT	10-mers; Guanine determination by N^7-G methylated DNA MS^n.	38
CID, PSD	Q-IT, MALDI-TOF	Metal ion adduct and photomodified DNA analysis.	39–42
CID	ESI-IT-MS^n	2- and 3-mer	43

[a] Q or q, quadrupole; IT, ion trap; EB, magnetic sector; h, hexapole; CID, collision-induced dissociation; NS, nozzle skimmer; PSD, postsource decay; ICR, ion cyclotron resonance; IRMPD, infrared multiphoton dissociation; BIRD, blackbody infrared radiative dissociation; ssDNA, single-stranded DNA; dsDNA, double-stranded DNA.

a-B and the complementary w sequence ions. Identification of such complementary fragments greatly simplifies spectral interpretation.

McLuckey determined the order of nucleobase (anion or neutral) loss under quadrupole ion trap conditions for multiply charged single-stranded oligodeoxyribonucleotides to be A^-, G^- or T^- and C^-.[27,47] Vrkic et al. investigated singly charged tri- and tetranucleotides and found the neutral base loss order to be A, T, G, and C.[45] Nucleobase loss from the 3′ is not favored. The subsequent decomposition step is the cleavage of the 3′ C-O bond of the sugar from which the base was lost yielding in a-B- and w-type ions. Consecutive base loss from both end fragments is likely, although w ions at lower charge states favor greatly the 5′ PO_3^- loss as well (w→y). Base loss from either-end fragments likely to be followed by a second strand cleavage to yield internal fragment ions. Other sequence ions such as c, x, and z are also observed.

In triple quadrupole collision cells, parent ions tend to undergo multiple collisions compared to the single collision process in a quadrupole ion trap (Q-IT). Consequently, triple quadrupole and other mass analyzers that utilize a quadrupole based collision cell do not have as strong a preference for base specific dissociation as do quadrupole ion traps as it was discovered by Barry et al.[14] or Boschenok and Sheil.[35] Crain et al. summarized the efforts on tandem mass spectrometry for sequence characterization and stated that base loss is not a prerequisite for backbone cleavage in fast time frame activation and the beam-type collisional activation of the triple quadrupole instrument is less selective and allows for multiple competitive dissociation reactions, thus giving more extensive structural information than Q-IT and FT-ICR.[34]

Little and coworkers investigated high-pressure nozzle skimmer and low-pressure collision-activated dissociation (most internal fragment ions)[20] and IRMPD (a-B- and w-series) of ESI anions and concluded that NS and IRMPD give similar amounts of sequence information for smaller DNA oligomers (up to 25-mers) but induce different yet complementary fragmentation patterns for a 50-mer.[23]

Fourier transform ion cyclotron mass spectrometry can be equipped with a range of unique fragmentation techniques not available to other analyzers. The nucleobase loss in ESI-FT product ion spectra for 50-mer oligodeoxynucleotides follows the order of A, G and C, and very few T's. Klassen et al. found that 7-mer DNAs fragmented in a BIRD-FT-MS also confirmed that the favored fragmentation path is neutral base loss followed by cleavage of the 3′-phosphodiester bond (a-B and w ions); however, no thymine loss is observed. The activation energy (E_a) for adenine neutral loss is 1.05 eV, for C and G E_a = 1.32 eV.[48] Dissociation kinetics is dependent on the internal energy of the precursor ions, hence temperature and charge state affect the preferential nucleobase loss. At lower internal energies (effective temperature <475 K) adenine loss is predominant, while at higher (effective temperature >540 K) C and G loss dominate.

Investigations by Wang et al. and Wan and coworkers on T-rich short deoxyoligonucleotides (4-, 5-, 6-, 8-, and 10-mers) under ESI and MALDI with four different activation types (low-energy LCQ IT, ESI-source CID, high-energy ESI-MS/MS, and PSD of MALDI) revealed that T base loss is least, G loss is most facile, while C and A bases are intermediate.[49–51] This observation is in accord with the proton affinities of the nucleobases (G > C > A > T).[52] It has also been concluded that the proton transfer from a phosphate to the leaving nucleobase plays an important role in the dissociation in the relatively low charge state oligodeoxyribonucleotides.

DNA duplexes have been investigated in ESI mass spectrometry[37,53–60] and under MS/MS with a hybrid quadrupole-TOF instrument.[60,61] More complicated DNA complexes, such as quadruplexes,[62] drugs, metals, and proteins have been widely explored.[63–65]

RNA MS/MS Sequencing in Negative Polarity

RNA gas phase fragmentation is significantly affected by the presence of the 2′-hydroxyl group, which has a stabilizing effect on the N-glycosidic bond presumably due its electronic effect, and hinders the nucleobase 1′-2′ *trans* elimination.[66] The nucleobase loss in the mechanism of RNA gas phase fragmentation does not play as an important role as for DNA. RNA sequence ions are

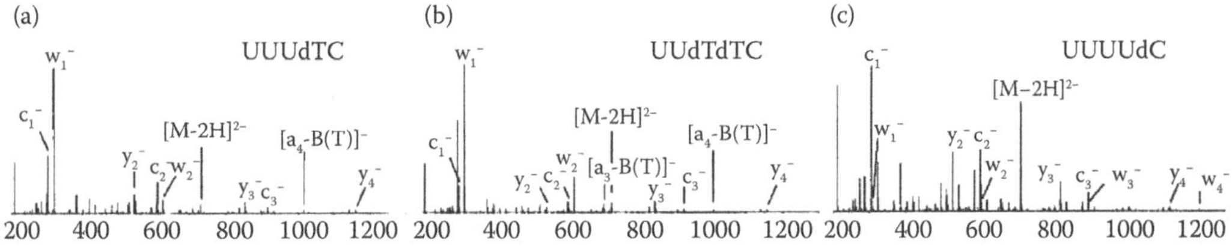

FIGURE 5.8 Product ion spectra of the mixed-sequence RNA/DNA pentanucleotides (a) UUUdTC, (b) UUdTdTC, and (c) UUUUdC. Dissociation of doubly charged precursor ions with a collision energy of 30 eV. The presence of a deoxyribonucleotide within the RNA sequence is indicated by abundant w_1^- and w_2^- fragment ions. Dissociation of the UUUUdC pentanucleotide with a 3′-terminal deoxycytidine does not generate an indicative w-type fragment ion. (Reprinted from Schurch, S., et al., *Journal of the American Society for Mass Spectrometry*, 13, 936–945, Copyright 2002, with permission from Elsevier.)

generated by cleavage around the phosphate diester bond such as 5′-O-P scission resulting in y and c sequence ions. In general, RNA fragmentation needs more activation energy than that of DNA. This phenomenon is demonstrated with DNA-RNA chimeric oligonucleotides upon low-energy CID in a hybrid quadrupole-TOF mass spectrometer[26] in negative polarity; preferential DNA cleavage with a-B- and w-type fragmentation was observed, versus c and y sequence ions for the RNA positions (Figure 5.8).

Primary structure determination of nonprotein-coding RNAs (ncRNA, lincRNA, siRNA, pre-microRNA, riboswitches) is of particular interest for biomolecular research and pharmaceutical development. Most recent publication by Taucher et al. describes top-down *de novo* sequencing of a 34 nt riboswitch size RNA by collisionally activated dissociation. The key of success to extending the size limitations to 30-mers and over relies on the minimization of energy differences of individual ions in the precursor ion cloud. The authors utilized collisional cooling and selecting low charge state parent ions in order to mitigate the internal (vibrational and Coulombic) energy of ions and as a result of that, producing relatively simple spectra that could be used for *de novo* sequencing of a 34-mer RNA molecule.[67]

5.2.2.3 Electrospray Ionization—Positive Mode

Positive ion mass spectrum has been obtained for DNA and RNA from solutions containing nitrogen-containing organic bases.[68] The phenomenon of positively charged oligonucleotide formation in the gas phase is explained as the excess positively charged weak ammonium base ions, in complex with the oligonucleotide, can dissociate under mild CID conditions in the ion source leaving extra positive charges on the nucleobases. Thus, the ammonium base should have similar or slightly lower proton affinity than the nucleobases in the oligonucleotide. Experimentally, it has been observed that the signal intensity decreases as the proton affinity of the nitrogen-containing base increases.

A typical protocol for positive polarity oligonucleotide analysis utilizes 5–10 mM ammonium acetate, pH 6 10–40% MeOH, 0.2–0.25 mM 1,2-cyclohexanediamine *tetra*-acetic acid (CDTA) resulting in three to nine positive charges for 14- to 77-mer DNA oligonucleotides. Organic bases were used including aniline, imidazole, diethylamine, and triethylamine instead of ammonia.

Protonated oligonucleotide gas phase collision-induced dissociation is initiated by release of protonated nucleobases. This mechanism is largely driven by the proton affinities of the critical heteroatom of the nucleobases in the nucleotides plus the stability of released tautomeric forms of the nucleobases. In general, the experimental order of nucleobase loss follows the trend of the predicted proton affinities of the critical heteroatom in the nucleoside that is leading to the most stable tautomeric form of the nucleobases, which is, with the exception of dG, not the preferential protonation site,[46] i.e., preferred protonation sites for the nucleobases are G: N^7, C: N^3, A: N^1, T: O^4,

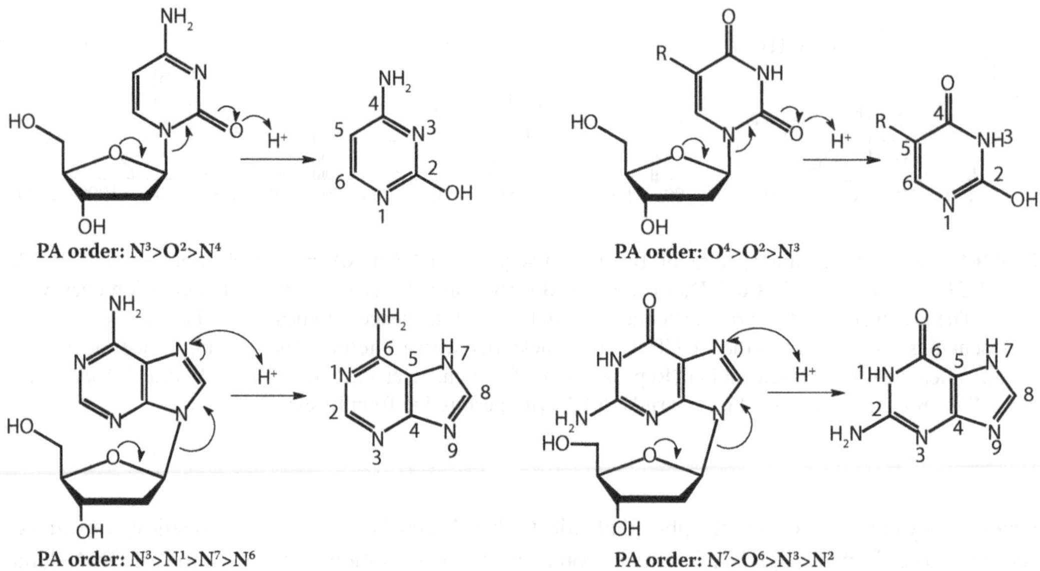

FIGURE 5.9 Positive ion ESI-MS of DNA. Recommended mechanisms for nucleobase loss initiated by base protonation either directly by the proton from ionization or by intramolecular zwitter-ion formation between the phosphodiester group and the nucleobase.

while protonation at G: N^7, C: O^2, A: N^7, T: O^2 results in the most stable tautomeric form of the product (Figure 5.9).

The proton affinities (PA) for nucleobases, nucleosides, and some amines are listed in Table 5.2. Nucleobase dissociation preference, typical sequence ions, and instrumentation are collected in Table 5.3 for positive mode tandem mass spectrometry of DNA.

5.2.2.4 MALDI—Positive Mode

RNA sequencing by characterization of enzymic digestion products by MALDI-MS has been successful up to 4000 Da.[77] However, the resolution and accuracy of MALDI beyond 4000 Da is limited

TABLE 5.2
Experimental PA Values

Nucleobase	PA (kcal/mol)[a,b]	dN	PA (kcal/mol)[a,b]	dN	PA[c] (kcal/mol)[d,e]	Amine	PA (kcal/mol)[f]
G	229.3/227.4	dG	237.9/234.4	dG (N^7)	226.5/234.4	Aniline	209.49
C	227.0/225.9	dA	237.0/233.6	dC (O^2)	226.4/234.9	Imidazole	223.50
A	225.3/224.2	dC	236.2/233.2	dA (N^7)	218.1/223.2	Diethylamine	225.91
T	210.5/209.0	dT	226.7/224.9	dT (O^2)	202.9/—	Triethylamine	232.19
$(CH_3O)_3PO$	212.9[a] (theoretical)						

[a] E. P. L. Hunter and S. G. Lias. 1998. *Journal of Physical and Chemical Reference Data* 27(3): 413–656.
[b] F. Greco et al. 1990. *Journal of the American Chemical Society* 112(25): 9092–9096.
[c] Computationally predicted proton affinity values for the given heteroatom.
[d] S. G. Lias, J. F. Liebman, and R. D. Levin. 1984. *Journal of Physical and Chemical Reference Data* 13(3): 695–808.
[e] T. Marino et al. 1994. *Theochem* 112(2–3): 185–195.
[f] J. Smets et al. 1996. *Chemical Physics Letters* 262(6): 789–796.

TABLE 5.3
Summary of Positive Polarity MS/MS of DNA[a]

Order of Base Loss	Base Loss Position Preference	Size	Fragment Ions	Instrumentation	Notes	Reference
N >> T	3′ > 5′ > internal[b]	2	FL-BH/w	ESI/Q-h-Q/CID	1	35
C ~ G ≥ A ≥≥ T	5′ > 3′ > internal	3, 4	a-B/w	ESI /Q-IT	2	46
C > G > A >> T	3′ > 5′ > internal	4	a-B/w	ESI/QQQ	3	73
C,G,A >> T	—	4–20	a-B/w	ESI/QQQ	4	74
N >> T	—	6–10	a-B/w	EBE-TOF	5	75
No T	—	4–7	w/d, a/z, d and z·	ECD	6	76
No T	—	4–7	a-B/w, -B and $-H_2O$	IRMPD	7	76

Notes: 1. Increasing the skimmer potential, thus the internal energy of the precursor ions, increased the relative intensities of the fragment ions without enhancing the sequence-specific fragmentation. Li^+ ion adducts yield more sequence ions than $(M+H)^+$ or $(M-H)^-$. $(M+Na)^+$ or $(M+K)^+$ give rise to few or no fragmentation.

2. T-rich sequences are cleaved by d/z-type fragmentation, explained by the low PA of T. This type of cleavage is initiated by the phosphate oxygen protonation.

3. T is more readily lost from 5′ than from the 3′ as seen for dinucleotides. Fragmentation at 3′ from T bases is absent or very low in abundance.

4. Dissociation 3′ to T nucleobases is disfavored, yielding to x and z ions exclusively. Greater abundance for high *m/z* ions is observed, compared to negative polarity experiments. Higher collision energy is required for positive ions to obtain similar extent of parent ion dissociation.

5. High collision energy ($E \sim 400$ eV) has been applied. Less selectivity toward base dissociation is observed. Thymine base loss and cleavage 3′ from a T nucleotide is absent because $(T+2H)^{2+}$ ion formation is disfavored. The authors identified the ion at $m/z = 81$ as a sugar derivative, supporting the theory that the sugar residue protonates more likely than the phosphate.

6. The electron-capture dissociation follows sequence dependent pathways for fragmentation.[76] Guanosines form radical w/d-type ions, and adenosines and cytosines result in even-electron w/d sequence ions; however, radical a/z-type ions are found in the ECD spectra of polydC oligodeoxyribonucleotides, while no even-electron a/z ions are found for polydGs. For mixed sequence 6-mer d(GCATGC), even-electron d and z· ions are also found. A novel type of ion is observed: d+H_2O of which structure is unclear. ECD of oligodeoxyribonucleotides is complicated and is not characterized thoroughly for application to primary structure determination.

7. Infrared multiphoton dissociation activation gives rise to ion types similar to those seen in CID. Abundant a-B/w fragment ions are assigned along with nucleobase and H_2O losses, which complicate spectral interpretation.

[a] N: any (other) nucleobase; ECD, electron capture dissociation.

[b] Except when T is the 3′ nucleobase.

by the low mass difference of C and U nucleotides (1 Da). A MALDI Q-q-TOF instrument has been used to evaluate the possibility to extend MALDI sequencing to longer RNAs by combination of enzymic digestion and collision-induced dissociation tandem mass spectrometry in an orthogonal time-of-flight mass analyzer. The dominating sequence ion types for RNA are the c and y ions, although nearly all types of cleavages along the phosphodiester backbone and of the N-glycosidic bonds (and combinations of these) have been observed. Neutral nucleobase loss has been observed in the order of C ~ G > A, whereas loss of uracil, either neutral or charged species, was not observed. A systematic investigation of the effect of increasing collision energy on the fragmentation revealed that protonated nucleobase loss requires higher energy than that of the neutral species. Furthermore, the higher collision energy resulted in more sequence ions, although more nucleobase losses (up to 3 in one ion) have been observed (Figure 5.10).[78]

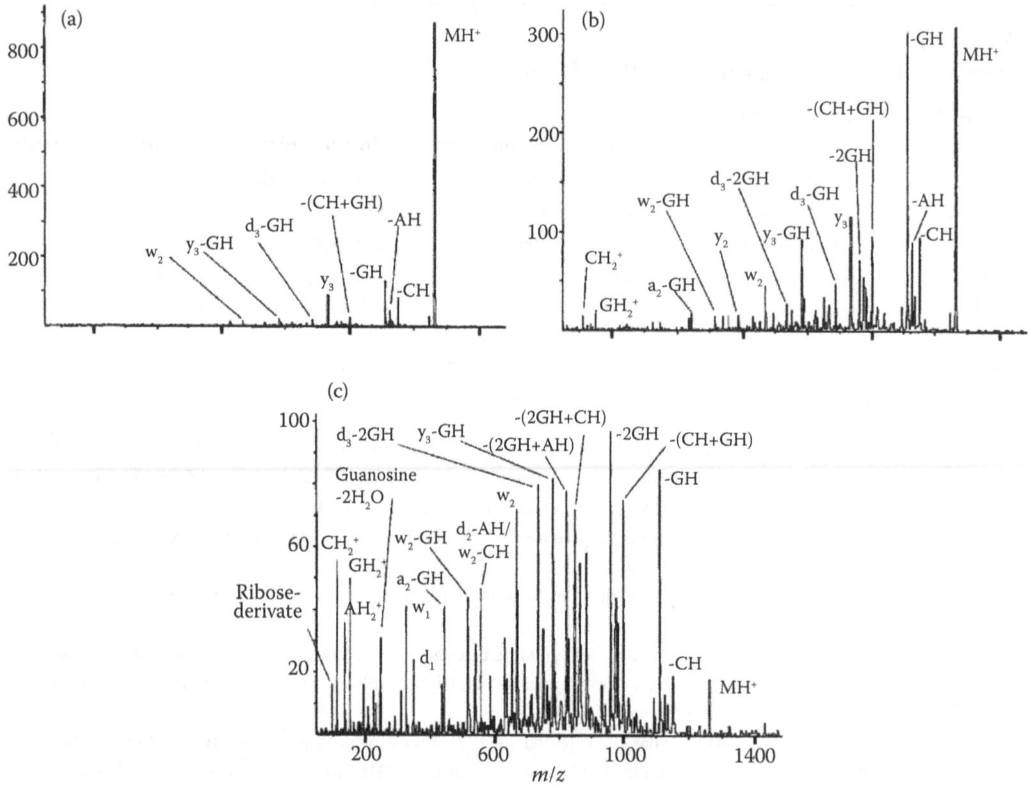

FIGURE 5.10 Effect of collision energy on the fragmentation pattern. The analyte was singly protonated RNA tetramer AGGC (m/z = 1263.26). The collision energies were (a) 40, (b) 50, and (c) 60 eV. (From Kirpekar, F., and T. N. Krogh, *Rapid Communications in Mass Spectrometry*, 15, 8–14, 2001. Copyright Wiley-VCH Verlag GmbH & Co. KGaA. Reproduced with permission.)

5.2.2.5 Software Tools for Spectral Interpretation

Except in the cases of very short oligonucleotides, very simple samples, or very few spectra, spectral interpretation can rapidly become the bottleneck in mass spectrometry based sequencing experiments. Again, taking the lead from proteomic applications, and particularly mass spectrometry-based *de novo* spectral interpretation algorithms, researchers have begun to develop informatics to aid in data interpretation for oligonucleotide sequence determination.

5.2.2.5.1 SOS

Computer software assisted *de novo* sequence determination for short (<20-mer) DNA and RNA oligomers is described by Rozenski and McCloskey.[79] The interactive, stand-alone Simple Oligonucleotide Sequencer (SOS) software works under user control on a residue-by-residue basis sequence determination. Sequence ladders for a-B, w, y, and d-H_2O ion series are constructed independently from a maximum of five nucleotides of DNA, RNA, or user defined modifications on the sugar, base, or the phosphate. The calculated ladder masses open a graphical window (ion panel) for each of the nucleotides in the mass spectrum, from which the user visually and manually selects the match by homology and overlap. The software is available for academic use from the authors (Figure 5.11).

5.2.2.5.2 Web-Based Tools

Web-based calculators (Nucleic Acids Masspec Toolbox) have been developed by Rozenski for calculations of oligonucleotide exact mass from the sequence, ESI series, CID fragment table, internal

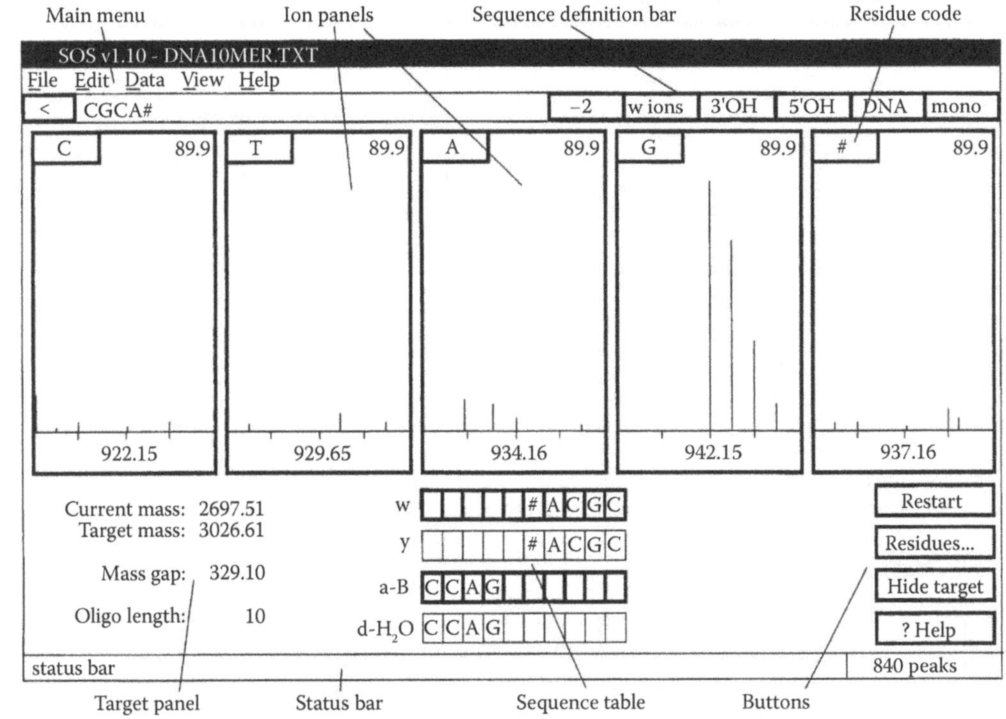

FIGURE 5.11 SOS graphic display generated during sequencing of the DNA oligonucleotide 5′-CCAGG#ACGC-3′, where # is an RNA residue, 2′-*O*-methylcytidine (rCm), which has been defined under residues. The ion panels display experimental ion patterns for candidates for the sixth residue from the 3′ end, for doubly charged w ions in the monoisotopic mass scale. Cumulative assignments from the w and a-B ion series at the point of processing account for 2697.51 Da, displayed in the target panel. From the user entered experimental molecular mass of 3026.61, the missing fifth nucleotide mass is calculated as 329.1 Da (the residue mass of dGp). If G is accepted as the next residue from ion panel data (also supported in the w_6^{-1} panels, not shown), the gap mass will be near zero. Further extension of the y ion series in the 3′-5′ direction by GG (not shown) is readily made, which provides limited sequence overlap in both directions, and leads to the full oligonucleotide candidate sequence 5′-CCAGG-rCm-ACGC-3′. (Reprinted from Rozenski, J., and J. A. McCloskey, *Journal of the American Society for Mass Spectrometry*, 13, 200–203. Copyright 2002, with permission from Elsevier.)

fragments, endo-(T_1, U_2, A), and exonuclease degradation for DNA, RNA, and modified bases (Mongo Oligo Mass Calculator v2.06). Additional calculators provide for oligonucleotide composition from exact mass (Oligo Composition Calculator v1.2) and isotope envelope from molecular formula (MolE-Molecular Mass Calculator v2.02) and elemental composition calculator from exact mass (v1.0) (http://library.med.utah.edu/masspec/).

Helpful resource from Rozenski and McCloskey[80] is the small subunit rRNA modification database at http://library.med.utah.edu/SSUmods/ and the Modomics database by Dunin-Horkawicz et al.[81,82]

5.2.2.5.3 COMPAS

A computer-aided comparative resequencing algorithm has been developed for polymorphism identification in genomic DNA for clinical, microbial, forensic and population genetics studies.

The method compares MS/MS data to fragment masses of a reference sequence. Genome-based DNA fragments of interest are polymerase chain reaction (PCR) amplified and analyzed in an ESI quadrupole ion trapping mass spectrometer. Tandem mass spectrometry (MS/MS) is performed on

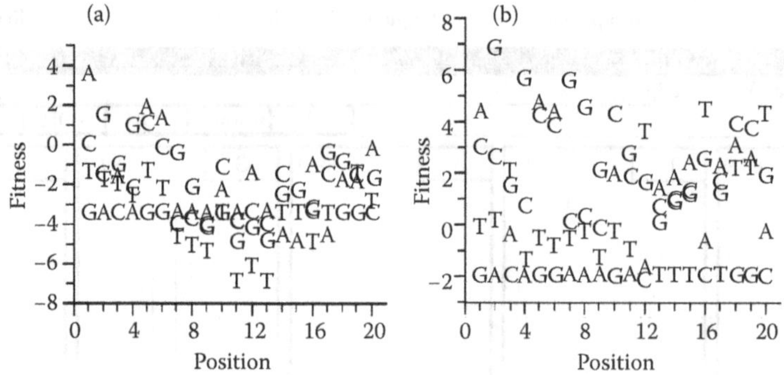

FIGURE 5.12 Sequence correlation diagrams of fitness versus position for a 20-mer oligodeoxynucleotide resulting from comparison of (a) reference sequence 1 (A in position 13) and (b) sequence 2 (T in position 13) to the MS/MS spectrum of sequence 2. Reference sequence yielded minimum FS values throughout the whole spectrum that proved the identity of sequence 2. (Reprinted with permission from Oberacher, H., et al., *Analytical Chemistry*, 74, 211–218. Copyright 2002, American Chemical Society.)

multiply charged oligonucleotide precursor ions applying selective collision-induced dissociation conditions (charge state, collision energy) to generate mainly a, a-B and w, w-B ion series. The sequence ladders in the mass spectrum are correlated with calculated mass values by the COMPAS software. The method consists of mass data table generation in an Excel table format for a reference sequence, and for mutations of the reference oligonucleotide at every position. The predicted values are compared to the MS/MS data and matching values are correlated with a so-called fitness (FS) function. Examples demonstrate that the fully automatable algorithm can be used to easily confirm the identity of an unknown sequence with a reference sequence as well as to detect point mutations in comparison to a given reference sequence (Figure 5.12).[83]

The COMPAS sequencing has been used for resequencing of SNP,[84] large oligonucleotide sequence verification (up to 80-mer)[32] with ESI-ion trap or ESI-QQ-TOF-MS/MS.[85] The COMPAS software is available for academic use from the authors or on-line at http://insertion.stanford.edu/comp_seq/comp_seq.html.

Other sequence analysis software with detailed description in the referenced papers are listed here by Ariadne (http://ariadne.riken.jp/),[86] SpecDiff,[87] and MS2Links.[88]

5.3 MALDI SEQUENCING

Some mass spectrometry based sequencing procedures are preceded by benchtop laboratory procedures such as enzymatic digestion, chemical degradation, or chain termination of polymerization to form a nested set of different length nucleic acid strands of the target oligonucleotide. MALDI-TOF mass spectrometry is popular for analysis of such complex sequencing mixtures because it gives rise to singly charged ions, hence simpler spectra. Also, MALDI is considered a less labor intensive and more robust ionization method than ESI. In the past, however, the resolution of MALDI-TOF mass spectra has been limited owing to a number of factors such as the oligonucleotide mixture complexity, applied matrix, and instrumentation. Alkali and alkali earth metal salt adducts decrease resolution.[89] Uneven energy distribution of ions, fragmentation during desorption, ionization, and metastable dissociation result in signal broadening and subsequent loss of detection sensitivity.[90] The mass detection range is limited owing to severely decreasing sensitivity and sequence specificity for higher masses. The low desorption and ionization ability of large DNAs paired with depurination (mostly G, -151 and A, -135) also complicate accurate mass determination or differentiation of large DNA fragments. Because RNA is less prone to depurination and alkali salt complexation, it is thus expected to be more suitable for MALDI-TOF-MS based sequencing. Reliability of RNA MALDI sequencing is only limited by resolving power and accuracy of the mass spectrometer to differenciate 1 Da mass difference between the U and C nucleotides especially at high mass ranges (>4000 Da).

Matrix-assisted laser desorption/ionization mass spectrometry for analysis and sequencing oligonucleotides has been made more practical by several significant developments. Alkali salts are exchanged on the target plate by application of organic and inorganic acid ammonium salts and/or ammonium form cation-exchange resins[91] as additives in the matrix. Peak broadening by ion energy distribution differences, fast fragmentation, and metastable decomposition in the field-drift region are minimized by delayed ion extraction (DE or time lag) MALDI[92] and reflector TOF. Implementation of delayed extraction MALDI mass spectrometry enabled DNA sequencing over 33 bases and routinely applied up to 50-mers, thanks to the resolution and sensitivity enhancement.[90] Depurination has been reduced by applying milder ionization conditions and matrices. Sanger

TABLE 5.4
MALDI Matrices for Oligonucleotide Analysis

Matrix	Acronym	Laser Wave-Length (nm)	Oligo Length	Reference
3,5-Dimethoxy-4-hydroxycinnamic acid (sinapinic acid)	SA			96, 97
2,5-Dihydroxybenzoic acid and 3,5-dimethoxy-4-hydroxycinnamic acid	DHB/SA	266–355		12, 98
2,5-Dihydroxybenzoic acid (gentisic acid)	DHB	266, 337, 355		99
α-Cyano-4-hydroxycinnamic acid	CHCA	337, 355		100
Succinic acid or urea		2.79μm, 2.94μm, 10.6μm		91
2-Aminobenzoic acid/nicotinic acid (20–50%)	AA/NA	266–355		91, 101
3-Hydroxy-4-methoxybenzaldehyde/methylsalicylic acid		266–355		102
Water/corroded copper			60-80	103
4-Hydroxy-3-methoxycinnamic acid (ferulic acid)	HMCA	266, 355, 337	10	104 51
3,5-Dihydroxybenzoic acid		355	20-80	105
2,4,6-Trihydroxy-acetophenone/diammonium sulphate, diammonium hydrogen citrate, diammonium-L-tartarate				106
3-Hydroxypicolinic acid	HPA	266, 337, 355	<500	107–109
Picolinic acid	PA	266		108
3-Aminopicolinic acid		266/355		110
3-Hydroxypicolinic acid/picolinic acid (1:8, 4:1)		266, 337/355		108, 111
6-aza-2-Thiothymine	ATT	377	20	112
2,4,6-Trihydroxy-acetophenone/2,3,4-trihydroxy-acetophenone (2:1)		337		113
8-Hydroxyquinoline				114
N-(3-Indolylacetyl)-L-leucine	IAL	337	25	115
Liquid glycerol		2.94 μm	2180	116
3-Hydroxypicolinic acid/picolinic acid/ammonium fluoride		337		117
4-Nitrophenol		355	1000	118
5-Methoxysalicylic acid/spermine		337		119
2-Aminobenzoic acid/ammonium fluoride	ABA/NH_4F		8	120

Source: Reprinted from Wu, J. and S. A. McLuckey, *International Journal of Mass Spectrometry* 237, 197–241. Copyright 2004, with permission of Elsevier.

sequencing mixture complexity is simplified by affinity purification, applying biotin on the primer or on the ddNTPs followed by filtration of the oligonucleotides of interest by streptavidin coated magnetic beads. If suitable, DNA is transcribed to RNA or vice versa to trade off lower depurination and salt formation for increased ambiguity in C/U differentiation at higher masses.

Molecular weight determination by MALDI-MS up to 461 nt long RNAs was reported compared to 150 nt for DNA,[93] which is in the range of Sanger sequencing (600 nt), although the expected application of MALDI-MS on Sanger sequencing mixtures of that size[94] has not yet been implemented.

The success of MALDI-MS for oligonucleotide analysis relies on the proper combination of matrix material and laser wavelength.[95] Suitable matrices for oligo MALDI are listed in Table 5.4.

5.3.1 Direct MALDI Sequencing

In contrast to the MS/MS based approach utilized with ESI-MS, direct MALDI sequencing of DNA and RNA is based on fragmentation during ionization with single stage mass analysis. Nucleic acid fragmentation in the MALDI source is classified into four categories: prompt, fast, early metastable, and metastable fragmentation,[95,121] depending on the time frame in which the metastable ion decomposes compared to the ion acceleration time frame. Prompt fragment ions are formed during ion generation, thus detected as fully accelerated ions. Fast fragmented ions slightly broaden peak shapes at the higher mass side that might be compensated by delayed ion extraction[92] or in reflectron-TOF (R-TOF). Early metastable fragmentation occurs during acceleration, thus the detected ions are dispersed in a wide spectral region and is responsible for limitations of detection of large oligonucleotides (>50-mers). Metastable fragmentation also occurs in the field-free drift region. These ions are detected concurrently with their precursor ions in linear TOF but time-dispersed in R-TOF. This characteristic is exploited when these ions are detected as sharp peaks in post-source decay (PSD) analysis.[122]

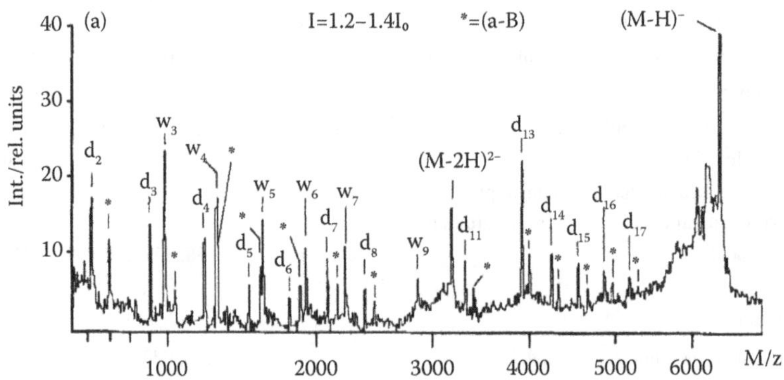

(b)

Simplified fragmentation pattern:

```
                            9    7 6 5 4 3              :w
3′-d4T-5′-5′-A-C-A-C-C-C-A-A-T-T-C-T-G-A-A-A-A-T-G-G-3′
      d:       2 3  4  5  6 7  8        11  13 14 15 16 17
   (a-B)
```

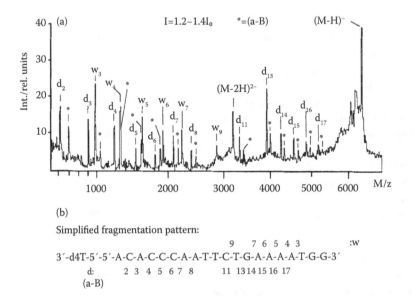

FIGURE 5.13 Direct MALDI sequencing. (a) Negative-ion IR-MALDI-RTOF mass spectrum of a 5′-modified antisense-DNA 21-mer obtained with succinic acid as matrix at increased laser irradiance ($I = 1.2–1.4I_0$). The found d-, a-B-, and w-ions are assigned; a-B-ions are indicated by asterisks. The modified nucleoside d4T (2′,3′-dideoxy-2′,3′-didehydrothymidine) has been added to the 5′-end of the antisense-DNA 20-mer by a 5′,5′-phosphate diester linkage. Total oligodeoxynucleotide load: ~10 pmol. Sum of 15 single-shot spectra. (b) Simplified fragmentation pattern derived from part a. (From Nordhoff, E., et al., *Journal of Mass Spectrometry*, 30, 99–112, 1995. Copyright Wiley-VCH Verlag GmbH & Co. KGaA. Reproduced with permission.)

As examples, direct sequencing by IR-MALDI-R-TOF in negative polarity with succinic acid matrix confirmed 14 bases of 21. Approximately 30% greater laser irridance has been applied compared to optimal value for intact molecule. The most abundant fragment ions fell in the w-type, along with a-B and d sequence ions. A, G, and C sites were cleaved at similar yields, while no fragmentation has been observed for Ts. Internal fragments were found also. Sequence determination based on d- and w-type ions was feasible (Figure 5.13).[121]

Positive ion MALDI direct sequencing has enhanced fragmentation compared to negative polarity. However, molecular ion signals are weaker and low-mass nucleobase-related ions are more abundant in positive mode than in negative. Ion types fall into the analogous wH_2^+ and $(aH_2\text{-}B)^+$ series.[123] Implication of UV-DE-MALDI-TOF at 266 nm laser wavelength using picolinic acid matrix resulted in direct sequence determination for a 11-mer DNA (Figure 5.14).[92]

Direct fragmentation/sequencing methods are summarized in Table 5.5.

5.3.2 Failure Analysis

Failure analysis for chemically synthesized oligonucleotides is in itself a valuable application for mass spectrometry but also provides sequence information based on the fact that efficiency of couplings in the stepwise chemical synthesis of oligonucleotides is never 100%, rather typically 95–99.9%; thus shortmers of the target sequence are formed. The failure oligos are stabilized by the capping step until the end of the synthesis. In the final cleavage step the full length product along with the failures are liberated. The short oligonucleotides show up in the crude product mixture at low concentration (0.1–5%). MALDI mass spectrometry of the crude product mixture of a synthetic oligonucleotide reveals the short oligonucleotide failure sequence ladder, from which each nucleotide in the final full length product can be determined by the *m/z* differences between adjacent peaks (Figure 5.15).[92,128]

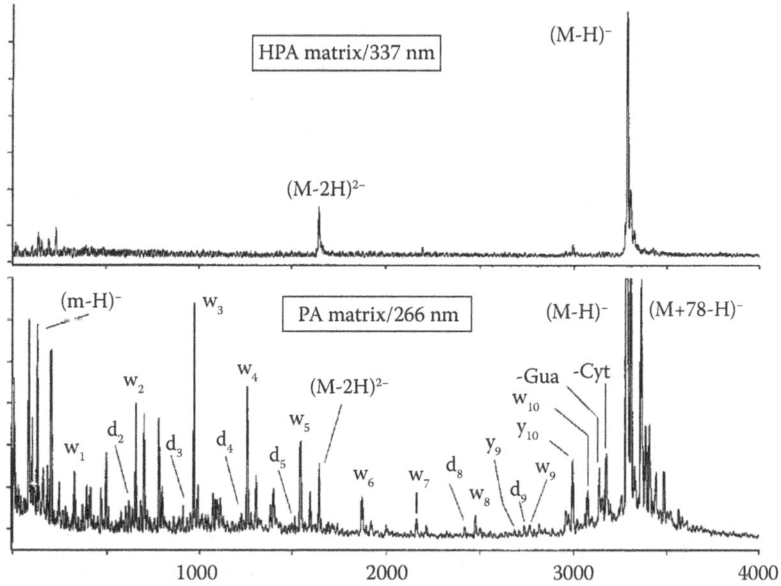

FIGURE 5.14 Sequencing of the 11-nucleotide deoxyribonucleotide CACACGCCAGT by fast fragmentation. Standard conditions (3-HPA matrix, 337-nm wavelength) generated stable ions without fragmentation. Picolinic acid (PA) matrix at 266-nm wavelength generates a complete series of fragments (w series) that permit sequence readout. $(m\text{-}H)^-$ denotes the most abundant matrix ion (m/z = 122). Instrument, 1.3 m linear. (Reprinted with permission from Juhasz, P., et al., *Analytical Chemistry*, 68, 941–946. Copyright 1996, American Chemical Society.)

TABLE 5.5
Summary of MALDI Methods for Oligonucleotide Direct Fragmentation/Sequencing[a]

Polarity	Bases	Degradation Type	Order of Base Loss	Instrument	Matrix	Reference
–	>2	BH	G,A > C >> T (U)	UV-MALDI	FA, DHB, CHCA, SA	1, 124
–	21	a-B/w, d, w-d (internal) Direct sequencing	A,G,C and no T; DNA>RNA>dT_n	IR-MALDI	SA	124, 125
+–			G >> T	UV-MALDI-FT-ICR-MS	DHB, AA/NA, HPA	123, 125
–	15	a-B/w Direct sequencing	G > C > A >> T; G ~ C > A and no T	UV-MALDI-L-TOF	DHB (355 nm)	126
–	11	w, d Direct sequencing		UV-DE-MALDI	PA (266 nm)	92
–	10	a-B/w Direct sequencing	G > C ~ A >> T	DE-RP-MALDI-TOF PSD and CID	HMCA (337 nm)	49, 51
–	4	a-B, d, w, y, $(w+HPO_3)^-$		CAD and SORI-FT-ICR-MS	HPA	127
+	25	w, d (PO, MP), a (PS, ON), z	G, C, A, T	DE-MALDI-TOF	HPA, IAL (337 nm)	115
+	8	PSD: a-B/w, z-B/d Fast fr.:w, b, y, d	A >> T	UV-MALDI-TOF	ABA/NH_4F	120

[a] PO, phosphate; MP, methylphosphonate; PS, phosphorothioate; ODN, oligodeoxyribonucleotide; ON, oligoribonucleotide; CAD or CID, collisionally activated dissociation; SORI, sustained off-resonance irridation.

5.3.3 Degradation

Degradation sequencing techniques have important applications in the study of DNA-protein interactions (footprinting), nucleic acid structure, and epigenetic modifications to DNA. Natural or just slightly modified nucleic acids can be degraded by enzymatic digestion. Nucleobase specific chemical degradations are initiated by reactions on one of the heterocyclic nucleobase rings, followed by elimination of the nucleobase and cleavage after and before the abasic sugar moiety. Nonspecific

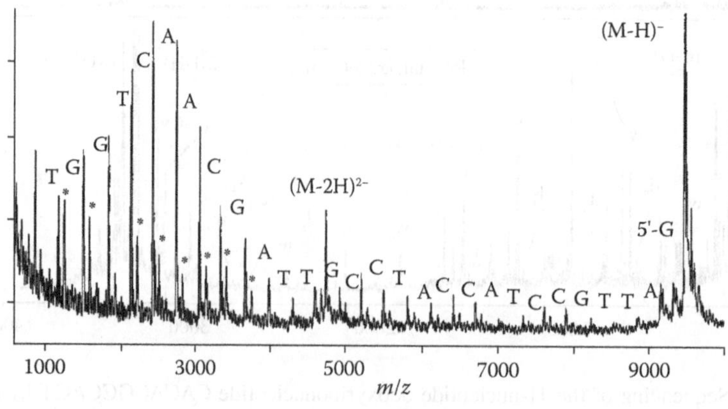

FIGURE 5.15 DE-MALDI mass spectrum of a crude synthetic DNA 31-mer (M_r = 9486.2 calculated). Mass measurements on the failure products define the sequence up the 3′ trinucleotide. Asterisks indicate +80 Da satellite peaks. Instrument, 1.3 m linear; matrix, 3-HPA. (Reprinted with permission from Juhasz, P., et al., *Analytical Chemistry*, 68, 941–946. Copyright 1996, American Chemical Society.)

scission of an oligoribonucleotide strand occurs in strong acid[129] or alkali;[2] however, oligodeoxyribonucleotides are resistant to aqueous bases. Phosphorothioates can be degraded using iodoethanol.[130] Finally, nucleic acids can be broken down by gas phase excitation fragmentation (tandem mass spectrometry) as well (Section 5.2).

5.3.3.1 Enzymatic Degradation

Enzymatic degradation of DNA[101,106,131] or RNA[132] results in distinct cleavage product mixtures, from which the oligonucleotide sequence can be inferred. Modifications in the natural chemical structure, such as phosphorothioate or 2′-OMe make DNAs and RNAs resistant to nucleases, thus these nucleotide analogs cannot be digested and directly sequenced by this method. There are, however, chemical and enzymatic implementations to overcome enzymic resistance of such modified oligonucleotides.[106,133]

A general approach to enzymatic degradation of DNA for MS analysis involves the use of two complementary enzymes, which digest the oligonucleotide from opposite ends of the sequence, creating overlapping sequence coverage.

5.3.3.1.1 Exonucleases CSP and SVP

Pieles and coworkers described sequencing of short DNAs with exonucleases.[106] As an example, a DNA 12-mer, containing only one 2′-O-methyl adenosine at position 5, has been digested with calf spleen phosphodiesterase (CSP), which digests DNA from the 5′ to the 3′ end. The cleavage was quenched after 30 min by mixing the digest reaction mixture with diammonium citrate or tartarate containing 2,4,6-trihydroxy acetophenone matrix solution, and deposited and evaporated on to the MALDI target plate. The first five y-type fragments have been observed in the negative ion mass spectrum. CSP enzyme did not cleave beyond the 2′-OMe-A nucleotide.

Snake venom phosphodiesterase (SVP), however, an enzyme that cleaves DNA in the 3′ to 5′ direction, did not stop digesting the oligonucleotide at the 2′ modified adenosine. Nine fragments have been observed, hence by combination of CSP and SVP digestion results, full sequence verification has been possible (Table 5.6).

TABLE 5.6
Negative Ion MALDI-TOF MS Results from Partial Digestion of DNA 12-mer Carrying One 2′-O-Methyl Adenosine (X) in Position 5[a]

Enzyme	5′-d(GCTTXCTCGAGT)↓	Fragment Type	Calculated Neutral Exact Mass	Mass Found[106]
CSP	GCTTXCTCGAGT	y_1	3664.65	3665.8
CSP	CTTXCTCGAGT	y_2	3635.59	3336.6
CSP	TTXCTCGAGT	y_3	3046.55	3046.9
CSP	TXCTCGAGT	y_4	2742.50	2742.8
CSP	XCTCGAGT	y_5	2438.45	2438.3
SVP	GCTTXCTCGAGT	b_1	3664.65	3666.0
SVP	GCTTXCTCGAG	b_2	3360.60	3362.0
SVP	GCTTXCTCGA	b_3	3031.55	3032.8
SVP	GCTTXCTCG	b_4	2718.49	2719.3
SVP	GCTTXCTC	b_5	2389.44	2389.8
SVP	GCTTXCT	b_6	2100.39	2100.5
SVP	GCTTXC	b_7	1796.34	1796.2
SVP	GCTTX	b_8	1507.30	1506.9
SVP	GCTT	b_9	1164.23	1163.9

[a] CSP, calf spleen phosphodiesterase; SVP, snake venom phosphodiesterase; digestion time, 30 min.

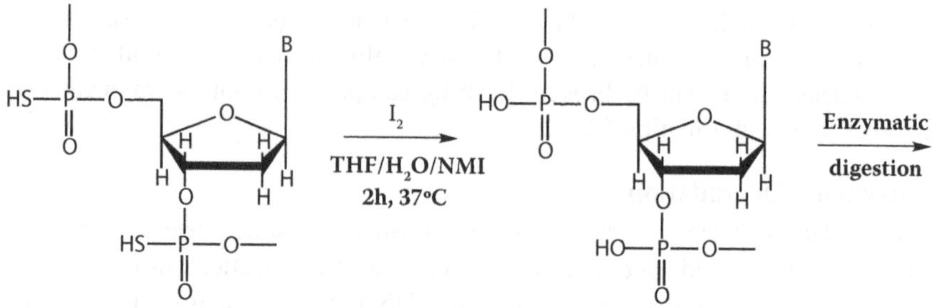

FIGURE 5.16 Phosphorothioate DNA sequencing by combination of oxidation by I_2 and nucleolytic degradation.

5.3.3.1.2 Phosphorothioate Oxidation by Iodine Followed by Enzymatic Digestion

Phosphorothioate modification is widespread for stabilizing antisense deoxyribonucleotides in biological media. Phosphorothioates are resistant to exonuclease cleavage and only partial cleavage is obtained with SVP at 55°C overnight treatment. Nevertheless, phosphorothioates can be converted to regular phosphates by chemical oxidation. Iodine solution (10 mg/mL in THF/water/N-methylimidazole, 16:4:1) oxidizes 94% of thiol groups to hydroxyl in 2 hours at 37°C (Figure 5.16). Oxidation of phosphorothioate oligonucleotides to their intact phosphodiester analogs *prior* to enzymolysis facilitates cleavage and reduces overall oligomer sequencing time. The DNA, comprising mostly regular phosphate diester linkages, can be digested by SVP and CSP in time-dependent experiments. Full sequence determination was achieved on a 21-mer phosphorothioate by oxidation and combining 3′ and 5′ exonuclease digest mixture by negative ion MALDI MS.[133]

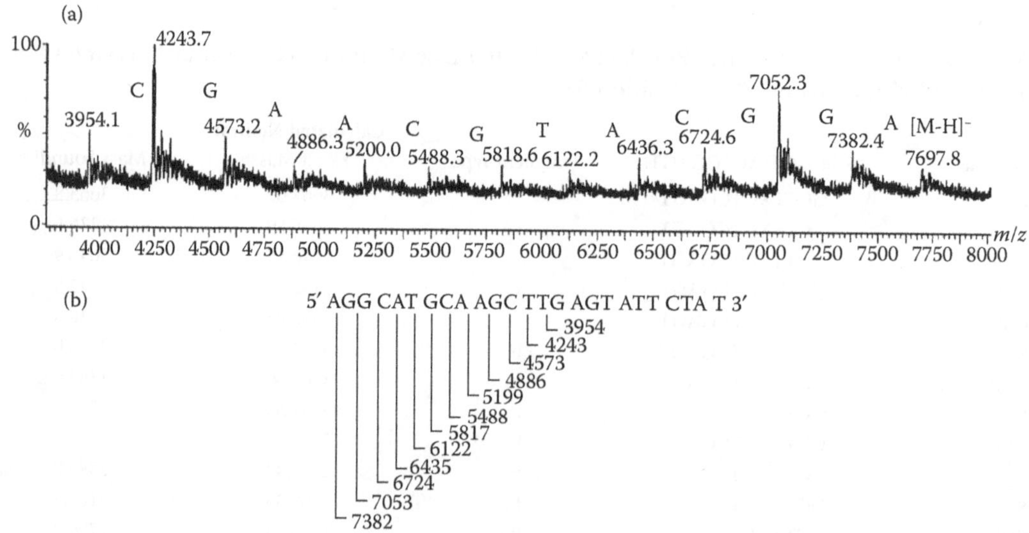

FIGURE 5.17 (a) Negative ion UV-MALDI mass spectrum of a synthetic 25-mer DNA (average molecular weight 7696 Da) subjected to BSP digestion for 1 min. (b) Schematic representation of observed cleavages. (From Tolson, D. A., and N. H. Nicholson, *Nucleic Acids Research*, 26, 446–451, 1998. With permission from Oxford University Press.)

5.3.3.1.3 BSP and SVP

Tolson et al.[132] used the combination of bovine spleen phosphodiesterase (BSP) and SVP. BSP digestion of a 25-mer DNA for 1 min results in a sequence ladder with y fragments. From the differences between successive peaks in the negative ion UV-MALDI-TOF mass spectrum, the sequence from the 5′-end can be determined (Figure 5.17).

DNA sequencing by MALDI-TOF is not limited by resolution, because the least mass difference between the nucleotides dA and dT is 9 Da, which is readily resolved by the MALDI-TOF instrument even at high masses. Continuing the digest for longer time periods, further sequence information could have been obtained. In principle, while full sequence information could be obtained from the BSP digest alone, smaller fragments (<1000 Da) are not readily observed, because of either incomplete digestion or interference from the matrix.

5.3.3.1.4 Enzymatic RNA Sequencing[2,106,131,134]

The principle of RNA sequencing using exonucleases is identical to what has been reported for DNA. Tolson et al. combined SVP digestion and MALDI MS for RNA sequence determination. SVP produced b-type ladder sequences (3′-OH shortmers) (Figure 5.18).

Reliable differentiation between U and C was not feasible, owing to the poor resolution of MALDI at higher masses that implicated the application of base specific endonucleases.

5.3.3.1.5 Endonucleases

Several ribonucleases with a high degree of base specificity have been employed to yield uniform cleavage patterns. In contrast to the enzymatic degradation approach for DNA in which complementary enzymes are used to generate sequence information from each end of the DNA, combinations of enzymes used for RNA sequencing yield relatively small, but unique fragments that can be overlapped to reconstruct the original sequence. RNA secondary structure can strongly affect the enzymatic cleavage pattern,[135,136] thus denaturing conditions, which do not disturb activity and

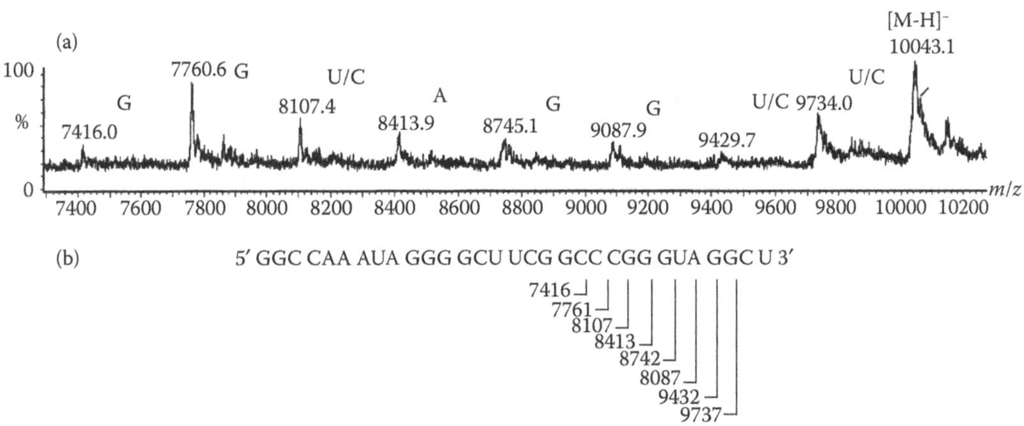

FIGURE 5.18 Negative ion UV-MALDI-TOF mass spectrum of a synthetic 31-mer RNA (average molecular weight 10,044 Da) subjected to SVP digestion for 2 min. A ladder sequence with fragments containing the original 5′-terminus is generated. Sequence information can be derived from mass differences between successive peaks. However, because the U and C residues differ in mass by only 1 Da, the mass of fragments containing these residues cannot be determined with the necessary accuracy, and the resulting assignments therefore contain partial ambiguities where the U and C residues are described as "U/C" as shown. (From Tolson, D. A., and N. H. Nicholson, *Nucleic Acids Research*, 26, 446–451, 1998. With permission from Oxford University Press.)

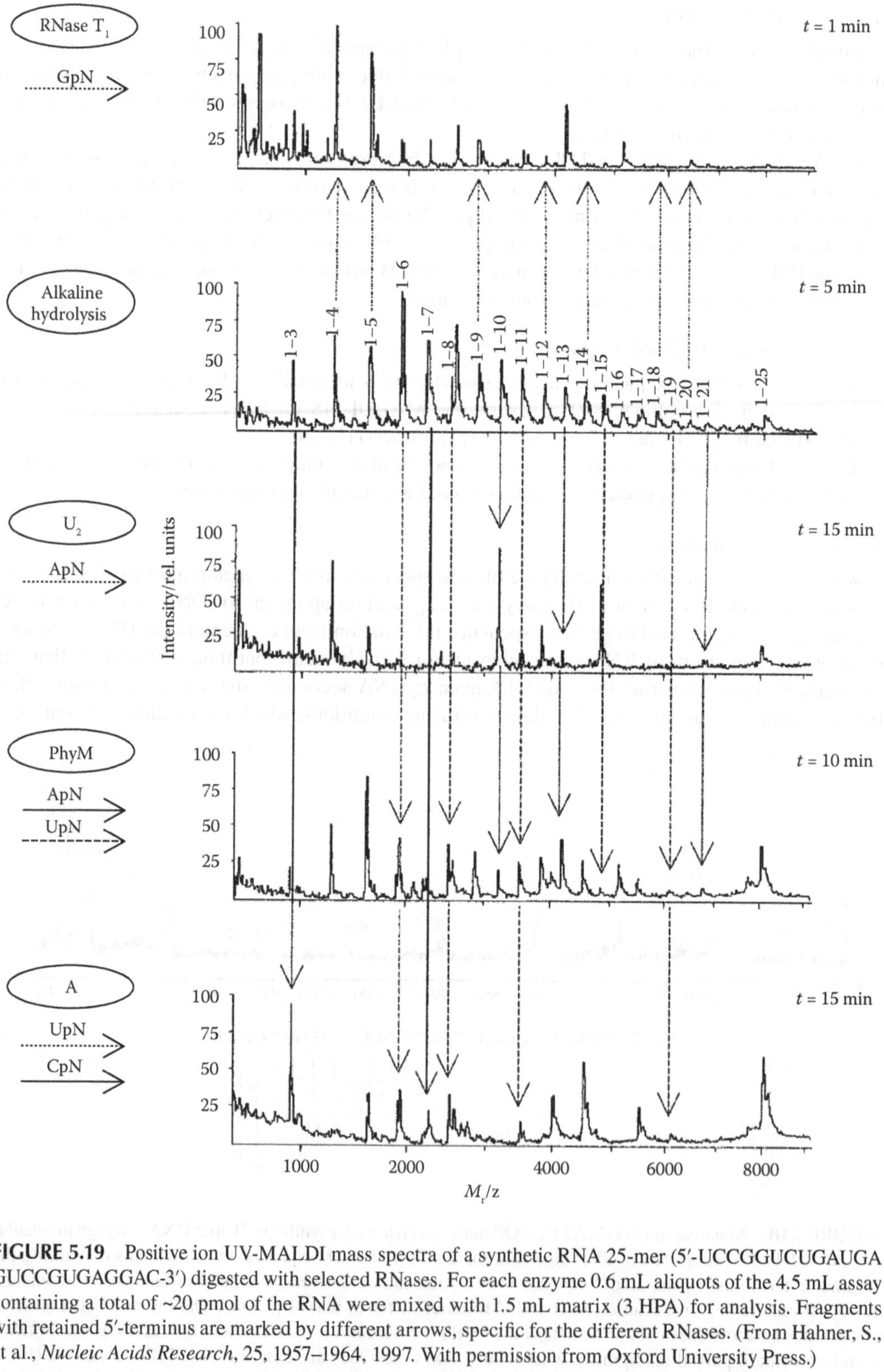

FIGURE 5.19 Positive ion UV-MALDI mass spectra of a synthetic RNA 25-mer (5′-UCCGGUCUGAUGA GUCCGUGAGGAC-3′) digested with selected RNases. For each enzyme 0.6 mL aliquots of the 4.5 mL assay containing a total of ~20 pmol of the RNA were mixed with 1.5 mL matrix (3 HPA) for analysis. Fragments with retained 5′-terminus are marked by different arrows, specific for the different RNases. (From Hahner, S., et al., *Nucleic Acids Research*, 25, 1957–1964, 1997. With permission from Oxford University Press.)

specificity are recommended.[137–140] Fragments are generated at specific cleavage sites both by partial as well as complete digestion.

5.3.3.1.6 CL3 and Phy M

Individual endonucleases with known cleavage specificity can be used in specialized cases to confirm sequence amibiguity. For example, endonucleases CL3 and Phy M (CL3: C specific, but cleaves at U also if in a sequence of UAG; Phy M is U, G, and A specific, both resulting in 2′-3′ cyclic phosphates; i.e., CpN→Ccp for CL3) were applied to determine uncertain positions in the RNA sequence. CL3 cleaved an example 16-mer RNA at two positions in 1 min: (UCC | AAA UUU GGU | AGG A) resulting in two observed digest products: UCC AAA UUU GGU 2′-3′cyclic phosphate (cp) (c_{12}) and AAA UUU GGUcp (y_3–c_{12}). In a similar example, Phy M cleaved a 23-mer RNA (GGG UCC AAA UUU U|G|G U|AG GAC CC) in 2 min to four assignable species: 3 y-type: GG UAG GAC CC, G UAG GAC CC, AG GAC CC and 1 c-type fragment: GGG UCC AAA UUU Ucp.[132]

5.3.3.1.7 RNase T_1, U_2, Phy M, and A

Hahner et al. used four different endonucleases and alkaline degradation for RNA sequencing and concluded that 5′ fragments in MALDI mass spectra from the digest mixtures are most useful for sequence determination. Specificity of the enzymes was not always perfect and 3′ fragments, internal fragments, and complexity of the digest mix disturbed the analysis (Figure 5.19 and Table 5.7). The 5′-biotinyl RNA has been used to simplify digest mixture by straptavidin coated magnetic beads with limited improvement.[2]

Kirpekar and coworkers sequenced an *in vitro* RNA 54-mer, a transcript from a DNA plasmid digest template by 3′ exonuclease with positive ion MALDI-MS. Poor mass resolution was observed due to 5′ phosphate heterogeneity and template independent, unspecific elongation by RNA polymerase. The 5′ phosphate heterogeneity is explained by the fact, that 5′-triphosphates (natural products of enzymatic RNA polymerization) can hydrolize to diphosphates or to phosphates during workup or during ionization in MALDI. Mass resolution was significantly improved by removing all 5′ phosphates with calf intestine phosphatase (CIP). Finally, a mixture of 54-mer, contaminated by nonspecific 55-mers, has been digested in a time course experiment with 3′-exonuclease from *Crotalus durissus* (South American rattlesnake). Shortmer peaks were observed after 1 min digestion (47-mer through 54-mer). Surprisingly enough, a broad distribution of all possible species from 4 to 47 nucleotides were found after 20 min, at relatively balanced intensities. Certain regions in the RNA appeared to attenuate the action of the 3′-exonuclease. Such regions were at position 46–48 and 35–37 from the 5′ end. The apparent broad distribution of exonuclease products is largely due to the phenomenon that smaller molecular ions give higher signal intensities than larger ones in MALDI-MS. The full length 54-mer RNA mass has been determined (17460.1 versus 17458.7 calculated), but this accuracy is insufficient for differentiation of bases C and U (Figure 5.20).[94]

Faulstich et al. described a potential solution to the major issue of low resolution of MALDI to differentiate C and U (1 Da mass unit difference). An 8-mer and a 22-mer RNA were digested with 5′-3′ exonuclease from bovine spleen. This enzyme has attenuated activity to cleave at cytidines over uridines resulting in significantly higher MS signal intensity for oligonucleotides 5′-terminated with cytidines compared to 5′-terminated uridines. The mass differences between adjacent peaks in the time-dependent negative ion mass spectra were used to determine A, G, and the pyrimidine nucleobases. C and U bases were differntiated by comparing their peak intensities. Significantly higher intensity peaks determined C positions. Also, chicken liver endonuclease (CL3) was used to cleave preferentially at cytidines, another verification for the C bases.[141] Table 5.8 summarizes the enzymes for DNA/RNA digestion for MS sequencing.

5.3.3.1.8 RNA Mass Mapping[143]

An emerging approach to sequence determination of RNA is the characterization of RNA by nucleotide specific endonuclease (RNase T_1, G specific) digestion followed by MALDI and ESI mass

TABLE 5.7
List of the Expected Observed 5′-Fragments for Digests of the Synthetic RNA 25-mer with the RNases T_1, U_2, Phy M, and A (Compare Also Spectra of Figure 5.17)[2]

Enzyme	Sequence UCCGGUCUGAUGAGUCCGUGAGGAC	M_r Calculated	M_r Measured
RNase T_1	UCCGGUCUGAUGAGUCCGUGAGG	7458.5	7456.3
	UCCGGUCUGAUGAGUCCGUGAG	7113.3	7112.0
	UCCGGUCUGAUGAGUCCGUG	6438.8	6437.9
	UCCGGUCUGAUGAGUCCG	5787.5	5788.5
	UCCGGUCUGAUGAG	4525.7	4524.1
	UCCGGUCUGAUG	3851.3	3851.0
	UCCGGUCUG	2870.7	2870.2
	UCCGG	1608.0	1608.0
	UCCG	1262.8	1262.4
RNase U_2	UCCGGUCUGAUGAGUCCGUGAGGA	7787.7	—
	UCCGGUCUGAUGAGUCCGUGA	6768.1	6768.7
	UCCGGUCUGAUGA	4180.5	4179.7
	UCCGGUCUGA	3199.9	3198.8
RNase Phy M	UCCGGUCUGAUGAGUCCGUGAGGA	7787.7	—
	UCCGGUCUGAUGAGUCCGUGA	6768.1	6769.2
	UCCGGUCUGAUGAGUCCGU	6093.7	6093.1
	UCCGGUCUGAUGAGU	4831.9	4830.4
	UCCGGUCUGAUGA	4180.5	4179.8
	UCCGGUCUGAU	3506.1	3505.9
	UCCGGUCUGA	3199.9	3199.5
	UCCGGUCU	2525.5	2525.5
	UCCGGU	1914.2	1914.0
RNase A	UCCGGUCUGAUGAGUCCGU	6093.7	6094.7
	UCCGGUCUGAUGAGUCC	5442.3	—
	UCCGGUCUGAUGAGUC	5137.1	—
	UCCGGUCUGAUGAGU	4831.9	—
	UCCGGUCUGAU	3506.1	3506.2
	UCCGGUCU	2525.5	2525.8
	UCCGGUC	2219.4	2219.7
	UCCGGU	1914.2	1913.6
	UCC	917.6	916.8

spectrometry and a fragment mass fingerprint search of sequence databases. Similarly to peptide mass fingerprinting (PMF),[149–151] RNAs are amenable to identification by comparison of accurate mass fingerprint of enzymatic RNA digest to *in silico* digest of all entries in a genomic database. Identification issues are discussed thoroughly in recent reviews on the application of mass spectrometry to nucleotide-specific digestion mixtures[152,153] and on rRNA with ESI-MS.[154]

RNA characterization by specific cleavage and mass spectrometry, in cases when the genetic origin of the RNA is known, has been demonstrated in the literature.[2,155] Posttranscriptional modifications have also been studied by MALDI-TOF.[77] Identification of genomic DNA has been demonstrated following the same concept, namely PCR amplification of the DNA in question and transcription to RNA, which is then cleaved to fragments by base-specific nucleolytic digestion,

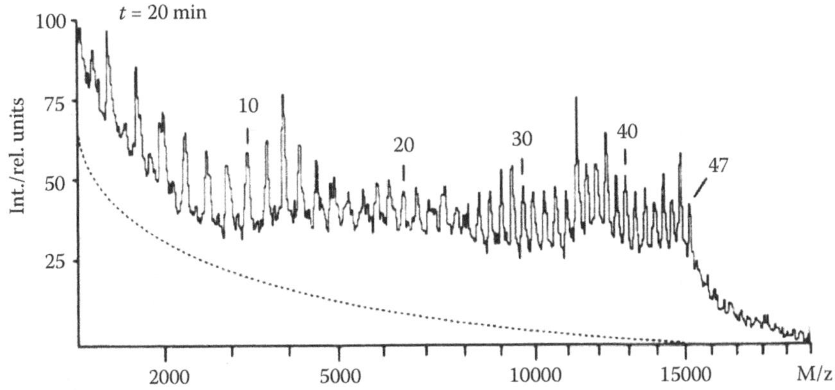

FIGURE 5.20 The 3′-exonuclease digests of a 54/55 nt CIP treated *in vitro* transcript. The numbers indicate the length of the digestion products in nucleotides. Positive ion UV-MALDI mass spectra of samples after 20 min incubation. The broken line indicates the background signal intensity. (From Kirpekar, F., et al., *Nucleic Acids Research*, 22, 3866–3870, 1994. With permission from Oxford University Press.)

and finally visualization of the pattern by mass spectrometry. Such examples include bacterial species identification,[156,157] single nucleotide polymorphisms (SNP),[158,159] short tandem repeat polymorphisms (STRP),[160] and tRNA detection.[155]

Interpretation of the raw MS data into meaningful RNA sequence information requires specialized algorithmic processing. RNA identification utilizing the mass fingerprint concept relies on bioinformatics tools. A demonstration of one approach, the RNA mass mapping (RMM) stand-alone platform[143] for correct location of DNA origin, is described below.

The MS mass list and the genome sequence database are imported to RMM. The reverse complement of the genome sequence is added to the sequence database. Genome positions that match the RNA fragment mass values are found and scored by a probability function. Finally, all RNA sequence matches are scored and the results are reported.

In the example illustrated in Figure 5.21, the DNA template was successfully identified for unknown RNA from picomoles of sample prokaryotic RNA with appropriate database and database search platform (RMM) in one day.[143]

5.3.3.2 Chemical Degradation

Oligonucleotides are prone to nucleobase specific degradation induced by chemical reagents. Similarly to endonucleolytic digestion, chemical degradation cleaves the oligonucleotides at a specific nucleobase.[136,161] The general mechanism of base specific degradation is depicted in Figure 5.22, including two reaction steps, one of which a chemical reagent reacts specifically with one or two kinds of nucleobases and secondly the oligonucleotide falls apart in a second reaction step at the modified base in a second chemical reaction. The illustrative example of nucleobase specific chemical degradation of DNA is the Maxam-Gilbert sequencing.[161]

The single-stranded, P^{32} labeled radioactive DNA is partially cleaved by a chemical procedure at each repetition of a nuclobase. The radioactive labeling is not applied if the fragments are detected by MS. Four aliquots of the DNA sample are degraded in four different chemical reaction mixtures.

1. Purines are partially methylated with dimethyl sulphate (DMS) and the charged nucleobases are hydrolyzed by heating (90°C, 15 min, pH 7). Somewhat depurinated DNA is then heated in NaOH (0.1 M, 90°C, 30 min) to get selective cleavage at G-s and weak cleavages at A-s. Alternatively aqueous piperidine (1 M, 90°C, 30 min) could be used in one step to depurinate and hydrolyze at methylated G-s.

TABLE 5.8
List of DNA and RNA Nucleases and Conditions of Use[a]

Enzyme	Type	Nucleic Acid Type	Source	Conditions[b]	Notes	Reference
Calf spleen phosphodiesterase (CSP)	5′-exonuclease	DNA	Calf spleen			90, 106
Bovine spleen phosphodiesterase (BSP)	5′-exonuclease	DNA/RNA	Bovine spleen		y fragments in 1 min	90
Phosphodiesterase I	5′-exonuclease	DNA/RNA	*Crotalus adamanteus*	0.003; 15 mM $MgCl_2$/50 mM Tris, pH 8.9, 45°C, 2 hours	Nonspecific	9
Snake venom phosphodiesterase (SVP)	3′-exonuclease	DNA/RNA	Snake venom		B fragments[132]	90, 106
Phosphodiesterase	3′-exonuclease	RNA	*Crotalus durissus*			94, 141
Custavin	Endonuclease	RNA	*Cucumis sativus L.*	0.05 ng; 10 mM Tris-HCl, pH 6.5, 37°C, 30 min[142]	C specific	142
RNase CL3	Endonuclease	RNA	Chicken liver	0.01; 10 mM Tris-HCl, pH 6.5, 37°C, 30 min[140]	C specific[140] U, G, or A specific in 1 min[132] C preference[141]	
RNase T_1	Endonuclease	RNA	*Aspergyllus oryzae*	0.2; 2 0mM Tris-HCl, pH 5.7, 37°C, 5 min[137]	G specific	134, 139, 143, 144
RNase U_2	Endonuclease	RNA	*Ustilago sphaerogena*	0.01; 20 mM DAC, pH 5, 37°C, 15 min[137]	A specific	134, 139, 143, 144
RNAse A	Endonuclease	RNA	Bovine pancreas	4×10^{-9}; 10 mM Tris-HCl, pH 7.5, 37°C, 30 min[145]	Pyrimidine nucleoside specific	139, 143, 144

RNase Phy M	Endonuclease	RNA	*Physarum polycephalum*	20; 20 mM DAC, pH 5, 50°C, 15 min[138]	U, G, or A specific in 2 min cleavage	2, 132
RNase T_2	Endonuclease	RNA		10; Thermosequenase buffer (Amersham Biosciences), 37°C, 45 min[146]	No specificity	147
Mung bean nuclease I	Endonuclease	DNA/RNA	Mung bean sprouts	50 mM NaOAc, 30 mM NaCl, 1 mM $ZnSO_4$, pH 5, 30°C	Single strand specific	148
Endonuclease S1	Endonuclease	DNA/RNA	*Aspergillus oryzae*	33 mM NH_4Ac, 50 mM NH_4Cl, 0.03 mM $ZnSO_4$, pH 4.5	Single strand specific	148
Nuclease P_1	Endonuclease	DNA/RNA	*Penicillium citrinum*	0.07; 20 mM NaOAc/50 mM NaCl/5 mM $ZnCl_2$, pH 5.3, 70°C, 2 hours	Single strand specific	9
Alkaline phosphatase	Hydrolase	Nucleotide	*Crotalus adamanteus*	15 mM $MgCl_2$/50 mM Tris, pH 8.9, 45°C, 2 hours	Removes phosphate from nucleotides	9

[a] Endonuclease cleavage products are cyclic phosphates that hydrolyze to phosphates by the same enzyme in a second, slower step.[2]

[b] Unit RNase/μg RNA; buffer, pH, temperature, incubation time for max number of fragments.

(a)

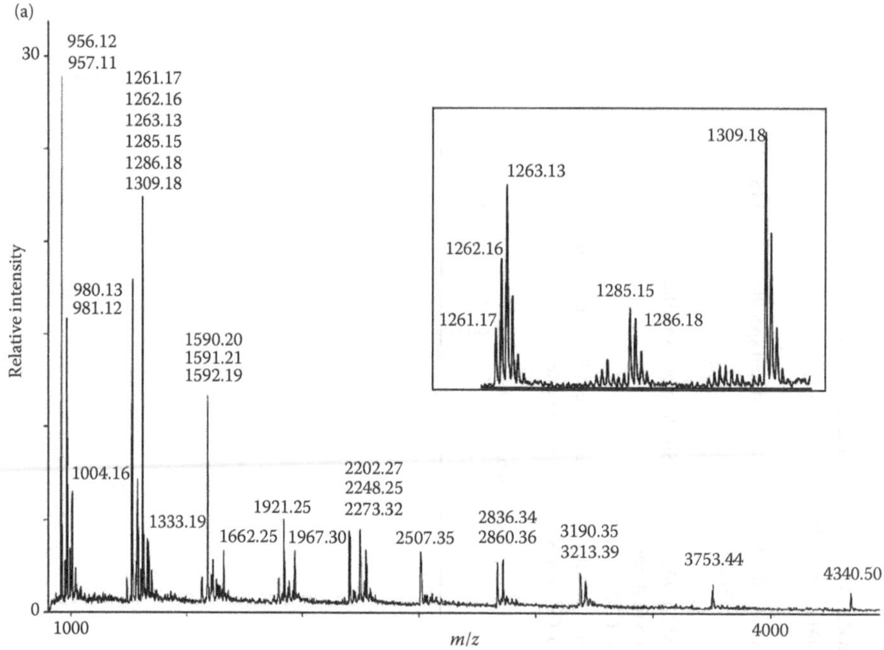

(b)

ACTGGGGCGGTACGCGCTCGAAAAGATATCGAGCGCGCCCTATGG***TCATCTCAG***CCGGGACAGAGACC
CGCGAAGAGTGCAAGAGCAAAAGATGACTTGACAGTGTTCTTCCCAACGAGGAACGCTGACGCGAAA
GCGTGGTCTAGCGAACCAATTAGCCTGCTTGATGCGGGCAATTGATGACAGAAAAGCTACCCTAGGGA
TAACAGAGTCGTCACTCGCAAGAGCACATATCGACCGAGTGGCTTGCTACCTCGATGTCGGTTCCCTC
CATCCTGCCCGTGCAGAAGCGGGCAAGGGTGAGG***TTG***TTCGCCTATTAAAGGAGGTCGTGAGCTGGG***T***
TTAGACCGTCGTGAGACAGGTCGGCTGCTA

FIGURE 5.21 Mass spectrometry data and search result for a *Haloarcula marismortui* 23S rRNA subfragment. (a) Mass spectrum of *H. marismortui* 23S rRNA subfragment (around positions 2323–2630) digested with RNase T_1. Assigned masses are from singly protonated digestion products; these masses were used in the subsequent genome search. Insert: zoom on peak clusters to illustrate the effect of digestion products with partially overlapping isotope distributions. (b) Top scoring genomic region with flanks for RNA mass mapping of *H. marismortui* 23S rRNA subfragments 2323–2630. Underlined: identified sequence. Grey highlight: RNase T_1 digestion fragments with masses in peak list. Bold italic: RNase T_1 digestion fragments with masses not present in peak list. (From Matthiesen, R. and F. Kirpekar, *Nucleic Acids Research*, 37, e48/1–e48/10, 2009. With permission from Oxford University Press.)

2. Strong A and weak G cleavage occurs if partially methylated DNA is depurinated at A-s at pH 1 and then hydrolyzed at pH 13 (Figure 5.23). An alternative for A determination is the strong A and weak C cleavage reaction: A and C rings are partially opened in NaOH (1.15 M, 90°C, 30 min) and DNA is hydrolyzed by piperidine (1 M, 90°C, 30 min) mostly at A-s and somewhat at C-s.
3. T and C equal cleavage: DNA is treated with hydrazine (16 M aqueous, 20°C, 15 min) to open pyrimidine bases partially, then hydrolyzed with piperidine (0.5 M, 90°C, 30 min).
4. Cytosine cleavage is achieved by hydrazine cleavage at high sodium chloride concentration (2–5 M) solution.

The Maxam–Gilbert degraded reaction mixtures can be used for sequence analysis via mass spectrometry after a short gel filtration and desalting.[162]

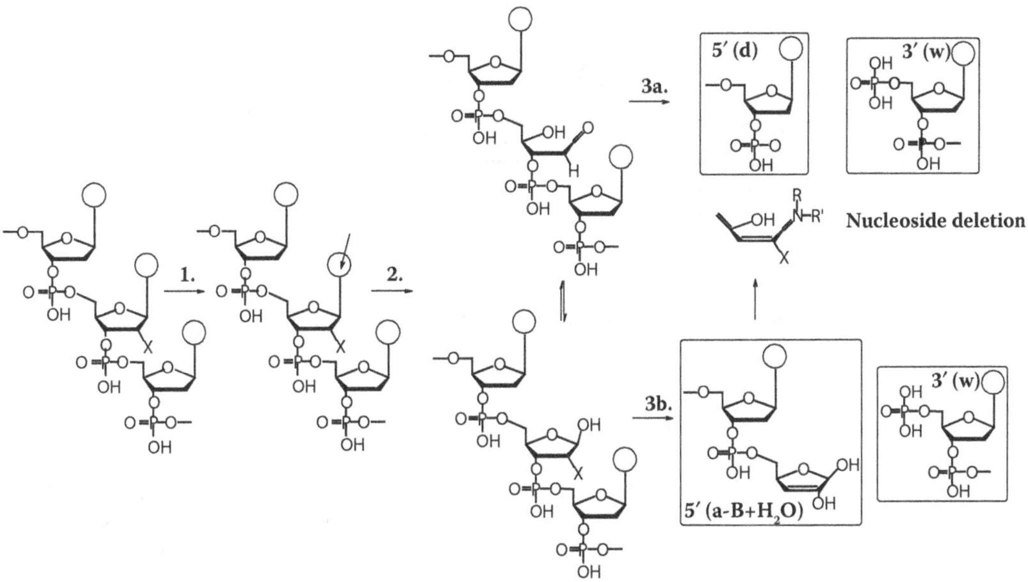

FIGURE 5.22 Nucleobase specific chemical degradation of oligonucleotides: 1, The oligonucleotide is modified by a nucleobase specific reagent. The glycosidic bond becomes sensitive; 2, The modified nucleobase is lost; and 3, strand scission occurs in a β–elimination reaction. DNA (X = H) tends to be fragmented by sugar ring opening (3a) and by deletion of an entire nucleoside. Ribose (X = OH, OMe) is less susceptible to ring opening than deoxyribose owing to the stabilizing effect of the 2′-oxygen, hence less reactive and cleaved in acid or base catalyzed β elimination (3b).

5.3.3.2.1 RNA

With slight modifications, chemical degradation methods for DNA are applicable to RNA.

1. RNA cleavage at guanines can be achieved if a partially methylated oligoribonucleotide guanine ring is reduced with $NaBH_4$ and depurinated. Mild aqueous aniline treatment results in chain cleavage.[163]
2. RNA degradation at adenines is induced by a ring opening reaction with diethyl-pyrocarbonate followed by the aniline scission.[164,165]

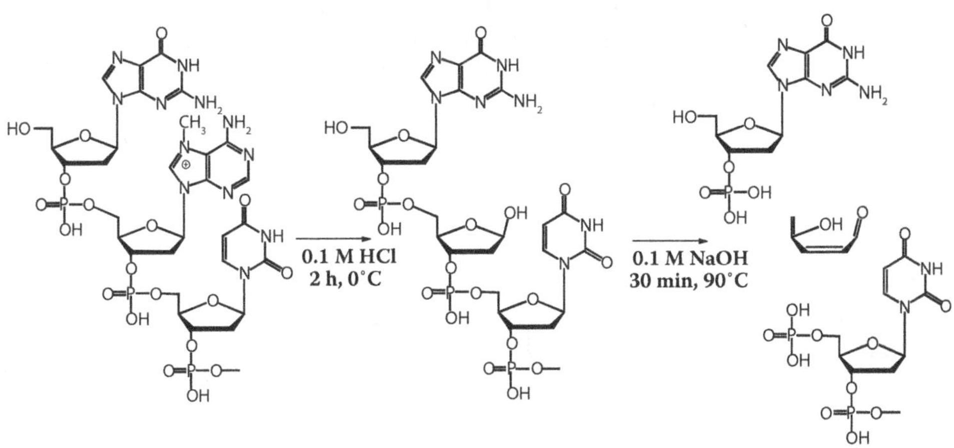

FIGURE 5.23 The Maxam–Gilbert "A" specific reaction. Note: adenine methylation can occur on ^{3}N and ^{6}N also.

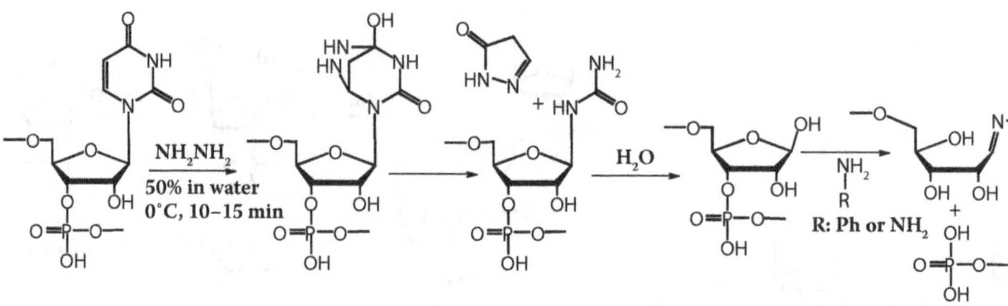

FIGURE 5.24 Uridine specific degradation reaction with hydrazine. Similar mechanism applies to C and T with other dinucleophiles.

3. Methylated cytosine ring is opened by hydrazine, followed by chain scission.[166,167]
4. Unprotonated hydrazine reacts with pyrimidine nucleobases in the U >> C > T order.[168] Uridine preferentially reacts with hydrazine both under aqueous or anhydrous conditions but sodium chloride suppresses hydrazinolysis of U in favor of C. Sodium chloride (2 M) in aqueous hydrazine was used for DNA and NaCl (3 M) in hydrazine for RNA C selective scission.[136,169]
5. Aqueous hydrazine specifically degrades the U ring, causing 3′O-P scission in buffered aniline solution (Figure 5.24).[132]
6. A C-specific chemical degradation protocol is described[135] in which cytidine is degraded by hydroxylamine, then followed by strand scission with aniline.[2,9]

5.3.3.2.2 Phosphorothioates

Polo et al. reported that phosphorothioate oligonucleotides can be uniquely cleaved with iodoethanol resulting in 5′ thiophosphate (w-type) and 3′ hydroxyl (b-type) fragments from which bidirectional sequencing is possible. The protocol will cleave specifically only at thioate sites, while the regular phosphate diesters remain uncleaved (Figure 5.25).[130] A collection of base specific chemical reagents for chemical degradation of oligonucleotides is found in Table 5.9.[170]

5.3.3.2.3 Alkaline Hydrolysis

DNA itself is quite stable to alkaline hydrolysis, but methylphosphonate DNA 15-mer was partially hydrolized by 3% NH_4OH (50°C, 1 hour) at 1 mg/mL concentration and analyzed by positive ion MALDI-MS. Increased relative intensities of 3′-hydroxyl (b-type) and 5′-phosphate (w-type) fragments were observed, compared to a failure analysis of the synthetic crude mixture.[128]

Confirmation of the structures of novel, chemically synthesized antisense DNA oligonucleotides having alternating methylphosphonate/phosphodiester internucleotide linkages was accomplished by base catalyzed hydrolysis followed by MALDI mass spectrometry.[148]

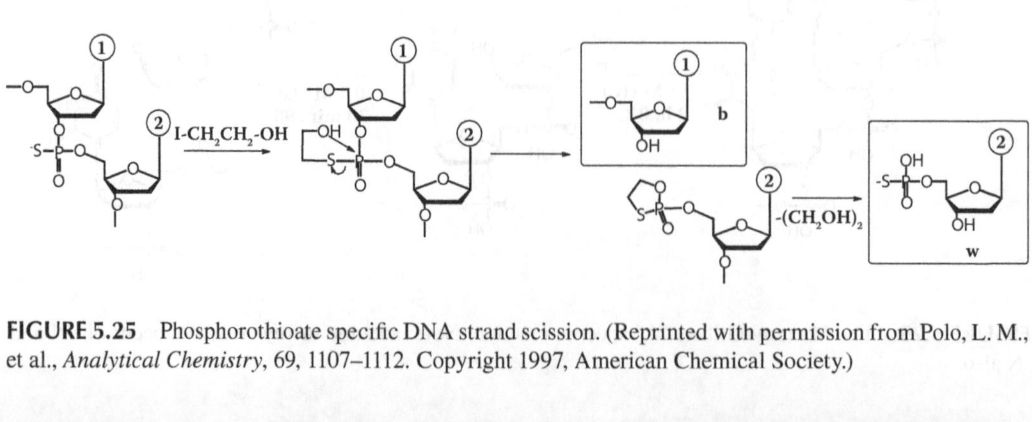

FIGURE 5.25 Phosphorothioate specific DNA strand scission. (Reprinted with permission from Polo, L. M., et al., *Analytical Chemistry*, 69, 1107–1112. Copyright 1997, American Chemical Society.)

TABLE 5.9
Base Specific Chemical Degradation with Base Preferences, Reaction Conditions, and References

Cleavage	Reagent	Reference
A	K_2PdCl_4, pH 2.0	171
A > G	DMS/0.1 M HCl, 0°C→0.1 M NaOH, 90°C	161
A + G	0.1% DEP, pH 5, 90°C (acetate buffer)	172
A + G	60–80% formic acid	173, 174
A + G	pH 4, 80°C (citrate buffer)	175
A + G	2–3% diphenylamine in 66% formic acid	176, 177
A > C	NaOH	178
G	DMS	178
G	Methylene blue	179
G	0.1% Methylene blue, visible light	179, 180
G	4% DMS, pH 3.5 (formate buffer)	174, 181
G	0.5% DMS, pH 8 (cacodylate buffer)	177
G > A	DMS→90°C pH 7→0.1M alkali 90°C or Δ aq. piperidine	161, 178
G > T	1 M methylamine, UV irridation	182, 183
G >> C	0.3% DEP, pH 8, 90°C (cacodylate buffer)	172
G + A	H^+	178
C	1 M NH_2NH_2/2–5 M NaCl, 20°C	161, 178
C	NH_2NH_2-H_2O (3:1), 5 M NH_2NH_2*AcOH	184
C	3 M NH_2OH-HCl, pH 6	175, 184
C	2–3 M H_2O_2, pH 8.3 or 7.4 (carbonate buffer)	185
C + T	1 M NH_2NH_2, 20°C	161, 178
C + T	NH_2NH_2-H_2O (7:4)	177
T	OsO_4	179
T	1 M spermine, UV irridation	182, 183
T	0.5 M $NaBH_4$, pH 8–10	186
T >> C	2–3 M H_2O_2, pH 9.6 (carbonate buffer)	185, 186
T >> G,C	0.1 mM KMnO4	174, 175, 184
T > G >> A,C	1 M cyclohexylamine, UV irridation	187

Source: Franca, L. T. C., et al., *Quarterly Reviews of Biophysics*, 35, 169–200, 2002. With permission from Cambridge University Press.

RNA readily undergoes alkaline hydrolysis, in contrast to DNA. Partial alkaline hydrolysis of RNA was performed with NH_4OH (1:1 mix of 25% NH_4OH and 5–10 µM RNA sample at 60°C, 5 min). The RNA (5′-UCCGGUCUGAUGAGUCCGUGAGGAC-3′) was uniformly cleaved at each position between nucleotides 3 and 21. The positive ion MALDI-MS spectrum of the alkaline hydrolysis of RNA was used as a standard (nucleobase position tag) to determine the sequence by combination of different base specific endonucleolytic cleavages.[2,137,139]

5.3.3.2.4 Acidic Hydrolysis

DNA sequencing by acidic hydrolysis is inappropriate owing to the inherited sensitivity of oligodeoxyribonucleotides to depurination as shown below by Nordhoff et al.[121] The authors incubated a 50 pmol sample of a 19-mer (CCC AAT TGA CCA ACT CTG G) in 10 µL 0.1 M HCl for 1.5 hours in which time frame, all the purine nucleobases hydrolyzed (were replaced by hydroxyls) and the resulting oligonucleotide was cleaved on each side of the depurinated nucleoside. Assignation of nucleobases to the abasic moieties was not possible; hence, the sequence was only partially determined (Table 5.10).

Bahr et al. reported acidic hydrolysis of RNA, an RNA duplex, and phosphorothioate RNA followed by high-resolution MALDI-MS and MS/MS. Acidic hydrolysis of RNA is a convenient way for full sequencing oligoribonucleotides up to 24-mers, the typical length range of siRNAs.[129] The low pH hydrolysis of RNA is not base specific; usually, all phosphodiester bonds were cleaved. Nucleobase losses and internal fragmentation resulting from multiple cleavages are usually not observed under the recommended conditions. The mechanism for RNA cleavage[188] at low pH (~pH 1) is shown on Figure 5.26.

The general procedure for regular RNA is as follows: RNA solution (20 μM, 4 μL) was mixed with TFA (3.75%, 1 μL, ~pH 1) and gently shaken for 20 min at 37°C for hydrolysis, but, 5 min at 60°C led to comparable results. TFA was preferred as acid, because it is available at high purity and facilitates a good crystallization with the HPA matrix. Phosphorothioate RNA hydrolyzes slower than DNA, thus 15 min at 60°C is recommended. MALDI-MS: After hydrolysis, 0.5 μL of each mixture was mixed with 2 μL matrix solution directly on the MALDI target and dried in a stream of air. The matrix used was 3-hydroxypicolinic acid (45 mg/mL in water) containing 5 mg/mL diammoniumhydrogencitrate.

Mass spectrometric analysis of RNA is limited by the U/C (1 Da) difference especially at higher mass ranges. High-resolution reflector TOF or orbitrap instruments are recommended for reliably differentiating Pyrimidines.

The authors give examples for single-strand siRNA full sequence determinations. The terminal dimers and trimers sometimes difficult to sequence, because of the high noise in the low mass regions, mostly due to the matrix adducts and because siRNAs consist of dTdT starts at the 3′ end. The unsequenced shortmers were determined by either high-accuracy MS from which composition could be deduced or if composition alone did not facilitate dependable sequence information, then tandem mass spectrometry (MS/MS) was applied.

Double-stranded siRNAs were also digested by acidic hydrolysis. The strands were separated and cleaved similarly to single strands during the procedure. High-accuracy mass determination of the full length single strands followed by search for mass differences between sequence ladder

TABLE 5.10
Partial Sequencing of 19-mer DNA by Acidic Hydrolysis IR-MALDI-TOF with Succinic Acid Matrix

Sequence[a]	Sequence Ions	Found Masses
CCC	d_3, $(a\text{-}B)_4$, $(a\text{-}B)_4+H_2O$	885.3, 983.4, 1001.0
CCCa	d_4, d_4+H_2O	1081.5, 1198.2
CCCaa	$(a\text{-}B)_5$	1179.7
CCCaaTT	d_7	1983.8
CCCaaTTa	$(a\text{-}B)_8$, $(a\text{-}B)_8+H_2O$, d_8	1885.4, 2002.9, 2081.9
CCCaaTTaa	$a\text{-}B_9$	2180.1
CCCaaTTaaCCa	c_{11}	2836.7
CCCaaTTaaCCaa	d_{12}	3052.8
CCCaaTTaaCCaaCTCTa	$(a\text{-}B)_{18}+H_2O$ (+2A, +AG, +2G)	4552.4, (4787.7, 4803.5, 4920.0)
CTCTaa	y_6	1597.2
aCTCTaa	y_7	1792.9
CCaaCTCTaa	y_{10}	2567.2
aCCaaCTCTaa	y_{11}	2763.2

[a] Here a, depurinated nucleoside (glycosidic OH).[121]

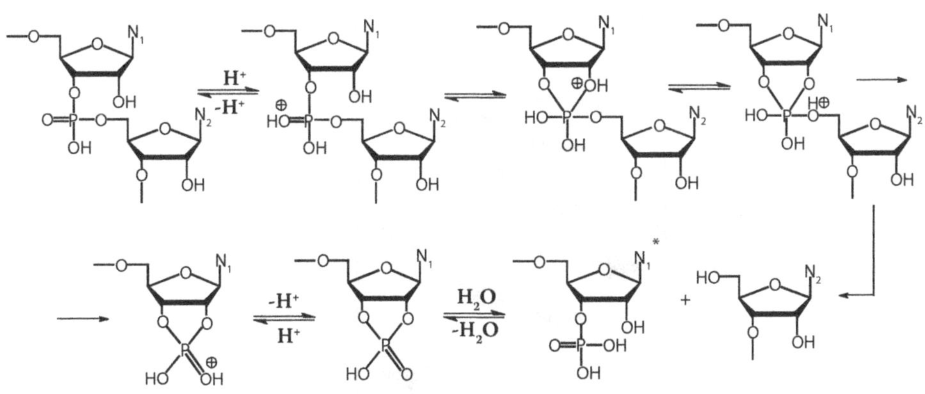

FIGURE 5.26 Mechanism for RNA acid hydrolysis. *2′-phosphate could also be formed, but irrelevant for mass spectrometry, being isobaric with the shown 3′-phosphate.[188] The general procedure for regular RNA is as follows: RNA solution (20 μM, 4 μL) was mixed with TFA (3.75%, 1 μL, ~pH 1) and gently shaken for 20 min at 37°C for hydrolysis, but, 5 min at 60°C led to comparable results. TFA was preferred as acid, since it is available at high purity and facilitates a good crystallization with the HPA matrix. Phosphorothioate RNA hydrolyzes slower than DNA, thus 15 min at 60°C is recommended. MALDI-MS: After hydrolysis, 0.5 μL of each mixture was mixed with 2 μL matrix solution directly on the MALDI target and dried in a stream of air. The matrix used was 3-hydroxypicolinic acid (45 mg/mL in water) containing 5 mg/mL diammoniumhydrogencitrate. Mass spectrometric analysis of RNA is limited by the U/C (1 Da) difference especially at higher mass ranges. High resolution reflector TOF or orbitrap instruments are recommended for reliably differentiating pyrimidines.[129]

peaks in the high-resolution MALDI-MS gave rise to the determination of consecutive nucleotides and sequence determination.

Acidic hydrolysis of phosphorothioate RNA resulted in d-type 5′ fragments (thiophosphate on 3′) that hydrolyze to 3′ phosphate (5′-NsNs . . . Np-3′). Such mass difference (-16) has to be considered when interpreting the results.[129]

5.3.4 Sanger DNA Sequencing by MALDI

Classical sequencing by the Sanger method is based on enzymatic polymerization to a template DNA (that is, the target of sequencing) initiated by a primer DNA and in the presence of modified nucleotides that terminate elongation (dideoxy nucleotide triphosphates, i.e., ddNTPs).[189–192] The sequencing mixture contains only limited amount of ddNTPs compared to regular dNTPs, such that termination of strand extension occurs only at low percentage; thus a pool of DNA sequences of all possible lengths is formed. The reaction was originally performed with radioactive tags and later by fluorescent dye labeled primers in four different reaction vessels for each of the four dNTPs. The newly synthesized and labeled DNA fragments are denatured and separated by size on a denaturing polyacrylamide-urea gel on four individual lanes for the four nucleobases. The DNA bands on the gel were visualized and the sequence is read out (Figure 5.27).

A widely used method, more suitable for high-throughput sequencing, uses a single extension-termination reaction done in the presence of four different fluorescently labeled ddNTPs. A single capillary electrophoresis run separates the differentially terminated oligos, which are identified for sequence read-out using laser induced fluorescence.

As an alternative to this classical approach, sequence determination can also be performed by primer extension and analysis of the chain termination mixture by mass spectrometry. MS is used to separate the different length DNA fragments and has the advantage that no labeling of the primers or the ddNTPs is necessary. A disadvantage of mass analysis of the chain termination

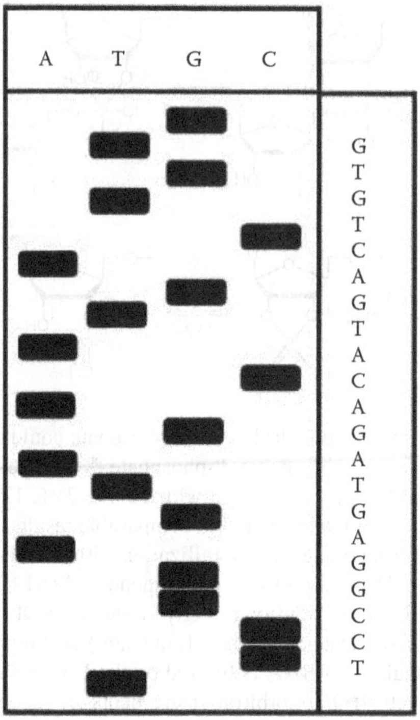

FIGURE 5.27 Idealized drawing of a gel electropherogram.

polymerization mixture is its limited sensitivity compared to radio- or fluorescent-labeled DNA detection by electrophoresis.

The classical Sanger termination method using four different reaction vessels, each containing a set of oligonucleotides terminated at a particular nucleobase, could also be analyzed by MALDI. The four mass spectra obtained from the four vessels would theoretically have DNAs at different lengths with the same 3′ terminal sequence. The sequence can be reconstructed from the combination of the four spectra results.

The first example of MS analysis of the terminating sequencing mixture, utilizing synthetic DNA sequences that would theoretically be found in a Sanger reaction mixture was described by Fitzgerald et al. A 24-mer range was identified (17- to 41-mers) at 1 pmol scale with this "mock DNA sequencing" method.[190]

Initial problems encountered included low sensitivity, poor ionization of DNA, and counterion heterogeneity. However, in later work, sensitivity was lowered to femtomoles[191] and resolution of small fragments increased to >500,[193] mostly by correcting for the velocity distribution of ions formed in the ion source (DE-MALDI[194]). Application of delayed extraction–matrix-assisted laser desorption/ionization on DNA sequencing to a high yield Sanger sequencing protocol has been described by Roskey et al.[189] and was appropriate for sequencing 40- and 50-mer DNA using a 13-mer primer.

Two interesting MALDI sequencing experiments were desribed by Köster et al.[192] Known 39- and 78-mer DNA templates, labeled with biotin on the 3′-end were immobilized on streptavidin coated magnetic beads. A 14-mer primer was annealed to the 3′-end of the 39- and 78-mer immobilized templates and then elongated and terminated in four base-specific Sanger sequencing reactions. Purine triphosphates were stabilized for MS by using N^7-deaza analogs, to reduce depurination and fragmentation during ionization/desorption. The Sanger mixture components were washed off, and the duplex oligonucleotides' phosphate backbone counterions were ion exchanged with excess ammonium salts, owing to the solid phase nature of the methodology. Finally, the immobilized nucleic acids were directly mixed with the matrix solution and analyzed by MALDI-MS. Only the annealed and extended primer ladder oligonucleotides were denatured and observed in the MS. The authors noticed an approximately 50% in-chain termination in the extension process, perhaps because of a special, sequence dependent tertiary structure formation. Also, some depurination and a full length+1 elongation product complicated the interpretation of results.

A potential *de novo* high-throughput version of the Köster solid phase sequencing was described as well. An 18-mer biotinylated DNA (GTG ATG CGT CGG ATC ATC-Bio) was immobilized on streptavidin coated magnetic beans. A 5′-fluorescein labeled primer 23-mer (F-GAT GAT CCG ACG CAT CAC-AGC TC) was annealed to the 18-mer, leaving a 5-base overhang that was used to capture a 15-mer template sequence (TCG GTT CCA AGA GCT). The primer was then elongated in the presence of ddNTPs, conditioned and directly MALDI-TOF MS analyzed successfully, although a relatively low intensity of peaks were observed.

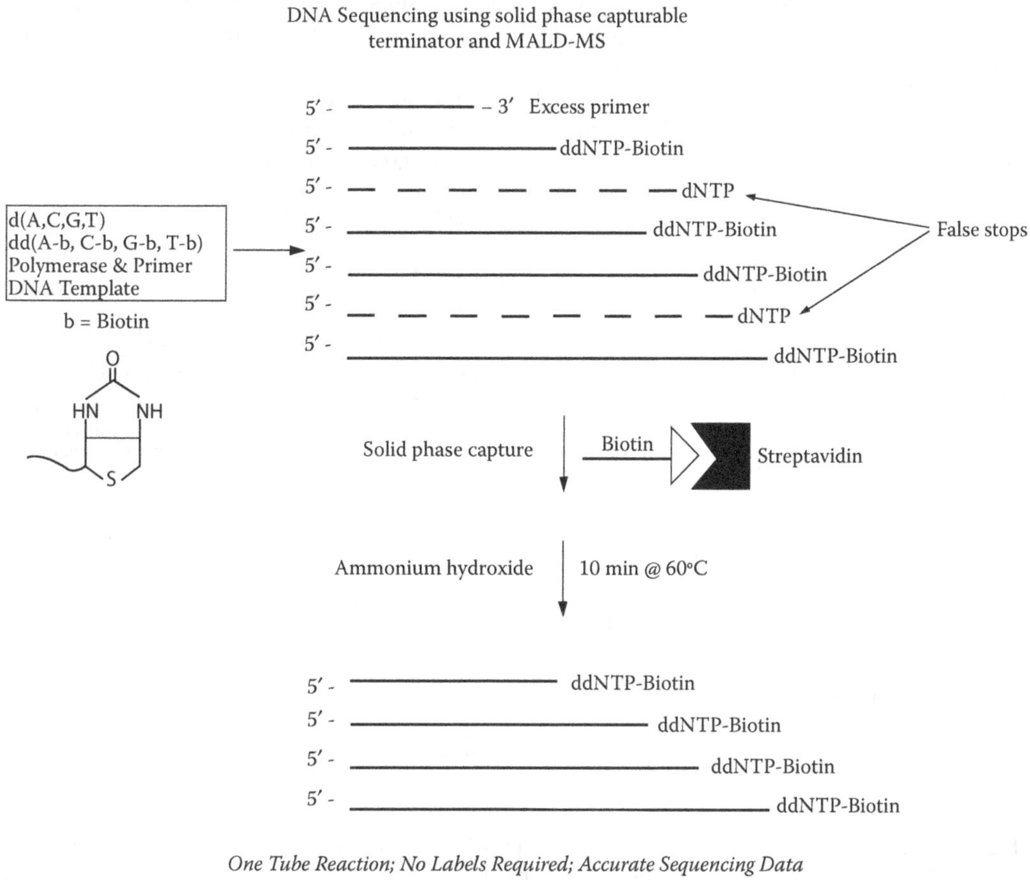

FIGURE 5.28 Schematic representation of SPC sequencing. Sanger sequencing reactions are carried out with biotinylated dideoxynucleotides. After solid phase capture of the correctly terminated DNA sequencing fragments, excess primer, incorrectly terminated DNA fragments, and other impurities are removed. The pure DNA products are then released from the solid phase for analysis by mass spectrometry. (Reprinted from Edwards, J. R., et al., *Mutation Research: Fundamental and Molecular Mechanisms of Mutagenesis*, 573, 3–12. Copyright 2005, with permission from Elsevier.)

```
3′-Bio-CTA CTA GGC TGC GTA GTG TCG GTT CCA AGA GCT
  5′-F-GAT GAT CCG ACG CAT CAC AGC TC
  5′-F-GAT GAT CCG ACG CAT CAC AGC TCA
  5′-F-GAT GAT CCG ACG CAT CAC AGC TCA G
  5′-F-GAT GAT CCG ACG CAT CAC AGC TCA GG
  5′-F-GAT GAT CCG ACG CAT CAC AGC TCA GGT
  etc.
```

Primer elongation on immobilized template by Sequenase™. Template sequence in bold and underlined. Biotinylated sequence was immobilized, fluorescein labeled primer hybridized, to which the target sequence was hybridized on a 5-base overhang. Elongation of the primer in the presence of ddNTPs resulted in partially terminated sequences that were MALDI-MS analyzed. Bio, biotin; F, fluorescein.

A major hurdle of MALDI-TOF MS analysis of the Sanger terminating mixture is low sensitivity and peak broadening due to the complexity of the sequencing mixture that contains the primer at high concentration, compared to the sequence ladder fragments as well as the enzyme, the triphosphates (NTPs), and dideoxy nucleotides (ddNTPs) or dye or tagged ddNTPs, alkali, and alkali earth metals that are bound to DNA. The interpretation of the results are further complicated by enzyme errors, when falsely stopped DNA sequencing fragments are generated, differing by only 16 Da from the dideoxy terminated fragments.

Edwards et al. published a solution to implement the hurdles desribed above, in which biotin labeled ddNTP-s (alkinylated N^7-deaza purines or alkin- and alkenylated 5-methyl pyrimidines)[195] are used. Short reads (5–20 bps) in question, that heterozygous by deletion or insertion mutations

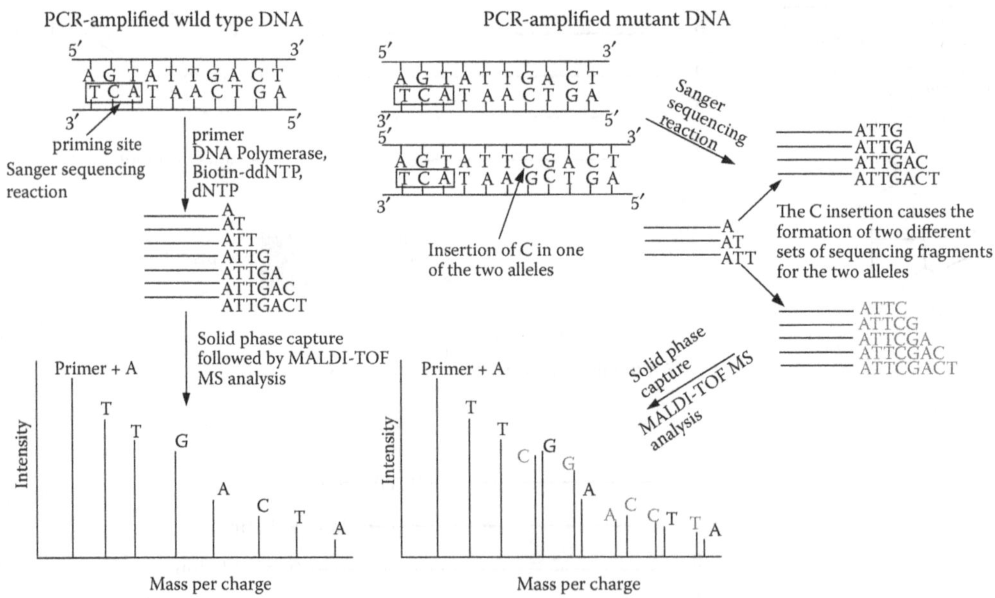

FIGURE 5.29 Schematic representation of SPC sequencing used for frame shift mutation detection. Wild-type DNA has the same sequence for both alleles leading to the production of a single peak at every position in the mass spectrum that will correspond to the nucleotide existing at that position in the sequence (left panel). However, mutant DNA with an insertion or deletion (insertion shown here as an example) in one allele can have two different bases coexisting at the same sequence position. This leads to the formation of two distinct mass peaks at each position in the spectrum (right panel). By calculating the corresponding mass differences, the sequences of both alleles can be simultaneously read and the mutation site can be accurately characterized. (Reprinted from Edwards, J. R., et al., *Mutation Research: Fundamental and Molecular Mechanisms of Mutagenesis*, 573, 3–12. Copyright 2005, with permission from Elsevier.)

such as 185delAG and 5382insC in the BRCA1 genes can be accurately determined, while the electrophoretic method fails. The principal advantage of the solid phase capturable dideoxynucleotides (SPC-ddNTPs) sequencing is that highly accurate determination of different bases that coexist at a single locus along the DNA as a direct consequence of substitution or frameshift mutation is possible as compared to other sequencing methods that measure fluorescent or radioactive signals emmited from the labeled DNA (Figures 5.28 and 5.29).[195]

Kwon et al. reported transcriptional DNA sequencing of ribonucleotides up to 56nt long using RNA ladder transcripts of DNA and MALDI-TOF analysis. A nicked form of phage SP6 polymerase substantially enhanced the yields and increased the effective sensitivity of the mass spectrometric analysis. The small difference between C and U nucleotides were increased by the use of α-thiouridine-triphosphate. Homozygous and heterozygous samples were unambiguously differentiated with the described DNA sequencing and genotyping method using these RNA transcript ladders.[196]

5.3.4.1 Minisequencing

Genotyping of single nucleotide polymorphism is of high interest for biomedical and forensic research and applications. Accurate, high-throughput and low-cost methods are under investigation.[197–211]

Mengel-Jorgensen et al. reported a single-base extension (SBE) assay for SNP detection using biotin labeled different mass ddNTPs, DNA/RNA chimeric primers, endonucleolytic cleavage of primers after extension, and monomeric avidin triethylamine purification (MATP) in combination with MALDI mass determination of the single-base extended primers.[146] The main goal of this technique was multiplexed SNP determination at low cost. The schematic description of the process is summarized in Figure 5.30.

Determination of SBE by mass spectrometry is based on the following: the DNA region of interest from a biological sample (e.g., blood) is PCR amplified. Primer DNA is designed (15–40 bases long) immediately upstream of the SNP and extended by a ddNTP. Finally, the extended primer mass is determined from which the SNP base can be determined.

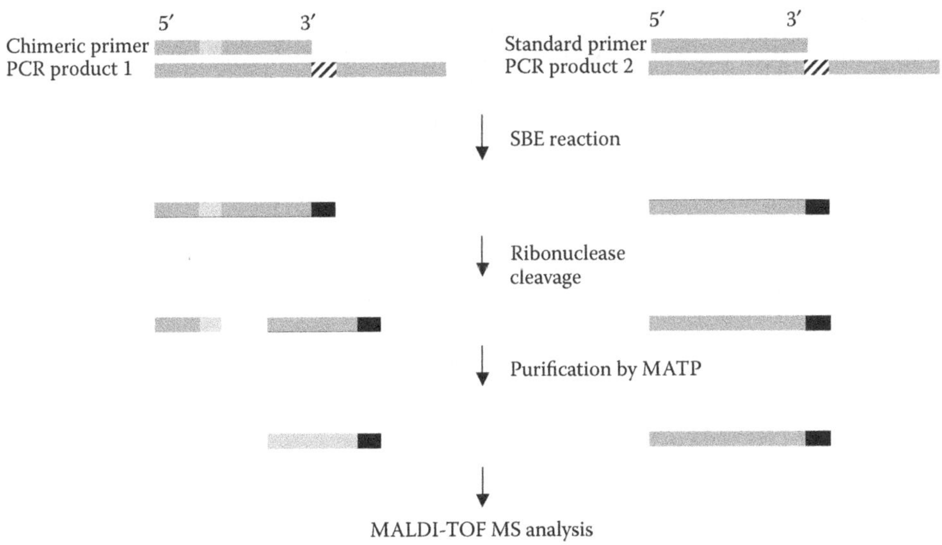

FIGURE 5.30 Principle of the SBE assay. A chimeric primer is used to detect the SNP on PCR product 1 and a standard primer is used to detect the SNP on PCR product 2. The positions of the SNPs are shown as hatched boxes. Deoxyribonucleotide strands are shown in dark gray, the position of the ribonucleotide in the chimeric primer is shown in light gray, and the biotin-labeled dideoxyribonucleotides are shown in black. (Reprinted with permission from Mengel-Jorgensen, J., et al., *Analytical Chemistry*, 77, 5229–5235. Copyright 2005, American Chemical Society.)

The number of SNPs to be determined by a single SNP assay is between 5 and 50, which makes the design of primers and determination of the SNP by MALDI-MS complicated or impossible for larger numbers of multiplexing. The difficulty relies on the complexity of the mass spectra due to the large number of DNAs, overlapping peaks in the spectra and low resolution at higher mass regions (>5000) to differentiate a single A/T (9 Da) mutation.

The solution for the described difficulties were shown below. The higher resolution and accuracy MALDI mass region of 5–10 bases (1500–3000 Da) can be used if the extension primers are cleaved 4–9 bases upstream of the SNP site. This is possible using chimeric DNA primers with only one RNA base at position 4–9 from 3′-terminus and an RNase cleavage after the extension step. A combination of nucleases was found to be effective to cleave any RNA bases in such chimeras: 0.5 μg of RNase A, 500 units of RNase T_1, and 2–8 units of RNase T_2. Not only chimeric but regular primers were used to extend the multiplexicity. To increase the mass difference between A and T (9 Da), different sizes of biotin labeled ddNTPs were used for extension, of which the smallest mass difference was 16 Da (A/G). The biotin labeled ddNTPs incorporate in the primer during the assay; thus the reaction mixture complexity can be simplified by the monomeric avidin triethylamine purification (MATP) technique. As a result, a cheaper and multiplex MALDI MS based SNP determination assay has been discovered.

5.3.4.2 Sequencing (*de novo*) of Highly Modified Oligonucleotides[8,9]

Sequence verification of oligonucleotide drug candidates has become a regulatory requirement in API manufacturing. The difficulty of sequencing highly modified oligo APIs comprised of their complex chemical structure: none of the above described methods can be applied as described.

TABLE 5.11
Fragmentation Preferences and Reaction Conditions of Sequence Verification and *de novo* Sequencing of Highly Modified Oligonucleotides[a]

Cleavage Passenger[8]	Cleavage Guide[8]	Cleavage *de novo*[9]	Reagent	Note
dG > dA	mA > mG > fC	dG, dA	DMS (10 μL), 15–30 min, 45°C	
fU, fC, iB	fU, fC, mA > mG	fC, fU>dA, dT	NaOH (0.25 M, 100 μL), to EtOH (50 μL), 90–120 min, 65°C	
dG, dA, fU	mA, (rU, rG) > mG	dG, dA	DEPC (10 μL), NaOAc buffer (50 mM, 150 μL) pH 5.2, 15–30 min, 45°C	No cleavage at fC.
fU, fC > dG	fU, fC, (rU) > mG	dG>dA, fC	$NH_2OH \times HCl$ (4 M, 20 μL, pH 6 with DEA), 30 min, 45°C	No cleavage at mA, one mG cleavage in a sequence of fCs and fU.
fU > dG	fU, rU, rG, rA	fU, fC, dA, dG	Piperidine (100 μL, 1 M), 30 min, 85°C	No cleavage at mG, mA
dG, dA	—	dG>dA	Aniline/AcOH/water (100 μL 10:6:50) , 30 min, 85°C	No cleavage at fC, fU (except one case), one dG and one dA cleavage missed.
—	rU, rG, rA	—	Formic acid (150 μL, 98%), 5 min, 70°C	No cleavage at fC, fU, mA, mG.

[a] Reagents are added to an oligonucleotide solution (50 μL, 100 μM) except DEPC, piperidine, and aniline, where oligonucleotide solution was concentrated to dryness and reagents applied on the dry material. Reactions were quenched after the given reaction time, precipitated with 2-propanol (800–1100 μL), vortexed and centrifuged (20–30 min, 4°C, 14 k RPM), supernatant removed, and the residuals were washed with ethanol (75%, 1 mL) and centrifuged. Ethanol wash was removed and residuals concentrated to dryness. DEPC, DMS, and hydroxylamine degradation reactions were followed by piperidine hydrolysis (1 M for 30 min at 85°C).

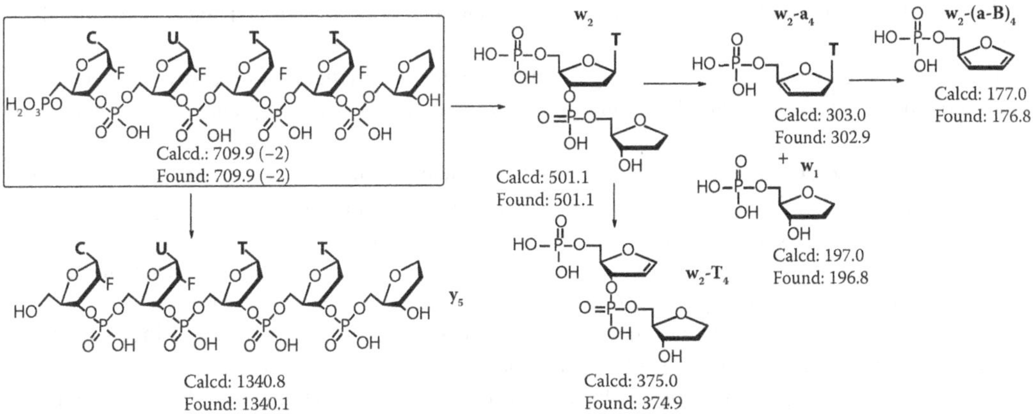

FIGURE 5.31 Tandem IT-MS on the 3′-terminus of the modified oligonucleotide (passenger) revealed its structural features, although full sequence verification in this specific example is not achieved.

Especially, in the case of *de novo* sequence determination of a mixed sequence oligo drug, only a combination of methodologies can lead to success.

Farand et al. recently reported a robust methodology for sequence determination of 21–23 nt long phosphodiester oligonucleotides containing DNA, RNA, 2′-F (U and C), 2′-OMe (A, G, U) and abasic nucleotides by combination of enzymatic cleavage (nuclease P_1, phosphodiesterase I, alkaline phosphatase), chemical degradation (DEPC, DMS/piperidine, hydroxylamine/piperidine, aniline, or piperidine), acidic and alkaline hydrolysis, ESI-TOF, and tandem ion trap (IT) mass spectrometry (Table 5.11).

Cleavage products were w or y and d or b type. Cleavage site is defined as 3′ of the listed nucleoside leaving either 3′-OH or phosphate on the 5′-terminus (in which case the remainder 3′-terminus had a 5′-phosphate or OH, respectively). According to the mechanism of piperidine degradation (1 M aqueous for 30 min at 85°C) (Figure 5.22, path 3a), a secondary cleavage may also occur, resulting in a nucleoside loss and two phosphate terminated fragments.

The 5′ dimer, 3′ trimers not found, those were sequenced by tandem IT-MS. As an example the 3′-terminus of the passenger nucleotide were isolated (m/z = 709.2, charge state -2) and fragmented by amplitude variation of the ion trap (Figure 5.31). First, the 5′ phosphate was lost (m = 1340.1) followed by strand scission between thymidines at increased amplitude (m = 501.1). The 5′-phosphate dimer p-T-iB (iB: abasic deoxynucleoside) lost the thymine base (m = 374.9) and broke to the eliminated p-T (a type, m = 302.9) and p-iB (196.8) and also p-Fu (Fu: hydroxymethyl furane) (a-B type, m = 176.8). Tandem mass spectrometry sequencing and MS analysis of the chemically degraded passenger and guide oligonucleotides gave rise to full sequence determination.

The same authors presented a *de novo* sequencing of a model oligonucleotide API. The methodology consists of (1) exact mass determination of the full length compound, (2) nucleoside composition analysis by enzymatic cleavage and ultra-performance liquid chromatography (UPLC), (3) partial hydrolysis with NH_2OH/piperidine and consecutive mass spectrometry for determination of the 5′-terminal nucleotides (using average masses from deconvolution), (4) mass ladder examination of seven different degradations by MS to figure out the rest of the base order, and (5) comparison of predicted sequence with UPLC nucleoside composition analysis results was used to ascertain the identity of nucleosides that have the same mass (rA-dG).[9]

5.4 SUMMARY

Sequencing of oligonucleotides by mass spectrometry has its unique role besides the traditional large-scale sequencing techniques in the areas of sequencing short and modified or mutant DNAs, highly structured and/or modified short RNAs and oligonucleotide-based pharmaceuticals and drug candidates.

Classical categories of MS sequencings include sequence verification or confirmation and *de novo* sequencing, classes that are differentiated by the availability of the sequence information *prior* to the analysis.

Three mild ionization sources are reported for oligonucleotide mass spectrometry (FAB, MALDI, and ESI), of which two have been discussed in detail in this chapter focusing on sequencing applications. Although the resolution, accuracy, and sensitivity is highly dependent on the ionization method, the instrumental configuration of the mass spectrometer including the mass analyzer, collision cell, and detector has major effects on those performance attributes as well.

Many times the available ionization source and instrument configuration determines what techniques can be applied; in other cases, the sequencing approach defines what kind of instrumentation is most suited for the intended purpose. One aim of this review is to help the user determine whether MALDI or ESI ionization is best for an actual oligonucleotide sequencing problem.

Tandem mass spectrometry (MS/MS) is an attractive choice for sequence verification of short oligonucleotides (Section 5.2). Separation methods for oligonucleotide LC-MS have improved significantly in the past decade (Section 5.2.2.1) and have aided in the successful development of MS/MS techniques. Multistage MS is usually performed using electrospray ionization of negatively charged phosphodiester oligonucleotides in aqueous solutions, hence in negative polarity (Section 5.2.2.2), although positive mode ESI (Section 5.2.2.3) and MALDI-MS/MS in positive mode (Section 5.2.2.4) has also been described. MS/MS sequencing is typically applied for 20-mers but successful software supported resequencing is reported for longer DNAs up to 80-mers (Section 5.2.2.5) MS/MS sequence verification has its major role in the quality control of manufacturing of oligonucleotide-based active pharmaceutical ingredients because sequence verification is a regulatory requirement for such medicinal candidates or products (Section 5.4). While full sequence verification by MS/MS is routinely performed for ~20-mers, successful sequencing of longer oligonucleotides (30- to 40-mers) is occasionally possible, however, highly dependent on the sequence, modifications in the molecule, purity, concentration of the sample, instrumental attributes (resolution and accuracy), sequencing technique, and data interpretation. Recent advancement in simplification of the tandem mass spectrum by mitigating the internal energy of parent ions in beam-type MS/MS gave rise to *de novo* sequencing of an RNA 34-mer (Section 5.2.2.2). Software support is available for interpretation of the MS/MS data of *de novo* sequencing for research purposes for limited number of nucleotides only. Software-aided MS/MS resequencing of 80-mer DNA is reported that also underlines the importance of computational tools in the mass spectrometry sequencing discipline (Section 5.2.2.5).

MALDI ionization produces predominantly singly charged ions; thus the spectrum is simpler than that of an oligonucleotide ionized with ESI (Section 5.3), where multiply charged ions complicate the spectrum. Sample introduction into the mass spectrometer is easier in case of MALDI, but it is disadvantageous that separation techniques cannot be readily "hyphenated," as it can be easily done with LC-ESI-MS configurations.

The simplest MALDI sequencing methods include direct sequencing (Section 5.3.1) and failure analysis of synthetic oligonucleotides (Section 5.3.2). Direct MALDI sequencing is a technique in which an appropriately chosen matrix and laser wavelength combination (Table 5.4), or postsource decay, is applied to the ion cloud, causing fragmentation from which the sequence is determined. Sequencing by failure analysis relies on a ladder created by shortmer sequences that are present in synthetic oligonucleotides. The sequence is read out from the differences of the successive peaks in the MALDI mass spectrum.

This sequence ladder can also be generated by different degradation techniques. Enzymatic degradation (Section 5.3.3.1) by exonucleases (CSP, BSP, and SVP) on DNA and RNA produces 5′ or 3′ sequence ladders. Often, the enzymatic digestion mixtures can be quenched by the MALDI matrix components and directly applied onto the target plate of the MALDI spectrometer. Specific endonucleases produce overlapping RNA fragments that lead to partial sequence information. Appropriate combination of exo- and endonucleases results in successful full *de novo* sequence determination of RNAs.

Phosphorothioate oligodeoxynucleotides are resistant to direct enzymatic digestion such that direct treatment with nucleases even at elevated temperature and for elongated times give rise to only partial cleavage. However, a phosphorothioate DNA can be converted to regular phosphate followed by cleavage with exonucleases, extending the applicability of enzymatic sequencing onto phosphorothioates as well. A collection of nucleases applied for MS sequencing of oligonucleotides is listed in Table 5.8.

Chemical degradation has also been used to create MALDI sequencing ladders. Phosphorothioate oligonucleotides can be specifically fragmented at thiophosphate diester bonds by a chemical reaction with iodoethanol. Nucleobase selective/specific chemical degradations (Section 5.3.3.2) are summarized in Table 5.9, many of which have been applied to mass spectrometric sequencing of DNA and RNA. Nonspecific chemical cleavage of RNA by acid or base followed by MALDI-MS is a rugged method for sequence analysis, although not widespread as much as the enzymatic techniques. DNA is highly stable to basic conditions and sensitive to depurination in slightly acidic media; thus, only partial sequence information was gained by acidic hydrolysis of a DNA oligonucleotide.

The analysis of the Sanger chain termination mixtures, traditionally done by (capillary) electrophoresis, has been performed by mass spectrometry (Section 5.3.4). The major hurdles for mass analysis of the Sanger mixture are the very low concentration of the sequence ladders in the terminating mixture and the complexity of the sequencing solution, which contain the primer, dideoxy- and regular nucleoside triphosphates, the transcription enzyme, and salts. The concentration issue can be addressed by amplification of the target DNA, and simplification of the reaction mixtures by using biotinylated primer or ddNTP followed by affinity filtration has shown to be successful in frame shift mutation analysis (Section 5.3.4). Multiplexed single base extension methodology and analysis by MS can be an appropriate tool in some applications for SNP determination (mini-sequencing, Section 5.3.4).

Similarly to peptide mass fingerprinting (PMF), RNA can be degraded by RNase T_1 and the cleavage mixture analyzed by MS and compared to gene banks by a special computer-aided search algorithm (Section 5.3.2.1).

A combination of composition determination by enzymatic degradation and UPLC analysis, acidic, basic, and nucleobase specific chemical cleavage followed by ESI-TOF and multistage ion trapping MS/MS was recommended for highly modified oligonucleotide sequencing. The rugged procedure was applied for sequence verification and *de novo* sequencing of oligonucleotide-based API-type molecules (21- to 23-mers) including DNA, RNA, 2′-F, 2′-OMe, and abasic moieties.

In summary, mass spectrometry is a versatile tool for sequence analysis of short oligonucleotides. There is a strong desire, especially from the industrial and regulatory institutions, for development of sequencing technologies suitable for long and complex oligonucleotides. In the rapidly developing research area of noncoding RNAs, sequence determination is also of increasing importance. The extremely high resolving power of novel separation techniques hyphenated to mass spectrometers, and a wide choice of fragmentation and degradation or chain termination methods are the key tools for successful full sequence verification or *de novo* sequencings. The development of more sensitive, more accurate, and extremely high resolution mass spectrometers along with the application of highly reliable, high-throughput, and robust MS sequencing methodologies will inevitably advance the oligonucleotide sequencing field.

ACKNOWLEDGMENTS

The author thanks Agilent Laboratories for all the support during the preparation of this review, especially to Alex Apffel, Joel Myerson, Doug Dellinger, Todd Kreutzian, and Travis Betz (Nucleic Acids Solutions Division) for the careful reading, corrections, and additions. The author appreciates the personal discussions with Professor Scott McLuckey, Finn Kirpekar, Herbert Oberacher, and Julie Farand at the 22nd ASMS Sanibel Conference on Mass Spectrometry, St. Pete Beach, FL, 2010. I also thank my wife, Andrea Miskolczi, for her supportive patience during the preparation of this manuscript.

REFERENCES

1. Kirpekar, F., et al. 1998. DNA sequence analysis by MALDI mass spectrometry. *Nucleic Acids Research* 26(11): 2554–2559.
2. Hahner, S., et al. 1997. Matrix-assisted laser desorption/ionization mass spectrometry (MALDI) of endonuclease digests of RNA. *Nucleic Acids Research* 25(10): 1957–1964.
3. Lane, D. J., et al. 1985. Rapid determination of 16S ribosomal RNA sequences for phylogenetic analyses. *Proceedings of the National Academy of Sciences of the United States of America* 82(20): 6955–6959.
4. Bauer, G. J. 1990. Rna sequencing using fluorescent-labeled dideoxynucleotides and automated fluorescence detection. *Nucleic Acids Research* 18(4): 879–884.
5. Kramer, F. R., and D. R. Mills. 1978. RNA sequencing with radioactive chain-terminating ribonucleotides. *Proceedings of the National Academy of Sciences of the United States of America* 75(11): 5334–5338.
6. Edwards, J. R., H. Ruparel, and J. Y. Ju. 2005. Mass-spectrometry DNA sequencing. *Mutation Research: Fundamental and Molecular Mechanisms of Mutagenesis* 573(1–2): 3–12.
7. Giessing, A. M. B., et al. 2009. Identification of 8-methyladenosine as the modification catalyzed by the radical SAM methyltransferase Cfr that confers antibiotic resistance in bacteria. *RNA: A Publication of the RNA Society* 15(2): 327–336.
8. Farand, J., and M. Beverly. 2008. Sequence confirmation of modified oligonucleotides using chemical degradation, electrospray ionization, time-of-flight, and tandem mass spectrometry. *Analytical Chemistry* 80(19): 7414–7421.
9. Farand, J., and F. Gosselin. 2009. De novo sequence determination of modified oligonucleotides. *Analytical Chemistry* 81(10): 3723–3730.
10. Hillenkamp, F., et al. 1990. Matrix assisted UV-laser desorption/ionization: a new approach to mass spectrometry of large biomolecules. In *Biological Mass Spectrometry, Proceedings of the 2nd International Symposium on Mass Spectrometry in the Health and Life Sciences*, 2nd, Elsevier, Amsterdam, Netherlands, pp. 49–60.
11. Tanaka, K., et al. 1988. Protein and polymer analyses up to m/z 100,000 by laser ionization time-of-flight mass spectrometry. *Rapid Communications in Mass Spectrometry* 2(8): 151–153.
12. Boernsen, K. O., M. Schaer, and H. M. Widmer. 1990. Matrix-assisted laser desorption and ionization mass spectrometry and its applications in chemistry. *Chimia* 44(12): 412–416.
13. Fenn, J. B., et al. 1989. Electrospray ionization for mass-spectrometry of large biomolecules. *Science* 246(4926): 64–71.
14. Barry, J. P., et al. 1995. Mass and sequence verification of modified oligonucleotides using electrospray tandem mass-spectrometry. *Journal of Mass Spectrometry* 30(7): 993–1006.
15. McLuckey, S. A., G. J. Van Berkel, and G. L. Glish. 1992. Tandem mass spectrometry of small, multiply charged oligonucleotides. *Journal of the American Society for Mass Spectrometry* 3(1): 60–70.
16. Comisarow, M. B., and A. G. Marshall. 1974. Fourier transform ion cyclotron resonance spectroscopy. *Chemical Physics Letters* 25(2): 282–283.
17. Makarov, A. 2000. Electrostatic axially harmonic orbital trapping: A high-performance technique of mass analysis. *Analytical Chemistry* 72(6): 1156–1162.
18. Electrospray Ionization. Wikipedia. http://en.wikipedia.org/wiki/Electrospray_ionization.
19. Wu, J., and S. A. McLuckey 2004. Gas-phase fragmentation of oligonucleotide ions. *International Journal of Mass Spectrometry* 237(2-3): 197–241.
20. Little, D. P., et al. 1994. Rapid sequencing of oligonucleotides by high-resolution mass spectrometry. *Journal of the American Chemical Society* 116(11): 4893–4897.
21. Loo, J. A., H. R. Udseth, and R. D. Smith. 1988. Collisional effects on the charge distribution of ions from large molecules, formed by electrospray-ionization mass spectrometry. *Rapid Communications in Mass Spectrometry* 2(10): 207–210.
22. Wolter, M. A., and J. W. Engels. 1995. Nanoelectrospray mass spectrometry/mass spectrometry for the analysis of modified oligoribonucleotides. *European Journal of Mass Spectrometry* 1(6): 583–590.

23. Little, D. P., and F. W. McLafferty. 1995. Sequencing 50-mer DNAs using electrospray tandem mass spectrometry and complementary fragmentation methods. *Journal of the American Chemical Society* 117(25): 6783–6784.
24. Little, D. P., et al. 1996. Sequence information from 42-108-mer DNAs (complete for a 50-mer) by tandem mass spectrometry. *Journal of the American Chemical Society* 118(39): 9352–9359.
25. Little, D. P., T. W. Thannhauser, and F. W. McLafferty. 1995. Verification of 50- to 100-mer DNA and RNA sequences with high-resolution mass spectrometry. *Proceedings of the National Academy of Sciences of the United States of America* 92(6): 2318–2322.
26. Schurch, S., E. Bernal-Mendez, and C. J. Leumann. 2002. Electrospray tandem mass spectrometry of mixed-sequence RNA/DNA oligonucleotides. *Journal of the American Society for Mass Spectrometry* 13(8): 936–945.
27. McLuckey, S. A., and S. Habibi-Goudarzi. 1993. Decompositions of multiply charged oligonucleotide anions. *Journal of the American Chemical Society* 115(25): 12085–12095.
28. McLuckey, S. A., and G. Vaidyanathan. 1997. Charge state effects in the decompositions of single-nucleobase oligonucleotide polyanions. *International Journal of Mass Spectrometry and Ion Processes* 162(1–3): 1–16.
29. Apffel, A., et al. 1997. Analysis of oligonucleotides by HPLC-electrospray ionization mass spectrometry. *Analytical Chemistry* 69(7): 1320–1325.
30. Huber, C. G., and H. Oberacher. 2001. Analysis of nucleic acids by on-line liquid chromatography-mass spectrometry. *Mass Spectrometry Reviews* 20(5): 310–343.
31. Premstaller, A., H. Oberacher, and C. G. Huber. 2000. High-performance liquid chromatography-electrospray ionization mass spectrometry of single- and double-stranded nucleic acids using monolithic capillary columns. *Analytical Chemistry* 72(18): 4386–4393.
32. Oberacher, H., et al. 2004. Applicability of tandem mass spectrometry to the automated comparative sequencing of long-chain oligonucleotides. *Journal of the American Society for Mass Spectrometry* 15(4): 510–522.
33. Gentil, E., and J. Banoub. 1996. Characterization and differentiation of isomeric self-complementary DNA oligomers by electrospray tandem mass spectrometry. *Journal of Mass Spectrometry* 31(1): 83–94.
34. Crain, P. F., et al. 1996. Characterization of posttranscriptional modification in nucleic acids by tandem mass spectrometry. In *Mass Spectrometry in the Biological Sciences*, 497–517. New York: Springer.
35. Boschenok, J., and M. M. Sheil. 1996. Electrospray tandem mass spectrometry of nucleotides. *Rapid Communications in Mass Spectrometry* 10(1): 144–149.
36. Aaserud, D. J., et al. 1997. DNA sequencing with blackbody infrared radiative dissociation of electrosprayed ions. *International Journal of Mass Spectrometry* 167: 705–712.
37. McLafferty, F. W., et al. 1997. Double stranded DNA sequencing by tandem mass spectrometry. *International Journal of Mass Spectrometry and Ion Processes* 165/166: 457–466.
38. Marzilli, L. A., et al. 1999. Oligonucleotide sequencing using guanine-specific methylation and electrospray ionization ion trap mass spectrometry. *Journal of Mass Spectrometry* 34(4): 276–280.
39. Wang, Y., J. S. Taylor, and M. L. Gross. 2001. Fragmentation of electrospray-produced oligodeoxynucleotide ions adducted to metal ions. *Journal of the American Society for Mass Spectrometry* 12(5): 550–556.
40. Wang, Y., J.-S. Taylor, and M. L. Gross. 1999. Nuclease P1 digestion combined with tandem mass spectrometry for the structure determination of DNA photoproducts. *Chemical Research in Toxicology* 12(11): 1077–1082.
41. Wang, Y., J.-S. Taylor, and M. L. Gross. 1999. Differentiation of isomeric photomodified oligodeoxynucleotides by fragmentation of ions produced by matrix-assisted laser desorption ionization and electrospray ionization. *Journal of the American Society for Mass Spectrometry* 10(4): 329–338.
42. Wang, Y., J.-S. Taylor, and M. L. Gross. 2001. Fragmentation of photomodified oligodeoxynucleotides adducted with metal ions in an electrospray-ionization ion-trap mass spectrometer. *Journal of the American Society for Mass Spectrometry* 12(11): 1174–1179.
43. Habibi-Goudarzi, S., and S. A. McLuckey. 1995. Ion trap collisional activation of the deprotonated deoxymononucleoside and deoxydinucleoside monophosphates. *Journal of the American Society for Mass Spectrometry* 6(2): 102–113.
44. Ho, Y., and P. Kebarle. 1997. Studies of the dissociation mechanisms of deprotonated mononucleotides by energy resolved collision-induced dissociation. *International Journal of Mass Spectrometry and Ion Processes* 165/166: 433–455.

45. Vrkic, A. K., R. A. J. O'Hair, and S. Foote. 2000. Fragmentation reactions of all 64 deprotonated trinucleotide and 16 mixed base tetranucleotide anions via tandem mass spectrometry in an ion trap. *Australian Journal of Chemistry* 53(4): 307–319.
46. Vrkic, A. K., et al. 2000. Fragmentation reactions of all 64 protonated trimer oligodeoxynucleotides and 16 mixed base tetramer oligodeoxynucleotides via tandem mass spectrometry in an ion trap. *International Journal of Mass Spectrometry* 194(2/3): 145–164.
47. McLuckey, S. A., G. Vaidyanathan, and S. Habibi-Goudarzi. 1995. Charged vs. neutral nucleobase loss from multiply charged oligoribonucleotide anions. *Journal of Mass Spectrometry* 30(9): 1222–1229.
48. Klassen, J. S., P. D. Schnier, and E. R. Williams. 1998. Blackbody infrared radiative dissociation of oligonucleotide anions. *Journal of the American Society for Mass Spectrometry* 9(11): 1117–1124.
49. Wang, Z., et al. 1998. Structure and fragmentation mechanisms of isomeric T-rich oligodeoxynucleotides: a comparison of four tandem mass spectrometric methods. *Journal of the American Society for Mass Spectrometry* 9(7): 683–691.
50. Wan, K. X., et al. 2001. Fragmentation mechanisms of oligodeoxynucleotides studied by H/D exchange and electrospray ionization tandem mass spectrometry. *Journal of the American Society for Mass Spectrometry* 12(2): 193–205.
51. Wan, K. X., and M. L. Gross. 2001. Fragmentation mechanisms of oligodeoxynucleotides: effects of replacing phosphates with methylphosphonates and thymines with other bases in T-rich sequences. *Journal of the American Society for Mass Spectrometry* 12(5): 580–589.
52. Greco, F., et al. 1990. Gas-phase proton affinity of deoxyribonucleosides and related nucleobases by fast atom bombardment tandem mass spectrometry. *Journal of the American Chemical Society* 112(25): 9092–9096.
53. Ganem, B., Y. T. Li, and J. D. Henion. 1993. Detection of oligonucleotide duplex forms by ion-spray mass spectrometry. *Tetrahedron Letters* 34(9): 1445–1448.
54. Bayer, E., et al. 1994. Analysis of double-stranded oligonucleotides by electrospray mass spectrometry. *Analytical Chemistry* 66(22): 3858–3863.
55. Doktycz, M. J., S. Habibi-Goudarzi, and S. A. McLuckey. 1994. Accumulation and storage of ionized duplex DNA molecules in a quadrupole ion trap. *Analytical Chemistry* 66(20): 3416–3422.
56. Aaserud, D. J., et al. 1996. Accurate base composition of double-strand DNA by mass spectrometry. *Journal of the American Society for Mass Spectrometry* 7(12): 1266–1269.
57. Greig, M. J., H.-J. Gaus, and R. H. Griffey. 1996. Negative ionization micro electrospray mass spectrometry of oligodeoxyribonucleotides and their complexes. *Rapid Communications in Mass Spectrometry* 10(1): 47–50.
58. Schnier, P. D., et al. 1998. Activation energies for dissociation of double strand oligonucleotide anions: evidence for Watson–Crick base pairing in vacuo. *Journal of the American Chemical Society* 120(37): 9605–9613.
59. Gabelica, V., and E. De Pauw. 2001. Comparison between solution-phase stability and gas-phase kinetic stability of oligodeoxynucleotide duplexes. *Journal of Mass Spectrometry* 36(4): 397–402.
60. Gabelica, V., and E. De Pauw. 2002. Comparison of the collision-induced dissociation of duplex DNA at different collision regimes: evidence for a multistep dissociation mechanism. *Journal of the American Society for Mass Spectrometry* 13(1): 91–98.
61. Gabelica, V., and E. De Pauw. 2002. Collision-induced dissociation of 16-mer DNA duplexes with various sequences: evidence for conservation of the double helix conformation in the gas phase. *International Journal of Mass Spectrometry* 219(1): 151–159.
62. Goodlett, D. R., et al. 1993. Direct observation of a DNA quadruplex by electrospray ionization mass spectrometry. *Biological Mass Spectrometry* 22(3): 181–183.
63. Muddiman, D. C., and R. D. Smith. 1998. Sequencing and characterization of larger oligonucleotides by electrospray ionization fourier transform ion cyclotron resonance mass spectrometry. *Reviews in Analytical Chemistry* 17(1): 1–68.
64. Beck, J. L., et al. 2001. Electrospray ionization mass spectrometry of oligonucleotide complexes with drugs, metals, and proteins. *Mass Spectrometry Reviews* 20(2): 61–87.
65. Hofstadler, S. A., and R. H. Griffey. 2001. Analysis of noncovalent complexes of DNA and RNA by mass spectrometry. *Chemical Reviews* 101(2): 377–390.
66. Tang, W., L. Zhu, and L. M. Smith. 1997. Controlling DNA fragmentation in MALDI-MS by chemical modification. *Analytical Chemistry* 69(3): 302–312.
67. Taucher, M., U. Rieder, and K. Breuker. 2010. Minimizing base loss and internal fragmentation in collisionally activated dissociation of multiply deprotonated RNA. *Journal of the American Society for Mass Spectrometry* 21(2): 278–285.

68. Sannes-Lowery, K. A., et al. 1997. Positive ion electrospray ionization mass spectrometry of oligonucleotides. *Journal of the American Society for Mass Spectrometry* 8(1): 90–95.
69. Hunter, E. P. L., and S. G. Lias. 1998. Evaluated gas phase basicities and proton affinities of molecules: an update. *Journal of Physical and Chemical Reference Data* 27(3): 413–656.
70. Lias, S. G., J. F. Liebman, and R. D. Levin. 1984. Evaluated gas phase basicities and proton affinities of molecules; heats of formation of protonated molecules. *Journal of Physical and Chemical Reference Data* 13(3): 695–808.
71. Marino, T., et al. 1994. Molecular orbital study of the protonation of dA, dG, dC and dT 2′-deoxyribonucleosides. *Theochem* 112(2–3): 185–195.
72. Smets, J., et al. 1996. Multiple site proton affinities of methylated nucleic acid bases. *Chemical Physics Letters* 262(6): 789–796.
73. Ni, J., M. A. A. Mathews, and J. A. McCloskey. 1997. Collision-induced dissociation of polyprotonated oligonucleotides produced by electrospray ionization. *Rapid Communications in Mass Spectrometry* 11(6): 535–540.
74. Wang, P., M. G. Bartlett, and L. B. Martin. 1997. Electrospray collision-induced dissociation mass spectra of positively charged oligonucleotides. *Rapid Communications in Mass Spectrometry* 11(8): 846–856.
75. Weimann, A., P. Iannitti-Tito, and M. M. Sheil. 2000. Characterization of product ions in high-energy tandem mass spectra of protonated oligonucleotides formed by electrospray ionization. *International Journal of Mass Spectrometry* 194(2/3): 269–288.
76. Hakansson, K., et al. 2003. Electron capture dissociation and infrared multiphoton dissociation of oligodeoxynucleotide dications. *Journal of the American Society for Mass Spectrometry* 14(1): 23–41.
77. Kirpekar, F., S. Douthwaite, and P. Roepstorff. 2000. Mapping posttranscriptional modifications in 5S ribosomal RNA by MALDI mass spectrometry. *RNA-A Publication of the RNA Society* 6(2): 296–306.
78. Kirpekar, F., and T. N. Krogh. 2001. RNA fragmentation studied in a matrix-assisted laser desorption/ionisation tandem quadrupole/orthogonal time-of-flight mass spectrometer. *Rapid Communications in Mass Spectrometry* 15(1): 8–14.
79. Rozenski, J., and J. A. McCloskey. 2002. SOS: A simple interactive program for ab initio oligonucleotide sequencing by mass spectrometry. *Journal of the American Society for Mass Spectrometry* 13(3): 200–203.
80. McCloskey, J. A., and J. Rozenski. 2005. The small subunit rRNA modification database. *Nucleic Acids Research* 33(Database issue): D135–D138.
81. Dunin-Horkawicz, S., et al. 2006. MODOMICS: A database of RNA modification pathways. *Nucleic Acids Research* 34(Database issue): D145–D149.
82. Czerwoniec, A., et al. 2009. MODOMICS: A database of RNA modification pathways. 2008 update. *Nucleic Acids Research* 37(Database issue): D118–D121.
83. Oberacher, H., B. Wellenzohn, and C. G. Huber. 2002. Comparative sequencing of nucleic acids by liquid chromatography-tandem mass spectrometry. *Analytical Chemistry* 74(1): 211–218.
84. Oberacher, H., et al. 2002. Re-sequencing of multiple single nucleotide polymorphisms by liquid chromatography-electrospray ionization mass spectrometry. *Nucleic Acids Research* 30(14): e67.
85. Oberacher, H., and F. Pitterl. 2009. On the use of ESI-QqTOF-MS/MS for the comparative sequencing of nucleic acids. *Biopolymers* 91(6): 401–409.
86. Nakayama, H., et al. 2009. Ariadne: A database search engine for identification and chemical analysis of RNA using tandem mass spectrometry data. *Nucleic Acids Research* 37(6): e47.
87. Muhammad, W. T., et al. 2003. Automated discrimination of polymerase chain reaction products with closely related sequences by software-based detection of characteristic peaks in product ion spectra. *Rapid Communications in Mass Spectrometry* 17(24): 2755–2762.
88. Kellersberger, K. A., et al. 2004. Top-down characterization of nucleic acids modified by structural probes using high-resolution tandem mass spectrometry and automated data interpretation. *Analytical Chemistry* 76(9): 2438–2445.
89. Huth-Fehre, T., et al. 1992. Matrix-assisted laser desorption mass spectrometry of oligodeoxythymidylic acids. *Rapid Communications in Mass Spectrometry* 6(3): 209–213.
90. Smirnov, I. P., et al. 1996. Sequencing oligonucleotides by exonuclease digestion and delayed extraction matrix-assisted laser desorption ionization time-of-flight mass spectrometry. *Analytical Biochemistry* 238(1): 19–25.
91. Nordhoff, E., et al. 1992. Matrix-assisted laser desorption ionization mass-spectrometry of nucleic-acids with wavelengths in the ultraviolet and infrared. *Rapid Communications in Mass Spectrometry* 6(12): 771–776.
92. Juhasz, P., et al. 1996. Applications of delayed extraction matrix-assisted laser desorption ionization time-of-flight mass spectrometry to oligonucleotide analysis. *Analytical Chemistry* 68(6): 941–946.

93. Tang, K., S. L. Allman, and C. H. Chen. 1993. Matrix-assisted laser desorption ionization of oligonucleotides with various matrixes. *Rapid Communications in Mass Spectrometry* 7(10): 943–948.
94. Kirpekar, F., et al., Matrix-assisted laser-desorption ionization mass-spectrometry of enzymatically synthesized Rna up to 150 Kda. *Nucleic Acids Research* 22(19): 3866–3870.
95. Nordhoff, E., F. Kirpekar, and P. Roepstorff. 1996. Mass spectrometry of nucleic acids. *Mass Spectrometry Reviews* 15(2): 67–138.
96. Beavis, R. C., and B. T. Chait. 1989. Matrix-assisted laser-desorption mass spectrometry using 355 nm radiation. *Rapid Communications in Mass Spectrometry* 3(12): 436–439.
97. Beavis, R. C., and B. T. Chait. 1989. Cinnamic acid derivatives as matrices for ultraviolet laser desorption mass spectrometry of proteins. *Rapid Communications in Mass Spectrometry* 3(12): 432–435.
98. Parr, G. R., M. C. Fitzgerald, and L. M. Smith. 1992. Matrix-assisted laser desorption/ionization mass spectrometry of synthetic oligodeoxyribonucleotides. *Rapid Communications in Mass Spectrometry* 6(6): 369–372.
99. Strupat, K., M. Karas, and F. Hillenkamp. 1991. 2,5-Dihydroxybenzoic acid: a new matrix for laser desorption-ionization mass spectrometry. *International Journal of Mass Spectrometry and Ion Processes* 111: 89–102.
100. Beavis, R. C., T. Chaudhary, and B. T. Chait. 1992. α-Cyano-4-hydroxycinnamic acid as a matrix for matrix-assisted laser desorption mass spectrometry. *Organic Mass Spectrometry* 27(2): 156–158.
101. Zhang, L. K., and M. L. Gross. 2000. Matrix-assisted laser desorption/ionization mass spectrometry methods for oligodeoxynucleotides: Improvements in matrix, detection limits, quantification, and sequencing. *Journal of the American Society for Mass Spectrometry* 11(10): 854–865.
102. Tang, K., S. L. Allman, and C. H. Chen. 1992. Mass-spectrometry of laser-desorbed oligonucleotides. *Rapid Communications in Mass Spectrometry* 6(6): 365–368.
103. Schieltz, D. M., et al. 1992. Mass spectrometry of DNA mixtures by laser ablation from frozen aqueous solution. *Rapid Communications in Mass Spectrometry* 6(10): 631–636.
104. Schneider, K., and B. T. Chait. 1993. Matrix-assisted laser desorption mass spectrometry of homopolymer oligodeoxyribonucleotides. Influence of base composition on the mass spectrometric response. *Organic Mass Spectrometry* 28(11): 1353–1361.
105. Chen, C. H., et al. 1993. Matrix- and substrate-assisted laser desorption for fast DNA sequencing. In *Proceedings of SPIE—The International Society for Optical Engineering* (Proceedings of Advances in DNA Sequencing Technology) 1891: 94–101.
106. Pieles, U., et al. 1993. Matrix-assisted laser desorption ionization time-of-flight mass spectrometry: a powerful tool for the mass and sequence analysis of natural and modified oligonucleotides. *Nucleic Acids Research* 21(14): 3191–3196.
107. Wu, K. J., A. Steding, and C. H. Becker. 1993. Matrix-assisted laser desorption time-of-flight mass-spectrometry of oligonucleotides using 3-hydroxypicolinic acid as an ultraviolet-sensitive matrix. *Rapid Communications in Mass Spectrometry* 7(2): 142–146.
108. Tang, K., et al. 1994. Detection of 500-nucleotide DNA by laser-desorption mass-spectrometry. *Rapid Communications in Mass Spectrometry* 8(9): 727–730.
109. Bai, J., et al. 1995. Procedures for detection of DNA by matrix-assisted laser desorption/ionization mass spectrometry using a modified nafion film substrate. *Rapid Communications in Mass Spectrometry* 9(12): 1172–1176.
110. Taranenko, N. I., et al. 1994. 3-Aminopicolinic acid as a matrix for laser desorption mass spectrometry of biopolymers. *Rapid Communications in Mass Spectrometry* 8(12): 1001–1006.
111. Liu, Y.-H., et al. 1995. Use of a nitrocellulose film substrate in matrix-assisted laser desorption/ionization mass spectrometry for DNA mapping and screening. *Analytical Chemistry* 67(19): 3482–3490.
112. Lecchi, P., H. M. T. Le, and L. K. Pannell. 1995. 6-Aza-2-thiothymine: a matrix for MALDI spectra of oligonucleotides. *Nucleic Acids Research* 23(7): 1276–1277.
113. Zhu, Y. F., et al. 1996. The study of 2,3,4-trihydroxyacetophenone and 2,4,6-trihydroxyacetophenone as matrixes for DNA detection in matrix-assisted laser desorption/ionization time-of-flight mass spectrometry. *Rapid Communications in Mass Spectrometry* 10(3): 383–388.
114. Hunter, J. M., et al. 1998. Volatile matrices for matrix-assisted laser desorption/ionization mass spectrometry. PCT Int. Appl. WO 9854751.
115. Wang, B. H., et al. 1997. Sequencing of modified oligonucleotides using in-source fragmentation and delayed pulsed ion extraction matrix-assisted laser desorption ionization time-of-flight mass spectrometry. *International Journal of Mass Spectrometry and Ion Processes* 169: 331–350.
116. Berkenkamp, S., F. Kirpekar, and F. Hillenkamp. 1998. Infrared MALDI mass spectrometry of large nucleic acids. *Science* 281(5374): 260–262.

117. Taranenko, N. I., et al. 1998. Matrix-assisted laser desorption/ionization for short tandem repeat loci. *Rapid Communications in Mass Spectrometry* 12(8): 413–418.
118. Lin, H., J. M. Hunter, and C. H. Becker. 1999. Laser desorption of DNA oligomers larger than one kilobase from cooled 4-nitrophenol. *Rapid Communications in Mass Spectrometry* 13(23): 2335–2340.
119. Distler, A. M., and J. Allison. 2001. 5-Methoxysalicylic acid and spermine: a new matrix for the matrix-assisted laser desorption/ionization mass spectrometry analysis of oligonucleotides. *Journal of the American Society for Mass Spectrometry* 12(4): 456–462.
120. Chan, T. W. D., Y. M. E. Fung, and Y. C. L. Li. 2002. A study of fast and metastable dissociations of adenine-thymine binary-base oligonucleotides by using positive-ion MALDI-TOF mass spectrometry. *Journal of the American Society for Mass Spectrometry* 13(9): 1052–1064.
121. Nordhoff, E., et al. 1995. Direct mass spectrometric sequencing of low-picomole amounts of oligodeoxynucleotides with up to 21 bases by matrix-assisted laser desorption/ionization mass spectrometry. *Journal of Mass Spectrometry* 30(1): 99–112.
122. Gross, J., et al. 2001. Metastable decay of negatively charged oligodeoxynucleotides analyzed with ultraviolet matrix-assisted laser desorption/ionization post-source decay and deuterium exchange. *Journal of the American Society for Mass Spectrometry* 12(2): 180–192.
123. Stemmler, E. A., et al. 1993. Matrix-assisted laser-desorption ionization Fourier-transform mass-spectrometry of oligodeoxyribonucleotides. *Rapid Communications in Mass Spectrometry* 7(9): 828–836.
124. Nordhoff, E., et al. 1993. Ion stability of nucleic-acids in infrared matrix-assisted laser-desorption ionization mass-spectrometry. *Nucleic Acids Research* 21(15): 3347–3357.
125. Stemmler, E. A., et al. 1995. Analysis of modified oligonucleotides by matrix-assisted laser desorption/ionization Fourier-transform mass-spectrometry. *Analytical Chemistry* 67(17): 2924–2930.
126. Zhu, L., et al. 1995. Oligodeoxynucleotide fragmentation in Maldi/Tof mass-spectrometry using 355-Nm radiation. *Journal of the American Chemical Society* 117(22): 6048–6056.
127. Hettich, R. L., and E. A. Stemmler. 1996. Investigation of oligonucleotide fragmentation with matrix-assisted laser desorption ionization Fourier-transform mass spectrometry and sustained off-resonance irradiation. *Rapid Communications in Mass Spectrometry* 10(3): 321–327.
128. Keough, T., et al. 1993. Antisense DNA oligonucleotides. 2. The use of matrix-assisted laser desorption ionization mass-spectrometry for the sequence verification of methylphosphonate oligodeoxyribonucleotides. *Rapid Communications in Mass Spectrometry* 7(3): 195–200.
129. Bahr, U., H. Aygun, and M. Karas. 2009. Sequencing of single and double stranded RNA oligonucleotides by acid hydrolysis and MALDI mass spectrometry. *Analytical Chemistry* 81(8): 3173–3179.
130. Polo, L. M., T. D. McCarley, and P. A. Limbach. 1997. Chemical sequencing of phosphorothioate oligonucleotides using matrix-assisted laser desorption/ionization time-of-flight mass spectrometry. *Analytical Chemistry* 69(6): 1107–1112.
131. Bentzley, C. M., et al. 1996. Oligonucleotide sequence and composition determined by matrix-assisted laser desorption/ionization. *Analytical Chemistry* 68(13): 2141–2146.
132. Tolson, D. A., and N. H. Nicholson. 1998. Sequencing RNA by a combination of exonuclease digestion and uridine specific chemical cleavage using MALDI-TOF. *Nucleic Acids Research* 26(2): 446–451.
133. Schuette, J. M., et al. 1995. Sequence analysis of phosphorothioate oligonucleotides via matrix-assisted laser desorption ionization time-of-flight mass spectrometry. *Journal of Pharmaceutical and Biomedical Analysis* 13(10): 1195–1203.
134. Kuchino, Y., and S. Nishimura. 1987. Enzymatic RNA sequencing. *Methods in Enzymology* 180: 154–163.
135. Waldmann, R., H. J. Gross, and G. Krupp. 1987. Protocol for rapid chemical RNA sequencing. *Nucleic Acids Research* 15(17): 7209.
136. Peattie, D. A. 1979. Direct chemical method for sequencing RNA. *Proceedings of the National Academy of Sciences of the United States of America* 76(4): 1760–1764.
137. Donis-Keller, H., A. M. Maxam, and W. Gilbert. 1977. Mapping adenines, guanines, and pyrimidines in RNA. *Nucleic Acids Research* 4(8): 2527–2538.
138. Donis-Keller, H. 1980. Phy M: an RNase activity specific for U and A residues useful in RNA sequence analysis. *Nucleic Acids Research* 8(14): 3133–3142.
139. Simoncsits, A., et al. 1977. New rapid gel sequencing method for RNA. *Nature* 269(5631): 833–836.
140. Boguski, M. S., P. A. Hieter, and C. C. Levy 1980. Identification of a cytidine-specific ribonuclease from chicken liver. *Journal of Biological Chemistry* 255(5): 2160–2163.
141. Faulstich, K., et al. 1997. A sequencing method for RNA oligonucleotides based on mass spectrometry. *Analytical Chemistry* 69(21): 4349–4353.

142. Rojo, M. A., et al. 1994. Cusativin, a new cytidine-specific ribonuclease accumulated in seeds of Cucumis-Sativus L. *Planta* 194(3): 328–338.
143. Matthiesen, R., and F. Kirpekar. 2009. Identification of RNA molecules by specific enzyme digestion and mass spectrometry: software for and implementation of RNA mass mapping. *Nucleic Acids Research* 37(6): e48/1–e48/10.
144. Gupta, R. C., and K. Randerath. 1977. Use of specific endonuclease cleavage in RNA sequencing. *Nucleic Acids Research* 4(6): 1957–1978.
145. Breslow, R., and R. Xu. 1993. Recognition and catalysis in nucleic acid chemistry. *Proceedings of the National Academy of Sciences of the United States of America* 90(4): 1201–1207.
146. Mengel-Jorgensen, J., et al. 2005. Typing of multiple single-nucleotide polymorphisms using ribonuclease cleavage of DNA/RNA chimeric single-base extension primers and detection by MALDI-TOF mass spectrometry. *Analytical Chemistry* 77(16): 5229–5235.
147. Deshpande, R. A., and V. Shankar. 2002. Ribonucleases from T2 family. *Critical Reviews in Microbiology* 28(2): 79–122.
148. Keough, T., et al. 1996. Detailed characterization of antisense DNA oligonucleotides. *Analytical Chemistry* 68(19): 3405–3412.
149. Henzel, W. J., et al. 1993. Identifying proteins from two-dimensional gels by molecular mass searching of peptide fragments in protein sequence databases. *Proceedings of the National Academy of Sciences of the United States of America* 90(11): 5011–5015.
150. Mann, M., P. Hoejrup, and P. Roepstorff. 1993. Use of mass spectrometric molecular weight information to identify proteins in sequence databases. *Biological Mass Spectrometry* 22(6): 338–345.
151. Pappin, D. J., P. Hojrup, and A. J. Bleasby. 1993. Rapid indentification of proteins by peptide-mass fingerprinting. *Current Biology* 3(6): 327–332.
152. Kirpekar, F. 2006. RNA mapping. In *The Encyclopedia of Mass Spectrometry. Biological Applications, Part B: Carbohydrates, Nucleic Acids and Other Biological Compounds* 3: 10–14.
153. Zhang, Z., et al. 2006. Microbial identification by mass cataloging. *BMC Bioinformatics* 7.
154. Kowalak, J. A., et al. 1993. A novel method for the determination of post-transcriptional modification in RNA by mass spectrometry. *Nucleic Acids Research* 21(19): 4577–4585.
155. Hossain, M., and P. A. Limbach. 2007. Mass spectrometry-based detection of transfer RNAs by their signature endonuclease digestion products. *RNA-A Publication of the RNA Society* 13(2): 295–303.
156. Von Wintzingerode, F., et al. 2002. Base-specific fragmentation of amplified 16S rRNA genes analyzed by mass spectrometry: a tool for rapid bacterial identification. *Proceedings of the National Academy of Sciences of the United States of America* 99(10): 7039–7044.
157. Lefmann, M., et al. 2004. Novel mass spectrometry-based tool for genotypic identification of mycobacteria. *Journal of Clinical Microbiology* 42(1): 339–346.
158. Hartmer, R., et al. 2003. RNase T1 mediated base-specific cleavage and MALDI-TOF MS for high-throughput comparative sequence analysis. *Nucleic Acids Research* 31(9): e47/1–e47/10.
159. Krebs, S., et al. 2003. RNaseCut: A MALDI mass spectrometry-based method for SNP discovery. *Nucleic Acids Research* 31(7): e37/1–e37/8.
160. Seichter, D., S. Krebs, and M. Foerster. 2004. Rapid and accurate characterization of short tandem repeats by MALDI-TOF analysis of endonuclease cleaved RNA transcripts. *Nucleic Acids Research* 32(2): e16/1–e16/10.
161. Maxam, A. M., and W. Gilbert. 1977. A new method for sequencing DNA. *Proceedings of the National Academy of Sciences of the United States of America* 74(2): 560–564.
162. Isola, N. R., et al. 1999. Chemical cleavage sequencing of DNA using matrix-assisted laser desorption/ionization time-of-flight mass spectrometry. *Analytical Chemistry* 71(13): 2266–2269.
163. Wintermeyer, W., and H. G. Zachau. 1975. Tertiary structure interactions of 7-methylguanosine in yeast tRNAPhe as studied by borohydride reduction. *FEBS Letters* 58(1): 306–309.
164. Vincze, A., et al. 1973. Reaction of diethyl pyrocarbonate with nucleic acid components. Bases and nucleosides derived from guanine, cytosine, and uracil. *Journal of the American Chemical Society* 95(8): 2677–2682.
165. Ehrenberg, L., I. Fedorcsak, and F. Solymosy. 1976. Diethyl pyrocarbonate in nucleic acid research. *Progress in Nucleic Acid Research and Molecular Biology* 16: 189–262.
166. Ehresmann, C., et al. 1987. Probing the structure of RNAs in solution. *Nucleic Acids Research* 15(22): 9109–9128.
167. Cashmore, A. R., and G. B. Petersen. 1978. The degradation of DNA by hydrazine: identification of 3-ureidopyrazole as a product of the hydrazinolysis of deoxycytidylic acid residues. *Nucleic Acids Research* 5(7): 2485–2491.

168. Verwoerd, D. W., and W. Zillig. 1963. A specific partial hydrolysis procedure for soluble RNA. *Biochimica et Biophysica Acta* 68: 484–486.
169. Levene, P. A., and L. W. Bass. 1926. The action of hydrazine hydrate on uridine. *Journal of Biological Chemistry* 71: 167–172.
170. Franca, L. T. C., E. Carrilho, and T. B. L. Kist. 2002. A review of DNA sequencing techniques. *Quarterly Reviews of Biophysics* 35(2): 169–200.
171. Iverson, B. L., and P. B. Dervan. 1987. Adenine specific DNA chemical sequencing reaction. *Nucleic Acids Research* 15(19): 7823–7830.
172. Krayev, A. S. 1987. The use of diethylpyrocarbonate for sequencing adenines and guanines in DNA. *FEBS Letters* 130(1): 19–22.
173. Ovchinnikov, Y. A., et al. 1979. Primary structure of an EcoRI fragment of lambda imm434 DNA containing regions cI-cro of phage 434 and cII-o of phage lambda. *Gene* 6(3): 235–249.
174. Rosenthal, A., et al. 1985. Solid-phase methods for sequencing of nucleic acids. I. Simultaneous sequencing of different oligodeoxyribonucleotides using a new, mechanically stable anion-exchange paper. *Nucleic Acids Research* 13(4): 1173–1184.
175. Hudspeth, M. E., et al. 1982. Location and structure of the var1 gene on yeast mitochondrial DNA: nucleotide sequence of the 40.0 allele. *Cell* 30(2): 617–626.
176. Korobko, V. G., and S. A. Grachev. 1977. Sequence determination in DNA by a modified chemical method. *Bioorganicheskaya Khimiya* 3(10): 1420–1422.
177. Banaszuk, A. M., et al. 1983. An efficient method for the sequence analysis of oligodeoxyribonucleotides. *Analytical Biochemistry* 128(2): 281–286.
178. Maxam, A. M., and W. Gilbert. 1980. Sequencing end-labeled DNA with base-specific chemical cleavages. *Methods in Enzymology* 65(Nucleic Acids, Pt. I): 499–560.
179. Friedmann, T., and D. M. Brown. 1978. Base-specific reactions useful for DNA sequencing methylene blue-sensitized photooxidation of guanine and osmium tetraoxide modification of thymine. *Nucleic Acids Research* 5(2): 615–622.
180. Stalker, D. M., W. R. Hiatt, and L. Comai. 1985. A single amino acid substitution in the enzyme 5-enolpyruvylshikimate-3-phosphate synthase confers resistance to the herbicide glyphosate. *Journal of Biological Chemistry* 260(8): 4724–4728.
181. Korobko, V. G., S. A. Grachev, and M. N. Kolosov. 1978. G-specific degradation of single-stranded DNA. Nucleotide sequence determination in restriction fragments of phage M13 DNA. *Bioorganicheskaya Khimiya* 4(9): 1281–1283.
182. Sugiyama, H., et al. 1983. A new, convenient method for determining T residues in chemical DNA sequencing by using photoreaction with spermine. *Nucleic Acids Symposium Series* 12: 103–106.
183. Saito, I., et al. 1984. A new procedure for determining thymine residues in DNA sequencing. Photoinduced cleavage of DNA fragments in the presence of spermine. *Nucleic Acids Research* 12(6): 2879–2885.
184. Rubin, C. M., and C. W. Schmid. 1980. Pyrimidine-specific chemical reactions useful for DNA sequencing. *Nucleic Acids Research* 8(20): 4613–4619.
185. Sverdlov, E. D., and N. F. Kalinina. 1983. DNA interaction with hydrogen peroxide. Method for determining pyrimidine bases in DNA. *Bioorganicheskaya Khimiya* 9(12): 1696–1698.
186. Sverdlov, E. D., and N. F. Kalinina. 1984. Chemical modifications of double-stranded DNA as a method for detection of exposed individual base pairs. *Doklady Akademii Nauk SSSR* 274(6): 1508–1510, 2 plates.
187. Simoncsits, A., and I. Torok. 1982. A photoinduced cleavage of DNA useful for determining T residues. *Nucleic Acids Research* 10(24): 7959–7964.
188. Oivanen, M., S. Kuusela, and H. Loennberg. 1998. Kinetics and mechanisms for the cleavage and isomerization of the phosphodiester bonds of RNA by Bronsted acids and bases. *Chemical Reviews* 98(3): 961–990.
189. Roskey, M. T., et al. 1996. DNA sequencing by delayed extraction matrix-assisted laser desorption/ionization time of flight mass spectrometry. *Proceedings of the National Academy of Sciences of the United States of America* 93(10): 4724–4729.
190. Fitzgerald, M. C., L. Zhu, and L. M. Smith. 1993. The analysis of mock DNA-sequencing reactions using matrix-assisted laser-desorption ionization mass-spectrometry. *Rapid Communications in Mass Spectrometry* 7(10): 895–897.
191. Shaler, T. A., et al. 1995. Analysis of enzymatic DNA-sequencing reactions by matrix-assisted laser-desorption ionization time-of-flight mass-spectrometry. *Rapid Communications in Mass Spectrometry* 9(10): 942–947.
192. Koster, H., et al. 1996. A strategy for rapid and efficient DNA sequencing by mass spectrometry. *Nature Biotechnology* 14(9): 1123–1128.

193. Christian, N. P., et al. 1995. High resolution matrix-assisted laser desorption/ionization time-of-flight analysis of single-stranded DNA of 27 to 68 nucleotides in length. *Rapid Communications in Mass Spectrometry* 9(11): 1061–1066.
194. Colby, S. M., T. B. King, and J. Reilly. 1994. Improving the resolution of matrix-assisted laser desorption/ionization time-of-flight mass spectrometry by exploiting the correlation between ion position and velocity. *Rapid Communications in Mass Spectrometry* 8(11): 865–868.
195. Edwards, J. R., Y. Itagaki, and J. Y. Ju. 2001. DNA sequencing using biotinylated dideoxynucleotides and mass spectrometry. *Nucleic Acids Research* 29(21): e104.
196. Kwon, Y.-S., et al. 2001. DNA sequencing and genotyping by transcriptional synthesis of chain-terminated RNA ladders and MALDI-TOF mass spectrometry. *Nucleic Acids Research* 29(3): e 11/1–e 11/6.
197. Sun, X., H. Ding, and B. Guo. 2000. A new MALDI-TOF based mini-sequencing assay for genotyping of SNPS. *Nucleic Acids Research* 28(12): e68, ii–viii.
198. Little, D. P., et al. 1997. Identification of apolipoprotein E polymorphisms using temperature cycled primer oligo base extension and mass spectrometry. *European Journal of Clinical Chemistry and Clinical Biochemistry* 35(7): 545–548.
199. Little, D. P., et al. 1997. Detection of RET proto-oncogene codon 634 mutations using mass spectrometry. *Journal of Molecular Medicine* 75(10): 745–750.
200. Braun, A., D. Little, and H. Koster. 1997. Detecting CFTR gene mutations by using primer oligo base extension and mass spectrometry. *Clinical Chemistry* 43(7): 1151–1158.
201. Little, D. P., et al. 1997. Mass spectrometry from miniaturized arrays for full comparative DNA analysis. *Nature Medicine* 3(12): 1413–1416.
202. Braun, A., et al. 1997. Improved analysis of microsatellites using mass spectrometry. *Genomics* 46(1): 18–23.
203. Blondal, T., et al. 2003. A novel MALDI-TOF based methodology for genotyping single nucleotide polymorphisms. *Nucleic Acids Research* 31(24): e155/1–e155/7.
204. Liu, Y., X. Sun, and B. Guo. 2003. Matrix-assisted laser desorption/ionization time-of-flight analysis of low-concentration oligonucleotides and mini-sequencing products. *Rapid Communications in Mass Spectrometry* 17(20): 2354–2360.
205. Luo, C., L. Deng, and C. Zeng. 2004. High throughput SNP genotyping with two mini-sequencing assays. *Acta Biochimica et Biophysica Sinica* 36(6): 379–384.
206. Jehan, T., and S. Lakhanpaul. 2006. Single nucleotide polymorphism (SNP)-methods and applications in plant genetics: a review. *Indian Journal of Biotechnology* 5(4): 435–459.
207. Saxon, A., D.-D. Sanchez, and F. D. Gilliland. 2005. Oligonucleotide primers and probes for genotyping GSTM1 gene and GSTP1 gene as well as determining susceptibility of an individual to allergen induced hypersensitivity. 2006 (University of California, USA). U.S. Patent Application. 2005-31822, 2006154261.
208. Sun, X., and B. Guo. 2006. Genotyping single-nucleotide polymorphisms by matrix-assisted laser desorption/ionization time-of-flight-based mini-sequencing. *Methods in Molecular Medicine* 128 (Cardiovascular Disease, Volume 1): 225–230.
209. Afanas'ev, M. V., et al. 2007. Use of a mini-sequencing test, followed by the MALDI-TOF mass-spectrometric analysis to evaluate rifampicin and isoniazid resistance in Mycobacterium tuberculosis circulation in the Russian Federation. *Problemy Tuberkuleza i Boleznei Legkikh* 2007(7): 37–42.
210. Malakhova, M. V., et al. 2009. Hepatitis B virus genetic typing using mass-spectrometry. *Bulletin of Experimental Biology and Medicine* 147(2): 220–225.
211. Zerikly, M., and G. L. Challis. 2009. Strategies for the discovery of new natural products by genome mining. *Chembiochem* 10(4): 625–633.

6 T_m Analysis of Oligonucleotides

Huihe Zhu
Girindus America, Inc.
G. Susan Srivatsa
ElixinPharma

CONTENTS

6.1 INTRODUCTION

In 1953, Watson and Crick discovered that the DNA helix is formed by two single nucleic acid strands binding noncovalently via base-pairing interaction. The thermodynamic stability of the two single strands to form a resultant DNA duplex is a function of the hydrogen bonding between pyrimidine and purine bases and the stacking of the aromatic rings. Other factors contributing to the strength of this interaction are salt concentration, strand concentration, sequence, and chemical makeup of the single strands (RNA, DNA, or otherwise modified sugars, backbones, or heterocyclic bases). In nature, the balance between a duplex and denatured state is modulated by enzymes that are responsible for the maintenance and replication of cellular function. Enzymes involved in DNA replication, transcription, recombination, and RNA processing reduce the forces necessary to denature the secondary and tertiary structures of nucleic acids, thereby facilitating performance of their respective functions. The study of these enzymes and their function is extensive. Within the field of molecular biology, understanding their thermodynamic properties is essential for successful polymerase chain reaction (PCR) primer and probe design. Likewise, thermodynamics of the binding of oligonucleotides to their molecular targets is central to the development of potent therapeutic agents.

Historically, empirical experiments that characterized the hybridization of oligonucleotide drug candidates to complementary RNA targets were used to guide the discovery of antisense therapeutics. Extensive literature exists describing the thermodynamic stability and binding affinity of synthetic and naturally occurring oligonucleotides with their RNA targets. This understanding has subsequently been applied to develop other classes of therapeutic oligonucleotides such as siRNA, microRNA, aptamers, and decoys.

Thermodynamic parameters can be determined using a thermal denaturing process, known as "melting," which breaks the hydrogen bonding and stacking interaction between two complementary strands to form two single strands (Figure 6.1). The temperature at which half the population is denatured and half is in a duplex state is defined as the melting temperature (T_m). Analysis of T_m values under different experimental conditions allows one to extract these individual thermodynamic parameters.[1–3]

Melting temperature of oligonucleotide duplexes can be measured by different methods, such as ultraviolet spectroscopy (UV), differential scanning calorimetry (DSC), nuclear magnetic resonance spectroscopy (NMR), circular dichroism spectroscopy (CD), and Raman spectroscopy. Among these alternatives, ultraviolet spectroscopy is one of the oldest, having been used since the 1950s. It is still the simplest and most sensitive method for thermodynamic characterization of the stability of nucleic acids. In addition to the thermodynamics of duplexes and other nucleic acid structures, extensive studies of the hybridization kinetics have also been performed. It is not our intention to provide an extensive review of the literature on this topic but rather to provide a practical overview of T_m analysis and its applications in the oligonucleotide laboratory.

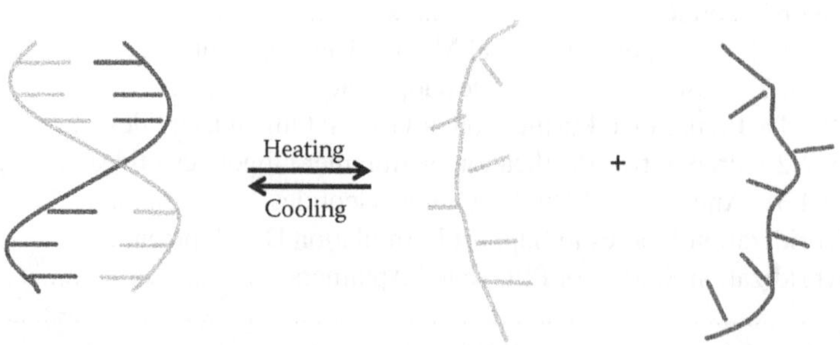

FIGURE 6.1 Temperature-controlled association and dissociation of duplexes.

6.2 UV ABSORBANCE OF OLIGONUCLEOTIDES

Oligonucleotides are composed of nucleoside monomers that are connected by phosphodiester bonds. The ultraviolet absorption characteristics of the individual nucleobases determine the overall ultraviolet absorption of the oligonucleotide strand. The aromatic heterocyclic purine or pyrimidine bases absorb strongly with maxima near 260 nm. Thus oligonucleotides also have maximum absorbance near 260 nm. Oligonucleotides in solution have a different UV profile from single nucleobases, as a result of π-π interactions from base stacking that affect the transition dipoles of the bases.[4] Typically, the total UV absorption of oligonucleotides is lower than that of the constituent monomer nucleosides. This phenomenon is known as the hypochromic effect, or hypochromicity (which means "less color").

In the case of double strands, formation of ordered structures such as a double-helical duplex results in greater hypochromicity owing to increased stacking relative to the single strands. During denaturation of the double strand to single strands, the ordered stacking structure of oligonucleotides is disturbed, resulting in a corresponding increase in UV absorption. This phenomenon is referred to as hyperchromicity (which means "more color").

The change in structure during denaturation of the double strand to the corresponding single strands can be monitored by the change in intensity of UV absorption as a function of temperature, as shown in Figure 6.2.

During the denaturing process, the magnitude of the hyperchromic effect typically ranges from 20 to 40%, depending on the length of the duplex and base composition.[5] The hyperchromicity is calculated using Equation 6.1; A_{ss} and A_d are the UV absorbance at high (single strands) and low (duplex) temperature, respectively.

$$\%\ \text{hyperchromicity} = 100\,(A_{ss} - A_d)\,/A_d \tag{6.1}$$

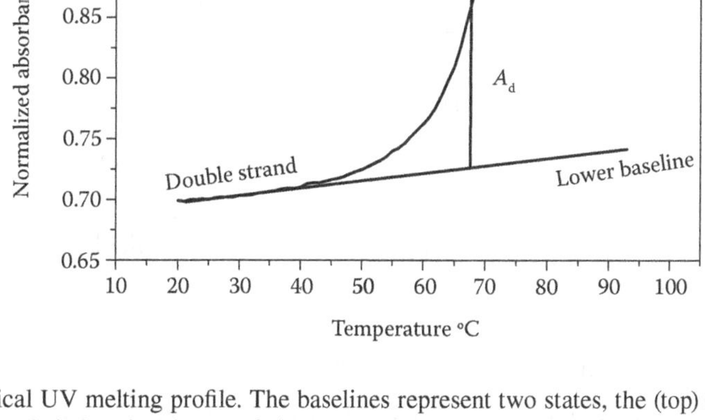

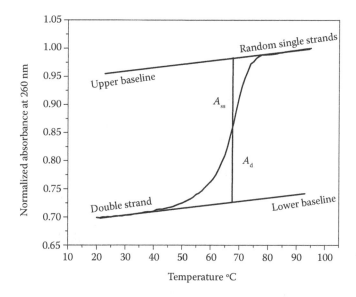

FIGURE 6.2 Typical UV melting profile. The baselines represent two states, the (top) single-stranded and (bottom) double-stranded. A_{ss}, absorbance of single strands; A_d, absorbance of duplex. The ratio of fractions of associated structure, f, at any temperature, can be calculated from $A_d/(A_{ss} + A_d)$. T_m is the point at which $A_{ss} = A_d$; in this case, $T_m = 67.4°C$. The data are for 21-base siRNA duplex at 8 μM total strand concentration dissolved in a 10 mM phosphate, pH7 buffer containing 25 mM NaCl and 1 mM EDTA.

The hyperchromic and hypochromic phenomena observed during melting and formation of DNA duplex was reported by Marmur and Doty in 1959.[6] Usually, denaturation of the target oligonucleotide duplex is monitored by UV absorbance as a function of temperature at an absorption wavelength of 260 nm. Plotting absorbance versus temperature results in a so-called melting curve. For sharp transitions, the T_m is identical to the maximum of the first derivative of the melting curve. Thermodynamic parameters can be derived from UV melting data using van't Hoff analysis, assuming a two-state model.[1,7,8] While this chapter primarily focuses on the UV analysis of thermal melting of oligonucleotides, equations analogous to Equation 6.1 may apply to T_m measurements made by other methods.

6.3 THERMODYNAMIC PARAMETERS OF OLIGONUCLEOTIDES

The affinity of an antisense candidate to a target complementary RNA strand can be evaluated from thermodynamic data obtained from UV melting experiments. Higher melting temperature or increased melting temperature (ΔT_m) of a candidate as compared to a control strand is not necessarily indicative of higher affinity. It is the changed free energy ($\Delta\Delta G_{37}^{o}$) that reflects the effect of an increase or decrease in affinity.[9] An approach to obtain thermodynamic parameters from UV-melting experiments is discussed below. For non self-complementary duplexes, the equilibrium is described by Equation 6.2, in which S_1, S_2, and D represent the two complementary single strands and duplex, respectively. The fraction of duplex (f) is defined in Equation 6.3 as a ratio of duplex concentration [D] and total strand concentration (C_T) and is calculated from the absorbance owing to the duplex (A_d) and single strand (A_{ss}). The association constant K at each temperature is defined as a ratio of duplex concentration [D] to single-strand concentration [S] and is calculated from Equation 6.4 using f and C_T. At the melting temperature, T_m, $f = 0.5$, which allows one to derive Equation 6.6, the van't Hoff equation for association of non self-complementary single strands. Multiple experiments at different concentrations yield a plot of $1/T_m$ versus $\ln C_T$; from this plot $\Delta H°$ and $\Delta S°$ are determined and used to calculate $\Delta G°$.

$$S_1 + S_2 \leftrightarrow D \tag{6.2}$$

$$f = 2\,[D]/C_T = (A_d)/(A_d + A_{ss}) \tag{6.3}$$

$$K = [D]/[S_1]\,[S_2] = 2f/[(1-f)^2\, C_T] \tag{6.4}$$

$$\Delta G° = -RT \ln K = \Delta H° - T\,\Delta S° \tag{6.5}$$

$$1/T_m = (R/\Delta H°)\ln C_T + (\Delta S° - R\ln 4)/\Delta H° \tag{6.6}$$

Thermodynamic parameters for other oligonucleotide systems, such as association of self-complementary duplexes or intramolecular hairpins, can also be calculated in this manner using similar van't Hoff relations.[1,7,8]

6.4 KINETICS OF HYBRIDIZATION PROCESS

It is important to keep in mind that kinetics also plays a critical role in hybridization. For short antisense oligonucleotides that hybridize to their unstructured complementary strand, the association rates are relatively rapid at 10^6–10^7 $M^{-1}\,s^{-1}$ and are not affected by length or sequence. Hybridization to structured regions varies widely in the range of 10^1–10^7 M^{-1} s^{-1}. However, dissociation rates appear to be primarily dependent upon the length and sequence of the oligonucleotide and may be unaffected by the structure of the target.[10]

UV melting profiles can be used to provide information about kinetics of the association and dissociation rates by monitoring both the heating and cooling profiles. Identical absorbance *versus* temperature profiles indicate that the association and dissociation are fast relative to the temperature gradient used. When heating and cooling profiles are not superimposable (a situation known as hysteresis), it may be due to slow kinetics or formation of other secondary structures.[11] In this case, care must be exercised to accurately determine the melting temperature. The temperature gradient should be appropriately lowered to adequately deal with the slower kinetics of the transition. If hysteresis occurs, neither heating nor cooling profile should be used, because neither represents true equilibrium curves. Mergny and coworkers have demonstrated that from heating/cooling cycles, the kinetic constants of association k_{on} and dissociation k_{off} at each temperature can be extracted.[12] Further analysis yields the activation energies of association and dissociation that can be used to calculate a theoretical equilibrium melting curve for comparison with that determined experimentally.

6.5 PREDICTING MELTING TEMPERATURE

The stability of oligonucleotide duplexes, and thus T_m, is closely related to the base composition of the strands. A higher percentage of guanosine-cytidine (G-C) base pairs relative to adenosine-thymidine base pairs (A-C) generally results in greater stability of duplexes.[6] An original use of T_m analysis was to determine the GC content of large DNA molecules.[13] Correspondingly, the T_m for oligonucleotides shorter than 20 base pairs can be estimated from the base content according[14]

$$T_m = 4°\text{C (number of C + number of G)} + 2°\text{C (number of A + number of T)} \quad (6.7)$$

This "4 + 2 rule" was developed from immobilized membrane experiments, so the T_m of an oligonucleotide duplex in solution is often much lower than the calculated T_m. Also, Wells and coworkers have found that DNAs with the same base composition but different nucleotide sequences (sequence isomers) do not show identical helix-coil transition profiles. Some sequence isomeric DNAs can have T_m values differing by as much as 9°C.[15]

Many research groups have developed improved models to correct the above 4 + 2 rule to more accurately predict the T_m for any sequence. The nearest-neighbor (N-N) model was pioneered by Zimm and Tinoco to study the effect of identity and orientation of neighboring bases to the stability of DNA duplex.[16,17] Breslauer reported the thermodynamic profiles for all 10 nearest-neighbor interactions possible in a Watson–Crick DNA duplex structure using DSC and UV melting analysis of a series of self-complementary oligonucleotide sequences.[18] With the advent of chemical synthesis of RNA oligonucleotides, thermodynamic parameters became available for RNA duplexes as well[3] (Table 6.1). The T_m value for a given duplex in 1 M NaCl may be predicted by using the parameters specified in Table 6.1 and Equation 6.6.[2,3]

The addition of metal ions to duplex solutions can increase the stability of duplexes by reducing the repulsion between negatively charged phosphate groups. It has been found that the melting temperature of DNA increases linearly with the logarithm of alkali-metal cation concentration.[19,20] Therefore, it is important to take into account the salt concentration when estimating melting temperatures. In the nearest-neighbor model, the salt concentration used for $\Delta H°$ and $\Delta S°$ (Table 6.1) are obtained from melting experiments performed in 1 M Na^+ buffer, while in PCR reactions, 50 mM of monovalent salt concentration is normally used. To compensate for this difference, Owczarzy proposed a corrected nearest-neighbor method[21] (Equation 6.8), in which f_{GC} is the fraction of GC base pair in the oligonucleotide.

TABLE 6.1
Thermodynamic Parameters for All Nearest-Neighbors Used for Calculating Duplex T_m Values and Thermodynamic Parameters[2,3]

	DNA			RNA		
Sequences	ΔH° (kcal/mol)	ΔS° (cal/K mol)	ΔG°$_{37}$ (kcal/mol)	ΔH° (kcal/mol)	ΔS° (cal/K mol)	ΔG°37 (kcal/mol)
AA/TT	-7.9	-22.2	-1.00	-6.82	-19.0	-0.93
AT/TA	-7.2	-20.4	-0.88	-9.38	-26.7	-1.10
TA/AT	-7.2	-21.3	-0.58	-7.69	-20.5	-1.33
CA/GT	-8.5	-22.7	-1.45	-10.44	-26.9	-2.11
GT/CA	-8.4	-22.4	-1.44	-11.40	-29.5	-2.24
CT/GA	-7.8	-21.0	-1.28	-10.48	-27.1	-2.08
GA/CT	-8.2	-22.2	-1.30	-12.44	-32.5	-2.35
CG/GC	-10.6	-27.2	-2.17	-10.64	-26.7	-2.36
GC/CG	-9.8	-24.4	-2.24	-14.88	-36.9	-3.42
GG/CC	-8.0	-19.9	-1.84	-13.39	-32.7	3.26

$$\frac{1}{T_m(\mathrm{Na}^+)} = \frac{1}{T_m(1\,\mathrm{M\,Na}^+)} + [(4.29 f_{GC} - 3.95)\ln[\mathrm{Na}^+] + 0.940\ln^2[\mathrm{Na}^+]] \times 10^{-5} \quad (6.8)$$

A direct T_m calculator based on the equation above can be found on the Web.[22]

6.6 EXPERIMENTALLY DETERMINING MELTING TEMPERATURE

6.6.1 Configuration of Instruments

To precisely measure a UV melting curve, a photometer with high accuracy and stability of absorbance reading is required. In some cases, a wavelength other than 260 nm may be used to increase sensitivity, such as in quadruplexes (also called tetraplexes) that have a UV maximum at 295 nm.[23] A wavelength at which the duplex has no absorbance can be used as a baseline or to detect artifacts

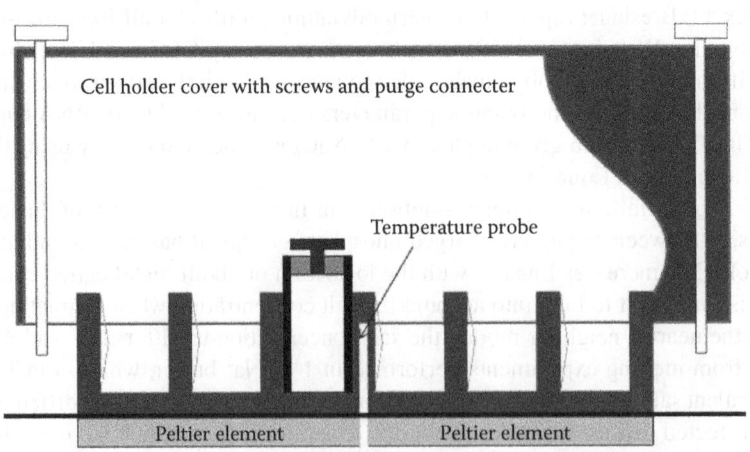

FIGURE 6.3 Micro auto 6 T_m cell holder from Beckman DU series spectrophotometer.

owing to instrument variations, air bubble formation, and other problems. The UV spectrophotometer is typically equipped with a Peltier temperature controller to allow programming of different heating and cooling rates. Such controllers allow temperature-change rates as low as 0.06°C/min for highly accurate measurements.[24] An integrated temperature sensor is essential for accurately measuring the temperature of the cell. The temperature inside the cell can also be measured with an external immersion temperature probe. For subambient temperatures, a nitrogen purge feature is needed to prevent condensation on the cell walls. A multicell holder can be employed to measure small volumes of multiple samples. This device reduces sample size requirements and saves time by allowing multiple samples to be analyzed under identical conditions. Figure 6.3 shows the design of a cell holder from Beckman Instruments.[25]

6.6.2 Concentration of Oligonucleotides

Generally, a higher concentration of oligonucleotide will drive the equilibrium toward duplex formation, resulting in higher observed T_m values. However, a high concentration of sample can cause the UV absorption measurement to deviate from the linear range. Therefore, when preparing oligonucleotide samples, an absorbance maximum of 1.5 or less is suggested for analysis. To measure more concentrated samples, UV cuvettes with shorter light path length, such as 0.1 cm, can be used. Alternately, one can use other methods such as DSC or NMR for these situations.

To measure the melting temperature of a duplex formed from non self-complementary single strands, it is critical to mix the two strands in equimolar concentrations. As discussed previously, base stacking of the single strands and the resulting hypochromism can lower the UV absorption relative to the constituent bases. Therefore, it is best to determine the concentration of each single strand at high temperature, such as 90°C, so that potential self-secondary structure is minimized and the concentration determination is most accurate.[26] Alternatively, one can factor hypochromism into the calculation for lower-temperature measurements.[27] The concentration of each single strand can be determined from the absorbance using extinction coefficients derived from the constituent nucleosides (see Chapter 12).

Once the concentration of each single strand has been determined, solutions of equal concentration are prepared with the desired buffer and ionic strength. After sample solutions have been prepared and mixed, the mixture is transferred to a cuvette and sealed, either with a Teflon cap or a layer of silicone oil, to limit evaporation. Heating the sample for 3–5 min at 90°C prior to beginning of the experiment can help remove dissolved gas from the samples and avoid bubble formation during measurements. This heating step followed by cooling the sample to room temperature can also ensure that the two single strands anneal into a perfect duplex. Incomplete annealing can result in inaccurate T_m readings due to incomplete or imperfect duplex formation. The melting temperature is typically measured with a heating rate of 1°C/min or slower. Most instruments allow programming of both heating and cooling rates to monitor reversibility of duplex formation. In general, a slower rate results in a more accurate equilibrium-temperature measurement. Start and end temperatures are chosen such that plenty of baseline data is collected on each side of the denaturation transition. Sampling rate is adjusted to ensure enough data points for accurate representation of the melting curve.

The impact of sample purity on hybridization has been studied by comparing the apparent stability of an siRNA duplex formed from purified and unpurified (crude) single strands. The hybridization of crude antisense strand to purified complement strand has also been investigated. The melt curves in Figure 6.4 show that hybridization of purified siRNA results in a well-resolved two state transition, while the crude single strands do not. The observed absorbance *versus* temperature plot of unpurified material shows a broad, flat melt profile. At the same sample concentrations, the purified strands have greater hyperchroism than the crude strands, indicating the disruption of a well-ordered structure in the case of the purified strands upon heating. For the unpurified crude sample, the observed flat T_m profile indicates the contribution of failure (or non full length) sequences from

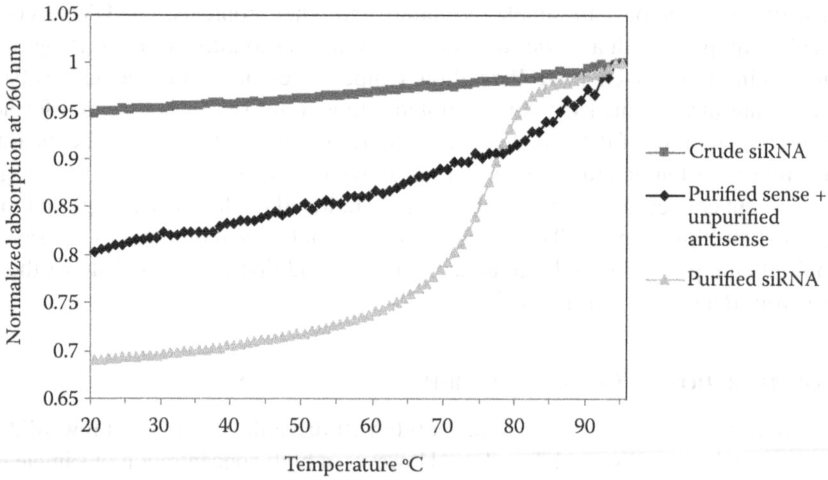

FIGURE 6.4 Relationship between sample purity and melting transition of a siRNA with 33% of GC content and 2′-OMe modification. Each sample contains 4 μM complement strand and 4 μM antisense strand, 1 mM EDTA, 10 mM phosphate, and 0.1 M sodium salt.

synthesis to UV melting. These failure sequences inhibit proper full length duplex formation, resulting in a series of imperfect duplexes melting at different temperatures thereby masking the transition of the perfect full length duplex. As expected, the T_m profile of annealing the purified antisense strand with the unpurified complement strand falls between all-crude and all-purified duplexes.

6.6.3 Buffer Selection

The optimum buffer system for melting measurements should have a temperature-independent p*K*a. For example, the commonly used Tris buffer is undesirable owing to its large temperature-dependent p*K*a.[26] It is also desirable that the selected buffer has no UV absorbance at or near the wavelength of the oligonucleotides being measured. It is important to confirm the chemical stability of the oligonucleotides in the selected buffer and pH. Cacodylate buffer, pH 5–7, has been used due to its compatibility with Mg^{2+} and its antibacterial properties. Phosphate buffer near pH 7 has been the buffer of choice for many studies intended to simulate simple pharmaceutical formulations. SantaLucia has summarized the merits of different buffer systems[26] for thermal melt analysis. To prevent potential hydrolysis of RNA strands by divalent metal ions, EDTA can be added. pH control is critical to avoid variations due to factors such as protonation or deprotonation of bases, or chemical degradation of the oligonucleotides.[13,28]

As discussed previously, the ionic strength can affect the stability of oligonucleotide duplexes due to the screening of neighboring phosphate charges by positive counterions.[28] The dependence of the melt profile on the salt concentration of a 2′OMe-siRNA duplex solution with 33% GC content is shown in Figure 6.5. A clear two-state transition is evident for a 4 μM duplex in 25 mM NaCl, and the T_m increases with increasing salt concentration. However, there is no clear melt transition observed at the same concentration when the sample is dissolved in water with no added salt.

It is possible to achieve high melting temperatures and sharpened transitions even in the absence of added salt by significantly increasing the concentration of the oligonucleotide strands (for a more detailed discussion, see Section 6.9.4.3 and Figure 6.9). For certain oligonucleotides, the nature of the metal ions, in addition to the strand concentration, can also impact the stability of a folded structure. For example, potassium ions increase the stability of quadruplex better than sodium ions in G-quartet sequences.[29,30]

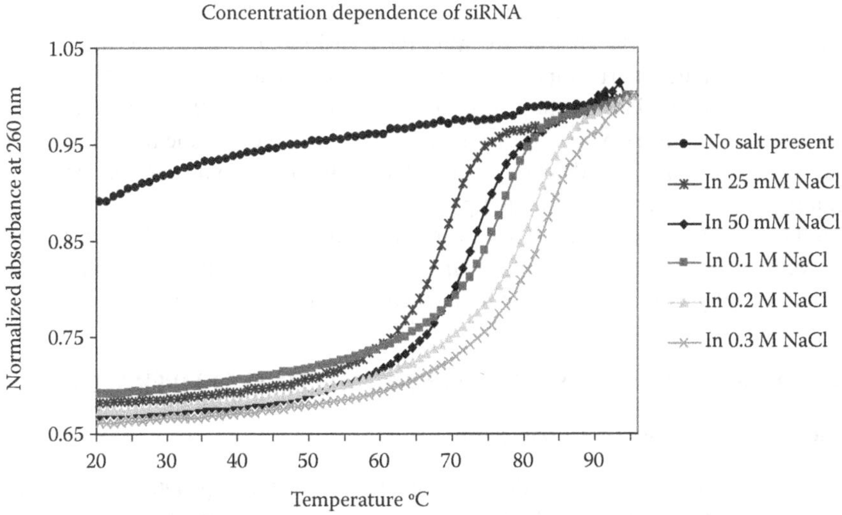

FIGURE 6.5 Salt-concentration dependence of melting profile of a 2′OMe-siRNA with 33% GC content. Each sample contains 4 μM oligonucleotide duplex, 1 mM EDTA, 10 mM phosphate at various concentrations of NaCl from 25 to 300 mM. Sample in solid circle is dissolved in water.

6.6.4 T_m Determination

T_m values can be obtained by different analyses of the melt curve, such as first derivative, two-point average, or nonlinear fit. Differences among the three methods have been discussed by Simonian.[25] The first derivative (as shown in Figure 6.6) has frequently been used to find an inflection point and is provided by most T_m analysis software programs. However, the quality of the T_m determination by this method will be affected if the raw data are noisy. As shown in Figure 6.2, the two-point average method determines T_m as the temperature at the midpoint between two parallel lines drawn from

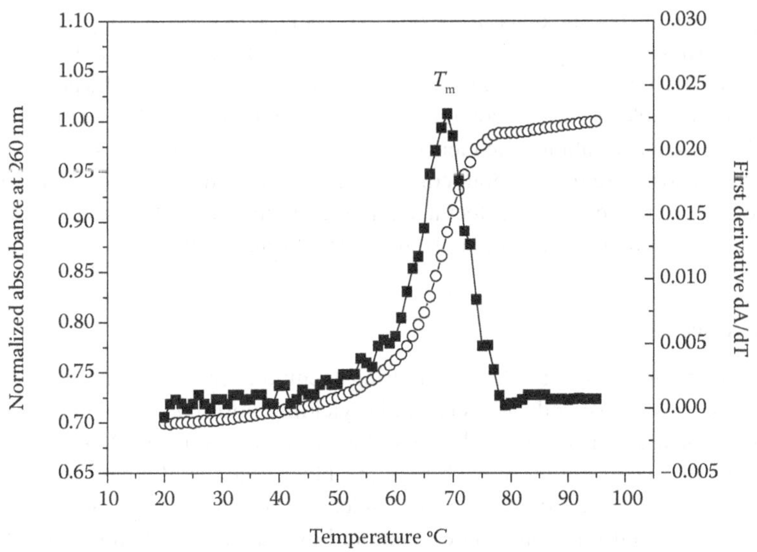

FIGURE 6.6 T_m determined by first derivative for sample in Figure 6.2. T_m determined by first derivative method is 68.9°C and by two-point average method is 67.4°C.

lower and upper temperatures; or, the temperature at which $A_{ss} = A_d$. The lower and upper baselines are a linear fit of absorbance data before and after the transition. Proper use of this method requires one to choose a low enough start temperature and high enough end temperature to ensure collection of sufficient data on both sides of the transition. However, because the method involves subjective judgments, the data could vary for different operators. If the raw melting data are noisy or if acquired data points cannot provide enough of a plateau at one or both states, a nonlinear curve fit is a useful tool to determine the melting temperature.

One should be aware that different processing methods can result in different T_m values. If no processing software is provided with the instrument, the raw data can be exported for further analysis using commercially available software such as Origin™ and MELTWIN™.[31,32]

6.7 OTHER TECHNIQUES OF MEASURING MELTING TEMPERATURE

6.7.1 Differential Scanning Calorimetry

As discussed earlier, thermodynamic data derived from UV melting data are dependent on the model used. Differential scanning calorimetry (DSC)[33–35] offers a direct measurement of change in the heat of biomolecules over a temperature gradient and is thus independent of such models. Compared to the UV method, DSC measurements require larger sample sizes. However, this can be a significant advantage in assessing the thermodynamic properties of more concentrated oligonucleotide solutions such as drug product formulations that are usually in the range of 1–150 mg/mL. An example where DSC is used to better understand the annealing properties of a siRNA in a manufacturing environment is presented in Section 6.9.4.3. DSC and UV used together can complement each other and provide validation of the experimentally determined thermodynamic parameters.

6.7.2 Circular Dichroism

Circular dichroism (CD) spectroscopy measures the differential absorption of left- and right-handed circular polarized light caused by optically active chiral molecules. CD has been used extensively to gain secondary and polymorphic structural information of biomolecules.[36,37] The advantage of determining melting temperatures using CD is the ability to observe the dynamic loss of ordered structure of duplexes during the denaturing process. Using CD spectroscopy, different conformations adopted by r(purine) · d(pyrimidine) hybrid or d(purine) · r(pyrimidine) hybrid from B-DNA and A-RNA have been determined.[38] CD spectra also allow an assessment of whether the individual strands are in self-complexes, which could compete with duplex or triplex formation.[39] It also has been reported from concentration dependence studies that some strands dissociate from the initial duplex to a hairpin structure to a random coil through two well-defined transitions as detected by CD, while the UV method can only detect a single transition.[40] Therefore, temperature-dependent CD can be used as a secondary resource to study thermal transitions.

6.7.3 Raman Spectroscopy

Raman spectroscopy is a powerful light scattering technique in which one measures the wavelength and intensity of inelastically scattered light, yielding information about the vibrational and rotational energy of a given molecule. Because the Raman spectra of DNA have well-resolved vibrational bands, the contribution of each building block, such as the base, sugar, or phosphate moiety, to the stability of the oligonucleotide duplex can be investigated at local conformation. The change of a local conformation can be monitored by studying the temperature dependency of a particular vibrational band. Studies of poly(dA) · poly(dT) and poly (dA-dT) · poly(dA-dT) have shown Raman to be a useful methodology to distinguish between the thermodynamic contributions of base stacking, base pairing, and backbone conformational ordering in the molecular mechanism of B-form DNA.[41]

6.7.4 NMR Spectroscopy

NMR spectroscopy has been used widely for the study of hybridization.[42–44] The melting of a model aptamer by monitoring the imino protons involved in hydrogen bonding has been described (Chapter 13). Similarly, for siRNA, H-NMR spectroscopy has been used to record the resonances of imino protons of U and G as a way of establishing duplex structure. For oligonucleotides in the single-strand form, the imino protons rapidly exchange with the deuterium atoms in the solvent, and thus no signals are observed in the ^{1}H-NMR spectrum. When the imino protons are involved in base pair forming hydrogen bonds, as in the case of duplex structure, exchange with deuterium is slow and distinct resonances are observed. As a result, the disappearance of the imino proton resonances as a function of temperature reflects the denaturing of the duplex to single strands.[45] Similar to DSC and CD, NMR offers the added advantage of evaluating the thermal stability of nucleic acids at much higher concentrations than are available for UV-based methods.

6.8 RESEARCH APPLICATIONS OF HYBRIDIZATION TECHNIQUES

Self-recognition of complementary oligonucleotide strands has been exploited for a variety of applications. To mimic the role of a cell's transportation system, researchers have designed a model system in which a DNA strand can walk on a track autonomously ("DNA walker") by forming successive base pairs that are synchronized with two "fuel strands."[46] By selecting gold particles of a certain size and/or particular DNA structures, controlled structures of nanotubes have been engineered through self-assembly of DNA strands.[47] In another study, quadruplex structures were immobilized on silicon cantilevers and a conformational change was induced by adding a complementary oligonucleotide strand. This allowed monitoring of an increased "open to close" response that has potential for use in nanoscale machinery and mechanical-biosensor applications.[48] Unique distance and helix features of hybridized duplex structure of oligonucleotides allowed synthetic chemists to use short DNA duplexes as templates for organic synthesis to induce stereoselectivity,[49] control reactivity,[50] and direct ordered multistep synthesis in a single solution.[51] A biosensor that can distinguish genetically modified food from nongenetically modified food via the specific hybridization ability of immobilized oligonucleotide sequence to target 35 S promoter or nopaline synthase terminator has been reported.[52] DNA-DNA hybridization techniques have also been used to compare genetic similarity between bacterial species.[53]

6.9 APPLICATIONS OF HYBRIDIZATION STUDIES TO SUPPORT DEVELOPMENT OF OLIGONUCLEOTIDE THERAPEUTICS

Hybridization studies, by virtue of their simplicity and versatility, have been used in a wide variety of applications in support of therapeutic development of oligonucleotides. In a development setting, the most obvious and common use of a T_m is to confirm the identity of a given oligonucleotide sequence. It is often listed as an identity test in certificates of analysis of batches of drug substance or drug products intended for clinical use (Chapter 19). It can also be used as a surrogate test for confirming the biological activity of oligonucleotides.

In addition, there is a myriad of other applications of T_m that aid the development scientist in making strategic decisions during chemical and pharmaceutical development. A select few examples are described here to highlight the sheer versatility of the T_m test. These examples include the use of T_m test in purification process development, process development for annealing of single strands to form duplexes, analytical method development, formulation development, and characterization of aptamers.

6.9.1 Use of T_m to Confirm Identity

6.9.1.1 Sequence Identity of Single-Stranded Oligonucleotides

Oligonucleotides exhibit a strong and specific hybridization to their complementary sequence. A sharp transition is observed in the absorbance versus temperature plot as the Watson–Crick duplex converts to single strands. On the basis of this simple principle, hybridization studies have been used as an indirect way to assure identity and sequence authenticity of synthetic oligonucleotides. The melting temperature of a duplex is sensitive to sequence modifications and base deletions.[54,55] For phosphorothioate-modified oligonucleotides of approximately 20 base units in length, the duplex is destabilized by approximately 0.5°C per modification. Sequences with a single 2′-*O*-Methyl modification have variations in T_m values from −0.3 to +1.1°C.[9] Thus, in a development setting, recording of T_m of a given sample against a well-characterized reference standard serves as an excellent means of confirming sequence authenticity of a given sample, when combined with other identity tests such as molecular weight by mass spectrometry. The sequence of the well-characterized reference standard is generally established through independent means such as chemical or enzymatic degradation followed by mass spectrometry (Chapter 5).

6.9.1.2 Sequence Identity of Double-Stranded Oligonucleotides Including siRNA

Recently, there has been considerable interest in exploiting RNA interference for clinical applications. The therapeutic agents are short interfering RNA strands, siRNAs, which are duplexes formed by hybridization of two complementary RNA strands. During manufacturing, the two individual single strand sequences are typically synthesized using commercially available solid phase synthesizers, purified using preparative anion-exchange chromatography and subsequently annealed to yield the desired siRNA duplex.

A direct method for demonstrating the presence of the correct duplex is measurement of the melting temperature T_m. As with single strands, the T_m of an siRNA duplex is very sensitive to base deletions and modifications[55] and hence can serve as a simple and effective test to confirm the presence of correct duplex in a sample, particularly when run against a well-characterized reference standard in a buffer solution that is representative of the clinical formulations.

The presence of the correct duplex can also be confirmed by performing mass spectrometric analysis to show the presence of the correct two single strands or by proton NMR to monitor the imino protons involved in hydrogen bonding in the duplex (Chapter 13).

6.9.1.3 Confirmation of Presence of Oligonucleotide Duplex in Formulated Drug Product

Demonstration of the presence of siRNA or DNA decoys in the duplex form is a key requirement for the quality control analysis of drug product formulations. T_m represents a simple and easy technique to meet this regulatory requirement.

Oligonucleotides have relatively strong molar extinction coefficients, on the order of 10^5 $M^{-1}cm^{-1}$. In order to keep the absorbance values within the measurable absorbance range, the samples are diluted significantly in the formulation buffer prior to T_m analysis. For DNA duplexes or siRNA, the concentrations of samples analyzed are often several orders of magnitude lower than the concentrations of the drug product used in clinical dosage forms. As duplex formation is favored with increasing concentration, demonstrating the presence of stable duplex under conditions of T_m analysis represents a worst-case scenario with respect to duplex stability and provides a higher degree of assurance of the presence of duplex in the drug product.

Other techniques for demonstrating the presence of intact duplex in a formulated drug product include nondenaturing chromatographic methods such as SEC (Chapter 3), AX-HPLC (Chapter 2) and IP-RP-HPLC (Chapter 1).

6.9.2 Melting Temperature as a Surrogate Activity Assay for Antisense Oligonucleotides

Antisense technology represents one of the most common approaches for the application of oligonucleotides as therapeutic agents for the treatment of human disease.[56] Vitravene™ was the first oligonucleotide drug to be approved by the US FDA and European authorities for the treatment of CMV retinitis. Since then, a significant number of antisense oligonucleotides have advanced to clinical development for treatment of a variety of indications.

Antisense oligonucleotides are single stranded DNA usually 20–21 base units in length or LNA/DNA chimera of shorter chain lengths. The mechanism of action involves binding of the oligonucleotide drug to the complementary messenger RNA thereby inhibiting the production of harmful or overexpressed proteins. Hybridization studies of the oligonucleotide drug substance with a synthetically made DNA or RNA complementary strand represents a reasonable simulation of antisense activity. The mechanism of binding to the synthetic complementary sequence is functionally similar to binding to the messenger RNA complement. The test itself is easy to perform and is likely to be much more reproducible than bioassays.

In the case of Vitravene, also known as ISIS 2922, select nucleosides were substituted in one or two positions to determine the effect of mismatched base pairs on the T_m. The biological activity of these sequences toward the inhibition of IE55 expression in transformed U373 cells was also reported.[57] The sequences of interest are shown below. Base substitutions are designated by bold, underlined letters.

ISIS 2922:	(5′-GCGTTTGCTCTTCTTCTTGCG-3′)
Single base substitution:	(5′-GCGTTTGCTC<u>**C**</u>TCTTCTTGCG-3′)
Double base substitution:	(5′-GCGTTT<u>**T**</u>CTCTTCT<u>**G**</u>CTTGCG-3′)
RNA complement:	(5′-CGCAAGAAGAAGAGCAAACGC-3′)

The effect of a single base substitution of a C for a T resulted in a significant 5.9°C drop in T_m from 54.7°C to 48.8°C. There was a corresponding, more than twofold decrease in EC_{50} from 34 to 82 nM in the IE55 expression assay. The effect of a double base switch was even more dramatic, with a 17.5°C drop in T_m to 37.2°C and a corresponding decrease in EC_{50} values to 124 nM.

This example clearly demonstrates that the T_m test is exquisitely sensitive to changes in base sequence. It should be noted, however, that the T_m values are less sensitive to sequence mismatches at the terminal base. This test, when performed against a well-characterized reference standard whose biological activity has been previously documented, can serve as a robust alternative to antisense activity assays for purposes of quality control in a development setting.

6.9.3 Use of T_m to Support Analytical Method Development

Stability of siRNA duplexes are often evaluated by both nondenaturing and denaturing HPLC methods. Nondenaturing methods provide information about the duplex/single-strand equilibrium while denaturing methods address the chemical stability of the two single strands that make up the duplex (Chapter 15). During the development of denaturing methods, it has been extremely useful to record the T_m of the duplex in the intended mobile phase to get an assessment of the stability of the duplex. The higher the T_m, the stronger the duplex and more difficult the HPLC separation to denature/resolve the two strands. An example of this application is shown in Table 6.2, in which the melting temperatures of a 21-base unmodified siRNA duplex in various denaturing RP-HPLC mobile phases are reported. From the data shown, it is clear that the presence of ion-pairing reagents such as triethlyammonium acetate (TEAA) or hexylammonium acetate (HAA) serves to destabilize the duplex and it is possible to run the HPLC separation at much lower temperatures to achieve similar chromatographic resolution for the two strands.

TABLE 6.2
Melting Temperature of a 20-Base siRNA Duplex in RP-HPLC Mobile Phase Compositions

siRNA Duplex Diluted in RP-HPLC Mobile Phase	T_m, °C
100 mM triethylammonium acetate (TEAA) w/ 5% acetonitrile, pH 7.0	45.5
100 mM hexylammonium acetate (HAA) w/ 30% acetonitrile, pH 7.4	39.5
25 mM Tris + 0.5 M LiCl w/ 10% acetonitrile, pH 8.2	74.0
20 mM sodium phosphate + 0.5 M sodium bromide w/ 10% acetonitrile, pH 8.0	70.0

Source: Data courtesy of Agilent Technologies, Nucleic Acids Solutions Division.

6.9.4 Use of T_m to Support Process Development

6.9.4.1 Preparative Purification of G-Rich Oligonucleotides

There are multiple reports in the literature about stable secondary structures adopted by G-rich oligonucleotides.[58] Four consecutive guanines can form a guanine-tetraplex (also known as quadruplex) in both the solution[59,60] and crystalline states.[60–62] When one guanine is replaced with an adenine or a uridine, an eight-stranded helical fragment[63] and a bulged tetraplex[64] have been observed. As part of planning for purification of a G-rich oligonucleotide, it can be very informative to first record the thermal melting curves to determine if they have a tendency to form stable secondary structures.

In the following example, the antisense strand of an unmodified siRNA containing three G's in a row was found to form a well-defined aggregate in solution. The sequence of interest is presented below.[65,66]

3′-UU **GGG** A GUU UUU GUU CAA ACG-5′

The T_m of the antisense strand was measured in phosphate-buffered saline at pH 7.4. There was a clean transition with a melting temperature of 59°C. Further analysis by size exclusion

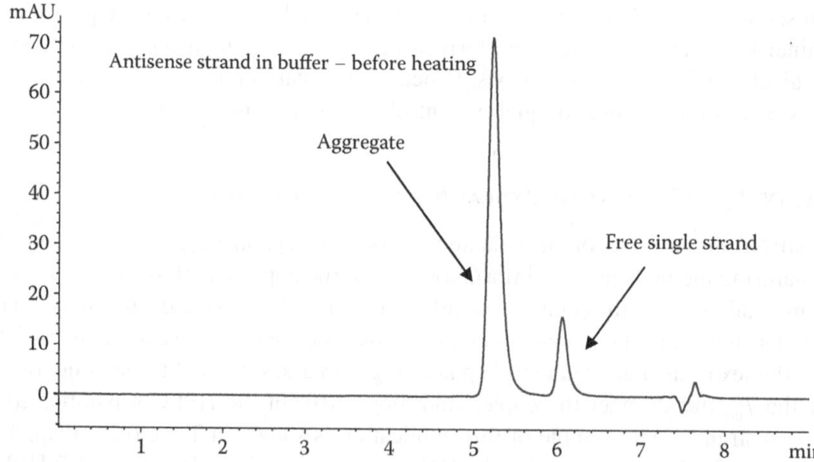

FIGURE 6.7 Size exclusion chromatography analysis of G-rich RNA strand analyzed without heating. Sample was prepared in 20 mM phosphate + 1 M NaBr + 10% ACN at pH 8. Sample was analyzed using an Agilent GF250 column at 10°C using 100 mM TEAA + 0.05% Tween 20 at pH 8.3 and a column flow rate of 0.4 mL/min. (Courtesy of TransDerm, Inc.)

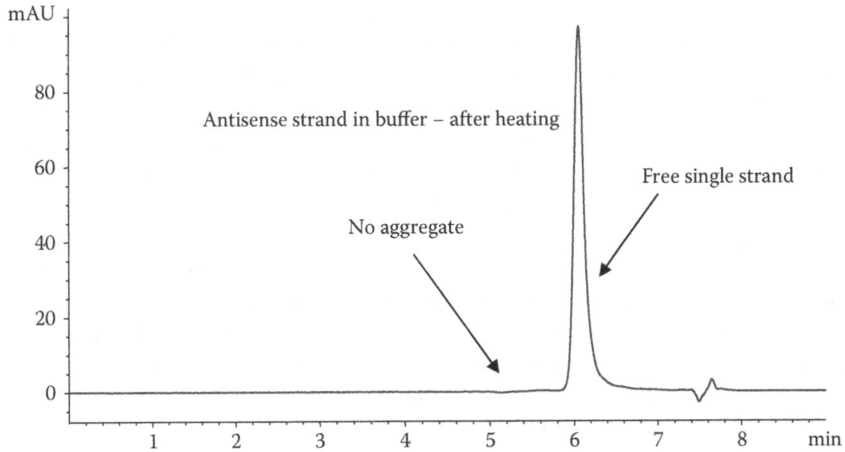

FIGURE 6.8 Size exclusion chromatography analysis of G-rich RNA strand analyzed after heating to 80°C for 10 min and cooling to ambient conditions. Sample was prepared in 20 mM phosphate + 1 M NaBr + 10% ACN at pH 8. Sample was analyzed using an Agilent GF250 column at 10°C using 100 mM TEAA + 0.05% Tween 20 at pH 8.3 and a column flow rate of 0.4 mL/min. (Courtesy of TransDerm, Inc.)

chromatography (SEC) revealed that the sample was predominantly in a tetraplex form as evidenced by a large peak corresponding to 25 kD (see Figure 6.7). Reanalysis after heating the sample to 80°C for 10 min and cooling to room temperature showed successful denaturing of the G-rich antisense strand. The chromatogram for the subsequent analysis shown in Figure 6.8 confirms that the sample is primarily in the denatured single-strand form and amenable to purification.

On the basis of these observations, the preparative chromatographic purification conditions were modified such that the crude samples were heated prior to purification and the purification conditions were selected to maintain highly denaturing conditions.

6.9.4.2 Preparative Purification of Aptamer Oligonucleotides

Aptamers are single-stranded RNA or DNA molecules that adopt well-defined three-dimensional conformations and function similarly to monoclonal antibodies. They selectively bind to target proteins with high affinity, thereby inhibiting their action. By their very nature, the conformation of aptamers can be sensitive to temperature. The manufacturing process for aptamers typically involves chemical synthesis of the oligonucleotide by solid phase synthesis followed by preparative anion-exchange purification. In order to ensure adequate purification of the desired oligonucleotide, it is necessary that the purification be conducted under denaturing conditions allowing for maximum resolution of the full length aptamer from the process related failure sequences. In the case of the example shown later in Section 6.9.6, UV thermal melting experiments showed that the aptamer had a T_m of 64°C (Table 6.4). On the basis of this information, an appropriately high temperature was selected to fully denature the aptamer during purification. Alternately, it is possible to select denaturing mobile phase conditions that achieve a lower T_m value for the aptamer, thereby allowing for successful purification at a reduced temperature.

6.9.4.3 Annealing of Duplex Oligonucleotides

During manufacture of duplex oligonucleotides, the annealing step involves bringing together the two single strands in an equimolar ratio in a suitable medium to form the duplex. The purified single-strand oligonucleotides are usually desalted prior to the annealing step. A logical question arises as to whether the presence of salt is necessary for optimal duplex formation. If the annealing step is carried out in a salt solution, as is usually done in a research environment, then the resultant duplex drug substance needs to be desalted prior to lyophilization. From a manufacturing point of view, it would be more efficient to perform the annealing step in water perhaps at a high oligonucleotide concentration, thereby avoiding an additional desalting step. Simple hybridization studies are extremely useful to assess whether the use of high concentration of oligonucleotide is sufficient for effective duplex formation in the absence of salt. This is illustrated by the following example in which the effect of salt and oligonucleotide concentration on the melting profile of an RNA duplex was evaluated. At the oligonucleotide concentrations of interest (1–100 mg/mL) the absorbance of these solutions was well beyond the range of UV analysis. Therefore, DSC was utilized to record the thermal melting profiles.

To study the effect of salt, melting profiles of a 10 mg/mL duplex of the 21-base siRNA were recorded in 1, 10, and 100 mM NaCl solutions. As expected (see Figure 6.9, left), the melt temperature of the duplex increases with increasing salt concentration, with the 100 mM NaCl solution exhibiting a T_m of 82°C. The effect of oligonucleotide concentration (1–100 mg/mL) on the melting profiles in the absence of added salt was also evaluated. The data confirmed expectations that the stability of the duplex was increased with increasing oligonucleotide concentration. The melt temperature of the duplex at 100 mg/mL in water was 86°C, somewhat more stable than of a 10 mg/mL duplex in 100 mM NaCl solution, indicating that the annealing step for this siRNA duplex could successfully be performed in water in the absence of added salt as long as the duplex concentration is maintained at approximately 100 mg/mL.

To further characterize this system, the effect of annealing medium during drug substance manufacturing on the final thermal melt profiles of a simulated drug product formulation was evaluated. Two samples were generated: (1) a 10 mg/mL duplex was annealed in 100 mM NaCl and (2) a 100 mg/mL duplex was annealed in water with no added salt; the sample was lyophilized and then dissolved at room temperature in 100 mM NaCl solution to give the same concentration as sample 1. This procedure yields two samples with the same oligonucleotide concentration and ionic strength, which is similar to the ionic strength of a typical parenteral formulation. The results of their UV thermal melting profiles are shown in Figure 6.10. It is clear that this siRNA duplex, whether annealed in water at high oligonucleotide concentration or in 100 mM NaCl, when dissolved in a simulated formulation, yielded similar thermal melting curves. Both samples had similar T_m values of 73.0° and 73.5°C, further providing assurance that this siRNA can be successfully annealed in water in the absence of salt during drug substance manufacturing.

6.9.5 Hybridization Studies to Support Formulation Development

The formulation of oligonucleotide duplexes for clinical applications requires the use of appropriate concentrations of buffers and salts to stabilize the duplex. In most instances, the concentrations of the duplexes are high enough (in the order of several mg/mL) for optimal duplex formation. Most formulations intended for parenteral injection include buffer salts and/or sodium chloride for chemical stability and isotonicity. An added benefit of the presence of these salts is the stabilization of the oligonucleotide duplex.

In the case of subcutaneous injections that are limited to low injection volumes, in order to deliver the desired dose the concentration of the oligonucleotide duplex required may be high, in the range of 100–150 mg/mL. In these situations, formulating in salt or buffer solution is not possible as the duplex in a water solution in itself may have osmolality values at or exceeding isotonicity. In such cases, it is very important to ensure the stability of the duplex at these high concentrations in the absence of added salt. Fortunately, the siRNA and DNA duplexes have sodium as the counterion, and it appears that this feature, coupled with the advantage of a high concentration, has been sufficient for stabilizing the duplex form in these predominantly water-based formulations (see Section 6.9.4.3 and Figure 6.9).

The case of very low concentrations of duplex needed for clinical applications poses a special challenge. This challenge is amplified for shorter chain duplexes, which have relatively low T_m values. In such cases, hybridization studies are essential to fully understand the stability of the duplex

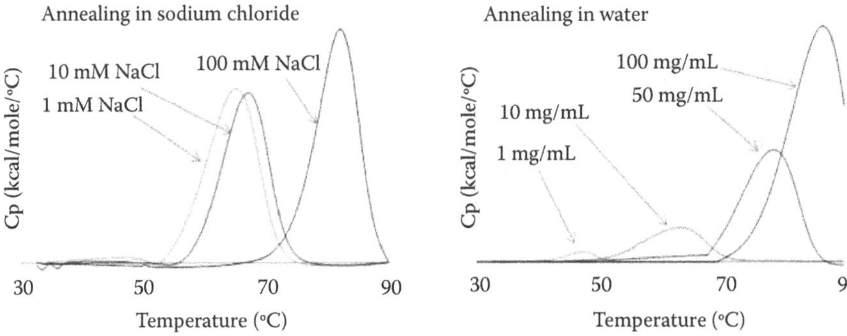

10mg/mL Duplex	
NaCl (mM)	T_m (°C)
1	65
10	67
100	82

Duplex (mg/mL)	T_m (°C)
1	47
10	63
50	78
100	86

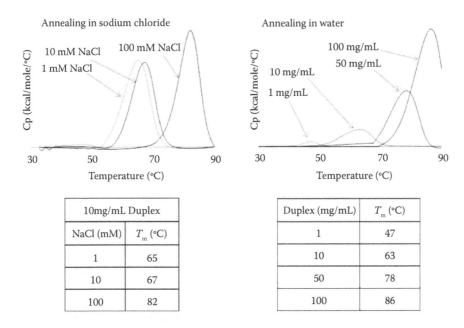

FIGURE 6.9 DSC analysis of a 21-base siRNA duplex. (Left) 10 mg/mL duplex in 1, 10, and 100 mM of NaCl solutions. (Right) 1, 10, 50, and 100 mg/mL duplex in water with no added salt. (Data courtesy of Alnylam Pharmaceuticals, Inc.)

for the intended application. In the example below, we discuss the case of a short chain DNA duplex of 10 base pairs, formed by hybridization of two 14-nucleotide single strands (strands 1 and 2). The duplex needed to be formulated at a relatively low concentration of 40 μM, not an optimum concentration for duplex stability. To complicate matters further, one of the strands, purine-rich strand 2, has the potential to form a self-complementary duplex across 7 base pairs.

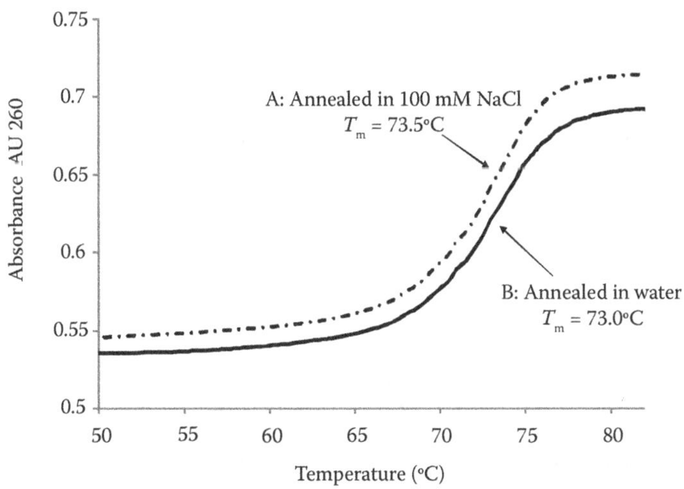

FIGURE 6.10 UV melting curves of a 21-base unmodified siRNA duplex. Sample A: 10 mg/mL duplex annealed in 100 mM NaCl; lyophilized and redissolved in 100 mM NaCl; Sample B: 100 mg/mL duplex annealed in water, lyophilized and redissolved in 100 mM NaCl. (Data courtesy of Alnylam Pharmaceuticals, Inc.)

TABLE 6.3
Thermodynamic Parameters for Duplex Formation of a DNA Duplex

Sample	Buffer[a]	ΔG_{37} (kcal/mol)	ΔH (kcal/mol)	ΔS (cal/K*mol)	T_m (40 µM each strand) (°C)
Strand 1	NS	0	0	0	N/A
Strand 2	NS	−19.0 ± 6.3	−42.7 ± 20	−5.74 ± 0.9	29.3 ± 0.5
Duplex	NS	−68.9 ± 4.9	−196.4 ± 15	−7.98 ± 0.1	45.1 ± 0.5
Duplex	PBS	−64.9 ± 2.6	−184.1 ± 8	−7.75 ± 0.03	44.6 ± 0.5

[a] NS, normal saline; PBS, phosphate-buffered saline.

The objective of the study was to evaluate the thermodynamic stability of this duplex in two different formulations (normal saline, NS; phosphate buffered saline, PBS) and if possible, perform modeling studies to predict the amount of duplex in the final formulation under conditions of clinical use.

Melting curves were measured by heating the oligonucleotide samples to 85°C for 5 min and then cooling to −1.6°C over a period of 25 min just prior to a melting experiment. Samples were heated at a constant rate of 0.8°C per minute, with data collection beginning at 0°C and ending at 95°C. The duplex to coil transition was monitored by measuring the absorbance at 260 nm. Air was used for the reference light beam. Single strands were evaluated at four concentrations. At least seven concentrations of duplex samples were analyzed for thermodynamic parameters using the program MELTWIN v2.1.[67]

Parameters for the individual single strands and the duplex are presented in Table 6.3. The thermodynamic parameters for the duplex in the two formulations are similar within experimental error. The T_m for 40 µM duplex was approximately 45°C in both formulations. The pyrimidine-rich strand 1 had no structure as evidenced by the lack of any transitions in the UV melt curves. The purine-rich strand 2 did form a self-complementary homoduplex in normal saline at a T_m of 29°C.

Using the thermodynamic parameters in normal saline, the equilibrium fraction of strand 2 in a heteroduplex with strand 1 and in a homoduplex with another strand 2 was simulated.[68] The

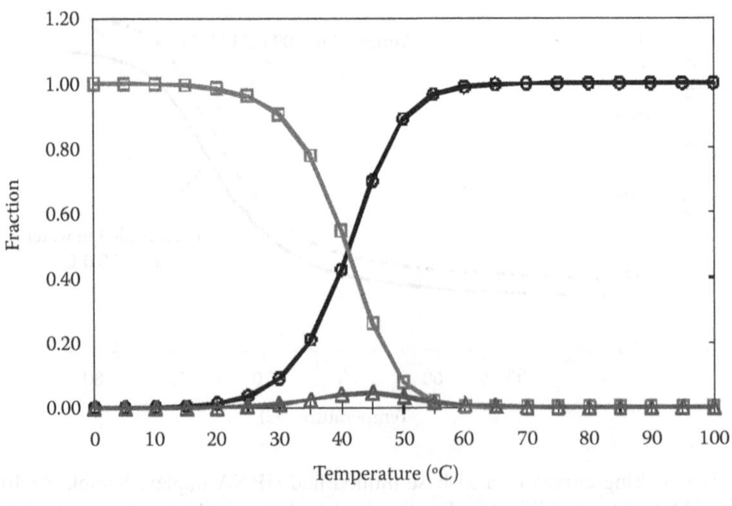

FIGURE 6.11 Plot of calculated fraction oligonucleotide in heteroduplex (squares), strand 2 single strand (diamonds), and strand 2 homoduplex (triangles) as a function of temperature in normal saline.

TABLE 6.4
Effect of PEGylation on the T_m of an Aptamer

Sample	ΔG_{37} (kcal/mol)	ΔH (kcal/mol)	ΔS (cal/K*mol)	T_m (°C)
Non-PEGylated aptamer	−3.03 ± 0.04	−37.8 ± 0.3	−112.2 ± 1.1	64.0
PEGylated aptamer	−2.90 ± 0.05	−41.8 ± 0.4	−125.4 ± 1.3	60.1

Source: Data courtesy of Regado Biosciences.

results are shown in Figure 6.11. At 40 μM strand concentration, less than 1% of strand 2 was in the homoduplex form at room temperature. The maximum fraction of homoduplex was less than 5% and occurred at 45°C.

In PBS, at 40 μM strand concentration, the sample was 99% duplex at room temperature (not shown). The purine-rich strand formed a self-complementary duplex but accounted for less than 1% of the sample. On the basis of the results of these studies, a PBS based formulation was successfully moved forward to support clinical trials.

6.9.6 Hybridization Studies of PEGylated Aptamers

T_m studies of aptamers can be an indication of the presence of correct secondary structure, an essential element of aptamer activity. Many aptamers in development are PEGylated to improve their pharmacokinetic profile. In these cases, thermal melting may be used to demonstrate that the PEGylation has not altered the secondary structure and the corresponding activity of the aptamer.

In the following example, hybridization studies of a novel PEGylated aptamer and its complementary strand are presented. The PEGylated aptamer is the active drug. Binding of the complement to the aptamer triggers a structural change in the aptamer thereby negating the aptamer's therapeutic effect. The hybridization experiments were performed to demonstrate that the PEGylation had no impact on the thermal melting. In addition, these studies were used to demonstrate the reproducibility of the manufacturing process of the aptamer. The RNA aptamer, 31 nucleotides in length, is stabilized against endonuclease degradation by the presence of 2′-fluoro and 2′-O-methyl sugar-containing residues and stabilized against exonuclease degradation by a 3′-inverted deoxythymidine cap. The nucleic acid portion of the aptamer is conjugated to a 40 kDa polyethylene glycol (PEG) carrier to enhance its bioavailability. The complement to the aptamer is a 2′-O-methyl RNA oligonucleotide, 15 nucleotides in length that is complementary to the 5′-half of the aptamer.

The UV thermal denaturation curves of the aptamer, the complement, and the aptamer-complement duplex were recorded with a Beckman DU-650 UV-Vis spectrophotometer at 260 nm in phosphate-buffered saline solution at pH 7.38 according to a typical protocol. The samples were incubated for 5 min at 85°C then cooled prior to data collection. Sample temperature was increased

TABLE 6.5
Thermodynamic Parameters for Aptamer: Complement Duplex as a Function of PEGylation

Sample	ΔG_{37} (kcal/mol)	ΔH (kcal/mol)	ΔS (cal/K*mol)	T_m (°C)
Complement annealed to non-PEGylated aptamer	−30.47 ± 0.17	−155.4 ± 1.3	−402.7 ± 3.6	93.5
Complement annealed to PEGylated aptamer	−30.71 ± 1.06	−160.2 ± 7.7	−417.5 ± 21.4	92.1

Source: Data courtesy of Regado Biosciences.

TABLE 6.6
Thermodynamic Parameters for Complement against Various Batches of Aptamer

Complement Lot	Aptamer Lot	ΔG_{37} (kcal/mol)	ΔH (kcal/mol)	ΔS (cal/K*mol)	T_m (°C)
1	1	−30.71 ± 1.06	−160.2 ± 7.7	−417.5 ± 21.4	92.1
	2	−30.78 ± 0.16	−160.2 ± 1.8	−417.4 ± 5.2	92.3
	3	−29.94 ± 0.34	−155.3 ± 2.4	−404.0 ± 6.5	92.1

Source: Data courtesy of Regado Biosciences.

from 10° to 100°C at the rate of 1°C per minute during data collection. The thermodynamic parameters were derived using two methods: (1) averages of the fits of six melting curves and (2) the $1/T_m$ versus ln(*Ct*) van't Hoff plot. The results of both methods and their error-weighted average are presented in Tables 6.4 through 6.6.

The effect of PEGylation on the thermodynamic properties of the aptamer is described in Table 6.4. The results show that PEGylation results in a small drop in the effective T_m from 64.0° to 60.1°C. At physiological temperatures of 37°C, these differences are considered inconsequential. The corresponding activity assays of the two samples demonstrated no significant differences as a result of PEGylation.

The complement had no stable secondary structures in solution as no clear transitions were observed in the melting curves. The thermodynamic parameters for the stability of aptamer: complement duplex were determined in phosphate buffered saline similar to the intended clinical formulation. In addition, the effect of PEGylation on the thermodynamic parameters of the aptamer:complement duplex was also determined. The T_m for the aptamer:complement duplex was found to be above 90°C indicating excellent stability of the duplex. The data presented in Table 6.5 demonstrate that the complement in the presence of aptamer, whether in the PEGylated or non-PEGylated form, will preferentially form the duplex at physiological temperatures. Furthermore, the thermodynamic parameters are equivalent to within experimental error for the duplex formed by the complement against the non-PEGylated or PEGylated aptamer.

Reproducibility of the manufacturing process was evaluated by assessing the thermodynamic parameters of the aptamer-complement binding for a given lot of complement against three different lots of PEGylated aptamer. The data presented in Table 6.6 show that T_m and other thermodynamic parameters are equivalent within experimental error for the various batches of aptamer.

As demonstrated by this example, hybridization studies have proven to be a simple yet extremely effective tool for supporting the chemical development of this novel aptamer-complement system.

As we continue to gain a better understanding of the sensitivity of T_m values to changes in aptamer structure, it is conceivable that one might consider the use of T_m, run against a well-characterized reference standard as a surrogate activity test of aptamers. This would require well-controlled experiments that document changes to aptamer sequence and structure and their impact on biological activity. A physical-chemical system that is suitably validated with respect to responsiveness to subtle changes in aptamer structure may be a more robust alternative to the currently used bioassays that tend to have a high degree of variability.

6.10 SUMMARY

Hybridization studies represent a simple yet highly versatile technique for the thermodynamic characterization of oligonucleotides. T_m profiles have been used extensively in research applications, leading to better understanding of cellular processes, development of novel chemical modifications, characterization of naturally occurring modified nucleic acids, and drug discovery applications. In a development setting, the applications range from the use of T_m for quality control testing to support of formulation, analytical, and process development. For antisense oligonucleotides, T_m can serve as a surrogate activity assay; for aptamers it can provide insight into secondary structures essential for biological activity. The sheer diversity of applications makes the T_m measurement a simple and yet powerful tool for the analysis and understanding of oligonucleotides in a research and development setting.

REFERENCES

1. Breslauer, K. J. 1995. Extracting thermodynamic data from equilibrium melting curves for oligonucleotide order-disorder transitions. *Methods Enzymology* 259: 221–242.
2. SantaLucia, J., Jr. 1998. A unified view of polymer, dumbbell, and oligonucleotide DNA nearest-neighbor thermodynamics. *Proceedings of the National Academy of Sciences United States of America* 95(4): 1460–1465.
3. Xia, T., et al. 1998. Thermodynamic parameters for an expanded nearest-neighbor model for formation of RNA duplexes with Watson–Crick base pairs. *Biochemistry* 37(42): 14719–14735.
4. Blackburn, G. M., and M. G. Gait. 1997. *Nucleic Acids in Chemistry and Biology*. 2nd ed. New York: IRL Press.
5. Owczarzy, R. 1993. Predictions of short DNA duplex thermodynamics and evaluation of next nearest neighbor interactions. In *Chemistry Department*, p. 188. Chicago: University of Illinois at Chicago.
6. Marmur, J., and P. Doty. 1959. Heterogeneity in deoxyribonucleic acids: I. Dependence on composition of the configurational stability of deoxyribonucleic acids. *Nature* 183(4673): 1427–1429.
7. Puglisi, J. D., and I. Tinoco, Jr. 1989. Absorbance melting curves of RNA. *Methods Enzymology* 180: 304–325.
8. SantaLucia, J., Jr., and D. H. Turner. 1997. Measuring the thermodynamics of RNA secondary structure formation. *Biopolymers* 44(3): 309–319.
9. Freier, S. 1993. Hybridization: Considerations affecting antisense drugs. In *Antisense Research and Applications*, S. T. Crooke and B. Lebleu, eds., 67–82. Boca Raton, FL: CRC.
10. Heidenreich, O., and F. Eckstein. 1992. Hammerhead ribozyme-mediated cleavage of the long terminal repeat RNA of human immunodeficiency virus type 1. *Journal of Biological Chemistry* 267(3): 1904–1909.
11. Mergny, J. L., and L. Lacroix. 2003. Analysis of thermal melting curves. *Oligonucleotides* 13(6): 515–537.
12. Mergny, J. L., and L. Lacroix. 1998. Kinetics and thermodynamics of i-DNA formation: phosphodiester versus modified oligodeoxynucleotides. *Nucleic Acids Research* 26(21): 4797–4803.
13. Marmur, J., and P. Doty. 1962. Determination of the base composition of deoxyribonucleic acid from its thermal denaturation temperature. *Journal of Molecular Biology* 5: 109–118.
14. Wallace, R. B., et al. 1979. Hybridization of synthetic oligodeoxyribonucleotides to phi chi 174 DNA: the effect of single base pair mismatch. *Nucleic Acids Research* 6(11): 3543–3557.
15. Wells, R. D., et al. 1970. Physicochemical studies on polydeoxyribonucleotides containing defined repeating nucleotide sequences. *Journal of Molecular Biology* 54(3): 465–497.
16. Crothers, D. M., and B. H. Zimm. 1964. Theory of the melting transition of synthetic polynucleotides: evaluation of the stacking free energy. *Journal of Molecular Biology* 9: 1–9.
17. Devoe, H., and I. Tinoco, Jr. 1962. The stability of helical polynucleotides: base contributions. *Journal of Molecular Biology* 4: 500–517.
18. Breslauer, K. J., et al. 1986. Predicting DNA duplex stability from the base sequence. *Proceedings of the National Academy of Sciences of the United States of America* 83(11): 3746–3750.
19. Frank-Kamenetskii, F. 1971. Simplification of the empirical relationship between melting temperature of DNA, its GC content and concentration of sodium ions in solution. *Biopolymers* 10(12): 2623–2624.
20. Owen, R. J., L. R. Hill, and S. P. Lapage. 1969. Determination of DNA base compositions from melting profiles in dilute buffers. *Biopolymers* 7(4): 503–516.
21. Owczarzy, R., et al. 2004. Effects of sodium ions on DNA duplex oligomers: improved predictions of melting temperatures. *Biochemistry* 43(12): 3537–3554.
22. Integrated DNA Technologies. 2010. http://www.idtdna.com.
23. Mergny, J. L., A. T. Phan, and L. Lacroix. 1998. Following G-quartet formation by UV-spectroscopy. *FEBS Letters* 435(1): 74–78.
24. Cary 100/300 UV-Vis Specteophotometers. Available from Varian Instruments.
25. Simonian, M. H. Experimental micro T_m analysis with DU series UV/Visible spectrophotometers. Available from Beckman Coulter.

26. Santalucia, J., Jr. 2000. The use of spectroscopic techniques in the study of DNA stability. In *Spectrophotometry and Spectrofluorimetry: A Practical Approach*, M. G. Gore, ed., 329–356. New York: Oxford University Press.
27. Tataurov, A. V., Y. You, and R. Owczarzy. 2008. Predicting ultraviolet spectrum of single stranded and double stranded deoxyribonucleic acids. *Biophysical Chemistry* 133(1–3): 66–70.
28. Schildkraut, C. 1965. Dependence of the melting temperature of DNA on salt concentration. *Biopolymers* 3(2): 195–208.
29. Risitano, A., and K. R. Fox. 2004. Influence of loop size on the stability of intramolecular DNA quadruplexes. *Nucleic Acids Research* 32(8): 2598–2606.
30. Ross, W. S., and C. C. Hardin. 1994. Ion-induced stabilization of the G-DNA quadruplex: free energy perturbation studies. *Journal of the American Chemical Society* 116(14): 6070–6080.
31. OriginLab. 2010. http://www.originlab.com.
32. MeltWin. 2010. MeltWin® 3.5. http://www.meltwin.com.
33. Breslauer, K. J. 1994. Extracting thermodynamic data from equilibrium melting curves for oligonucleotide order-disorder transitions. *Methods in Molecular Biology* 26: 347–372.
34. Plum, G. E., and K. J. Breslauer. 1995. Calorimetry of proteins and nucleic acids. *Current Opinion in Structural Biology* 5(5): 682–690.
35. Spink, C. H. 2008. Differential scanning calorimetry. *Methods in Cell Biology* 84: 115–141.
36. Gray, D. M., R. L. Ratliff, and M. R. Vaughan. 1992. Circular dichroism spectroscopy of DNA. In *Methods in Enzymology*, M. J. L. David and E. D. James, eds., 389–406. San Diego, CA: Academic Press.
37. Kypr, J., et al. 2009. Circular dichroism and conformational polymorphism of DNA. *Nucleic Acids Research* 37(6): 1713–1725.
38. Hung, S.-H., et al. 1994. Evidence from CD spectra that d(purine)-r(pyrimidine) and r(purine)-d(pyrimidine) hybrids are in different structural classes. *Nucleic Acids Research* 22(20): 4326–4334.
39. Gray, D. M., S. H. Hung, and K. H. Johnson. 1995. Absorption and circular dichroism spectroscopy of nucleic acid duplexes and triplexes. *Methods Enzymology* 246: 19–34.
40. Davis, T. M., et al. 1998. Melting of a DNA hairpin without hyperchromism. *Biochemistry* 37(19): 6975–6978.
41. Movileanu, L., J. M. Benevides, and G. J. Thomas, Jr. 2002. Determination of base and backbone contributions to the thermodynamics of premelting and melting transitions in B DNA. *Nucleic Acids Research* 30(17): 3767–3777.
42. Lane, A. N. 1989. N.m.r. assignments and temperature-dependent conformational transitions of a mutant trp operator-promoter in solution. *Biochemistry Journal* 259(3): 715–724.
43. Jaroszewski, J. W., et al. 1996. NMR investigations of duplex stability of phosphorothioate and phosphorodithioate DNA analogues modified in both strands. *Nucleic Acids Research* 24(5): 829–834.
44. Graber, D., H. Moroder, and R. Micura. 2008. 19F NMR spectroscopy for the analysis of RNA secondary structure populations. *Journal of the American Chemical Society* 130(51): 17230–17231.
45. Furtig, B., et al. 2003. NMR spectroscopy of RNA. *Chembiochem* 4(10): 936–962.
46. Omabegho, T., R. Sha, and N. C. Seeman. 2009. A bipedal DNA Brownian motor with coordinated legs. *Science* 324(5923): 67–71.
47. Sharma, J., et al. 2009. Control of self-assembly of DNA tubules through integration of gold nanoparticles. *Science* 323(5910): 112–116.
48. Shu, W., et al. 2005. DNA molecular motor driven micromechanical cantilever arrays. *Journal of the American Chemical Society* 127(48): 17054–17060.
49. Li, X., and D. R. Liu. 2003. Stereoselectivity in DNA-templated organic synthesis and its origins. *Journal of the American Chemical Society* 125(34): 10188–10189.
50. Li, X., and D. R. Liu. 2004. DNA-templated organic synthesis: nature's strategy for controlling chemical reactivity applied to synthetic molecules. *Angewandte Chemie (International Edition English)* 43(37): 4848–4870.
51. Snyder, T. M., and D. R. Liu. 2005. Ordered multistep synthesis in a single solution directed by DNA templates. *Angewandte Chemie* (International Edition English) 44(45): 7379–7382.
52. Tichoniuk, M., M. Ligaj, and M. Filipiak. 2008. Application of DNA hybridization biosensor as a screening method for the detection of genetically modified food components. *Sensors* 8(4): 2118–2135.
53. Goris, J., et al. 2007. DNA-DNA hybridization values and their relationship to whole-genome sequence similarities. *International Journal of Systematic and Evolutionary Microbiology* 57(1): 81–91.
54. Chen, D., et al. 1999. Analysis of internal (n–1)mer deletion sequences in synthetic oligodeoxyribonucleotides by hybridization to an immobilized probe array. *Nucleic Acids Research* 27(2): 389–395.

55. Naiser, T., et al. 2008. Impact of point-mutations on the hybridization affinity of surface-bound DNA/DNA and RNA/DNA oligonucleotide-duplexes: comparison of single base mismatches and base bulges. *BMC Biotechnology* 8: 48.
56. Crooke, S. T., and B. Lebleu. 1993. *Antisense Research and Applications*. Boca Raton, FL: CRC.
57. Anderson, K. P., et al. 1996. Inhibition of human cytomegalovirus immediate-early gene expression by an antisense oligonucleotide complementary to immediate-early RNA. *Antimicrobial Agents Chemotherapy* 40(9): 2004–2011.
58. Patel, D. J., et al. 1999. Structures of guanine-rich and cytosine-rich quadruplexes formed in vitro by telomeric, centromeric, and triplet repeat disease DNA sequences. In *Oxford Handbook of Nucleic Acid Structure*, ed. S. Neidle, 389–454. New York: Oxford University Press.
59. Cheong, C., and P. B. Moore. 1992. Solution structure of an unusually stable RNA tetraplex containing G- and U-quartet structures. *Biochemistry* 31(36): 8406–8414.
60. Kim, J., C. Cheong, and P. B. Moore. 1991. Tetramerization of an RNA oligonucleotide containing a GGGG sequence. *Nature* 351(6324): 331–332.
61. Laughlan, G., et al. 1994. The high-resolution crystal structure of a parallel-stranded guanine tetraplex. *Science* 265(5171): 520–524.
62. Phillips, K., et al. 1997. The crystal structure of a parallel-stranded guanine tetraplex at 0.95 A resolution. *Journal of Molecular Biology* 273(1): 171–182.
63. Pan, B., et al. 2003. An eight-stranded helical fragment in RNA crystal structure: Implications for tetraplex interaction. *Structure* 11(7): 825–831.
64. Pan, B., et al. 2003. Crystal structure of a bulged RNA tetraplex at 1.1 a resolution: implications for a novel binding site in RNA tetraplex. *Structure* 11(11): 1423–1430.
65. Hickerson, R. P., et al. 2008. Single-nucleotide-specific siRNA targeting in a dominant-negative skin model. *Journal of Investigative Dermatology* 128(3): 594–605.
66. Leachman, S. A., et al. 2010. First-in-human mutation-targeted siRNA phase Ib trial of an inherited skin disorder. *Molecular Therapy* 18(2): 442–446.
67. McDowell, J. A., and D. H. Turner. 1996. Investigation of the structural basis for thermodynamic stabilities of tandem GU mismatches: solution structure of (rGAGGUCUC)2 by two-dimensional NMR and simulated annealing. *Biochemistry* 35(45): 14077–14089.
68. SantaLucia, J., Jr., H. T. Allawi, and P. A. Seneviratne. 1996. Improved nearest-neighbor parameters for predicting DNA duplex stability. *Biochemistry* 35(11): 3555–3562.

7 Purity and Content Analysis of Oligonucleotides by Capillary Gel Electrophoresis

Judy Carmody
Avatar Pharmaceutical Services, Inc.
Bernhard Noll
Roche Kulmbach GmbH

CONTENTS

7.1 INTRODUCTION

Chromatography and electrophoresis are the most common techniques available for the separation of mixtures. For the analysis of large molecules, capillary gel electrophoresis (CGE) was not introduced until the late 1980s, well after most chromatographic techniques. Guttman et al.,[1] Cohen et al.,[2] and Kasper et al.,[3] all independently, combined the principles of slab gel electrophoresis with the column diameters of capillary zone electrophoresis (CZE) to improve the resolution, speed, and efficiencies demonstrated by slab gels, thereby establishing CGE as the preferred separation technique for large molecules.

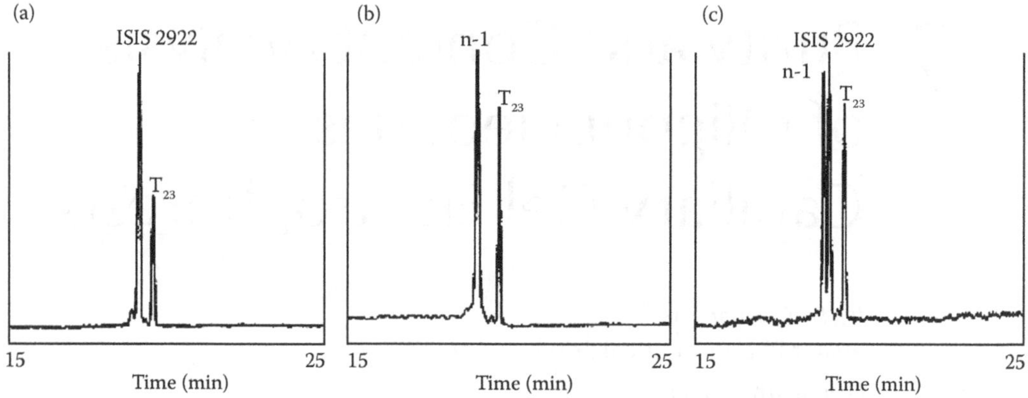

FIGURE 7.1 Capillary gel electropherograms of (a) ISIS 2922 and (b) n-1 deletion sequence of ISIS 2922 and (c) a mixture of (a) and (b) demonstrating length-based resolution. Denaturing conditions with detection by UV at 254 nm. (Reprinted from G. S. Srivatsa, et al., *Journal of Chromatography A*, 680, 469–477. Copyright 1994, with permission from Elsevier.)

CGE found applications in the separation of relatively short single-stranded oligonucleotides because of the desire to separate these species by a single base unit (i.e., sequencing)[4] or to separate identical length species containing different sequences (e.g., probes, primers, etc.).[5] However, early on, CGE was primarily used as a research tool. It wasn't until the mid-1990s when the need for regulated, validated assays pushed CGE from the research labs into the drug development setting. The approval of Vitravene™ (formerly known as ISIS 2922) represented the first NDA to use quantitative CGE (QCGE) for purity and content determination for both the active pharmaceutical ingredient (API) and formulated drug product (FDP).[6–8] This methodology obtained length-based resolution using denaturing conditions (Figure 7.1). It was successfully used for release and stability monitoring and was approved by both U.S. and European authorities. Owing to the propensity of oligonucleotides to form secondary structures, the separation of a single-stranded parent

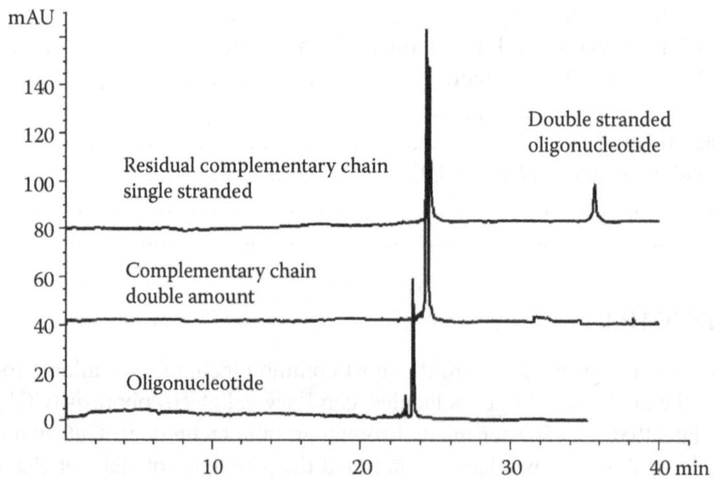

FIGURE 7.2 Separation of single- and double-stranded oligonucleotides in a single run using a 27 (w/v)% Polyethylenglykol 35000 (PEG) in 200 mM BisTris buffer solution, 20 (w/v)% Acetonitrile. (From Oligonucleotide analysis with the Agilent Capillary Electrophoresis System. Copyright Agilent Technologies, Inc. 2001. Reproduced with permission courtesy of Agilent Technologies, Inc.)

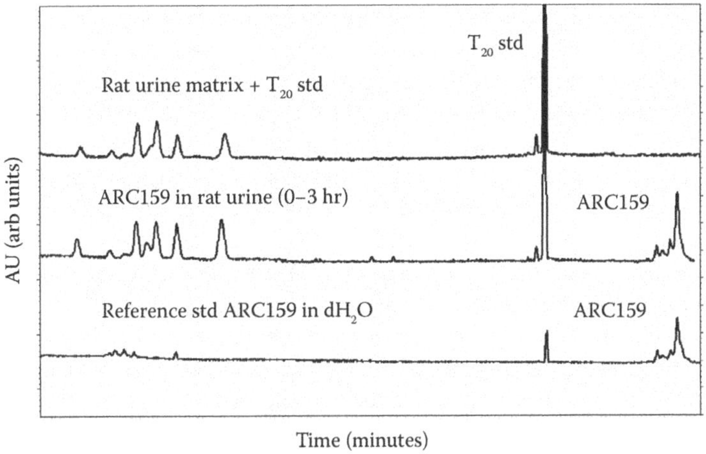

FIGURE 7.3 Intact fully 2′-O-Me composition aptamer ARC159 was detected by capillary gel electrophoresis in urine samples from Sprague-Dawley rats (0–3 hour collection interval). Samples were spiked with T_{20} internal standard and incubated in a pH 8.0 Tris-Cl, EDTA buffer containing 0.5% SDS and 500 μg/mL proteinase K at 65°C for 4 hours. Detection was performed at 260 nm using a 10% polyacrylamide gel-filled capillary at 25°C. (With kind permission from Springer Science+Business Media, Healy J. M., et al., *Pharmaceutical Research*, 21, 2234–2246, 2004.)

compound from shorter single-stranded nucleic acids usually requires denaturation[9–13] and for these applications CGE is usually carried out using a buffer system including a denaturing agent.

In native applications, the CGE (or CZE) can be used to verify certain base compositions when coupled with mass spectrometry (MS) detection, separate duplexes from single strands, as demonstrated by Aglient[14] (Figure 7.2), identify structured single strands or, as Healy et al. demonstrated, detection of intact aptamers (Figure 7.3).[15] In this paper, resolution of aptamers from chain-shortened metabolites was achieved. For longer duplex oligonucleotides, CGE has the capability to analyze and identify restriction fragments or polymerase chain reaction (PCR) products. It is the unique combination of the applied high field strengths and the sieving media that allows CGE to excel in the separation of oligonucleotides with great sensitivity and very little sample requirements.

7.2 ADVANTAGEOUS STRUCTURE

In nature, polynucleotides exist in two distinct species, namely ribonucleic acid (RNA) and deoxyribonucleic acid (DNA). These molecules form the genetic code (DNA) and the transcription products (RNA) of all living organisms. Polynucleotides typically consist of a linear backbone composed of sugar moieties of the nucleotides and the connecting phosphate groups. They can be present in single-stranded (unpaired) form or, in the case of complementary base sequences, allow the formation of hydrogen bonds between the respective purine and pyrimidine bases of two strands (Watson–Crick base pairing) resulting in a double strand. The three-dimensional, helical structure of the two congruent strands was first published in 1953 by James D. Watson and Francis Crick and is commonly referred to as the "double-helix." The individual bases of the nucleotide building blocks of the strand form the base sequence of the polynucleotide molecule. The main chemical difference between RNA and DNA is the lack of the OH group at the 2′ position of the ribose sugar and an additional methyl group at one of the pyrimidine bases for the DNA. Because neither of these modifications change the overall charge of the molecules, they have little influence on the electrophoretic separation. There is a minor effect of base composition and base sequence, but this is negligible compared to the effect of the difference in charge.

At neutral or moderately basic pH values, the naturally occurring unmodified phosphate groups (and most common synthetic analogues) carry one negative charge per phosphate group, resulting in an overall negatively charged molecule. Consequently, such polynucleotides will migrate toward the anode under electrophoresis. At pH values above approximately 10.5 the imino-proton at the guanine and thymine bases can dissociate, whereas at pH values below 6.5 the N3 nitrogen of the cytosin base and at approximately pH 4.5 the N1 nitrogen of the adenine can be proteated. These events lead to addition or loss of charges on the molecule and therefore affect separation. However, in the pH range of approximately 7–10, the charge-to-size ratio is virtually constant for all single-stranded polynucleotides and for all double-stranded DNA molecules, because it depends solely on the number of charges of the phosphate linkages in the backbone.

In the electrical field, the electrophoretic mobility of a molecule is directly proportional to the net charge of the molecule and the friction imposed on the molecule by the medium. Friction is dependent on particle size and shape as well as viscosity of the medium. An inherent property of polynucleotides is the linear charge density, i.e., a molecule containing 5 nucleotides has the same charge-to-mass ratio as one containing 50 nucleotides. As a result, oligonucleotide mobility in free solution is independent of molecular weight. Employing a similar separation mechanism as in slab gel systems (PAGE), this can be overcome by using sieving matrices or "gels." Therefore, in electrophoretic separation of oligonucleotides the gel matrix adds selectivity for separation by molecule size.

These principles form the basis of separation of polynucleotides in PAGE and CGE. However, CGE offers a number of advantages over PAGE for the separation of single and double-stranded DNA and RNA molecules in qualitative and quantitative applications for research, development, purification, and quality control. For example, employment of gel-filled capillaries increased speed up to 14-fold over slab gel techniques because of the much higher field strengths ~500 V/cm that could be applied in 50–100 μm fused silica capillaries. This improvement was achievable because of the improved heat dissipation as compared to the slab gel approach. The large surface of the capillary in relation to the low volume of the gel solution results in a 1000-fold reduction in the surface-to-gel ratio in the CGE technique. Furthermore, in CGE the capillary temperature can easily be held constant by tempering with an external liquid. As a consequence, band broadening is only caused by longitudinal diffusion and not by thermal convection allowing separations close to

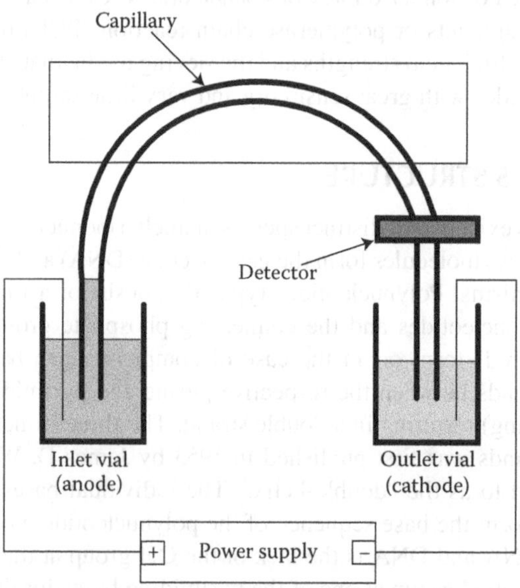

FIGURE 7.4 Depiction of a capillary electrophoresis apparatus. In normal mode, the injection is made from inlet vial or the anode. Oligonucleotides are typically analyzed using "reverse polarity" mode. In reverse mode, the injection occurs at the cathode and detection occurs at the anode.

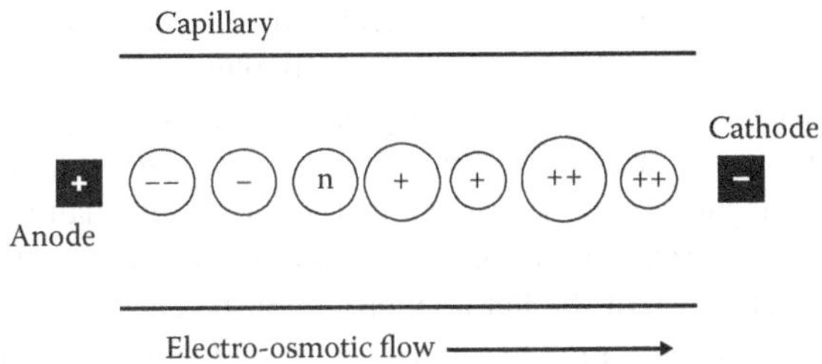

FIGURE 7.5 Representation of the flow of ions through the capillary during electrophoretic separation. Ions with higher charge and smaller size move more rapidly to the electrode and elute first. Larger ions with the same charge elute next before same size ions with lesser charge.

the theoretical limit. Analysis of large molecules with low diffusion coefficients, like DNA or RNA oligonucleotides, by capillary separations at high voltages results in a high theoretical plate number and make CGE especially suited for the separation of these macromolecules.

7.3 APPARATUS

The capillary electrophoresis apparatus, shown in Figure 7.4, consists of a power supply and a glass column filled with run buffer or electrolyte. Many systems have a thermostated capillary compartment to provide cooling for dissipation of heat from the capillary or to control the capillary temperature. The ends of the column are submerged in vials containing the same buffer. The buffer vials are often referred to as the inlet and outlet vials or reservoirs. Electrodes connected to the power supply extend into the vials containing the run buffer to complete the circuit. When the electric field is applied across the conductive liquid medium, the ions in the sample migrate through the column at different rates and in different directions based on their charge. The "normal" polarity in CE is considered to be the positive electrode (anode) at the inlet of the capillary and the negative electrode (cathode) at the outlet of the capillary. Positively charged ions migrate toward the cathode while negatively charged ions migrate toward the anode. Because of their negative charge, however, oligonucleotides are typically analyzed using "reverse polarity" mode. In this mode, the injection occurs at the cathode and detection occurs at the anode. As stated previously, the rates and directions of migration depend on the sizes of the ions and the magnitudes and signs of their charges (Figure 7.5). Smaller ions will elute earlier than larger ions with the same charge. If ions are of equal size, the one with the greater charge will travel faster than one of lesser charge. Neutral ions are not affected by the electric field; however, they do move through the capillary due to electro-osmosis as does the buffer. The movement of the buffer toward the negative electrode is termed electro-osmotic flow (EOF) and plays a fundamental role in capillary electrophoresis because the buffer components carry the sample ions with them.

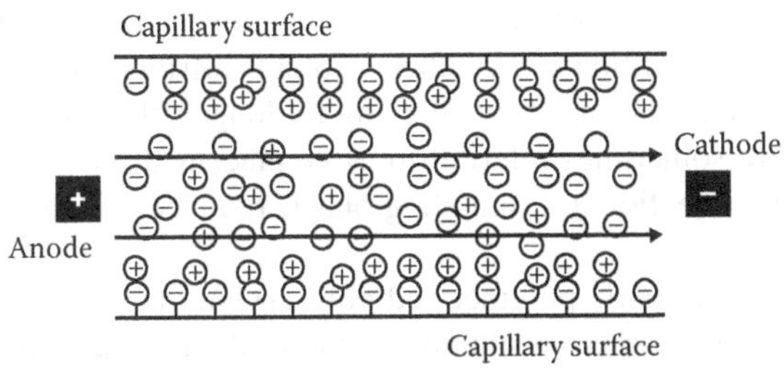

FIGURE 7.6 Electro-osmotic flow (EOF) is an electrical double-layer created at the capillary silica/solution surface.

The EOF is caused by the formation of an electrical double layer at the border between the capillary silica surface and the solution. Because of partial dissolution of the SiO_2 at the solution silica interface, the silica surface will carry negative charges. Positive charges present in the solution associate with the wall and move toward the negative electrode, resulting in a redirection of movement of the solvent toward the negative electrode. Cations migrate faster than the electro-osmotic flow, neutral ions move at the same rate and anions travel slower; so they reach the negative electrode in that order (Figure 7.6). Because of the flat flow profile of the electro-osmotic flow, this does not contribute to peak broadening; because negatively charged DNA/RNA molecules travel toward the positive pole, migration speed could be significantly reduced by electro-osmotic flow or even reversed. Therefore, for the separation of oligonucleotides, neutral coated capillaries are commonly used to abolish the electro-osmotic flow in these applications.

7.4 CAPILLARY GEL ELECTROPHORESIS

As stated earlier, nucleic acid molecules have a constant charge-to-size ratio and therefore are not successfully separated using free zone electrophoresis. However, when a "sieving gel" is employed, separation of these molecules can be achieved with high sensitivity. When employed with a capillary, this technique uses the same apparatus described above with the exception that the capillary is filled with a gel. A polyacrylamide/bisacrylamide cross-linked polymer, a linear, non-cross-linked polyacrylamide gel or hydroxypropyl methyl cellulose are typically used depending on the application and size of the molecule. Polyacrylamide has been widely used in slab gel electrophoresis and also in CGE by extension. The mechanism for separation is created through the formation of pores within the gels during cross-linking and as the charged solutes migrate through the capillary they are separated by a sieving mechanism based on size similar to size exclusion chromatography. Unlike size exclusion chromatography, in CGE, small molecules migrate through the pores faster and elute first then larger molecules elute as they are retained by the gel.[16]

7.4.1 Gels

The use of polymers as sieving matrixes was first considered by Bode[17,18] in 1977 when the migration of macromolecules through "dynamic pores" was observed. This proposal was further expanded by De Gennes in his work.[19] From this time, the most popular polymer matrixes used were agarose and polyacrylamides although several other gels were used for a wide range of applications and are well documented by Righetti and Gelfi.[20] Slab gel electrophoresis used cross-linked polyacrylamide gels for the separation of oligonucleotides with up to ~1000 bases. Agarose gels were used for the separation of oligonucleotides greater than 1000 bases in length.

In CGE the gels can be polymerized and covalently bound to the wall of the fused silica capillary or polymerized and injected either electrokinetically or by pressure into the capillary. Non-cross-linked linear polymer networks are primarily used as they can operate at higher temperatures, even at high field strengths, with greater resistance to breaking down.[21] Cross-linked gels do not perform as well under such extreme conditions.[22] Another favorable attribute of the linear polymers is that they can be readily replaced in the capillary allowing for longer lifetime of the system and removal of contaminates. They are sometimes referred to as "replacement gels," and if pressure injection is used over electrokinetic injection, the sampling bias is reduced and quantitative analysis can be achieved.[23]

Linear polymer gels are typically polymerized polyacrylamides. Addition of denaturing agents or organic solvents can improve peak resolution of the CGE. For denaturing gels, the most common additives to the gel and buffer include urea and organic solvents like formamide (FA)[24,25] and DMSO.[26–28] DMSO is a stronger denaturing agent than FA.[29,30] However, many single-stranded nucleic acids, especially when G- or GC-rich, are able to form stable secondary structures even in the presence of denaturing agents, causing decreased peak resolution between parent compound and chain shortened fragments, thereby leading to failure of CGE separations.[31–33] Currently, for the separation of single- and double-stranded DNA, neutrally coated fused silica capillaries filled with replaceable linear polymer in combination with a Tris/borate/EDTA buffer system (e.g., 100 mM Tris-boric acid, 2 mM EDTA, pH 8.5) are most commonly used.

7.4.2 Capillaries

CGE applications generally use fused silica capillaries with an internal diameter of either 50 or 100 μm in lengths of up to 60 cm as measured from the inlet buffer vial to the detector window. For denaturing applications, a neutrally coated capillary is used to control EOF and enhance peak efficiency. The length of the capillary from the detector window to the outlet buffer vial may be up to 20 cm in length depending on the instrument and application. Voltages applied across the capillaries depend on the total length of the capillary (from inlet to outlet buffer vials) and are measured in volts/cm and may be as high as 800. Peak spacing per base is a linear function of capillary length when all other variables are held constant,[34] and therefore separation efficiency is directly related to capillary length. However, the gain in resolution is achieved at the cost of additional separation time, which increases proportionally.

Temperature is a critical parameter in the separation of ODN, because many secondary structures can be resolved at elevated temperatures. Elevated temperatures were found helpful in CGE of long DNA fragments by reducing separation time and by resolving compressions[35–41] as well as by improving selectivity.[42,43] The reported temperature optimum for separations of DNA fragments or GC-rich ODN lies between 60° and 70°C.[27,28,44]

Increasing capillary temperature can improve separation efficiency; however, FA is prone to thermal decomposition at elevated temperatures.[32] The thermal stability of DMSO is much higher;[45] however, temperatures above 70°C are seldom used, because urea is thermally unstable[46] and the products of thermal decomposition of urea have a deleterious effect on separation performance.[28] In addition, elevated capillary temperatures may lead to increased fluctuations in the baseline as well as the current.

7.4.3 Injection Modes

Unlike other modes of chromatography, in CGE there is no continuous flow of mobile phase through the column to deliver the sample. Further, actual "injection" sample volumes required for a CGE separation are considerably less (nanoliters as opposed to microliters) than other forms of chromatography. Introduction of the sample occurs through one of several different ways including hydrostatic, hydrodynamic (pressure or vacuum), split flow, and electrokinetic. Hydrodynamic and electrokinetic injections are the two most commonly used techniques. Introduction of the buffer and the gel, for replaceable gel applications, is also performed similarly.

7.4.3.1 Hydrostatic Injection

The hydrostatic injection approach is used mostly with CE-MS. It is performed by raising the inlet end of the capillary or lowering the outlet end of the capillary. As the detector is usually fixed, typically the inlet end is raised. This injection approach has been used successfully for CE-MS by Lee et al.[47–49] Of note, if hydrostatic injection is not the injection mode employed, the ends of the capillaries need to be placed at the same height to prevent siphoning during the analysis.

7.4.3.2 Hydrodynamic Injection

Hydrodynamic injection can be performed in one of three ways: vacuum, pressure, or split flow. With any of these injections the amount of material introduced is dependent on the inner diameter of the column and the viscosity of the material being injected.

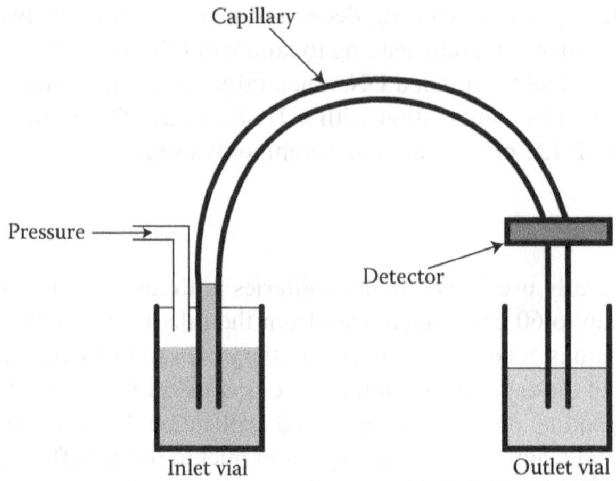

FIGURE 7.7 During a hydrodynamic pressure injection, the inlet end of the capillary and cathode (reverse polarity) are placed in the sample vial and a pressure is placed on the sample, forcing it into the capillary.

7.4.3.2.1 Pressure

Pressure injections are performed by placing the inlet end of the capillary into the inlet vial and applying a pressure to the vial (Figure 7.7). When the sample vial is pressurized, the sample or gel is forced into the capillary. The volume injected is dependent on the duration and magnitude of the pressurization, solution viscosity, and capillary dimensions. After injection, an electric field is applied and the separation is initiated.

7.4.3.2.2 Vacuum

Injections can also be made by applying a vacuum to the outlet vial while the capillary inlet is in the inlet vial (Figure 7.8). The material is pulled into the capillary. The volume of material injected depends on the size and duration of the vacuum applied, solution viscosity, and capillary dimensions. After injection, an electric field is applied and the separation is initiated.

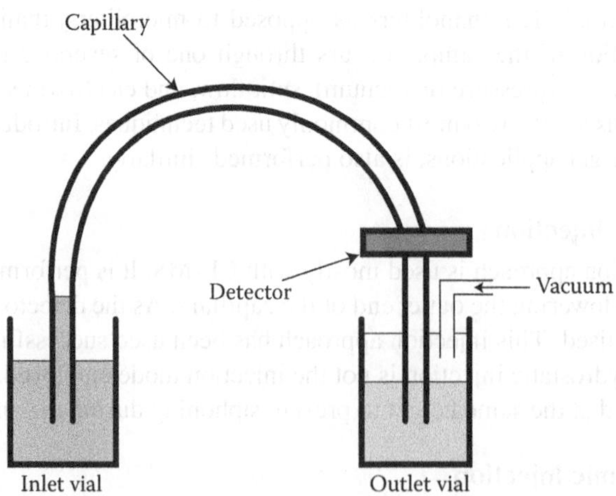

FIGURE 7.8 During a vacuum injection, the inlet end of the capillary and cathode (reverse polarity) are placed in the sample vial and a vacuum is applied on the outlet vial pulling the sample, into the capillary.

7.4.3.2.3 Split Flow

The split flow injection uses an HPLC type syringe and an injection block that is connected to the capillary and split-vent tubing. The injection occurs through hydrodynamic pressure being applied by the syringe. As a result of this pressure, the sample is introduced into the capillary and the vent tubing. The advantage to this type of injection is that it does not require constant pressure across the capillary; however, the amount of sample lost during the injection is often ≥99%.

7.4.3.2.4 Considerations

To minimize siphoning or band broadening during pressure or vacuum injections, the liquid levels in the inlet and outlet vials should be kept the same. This is also true of each end of the column; they should be at the same height. Siphoning may cause irreproducibility in the injections producing errors in quantitation. Further, any variability or loss in pressure or vacuum during injection will deleteriously impact injection reproducibility, so a constant pressure or vacuum is required for reproducible results.

Hydrodynamic injections give reproducible results as long as the pressure or vacuum, solution viscosity, and capillary dimensions remain constant and the instrumentation permits. Because the volume injected is dependent on the viscosity of the solution, any changes in the temperature of the solution will affect the reproducibility of the injection. Most instruments today have addressed this issue with temperature-controlled sample compartments and capillaries.

In CGE, hydrodynamic injections are not typically used as the action of introducing pressure or a vacuum may extrude the gels from the capillary. Further, with vacuum injections, the gels may provide too much resistance for consistent volume delivery. As such, electrokinetic injections are typically used with CGE.

7.4.3.3 Electrokinetic

Electrokinetic injections are performed by placing the inlet end of the capillary into the vial from which the injection is to be made, the outlet end of the capillary into a buffer vial, and applying a voltage across the circuit for a given period of time (Figure 7.9). After the solution is delivered,

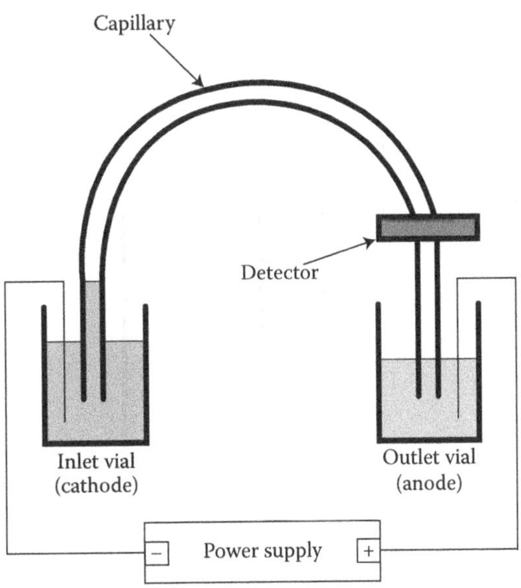

FIGURE 7.9 During an electrokinetic injection, the capillary inlet and the cathode are placed in the sample (inlet) vial and a voltage is placed across the system creating migration of the ions into and through the capillary pass the detector to the anode.

the inlet end of the capillary and the anode, and the outlet end of the capillary and the cathode are placed into buffer vials and a voltage is applied. In normal operating mode the inlet end of the capillary and the anode is in the vial from which the injection is to be made. In reverse operating mode the cathode is placed in the vial from which the injection is to be made and the polarity of the voltage reversed.

In normal mode neutral molecules are pulled into the capillary by the electro-osomotic flow. Ionic molecules are injected as a result of the electro-osmotic flow and their electrophoretic mobility. If the capillary has been treated, as is usually the case with CGE analysis of nucleic acids, to eliminate electro-osmotic flow effects and the cathode is at the detector end of the capillary, only cations are injected. The anions in solution are attracted to the anode that is in the inlet vial. As nucleic acids are negatively charged, reverse polarity is used.

7.4.3.3.1 Sampling Bias

Although electrokinetic mode of injection is primarily used for CGE as hydrodynamic modes of injection can extrude the gels from the capillaries, a sampling bias exists.[50] It was demonstrated by Huang et al. that larger quantities of solutes with higher electrophoretic mobilities are injected during electrokinetic modes of injection than solutes with lower electrophoretic mobilities (see Figure 7.10). For example, if there are equal amounts of anionic, cationic, and neutral species in a sample solution, once the electric field is applied the most highly charged cationic species have the highest mobility and will therefore get introduced into the capillary first in greater abundance. The next highly charged cations will follow then neutral ions and finally the singly and doubly charged anions in that order. The amount injected of each type is dependent on the mobility of each type (see Figure 7.10). After the injection, the separation is initiated and the ions migrate as shown in Figure 7.10 in their respective zones. As shown in Figure 7.11, because there are different amounts of each type of ion in each band, or zone, smaller intensity peaks are obtained for the later eluting species having the lesser electrophoretic mobility. A hydrodynamic injection is shown below the electrophoretic injection for comparison. As hydrodynamic injection does not involve application of a voltage, it does not demonstrate a sampling bias; all of the ionic species are represented equally, again, assuming an equal amount of each ionic species. A bias of up to 56% has been reported.[50]

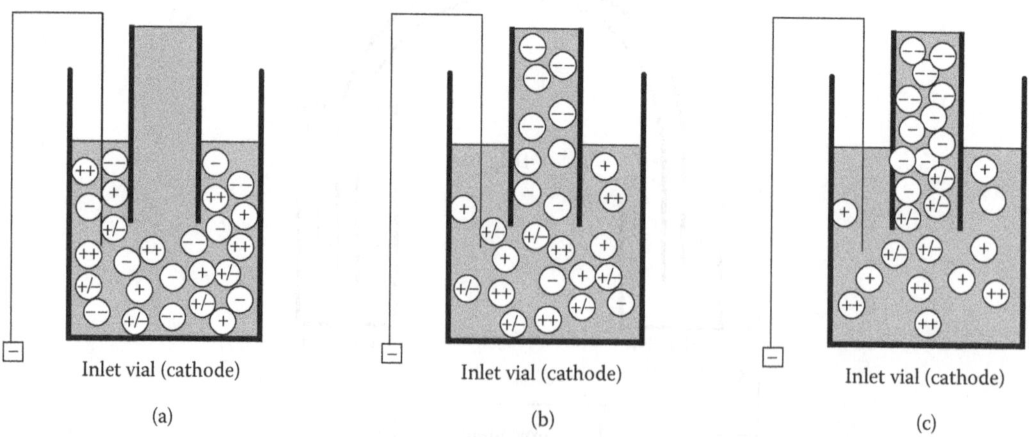

FIGURE 7.10 Sampling bias. (a) Typical solution prior to electrokinetic injection; ions are equally dispersed. (b) Voltage is applied; migration begins. Smaller more highly charged species migrate faster; so more of this type are introduced. (c) Unequal amounts of each component are introduced into the capillary and migrate as such when the voltage is applied during separation.

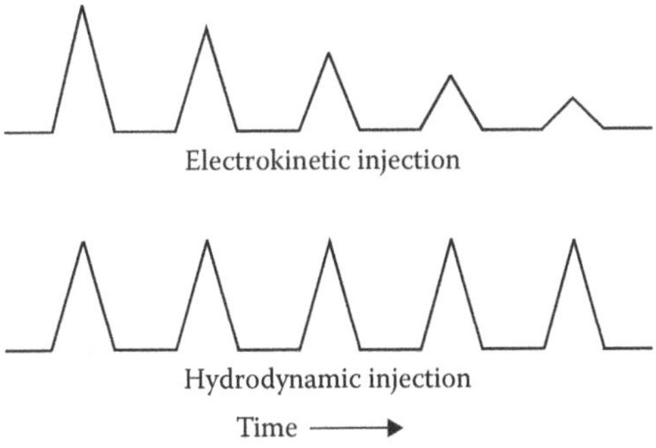

FIGURE 7.11 Depiction of electrophoretic separation with a sample containing equal amounts of cations, anions, and neutral species employing electrokinetic injection. As more highly charged species are introduced into the capillary and travel faster, the first peak demonstrates a greater response than subsequent lesser charged species or those having a larger size-to-charge ratio resulting in the top electropherogram. When hydrodynamic injection is employed, all components of the sample solution are introduced into the capillary equally and demonstrate equal responses as shown in the bottom electropherogram.

To correct for this sampling bias, Huang et al. divided the area response of each solute by a bias factor consisting of a ratio of the migration times of the solutes.[50] The sampling bias caused by differing electrophoretic mobilities is easily corrected if the samples are diluted in the same diluent and are of the same approximate concentration. As stated previously, the electro-osmotic flow of the solution and the electrophoretic mobility of the ionic species are affected by the diluent composition and concentration. So if there are variations in the diluent composition, such as salt type or concentration, the peak area responses will change significantly for electrokinetic injections as compared to hydrodynamic injections (Figure 7.11). One way to correct or minimize sampling bias is to dilute all samples with a large volume of running buffer making each sample more similar. However, this may result in dilution of the samples below the limit of detection or quantitation of the method. Others[51] have added a small amount of a concentrated, nondetected ion to the samples to yield samples with similar conductivities to reduce sampling bias as a result of electrokinetic injections. Another approach to reduce sampling bias is through the use of internal standards.[52,53]

Another issue with electrokinetic mode of injection is the depletion of the sample. With electrokinetic injections a larger amount of high-mobility solutes are removed upon injection as compared to lessor mobility solutes. If the samples are dilute, then multiple injections from the same vial can deplete the high-mobility solutes, thereby changing the makeup of the samples. It is recommended that one injection be performed from each vial.

7.4.4 Sample Size/Concentration Considerations

The length of the injected sample "band" is more important than its actual volume. The band should be as small as possible to reduce or minimize band spread resulting in loss of efficiency and resolution. Terabe et al.[54] demonstrated that the best efficiency was achieved with a sample band of <0.8 mm in width when using a 50 μm ID, 50 cm length column. This points to a sample band of <1% of the total capillary length to achieve optimal efficiency and resolution.[55] In general, the smallest volume that yields detectable peaks should be injected for maximum efficiency.

Of course, sample concentration also impacts efficiency and resolution. Mikkers et al.[56] demonstrated that the best peak shape and resolution was achieved when the sample concentration was on average 100 times less than the concentration of the run buffer. They showed that the presence of high concentrations of solute ions may distort the electric field in the capillary, causing distortion of peak shapes.

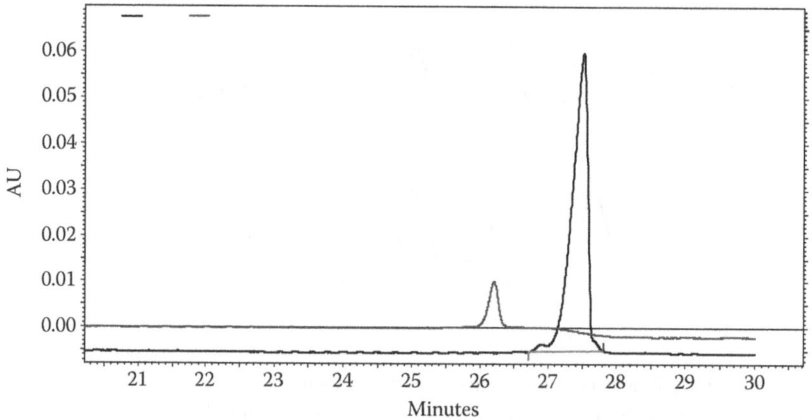

FIGURE 7.12 Top electropherogram represents a 26-mer single strand DNA prepared in PBS, while the bottom electropherogram is the same DNA prepared in water at the same concentration and injection volume.

7.4.5 Sample Preparation

To avoid any interference or obstruction from small particles, samples, gels, and buffers should be filtered or centrifuged prior to introduction into the capillary. Oligonucleotides should be diluted in the run buffer or a diluent that is less viscous than the run buffer. For analyses employing electrokinetic injection modes, sample/standard solutions should be prepared in water since all cations will enter the capillary and compete with each other reducing sensitivity. Figure 7.12 demonstrates how significantly the response is reduced by the presence of salt cations in the sample, further validating the need for desalting before injection.

7.4.6 Detector Selection

7.4.6.1 UV-Vis and Laser-Induced Fluorescence

Selection of the detector is based upon the oligonucleotide being analyzed. As oligonucleotides are chromophores, most single-stranded and duplex oligonucleotides are analyzed using UV-Vis detection using a wavelength, typically, of 254 nm. However, CGE with laser-induced fluorescence (LIF) detection has found great utility in, specifically, PCR and plasmid DNA analysis as it is 10,000 times more sensitive than UV.[57] Although LIF detection can be affected several ways, the most popular employs intercalating dyes and an Argon LIF detector with an excitation wavelength of 488 nm and an emission wavelength of 530 nm. Ethidium bromide (EtBr) is one of the most common dyes, but monomeric and dimeric intercalators such as oxazole yellow (YO) and its homodimer (YOYO); tiazole orange-thiazole blue heterodimer (TOTAB) and others have been used. Most detection and derivatization approaches are covered in a review article by Szulc et al.[58] These modes of detection are relatively stable, easy to use, and allow on-line column detection, which minimizes band spreading and has direct, simple capillary-to-detector connections.

7.4.6.2 Mass Spectrometry

Separation and detection by CE, despite this technique's advantages in analyzing length and purity of oligonucleotides, are often not sufficient for absolute identity determination of a given base sequence. For example, the verification of a certain base composition at a given length requires the determination of the molecular mass of an oligonucleotide. Molecular masses are determined by mass spectrometry (MS). During the past decade, MS has become an important tool in the analysis of oligonucleotides and can provide a basis for elucidation of length and sequence of

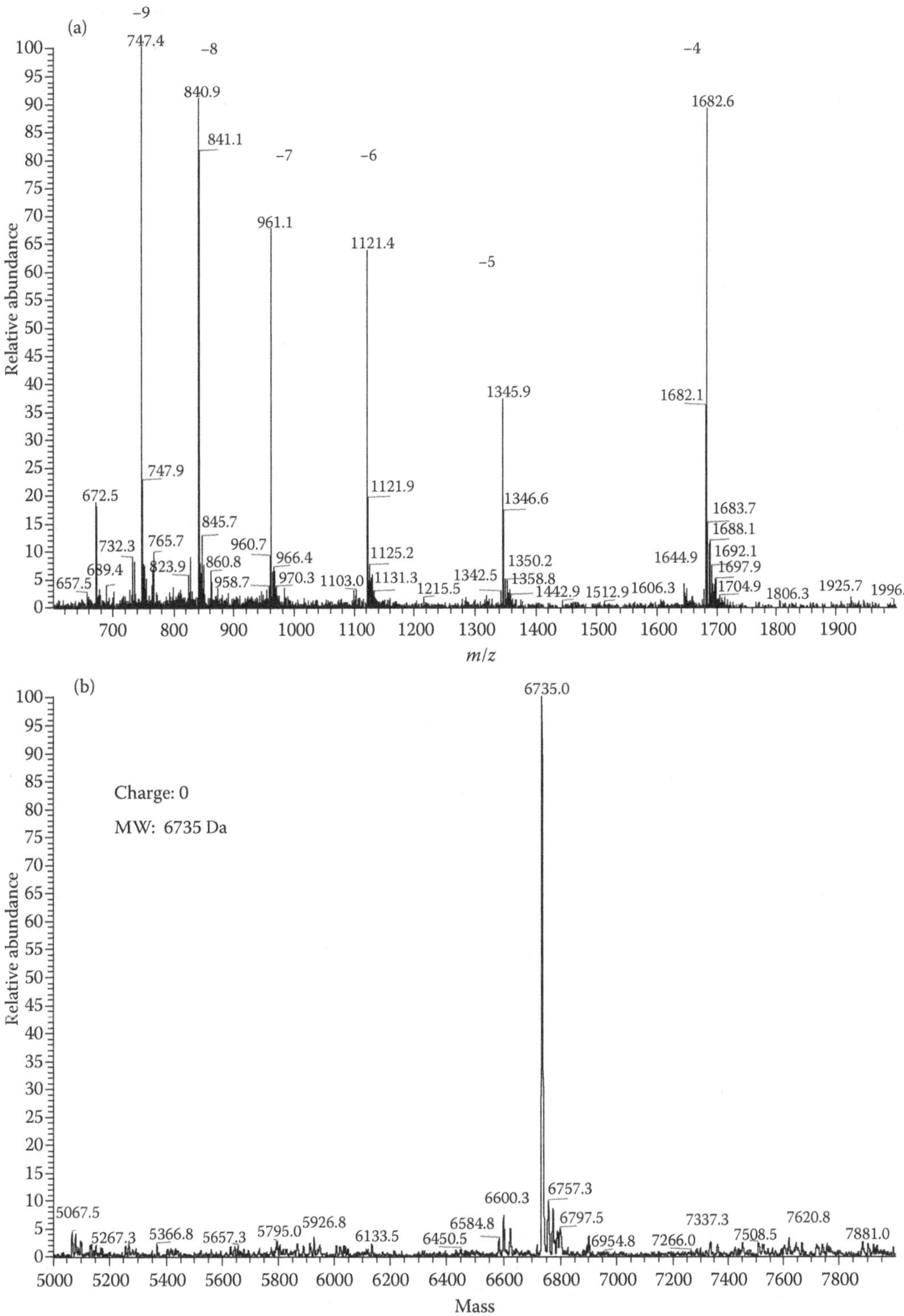

FIGURE 7.13 Mass analysis of oligonucleotides. (a) Raw mass data spectrum measured in negative ion mode. Intensity is plotted over *m*/*z* values. Negative numbers indicate charge states of the detected ion signals. (–4 = 1682.6; –5 = 1345.9; –6 = 1121.4; –7 = 961.1; –8 = 840.9; –9 = 747.4) (b) Deconvoluted mass spectrum. Main signal at a molecular weight of 6735 Daltons.

oligonucleotides.[59–61] However, until recently, only a few reports were found presenting MS analysis of oligonucleotides in combination with CE techniques.[59,62–65] Owing to its gentle properties on large biomolecules, the preferred ionization technique used for the analysis of oligonucleotides is electrospray ionization (ESI). During ESI ions are formed directly from a liquid solution, and therefore ESI-MS can be directly interfaced to CE.[65–68] During the ionization process of ESI, multiply charged oligonucleotide ions are produced in the ion source. The resulting mass spectrum contains an envelope of peaks that correspond to ions with various charge states of a relatively low mass-to-charge ratio ($m/z < 2500$) (Figure 7.13a). Computer algorithms can transform this spectrum to a zero charge spectrum to yield the molecular mass of the oligonucleotide, called a deconvoluted spectrum (Figure 7.13b).[69,70] The coupling of CE and MS for the analysis of oligonucleotides combines the advantages of CE with the ability of the mass spectrometer to provide sensitivity and selectivity for the target oligonucleotide, sequence variations, and fragments thereof.

A number of reports have addressed the combination of CGE with MS detection. Barry et al.[62] and Harsch et al.[63] coupled CE in a fused silica capillary coated with poly-(vinyl alcohol) and filled with a poly-(N-vinylpyrrolidone) matrix to negative ion ESI-MS for the analysis of short modified oligonucleotides. Freudemann et al.[64] and von Brocke et al.[65] reported the on-line coupling of CGE with negative ion ESI-MS for oligonucleotide analysis (up to 20 bases in length) using an entangled polymer solution, bis-Tris-borate buffer, and coated capillaries.

In addition to CGE-MS techniques, applications using capillary zone electrophoresis (CZE) in combination with MS have been reported. CZE is by far the most frequent CE-mode applied in combination with MS detection, mainly due to the fact that volatile electrolytes, as required for sensitive MS detection, are easily available for a broad pH range (<2 up to >12). The separation in CZE is based on different migration velocities (i.e., charge to size) and has been applied to a wide variety of analytes including oligonucleotides.[71–74] Schrader et al.[66–68] used CZE-negative ion ESI-MS and ESI-MS/MS in an ammonium carbonate buffer for the detection of oligonucleotides. Deforce et al.[59] investigated an on-line CZE-negative ion ESI-quadrupole time of flight (Q-TOF)-MS and reported single step desalting, separation, and characterization of oligos up to 120 bases in length using a 25 mM ammonium carbonate buffer supplemented with 0.2 mM CDTA (pH 9.7).[75] A quality control method for the characterization of oligonucleotides by CZE-ESI-Q-TOF-MS has been reported by the same group.[76]

One of the main challenges of the combination of CE with MS is the high affinity of cations such as sodium and potassium in the sample solution to the polyanionic backbone of the oligonucleotide, which leads to the formation of adducts during the ionization process. As a result, the analyte ions are dispensed among multiple adduct ions, leading to highly complex spectra and low sensitivity of the mass measurements. Effective removal of these cations is required to obtain satisfactory sensitivity and peak resolution in the mass spectra. One approach to remove the sodium and potassium ions associated to the oligonucleotide is to use an excess of ammonium ions as competing agent. Because of their lower affinity to the negative charges at the sugar-phosphate backbone the ammonium ions can dissociate from the analyte during the electrospray process thereby avoiding adduct formation.[77–80]

Another approach to remove unwanted cations from the sample is through the use of a chelating agent, such as trans-1,2-diaminocyclohexane-N,N,N′,N′-tetraacetic acid (CDTA) as it has been described first by Limbach et al.[81] and has been further used by other groups.[59]

Another challenge of CE is the limited volumes of sample that are typically loaded into a capillary. When the sample is not preconcentrated during loading, low sample volumes can result in reduced sensitivity. Several methods for preconcentration of the sample during the loading process into the capillary have been described. Sample stacking with transiently reversed polarity, originally proposed by Burgi and Chien,[82] can accomplish a signal enhancement of a factor of several hundred as compared to the classical use of the CZE. This technique removes the sample buffer prior to separation of the analytes by applying a high voltage with reversed polarity immediately after loading the sample. A prerequisite is that the sample is dissolved in a solution of lower conductivity than that of the CZE buffer. Because the nature of the stacking process, the method lends itself particularly well for the analysis of negatively charged species as oligonucleotides.[59,62,83,84] Using the sample stacking technique the unequivocal identification of low amounts (approximately 100 pmol/μL) of oligonucleotides of 24 base lengths was feasible.[76]

Recently, Feng et al.[85] described a CZE-MS application using a pressure assisted electrokinetic injection (PAEKI) technique, which uses external pressure to counterbalance the electro-osmotic flow (EOF) during sample introduction. Using PAEKI, single- and double-stranded oligonucleotides of up to 20-mer length could be loaded onto the capillary with a concentration enhancement of 300–800 times, allowing the identification of oligonucleotides at a low micromolar concentration. The dynamic linear range of the method was described to be in the range of approximately 2 orders of magnitude.

Combinations of CGE or CZE with mass spectrometry have been reported to be rapid, low cost, and reliable methods suitable for the quality control of oligonucleotides after synthesis. A comprehensive review of the different techniques can be found in a review article by Deforce et al.[86]

7.5 APPLICATIONS

Capillary gel electrophoresis has found multiple applications within the realm of oligonucleotide analysis in denaturing and nondenaturing applications and purity and content applications. Denaturing gel-filled capillaries are utilized mainly to separate short (up to several hundreds of bases) linear single-stranded DNA or RNA molecules based primarily on length while nondenaturing gels are used to separate longer double-stranded or structured polynucleotides (conjugates, plasmids, etc.).

7.5.1 Denaturing CGE

Applications of denaturing CGE include the purity determination of unmodified and modified DNA and RNA oligonucleotides in the length range of 10–60 bases for therapeutic molecules or several hundred bases in sequencing applications. Purity determination by CGE has almost become commonplace as evidenced by the increasing presence of application notes, booklets, and publications

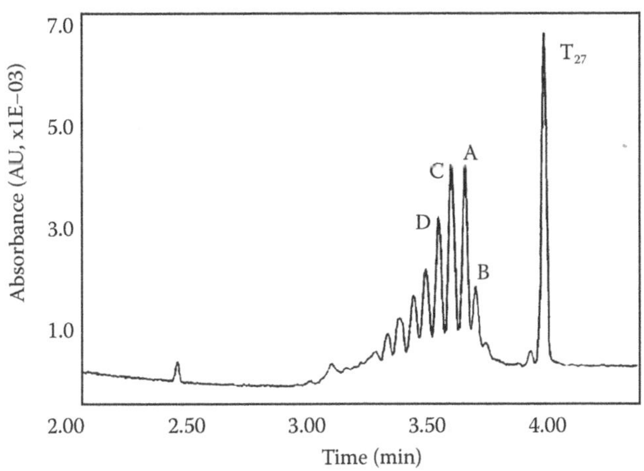

FIGURE 7.14 A capillary gel electropherogram metabolite profile resulting from analysis of a liver sample taken at 24 hours after intravenous administration. (A) intact CGP 69846A (20-base phosphorothioate oligonucleotide); (B) N+1 metabolite (seen only in tissue); (C, D) 1 and 2 base deletions, respectively; T_{27}, 27-base phosphorothioate oligonucleotide quantitative internal standard. (From Geary, R. S., et al., *Drug Metabolism and Disposition*, 25, 1272–1281, 1997. With permission.)

available from manufacturers, vendors, and others. These items have become valuable tools in the development of new approaches for the separation of more complex oligonucleotides and the separation of oligonucleotides in more complex matrices.

As CGE purity applications are on the rise so are quantitative applications. Several quantitative applications of CGE for the analysis of antisense ODN, immunomodulatory ODN or RNA have been described within the past 15 years. Leeds and others[6,87–97] have used CGE extensively for the quantitation of therapeutic oligonucleotides and their chain-shortened metabolites in bioanalytical applications (Figure 7.14). Chen et al.[98] demonstrated perhaps the first use of CGE for the quantitation of free and encapsulated oligonucleotides in liposome formulated drug products. Using eCAP™ ssDNA R-100 Gel and eCAP™ DNA capillary under denaturing conditions, they were able to quantitatively determine the amounts of free and encapsulated active pharmaceutical ingredient in the liposome to ensure consistent formulation behavior.

7.5.2 Nondenaturing or Native CGE

Another attractive use of CGE is under nondenaturing or native conditions. For nondenaturing gels, linear polyacrylamide monomers, or alkylcelluloses (e.g., hydroxypropylmethyl cellulose) are polymerized without urea, DMSO, or formamide. Parameters like duplex content or conjugation efficiency can be determined using native CGE in addition to the determination of secondary structure differences and the separation of double-stranded DNA such as restriction fragments.[23] However, when attempting to separate double-stranded DNA, lower concentrations of the gels are used as the resolution of larger molecules increases with increasing polymer chain length and decreasing concentration.[13,34,99] Applications involving the purity and size determination of duplex oligonucleotides and evaluation of the parameters that affect the determination of these items are detailed in an Agilent application by Cavender and Heiger.[100] In this paper, they show the effects of applied voltage, capillary length, and gel concentration on the separation of 15 pGEM DNA fragments that range from 36 to 2645 bp. Size determination was performed using Ferguson plots.

Other nondenaturing applications include quantification of plasmid structures using neutrally coated fused silica capillaries filled with replaceable 0.1% HPMC gels in combination with a Tris/

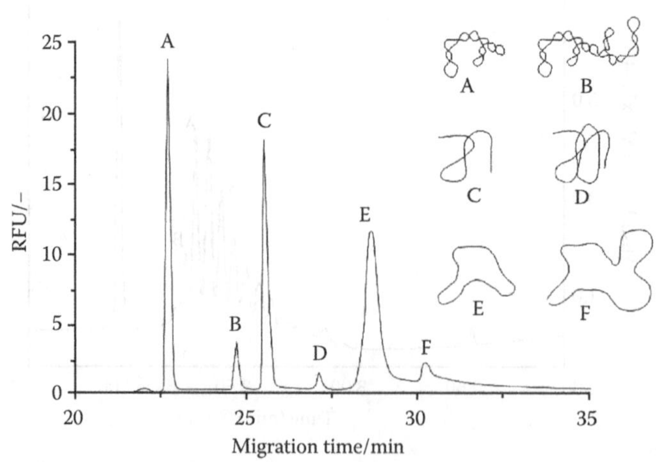

FIGURE 7.15 CGE analysis of different plasmid topologies of pUC19 plasmid DNA (2.7 kb) using 0.1% HPMC, DB-17 coated capillary, and TRIS/Borate/EDTA buffer with LIF detection. (A,B) CCC forms (monomer and dimer); (C,D) Linear forms (monomer and dimer); (E,F): OC forms (monomer and dimer). (Courtesy of M. Schleef, PlasmidFactory GmbH & Co., Bielefeld, Germany. With permission.)

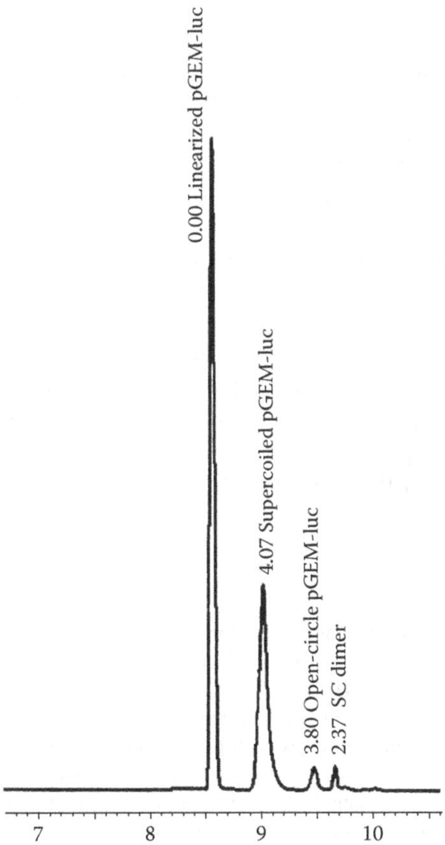

FIGURE 7.16 Capillary gel electrophoresis analysis of a 1:1 mixture of linearized and untreated pGEM-luc plasmid using eCAP™ dsDNA 1000 gel buffer diluted to 0.1X in TBE, eCAP™ LIFluor enhance stain, and eCAP™ DNA capillary 30 cm to detector. Detection employed an Argon laser at 488 nm excitation and 520 nm emission wavelengths. (Courtesy of Beckman Coulter, Inc. http://www.beckman.com, Publication Number P-12584A.)

borate/EDTA buffer-system (e.g., 100 mM Tris-boric acid, 2 mM EDTA, pH 8–8.5). Samples are typically prestained with YOYO-1 intercalating dye and analyzed using an Argon laser. All plasmid structures can be separated as demonstrated by Schmidt et al.[101] in Figure 7.15. The use of non-cross-linked polyacrylamide (PAA) was presented by Hebenbrock et al.[102] for studying plasmid DNA incorporated into the *Escherichia coli* hot cell. A 4% PAA gel was recommended for use as it demonstrated lower viscosity allowing for replacement. More recently, Boardman and Dewald[103] demonstrated the use of eCAP™ dsDNA 1000 Gel with eCAP™ enhanced dye and eCAP™ DNA capillary for the separation of plasmid isoforms in 4.9 kb pGEM-luc plasmid. Baseline resolution of supercoiled, linearized, open circular, and supercoiled dimer were achieved within 15 min (Figure 7.16).

Other CGE applications of note include viral load quantitation, DNA-protein interactions assessments, and genotyping. Wei and coworkers[104] adapted CGE-LIF to assess the quantification of viral load in patient specimens. Disease progression and drug therapy efficacy were evaluated. DNA-protein interactions were investigated by Xian et al.[105] using CGE. In a related note, sodium acetate content in antisense oligonucleotides was determined using CZE with reverse detection by Chen et al.[106] And finally, CGE-LIF has been successfully applied to identify humans through the analysis of genetic markers as shown in Figure 9 of Ref. 107.

7.6 SUMMARY

In this chapter we presented and gave background to different applications and assay conditions for the analysis of oligonucleotides by capillary electrophoresis. The reviewed methods can be applicable for single-stranded, double-stranded, structured, or conjugated DNA or RNA molecules for therapeutic or diagnostic use in a quality control or drug development setting.

In the quality control of oligonucleotide diagnostics, cost, speed, and ease of use are of importance and can favor the use of a CE application over other high-resolution analytical methods. The development of oligonucleotide therapeutics requires generation and implementation of reliable, reproducible validated assays to ensure high-quality purity, and content determination to withstand the rigors of regulatory scrutiny. In recent years there has been significant progress in the development of reliable, robust capillary gel electrophoresis assays for not only API's but also increasing complex drug product formulations as evidenced by the increase in papers published on the subject. As such, CE has proven to be a powerful tool in continuing the progress of oligonucleotide therapeutics through the drug development pipeline, as well as oligonucleotides for diagnostic use, and is expected to continue to play a critical role in the future.

REFERENCES

1. Guttman, A., A. Paulus, A. S. Cohen, and B. L. Karger. 1988. High-performance capillary gel electrophoresis: high resolution and micropreparative applications. In *Electrophoresis '88*, ed. C. Schafer-Nielsen, 151–159. Weinheim, Germany: VCH.
2. Cohen, A. S., D. R. Najarian, A. Paulus, A. Guttman, J. A. Smith, and B. L. Karger. 1988. Rapid separation and purification of oligonucleotides by high-performance capillary gel electrophoresis. *Proceedings of the National Academy of Sciences of the United States of America* 85: 9660–9663.
3. Kasper, T. J., M. Melera, P. Gozel, and R. G. Brownlee. 1988. Separation and detection of DNA by capillary electrophoresis. *Journal of Chromatography* 458: 303–312.
4. Schwartz, H., and A. Guttman. 1995. *Separation of DNA by Capillary Electrophoresis*, Fullerton, CA: Beckman Instruments.
5. El Rassi, Z., ed. 1995. *Carbohydrate Analysis.* Amsterdam: Elsevier.
6. Srivatsa, G. S., M. Batt, J. S. Schuette, R. Carlson, J. Fitchett, C. Lee, and D. L. Cole. 1994. Assay of phosphorothioate oligonucleotides in pharmaceutical formulations by capillary gel electrophoresis (QCGE). *Journal of Chromatography A* 680: 469–477.
7. Srivatsa, G. S., S. Winters, and R. Pourmand. 2001. Use of capillary gel electrophoresis for the analysis of phosphorothioate oligonucleotides. In *Capillary Electrophoresis of Nucleic Acids*. New York: Humana Press.
8. Srivatsa, G. S., P. Klopchin, M. Batt, M. Feldman, R. H. Carlson, and D. L. Cole. 1997. Selectivity of anion exchange chromatography and capillary gel electrophoresis for the analysis of phosphorothioate oligonucleotides, *Journal of Pharmaceutical and Biomedical Analysis* 16: 619–630.
9. Todorov, T. I., Y. Yamaguchi, and M. D. Morris. 2003. Effect of urea on the polymer buffer solutions used for the electrophoretic separations of nucleic acids. *Analytical Chemistry* 75(8):1837–1843.
10. Todorov, T. I., and M. D. Morris. 2002. Comparison of RNA single-stranded DNA and double-stranded DNA behavior during capillary electrophoresis in semidilute polymer solutions. *Electrophoresis* 23: 1033–1044.
11. Van der Schans, M. J., A. W. H. M. Kuypers, A. D. Kloosterman, H. J. T. Janssenm, and F. M. Everaerts. 1997. Comparison of resolution of double-stranded and single-stranded DNA in capillary electrophoresis. *Journal of Chromatography A* 772: 255–264.
12. Heller, C. 1999. Separation of double-stranded and single-stranded DNA in polymer solutions: I. Mobility and separation mechanism. *Electrophoresis* 1962–1976.
13. Heller, C. 1999. Separation of double-stranded and single-stranded DNA in polymer solutions: II. Separation, peak width and resolution. *Electrophoresis* 1978–1986.
14. Agilent Technologies. 2001. Oligonucleotide analysis with the Agilent Capillary Electrophoresis System, Publication Number 5988-4303EN, http://www.agilent.com.
15. Healy, J. M., S. D. Lewis, M. Kurz, R. M. Boomer, K. M. Thompson, C. Wilson, and T. G. McCauley. 2004. Pharmacokinetics and biodistribution of novel aptamer compositions. *Pharmaceutical Research* 21(12): 2234–2246.

16. Landers, J. P. 1993. Capillary electrophoresis: pioneering new approaches for biomolecular analysis. *Trends in Biochemistry Sciences* 18: 409–414.
17. Bode, H. J. 1977. The use of liquid polyacrylamide in electrophoresis. I. Mixed gels composed of agar-agar and liquid polyacrylamide. *Analytical Biochemistry* 83: 204–210.
18. Bode, H. J. 1977. The use of liquid polyacrylamide in electrophoresis. II. Relationship between gel viscosity and molecular sieving. *Analytical Biochemistry* 83: 364–371.
19. DeGennes, P. G. 1979. *Scaling Concepts in Polymer Physics*. Ithaca, NY: Cornell University Press.
20. Righetti, P. G., and C. Gelfi. Capillary electrophoresis of DNA. In *Capillary Electrophoresis in Analytical Biotechnology*, ed. P. G. Righetti, 431–476. Boca Raton, FL: CRC Press.
21. Guttman, A., and N. Cooke. 1991. Effect of the temperature on the separation of DNA restriction fragments in capillary gel electrophoresis. *Journal of Chromatography* 559: 285–294.
22. Tanaka, T. 1981. Gels. *Scientific American* 244: 124–138.
23. Guttman, A., and N. Cooke. 1991. Capillary gel affinity electrophoresis of DNA fragments. *Analytical Chemistry* 63: 2038–2042.
24. Rocheleau, M. J., R. J. Grey, D. Y. Chen, H. R. Harke, and N. J. Dovichi. 1992. Formamide modified polyacrylamide gels for DNA sequencing by capillary gel electrophoresis. *Electrophoresis* 13: 484–486.
25. Chen, S. H., and R. T. Tzeng. 1999. Polymer solution-filled column for the analysis of antisense phosphorothioates by capillary electrophoresis. *Electrophoresis* 20: 547–554.
26. Zimmer, C., and G. Luck. 1972. Stability and dissociation of the DNA complexes with distamycin A and netropsin in the presence of organic solvents, urea and high salt concentration. *Biochimica Biophysica Acta* 287: 376–385.
27. Salas-Solano, O., E. Carrilho, L. Kotler, A. W. Miller, W. Goetzinger, Z. Sosic, and B. L. Karger. 1998. Routine DNA sequencing of 1000 bases in less than one hour by capillary electrophoresis with replaceable linear polyacrylamide solutions. *Analytical Chemistry* 70: 3996–4003.
28. Kotler, L., H. He, A. W. Miller, and B. L. Karger. 2002. DNA sequencing of close to 1000 bases in 40 minutes by capillary electrophoresis using dimethyl sulfoxide and urea as denaturants in replaceable linear polyacrylamide solutions. *Electrophoresis* 23: 3062–3070.
29. Escara, J. F., and J. R. Hutton. 1980. Thermal stability and renaturation of DNA in dimethylsulphoxide solutions: acceleration of renaturation rate. *Biopolymers* 19: 1315–1327.
30. Hutton, J. R. 1977. Renaturation kinetics and thermal stability of DNA in aqueous solutions of formamide and urea. *Nucleic Acids Research* 4: 3537–3555.
31. Bowling, J. M., K. L. Bruner, J. L. Cmarik, and C. Tibbetts. 1991. Neighboring nucleotide interactions during DNA sequencing gel electrophoresis. *Nucleic Acids Research* 19: 3089–3097.
32. Konrad, K. D., and S. L. Pentoney, Jr. 1993. Contribution of secondary structure to DNA mobility in capillary gels. *Electrophoresis* 14: 502–508.
33. Satow, T., T. Akiyama, A. Machida, Y. Utagawa, and H. Kobayashi. 1993. Simultaneous determination of the migration coefficient of each base in heterogeneous oligo-DNA by gel filled capillary electrophoresis. *Journal of Chromatography A* 652: 23–30.
34. Heller, C. 2001. Principles of DNA separation with capillary electrophoresis. *Electrophoresis* 22: 629–643.
35. Rocheleau, M. J., R. J. Grey, D. Y. Chen, H. R. Harke, and N. J. Dovichi. 1992. Formamide modified polyacrylamide gels for DNA sequencing by capillary gel electrophoresis. *Electrophoresis* 484–486.
36. Fang, Y., J. Z. Zhang, J. Y. Hou, H. Lu, and N. J. Dovichi. 1996. Activation energy of the separation of DNA sequencing fragments in denaturing noncross-linked polyacrylamide by capillary electrophoresis. *Electrophoresis* 17: 1436–1442.
37. Chang, H. T., and E. S. Yeung. 1995. Dynamic control to improve the separation performance in capillary electrophoresis. *Electrophoresis* 16: 2069–2073.
38. Zhang, J., Y. Fang, J. Y. Hou, H. J. Ren, R. Jiang, P. Roos, and N. J. Dovichi. 1995. Use of non-cross-linked polyacrylamide for four-color DNA sequencing by capillary electrophoresis separation of fragments up to 640 bases in length in two hours. *Analytical Chemistry* 67: 4589–4593.
39. Guldberg, P., K. F. Henriksen, and F. Guttler. 1994. Constant denaturant gel electrophoresis without formamide. *Biotechniques* 16: 786–788.
40. Swerdlow, H., J. Z. Zhang, D. Y. Chen, H. R. Harke, R. Grey, S. L. Wu, N. J. Dovichi, and C. Fuller. 1991. Three DNA sequencing methods using capillary gel electrophoresis and laser-induced fluorescence. *Analytical Chemistry* 63: 2835–2841.
41. Kabatek, Z., K. Kleparnik, and B. Gas. 2001. *Journal of Chromatography A* 305–310.

42. Lu, H., E. Arriaga, D. Y. Chen, D. Figeys, and N. J. Dovichi. 1994. Activation energy of single stranded DNA moving through cross-linked polyacrylamide gels at 300 V/cm: effect of temperature on sequencing rate in high-electric-field capillary gel electrophoresis. *Journal of Chromatography A* 680: 503–510.
43. Kleparnik, K., F. Foret, J. Berka, W. Goetzinger, A. W. Miller, and B. L. Karger. 1996. The use of elevated column temperature to extend DNA sequencing read lengths in capillary electrophoresis with replaceable polymer matrices. *Electrophoresis* 17: 1860–1866.
44. Zhou, H., Y. Zhang, and Z. Ou-Yang. 2000. Elastic property of single double-stranded DNA molecules: theoretical study and comparison with experiments. *Physical Review E: Statistical Physics, Plasmas, Fluids, and Related Interdisciplinary Topics* 62: 1045–1058.
45. Head, D. L., and C. G. McCarty. 1973. *Tetrahedron Letters* 1405–1408.
46. Nachbaur, E., E. Baumgartner, and J. Schober. 1981. *Proc. Eur. Symp. Therm. Anal.* 417–421.
47. Lee, E. D., W. Muck, J. D. Henion, and T. R. Covey. 1989. Liquid junction coupling for capillary zone electrophoresis-ion spray mass spectrometry. *Biomedical and Environmental Mass Spectrometry* 18: 844–850.
48. Lee, E. D., W. Muck, J. D. Henion, and T. R. Covey. 1989. On-line capillary zone electrophoresis-ion-spray tandem mass spectrometry for the determination of dynorphins. *Journal of Chromatography,* 458: 313–321.
49. Lee, E. D., W. Muck, J. D. Henion, and T. R. Covey. 1989. Capillary zone electrophoresis—tandem mass spectrometry for the determination of sulfonated azo dyes. *Biomedical and Environmental Mass Spectrometry* 18: 253–257.
50. Xiaohua, H., M. J. Gordon, and R. N. Zare. 1988. Bias in quantitative capillary zone electrophoresis caused by electrokinetic sample injection. *Analytical Chemistry* 60: 375–377.
51. Jandik, P., and W. R. Jones. 1991. Optimization of detection sensitivity in the capillary electrophoresis of inorganic anions. *Journal of Chromatography* 546: 431–443.
52. Dose, E. V., and G. A. Guiochon. 1991. Internal standardisation technique for capillary zone electrophoresis. *Analytical Chemistry* 63: 1154–1158.
53. Altria, K. D. 2002. Improved performance in capillary electrophoresis using internal standards. *LC GC Europe* 15: 588–594.
54. Terabe, S., K. Otsuda, and T. Ando. 1989. Band broadening in electrokinetic chromatography with micellar solutions and open-tubular capillaries. *Analytical Chemistry* 61: 251–260.
55. Albin, M., P. D. Grossman, and S. E. Moring. 1993. Sensitivity enhancement for capillary electrophoresis. *Analytical Chemistry* 65: 489A–497A.
56. Mikkers, F. E. P., F. M. Everaerts, and Th. P. E. M. Verheggen. 1979. High-performance zone electrophoresis. *Journal of Chromatography A* 169: 11–20.
57. Zhu, H., S. M. Clark, S. C. Benson, H. S. Rye, A. N. Glazer, and R. A. Mathies. 1994. High-sensitivity CZE of double-stranded DNA fragments using monomeric and dimeric fluorescent intercalating dyes. *Analytical Chemistry* 66: 1941–1948.
58. Szulc, M. E., and I. S. Krull. 1994. Improved detection and derivatization in capillary electrophoresis. *J. Chromatography A* 649: 231–245.
59. Deforce, D. L. D., J. Raymackers, L. Meheus, F. Van Wijnendaele, A. De Leenheer, and E. G. Van den Eeckhout. 1998. Characterization of DNA oligonucleotides by coupling of capillary zone electrophoresis to electrospray ionization Q-TOF mass spectrometry. *Analytical Chemistry* 70(14): 3060–3068.
60. Huber, C. G., and A. Krajete. 1999. Analysis of nucleic acids by capillary ion-pair reversed-phase HPLC coupled to negative-ion electrospray ionization mass spectrometry. *Analytical Chemistry* 71(17): 3730–3739.
61. Huber, C. G., and A. Krajete. 2000. Comparison of direct infusion and on-line liquid chromatography/electrospray ionization mass spectrometry for the analysis of nucleic acids. *Journal of Mass Spectrometry* 35(7): 870–877.
62. Barry, J. P., J. Muth, S.-J. Law, B. L. Karger, and P. Vouros. 1996. Analysis of modified oligonucleotides by capillary electrophoresis in a polyvinylpyrrolidone matrix coupled with electrospray mass spectrometry. *Journal of Chromatography A* 732(1): 159–166.
63. Harsch, A., and P. Vouros. 1998. Interfacing of CE in a PVP matrix to ion trap mass spectrometry: analysis of isomeric and structurally related (N-acetylamino)fluorene-modified oligonucleotides. *Analytical Chemistry* 70(14): 3021–3027.
64. Freudemann, T. T., A. von Brocke, and E. Bayer. 2001. On-line coupling of capillary gel electrophoresis with electrospray mass spectrometry for oligonucleotide analysis. *Analytical Chemistry* 73(11): 2587–2593.

65. von Brocke, A., T. Freudemann, and E. Bayer. 2003. Performance of capillary gel electrophoretic analysis of oligonucleotides coupled on-line with electrospray mass spectrometry. *Journal of Chromatography A* 991(1): 129–141.
66. Janning, P., W. Schrader, and M. Linscheid. 1994. A new mass spectrometric approach to detect modifications in DNA. *Rapid Communications in Mass Spectrometry* 8(12): 1035–1340.
67. Schrader, W., and D. Linscheid. 1995. Determination of styrene oxide adducts in DNA and DNA components. *Journal of Chromatography A* 717(1–2): 117–125.
68. Schrader, W., and M. Linscheid. 1997. Styrene oxide DNA adducts: in vitro reaction and sensitive detection of modified oligonucleotides using capillary zone electrophoresis interfaced to electrospray mass spectrometry. *Archives of Toxicology* 71(9): 588–595.
69. Doktycz, M. J., G. B. Hurst, S. Habibi-Goudarzi, et al. 1995. Analysis of polymerase chain reaction-amplified DNA products by mass spectrometry using matrix-assisted laser desorption and electrospray: current status. *Analytical Biochemistry* 230(2): 205–214.
70. Walters, J. J., K. F. Fox, and A. Fox. 2002. Mass spectrometry and tandem mass spectrometry, alone or after liquid chromatography, for analysis of polymerase chain reaction products in the detection of genomic variation. *Journal of Chromatography B* 782(1–2): 57–66.
71. McKeon, J., and M. G. Khaledi. 2003. Evaluation of liposomal delivery of antisense oligonucleotide by capillary electrophoresis with laser-induced fluorescence detection. *Journal of Chromatography A* 1004(1–2): 39–46.
72. Zhu, J., and Y.-L. Feng. 2005. Size exclusive capillary electrophoresis separation of DNA oligonucleotides in small size linear polyacrylamide polymer solution. *Journal of Chromatography A* 1081(1): 19–23.
73. Murakami, Y., and M. Maeda. 2006. Separation of single-stranded DNAs using DNA conjugates having different migration properties in capillary electrophoresis. *Journal of Chromatography A* 1106(1–2): 118–123.
74. Wu, J.-F., L.-X. Chen, G.-A. Luo, Y.-S. Liu, Y.-M. Wang, and Z.-H. Jiang. 2006. Interaction study between double-stranded DNA and berberine using capillary zone electrophoresis. *Journal of Chromatography B* 833(2): 158–164.
75. Willems, A., D. L. Deforce, and J. Van Bocxlaer. 2008. Analysis of oligonucleotides using capillary zone electrophoresis and electrospray mass spectrometry. In *Methods in Molecular Biology,* vol. 384, 401–414. New York, Springer.
76. Willems, A. V., D. L. Deforce, and C. H. Van Peteghem, and J. F. Van Bocxlaer. 2005. Development of a quality control method for the characterization of oligonucleotides by capillary zone electrophoresis-electrospray ionization-quadrupole time of flight-mass spectrometry. *Electrophoresis* 26(7–8): 1412–1423.
77. Liu, C., Q. Wu, A. C. Harms, and R. D. Smith. 1996. On-line microdialysis sample cleanup for electrospray ionization mass spectrometry of nucleic acid samples. *Analytical Chemistry* 68(18): 3295–3299.
78. Huber, C. G., and A. Krajete. 2000. Comparison of direct infusion and on-line liquid chromatography/electrospray ionization mass spectrometry for the analysis of nucleic acids. *Journal of Mass Spectrometry* 35(7): 870–877.
79. Ragas, J. A., T. A. Simmons, and P. A. Limbach. 2000. A comparative study on methods of optimal sample preparation for the analysis of oligonucleotides by matrix-assisted laser desorption/ionization mass spectrometry. *Analyst* 125(4): 575–581.
80. Huber, C. G., and A. Krajete. 2000. Comparison of direct infusion and on-line liquid chromatography/electrospray ionization mass spectrometry for the analysis of nucleic acids. *Journal of Mass Spectrometry* 35(7): 870–877.
81. Limbach, P. A., P. F. Crain, and J. A. McCloskey. 1995. Molecular mass measurement of intact ribonucleic acids via electrospray ionization quadrupole mass spectrometry. *Journal of the American Society for Mass Spectrometry* 6(1): 27–39.
82. Chien, R. L., and D. S. Burgi. 1992. Sample stacking of an extremely large injection volume in high-performance capillary electrophoresis. *Analytical Chemistry* 64(9): 1046–1050.
83. Wolf, S. M., and P. Vouros. 1995. Incorporation of sample stacking techniques into the capillary electrophoresis CF-FAB mass spectrometric analysis of DNA adducts. *Analytical Chemistry* 67(5): 891–900.
84. Osbourn, D. M., D. J. Weiss, and C. E. Lunte. 2000. On-line preconcentration methods for capillary electrophoresis. *Electrophoresis* 21(14): 2768–2779.
85. Feng, Y. L., H. Lian, and J. Zhu. 2007. Application of pressure assisted electrokinetic injection technique in the measurements of DNA oligonucleotides and their adducts using capillary electrophoresis-mass spectrometry. *Journal of Chromatography A* 1148(2): 244–249.
86. Willems, A. V., D. L. Deforce, C. H. Van Peteghem, and J. F. Van Bocxlaer. 2005. Analysis of nucleic acid constituents by on-line capillary electrophoresis-mass spectrometry. *Electrophoresis* 26(7–8): 1221–1253.

87. Hardiman, T., J. C. Ewald, K. Lemuth, M. Reuss, and M. Siemann-Herzberg. 2008. Quantification of rRNA in Escherichia coli using capillary gel electrophoresis with laser-induced fluorescence detection. *Analytical Biochemistry* 374(1): 79–86.
88. Soucy, N. V., J. P. Riley, M. V. Templin, R. Geary, A. de Peyster, and A. A. Levin. 2006. Maternal and fetal distribution of a phosphorothioate oligonucleotide in rats after intravenous infusion. *Birth Defects Research, Part B: Developmental and Reproductive Toxicology* 77(1): 22–28.
89. Leeds, J. M., M. J. Graham, L. Truong, and L. L. Cummins. 1996. Quantitation of phosphorothioate oligonucleotide in human plasma. *Analytical Biochemistry* 36–43.
90. Geary, R. S., J. M. Leeds, J. Fitchett, T. Burckin, L. Truong, C. Spainhour, M. Creek, and A. A. Levin. 1997. Pharmacokinetics and metabolism in mice of a phosphorothioate oligonucleotide antisense inhibitor of C-*RAF*-1 kinase expression. *Drug Metabolism and Disposition* 25(11): 1272–1281.
91. Chen, S.-H., and J. M. Gallo. 1998. Use of capillary electrophoresis methods to characterize the pharmacokinetics of antisense drugs. *Electrophoresis* 19(16–17): 2861–2869.
92. Cohen, A. S., A. J. Bourque, B. H. Wang, D. L. Smisek, and A. A. Belenky. 1997. A nonradioisotope approach to study the in vivo metabolism of phosphorothioate oligonucleotides. *Antisense and Nucleic Acid Drug Development* 7(1): 13–22.
93. DeDionisio, L. A., and D. H. Lloyd. 1996. Capillary gel electrophoresis and antisense therapeutics. Analysis of DNA analogs. *Journal of Chromatography A* 735(1–2): 191–208.
94. Palm, A. K., and G. Marko-Varga. 2004. On-column electroextraction and separation of antisense oligonucleotides in human plasma by capillary gel electrophoresis. *Journal of Pharmaceutical and Biomedical Analysis* 415–423.
95. Yu, R. Z., R. S. Geary, D. K. Monteith, J. Matson, L. Truong, J. Fitchett, and A. A. Levin. 2004. Tissue disposition of a 2′-O-(2-methoxy) ethyl modified antisense oligonucleotides in monkeys. *Journal of Pharmaceutical Science* 48–59.
96. Yu, R. Z., R. S. Geary, and A. A. Levin. 2004. Application of novel quantitative bioanalytical methods for pharmacokinetic and pharmacokinetic/pharmacodynamic assessments of antisense oligonucleotides. *Current Opinion in Drug Discovery and Development* 195–203.
97. Noll, B. O., M. J. McCluskie, T. Sniatala, et al. 2005. Biodistribution and metabolism of immunostimulatory oligodeoxynucleotide CPG 7909 in mouse and rat tissues following subcutaneous administration. *Biochemical Pharmacology* 69(6): 981–991.
98. Chen, D., D. Cole, and G. S. Srivatsa. 2000. Determination of free and encapsulated oligonucleotides in liposome formulated drug product. *Journal of Pharmaceutical and Biomedical Analysis* 22(5): 791–801.
99. Barron, A. E., W. M. Sunada, and H. W. Blarch. 1996. Capillary electrophoresis of DNA in uncrosslinked polymer solutions: Evidence for a new mechanism of DNA separation. *Biotechnology and Bioengineering* 52: 259–270.
100. Agilent Technologies. 2001. Double-stranded DNA analysis with the Agilent Capillary Electrophoresis System, Publication Number 5988-4304EN. http://www.agilent.com.
101. Schmidt, T., K. Friehs, and E. Flaschel. 2001. Structures of a plasmid DNA. In *Plasmids for Therapy and Vaccination*, ed. M. Schleef, 29–44. Weinheim: Wiley-VCH, Weinheim.
102. Hebrenbrock, K., K. Schügerl, and R. Freitag. 1993. Analysis of plasmid-DNA and cell protein of recombinant E. coli by gel CZE. *Electrophoresis* 14: 753–758.
103. Boardman, C. L., and J. A. Dewald. Plasmid purity and heterogeneity analysis by capillary electrophoresis. Publication Number P-12584A. http://www.beckmancoulter.com.
104. Wei, L., D.-S. Yuan, and J.-M. Andrieu. 1994. Multi-target PCR analysis by capillary electrophoresis and laser-induced fluorescence. *Nature* 368: 269–271.
105. Xian, J., M. G. Harrington, and E. H. Davidson. 1996. DNA-protein binding assays from a single sea urchin egg: a high-sensitivity capillary electrophoresis method. *Proceedings of the National Academy of Sciences of the United States of America* 93(1): 86–90.
106. Chen, D., P. Klopchin, J. Parsons, G. S. Srivatsa. 1997. Determination of sodium acetate in antisense oligonucleotides by capillary zone electrophoresis. *Journal of Liquid Chromatography and Related Technologies* 20(8): 1185–1195.
107. Chapman, J., and J. Hobbs. 1999. Putting capillary electrophoresis to work. *LC-GC* 17(3): 86–99.

8 Bioanalysis of Therapeutic Oligonucleotides Using Hybridization-Based Immunoassay Techniques

Helen Legakis and Sandra Carriero
Charles River Laboratories Preclinal Services Montreal Inc.

CONTENTS

8.1 INTRODUCTION AND OVERVIEW

Over the past two decades, synthetic oligonucleotides have become a prime area of focus in biopharmaceutical drug discovery and development pipelines for the potential treatment of a broad range of disease indications including cancer, diabetes, viral infections, cardiovascular disease, as well as many inflammatory and degenerative disorders. Numerous oligonucleotide-based therapeutics are actively being investigated in both preclinical and clinical settings, with a first generation phosphorothioate antisense oligonucleotide (ASO), Vitravene™, approved by the Food and Drug Administration (FDA) in 1998 for the antiviral treatment of cytomegalovirus retinitis infection.[1] Since that time, there has been significant progress in the oligonucleotide field with the development of different classes of oligonucleotide-based therapeutic candidates, encompassing a number of different chemical modifications, drug delivery modalities and proposed mechanisms of action. Investigational oligonucleotide therapeutics, which have and continue to show great promise in the drug development pipeline, include the antisense oligonucleotides (ASOs),[2] small interfering RNAs (siRNAs),[3–5] immunomodulatory oligonucleotides (IMOs),[6] and aptamers.[7] Other lesser investigated classes include decoy oligonucleotides[8,9] and ribozymes/DNAzymes.[10,11] The potential applications of many of these oligonucleotide classes as viable drugs in the treatment of human diseases remains in its infancy. Nonetheless, remarkable progress has been made to improve upon the stability, efficacy, safety, potency, and delivery of oligonucleotide therapeutics, much of which has been gained from the exceptional knowledge acquired through second-generation antisense oligonucleotide research programs.

Predominantly, oligonucleotide drugs are parenterally administered by either intravenous injection/infusion or subcutaneous injection for the treatment of systemic disorders; however, local administration (i.e., inhalation, intracerebral injection, or ocular injection) has also been used to circumvent issues associated with nontargeted delivery and minimize potential systemic toxicity. Chemical modification and drug delivery continue to be prime research areas of focus because they can specifically improve upon the greatest hurdles faced with oligonucleotide therapeutics, namely site-specific targeting (i.e., tissue/cellular uptake), efficacy, off-target, and side effects. Chemical modifications to improve biostability increase plasma half-life such that distribution to tissues can occur and enhance target nucleic acid affinity have included 3′- and/or 5′-terminal (e.g., fatty acid, cholesterol), phosphate-backbone (e.g., phosphorothioate, methylphosphonate), sugar (e.g., 2′-fluoro, 2′-O-methyl), and heterocyclic base modifications.[12–15] Advancements in oligonucleotide delivery to boost cellular/tissue uptake comprise of chemical conjugation (e.g., cholesterol, PEG), or use of encapsulation agents such as liposomes (e.g., cationic lipids, PEG-lipids), cell penetrating peptides,[16] and nanoparticles, to name just a few.[17]

Highly responsive bioanalytical techniques are required to support both preclinical and clinical programs in order to study the toxicokinetics, pharmacokinetics, elimination, and biodistribution of these classes of potential therapeutic compounds. Accurate and precise quantitation of oligonucleotide therapeutics is critical in the establishment of exposure-response and exposure-toxicity relationships, both of which yield vital data for predicting doses in man and designing the appropriate clinical regimen. With the emerging importance of these therapeutic compounds as viable pharmaceuticals, development of novel methods for the fast, sensitive, and reliable quantitation of oligonucleotides in complex biological matrices, such as plasma, urine, and solid tissues (typically kidney, liver, spleen, and brain) has increased in demand. Other target-specific organs such as lung, skin, and colon have also been investigated rendering bioanalytical method development all the more challenging. Radio-labeled oligonucleotides[18] and conventional chromatographic techniques such as capillary gel electrophoresis (CGE),[19–22] ion-pair,[23] and anion-exchange[24,25] high-performance liquid chromatography (HPLC) and liquid chromatography/mass spectrometry (LC/MS)[26–28] have been extensively used for measuring such macromolecules in biological matrices. For the most part, their insufficient sensitivity and/or selectivity toward truncated sequences have compromised their ability to fully characterize the pharmacokinetic properties of therapeutic oligonucleotides in plasma following systemic exposure, particularly during the terminal elimination phase, when near complete biodistribution has occurred and very low levels of circulating drug are present in plasma.[29] Such methods would also yield inadequate sensitivity when local administration is performed, or when low doses of test drug are administered, predominantly in the case of the more highly potent siRNAs and IMOs. In addition, exhaustive sample cleanup procedures via liquid-liquid and/or solid phase extraction procedures are required for chromatographic and CGE-based methods, which limit method sensitivity and reduce sample throughput. These particular methods are also quite sensitive to matrix effects that can complicate chromatographs/electropherograms and decrease assay ruggedness. Nonetheless, they remain powerful tools for identifying and quantifying metabolites in the presence of the parent drug.

The increasing demand for ultra-sensitive bioanalytical assays has made hybridization based enzyme-linked immunosorbent assays (ELISAs) a highly desirable tool for oligonucleotide drug development programs. Hybridization-based ELISAs are bioanalytical assays readily used for the quantitative measurement of a variety of oligonucleotide therapeutics from toxicokinetic, pharmacokinetic, and animal biodistribution studies in support of preclinical and clinical drug development programs for a wide range of disease indications.[30] The hybridization-based ELISAs are ligand-binding assays specific for oligonucleotide analyte quantitation and are an extension of the well-known ELISA immunoassays for antigen and/or antibody analysis. This particular type of assay exploits the Watson–Crick base-pairing interactions of an oligonucleotide analyte and one or more complementary oligonucleotide probes. Much like a typical bioanalytical ELISA immunoassay, the fundamental steps in a typical hybridization-based ELISA are (1) pre-annealing the oligonucleotide test sample (in biological matrix) to its complementary capture probe or precoating a microtitre plate with capture probe for annealing to the test oligonucleotide, (2) washing the plate to remove any excess probe/reagents and eradicate any nonspecifically bound proteins, (3) binding of the complementary detection probe to the free oligonucleotide sequences, and (4) adding an enzyme conjugate that binds the free label on the detection probe. The enzyme subsequently reacts with a substrate in a colorimetric or fluorescent reaction producing a detectable signal that can be extrapolated back to the concentration of analyte present in the initial sample. To date, the hybridization-based immunoassay methods yield the best reported assay sensitivity and sample throughput as compared with other bioanalytical methods used for oligonucleotide quantitation. On the other hand, improving selectivity for parent oligonucleotide remains a challenge that continues to be explored in designing new hybridization-based methods. This has resulted in several types of hybridization-based ELISA methods being developed for oligonucleotide measurement. Published and routinely used hybridization methods for quantitation of all classes of oligonucleotide constructs are thoroughly discussed in this chapter. As with other bioanalytical tools, each distinct type of hybridization assay has its own inherent advantages and limitations depending on the oligonucleotide chemistry, length, stability, class, or delivery vehicle used. These considerations will be discussed in the upcoming sections.

8.2 HYBRIDIZATION TECHNIQUES AND INSTRUMENTATION—GENERAL CONSIDERATIONS

Hybridization-based immunoassays provide high selectivity and assay sensitivity with minimal sample matrix effects and generally little to no sample cleanup. These assays are typically performed on 96-well microtitre plates, allowing for the analysis of multiple samples in tandem on one plate using small sample volumes (5–100 μL per measurement). This is particularly important when sample volume is limited such as in the case of small rodent or pediatric studies. With the exception of some complex matrices, such as tissues, sample cleanup prior to analysis is not required, thereby rendering these assays less time consuming.

Prior to embarking upon analysis by a hybridization method, the key characteristics of the oligonucleotide drug must be thoroughly considered. How long is the oligonucleotide construct? Is it single or double stranded? Is the drug formulated in a delivery vehicle and, if so, how will the oligonucleotide be released into the assay milieu for accurate quantitation? Are the termini free for ensuing reactions with other oligonucleotide constructs in an assay? What types of chemical modifications exist and will they affect duplex stability significantly (i.e., effect on thermal melting temperature; T_m)? Understanding the nature of the oligonucleotide drug will allow one to select the best assay format and design the most suitable probes.

Single-stranded analytes can directly hybridize to their complementary probe, while double-stranded therapeutic oligonucleotides such as siRNA, or oligonucleotides which contain self-hybridizing regions (e.g., aptamers), must first be thermally denatured to expose nucleobases and allow for hybridization with the complementary probes. In the case of duplex siRNA, the generally adopted approach has been to design complementary probes to the pharmacologically active strand of the duplex (i.e., the antisense strand) that is used by the RNA-induced silencing complex (RISC) to degrade the target mRNA.[31] When a thermal denaturation step is applied to unprocessed blood plasma samples, care must be taken to minimize the denaturation temperature (i.e., below 70°C) or reduce the proportion of plasma in solution when heating at higher temperatures. This is to avoid precipitation of plasma components (e.g., fibrin) into the hybridization mixture, thereby making the sample difficult to handle and introducing nonspecific interference in the assay.

The oligonucleotide analyte is first immobilized to a microtitre plate by way of a complementary capture probe and quantified following hybridization with a complementary detection probe. Nucleobase pairing between probe and analyte can take place either in solution (i.e., homogenous assay) or on a solid phase support using a pre-capture probe coated microplate (i.e., heterogeneous assay). Immobilization labels that have been applied to published and routinely used hybridization methods include biotin and primary amines. A biotin-labeled capture probe, for example, can be immobilized to a precoated avidin plate surface. Alternatively, a probe containing a primary amine can covalently bind to a commercial DNA-Bind™ plate surface, which has been precoated with N-oxysuccinimide (NOS). Biotinylated probes are the preferred immobilization label. The binding between biotin-streptavidin is regarded as the strongest, noncovalent, biological interaction, with a dissociation constant in the femtomolar range.[32] This interaction is resistant to changes in pH and temperature. A biotinylated probe can therefore be tethered to a streptavidin-coated plate surface very rapidly with high affinity and specificity after a relatively short incubation period (30 min to 1 hour). In contrast, coupling of an aminated probe to a DNA-Bind surface requires overnight incubation (12–24 hours). In order to minimize competitive amide bond formation with primary amines on nucleotide bases to the NOS-coated plate surface, incubation with the aminated oligonucleotide probe is generally perfomed at low temperatures (approximately 4°C), and is followed by blocking with a bovine serum albumin solution (3% w/v). Detection of the amount of surface bound drug is measured through a label that is either directly or indirectly bound to the terminus of the detecting probe. A direct label may include a fluorescent or radiolabeled tag, while an indirect label requires a secondary high-affinity binding motif (e.g., antibody or protein) that is conjugated to a reporter enzyme. Indirect labels are preferred over direct labels owing to the fact that the latter may lose significant activity upon storage, thereby reducing the sensitivity of the assay.

The decision as to which substrate is best for any type of assay depends on the sensitivity desired, the timing requirements, and the detection device to be used. The conjugated enzyme reacts with a substrate either through enzymatic oxidation or hydrolysis, which catalyzes the formation of a measurable product, typically in a colorimetric or chemifluorescent reaction. Both horseradish peroxidase (HRP) and alkaline phosphatase (AP) are commonly used enzymes as there are a variety of commercially available substrates for the generation of either colorimetric or fluorescent products using an endpoint analysis. The HRP enzyme system is less expensive compared to AP; however, it cannot be used in combination with sodium azide, which might be used to reduce microbial contamination in many biological buffers as it inactivates peroxidase activity even at low concentrations. Although slightly more expensive than HRP, AP is considered more stable. Methods utilizing AP cannot use standard assay phosphate buffered saline (PBS) solutions, as the inorganic phosphate can act as a pseudosubstrate for AP and effectively reduce its specific activity with the substrate. Fluorogenic substrates have a tighter binding affinity for their enzyme target compared to colorimetric substrates. As such, fluorescent-based assays can provide a wider dynamic range, enhanced sensitivity, and overall precision compared to methods that utilize colorimetric detection. This enhancement in sensitivity has been reported to be two- to fivefold when compared with assays that utilize TMB as substrate.[33]

Clear plates made from polystyrene offer the most well-defined binding characteristics and are ideal for colorimetric assays. However, for fluorescent assays, opaque black plates are preferred as they reduce background due to autofluorescence, decrease well-to-well cross talk, and enhance sensitivity by reducing light scatter. Typically, employed detection enzymes along with a list of various substrates are included in Table 8.1. Customarily, the generated absorbance or fluorescent signal is directly proportional to the amount of surface bound analyte, with the exception of competitive hybridization assays, whereby signal intensity is indirectly proportional to the amount of analyte present in a sample (see Section 8.2.4). In the case of double-stranded oligonucleotides (e.g., siRNA) when only one strand is measured in the assay, the detected signal can be related back proportionally to the amount of double-stranded analyte on a molar basis (based on a 1:1 ratio of each sense and antisense strand).

TABLE 8.1
Commonly Used Detection Enzyme Conjugates and Substrates in Hybridization-Based ELISAs[a]

Enzyme Conjugate	Binding Label	Substrate	Signal
Streptavidin-HRP	Biotin	TMB	Absorbance
		ABTS	Absorbance
		OPD	Absorbance
		QuantaBlu™	Fluorescence
		Amplex® Red	Fluorescence
		HPA	Fluorescence
Anti-Digoxigenin-AP	Digoxigenin	p-NPP	Absorbance
		Attophos®	Fluorescence
		4-MUP	Fluorescence

[a] ABTS = 2,2′-azino-di [3-ethylbenzthiazoline sulfonate]; AP = alkaline phosphatase; HPA = p-hydroxyphenylacetic acid; HRP = horseradish peroxidase; 4-MUP = 4-Methylumbelliferyl phosphate; p-NPP = p-nitrophenyl phosphate; OPD = O-phenylenediamine dihydrochloride, TMB = 3,3′,5,5′-tetramethlybenzidine.

The sensitivity or lower limit of quantitation (LLOQ) of the hybridization assay can be optimized, depending on the requirements of a study, particularly when minimal systemic exposure is expected (e.g., inhalation or local injection as routes of administration), or if little sample is available. Various conditions such as probe and enzyme conjugate titer can be optimized to enhance hybridization assay complex formation. Such optimization will directly impact upon the signal-to-noise ratio in the assay, whereby the noise is derived mainly from nonspecifically bound proteins in the plasma or tissue matrices within the hybridization mixture. The hybridization reaction depends on the ability of the oligonucleotide analyte to bind with its complementary probe in a milieu that is slightly below their duplex thermal melting temperature.[34] The introduction of RNA analogues in the oligonucleotide probe sequence can enhance the hybridization assay sensitivity and selectivity for its target analyte by increasing the hybridization assay kinetics.[35] Nucleic acid analogues such as locked nucleic acid (LNA) contain a rigid bicyclic structure (C2′–C4′ methylene bridge) that confers stability to the hybridized probe duplex by improving nucleobase stacking interactions, increasing the duplex T_m and enhancing specificity to the complementary test analyte.[36] The enhanced probe specificity minimizes competitive re-annealing of double-stranded drugs, such as siRNA and decoy oligonucleotides, thus enhancing hybridization specificity between the quantifiable strand and its complementary probe.

Depending on the stringency of the hybridization conditions, duplexes can be formed between perfectly matched sequences, and also between mismatched base pairs. The hybridization stability of the latter is greatly reduced with every mismatched nucleotide (T_m decreases approximately 1°C per percent base mismatch). Stringency of hybridization can be enhanced by increasing the hybridization temperature, thereby reducing duplexes formed as a result of mismatched base pairing. Washing the microtitre plate with a physiologic buffer compatible with the detection system such as Tris-buffered saline (TBS) or phosphate-buffered saline (not compatible with phosphatase detection enzymes) in between hybridization steps can also remove any nonspecifically associated material. The stringency of the wash can be increased by lowering the salt concentration and increasing the wash temperature. This, in turn, improves the sensitivity of the assay by lowering the background signal.[34] The addition of detergents such as Tween-20 is also beneficial as they aid in the removal of any loosely bound protein and also acts as a hydrophobic blocking reagent to block sites on the surface that may become available due to protein desorption during wash steps. A concentration of 0.01–0.03% is recommended. The goal is to remove loosely bound protein without stripping off the specifically bound interactions or inactivating enzymes, which could occur if detergent concentrations are too high. To enhance precision, equal volumes of wash solutions should be dispensed into each well of the microplate with equal force. Automated washes are preferred over uncontrolled manual washing. The volume of wash solution should be high enough to cover entire surface area coated with oligonucleotide complex. Optimization of the number of wash cycles should be determined through experimentation to ensure removal of residual unbound protein without compromising hybridization efficiency. Less than three washes generally leaves residual unbound protein in wells while more than five washes risks unwanted protein desorption from blocked sites. Stringent and optimized washing is also necessary to remove any residual excess probe thereby minimizing background readings.

One-step hybridization assays without washing in between hybridization steps are generally avoided due to a high-concentration hook or prozone phenomenon. This phenomenon is inherent to assays where the reagent (i.e., probe) is concentration limited in comparison to the analyte which is in excess in the microplate.[37] This is apparent when increasing drug concentrations are measured with sample dilution in comparison to the same sample when analyzed neat (or undiluted). The addition of wash steps in between binding events can therefore improve the accuracy of the assay. The capture probe titer is also particularly important, given that under high concentrations of probe, steric hindrance at the level of the detection system can also result in a lower signal. Excess reference material and the inability of the detecting conjugate to effectively bind under these conditions can yield a prozone-like effect, which can be removed by reducing the total number of surface-bound probes on a plate, thus allowing the detection system to proceed more efficiently. In designing complementary probes, it is recommended to include a suitable linker arm (generally at least a C_{12} or C_{18} spacer arm) between the plate binding moiety and the first nucleotides of the probe in order to minimize steric interactions with the plate and reduce any conformational changes. A suitable spacer arm at the detecting end of the probe is also recommended so as to improve binding to the bulky enzyme conjugate.

Hybridization assays are easily amenable to automation providing high-throughput capabilities. They require low sample volume for oligonucleotide quantitation. The complementary probe can bind with high affinity with minimal interference and specificity, generally without the need for sample extraction. This has enabled the ability to develop ultra-sensitive hybridization assays with good accuracy and precision without the addition of an internal standard.

Depending on the particular assay format developed for bioanalysis, cross-reactivity of truncated sequences (i.e., metabolites produced by exo- and/or endonuclease hydrolysis) of an oligonucleotide analyte can be measured in addition to full length material and can result in an overestimation of the parent oligonucleotide being measured. As such, the specificity of the assay used for quantifying metabolites in the presence of full length parent oligonucleotide should be measured as part of the method development process. The general and critical features of hybridization-based ELISAs are discussed throughout this chapter, along with a comparison of the relative advantages and limitations of each assay format. The various hybridization-ELISA assay formats which are routinely used for oligonucleotide drug quantitation are detailed in Figures 8.1 to 8.4.

8.2.1 Ligation-Dependent Hybridization ELISA

This assay format has a reported sensitivity in the picomolar range, making it an ideal method to quantitatively determine oligonucleotides in biological matrices, particularly at the terminal elimination phase, thus making it easy to correlate tissue concentrations in support of preclinical and

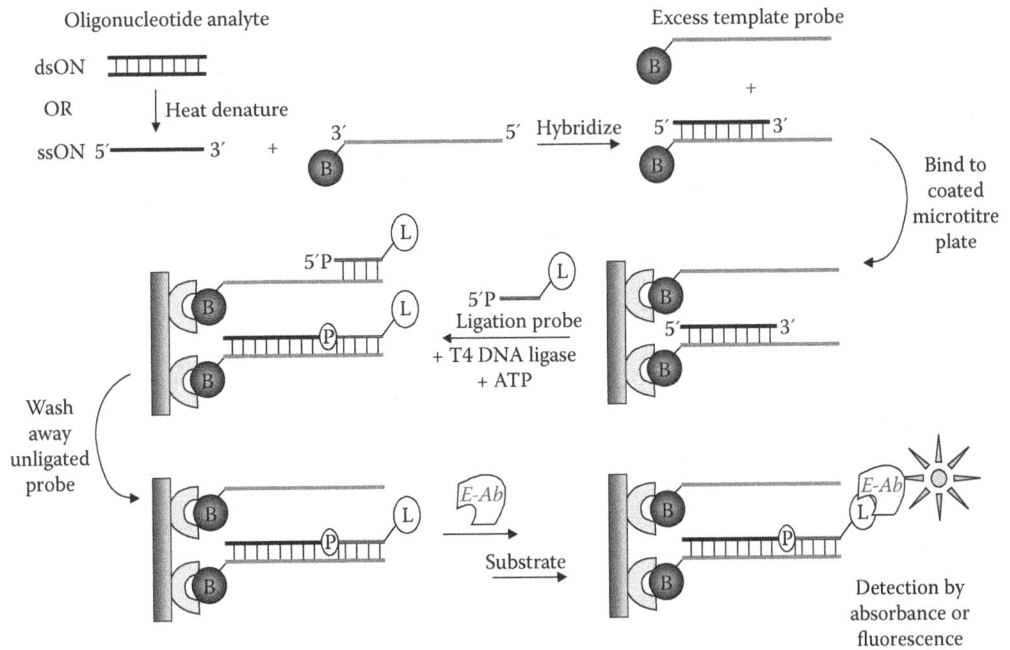

FIGURE 8.1 Representation of the heterogeneous noncompetitive ELISA/ligation assay for quantitation of single- or double-stranded oligonucleotide analytes. Ligation will only occur in the presence of an intact 3′-OH group on the analyte juxtaposed beside the 5′-phosphate group on the ligation probe. B, plate binding motif on template probe; L, indirect label motif on label probe; P, phosphate; circled P, new phosphodiester bond; E-Ab, enzyme-antibody conjugate used to catalyze substrate reaction.

clinical studies in support of regulatory submissions. The ligation-dependent hybridization assay, developed and patented by Isis Pharmaceuticals, is highly selective for parent analyte owing to its assay format which requires an intact 3′-end of the analyte for full ligation to occur.[38,39] The 5′-truncated sequences will, however, be detected in the assay. Nonetheless, metabolite formation *in vivo* generally occurs predominantly from the 3′-end via 3′→5′ exonuclease hydrolysis.[40,41] This assay format is easily amenable to quantitation of most classes of oligonucleotides including first- and second-generation ASO compounds, such as siRNA, and LNA, and is limited only by the availability of a 3′-hydroxyl group as a substrate for the T4 DNA ligase enzyme used in the reaction. It is not an ideal method for bioanalysis of aptamers given the post-SELEX stabilization performed on the molecules, which generally include the addition of an inverted deoxythymidine (idT) cap through a terminal 3′-3′ linkage or other 3′-modifications such as low and high molecular weight PEG or biotinylation.[42] This also includes synthetic CpG oligonucleotides that require accessible 5′-ends to elicit potent immunostimulatory effects. Following synthesis on solid support, two identical oligonucleotide segments are associated via their 3′-ends using a diDMT-glyceryl-linker, making them inaccessible for full ligation reaction and detection.[43]

The method is a heterogeneous noncompetitive assay and involves a hybridization/ligation reaction between complementary oligonucleotides, with an enzyme-linked immunoassay (ELISA) used for detection (Figure 8.1). The oligonucleotide test article is first denatured, as required, and hybridized to a complementary probe, generally referred to as the template probe. The template probe consists of an oligodeoxynucleotide sequence complementary to the single strand of the analyte and also contains a 9-deoxynucleotide generic overhang on its 5′-terminus and a surface capturing label on its 3′-end. The resulting double-stranded complex formed from the hybridization of the

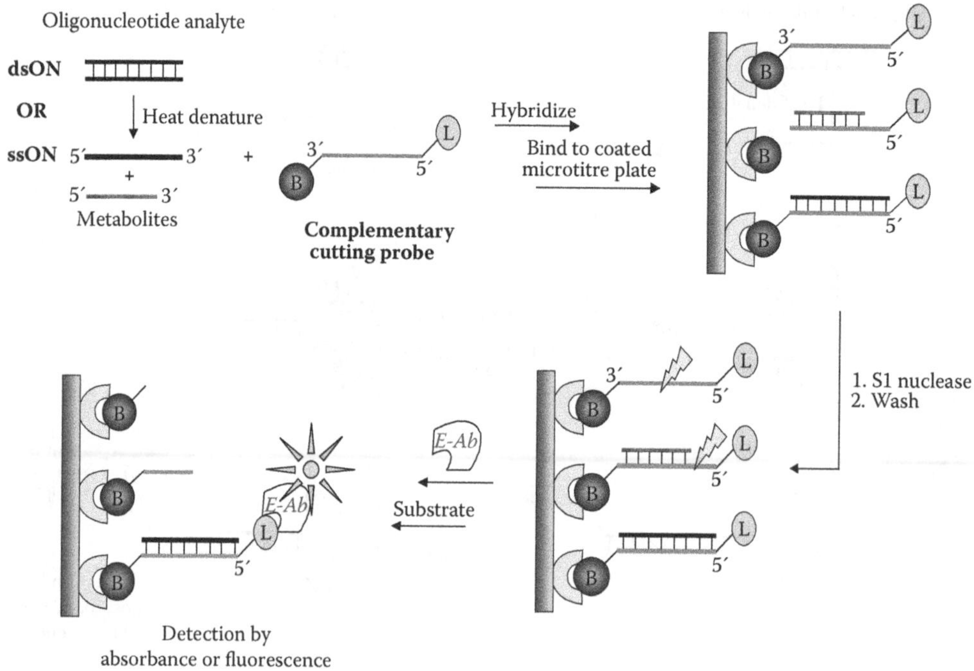

FIGURE 8.2 Representation of the noncompetitive nuclease-dependent hybridization assay for quantitation of single- or double-stranded oligonucleotide analytes. S1 Nuclease will nonspecifically cleave exposed single-stranded regions within the oligonucleotide thereby eradicating the label moiety. B, plate binding motif on cutting probe; L, indirect label motif on cutting probe; E-Ab, enzyme-antibody conjugate used to catalyze substrate reaction.

template probe with the single strand of analyte is captured onto the surface of a coated microtitre plate. This is followed by the addition of a 9-nucleotide ligation/signaling probe, complementary to the 5′-overhang of the template probe.

The ligation probe consists of an indirect label on its 3′-end and a 5′-phosphate group. The ligation reaction ensues on solid support by way of T4 DNA ligase which serves to bind the 5′-phosphate of the ligation probe onto the juxtaposed and intact 3′-end of the oligonucleotide test article, thereby producing a new phosphodiester linkage. Given that the binding affinity of the ligated test article is much higher than that of the ligation probe:template probe, any residual labeled ligation probe is effectively eliminated by stringent washing of the plate while leaving the ligated product intact. Measurement of the amount of incorporated indirect label is performed by subsequent reaction with a conjugated enzyme and substrate, thereby catalyzing the formation of a colorimetric or fluorescent product that is quantified on a microplate reader. In the initial report by Yu et al., a biotin label was used to tether the 3′-end of a template probe to a streptavidin coated microtitre plate and a digoxigenin-labeled ligation probe used for detection. This was followed by reaction with an anti-digoxigenin antibody conjugated to AP, thereby catalyzing the fluorescence of an AttoPhos substrate. An LLOQ of 50 pM was achieved for measurement of a phosphorothioate antisense oligonucleotide, with a 40-fold dynamic curve range. The assay was selective for full length antisense oligonucleotide with minimal observed cross-reactivity of truncated sequences generated from the 3′-end (<0.22%) as a result of the efficiency of T4 DNA ligase being very low in the presence of a gap as small as 1 nucleotide.[44] On the other hand, the quantitation of chain-shortened metabolites generated from the 3′-end is limited with this assay format given that the T4 DNA ligase enzyme

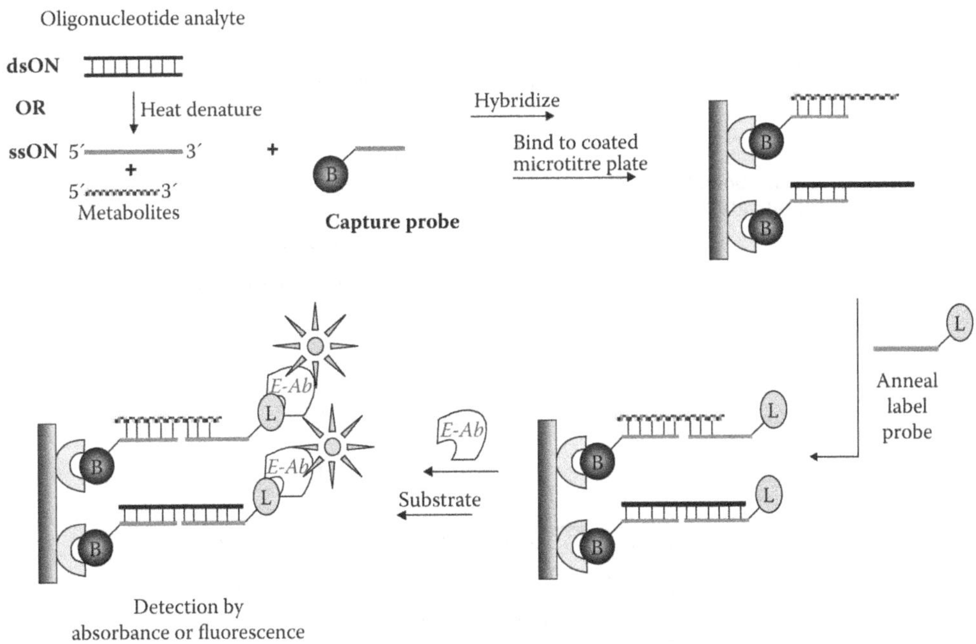

FIGURE 8.3 Representation of the noncompetitive dual-hybridization assay for quantitation of single- or double-stranded oligonucleotide analyte. Depending on the thermal stability of the capture and label probes, short-chained metabolite species may be quantified in the assay B, plate binding motif on capture probe; L, indirect label motif on label probe; E-Ab, enzyme-antibody conjugate used to catalyze substrate reaction.

depends on the proximity of the 3′-hydroxyl group of the analyte and the 5′-phosphate group of the ligation probe. It is possible to customize the assay to detect truncated sequences generated at the 3′-end by designing template probes which are complementary to identified metabolite sequences. In contrast, any metabolites generated from the 5′-end can participate with the assay because ligation can still occur as the 3′-end is intact, and may potentially result in an overestimation of parent drug.[45]

Assay sensitivity can be optimized through the introduction of LNA nucleotides within the region of the template probe that is complementary to the test oligonucleotide. Critical design must be used when incorporating LNA into the template probe to avoid self-annealing as a result of the enhanced hybridization and exceptional biostability of LNA. As such, for template probes with complementary regions greater than 15-nucleotides, it is recommended that the percentage of total LNA be reduced, and interspersed LNA/DNA bases be incorporated. The OligoDesign software provided by Exiqon can calculate the potential for self-hybridization within the oligonucleotide strand and is accessible for free at the following Web site:[45] http://lnatools.com/.

8.2.2 Nuclease-Dependent Hybridization ELISA

This assay format, also developed and patented by Isis Pharmaceuticals, accomplishes both immobilization and signaling in one single step by using a complementary probe (i.e., cutting probe) that spans the entire sequence of the oligonucleotide analyte (Figure 8.2).[46] The cutting probe is designed to incorporate a surface-capturing label at the 3′-end and a detectable marker at the 5′-terminus. The cutting probe and oligonucleotide test article of interest are pre-annealed and then tethered to a

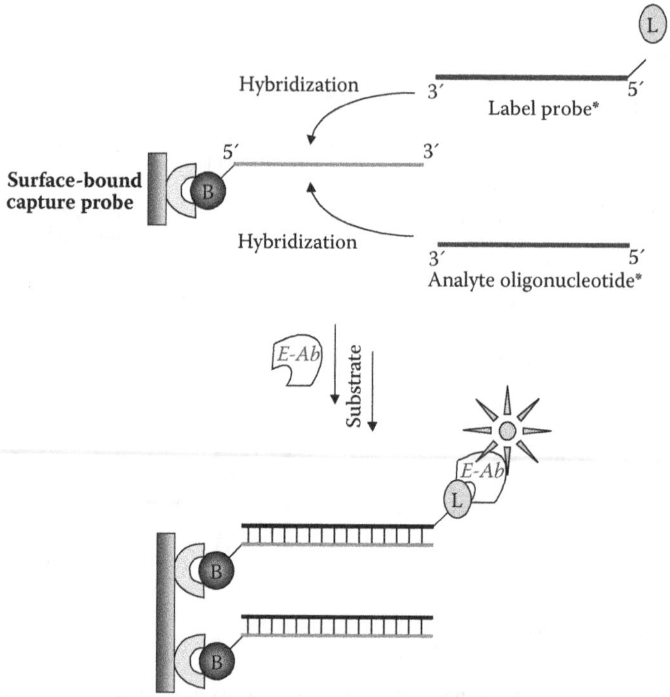

FIGURE 8.4 Representation of the competitive hybridization assay for quantitation of single oligonucleotide analytes. The sequences of the test analyte and the label probe are both directly complementary to the capture probe. As the concentration of test analyte increases, in relation to the label probe, the amount of surface captured label probe will decrease. This results in a signal that is inversely proportional to the amount of analyte present in the biological sample. B, Plate binding motif on capture probe; L, indirect label motif on label probe; E-Ab, enzyme-antibody conjugate used to catalyze substrate reaction.

coated 96-well plate by way of the label on the cutting probe's 3′-end. Once the hybridized duplex is captured to the plate, detection can occur on the opposing end of the same probe. This assay format is referred to as the nuclease-dependent hybridization ELISA owing to the fact that an S1 nuclease is utilized to enhance its selectivity for full length analyte. Any un-hybridized probe or weakly associated complexes resulting from mismatched base pairing can be cleaved by this single-strand–specific nuclease enzyme.[47] Nonspecific enzymatic cleavage of double-stranded material can occur at excess concentrations of enzyme. An optimal enzyme titer should therefore be established to avoid nonspecific cleavage of hybridized probe and optimal selectivity for parent oligonucleotide. Theoretically, the nuclease-dependent hybridization assay can be used to quantify all classes of oligonucleotides provided that the complementary probe designed includes nucleotides that are substrates for the S1 nuclease enzyme.

The key advantage of the cutting assay is that full length analyte remains intact and is detected following endonuclease digestion and any unbound probe or nonfully matched duplexes are degraded. Therefore, the signal generated is proportional to the amount of parent analyte hybridized to probe. However, in practice, truncated sequences can also participate with this assay owing to the fact that the enzyme only partially degrades nicked hybrids.[48] In order to circumvent this, Wei X et al. developed a two-step cutting method to improve the assays selectivity for full length analyte. This method utilized a cutting probe consisting of a 9-nucleotide overhang on its 5′-end and a biotin label on its 3′-end. Hybridization proceeded in solution, followed by immobilization of the duplex to a neutravidin-coated microtitre plate. A detection probe, complementary to the 5′-overhang of the cutting probe, contained a digoxigenin label on its 3′-end for detection. Similar to the concept of a ligation-dependent hybridization assay, the detection probe contained a 5′-phosphate group that could be ligated to an intact 3′-hydroxyl end of an analyte in the presence of T4 DNA ligase and ATP. Following incubation with S1 nuclease and a stringent wash, any unoccupied cutting probe or hybridized truncated sequences would be removed. The selectivity of the assay for full length analyte was improved by reverting to the two-step method as the cross-reactivity of 3′N-1 sequence decreased from 26 to 5.5%.[49]

8.2.3 Dual-Probe Hybridization ELISA

The dual-probe hybridization assay, also referred to as the "sandwich" hybridization assay, employs the most straightforward methodology of all the hybridization assays. It involves a hybridization reaction between a capture probe designed to target one portion of an oligonucleotide analyte and a detection probe that is complementary to the remaining portion of the analyte (Figure 8.3). Other than aptameric oligonucleotides which are commonly greater than 30 nucleotides in length, therapeutic oligonucleotides generally range from 16 to 30 nucleotides. As a result, the capture and detection probes are relatively short in length and require LNA modifications in their nucleic acid sequence to enhance the stability and hybridization specificity.[50] The probe sequences must first be verified for their potential to self-anneal to one another, as this would interfere with hybridization to the oligonucleotide analyte. The OligoDesign software from Exiqon can calculate the potential for self-hybridization and secondary structures between two oligonucleotide strands (http://lnatools.com/).[51] Depending on the plate surface chemistry being utilized, the plate can either be precoated with the capturing probe (e.g., DNA-Bind plates) or the capturing probe allowed to pre-anneal to the test oligonucleotide prior to plate binding (e.g., neutravidin/streptavidin coated plates). Another complementary oligonucleotide (i.e., detection probe) is able to hybridize to the unoccupied end of the oligonucleotide to form the "sandwich complex" (i.e., capture probe/analyte/detection probe). Measurement of the amount of surface-bound analyte is performed via detection of signaling product generated by the enzyme-labeled detection probe. Fluorescence or absorbance intensity is measured using a plate reader.

The dual-hybridization assay is known for its sensitivity, reported in the femtomolar range, particularly when LNA probes are used. However, its inability to discriminate between full length or chain-shortened analyte limits its extensive use. The measured levels of oligonucleotide are therefore an overestimation of parent or full length analyte making it an ideal bioanalytical tool for quantitation of total oligonucleotide content. This format is well suited for oligonucleotide programs in which the metabolites also exhibit biological activity. As a result of the methodology only being reliant upon straightforward Watson–Crick base pairing between complementary oligonucleotides, it is considered the method of choice when analyzing large oligonucleotide constructs (e.g., aptamers) or highly modified oligonucleotides with complex chemistries.

8.2.4 Competitive Hybridization ELISA

The competitive hybridization assay involves an antisense oligonucleotide analyte and an antisense analogue labeled on either its 3′- or 5′-end (i.e., detection probe), competing for hybridization to a complementary sense oligonucleotide sequence (i.e., capture probe) either in solution or on solid support by immobilizing the bound-capture probe to a microtitre plate (Figure 8.4). Once hybridization to the capture probe has ensued, detection can be achieved with a direct or indirect label following addition of a substrate. The label probe and complementary probe are introduced at fixed concentrations in the reaction. As the concentration of oligonucleotide analyte increases, the amount of competitively bound labeled analogue decreases. As a result, the detector response yields a signal profile (i.e., calibration curve) that is inversely proportional to the amount of oligonucleotide analyte present in a sample. The lower the concentration of the complementary target or label probe, the greater the sensitivity of the method. Although this assay format offers ultra-sensitivity, similar to the dual-hybridization method, both full length and n-truncated sequences can participate with the assay, as both 3′- and 5′-ends of the analyte do not need to be intact for detection.[52]

While the competitive assay format is not used as extensively as the ligation and dual-hybridization assay formats, it is generally employed for measurement of short nucleotide sequences or those oligonucleotide constructs that are not amenable to ligation as a result of having modified 3′-termini. Oligonucleotide therapeutics generally <15 nucleotides cannot be readily analyzed with the dual-ligation method as the corresponding capture and detection probes would be thermally unstable and lack optimal hybridization specificity as a result of their short nucleotide length. In addition, the competitive approach limits the dynamic range of the assay by as much as 25-fold compared with noncompetitive methods. This, in turn, would increase the number of sample dilutions that might be required in order to obtain a reading that is within the calibration range.

8.3 EXPERIMENTAL SETUP

8.3.1 Bioanalysis Using a Nonextraction Procedure

Biological study samples can be measured for oligonucleotide analyte by using a calibration curve prepared with a known amount of reference standard in blank biological matrix. The standard curve regression model should accurately reflect the concentration-response relationship. The unknown concentration of an analyte can be extrapolated from the calibration curve by finding the concentration that produces the same response as that obtained from the unknown sample. The dose-response relationship between standards is generally sigmoidal in shape for immunoassays rather than a linear function. The most commonly used curve fitting equations include the symmetric regression model, four-parameter logistic (4PL) model, or the asymmetric function, five-parameter logistic (5PL) model. The latter curve fit includes a fifth parameter that permits asymmetry to be modeled at the upper and lower asymptotes of the curve, ultimately improving the accuracy of concentration measurement.[53] Quality control (QC) samples (spiked with known amounts of reference standard at the low, mid and high portions of the curve) are loaded onto the 96-well microtitre plate to ensure that the calibration curve is under control and assays meet the predefined acceptance criteria (see Section 8.3.3). These samples should be prepared to accurately mimic the biological study samples and confirm that the calibration curve is acceptable for quantitation of biological study samples. This is particularly important in instances when a surrogate matrix or buffer (which differs from the biological matrix being assessed) is utilized to prepare the calibration curve. Dilution controls (diluted at an equivalent fold with blank matrix as the corresponding study samples) can also be employed to assess the dilution performance of the assay.

Following the administration of a therapeutic oligonucleotide *in vivo*, minimal, or no sample extraction is required for its bioanalysis using hybridization assays in nonviscous matrices such as plasma or urine. Plasma is a complex matrix containing a variety of proteins capable of binding oligonucleotides. Both phosphorothioate (PS) and 2′-O-methoxy ethyl (2′MOE) modifications used in second-generation ASOs are well documented in associating with proteins to improve their systemic biodistribution.[54] In addition, nonspecifically associated proteins can interfere with hybridization and detection events. Heating the reaction mixture to approximately 90°C can denature proteins associated with the analyte. Methods that require heat denaturation of double-stranded oligonucleotides in order to participate with the hybridization assay will also in turn denature nonspecifically associated plasma proteins, thereby enhancing the hybridization specificity with a complementary probe upon renaturation. Urine is generally void of proteins and can be analyzed without extraction. This biological matrix does, however, vary in pH between individuals depending on their acid-base status.[55] Urine samples may require buffering prior to analysis to minimize sample variability. For matrices that are limited in availability (e.g., vitreous humor and bronchoalveolar lavage fluid), a surrogate matrix (e.g., plasma or buffer) can be used to prepare the calibration standards. These samples themselves may require a pretreatment step such as a minimum fold dilution with the surrogate matrix to remove any potential interference between the two media.

Therapeutic oligonucleotides that require a vehicle for delivery to an intracellular target must first be liberated from this encapsulation for quantitation with a hybridization assay. Unmodified siRNA, for example, are susceptible to enzymatic degradation by endogenous nucleases following systemic administration and need to be encapsulated to reach their intracellular target.[56,57] Depending on the delivery vehicle used, simply heating the reaction mixture to denature the siRNA can also disrupt the lipid membrane, rendering the oligonucleotide free to bind to its complementary probe. In some cases, manual disruption alone (e.g., vortexing at high speed) may be sufficient enough to release the oligonucleotide. An anionic detergent such as Triton-X 100 can be added to solubilize most lipid bilayers, and release the oligonucleotide from formulation without the requirement of additional extraction. The release efficiency of the method can be measured by extrapolating formulated samples from a calibration curve prepared with free analyte and verifying the percent recovery. An accurate measurement of siRNA extrapolated from a free or "naked" curve can confirm that the siRNA is completely released. When preparing samples in biological matrix with free oligonucleotide analyte or any test oligonucleotide susceptible to nuclease degradation, care must be taken to avoid potential degradation. In such cases, heat-treated or stabilized matrices may be utilized.

8.3.2 Bioanalysis Using an Extraction Procedure

The quantitation of oligonucleotides in tissue first requires the mincing and homogenization of the tissue in a liquid medium in order to attain a consistent amount of tissue in each sample, thereby ensuring that an accurate amount of oligonucleotide drug is being quantified. Tissues are minced with a scalpel blade, followed by homogenization with a saline buffer. Typically, 200 mg of tissue is homogenized per every milliliter of homogenization buffer; however, amounts can be varied, depending on the tissue type and amounts available. Oligonucleotide purification using liquid-liquid extraction with organic solvents such as phenol, chloroform, and isoamyl alcohol remove organic-soluble impurities, while the aqueous phase containing the purified oligonucleotide drug can be evaporated and reconstituted prior to analysis. Concentrated ammonium hydroxide is used to extract ASOs from tissue and disrupt proteins associated with their PS backbone. In contrast, this basic solution can degrade oligonucleosides, such as siRNA, via alkaline hydrolysis, and therefore water must be used instead as the aqueous phase for extraction of ribonucleic acid (RNA).[58]

More recently, a simple sonication procedure has been adopted for the analysis of tissue samples (Charles River Laboratories, data not published). This method utilizes both sonication and a proteinase K enzyme to break down tissue membranes. This procedure is commonly used in purifying oligonucleotide from various tissue homogenates such as kidney and liver. The amount of homogenized tissue subjected to this procedure should be minimized (i.e., 10–40 mg of tissue per milliliter of homogenization buffer) to avoid interference. An ionic detergent such as sodium dodecyl sulfate (SDS) or a mild anionic detergent such as Triton X-100 can also be added to the hybridization mixture, or directly to the biological sample as part of the homogenization buffer, to disrupt nonspecific binding without compromising the Watson–Crick base pairing between analyte and probe. Alternatively, a liquid:liquid extraction with organic solvents may be necessary for the purification of analyte from much denser tissues such as skin and colon.

8.3.3 Method Development and Validation

Before a bioanalytical method can be used to support toxicology and clinical studies for a therapeutic oligonucleotide, the method must first be shown to be reliable and reproducible. Although no official FDA published guidance document exists for the validation of a bioanalytical method for macromolecules, White Papers jointly generated from the American Association of Pharmaceutical Scientists and U.S. FDA Bioanalytical Workshops exist to provide assistance for the validation of

TABLE 8.2
Method Validation Parameters for Hybridization-Based Assays

Validation Parameter	Reason Assessed	Experiment
Range of calibration	Establish the analyte concentration-response relationship and regression model for consistently fitting a curve to the calibration data	• Blank biological matrix spiked with increasing concentrations of the analyte analyzed in duplicate to obtain a mean result (a minimum of six nonzero calibrators that span the anticipated calibration curve range) • Concentration-response data fitted to the best regression model (generally 4-PL or 5-PL for ligand-binding assays) • Upper (ULOQ) and lower (LLOQ) limits of quantitation of the assay established using interassay precision and accuracy data
Interassay precision and accuracy	Verify the magnitude of random and systematic error associated with repeated measurement of the same homogenous (spiked) sample under specific conditions	• Quality control samples (prepared in identical study sample matrix) spiked at the LLOQ, low, mid, high, and ULOQ concentration levels analyzed n = 3 (in duplicate) on six independent occasions • Method accuracy determined from the percent deviation of the measured sample mean from the theoretical concentration • Precision expressed as %CV (SD/mean)
Selectivity	Verify ability of the assay to measure the analyte of interest in the presence of other constituents in the sample (i.e., nonspecific interference)	Ten different lots of blank biological matrix analyzed neat (unspiked) or spiked with reference standard at pre-determined concentrations
Specificity	Ability to bind the analyte of interest without cross-reactivity with variant forms of the analyte of other structurally related compounds that may be present in a study sample (e.g., truncated oligonucleotide metabolites)	Blank biological matrix spiked with predetermined concentrations of known metabolites to assess the relative cross-reactivity in the assay
Dilution linearity	Demonstrate that high concentrations of analyte (>ULOQ) can be accurately diluted to concentrations within validated range	• High concentration spiked sample in blank biological matrix diluted to multiple concentrations within the calibration curve range to evaluate linearity of dilution • Study sample (parallelism) by serially diluting an incurred sample with blank matrix to evaluate linearity of dilution
Prozone ("hook-effect")	Verify if signal suppression occurs as a result of high concentration of sample being present in a study sample	• A highly concentrated sample (spiked at least 100-fold above the established ULOQ) analyzed without dilution to evaluate prozone or hook effect • Established prozone effect may require multiple dilutions of study samples to ensure accurate results
Stability • In process: room temperature, 4°C, freeze-thaw • Long-term storage (–20° and/or –80°C)	Demonstrate acceptable stability of analyte in sample matrix under conditions of assay processing and storage	• Stability samples prepared at the low and high concentrations by spiking known amounts of analyte in blank biological matrix and subjecting validation samples to similar stability conditions as the corresponding study samples • Percent recovery compared to fresh quality control samples assessed • All stability sample handling should mimic study sample handling

ligand-binding methods, such as hybridization assays, to support investigational new drug applications (INDs).[59,60] The same validation guidelines are used for methods supporting clinical studies. The initial development of a hybridization assay must first address the requirements of the oligonucleotide drug program. Depending on the chemistry and class of oligonucleotide, delivery mechanism (i.e., free or formulated drug), and route of administration, the best assay format can be selected. The probe sequence, reagent concentrations, including probe concentrations, and incubation temperatures can be optimized to enhance the sensitivity and selectivity assay requirements. The reference standard should be well characterized and be identical to the analyte. A certificate of analysis containing batch information, purity, and stability should be provided for each reference standard.

The typical validation parameters for a bioanalytical method include the determination of selectivity, specificity, precision and accuracy, calibration curve range, prozone (or hook effect), and stability of the analyte in the defined biological matrix. These prevalidation parameters are evaluated to assess assay robustness and ensure easy transition of the bioanalytical method to validation in a GLP-compliant environment, thereby reducing the chances of assay re-optimization. Spiked samples should be prepared with a known amount of reference material in the same biological matrix as the anticipated study samples. A surrogate matrix can be used as a substitute in instances where the matrix is limited (e.g., vitreous humor, cerebrospinal fluid) or requires pre-assay extraction (e.g., tissues). In any case, QC samples should be prepared using the same unaltered matrix to evaluate the precision and accuracy of the assay when ever possible. Stability assessments are also conducted in QC samples prepared in the unaltered matrix to accurately reflect the storage conditions of the corresponding bioanalytical study samples. The typical validation parameters assessed for macromolecule ligand binding assays are described in Table 8.2.

The predefined acceptance criteria are established once the prestudy validation parameters are completed. If these criteria are not accurately defined, this may result in frequent assay failures. Once the bioanalytical method has been validated and approved for routine use, the precision and accuracy of the assay should be monitored regularly with the use of acceptance QC samples to ensure assay robustness. These parameters should also be closely examined during sample analysis. The reanalysis of incurred samples during bioanalysis can support assay reproducibility and has become a requirement from regulatory authorities such as the FDA. Incurred samples may behave differently from a blank sample prepared with a known amount of exogenous drug (i.e., spiked sample), in that it may contain additional components that a spiked sample may not. The outcome of incurred sample reanalysis (ISR) defines whether additional parameters need to be investigated or whether the assay can accurately quantify drug concentrations in biological study samples.[61]

8.4 CONCLUSIONS AND SUMMARY

The progress in the development of robust bioanalytical tools for oligonucleotide quantitation has resulted in the emergence of ultra-sensitive methods, selective for full length oligonucleotide analyte, with the ability to identify and measure metabolites. Routinely used chromatographic techniques such as CGE, HPLC, and LC/MS are capable in identifying and quantifying metabolite species in the presence of full length oligonucleotide analyte; yet their extensive sample clean up and limited assay sensitivity has compromised their ability to fully characterize the pharmacokinetic/pharmacodynamic properties of therapeutic oligonucleotides. Alternatively, hybridization-ELISA assays are exceptional in their sensitivity. However, their limited ability to select for parent oligonucleotide in the presence of truncated sequences has resulted in the utilization of a combination of bioanalytical techniques to fully characterize a therapeutic oligonucleotide following systemic administration. Various hybridization assay designs have emerged with enhanced selectivity for full length analyte quantitation, while maintaining the advantages that hybridization-ELISA assays offer. Depending on the type of hybridization assay format, a more specific design can be used. The ligation-based assay, for example, has enhanced selectivity for full length oligonucleotide analyte, particularly at the 3′-end. However, the cross-reactivity of truncated sequences generated from the 5′-end of the oligonucleotide analyte can still participate with this assay. To circumvent this issue, Tremblay GA designed a template probe, complementary to the entire sequence of the oligonucleotide analyte, containing a generic 9-nucleotide overhang on each end. Hybridization ensues in a heterogeneous solution containing oligonucleotide analyte, template probe, and a ligation probe containing a 5′-phosphate group and a biotinylated 3′-end. The biotinylated ligation probe can immobilize the hybridized complex to a streptavidin-coated plate surface. The exposed 5′-hydroxyl end of the analyte is subsequently phosphorylated using a polynucleotide kinase, followed with the addition of a secondary ligation probe containing a 3′-hydroxyl end and a digoxigenin label for detection. The addition of a T4 DNA ligase associates each end of the analyte via a phosphodiester linkage, and as a result, only intact analyte is detected.[62]

Optimization of hybridization-ELISA assays continues to make them exceptional bioanalytical tools for oligonucleotide measurement. As the growing interest in oligonucleotide therapeutics to treat disease targets intensifies, additional advances in oligonucleotide design and delivery will translate in the development of bioanalytical methods tailored for their quantitation. The enhanced bioavailability and specificity of therapeutic oligonucleotides have made them increasingly more potent, therefore requiring ultra-sensitive bioanalytical methods for their quantitation in biological fluids. Second-generation ASO technology and improved siRNA delivery using nanotechnology vehicles have enhanced the pharmacokinetic profiles of these oligonucleotide therapeutics in terms of plasma stability and cellular uptake. Multiple siRNA compounds can be combined in a liposomal formulation to enhance their therapeutic potential and are currently being evaluated in a clinical setting. The emergence of multiple therapeutic targets in preclinical and clinical development has increased the requirement of multiplexing technology coupled with ELISA-based assays, enabling the quantitation of multiple oligonucleotide therapies from a single sample. Simultaneous quantitation of up to 100 analytes can be achieved with Luminex™ xMAP-based technology. An oligonucleotide probe can be immobilized by covalent or affinity based noncovalent attachment to a microsphere that exhibits a discrete fluorescence-encoded intensity, which can be quantified by distinct analyte reporting.[63] Similarly, an electrochemiluminescent (ECL) assay developed by Mesoscale Discovery™, uses a 96-well plate format that allows noncovalent immobilization of up to 10 distinct biotin labeled oligonucleotide probes to a multispot avidin or streptavidin precoated electrode surface. A MSD Sulfo-TAG label composed of ruthenium (II) Tris-bipyridine-(4-methylsulfonate) can produce a quantifiable ECL signal following stimulation of the electrode surface.[64] With immunogenicity testing as a crucial part of biopharmaceutical development, multiplexing arrays for both oligonucleotide and cytokine quantitation could potentially assess both the pharmacokinetic properties as well as off-target effects, such as the immunogenic potential for a therapeutic oligonucleotide analyte.

With the emergence of more selective hybridization assay formats, advances in the application of multiplexing technology for the quantitation of full length analyte and identified metabolites using hybridization-based assays are becoming increasingly important. The expansion of multiplex hybridization assays will permit a more detailed pathway of both the therapeutic target and any potential biological off-target effects. The simplicity, sensitivity, minimal sample volume required, robustness, and accuracy of hybridization-ELISA assays will continue to make them desirable bioanalytical tools that will continue to adapt to the advances in oligonucleotide therapeutic discovery and development.

REFERENCES

1. Crooke, S. T. 2004. Progress in antisense technology. *Annual Review of Medicine* 55: 61–95.
2. Malik, R., and I. Roy. 2008. Design and development of antisense drugs. *Expert Opinion on Drug Discovery* 3(10): 1189–1207.

3. Pappas, T. C., A. G. Bader, B. F. Andruss, D. Brown, and L. P. Ford. 2008. Applying small RNA molecules to the directed treatment of human diseases: realizing the potential. *Expert Opinion on Therapeutic Targets* 12(1): 115–127.
4. Carthew, R. W., and E. J. Sontheimer. 2009. Origins and mechanisms of miRNAs and siRNAs. *Cell* 136(4): 642–655.
5. Castanotto, D., and J. J. Rossi. 2009. The promises and pitfalls of RNA-interference-based therapeutics. *Nature* 457(7228): 426–433.
6. Dorn, A., and S. Kippenberger. 2008. Clinical application of CpG-, non-CpG- and antisense oligodeoxynucleotides as immunomodulators. *Current Opinion in Molecular Therapeutics* 10(1): 10–20.
7. Kaur, G., and I. Roy. 2008. Therapeutic application of aptamers. *Expert Opinion on Investigational New Drugs* 17(1): 47–60.
8. Morishita, R., G. H. Gibbons, M. Horiuchi, K. E. Ellison, M. Nakama, L. Zhang, Y. Kaneda, T. Ogihara, and V. J. Dzau. 1995. A gene therapy strategy using a transcription factor decoy of the E2F binding site inhibits smooth muscle proliferation *in vivo*. *Proceedings of the National Academy of Sciences of the United States of America* 92(13): 5855–5859.
9. Sel, S., W. Henke, A. Dietrich, U. Herz, and H. Renz. 2006. Treatment of allergic asthma by targeting transcription factors using nucleic-acid based technologies. *Current Pharmaceutical Design* 12(25): 3291–3304.
10. Usman, N., and L. M. Blarr. 2000. Nuclease-resistant synthetic ribozymes: developing a new class of therapeutics. *Journal of Clinical Investigation* 106(10): 1197–1202.
11. Isaka, Y. 2007. DNAzymes as potential therapeutic molecules. *Current Opinion in Molecular Therapeutics* 9(2): 132–136.
12. Behlke, M. A. 2008. Chemical modification of siRNAs for *in vivo* use. *Oligonucleotides* 18(4): 305–320.
13. Urban, E., and C. R. Noe. 2003. Structural modifications of antisense oligonucleotides. *Farmaco* 58: 243–258.
14. Faria, M., and H. Ulrich. 2008. Sugar boost: when ribose modifications improve oligonucleotide performance. *Current Opinion in Molecular Therapeutics* 10: 168–175.
15. Prakash, T. P., and B. Bhat. 2007. 2′-Modified oligonucleotides for antisense therapeutics. *Current Topics in Medicinal Chemistry* 7: 641–649.
16. Abes, R., A. A. Arzumanov, H. M. Moulton, S. Abes, G. D. Ivanova, P. L. Iversen, M. J. Gait, and B. Lebleu. 2007. Cell-penetrating-peptide-based delivery of oligonucleotides: an overview. *Biochemical Society Transactions* 35(4): 775–779.
17. Zhao, X., F. Pan, C. M. Holt, A. L. Lewis, and J. R. Lu. 2009. Controlled delivery of antisense oligonucleotides: a brief review of current strategies. *Expert Opinion on Drug Delivery* 6(7): 673–686.
18. Cossum, P. A., H. Sasmor, D. Dellinger, L. Truong, L. Cummins, S. R. Owens, P. M. Markham, J. P. Shea, and S. Crooke. 1993. Disposition of the 14C-labeled phosphorothioate oligonucleotide ISIS 2105 after intravenous administration to rats. *Journal of Pharmacology and Experimental Therapeutics* 267(3): 1181–1190.
19. Leeds, J. M., M. J. Graham, L. Truong, and L. L. Cummins. 1996. Quantitation of phosphorothioate oligonucleotides in plasma. *Analytical Biochemistry* 235(1): 36–43.
20. Chen, S. H., and J. M. Gallo. 1998. Use of capillary electrophoresis methods to characterize the pharmacokinetics of antisense drugs. *Electrophoresis* 19: 2861–2869.
21. Reyderman, L., and S. Stavchansky. 1996. Determination of single-stranded oligodeoxynucleotides by capillary gel electrophoresis with laser-induced fluorescence and on column derivatization. *Journal of Chromatography A* 755(2): 271–280.
22. Willems, A. V., D. L. Deforce, C. H. Van Peteghem, and J. F. Van Bocxlaer. 2005. Analysis of nucleic acid constituents by on-line capillary electrophoresis-mass spectrometry. *Electrophoresis* 26(7–8): 1221–1253.
23. Bigelow, J. C., L. R. Chrin, L. A. Mathews, and J. J. McCormack. 1990. High-performance liquid chromatographic analysis of phosphorothioate analogues of oligodeoxynucleotides in biological fluids. *Journal of Chromatography B: Biomedical Sciences and Applications* 533: 133–140.
24. Chen, S. H., M. Qian, J. M. Brennan, and J. M. Gallo. 1997. Determination of antisense phophorothioate oligonucloetides and catabolites in biological fluids and tissue extracts using anion-exchange high-performance liquid chromatography and capillary gel electrophoresis. *Journal of Chromatography B: Biomedical Sciences and Applications* 692(1): 43–51.
25. Arora, V., D. C. Knapp, M. T. Reddy, D. D. Weller, and P. L. Iversen. 2003. Bioavailability and efficacy of antisense morpholino oligomers targeted to *c-myc* and cytochrome P-450 3A2 following oral administration in rats. *Journal of Pharmaceutical Sciences* 91(4): 1009–1018.

26. Huber, C. G., and H. Oberacher. 2001. Analysis of nucleic acids by on-line liquid chromatography-mass spectrometry. *Mass Spectrometry Reviews* 20(5): 310–343.
27. Lin, Z. J., W. Li, and G. Dai. 2007. Application of LC-MS for quantitative analysis and metabolite identification of therapeutic oligonucleotides. *Journal of Pharmaceutical and Biomedical Analysis* 44(2): 330–341.
28. Zhang, G., J. Lin, K. Srinivasan, O. Kavetskaia, and J. N. Duncan. 2007. Strategies for bioanalysis of an oligonucleotide class macromolecule from rat plasma using liquid chromatography-tandem mass spectrometry. *Analytical Chemistry* 79(9): 3416–3424.
29. Yu, R. Z., K. M. Lemonidis, M. J. Graham, J. E. Matson, R. M. Crook, D. L. Tribble, M. K. Wedel, A. A. Levin, and R. S. Geary. 2009. Cross-species comparison of in vivo PK/PD relationships for second-generation antisense oligonucleotides targeting apolipoprotein B-100. *Biochemical Pharmacology* 77(5): 910–919.
30. Leaman, D. W. 2008. Recent progress in oligonucleotide therapeutics: antisense to aptamers. *Expert Opinion on Drug Discovery* 3(9): 997–1009.
31. Schwarz, D. Z., G. Hutvágner, T. Du, Z. Xu, N. Aronin, and P. D. Zamore. 2003. Asymmetry in the assembly of the RNAi enzyme complex. *Cell* 115(2): 199–208.
32. Green, N. M. 1990. Avidin and streptavidin. *Methods in Enzymology* 184: 51–67.
33. Meng, Y., K. High, J. Antonello, M. W. Washabaugh, and Q. Zhao. Enhanced sensitivity and precision in an enzyme-linked immunosorbent assay with fluorogenic substrates compared with commonly used chromogenic substrates. *Analytical Biochemistry* 345: 227–236.
34. Blackburn, M. G., and M. J. Gait. 1996. *Nucleic Acids in Chemistry and Biology*. New York: Oxford University Press.
35. Braasch, D. A., and D. R. Corey. 2001. Locked nucleic acid (LNA): fine-tuning the recognition of DNA and RNA. *Chemistry and Biology* 8(1): 1–7.
36. You, Y., B. G. Moreira, M. A. Behlke, and R. Owczarzy. 2006. Design of LNA probes that improve mismatch discrimination. *Nucleic Acids Research* 34(8): e60.
37. Butch, A. W. 2000. Dilution protocols for detection of hook effects/prozone phenomenon. *Clinical Chemistry* 46(10): 1719–1720.
38. Yu, R. Z., B. Baker, A. Chappell, R. S. Geary, E. Cheung, and A. A. Levin. 2002. Development of an ultra-sensitive noncompetitive hybridization-ligation enzyme-linked immunosorbent assay for the determination of phosphorothioate oligodeoxynucleotide in plasma. *Analytical Biochemistry* 301(1): 19–25.
39. Baker, B. F., Z. Yu, and J. M. Leed. 2000. Method for quantitating oligonucleotides. U.S. Patent 6,355,438. Isis Pharmaceuticals Inc.
40. Eder, P. S., R. J. Devine, J. M. Dagle, and J. A. Walder. 1991. Substrate specificity and kinetics of degradation of antisense oligonucleotides by a 3′-exonuclease in plasma. *Antisense Research and Development* 1(2): 141–151.
41. Wòjcik, M., M. Cieślak, W. J. Stec, J. W. Goding, and M. Koziolkiewicz. 2007. Nucleotide pyrophosphatase/phosphodiesterase 1 is responsible for degradation of antisense phosphorothioate oligonucleotides. *Oligonucleotides* 17(1): 134–145.
42. Gamper, H. B., M. W. Reed, T. Cox, J. S. Virosco, A. D. Adams, A. A. Gall, J. K. Scholler, and R. B. Meyer, Jr. 1993. Facile preparation of nuclease resistant 3′ modified oligodeoxynucleotides. *Nucleic Acids Research* 21: 145–150.
43. Yu, D., E. R. Kandimalla, L. Bhagat, J. Y. Tang, Y. Cong, J. Tang, and S. Agrawal. 2002. 'Immunomers'-novel 3′-3′- linked CpG oligodeoxyribonucleotides as potent immunomodulatory agents. *Nucleic Acids Research* 30(20): 4460–4469.
44. Goffin, C., V. Bailly, and W. G. Verly. 1987. Nicks 3′ or 5′ to AP sites or to mis-paired bases, and one-nucleotide gaps can be sealed by T4 DNA ligase. *Nucleic Acids Research* 15(21): 8755–8771.
45. Tolstrup, N., P. S. Neilson, J. G. Kolberg, A. M. Frankel, H. Vissing, and S. Kauppinen. 2003. OligoDesign: optimal design of LNA (locked nucleic acid) oligonucleotide capture probes for gene expression profiling. *Nucleic Acids Research* 31(13): 3758–3762.
46. Yu, Z., B. F. Baker, and H. Wu. 2003. Nuclease-based method for detecting and quantifying oligonucleotides. WO/2002/059137. Isis Pharmaceuticals Inc.
47. Desai, N. A., and V. Shankar. 2003. Single-strand-specific nucleases. *FEMS Microbiology Reviews* 26(5): 457–491.
48. Shenk, T. E., C. Rhodes, P. W. J. Rigby, and P. Berg. 1975. Biochemical method for mapping mutational alterations in DNA with S1 nuclease: the location of deletions and temperature-sensitive mutations in simian virus 40. *Proceedings on the National Academy of Sciences* 72(3): 989–993.

49. Wei, X., G. Dai, G. Marcucci, Z. Liu, D. Hoyt, W. Blum, and K. K. Chan. 2006. A specific picomolar hybridization-based ELISA assay for the determination of phosphorothioate oligonucleotides in plasma and cellular matrices. *Pharmaceutical Research* 23(6): 1251–1264.
50. Efler, S. M., L. Zhang, B. O. Noll, E. Uhlmann, and H. L. Davis. 2005. Quantitation of oligonucleotides in human plasma with a novel hybridization assay offers greatly enhanced sensitivity over capillary gel electrophoresis. *Oligonucleotides* 15: 119–131.
51. Tolstrup, N., P. S. Neilson, J. G. Kolberg, A. M. Frankel, H. Vissing, and S. Kauppinen. 2003. OligoDesign: optimal design of LNA (locked nucleic acid) oligonucleotide capture probes for gene expression profiling. *Nucleic Acids Research* 31(13): 3758–3762.
52. Deverre, J. R., V. Boutet, E. Ezan, J. Grassi, and J. M. Grognet. 1997. A competitive enzyme hybridization assay for plasma determination of phosphodiester and phosphorothioate antisense oligonucleotides. *Nucleic Acids Research* 25(18): 3584–3589.
53. Gottschalk, P. G., and J. R. Dunn. 2005. The five-parameter logistic: A characterization and comparison with the four-parameter logistic. *Analytical Biochemistry* 343: 54–65.
54. Crooke, S. T. 2008. *Antisense Drug Technology Principles, Strategies, and Applications.* New York: CRC Press.
55. Pitts, R. F., W. D. Lotspeich, W. A. Schiess, J. L. Ayer, and P. Miner. 1948. The renal regulation of acid base balance in man I. The nature of the mechanism for acidifying the urine. *Journal of Clinical Investigation* 27(1): 48–56.
56. De Fougerolles, A. R. 2008. Delivery vehicles for small interfering RNA *in vivo. Human Gene Therapy* 19: 125–132.
57. Whitehead, K. A., R. Langer, and D. G. Anderson. 2009. Knocking down barriers: advances in siRNA delivery. *Nature Reviews* 8: 129–138.
58. Perreault, D. M., and E. V. Anslyn. 1997. Unifying the current data on the mechanism of cleavage—transesterification of RNA. *Angewandte Chemie (International Edition in English)* 36: 432–450.
59. U.S. Department of Health and Human Services, Food and Drug Administration, Center for Drug Evaluation and Research (CDER), Center for Veterinary Medicine (CVM). 2001. *Guidance for Industry: Bioanalytical Method Validation.*
60. Kelley, M., and B. DeSilva. 2007. Key elements of bioanalytical method validation for macromolecules. *AAPS Journal* 9(2): E156–E163.
61. Rocci, M. L., Jr., V. Devanarayan, D. B. Haughley, and P. Jardieu. 2007. Confirmatory reanalysis of incurred bioanalytical samples. *AAPS Journal* 9(3): E336–E343.
62. Tremblay, G. A., P. R. Oldfield, and A. J. Bartlett. 2009. A parent-specific hybridization assay for quantifying therapeutic oligonucleotides and siRNA in biological samples. Poster. Annual Meeting and Exposition of the American Association of Pharmaceutical Scientists (AAPS), Los Angeles, California, Nov. 8–12 (Patent Pending No. US 61/258,046).
63. Nolan, J. P., and F. Mandy. 2006. Multiplexed and Microparticle-based Analyses: Quantitative Tools for the Large-Scale Analysis of Biological Systems. *Cytometry* 69(5): 318–325.
64. Gugliermo-Viret, V., and P. Thullier. 2007. Comparison of an electrochemiluminescence assay in plate format over a colorimetric ELISA, for the detection of ricin B chain (RCA-B). *Journal of Immunological Methods* 328: 70–78.

9 Oligonucleotide Assay and Potency

Dennis P. Michaud
Avecia Biotechnology, Inc.

CONTENTS

9.1 INTRODUCTION

Assay is one of the fundamental tests for determining the active ingredient content of a sample. As a general term, assay has come to have various shades of meaning, from simply a synonym for an analytical test to complex extraction procedures to measure the therapeutic levels of a drug in biological matrices. Although there is little regulatory guidance on the definition of assay or how to measure it, assay is mentioned in ICH Harmonised Tripartite Guideline, "Validation of Analytical Procedures: Text and Methodology Q2 (R1),"[1] which outlines three major types of analytical tests:

1. *Identification tests* to ensure the identity of an analyte in a sample (e.g., molecular weight measurements and sequencing);
2. *Impurity testing* as either a quantitative test or a limit test to accurately reflect the purity characteristics in a sample (product- and process-related impurities in oligonucleotides);
3. *Assay* to quantitatively measure the major component(s) in a drug substance.

Assay has been used interchangeably with content, strength (the concentration of a drug substance), and potency (therapeutic activity as indicated by label claim). Assay has also been considered synonymous with purity. However, a distinction between these two terms is implied in ICH Q2(R1) and the term purity should be reserved for the control of impurities through specification. Assay, as defined in ICH Q2(R1), is the subject of this chapter.

9.2 COMPARISON WITH SMALL MOLECULE ASSAY

The standard references such as the U.S. Pharmacopeia (USP)[2] and European Pharmacopoeia (EP)[3] contain many examples of assay of drugs. The procedures generally require the chromatographic comparison of a sample preparation with a reference standard preparation of known assay. For example, using the peak areas and the weights of the sample (S) and reference standard (R), the assay can be calculated with the following typical equation:

$$\text{Assay}_\text{S} = \frac{\text{Area}_\text{S} \times Wt_\text{R}}{\text{Area}_\text{R} \times Wt_S} \times \text{Assay}_\text{R} \tag{9.1}$$

This equation is an example of the general equation,

$$Z_\text{S} = \frac{Q_\text{S}}{Q_\text{R}} \times Z_\text{R} \tag{9.2}$$

in which the ratio of some measurable quantities (Q) for a reference and sample are multiplied by a known attribute (Z) of the reference, such as assay, purity, or concentration to give the corresponding attribute for the sample.

This approach to assay calculation works well for small molecules where highly purified reference standards are generally available. However, with current manufacturing technology, similar standards of high purity for oligonucleotides are not available. Because of the iterative nature of oligonucleotide synthesis, a number of sequence-related failure sequence impurities are always present in oligonucleotides as well as nonoligonucleotidic components, such as the counterion and moisture. The purity of an oligonucleotide is dependent on the selectivity of the analytical method used, common methods being capillary gel electrophoresis, ion-exchange HPLC, and reversed phase ion-pair HPLC. In addition, the electrostatic nature of the sample, likely owing to the ionic phosphate backbone, as well as the tendency of oligonucleotides to be hygroscopic, can make accurate weighing of these materials difficult. These considerations can make the assay of oligonucleotides less straightforward than small molecule assay.

The approaches to assay determination outlined in this chapter are presented using a generic DNA phosphorothioate oligonucleotide.[4] However, the principles developed may be applied to other oligonucleotides as well.

9.3 POTENTIAL COMPONENTS IN AN OLIGONUCLEOTIDE SAMPLE

Oligonucleotide active pharmaceutical ingredients (API) potentially contain the following types of components/impurities:

1. Oligonucleotide product, sequence-related process impurities, degradation products
2. Moisture content and counterion (typically sodium)
3. Reagents, by-products from manufacturing, residual solvents, and metals

In good process control, the items in the third category are present at part per million levels and have no significant contribution to the assay calculation. Many of these components are removed during manufacture (starting materials and reagents, synthesis by-products) or are present at relatively low levels (residual solvents, metals). Thus the major components that could be considered in assay calculations for API are the oligonucleotide and its structurally related impurities, water, and

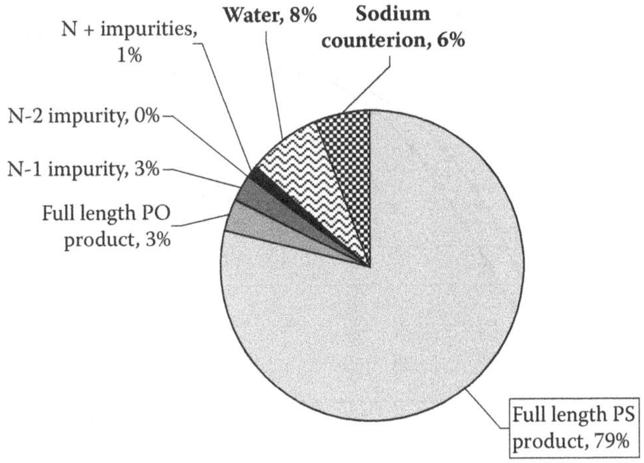

FIGURE 9.1 Typical components in a DNA phosphorothioate oligonucleotide. In a drug product, the excipients may also play a role in the assay calculations.

the sodium counterion. This is conveniently represented in the pie chart (Figure 9.1) for a generic DNA phosphorothioate oligonucleotide.

It is critical to define initially what constitutes the oligonucleotide API. Oligonucleotides are generally manufactured in the hydrated sodium salt form. Because the moisture content of an oligonucleotide sample is variable, many consider the anhydrous sodium salt form as the actual API. Still others consider the free acid form (the UV absorbing portion) of the oligonucleotide to be the API. Whichever form is chosen, the assay calculation will be with regard to that form and the assay value should be clearly labeled with the form addressed. Although there is no clear regulatory guidance in this matter, assay has usually been reported as the sodium salt on an anhydrous basis.[5] It is important to be consistent about reporting the form throughout a drug program. In this chapter, all three forms (hydrated sodium salt, anhydrous sodium salt, free acid) will be considered in the calculations.

9.4 HYGROSCOPICITY OF OLIGONUCLEOTIDES

Oligonucleotides have a tendency to take up or release water, depending on the surrounding relative humidity.[6] This hygroscopicity of oligonucleotides can be demonstrated by the careful measurement of weight gain or loss as a function of the relative humidity, as shown in Figure 9.2.

The contribution of moisture to the weight of a sample must be determined and accounted for in all analytical tests requiring w/w% calculations, such as extinction coefficient measurements, sodium determination, and assay calculations. Although a glove box may be used as a way to control humidity during handling of oligonucleotide samples, a particularly effective technique is to "condition" the sample at ambient temperature and humidity. A typical protocol for obtaining and using conditioned moisture values is as follows:

1. Measure initial moisture of bulk sample for Certificate of Analysis.
2. Condition sufficient sample at ambient temperature/relative humidity for 1–2 hours.
3. Measure the conditioned moisture of the sample using Karl Fischer (KF) analysis.
4. Prepare solutions for assay analysis and other applications requiring a moisture correction.
5. Measure the conditioned moisture of the sample using KF analysis.
6. Use average of the conditioned moisture values in calculations requiring a moisture correction.

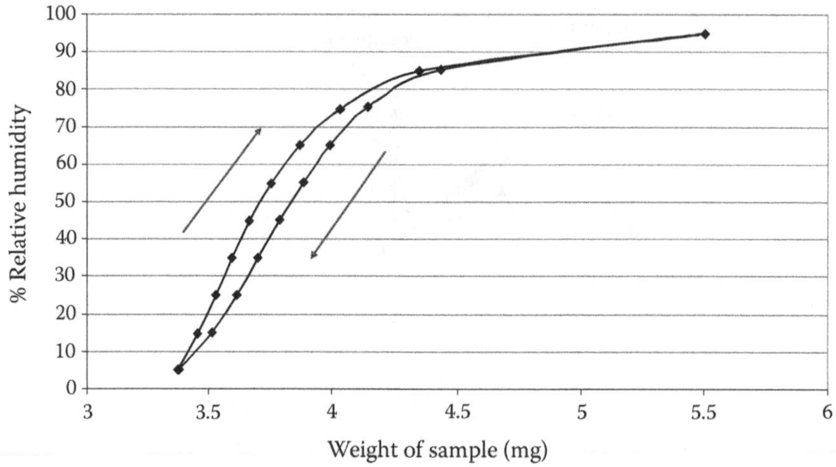

FIGURE 9.2 Hygroscopicity of a typical oligonucleotide.

Figure 9.3 illustrates how the moisture varies during the conditioning process of an oligonucleotide sample. In this example the sample initially contained 8% moisture. After one hour of conditioning at ambient temperature and relative humidity, the moisture of the sample increased to about 14% and maintained this hydration level for these conditions. Conditioned moisture treatment allows the oligonucleotide to be handled for analysis in a stable way. It is more important to have an accurate measure of the moisture content of an oligonucleotide than to attempt to lower the moisture value for sample preparation.

9.5 THE EXTINCTION COEFFICIENT

The extinction coefficient is an important constant in the quantitation of oligonucleotide solutions.[7] Because some assay methods require the use of an extinction coefficient, a brief introduction is provided here, although a more thorough treatment is provided in Chapter 12. The individual nucleosides strongly absorb UV light near 260 nm, and the oligo extinction coefficient has been

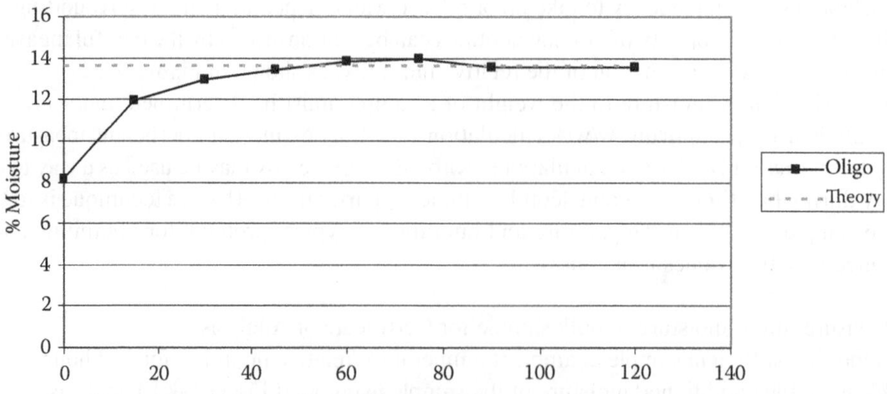

FIGURE 9.3 Conditioning an oligonucleotide for moisture content. The theory refers to the observation that at ambient temperature and humidity, many oligonucleotides absorb moisture to a percentage that corresponds to approximately three water molecules per base.[6]

approximated by a simple summation of the individual nucleotidic extinction coefficients.[8] The calculation was later refined using various approaches such as the nearest neighbor approach.[9,10] Additional empirical corrections have been employed in our laboratories to improve the nearest neighbor calculations and better approximate experimental values. In early phases of clinical development, theoretical values of extinction coefficient are sufficient for calculations. For the later clinical phases, an experimental determination of the extinction coefficient is strongly recommended. A typical experimental measurement would include:

1. Sample weights corrected for sodium and conditioned moisture (this is done because the extinction coefficient is a property of only the UV-absorbing component of the sample)
2. A stock solution prepared in water or preferably a suitable buffer
3. A series of dilutions made from the stock solution
4. UV measurement of the dilution series at 260 nm (or λ_{max} if desired)
5. Calculation of a regression line of absorbance versus molar concentration
6. The slope of the regression line is the extinction coefficient

This procedure yields experimental values for single-strand oligonucleotides that are in very good agreement with theoretical values calculated from the nearest neighbor approach with empirical corrections.

9.6 TYPES OF ASSAY METHODS

Assay methods for oligonucleotides may be categorized as:

1. Fractional methods
   Mass balance (gravimetric) methods
2. Comparative methods
   Using extinction coefficient as the reference
   Using external standards as the reference
3. Functional methods

Functional assay methods for oligonucleotides, where the biological or therapeutic activity are measured, are not discussed in this chapter.

There are two fundamental approaches to assay determination. Both of these approaches are variants of the general Equation 9.2. The first is the fractional approach, namely what portion of a whole comprises the substance of interest. This approach can be expressed as

$$\text{Assay} = \frac{\text{Wt}_{\text{product}}}{\text{Wt}_{\text{total}}} \times 100\% \tag{9.3}$$

The second approach to assay determination is a comparative approach. In this case the ratio of some measurable quantities (Q) for the reference and sample is multiplied by a known attribute of the reference, in this case the reference assay.

$$\text{Assay}_{\text{S}} = \frac{Q_{\text{S}}}{Q_{\text{R}}} \times \text{Assay}_{\text{R}} \tag{9.4}$$

From these two fundamental assay approaches, equations for assay of oligonucleotides can be derived.

9.6.1 Mass Balance Assays

Mass balance assays, also known as gravimetric assays, can be considered as a value of 100% corrected for selected measured components or impurities. Because this is a nonchromatographic method, the assay value is that of total or bulk oligonucleotide content.

$$\text{Assay} = \frac{\text{Wt}_{\text{oligo}}}{\text{Wt}_{\text{total}}} = 100\% - \text{H}_2\text{O}\% - \text{Na}\% - X\% \tag{9.5}$$

In Equation 9.5, "X%" refers to any other component that can be measured and contributes significantly to the assay such as extraneous salts or excipients in a drug product. In this and subsequent equations, "X%" will be considered zero unless required in the discussion. Essentially, the total weight of the sample is being corrected for any nonoligonucleotide components, Equation 9.6. (Note that in all subsequent equations, the percentage weight corrections are presented in decimal format for simplicity.)

$$\text{Wt}_{\text{oligo}} = \text{Wt}_{\text{total}} \times (1 - \text{H}_2\text{O} - \text{Na} - X) \tag{9.6}$$

Mass balance assay approaches would have greater utility if the result could be expressed as assay of oligonucleotide API rather than of total oligonucleotide content. Multiplication of the total content assay result by a purity value determined by HPLC to calculate API assay, for example, is a common practice and is justified in the following discussion.

HPLC is normally used to determine the purity of the oligonucleotide product (full length product [FLP]), and purity is reported as relative percent of the integrated peak area. However, one may question whether area percent purity is equivalent to weight/weight percent purity owing to the differences in extinction coefficients between the FLP and its various sequence failure impurities. The comparison of FLP and its impurities can be understood by examining Beer's law applied to multiple components.

In the standard Beer's law equation, the absorbance (A) is equal to the product of the molar extinction coefficient (ε), the molar concentration (c), and the cell path length (l).

$$A = \varepsilon c l \tag{9.7}$$

Beer's law is additive in the case of multiple components, under the same experimental conditions such as monitoring wavelength, temperature, and buffer composition.[11]

$$A_{\text{total}} = A_1 + A_2 + A_3 \tag{9.8}$$

$$A_{\text{total}} = \varepsilon_1 c_1 l + \varepsilon_2 c_2 l + \varepsilon_3 c_3 l \tag{9.9}$$

The concentration, normally expressed in moles per liter, may also be expanded to include a weight expression (g).

$$c = \frac{\text{moles}}{\text{L}} = \frac{g}{L \times MW} \tag{9.10}$$

If the concentration expressed as in Equation 9.10 is substituted into Beer's law and the terms rearranged, we obtain

$$A_{\text{total}} = \left(\frac{\varepsilon_1}{MW_1}\right)\left(\frac{g_1}{L}\right)l + \left(\frac{\varepsilon_2}{MW_2}\right)\left(\frac{g_2}{L}\right)l + \left(\frac{\varepsilon_3}{MW_3}\right)\left(\frac{g_3}{L}\right)l \tag{9.11}$$

TABLE 9.1
Conversion Factors for a Typical DNA Phosphorothioate Oligonucleotide and Its Impurities[a]

Sample	Extinction Coefficient (NN)	Molecular Weight (Na Salt)	Conversion Factor
FLP	141,818 /M/cm	6058	23.41 OD/mg
PO	141,818 /M/cm	6042	23.47 OD/mg
N-1	133,818 /M/cm	5716	23.41 OD/mg
N-2	127,909 /M/cm	5389	23.74 OD/mg
N-8	80,000 /M/cm	3346	23.91 OD/mg

[a] *Note:* OD is the absorbance of a 1 mL solution measured at 260 nm in a 1 cm path length cuvette.

which expresses the concentration in g/L. Thus the absorbances would be directly proportional to weights of each component in a solution provided that the ratio of extinction coefficient to molecular weight, called the conversion factor, is constant.

$$\left(\frac{\varepsilon_1}{MW_1}\right) \approx \left(\frac{\varepsilon_2}{MW_2}\right) \approx \left(\frac{\varepsilon_3}{MW_3}\right) \tag{9.12}$$

Whereas significantly different UV spectra are likely to be observed for impurities of small molecules, the assumption of a constant conversion factor, Equation 9.12, may be valid for oligonucleotides because (1) oligonucleotides consist of repeating units of simple nucleobases, (2) the major oligonucleotide impurities are shortmers and longmers also containing repeating units of the nucleobases, (3) the individual nucleobases have extinction coefficients in the same order of magnitude, and (4) all bases have absorption maxima around a single wavelength, i.e., 260 nm.

The principle of Equation 9.12 is demonstrated by calculating the conversion factors for a hypothetical DNA phosphorothioated oligonucleotide and four of its impurities, the monooxygenated (PO) impurity, and the shortmers, N-1, N-2, and N-8. Their extinction coefficients are calculated using the nearest neighbor approach, and the anhydrous sodium salt molecular weights are used in Table 9.1. The conversion factors calculated relate the UV absorption to a gravimetric quantity. Conversion factors for the other forms of oligonucleotide can also be calculated by using the appropriate molecular weight of that form.

Table 9.1 demonstrates that the conversion factor is relatively constant for the oligonucleotide and its impurities. Thus the conversion factor can be used to relate the UV absorption to a weight, and the area percent values for peaks in HPLC may also be interpreted as weight percent values.

On the basis of this rationale, we can calculate the weight of oligonucleotide API ($\text{Wt}_{\text{product}}$):

$$\text{Wt}_{\text{API}} = \text{Wt}_{\text{oligo}} \times \text{Purity} \tag{9.13}$$

The HPLC area percent purity results are combined with the gravimetric corrections to give an assay of oligonucleotide product (API):

$$\text{Assay}_{\text{API}} = \text{Purity}_{\text{API}} \times (1 - \text{H}_2\text{O} - \text{Na} - X) \tag{9.14}$$

The gravimetric correction terms in Equation 9.14 are for reporting the assay of the free acid form on an "as is" basis. The full list of correction terms for all forms and reporting modes are given in Table 9.2.

TABLE 9.2
Correction Terms for Gravimetric Assay for Total Oligonucleotide Content[a]

Form	"As Is" Basis	Anhydrous Basis
Hydrated	$1 - X$	N/A
Anhydrous Na salt	$1 - H_2O - X$	$1 - \frac{X}{1 - H_2O}$
Free acid	$1 - H_2O - Na^* - X$	$1 - \left(\frac{Na^* + X}{1 - H_2O}\right)$

[a] *Note:* The sodium value used in the calculation is the assay sodium value rather than the anhydrous value normally reported on certificates of analysis. $Na^* = Na_{anh}(1 - H_2O)$.

Mass Balance Assay Example:

Calculate the assay of API oligonucleotide as the free acid form on an "as is" basis and on an anhydrous basis.

1. Oligonucleotide A has a moisture content of 15.4% and an anhydrous sodium content of 5.77%.
2. Calculate the "as is" sodium content: 4.88% = 5.77% × (1 – 0.154).
3. Calculate the total oligo content by subtracting the moisture and sodium contributions: 0.7972 = 1 – 0.154 – 0.0488. Total oligonucleotide content is 79.72%.
4. With an HPLC purity of 87%, calculate the API content: 79.72% × 0.87 = 69.36% (API as free acid). The free acid API content on an "as is" basis for this sample is 69%.
5. Calculate the assay on an anhydrous basis: 87% × (1 – 0.0577) = 82%. The free acid API content on an anhydrous basis is 82%.

9.6.2 UV Spectrophotometric Assays

Comparative assays, as described in general by Equation 9.4, require the use of a reference.

In a UV-based assay, the theoretical conversion factor may be used as a reference. In this case $Assay_R = 1$ because the components of the theoretical conversion factor, extinction coefficient and molecular weight, are intrinsic constants of the molecule.

Therefore the UV assay is calculated as the ratio of conversion factors (OD/mg) for the sample and the theoretical value.

$$\text{Assay} = \frac{\left(\text{OD}/\text{mg}\right)_S}{\left(\text{OD}/\text{mg}\right)_R} \tag{9.15}$$

From the definitions of the conversion factors for sample and reference, this equation can be expanded as

$$\text{Assay} = \frac{\left(\text{Total OD}/\text{CorrWt}\right)}{\left(\text{ExtCoeff}/\text{MW}\right)} = \left(\frac{\text{Total OD}}{\text{ExtCoeff}}\right)\left(\frac{\text{MW}}{\text{CorrWt}}\right) \tag{9.16}$$

TABLE 9.3
Oligonucleotide Forms Defined by Choice of MW and Correction Factors[a]

	Molecular Weight		
Weight Correction	**Hydrated Na Salt**	**Anhydrous Na Salt**	**Free Acid**
mass	Hydrated salt as is	Anhydrous salt as is	Free acid as is
mass-H_2O	N/A	Anhydrous salt anhydrous basis	Free acid anhydrous basis
mass-H_2O-Na	N/A	N/A	N/A

[a] The molecular weight of the hydrated salt is estimated by assuming a hydration level of three water molecules per base.

In Equation 9.16, the following terms are defined from experiment:

$$\text{Total OD} = \text{Absorbance} \times \text{Vol.} \times \text{Dil.} \tag{9.17}$$

where Vol. is the initial volume of the sample solution and Dil. is the dilution factor required to lower the absorbance into the linear range of the spectrophotometer. Equation 9.18 defines the corrected sample weight.

$$\text{CorrWt} = \text{Wt} \times (1 - H_2O - \text{Na} - X) \tag{9.18}$$

Including the total OD and corrected weight terms into Equation 9.16 provides an assay formula for total oligonucleotide content. As with the mass balance assay, multiplication of the total assay by the HPLC purity will give the assay of the API:

$$\text{Assay}_{\text{API}} = \frac{\text{Purity}_{\text{API}} \times \text{Absorbance} \times \text{Vol.} \times \text{Dil.} \times MW}{\text{ExtCoeff} \times \text{Wt} \times \left(1 - H_2O - \text{Na} - X\right)} \tag{9.19}$$

Equation 9.19 can yield different assay values, depending on the form of the oligonucleotide under consideration. The choice of the molecular weight of each form and the degree of sample weight correction will determine the assay value for the particular form. This is summarized in Table 9.3.

Thus the first row in Table 9.3 shows different assay calculation options using the original uncorrected weight of sample and the appropriate molecular weight for the form of oligonucleotide desired.

9.6.3 HPLC Assays Using Reference Standards

A typical application of Equation 9.4 for calculating assay is conducting an HPLC analysis of the sample and comparing results with those of a well-characterized reference standard.

The measurable quantity (Q) in this case is the response factor (RF). The response factor, analogous to the UV conversion factor (CF = OD/mg), is the ratio of the total chromatographic response (A_{total}) and the mass injected on the column (M_{total}),

$$RF = \frac{A_{total}}{M_{total}} \tag{9.20}$$

The total mass is calculated by multiplying the sample concentration by the injection volume.

$$M_{total} = \frac{\text{CorrWt} \times \text{InjVol}}{\text{Volume} \times \text{Dilution}} \tag{9.21}$$

The total chromatographic response, Equation 9.22, is determined as the ratio of the main peak area and its purity multiplied by the flow rate. The flow rate term is included since the chromatographic peak area is a time-dependent absorbance measurement rather than a static reading as from a UV spectrophotometer.[12]

$$A_{total} = \frac{\text{Area}_{API} \times \text{Flow rate}}{\text{Purity}_{API}} \tag{9.22}$$

Combining the mass and response terms yields Equation 9.23, an expression for the response factor:

$$RF_{total} = \frac{\text{Area}_{API} \times \text{Flow rate} \times \text{Vol.} \times \text{Dil.}}{\text{Purity}_{API} \times \text{InjVol} \times \text{Wt} \times (1 - H_2O - \text{Na} - X)} \tag{9.23}$$

Removing the purity term provides the response factor for the API:

$$RF_{API} = \frac{\text{Area}_{API} \times \text{Flow rate} \times \text{Vol.} \times \text{Dil.}}{\text{InjVol} \times \text{Wt} \times \left(1 - H_2O - \text{Na} - X\right)} \tag{9.24}$$

A comparative HPLC assay, therefore, is the ratio of response factors of sample and reference multiplied by the known assay value for the reference.

$$\text{Assay}_S = \frac{RF_S}{RF_R} \times \text{Assay}_R \tag{9.25}$$

If solution preparation and instrument parameters are the same for both sample and reference, then some of the terms in the ratio of RF will cancel out. A reduced response factor with these terms eliminated can be stated:

$$RF_{reduced} = \frac{\text{Area}_{API}}{\text{Wt} \times \left(1 - H_2O - \text{Na} - X\right)} \tag{9.26}$$

Assay using HPLC data from both sample and reference standard is then calculated using Equation 9.27. This assay calculation requires a reference standard that has been previously qualified.

$$\text{Assay}_S = \frac{\text{Area}_S \times \text{Wt}_R \times (1 - H_2O_R - \text{Na}_R - X_R)}{\text{Area}_R \times \text{Wt}_S \times \left(1 - H_2O_S - \text{Na}_S - X_S\right)} \times \text{Assay}_R \tag{9.27}$$

This equation, which is very similar to Equation 9.1 used for small molecules, shows that the weight corrections can be applied to both reference and sample, depending on the form of the oligonucleotide and how the assay is reported. This equation is often used to establish sample conformity with the reference standard.

One of the concerns about assay determination using a reference standard is the moisture content and quality of the reference sample each time assay is performed. Variable moisture content may be addressed by preparing a larger volume of reference solution and storing aliquots frozen for later use. The stability of the reference standard under these conditions should be known.

9.6.4 HPLC Assays Using Extinction Coefficient

If the HPLC assay is performed using UV detection, then the extinction coefficient may serve as an alternate reference. Using the extinction coefficient as a reference has distinct advantages over use of a separate reference sample. In addition to eliminating the aforementioned moisture and stability concerns of the reference sample, time and effort are saved in preparation and analysis of the additional reference sample. It has been previously shown in the derivation of Equation 9.19 that extinction coefficient is an appropriate reference for UV assay, and a logical extension is the application of extinction coefficient as a reference in HPLC assay. In this case, the theoretical conversion factor can be used as a substitute for the reference response factor

$$\text{Assay} = \frac{\text{RF}_{\text{sample}}}{\text{RF}_{\text{reference}}} = \frac{\text{RF}_{\text{sample}}}{\left(\text{ExtCoeff}/\text{MW}\right)} \tag{9.28}$$

When the extinction coefficient and molecular weight for the API are combined with the response factor of the sample, Equation 9.24, the assay calculation becomes

$$\text{Assay}_{\text{API}} = \frac{\text{Area} \times \text{Flow rate} \times \text{Vol.} \times \text{Dil.} \times \text{MW} \times 10^{-6}}{\text{ExtCoeff} \times 60 \times \text{InjVol} \times \text{Wt} \times \left(1 - \text{H}_2\text{O} - \text{Na} - X\right)} \tag{9.29}$$

Equation 9.29 is derived assuming the HPLC peak area units are μA.s, where the term 60 is a time conversion and the 10^{-6} term scales the absorbance. These terms may be changed to accommodate the peak reporting conventions of different instruments. The equipment should be appropriately qualified, and the precision and accuracy of parameters such as flow rate and injection volume should be established.

9.7 COMPARISON OF ASSAY METHODS

As previously stated, assay values can vary considerably, depending on the form of the oligonucleotide selected, if API or total oligonucleotide content is desired, and if reporting is on an anhydrous basis. When these parameters are appropriately defined, then ideally the different assay approaches described in this chapter should provide similar assay information.

To test this, the assay approaches are compared using data for a 20 base DNA phosphorothioate oligonucleotide.[13] The comparison includes calculation results from all three forms of the oligonucleotide (hydrated sodium salt, anhydrous sodium salt, free acid), API versus total oligonucleotide content, and reporting the results on an "as is" basis or an anhydrous basis.

The data in Table 9.4 used for the comparison are from two analysts doing the analyses in triplicate (average data used when applicable).

TABLE 9.4
Data for Assay Comparison

Assay Category	Parameter	Value
Gravimetric calculations	Moisture content	16.94%
	Sodium (anhydrous value)	5.50%
UV data	Average weight	10.41 mg
	Average absorbance @ 260nm	0.958
	Volume	10 mL
	Dilution	20-fold
HPLC data (sample)	Moisture content	16.94%
	Sodium (anhydrous value)	5.50%
	Average weight	10.41 mg
	Average main peak area (µAu.s)	6,469,421
	Average purity	86.65%
HPLC data (reference)	Moisture content	17.39%
	Sodium (anhydrous value)	5.17%
	Average weight	10.36 mg
	Average main peak area (µAu.s)	6,919,534
	Average purity	91.55%

Table 9.5 contains other oligonucleotide data and instrument parameters needed in the comparison. The following several examples are assay calculations using the data in Tables 9.4 and 9.5 and the appropriate assay equation.

Examples of Mass Balance Assay Using Equation 9.14:

Calculate assay of the total oligo content as the free acid on an as is basis:

$$\text{Assay}_{\text{total}} = (1 - 0.1694 - 0.055(1 - 0.1694)) = 0.785$$

Calculate assay of API sodium salt on an as is basis:

$$\text{Assay}_{\text{API}} = 0.867 \times (1 - 0.1694) = 0.720$$

Calculate assay of API sodium salt on an anhydrous basis:

$$\text{Assay}_{\text{API}} = 0.867 \times (1 - 0.0) = 0.867$$

TABLE 9.5
Additional Data for Assay Comparison

Parameter	Value
Extinction coefficient	153,957
Molecular weight (hydrated salt, estimated)	8,179
Molecular weight (anhydrous sodium salt)	7,044
Molecular weight (free acid)	6,604
HPLC flow rate	0.40 mL/min
HPLC injection volume	0.005 mL
HPLC sample volume	10 mL
HPLC dilution factor (average)	1.925

Notice that in this case the assay for the sodium salt on an anhydrous basis is equivalent to the HPLC purity.

Examples of UV Assay Using Equation 9.19:

Calculate assay of API free acid on an as is basis:

$$\text{Assay} = \frac{0.8665 \times 0.958 \times 10 \times 20 \times 6604}{153957 \times 10.41 \times \left(1 - 0.0 - 0.0\right)} = 0.684$$

The molecular weight of the free acid is used, and no correction for the sample weight is needed for reporting on an as is basis.

Calculate assay of API sodium salt on an anhydrous basis:

$$\text{Assay} = \frac{0.8665 \times 0.958 \times 10 \times 20 \times 7044}{153957 \times 10.41 \times \left(1 - 0.1694\right)} = 0.878$$

In this case the molecular weight of the sodium salt is used, and the sample weight is corrected for moisture to report on an anhydrous basis.

Examples of HPLC Assay Using a Reference Standard and Equation 9.27

Calculate assay of API sodium salt on an anhydrous basis:

$$\text{Assay} = \frac{6469421 \times 10.36 \times (1 - 0.1739)}{6919534 \times 10.41 \times \left(1 - 0.1694\right)} \times 0.9155 = 0.847$$

Note that reporting the oligonucleotide assay as the sodium salt on an anhydrous basis requires a moisture correction for both sample and reference.

Calculate assay of API free acid on an anhydrous basis:

First, convert anhydrous sodium value of the reference standard to the assay sodium value:

$$Na_{assay} = 0.0517 \times (1 - 0.1739) = 0.0427$$

$$\text{Assay} = \frac{6469421 \times 10.36 \times (1 - 0.1739 - 0.0427)}{6919534 \times 10.41 \times \left(1 - 0.1694\right)} \times 0.9155 = 0.803$$

In this case, the reference standard weight corrections determine the oligo form and the sample weight corrections determine the reporting format (anhydrous basis).

Examples of HPLC Assay Using Extinction Coefficient as the Reference and Equation 9.29

Calculate assay of API sodium salt on an anhydrous basis:

$$\text{Assay} = \frac{6469421 \times 0.40 \times 10 \times 1.925 \times 7044 \times 10^{-6}}{153957 \times 60 \times 0.005 \times 10.41 \times \left(1 - 0.1694\right)} = 0.879$$

The molecular weight of the sodium salt was used and the sample weight was corrected for moisture.

Calculate assay of total oligo content free acid on an as is basis:

$$\text{Assay} = \frac{6469421 \times 0.40 \times 10 \times 1.925 \times 6604 \times 10^{-6}}{153957 \times 60 \times 0.8665 \times 0.005 \times 10.41 \times (1-0.0)} = 0.790$$

The free acid molecular weight is used as well as the HPLC purity to obtain total oligo content as the free acid. No sample weight correction is needed for reporting on an as is basis.

It is readily apparent from the treatment of the data by each of the derived equations that assay results can vary considerably. The data in Tables 9.6 through 9.9 are summarized in Figure 9.4.

Figure 9.4 shows that all four methods of assay determination are in good agreement. The figure also demonstrates that the wide variation in calculated assay is due to the different forms of oligonucleotide, whether total oligo content or API is desired, and if the data are reported on an anhydrous basis.

TABLE 9.6
Summary of Mass Balance Assay

	Total Oligo		FLP (API)	
Form	**As Is**	**Anhydrous**	**As Is**	**Anhydrous**
Hydrated salt	100.0%	N/A	86.7%	N/A
Sodium salt	83.1%	100.0%	72.0%	86.7%
Free acid	78.5%	94.5%	68.0%	81.9%

TABLE 9.7
Summary of UV Assay

	Total Oligo		FLP (API)	
Form	**As Is**	**Anhydrous**	**As Is**	**Anhydrous**
Hydrated salt	97.8%	N/A	84.7%	N/A
Sodium salt	84.2%	101.4%	73.0%	87.8%
Free acid	79.0%	95.1%	68.4%	82.4%

TABLE 9.8
Summary of HPLC Assay Using a Reference Standard

	Total Oligo		FLP (API)	
Form	**As Is**	**Anhydrous**	**As Is**	**Anhydrous**
Hydrated salt	N/A	N/A	85.2%	N/A
Sodium salt	N/A	N/A	70.4%	84.7%
Free acid	N/A	N/A	66.7%	80.3%

TABLE 9.9
Summary of HPLC Assay Using Extinction Coefficient as the Reference

	Total Oligo		FLP (API)	
Form	**As Is**	**Anhydrous**	**As Is**	**Anhydrous**
Hydrated salt	97.8%	N/A	84.7%	N/A
Sodium salt	84.2%	101.4%	73.0%	87.9%
Free acid	79.0%	95.1%	68.4%	82.4%

9.8 ASSAY CALCULATIONS IN THE PRESENCE OF EXCIPIENTS

In the formulation of drug products, the active ingredients are usually mixed with excipients such as binders and fillers. Assay is particularly important in the analysis of drug products because it determines the amount of active component for dosing, formulation studies, and other applications. Any extraneous materials present in an oligonucleotide sample should affect the assay calculation. In order to test the equations derived earlier for their ability to provide meaningful assay in the presence of excipients, a simulation was conducted. In this simulation, 13.05% by weight of an excipient was added to the sample. The excipient was assumed to make no contribution to the moisture or sodium content in the sample and was considered as non-UV absorbing. The moisture and sodium values are maintained in the same proportion with respect to the oligonucleotide. If the same sample weight is used, then the UV absorbance and HPLC peak area would proportionally decrease relative to the values in the absence of excipient. Sample purity values and reference data would not be affected.

The simulation uses the data in Table 9.10 and the supplemental data from Table 9.5. In this simulation the same sample weight is assumed, and the bold type values in Table 9.10 are proportionately estimated values based on the theoretical addition of an excipient.

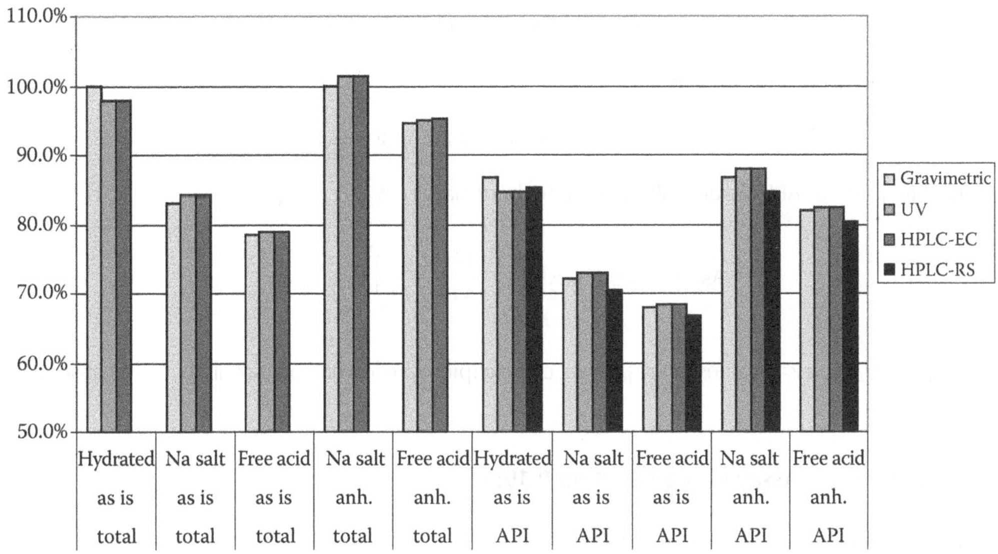

FIGURE 9.4 Comparison of assay methods (gravimetric are the mass balance results, UV are the UV assay results, HPLC-EC are HPLC assay results using the extinction coefficient as the reference, and HPLC-RS are HPLC assay results using an external reference standard).

TABLE 9.10
Data for Assay Comparison (Assuming 13.05% Excipient Present)

Category	Parameter	Value
Gravimetric calculations	**Moisture**	**14.73%**
	Sodium (assay basis)	**3.98%**
UV Data	Average weight	10.41 mg
	Average absorbance at 260 nm	**0.833**
	Volume	10 mL
	Dilution	20-fold
HPLC data (sample)	**Moisture**	**14.73%**
	Sodium (assay basis)	**3.98%**
	Average weight	10.41 mg
		5,626,362
	Average purity	86.65%
HPLC data (reference)	Moisture	17.39%
	Sodium	5.17%
	Average weight	10.36 mg
	Average main peak area	6,919,534
	Average purity	91.55%

[a] Bold type indicates proportionately estimated values based on the theoretical addition of an excipient.

Examples of Mass Balance Assay Using Equation 9.14:

In a formulation containing 13.05% by weight of excipient, calculate the assay for the total oligo content as the free acid on an as is basis:

$$\text{Assay}_{\text{total}} = (1 - 0.1473 - 0.0398 - 0.1305) = 0.682$$

Calculate assay of API sodium salt on an as is basis:

$$\text{Assay}_{\text{API}} = 0.867 \times (1 - 0.1473 - 0.1305) = 0.626$$

Calculate assay of API sodium salt on an anhydrous basis:

$$\text{Assay}_{\text{API}} = 0.867 \times \left(1 - \left(\frac{0.1305}{1 - 0.1473}\right)\right) = 0.734$$

As expected, the assay values in the presence of excipient are lower than calculated in the absence of excipient.

Examples of UV Assay Using Equation 9.19:

Calculate assay of API free acid on an as is basis:

$$\text{Assay} = \frac{0.8665 \times 0.833 \times 10 \times 20 \times 6604}{153957 \times 10.41 \times (1 - 0.0 - 0.0 - 0.0)} = 0.595$$

The molecular weight of the free acid was used, and no sample weight correction is needed for reporting on an as is basis.

Calculate assay of API sodium salt on an anhydrous basis:

$$\text{Assay} = \frac{0.8665 \times 0.833 \times 10 \times 20 \times 7044}{153957 \times 10.41 \times (1 - 0.1473 - 0.0 - 0.0)} = 0.744$$

In this case the molecular weight of the sodium salt is used, and the sample weight is corrected for moisture to report on an anhydrous basis.

HPLC Assay Using a Reference Standard and Equation 9.27:

Calculate assay of API sodium salt on an anhydrous basis:

$$\text{Assay} = \frac{5626362 \times 10.36 \times (1 - 0.1739)}{6919534 \times 10.41 \times (1 - 0.1473)} \times 0.9155 = 0.718$$

The reference standard has no excipient and its data are unchanged.

Calculate assay of API free acid on an anhydrous basis:

First, convert anhydrous sodium value of the reference standard to the assay sodium value:

$$Na_{assay} = 0.0517 \times (1 - 0.1739) = 0.0427$$

$$\text{Assay} = \frac{5626362 \times 10.36 \times (1 - 0.1739 - 0.0427)}{6919534 \times 10.41 \times (1 - 0.1473)} \times 0.9155 = 0.681$$

HPLC Assay Using Extinction Coefficient as the Reference and Equation 9.29:

Calculate assay of API sodium salt on an anhydrous basis:

$$\text{Assay} = \frac{5626362 \times 0.40 \times 10 \times 1.925 \times 7044 \times 10^{-6}}{153957 \times 60 \times 0.005 \times 10.41 \times (1 - 0.1473)} = 0.744$$

Calculate assay of total oligo content free acid on an as is basis:

$$\text{Assay} = \frac{5626362 \times 0.40 \times 10 \times 1.925 \times 6604 \times 10^{-6}}{153957 \times 60 \times 0.8665 \times 0.005 \times 10.41 \times (1 - 0.0)} = 0.687$$

The free acid molecular weight is used as well as the HPLC purity to obtain total oligo content as the free acid. No sample weight correction is needed for reporting on an as is basis.

In all calculations the presence of an excipient lowers the overall assay values. Figure 9.5 compares the assay method results with and without excipient.

If the excipient is accounted for in the weight corrections, then the assay values would be similar to the values calculated in the absence of excipient.

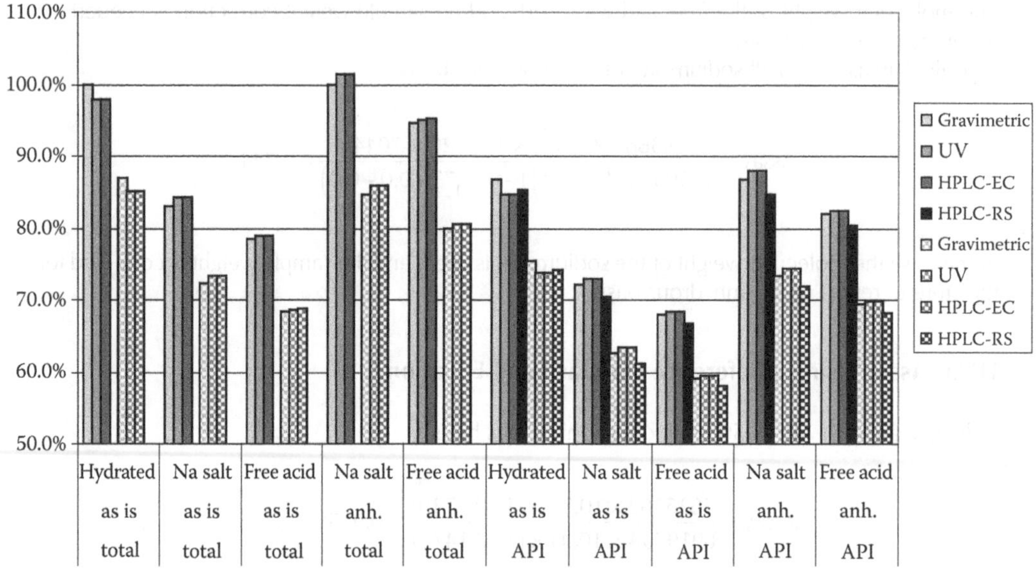

FIGURE 9.5 Comparison of assay methods (solid columns are with no excipient and checkered columns are with excipient, gravimetric are the mass balance results, UV are the UV assay results, HPLC-EC are HPLC assay results using the extinction coefficient as the reference, and HPLC-RS are HPLC assay results using an external reference standard).

9.9 ERROR ANALYSIS

As with any analytical result, confidence in the calculation of assay is dependent on the uncertainty of the analytical measurements and the tolerances of the analytical equipment used to generate the data. If systematic errors from the analyst or instrument calibration are controlled or minimized, then the remaining random errors are amenable to statistical analysis. Random error or uncertainty is calculated as the square root of the sum of the squares of each error.

TABLE 9.11
Summary of Mass Balance Assay in the Presence of Excipients

	Total Oligo		FLP (API)	
Form	**As Is**	**Anhydrous**	**As Is**	**Anhydrous**
Hydrated salt	87.0%	N/A	75.3%	N/A
Sodium salt	72.2%	84.7%	62.6%	73.4%
Free acid	68.3%	80.0%	59.1%	69.4%

TABLE 9.12
Summary of UV Assay in the Presence of Excipients

	Total Oligo		FLP (API)	
Form	**As Is**	**Anhydrous**	**As Is**	**Anhydrous**
Hydrated salt	85.0%	N/A	73.7%	N/A
Sodium salt	73.2%	85.9%	63.5%	74.4%
Free acid	68.7%	80.5%	59.5%	69.8%

TABLE 9.13
Summary of HPLC Assay in the Presence of Excipients Using a Reference Standard

	Total Oligo		FLP (API)	
Form	**As Is**	**Anhydrous**	**As Is**	**Anhydrous**
Hydrated salt	N/A	N/A	74.1%	N/A
Sodium salt	N/A	N/A	61.2%	71.8%
Free acid	N/A	N/A	58.0%	68.1%

TABLE 9.14
Summary of HPLC Assay in the Presence of Excipients Using Extinction Coefficient as the Reference

	Total Oligo		FLP (API)	
Form	**As Is**	**Anhydrous**	**As Is**	**Anhydrous**
Hydrated salt	85.1%	N/A	73.7%	N/A
Sodium salt	73.3%	85.9%	63.5%	74.4%
Free acid	68.7%	80.5%	59.5%	69.8%

TABLE 9.15
Actual and Estimated Tolerances for Factors Used in Assay Calculations

Parameter	Value	Tolerance or Uncertainty	Absolute Error (1SD)	Relative Error
Moisture	16.94%		0.17%	
Sodium	5.50%		0.05%	
HPLC purity	91.55%	0.4%		0.44%
UV absorbance	0.958	0.002		0.21%
Volume	10.00 mL	0.02 mL		0.20%
Dilution[a]	20	0.052		0.26%
Molecular weight	6604	0		0.00%
Extinction coefficient	153,957	1000		0.65%
Sample weight	10.41 mg	0.02 mg		0.19%
HPLC peak area	6,469,421	5000		0.08%
Flow rate	0.40 mL/min	0.011 mL/min		0.10%
Injection volume	0.005 mL	0.0001 mL		0.35%

[a] Dilution errors are dependent on the size of the glassware, i.e., a 100 mL volumetric flask has a smaller relative tolerance than a 2 mL volumetric flask.

Uncertainty in addition and subtraction operations is calculated from the absolute error of the measurement:

$$e_{tot} = \sqrt{e_1^2 + e_2^2 + e_3^2}$$

Uncertainty in multiplication and division operations is calculated from the relative error of the measurement:

$$\%e_{tot} = \sqrt{(\%e_1)^2 + (\%e_2)^2 + (\%e_3)^2}$$

An estimation of the random error associated with each measurement and a propagation of the uncertainty in the assay calculation were explored using the actual and estimated tolerances listed in Table 9.15. The uncertainty ranged from 0.4 to 2.9% depending on the equation used and various other experimental factors.

9.10 ASSAY OF MORE COMPLEX OLIGONUCLEOTIDES

Recently, more complex oligonucleotides have been investigated as therapeutic agents.[14] These include conjugation of the oligonucleotide to polyethylene glycol (PEG), aptamers in which the sequence is designed to induce internal folding or secondary structure, and duplexes in which two complementary single-strand oligonucleotides are held together through Watson–Crick interactions. These more advanced types of oligonucleotides pose some significant analytical challenges. Purity, for example, is not only dependent on the analytical method as mentioned previously but also on the additional structural dimensions of these molecules. Does the purity of an aptamer include the percent of unfolded oligonucleotide or aggregates as well as the standard related impurities? What does the purity value of a duplex indicate about impurities from each single strand, excess of one of the single strands, or duplex formation from the impurities themselves? Clearly, no single analytical method is capable of providing all the answers. Moreover, secondary structure formation and duplex annealing may be more dynamic processes than expected, and the analytical methods themselves may influence the result. Determining assay in this complicated scenario should be approached with caution. It may be more appropriate to report assay for these systems as total oligonucleotide content rather than API content, at least until the purity methods and the interpretation of results become well established.

Example of an Assay of PEGylated Oligonucleotide

In this example a 13,000 molecular weight oligonucleotide was conjugated to a 40 K PEG. The extinction coefficient for the molecule was determined to be 305,000/M/cm. A 50 mg sample dissolved in 10 mL water was diluted 30-fold and measured by UV at 260 nm immediately after solution preparation because viscosity and nonhomogeneity of the solution can result in lower precision.

The total oligonucleotide content for the PEGylated oligonucleotide free acid on an as is basis can be calculated:

$$\text{Assay} = \frac{0.932 \times 10 \times 30 \times 53000}{305000 \times 50 \times (1 - 0.0 - 0.0)} = 0.972$$

Initially, it may seem that this assay value is an appropriate value. However, the polydispersity of PEG can generate significant uncertainty in the final assay number. To reduce this uncertainty, many have chosen to report the assay result in terms of oligonucleotide content. Using the molecular weight of the unPEGylated oligonucleotide rather than a molecular weight average of the PEGylated oligonucleotide, the assay result is about 25% of the original PEGylated assay value.

$$\text{Assay} = \frac{0.932 \times 10 \times 30 \times 13000}{305000 \times 50 \times (1 - 0.0 - 0.0)} = 0.238$$

9.11 SUMMARY

The assay determination of an oligonucleotide is less straightforward than for small molecule drug products. Complicated by variable moisture content, ever-present product related impurities, lack of highly purified reference standards, and a multiplicity of analytical purity methods, assay calculations can nonetheless be approached in a systematic way. One of the most critical aspects of assay determination is defining the form of the oligonucleotide of interest and how it will be reported. Four approaches to assay calculation were described and shown to provide similar results. The extinction coefficient was shown to be a suitable alternative to a reference standard sample when a reference point is required in the assay.

REFERENCES

1. ICH Harmonised Tripartite Guideline. 2005. Validation of analytical procedures: Text and methodology Q2(R1). http://www.ich.org/LOB/media/MEDIA417.pdf.
2. U.S. Pharmacopeia. 2009. *United States Pharmacopeia and National Formulary* (USP32-NF 27). Rockville, MD: U.S. Pharmacopeial Convention.
3. *European Pharmacopoeia*. 2008. Strasbourg: European Directorate for the Quality of Medicines & HealthCare.
4. Portions of this chapter have been presented at TIDES (Michaud, D. P., 2009. Assay of oligonucleotides. Pre-conference Workshop Managing Analytical Development and Quality Control of Oligonucleotides, Las Vegas) and EuroTides (Michaud, D. P., 2009. Assay calculations for oligonucleotides. Pre-conference Workshop Managing Analytical Development and Quality Control of Oligonucleotides, Amsterdam).
5. Kambhampati, R. V. B. 2007. Points to consider for the submission of Chemistry, Manufacturing, and Controls (CMC) information in oligonucleotide-based therapeutic drug applications. DIA Industry and Health Authority Conference on Oligonucleotide-based Therapeutics. http://www.fda.gov/downloads/AboutFDA/CentersOffices/CDER/ucm103524.pdf.
6. Brown, T., and D. J. S. Brown. 1991. Modern machine-aided methods of oligodeoxyribonucleotide synthesis. In *Oligonucleotides and Analogues. A Practical Approach,* ed. F. Eckstein, 1–24. Oxford: IRL Press.
7. Schweitzer, M., and J. W. Engels. 1997. Analysis: analysis of oligonucleotides. In *Antisense—From Technology to Therapy*, ed. R. Schlingensiepen, W. Brysch, and K.-H. Schlingensiepen. Berlin: Blackwell Science.
8. Sambrook, J., E. F. Fritsch, and T. Maniatis. 1989. *Molecular Cloning: A Laboratory Manual*, 2nd ed. Cold Spring Harbor, NY: Cold Spring Harbor Laboratory Press.
9. Kallansrud, G., and B. Ward. 1996. A comparison of measured and calculated single- and double-stranded oligodeoxynucleotide extinction coefficients. *Analytical Biochemistry* 236: 134–138.
10. Cavaluzzi, M. J., and P. N. Borer. 2004. Revised UV extinction coefficients for nucleoside-5′-monophosphates and unpaired DNA and RNA. *Nucleic Acids Research* 32: 1–9.
11. Harris, D. C. 1991. *Quantitative Chemical Analysis*, 519. New York: W. H. Freeman.
12. Karger, B. L., L. R. Snyder, and C. Horvath. 1973. *An Introduction to Separation Science*, 71–73. New York: John Wiley.
13. Material for this study was kindly provided by Giuliani Pharma S.p.A., and permission was granted for use of the data in this chapter.
14. Sinha, N. D., and D. P. Michaud. 2007. Recent developments in the chemistry, analysis and control for the manufacture of therapeutic-grade synthetic oligonucleotides. *Current Opinion in Drug Discovery and Development* 10(6): 807–818.

10 Microbial Analysis
Endotoxin Testing

Barbara J. Markley
Associates of Cape Cod, Inc.

CONTENTS

10.1 INTRODUCTION

Limulus amebocyte lysate (LAL) has had a tremendous impact since its introduction in the early 1970s. It is now used to quantify endotoxin in parenteral drugs including oligonucleotides and medical devices. It is used in research on a wide variety of disciplines, from clinical to environmental applications. LAL is used in basic research on gram-negative bacterial wall structure, lipopolysaccharide (LPS) biosynthesis, and bioactivity. In 1983, the LAL test for gram-negative bacterial endotoxin became the first *in vitro* test to replace an *in vivo* test for a toxic substance when the U.S. Pharmacopeia (USP) began replacing the pyrogen test (rabbit fever test) with the bacterial endotoxins test (BET) in the water and radiopharmaceutical monographs. Today, the majority of the monographs specify the LAL test.

10.2 LAL HISTORY

Howell published his observations on the coagulation of the blood of *Limulus* in 1885.[1] Twenty-five years later, Loeb described the blood cells of the horseshoe crab.[2] Frederik Bang described the formation of blood clots in *Limulus* in the presence of a toxic marine bacterium in papers as early as 1953, but it was not until the mid 1960s that he and Jack Levin at the Marine Biological Laboratory in Woods Hole, MA, elucidated the role of endotoxin, demonstrated extracellular coagulation *in vitro*, and monitored the kinetics of coagulation by following the rate of increase in turbidity in the *in vitro* reaction mixture.[3]

As news of the clotting reaction spread, Stanley W. Watson at the Woods Hole Oceanographic Institution began to prepare his own lysate to determine whether or not his cytoplasmic membrane preparations from gram-negative marine bacteria were free of outer membrane contaminants. He demonstrated that lysate could quantify the amount of endotoxin per cell and showed that LAL was sensitive enough to detect the greater amount per cell when cells were grown in rich nutrient media than when grown in minimal media.

Throughout the 1970s, papers were published on the biochemistry of coagulation, the specificity of LAL for endotoxins, and the use of the test as an *in vitro* substitute for the USP pyrogen test. Researchers at the Food and Drug Administration (FDA) investigated the use of LAL for detecting endotoxins in radiopharmaceuticals and biologicals. Studies comparing the responses of rabbits to increasing doses of endotoxin were initiated to determine the endotoxin pyrogenic threshold dose. In 1973, in anticipation of the use of LAL for drug and device testing, the FDA published the first manufacturing standards for LAL.

Finding that his LAL preparations were in demand by some of the pharmaceutical companies, Dr. Watson founded Associates of Cape Cod, Inc. (ACC) in 1974. In 1977, the third major LAL method, the chromogenic method, was described by Nakamura et al. Also in 1977, the FDA published final manufacturing standards for LAL, allowed the use of the LAL test as an end-product release test for medical devices without prior approval, and granted the first license to manufacture LAL to ACC.[4]

Use of the LAL test accelerated dramatically as more and more companies began to monitor raw materials and in-process manufacturing or materials. Endotoxin limits were hotly debated. While the LAL test had been demonstrated to be much more sensitive to endotoxin than the rabbit fever test, many manufacturers were adamant that the test should only be used as a pass/fail test at the same level of detection as the rabbit. A Health Industries Manufacturers Association (HIMA) study determined that the threshold pyrogenic dose in the rabbit for the endotoxin standard in common use in the medical devices industry was 0.1 ng/mL in a dose of 10 mL per kg body weight. This level was not easily established because of extreme variation in response to endotoxin between rabbit colonies in the study. The endotoxin used in the HIMA study was manufactured by Difco. It was purified by the Westphal method from cells of *Escherichia coli* O55:B5. This standard preparation remained the standard for testing medical devices until 1987. Meanwhile, the Bureau of Biologics (now the Center for Biological Evaluation and Research, CBER), which was given the responsibility for monitoring the standards of manufacture and the licensing of LAL manufacturers, had their own endotoxin preparation, a Westphal extraction from cells of *Escherichia coli* O113:H10. In collaboration with USP, this endotoxin standard, EC-2, became USP's first endotoxin reference standard to support the new USP Bacterial Endotoxins Test (BET) that became official in 1980.

FDA distributed the first draft of a guideline for using the LAL test as an end-product test in 1980; after comments, a second draft was published in 1983. The final guideline was published in 1987. By 1987, formulations for end point and kinetic and end point chromogenic LAL methods were commercially available, so these methods were included in the 1987 guideline. LAL licensing for the kinetic chromogenic method followed later, so guidance was issued in 1991 for the use of the turbidimetric and chromogenic kinetic methods for testing drugs and biologicals.

LAL was initially used to assess the level of endotoxin in in-process materials even though the rabbit was still the official release test. The experiences with the LAL test ultimately led to the adoption of the LAL test for product release. In fact, approvals to release drugs and biological products seemed slow once the approval had been given to devices in 1977. By 1983, though, the approvals were being given and by 1987, release using LAL was very strongly encouraged in the industry. Today, the LAL test is so commonly performed that the rabbit test seems unusual. But, even today, the LAL test is still being revised. Revisions have been made to various pharmacopoeias that have bacterial endotoxins tests and these tests are being added to some that have not previously accepted the LAL test. The Japanese Pharmacopoeia (JP) led the JP, European Pharmacopoeia (EP), and U.S. Pharmacopeia (USP) in drafting a "harmonized" bacterial endotoxins test that became official (in USP and EP) January 1, 2001.

10.3 HORSESHOE CRABS

10.3.1 The Living Fossil

The horseshoe crab is often described as a "living fossil." Various sources suggest that there have been no substantial morphological changes to the horseshoe crab since the Devonian period 400 million years ago.[5] Fossils, found to be 300 million years old, are clearly recognizable as horseshoe crabs. Despite their name, though, horseshoe crabs are not crustaceans but are more closely related to spiders, scorpions, mites, and ticks than to true crabs.

Horseshoe crabs are found today on the eastern shores of the continents. *Limulus polyphemus* is found from the Gulf of Maine to the Florida Keys, and around the Gulf Coast of Florida. *Carcinoscorpius* and species of *Tachypleus* are found in Southeast Asia. Their ranges overlap, but they occupy distinct ecological niches.

10.3.2 Horseshoe Crab Blood

Hemolymph contains circulating, nucleated cells called amebocytes. The oxygen-carrying protein, hemocyanin, is located in the hemolymph; it is not cell bound like the equivalent protein, hemoglobin, in mammals. Hemocyanin also differs from hemoglobin because it contains copper rather than iron. When hemocyanin is oxidized, it turns blue.

The great majority of the bacteria in the horseshoe crab's ocean environment are gram-negative. Thus any injury to the horseshoe crab will expose its blood to these organisms and to endotoxin. Horseshoe crabs cannot respond to microbial infection by producing antibodies. Instead, they fight infection by the clotting of their blood. Coagulation causes the cells to be killed and the infection to be cleared. The clot that is formed does not contain covalent bonds. It is physically fragile and will break down if disturbed. The horseshoe crab can survive even massive clotting throughout its circulatory system. High levels of endotoxin are known to cause blood coagulation in mammals as well, but in mammals, disseminated intravascular coagulation leads to death.

10.4 DEFINITIONS

In order to get started, several definitions are needed to facilitate understanding the next chapters on endotoxins and LAL test methods. This section should also clarify the distinction between the terms pyrogen, endotoxin, and lipopolysaccharide.

10.4.1 Pyrogen

The first is pyrogen. The word comes from two Greek terms: "pyro," meaning "fire" or "heat," and "gen," meaning "production (of)." Thus a pyrogen is something that generates or causes a heat (or fire) fever, and pyrogenic means producing or produced by heat or fever. One of the first obvious indications of illness caused by infection is onset of fever. Fevers may result from other sources, but those that follow administration of parenteral drugs are attributed to contamination of the drug or drug delivery device with heat stable endotoxins from gram-negative bacteria.

10.4.2 Endotoxin

Endotoxins refer to heat-stable toxins located "within" a cell. In contrast, the secreted/excreted heat-labile toxins from bacteria and other microorganisms are referred to as exotoxins, i.e., toxins that are "outside" the cell. These are usually proteins.

10.4.3 Gram Reaction

Pyrogenic endotoxins are found only in the cells of gram-negative bacteria. Bacteria are divided for purposes of classification into at least two major groups on the basis of staining characteristics. The different staining properties are a consequence of the different structures of the cell envelope. The Gram stain quickly distinguishes the two groups: gram-positive cells retain the primary stain, and gram-negative cells do not. Gram-negative bacteria can be decolorized in the Gram stain procedure and counterstained pink to distinguish them from the darkly purple stained gram-positive bacteria.

10.4.4 Lipopolysaccharide

The search for the biochemical structure that was the source of the toxicity of endotoxins led to the discovery of a substance unique to gram-negative cells. The substance forms a macromolecular structure in the wall of the cells and is called lipopolysaccharide (LPS).

When the LAL test was first proposed as a substitute for the rabbit fever or pyrogen test, the LAL test was often referred to as a pyrogen test. This is not appropriate terminology because LAL is not used to measure a fever response; it can only be used to measure the concentration of endotoxin or LPS in solution. The concentration of endotoxin in solution determines whether or not the solution would be expected to be pyrogenic if administered intravenously or intrathecally. Therefore, the LAL test is a test for endotoxins.

LPS is the biochemically purified toxic molecule found in endotoxins. The term endotoxin is used broadly to mean the complexes of LPS and other substances that are closely associated with LPS in the wall of the cell. Even "purified" endotoxins may not be pure LPS.

10.4.5 *Limulus* Amebocyte Lysate

LAL is the aqueous extract obtained after lysis of the blood cells (amebocytes) of the horseshoe crab, *Limulus polyphemus*. This extract contains the amebocyte proteins that cause the coagulation of horseshoe crab blood. Horseshoe crab blood clots in response to wounds, but the causative agent that initiates the clotting reaction is lipopolysaccharide. The reaction is initiated in response to the LPS on the surface of the whole cell or to the LPS-containing wall fragments or endotoxins. The LAL test is the test using LAL reagent to detect and quantify gram-negative bacterial endotoxins. The form of the test may be specified according to the type of indicator that is used: gel clot, turbidimetric, or chromogenic.

10.5 ENDOTOXINS

This section will cover the pyrogenic substance detected by *Limulus* Amebocyte Lysate, which is gram-negative bacterial endotoxin. The LAL test is, of itself, very simple to perform, but to use it appropriately, an understanding of the biochemistry and bioactivities of endotoxins is essential.

Pyrogens may be exogenous or endogenous. An endogenous pyrogen, or factor responsible for direct activation of the fever response *in vivo*, may be activated by various pathways but always by some exogenous substance such as endotoxin. There are a number of types and sources of exogenous pyrogens. Examples are given in Table 10.1.

Typical cell membranes are phospholipid bilayers. Phospholipids (PL) are amphipatic; that is, they are water soluble at one end of the molecule and water insoluble at the other end. This structure results in forces such that molecules arrange themselves into a membrane spontaneously. The water insoluble or hydrophobic ("water fearing") ends are sequestered in a water-free environment by the water soluble or hydrophilic ("water loving") ends. The cytoplasmic membrane, or inner membrane as it is called in gram-negative bacteria, is illustrated in Figure 10.1.

In the figure, a phospholipid molecule is depicted as a small circle with two lines. The circle depicts the polar head of the molecule, which is hydrophilic, and the lines represent the nonpolar, hydrophobic, acyl, or hydrocarbon chains. The bilayer is formed by the alignment of the molecules with the hydrophobic ends of two molecules "facing" each other and the hydrophilic ends in the water phase. The rapid and spontaneous formation of a lipid bilayer in water is driven by the hydrophobic interactions. Water is excluded from the hydrocarbon tails as the tails become sequestered in the nonpolar interior of the bilayer. Van der Waals forces between the hydrocarbon tails favor close packing and further stabilize the structure.

As depicted in Figure 10.1, the cell envelope of the gram-negative bacterium includes the cytoplasmic membrane, the peptidoglycan in the periplasmic space, and the outer membrane. The outer membrane is unique to gram-negative bacteria.

It is the presence of this outer membrane that causes the differential staining reaction in the Gram stain procedure. Gram-positive bacteria are so-called because they retain the initial stain;

TABLE 10.1
Examples of Exogenous Pyrogens

Substances from microbial sources:		
	Bacteria:	Endotoxins (lipopolysaccharides)
		Peptidoglycans
		Lipoteichoic acids
		Exotoxins (proteins)
	Viruses:	Proteins
	Fungi:	"toxins"
	Algae:	"toxins"
Nonmicrobial sources:		
	Antigens (e.g., albumins, antibiotics)	
	Steroids	
	Poly I:C (synthetic double-stranded nucleic acid)	

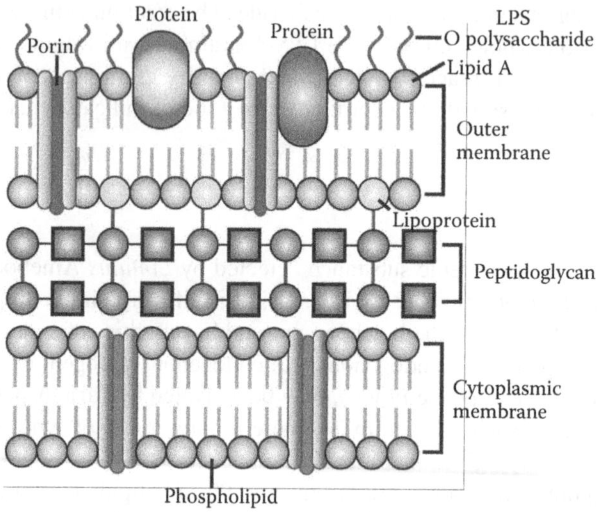

FIGURE 10.1 Gram-negative bacterial cell wall.

gram-negative bacteria, decolorized in the next step of the procedure, are counterstained for easy visualization. The differences in cell wall structure are illustrated in Figure 10.2. The region outside the cytoplasmic membrane in the gram-positive cell is a thick matrix of teichoic acid and peptidoglycan.

In gram-negative bacteria both cytoplasmic membrane of phospholipid leaflet and the inner leaflet of the outer membrane consist primarily of phospholipid. The outer leaflet of the outer membrane is not phospholipid. It is lipopolysaccharide (LPS). As illustrated in Figure 10.1, the molecules are larger than phospholipids, but they have the same amphiphilic nature, which allows them to substitute for phospholipids. Lipopolysaccharides are required for an outer membrane to exist. No mutants are known that will survive without some form of LPS.

Both membranes have proteins associated with the structure. Proteins are located on either side of the membrane and some traverse the membrane.

An LPS molecule has three distinct regions: lipid A, a core oligosaccharide (sometimes separated into an inner and outer core), and a repeating oligosaccharide as illustrated in Figure 10.3. Lipid A is a disaccharide with substituted fatty acids and ionizable groups such as phosphates. The core contains unusual heptoses and a sugar unique to LPS, the 3-deoxy-D-octulosonate (KDO). The repeating oligosaccharide, also called the O-antigen, is made up of a variable number of units of a few sugars that have a specific arrangement within the unit. The types of sugars and their arrangement are characteristic of the species and often the strain. The repeating oligosaccharide, core, and

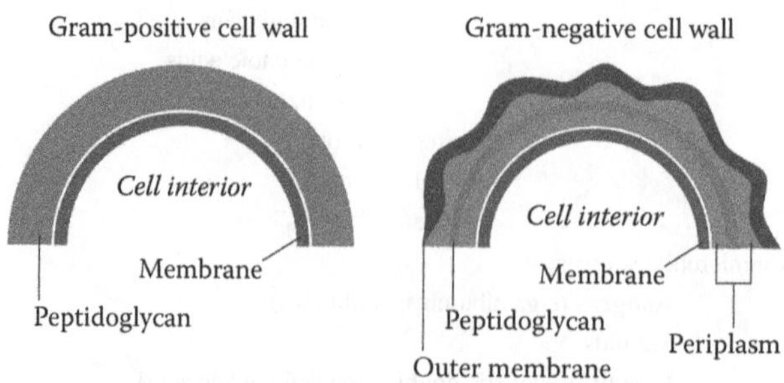

FIGURE 10.2 Cell wall cross section.

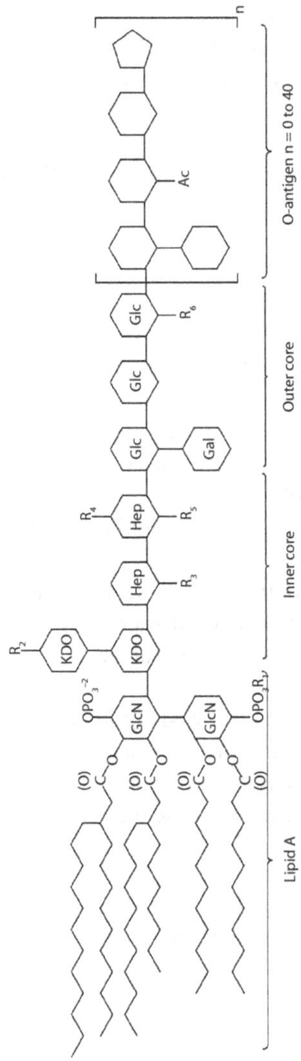

FIGURE 10.3 Schematic of a molecule of lipopolysaccharide. (Modified from Kadrmas J. L. and Raetz C. R. *Journal of Biological Chemistry,* 273, 2799–2807, 1998.)

the disaccharide of the lipid A are hydrophilic. The fatty acids confer the hydrophobic character of the lipid A region.

Lipid A is the most highly conserved portion of the LPS molecule. While minor differences exist, lipid A's from marine bacteria are the same as lipid A's from the human intestinal tract. Lipid A is the portion of the molecule that is recognized by LAL and that causes the majority of the toxic effects credited to endotoxins. Because LAL reacts to lipid A and lipid A is part of all endotoxins, LAL reacts to some degree with all endotoxins.

The sugars in the core region vary to some extent between the species of bacteria. The presence of KDO is not required. If KDO is present, the number of molecules may vary, typically from one to three. Before LAL, endotoxin concentration could be estimated from the amount of KDO by analytical chemistry, but this is not nearly as sensitive an assay as LAL.

The composition of the repeating oligosaccharide is quite diverse between types of gram-negative bacteria. The number of repeating units, and therefore the length of the repeating region, is heterogeneous even within the same cell. The molecular weight of the naturally occurring LPS depends on the presence and number of repeating units and can vary from 2500 to 25,000 Da. In water, the aggregated molecules form membranous particles that are millions of Daltons in molecular weight.

Phosphate groups and other ionizable groups may be present in the lipid A and core. In solutions with pH > 2, these groups will have a negative charge, giving endotoxin and the bacterial cell an overall negative charge. The electrostatic charge is decreased by cations bound to the anions. The charge must be neutralized to maintain a stable structure. Cations include metal ions, polyamines, and even charged proteins that are closely associated with LPS in the membrane.

10.5.1 Biological Activity

Endotoxins elicit biological responses in animals whether they are bound in the outer membrane of the intact cell, bound in membrane fragments in lysed wall particles, or purified in solution. These biological activities range from immunological activity to lethal toxic shock as shown in Table 10.2. The immunological activities depend heavily on LPS as an antigen and therefore on the repeating oligosaccharide units that tend to be accessible on the surface of the cell or the aggregation of endotoxin. The majority of the toxic responses to endotoxins are caused by the lipid A; consequently, most of the toxic activities of endotoxins are elicited when the bacterial cell is lysed to expose the membranous fragments.

The biological activities of endotoxins are diverse. Both the type and intensity of the mammalian response to endotoxin depends on the bacterial source, the purification state, and the background or type of solution in which the endotoxin is presented. As a biological assay, the LAL test is also influenced by the same factors.

Research by Greiseman and Hornick showed significant differences in human threshold pyrogenic doses between endotoxin preparations.[6] It was shown that for LPS from *Pseudomonas sp.*, 50–70 ng/kg was required to elicit a pyrogenic response, while administration of only 0.1–0.14 ng/kg of *Salmonella typhosa* LPS was required to cause the same response which is a 500-fold difference

TABLE 10.2
Biological Activities of Endotoxin/LPS

Pyrogenicity
Disseminated intravascular coagulation
Shock
Lethality
Leucopenia
Leucocytosis
Local Schwartzman reaction
Bone marrow necrosis
Embryonic bone resorption
Complement activation—classical pathway activated by lipid A alternate pathway by oligosaccharide
Depression of blood pressure
Platelet aggregation
Induction of plasminogen activator
Induction of nonspecific tolerance
Induction of endotoxin tolerance
Adjuvant activity
Abortion
Mitogenicity
Inhibition of mitosis
Macrophage activation
Induction of cytokine production—interleukins, tumor necrosis factor, interferon, immunoglobulins, etc.

in pyrogenicity. Naturally occurring endotoxins, which may be what a combination of whole bacterial cells and cell debris derived from the breakdown or disintegration of dead cells, are reportedly less pyrogenic than purified standard endotoxins of equivalent LAL reactivity. This may be because the pyrogenic response is primarily elicited by the form of the endotoxin in the cell debris.

Aggregation size and three-dimensional presentation affect the biological activity of LPS. Size is affected by the type of metal ion bound within the aggregate, but in typical solutions of purified endotoxins, these salts exchange relatively readily. Purified LPS can be prepared in such a way that only one type of cation is available. Pure endotoxin-triethylamine salts form relatively small aggregates and have a high pyrogenicity. At the other extreme, endotoxin-calcium salts have large aggregate sizes and low pyrogenicity. Other biological activities such as lethality and mitogenicity are more dependent on other factors, such as species of animal, than on endotoxin aggregate size.

The pH of the solution affects the structure by changing the charge on the head groups. Acidic solutions diminish the charge repulsion, and there is increased hydrogen bonding to give the largest membranous sheet structures. Basic solutions favor a loss of hydrogen bonding with an increase in the negative charge with subsequent disaggregation into tubular and micellar-like particles.

Substances that bind cations will cause a destabilization of the aggregate size. Thus, introduction of chelating agents may reduce the molecular weight of a purified endotoxin to 300,000 Da or less.

Surface active agents (like LPS) are amphipathic molecules and as such are capable of disrupting membranes. Molecules of LPS will interact with surfactants and become dispersed with respect to the LPS membrane.

Besides the naturally occurring differences in activity between endotoxins, there is also a difference in activity per unit weight that is a function of the ratio of lipid A to LPS in any preparation. The lipid A to LPS ratio is affected by differences in the nutritional background or growth of the cells from which the endotoxin is derived and by differences in the method of extraction. Consequently, equal masses of different endotoxin preparations may have very different activities in either a pyrogen or LAL test as was demonstrated by Greisman and Hornick.[6]

10.5.2 Development of Endotoxin Standards

As long as pyrogenicity was defined by the pyrogen test (rabbit fever test), the labeling of a material as "nonpyrogenic" was straightforward. The rabbits did or did not respond, and it was necessary to know the differences in activity between endotoxins to determine pyrogenicity. The LAL test quantified endotoxin, so the definition of nonpyrogenic had to be in terms of concentration. This required the establishment of a reference endotoxin and endotoxin limits against which the endotoxin concentration of the unknown could be compared.

The first reference standard endotoxin (RSE) referenced by USP (standard reference endotoxin) was the same as the endotoxin used by the Food and Drug Administration, Office of Biologics (later Center for Biologics Evaluation and Research [CBER]). It was derived from cells of *Escherichia coli* O113:H10:K negative and was referred to as EC-2. The amount of endotoxin in an unknown was measured using an LAL test in which the sensitivity of the reagent was calibrated against the RSE and the results were reported in ng/mL of RSE.

A control standard endotoxin (CSE) is any endotoxin preparation, other than RSE, used to calibrate the LAL test. In order to use a CSE instead of RSE, the relative activity of CSE to RSE had to be known. The sensitivity of the lot of LAL was determined in a parallel assay with RSE and CSE. The sensitivity with RSE was expressed in ng RSE/mL and sensitivity with CSE was expressed in ng CSE/mL. The potency of the CSE was then expressed as ng CSE/ng RSE, a potentially confusing and easily misinterpreted practice if the identity of endotoxin got separated from its weight.

Surveys of U.S. rabbit colonies were performed to relate the concentration of a purified endotoxin preparation to the pyrogenic response. From these studies, the threshold pyrogenic limit was determined to be 0.1 ng/mL in a 10 mL/kg dose. The threshold pyrogenic limit is the concentration of endotoxin that caused 50% of the rabbits tested to demonstrate a fever within a 95% confidence interval. The study was performed with the endotoxin preparation, derived from cells of *E. coli* O55:B5, that the then Office of Medical Devices subsequently retained as their LAL reference endotoxin. The two endotoxins were very similar and later were determined to have equivalent activity in a national study. Still, there was often a difference demonstrated between the activities of the two endotoxins within laboratories using different lots of LAL. This made the issue of "real" concentrations of endotoxin even more obscure because the activity of a ng of O55:B5 endotoxin did not always equal that of a ng of O113.

The Food and Drug Administration helped clarify the issue of which endotoxin preparation was which, CSE or RSE, when they defined EC-2 to have five endotoxin units (EU) per ng. By changing the units, it became possible to relate the intensity of an LAL response to the expected rabbit response relative to the reference standard endotoxin, EC-2. The response was now relative only to activity, more closely a function of the lipid A concentration, than to weight of LPS. Eventually, EC-5 (the replacement for EC-2) was accepted as the reference standard for medical devices as well as for drugs and biologics.

Today, a subsequent lot of endotoxin EC-6 has been accepted as an international standard. USP refers to this preparation as "lot G." FDA as EC-6, and the World Health Organization as the Second International Standard for endotoxin. The EP has accepted this preparation as the current BRP. USP and FDA still refer to the units of activity as EUs; the World Health Organization and EP refer to the units of activity as International Units (IUs).

The Japanese Pharmacopoeia (JP) recognizes a "reference" endotoxin other than the one currently used to establish the sensitivity of U.S. licensed gel-clot reagents or establish the suitability of licensed photometric LAL reagents. As such, the Japanese Primary Standard (JPS), derived from a completely different strain of *E. coli*, is technically a secondary standard that has been assayed with reference to the U.S. RSE (similar to CSE).

10.5.3 Endotoxin Limits

Once the practice of expressing endotoxin activities in EU/mL were established, the threshold pyrogenic limit could be given as 0.5 EU/mL in a dose of 10 mL/kg body weight, or, more simply, as 5 EU/kg body weight. This limit applied to parenterals administered by nonintrathecal routes.

10.6 THE *LIMULUS* AMEBOCYTE LYSATE TEST

The LAL test uses the horseshoe crab's host defense mechanism (blood clotting proteins and the coagulation cascade reaction) for responding to gram-negative bacterial infection. The proteins in the coagulation cascade are the basis of the LAL test.

Gram-negative bacteria exist in very large numbers in the ocean, a fact that was not appreciated until the advent of the acridine orange stain. Samples taken from the open ocean show the various shapes of the predominantly gram-negative, slowly metabolizing, bacteria. Acridine orange staining is an extremely useful technique to enumerate bacteria in a sample. It is more sensitive than plate count methods that rely on cell division to observe colony formation or on light microscopy where the small, slowly growing cells cannot be distinguished from particles present in seawater. Gram-negative bacteria number in the millions of cells per milliliter of sea water. Thus, if a horseshoe crab is wounded, it will be exposed to gram-negative bacteria and endotoxins in the water and the coagulation cascade will be initiated.

10.6.1 LAL Biochemistry

The names given to the horseshoe crab's clottable protein and the protein's major cleavage product are reminiscent of human blood clotting. The clottable protein was called coagulogen; the mammalian protein is fibrinogen. The major cleavage product was called coagulin; the mammalian fragment is called

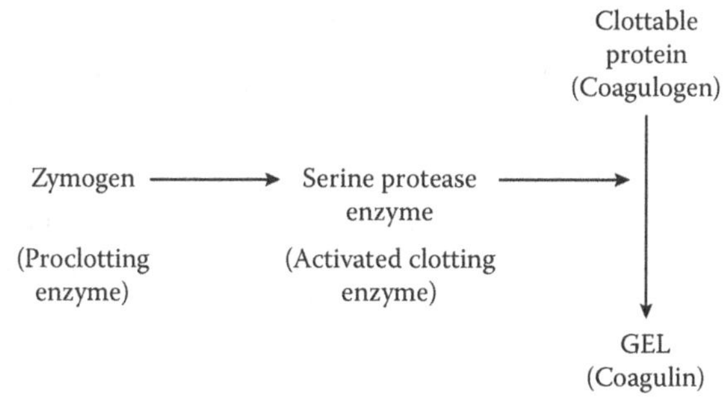

FIGURE 10.4 First concept of the LAL coagulation reaction (Modified from Levin, J., *Biomedical Applications of the Horseshoe Crab (Limulidae)*, ed. E. Cohen, 131–146. New York: Alan R. Liss, 1979.)

fibrin. Here, though, is where the similarity between horseshoe crab and mammalian blood clotting ends. Unlike fibrin, coagulin does not form covalent linkages between molecules when the clot is formed. This means that the horseshoe crab clot is physically fragile and easily disrupted, a fact that probably contributes to the ability of the horseshoe crab to survive the coagulation of its circulatory system.

Levin and Bang recognized that horseshoe crab coagulation was most likely a cascade of enzyme activations, and they proposed a simple model (Figure 10.4) showing those proteins that were known at the time.[7,8] In this model, endotoxin initiates a reaction with an unknown number of steps that ultimately results in the activation of the "clotting enzyme." Activated clotting enzyme cleaved the protein substrate, coagulogen, releasing the fragment, coagulin. Coagulin molecules coalesced by ionic interaction into a gel matrix. When this reaction occurred *in vitro*, one had the LAL "gel-clot" test.

Our current understanding of the LAL biochemical reaction is illustrated in Figure 10.5, which includes the intermediate activation steps and has assigned names to the enzymes that comprise the cascade. Figure 10.5 also shows an alternate pathway by which various β-glucan structures may initiate or augment the primary, endotoxin-initiated pathway.

The enzymes in the coagulation cascade are serine proteases. Serine proteases, as a class, require divalent cations to activate. The reaction between the activated clotting enzyme and coagulogen is extremely specific. Trypsin is a known serine protease able to activate the natural substrate (coagulogen) at the same site(s) as the activated clotting protein and release coagulin to form a gel clot. Consequently, trypsin is capable of causing a false positive LAL test; able to cause a gel-clot response in the absence of endotoxin.

The LAL test (or horseshoe crab blood coagulation) is extremely sensitive to endotoxin because of the cascade of enzyme activations. One molecule of endotoxin activates a molecule of factor C. A single molecule of activated factor C then activates as many molecules of factor B as it comes in contact with. As factor B is activated, it activates molecules of inactive clotting enzyme. Activated

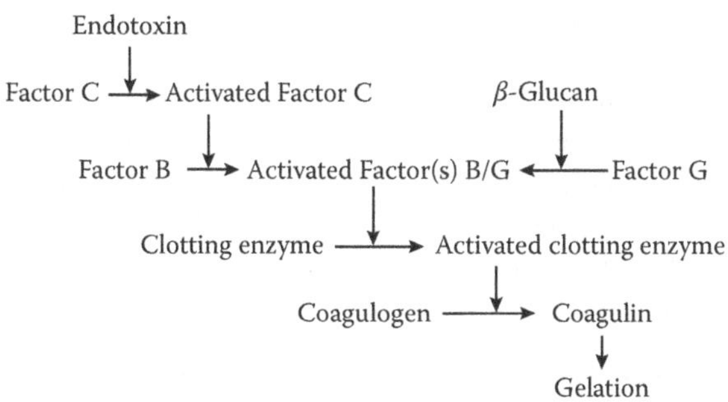

FIGURE 10.5 LAL coagulation cascade. (Modified from Iwanaga, S., et al., *Microbiology*, 29–32, 1985.)

clotting enzyme clips the coagulogen substrate at faster and faster rates as activated enzyme accumulates. While this may be an oversimplification of the biochemistry, the more steps there are in an activation cascade, the greater the magnitude of the final response. The effect of one molecule of endotoxin has been increased by orders of magnitude.

Levin and Bang showed that the turbidity or cloudiness of the *in vitro* LAL reaction mixture increased with time after addition of endotoxin.[9] The turbidity is caused by the increasing number and size of the coagulin matrices that become visible before the reaction mixtures gels completely. Levin and Bang showed that the rate of increase in turbidity was a function of endotoxin concentration. Turbidity became an alternate indicator of the LAL reaction and the test protocols were classified as turbidimetric tests.

It was also shown by Nakamura et al.[4] in 1977, that activated clotting enzyme would cleave the same synthetic substrate as activated factor X, one of the blood clotting enzymes in the human coagulation cascade. This synthetic peptide substrate could be used to measure the amount of activated clotting enzyme in the same way as it was used to measure the amount of activated factor X. The substrate has a chromogen, a *para*-nitroanalide (*p*NA), attached at the homologous amino acid cleavage site. In its bound form, *p*NA is colorless; when it is released as a chromophore by the action of the activated enzyme, it turns yellow. The rate of increase in intensity of yellow is a function of the rate of activation of the proteases in the cascade and therefore a function of endotoxin concentration. A chromogenic LAL test is one that uses the absorbance of the synthetic substrate as an indicator of the amount of activated clotting enzyme and the presence of endotoxin.

Several patented substrates that are more suitable indicators of clotting enzyme activation are available and in use in chromogenic formulations today. These substrates have been designed to more closely represent the amino acid sequence present in coagulogen. No matter how similar they are, though, the synthetic substrates are not as specific for activated clotting enzyme as the naturally occurring substrate, coagulogen. Other serine proteases, in addition to trypsin, are capable of cleaving the synthetic substrate and of causing a false positive response.

10.6.2 LAL Test Methods

While there have been modifications to the test procedures, there are three commonly used ways to read an LAL test: by gelation, by some measure of turbidity, or by some measure of intensity of color ("chromogenicity"). All three methods of performing the LAL test may be read after a fixed incubation period. Two of the methods, turbidimetric and chromogenic, may be read "continuously" in an incubating optical reader by collecting optical densities at frequent, known time intervals. We refer to the methods that read after a fixed time interval as being end point tests and to the ones that are read continuously with time as being kinetic tests.

10.6.2.1 Gel-Clot Test Method

The gel-clot method is the simplest and most widely used form of the LAL test. It has been used over the longest period of time and requires the fewest pieces of equipment. Because of this, the bacterial endotoxins test (BET), the U.S. Pharmacopeial or referee test is the gel-clot method.

Gel-clot tests are end point tests. A test is performed by mixing a small volume of test sample with a small volume of LAL reagent in a reaction tube. The mixture is incubated, typically for an hour at 37°C, and then read at the end of the incubation period for presence or absence of a firm gel clot that remains intact on inversion of the tube. When a gel clot remains intact on inversion of the test tube through 180 degrees, the test is scored as positive. If the gel clot has not formed, or has formed but fails to remain intact on inversion, then the test is scored as negative.

10.6.2.2 Definition of Sensitivity of the Gel-Clot Reagent

The least concentration of endotoxin to cause a positive test under standard conditions is defined to be the sensitivity of the reagent. Sensitivity is commonly referred to as lambda (λ) in LAL test protocols. The sensitivity of each lot of gel-clot reagent is determined by the manufacturer. If the reagent is licensed, its sensitivity will have been determined by the manufacturer using the U.S. reference standard endotoxin (RSE) obtained from CBER (currently EC-6). RSE is tested using the LAL reagent in question and with the reference lot of LAL reagent, which is also obtained from CBER. The sensitivity of the reference LAL has to be confirmed in order to use the results of the test with the manufacturer's lot.

Sensitivity of licensed LAL reagents is determined by testing serial dilutions of RSE, typically starting with 1 or 0.5 EU/mL and ending with a low enough concentration of RSE to define the end point. Table 10.3 illustrates such a titration. The end point is the lowest concentration of endotoxin in a series of decreasing concentrations to give a positive test. It must always be followed by a dilution that is negative to confirm that an end point has been reached.

Typically, LAL tests are performed with replicates and with negative controls. Negative controls ensure that test materials (including water) are not contaminated.

The end point in this series is the 0.125 EU/mL concentration: $\lambda = 0.125$ EU/mL.

10.6.2.3 Pass/Fail Tests by the Gel-Clot Method

Even though the gel-clot test can be read only as positive or negative, it is possible to use the test to perform either a pass/fail test or an assay on a sample with an unknown level of endotoxin. The pass/fail test is as its name implies: a positive reaction in the sample fails the test and a negative reaction passes. If the limit for a sample is 0.25 EU/mL, then the sample could be tested with an LAL of that sensitivity. If the sample were positive, then the sample would have 0.25 EU/mL or more endotoxin and the test would fail. If the sample were negative, then the sample would have less than 0.25 EU/mL and the sample would pass.

It would also be possible to perform a pass/fail test on the sample with a reagent that is more sensitive than the limit. For example, a reagent with a sensitivity of 0.125 EU/mL could be used. In this case, it would be necessary to dilute the sample 1:2 and test the dilution in order to have a pass/fail test at a limit of 0.25 EU/mL. If the 1:2 dilution were positive, it would mean that the dilution had 0.125 EU/mL or more endotoxin and the sample, which is twice as concentrated, would have 0.25 EU/mL or more endotoxin. The sample would fail.

This is a very important concept because it is the basis of the pass/fail pharmacopeial test for endotoxin. If the substance to be tested has an endotoxin limit specified in its monograph, then it would be appropriate to dilute the substance to some concentration that would allow a test of that dilution to return either a pass or fail result. It is easy to calculate the dilution for a pass/fail test at the limit. The dilution factor is simply the endotoxin limit for the substance in EU/mL divided by the sensitivity of the LAL reagent. As above, if the limit is 0.25 EU/mL and the LAL has a sensitivity of 0.125 EU/mL, then 0.25 EU/mL divided by 0.125 EU/mL is 2. The substance must be diluted 1:2 to perform a pass/fail test. This dilution is referred to as the maximum valid dilution (MVD) in the BET test and in other documents on methods or techniques.

10.6.2.4 Assay by the Gel-Clot Method

The gel-clot test can also be used to quantify endotoxin. The sample is assayed by titrating to an end point. This is accomplished by making a series of dilutions of the sample and then testing each

TABLE 10.3
Titration of Reference Standard Endotoxin (RSE)

Concentration of RSE (EU/mL)				
0.5	0.25	0.125	0.0625	0.03125
+	+	+	–	–

dilution until the last positive dilution is found. Typically, the assay is performed on serial twofold dilutions, but it may be performed on any geometrically increasing dilution series. Because the test is performed as a titration, the error about the end point is one tube. If the last positive test were at a 1:4 dilution, then within the error, the last positive tube might have been the 1:2 or 1:8 dilution. This error is referred to as plus or minus a tube. By convention, the titration is performed on twofold serial dilutions, so the error is expressed plus or minus a twofold dilution.

10.6.2.5 Convention for Expressing Dilutions

In microbiology and for LAL tests, the dilution is expressed as the number of parts in a total number of parts. Therefore a 1:2 dilution means one part in a total of 2 parts; it was made by mixing equal volumes of sample and diluent. A 1:10 dilution is achieved by mixing one part sample with 9 parts diluent. A 1:1 sample is one that has not been diluted.

10.6.2.6 Calculation of Sample Concentration

The concentration of endotoxin in a sample is calculated as follows:

$$\text{Concentration} = \lambda/\text{End Point Dilution} = \lambda \times \text{End Point Dilution Factor}$$

Refer to the example given in Table 10.4. The concentration of endotoxin may be calculated for each replicate. The following calculations are for replicate 1:

$$\text{Concentration} = 0.125\ \text{EU/mL}/0.25 = 0.5\ \text{EU/mL}$$

Or the equivalent expression,

$$\text{Concentration in replicate 1} = 0.125\ \text{EU/mL}/(1/4) = 0.125 \times 4 = 0.5\ \text{EU/mL}$$

10.6.2.7 Calculation of the Geometric Mean

Because assays are performed using dilution schemes that are geometric progressions (1, 2, 4, 8, 16, etc.), the average of the results derived from replicate tests must be calculated as the geometric mean (an arithmetic mean is only appropriate for arithmetic progressions such as 1, 2, 3, 4, 5, etc.). A geometric mean is the antilog of the mean of the logarithms of values.

$\lambda = 0.125$ EU/mL

TABLE 10.4
Illustration of Results from a Gel-Clot Assay and Calculation of the Geometric Mean Concentration

	Sample Dilution						
Replicate Number	**1:1**	**1:2**	**1:4**	**1:8**	**1:16**	**Concentration (EU/mL)**	**Log Concentration**
1	+	+	+	–	–	0.5	–0.301
2	+	+	+	–	–	0.5	–0.301
3	+	+	–	–	–	0.25	–0.602
4	+	+	+	–	–	0.5	–0.301
						Sum of the log of the concentrations:	–1.505
						Sum divided by the number of replicates:	–0.376
						Antilog = geometric mean:	0.42

The end points of replicates 1, 2, and 4 are the 1:4 dilutions; the end point of replicate 3 is the 1:2 dilution.

In an endotoxin assay, the end point of each replicate can be expressed in terms of endotoxin concentration. To calculate the geometric mean, first, find the logarithm of each of the concentrations. Add the logarithmic values and divide by the number of replicates to get the average logarithmic value. Calculate the antilog of the average logarithmic value to determine the geometric mean. It does not make any difference which base is used for the logarithmic calculation as long as it is used throughout the calculation. The example in Table 10.4 uses logarithms to the base 10. The example in Table 10.5 includes four replicates, but this should not be taken to mean that unknowns need to be assayed in quadruplicate. Test protocols typically include at least duplicates.

$\lambda = 0.125$ EU/mL

It is most convenient if the end point dilutions are expressed as decimal fractions before calculating their logarithms.

Results of an LAL assay are reported as the concentration obtained. If the GM concentration is 1.4 EU/mL, then this value is compared directly with the limit for the sample. If the limit is 2 EU/mL, then the sample passes. If the limit were 1 EU/mL, the sample fails. If the limit were 1.4 EU/mL, the sample would fail.

The range of error for a result of 1.4 EU/mL is 0.7–2.8 EU/mL. The error is used to assess how closely different tests agree with one another. The full range is often necessary to determine whether or not results obtained from different laboratories are the same. Within a laboratory and certainly between tests performed by the same analyst, results are even more reproducible and tend to vary within a twofold on only one side or the other, not on both sides of the end point concentration.

10.6.2.8 Geometric Mean Sensitivity

A geometric mean is also the appropriate calculation to make when performing RSE or CSE titrations to confirm LAL reagent sensitivity. The geometric mean is calculated on the sensitivities obtained for each replicate.

10.6.2.9 How to Choose a Suitable LAL Sensitivity

Ideally, one lot of LAL reagent should be used for all applications within a laboratory. This reduces the amount of work required to accept the lot and to validate products (see regulatory chapter).

TABLE 10.5
Alternative Method to Calculate Sample Concentration by Calculating the Geometric Mean of the Dilutions at the End Points First

	Sample Dilution						
Replicate Number	1:1	1:2	1:4	1:8	1:16	End Point Dilution	Log End Point Dilution
1	+	+	+	–	–	0.25	–0.602
2	+	+	+	–	–	0.25	–0.602
3	+	+	–	–	–	0.50	–0.301
4	+	+	+	–	–	0.25	–0.602
						Sum of the log of the end point dilutions:	–2.107
						Sum divided by the number of replicates:	–0.527
						Antilog = geometric mean:	0.3
						λ/geometric mean:	0.42

When sample formulations are complex, especially if they contain proteins, electrolytes, or are buffered out of the range required to perform the test, it is more likely that the more sensitive LAL reagents will be a better choice. If the formulations and types of products are quite diverse, then their interference characteristics are also likely to vary considerably and one would choose the most sensitive LAL available.

If interference is not an issue, then consideration should be given to the range of endotoxin limits represented by the products to be tested. While at first glance it may appear that fewer dilutions are required if a less sensitive reagent is used, most of the time it will be possible to achieve even large dilutions in a minimum of dilution steps. Many times, it is not the product, but the raw material or in-process material that requires the strictest limit. If, for example, the process water has the strictest limit, then it might define the minimum sensitivity for the LAL reagent used to test it and all the other materials.

In the final analysis, the choice of reagent sensitivity (and even the choice of LAL method) will depend on the way the reagent reacts with the sample or product.

10.7 PHOTOMETRIC METHODS

Photometric methods include both the turbidimetric and chromogenic forms of the LAL test. Whether read as end point or kinetic tests, both require an optical reader to obtain optical density values. Turbidimetric tests are typically read at shorter wavelengths to enhance sensitivity to small particles. Chromogenic tests are read at the wavelength that best absorbs the color of the leaving group. For the yellow *para*-nitroanaline (*p*NA), this is 405 nm. If *p*NA is derivatized to form an azo (magenta) dye, greatest absorbance is obtained 540 nm.

End point tests do not require collection of very many data points, so these tests may be performed with optical density readers that are commonly available and they may be analyzed without the assistance of a computer.

Kinetic tests involve the collection and analysis of many data points during the reaction period, so a computer with adequate data storage capacity is necessary. Kinetic tests must be performed in optical readers capable of incubating multiple samples. The reader must be capable of reading and storing optical density data at very frequent, set intervals; these are typically every 10–20 s.

For turbidimetric assays, the reader must be capable of obtaining optical density values from each reaction mixture without disturbing the integrity of the coagulin matrix. For most chromogenic assays, turbidity develops along with the increase in color intensity because these chromogenic formulations still contain coagulogen. Therefore, even though it is not as critical as for turbidimetric tests, it is still advisable to use an optical reader that is capable of reading samples without moving them.

10.7.1 Reaction Kinetics

The reaction kinetics (Figure 10.6) are similar whether the indicator of concentration of activated clotting enzyme is gelation, turbidity of coagulin matrices, or absorbance by *p*NA. The shape of the curve after mixing reagent with endotoxin is defined by a period of little or no change in the rate of increase in optical density. This is followed by a period of increasing optical density and then another period of constant optical density. The differences in the rates of increase in optical density are a function of the endotoxin concentration.

The rate of increase in optical density increases with increasing concentration of endotoxin. At low concentrations of endotoxin, the rates are clearly different between concentrations, but as the endotoxin concentrations become greater, the differences in rate between them become very similar and may be indistinguishable. Typically, LAL reaction rates are expressed in terms of the time it takes for the optical density to reach a predetermined optical density value or "threshold optical density." This time is referred to as an "onset time." The onset times become shorter with increasing concentrations of endotoxin. At high concentrations the difference between onset times becomes

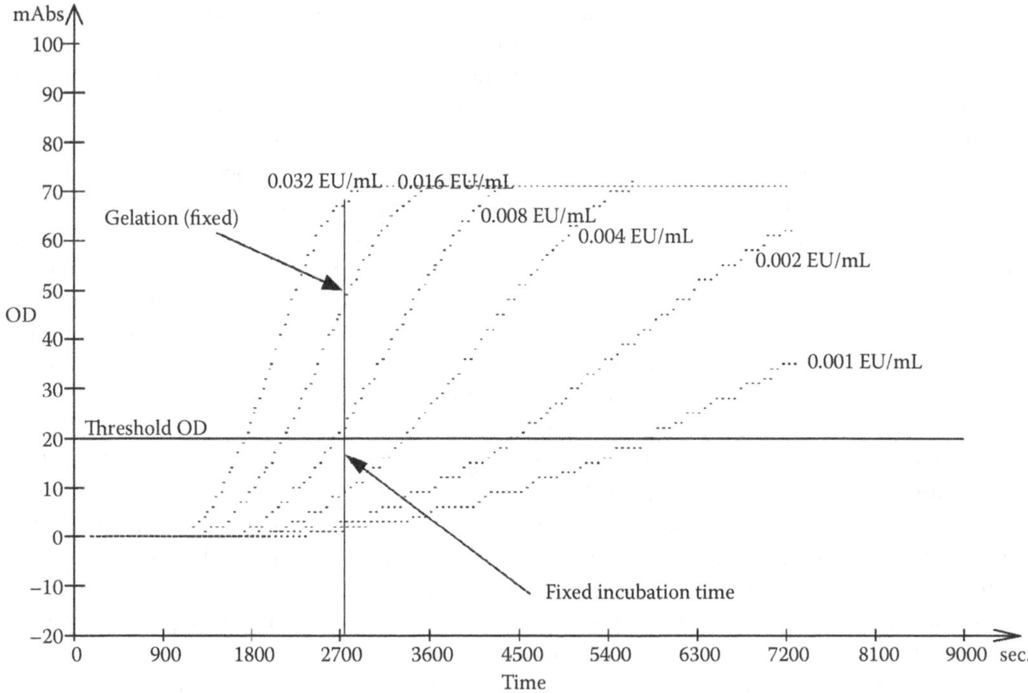

FIGURE 10.6 Kinetics of increase in activated clotting enzyme as measured by turbidity or coagulin matrices or absorbance of *p*NA.

so narrow that distinguishing between concentrations requires optical density values to be read at extremely precise and narrow intervals.

Each reaction is limited by the concentration of substrate (coagulogen or synthetic substrate) in the formulation. Once it is cleaved, there can be no more increase in optical density. The initial cascade step covers the period before any visible changes in optical density can be discerned as well as a lag that may result from initiation of the reaction itself.

10.7.2 End Point Test Protocols

There are six, twofold, serially decreasing concentrations of endotoxin represented in Figure 10.6. Gelation occurs toward the end of the increase in turbidity in formulations that do not include a synthetic substrate. The gel-clot method is an end point test because its presence or absence is read after a set incubation period. As the incubation period is lengthened, fewer and fewer concentrations of endotoxin reach the gelation point. Thus one way to increase the sensitivity of a gel-clot test is to increase the incubation time. For regulatory purposes, though, standard gel-clot procedures require a 1 hour incubation period.

End point tests are performed by photometric methods in a similar manner. The optical density of each sample is read at time zero (blank) and again after a set period of incubation. The reaction may be stopped by addition of acid (chromogenic) or read at the precise time without stopping the reaction (turbidimetric method). End point turbidimetric tests are most easily performed by staggering the addition of LAL reagent and performing the test on samples in separate reaction tubes.

If the formulation requires addition of substrate after preincubation for a set period of time (two-step chromogenic method), then the LAL reaction is allowed to progress for an initial period of incubation; substrate is added and incubated for a short interval; and the reaction is stopped with acid and read in the spectrophotometer. The timing of the initial LAL reaction and the timing of the substrate/enzyme reaction must be precise and exactly the same for each sample. In addition, it is a good idea to confirm with any new formulation that the recommended period of the substrate incubation is short enough that the optical density reading is obtained before the substrate is used up in the reaction. This is particularly important if the assay will be used to detect relatively high concentrations of endotoxin.

Once the values are read, a standard curve is constructed by linear regression of optical density versus endotoxin concentration. Only those optical density values that are obtained from reactions that are increasing in optical density at the time of the reading are valid to use in the construction of the standard curve. Values that are obtained from reactions that have not yet started (initial lag phase) or from reactions that have saturated (show no further increase in optical density with time) cannot be included. Because of this necessity, the range of concentrations that may be included on an end point standard curve is very narrow, typically within one log. Also, the test is technically easier to perform over a range of concentrations that have well-separated onset times, so end point tests are used typically to measure endotoxin concentrations that are less than 1 EU/mL.

By inspection of the LAL kinetics shown in Figure 10.6, the 0.016, 0.008, 0.004, and 0.002 EU/mL concentrations have progressed beyond the background and are not yet saturated. These three values would be included on the standard curve. The 0.032 and the 0.001 EU/mL concentrations are "borderline" and might be included. The problem is that one does not view the reaction kinetics when performing an end point assay, so the decision about whether or not to include a concentration on the standard curve must be made by inspection of the plot of the standard line. The FDA guideline recommends that the absolute value of the coefficient of correlation (*r*) be equal to or greater than 0.980. This correlation is obtainable even when the standard line exhibits the flattening off at either end that is characteristic of either saturation or failure to be significantly greater than the blank. Decisions about whether or not to retain the points are ultimately based on the values of the recovered endotoxin when the ODs of the standard concentrations are used in the line equation.

The reagent manufacturer will typically advise on appropriate incubation time. Alternately, a technician can test several incubation times to determine which time will allow inclusion of the concentrations desired on the standard curve.

10.7.3 Kinetic Tests

The advantage of kinetic methods over end point methods is that there is no set period of incubation. Once the reaction is initiated, a spectrophotometer collects optical density values until collection is terminated at any time chosen by the technician. The only requirement to quantify endotoxin is to obtain an onset time. Therefore, the range of concentrations that may be included in a standard curve is quite broad and limited only by the ability of the instrument to distinguish the times between concentrations and by the intrinsic ability of the reagent formulation to show a response.

Concentration of endotoxin is calculated from a standard curve constructed by linear regression of the log of the onset time on the log of the endotoxin concentration (Figure 10.7). Unfortunately, LAL reaction kinetics are not simple and do not lend themselves to linear regressions, even on logarithmic plots, that are in fact linear. When enough points are included on the standard curve, it is clear that there is some curvature even though the regression coefficient indicates satisfactory linearity. For this reason, and because onset times become closer and closer as endotoxin concentrations increase, broad range standard curves (4 or more log intervals) are typically used to estimate endotoxin concentrations and narrower ranges (1–3 log intervals) for more precise determinations.

The absolute value of the correlation coefficient for the calibration curve recommended in the FDA guideline is 0.980. This is easily obtained even over broad ranges of endotoxin concentrations.

10.7.4 Sensitivity of Photometric Methods

The sensitivity of a photometric method is defined as the least concentration included on the standard curve. The LAL reagent manufacturer will indicate a range over which the specific formulation

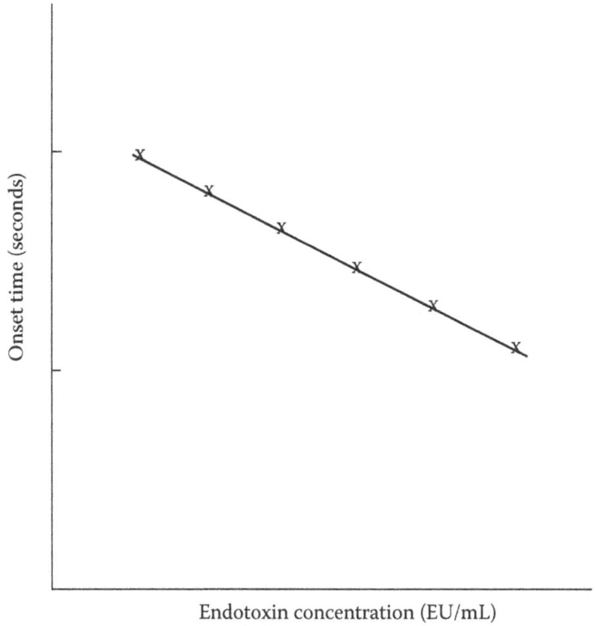

FIGURE 10.7 Representation of a kinetic photometric standard curve comprised of six endotoxin concentrations tested in triplicate.

may be used, but the actual sensitivity of the test is set by the technician. Sensitivity is influenced by the type of instrumentation, the length of the light path length reaction volume in the cuvette or microplate well, and the type of material (glass or plastic) used for the cuvette or microplate.

10.7.5 Selecting a Test Method

It is usually most convenient and easiest to begin a testing program with the gel-clot method. The equipment is minimal, and therefore it does not require a large capital outlay to start testing. Qualifying the test and the technician is relatively easy because the sensitivity of the reagent is established by the manufacturer. Results are simple to read because the test in each tube is either positive or negative. This test is the oldest form of the LAL test and typical responses to different types of substances are relatively well known. Because it is typically less sensitive than photometric methods, it also tends to be less sensitive to conditions that cause interference. The gel-clot method is currently the compendial method which means that controversial results are arbitrated by the gel-clot test.

Photometric methods provide tests with greater sensitivities. The greatest sensitivities are obtained by kinetic methods. Chromogenic methods may have sensitivities as great as 0.005 EU/mL and turbidimetric methods as great as 0.001 EU/mL. The greater sensitivities are especially useful for samples that cannot be tested by gel clot because of interference. These samples can be diluted further (greater MVD) to overcome the interference and thus be successfully tested by the photometric methods.

Characterization and validation of samples for photometric methods is typically more complex and time consuming than it is for the gel-clot method.[10] Technicians do require more skill to perform the tests reproducibly and more training to interpret test results.

Photometric methods require additional instrumentation to incubate the tests, collect, and analyze the data. Both the instrument and software need to be qualified (validated) if the tests will be used in a regulated environment. Three sources for information on this are: the USP chapter on validation, the International Conference on Harmonisation (ICH) validation documents, and the

FDA guidance on validation. This adds to the capital costs but may ultimately reduce the cost of labor once products are validated and if the test is sufficiently automated.

10.8 INTERFERENCE

The term, interference, is usually used to mean recovery of either more or less endotoxin than is actually present in the sample. Recovery of more endotoxin is referred to as enhancement. Recovery of less endotoxin is referred to as inhibition. False positive tests must be distinguished from interference. A false positive test gives a reaction in the absence of endotoxin. False negative tests are those reactions that should have given a response but do not react. These are generally considered to be extremes of inhibition.

Trypsin is a substance that will give a false positive in all LAL methods. Other serine proteases may be active in the chromogenic method as discussed above. β glucans enhance LAL reactions by activating the clotting enzyme in the absence of endotoxin. The β glucans also affect various formulations of LAL reagents to different extents.

Both the physical configuration of endotoxin and the biochemical state of the LAL proteins can cause either inhibition or enhancement. The aggregate size and the interaction of endotoxin with other substances in solution may hinder or facilitate the activation of the LAL cascade relative to the activation by endotoxin in water. The LAL reaction depends on the activity of the enzymes in the cascade. Any solution that affects the rate of the enzyme reactions will affect the sensitivity of the assay. A reduction in sensitivity is evident as inhibition, while an increase in sensitivity is enhancement.

A wide range of substances can interfere with the LAL test. Enzyme reactions have optimum pH's, temperatures, ionic conditions, etc. Proteins are denatured by high salt concentrations, alcohols, and surfactants. Substances that affect the structure of endotoxins may also affect the activity of the LAL enzymes. It is often difficult to determine the precise nature of interference, but that is less important than demonstrating that the interference can be overcome.

One source of interference that is easy to check and correct is the pH of the reaction mixture. In general, the reaction works best in the pH range 6–8. If the pH is out of the acceptable range, the test will fail. Either the gel will not form or the positive product controls will fail to be recovered.

It may be possible to distinguish between inactivation of LAL proteins and endotoxin stereochemistry as the cause of inhibition. If LAL proteins are denatured, no amount of added endotoxin will overcome the interference. On the other hand, if the endotoxin presentation is less than optimum, it will be possible to overcome inhibition with greater concentrations of endotoxin. This is why it is so important to choose a concentration for inhibition controls that will assure that the sensitivity of the test is confirmed.

Inhibition controls for the gel-clot method are prepared by adding endotoxin to the sample to achieve a concentration in the sample that is twice the sensitivity of the LAL (2λ). If this control fails to clot, then there is inhibition that exceeds the error of the test. In other methods, added endotoxin must be recovered within the error of the method, but the choice of the concentration of the added endotoxin is not as clearly defined as for gel clot. Concentrations are typically chosen from the middle area of the calibration curve. Controls for broad range calibration curves are less likely to assure lack of interference at the sensitivity of the test because the concentrations will be much greater than the sensitivity. For a range from 0.005 to 50 EU/mL, the added concentration is often 0.5 EU/mL, which is a hundred-fold greater than the intended sensitivity of the test.[11]

10.9 TESTING OLIGONUCLEOTIDES FOR THE PRESENCE OF ENDOTOXINS

Oligonucleotides present many challenges when testing for endotoxins. Certain groups of oligonucleotides are polyanionic in nature and present false positives or variable results with increasing dilutions in endotoxin testing if the anionic nature is not dealt with by either overcoming the interference with dilution, adjusting the pH, or neutralizing the charges.

Other challenges and interferences in endotoxin testing presented by oligonucleotides are due to the following:

- Materials that are used to chemically stabilize by specific modification of the oligonucleotide backbone
- Oligonucleotides that are loaded into polymeric nanocarriers resulting in polyion complex micelles
- Folded conformers and complexes
  - Hairpin duplex structures
  - Biomolecular hairpin triplex structures
  - Monomolecular triplex structures

For oligonucleotides that have micelles or folded structures, the micelles must be disrupted and folded structures must be separated before endotoxin testing otherwise endotoxin may be unavailable for detection since the LAL reagent cannot penetrate the micelles or folded structures. The following treatments can be used to address micelles and folded structures and allow accurate measurement of endotoxin:

- Adjust pH to disrupt the micelle being careful not to effect the endotoxin concentration. The pH should be measured with the lysate/sample mixture to ensure that assay is being conducted in the appropriate pH range specified by each manufacturer and adjusted if outside of the needed pH range.
- Surfactants can also be used to disrupt the micelle being careful not to effect the endotoxin concentration. Surfactants can also cause inhibition with the LAL cascade, which should be monitored carefully.
- Dissolve or dilute the oligonucleotide with small chain alcohols such as methanol, ethanol, or isopropanol. The smallest amount of alcohol should be considered because of inhibition with the LAL cascade from the alcohol. The best approach is to try several different percentages of alcohol to determine which one gives the most consistent results.
- Heat the oligonucleotide at a temperature to separate the strands being careful not to destroy any endotoxin present. A starting point might be 95°C for 2 min for some oligonucleotides but may be too high for some oligonucleotides. These oligonucleotides might be better treated at lower temperatures.

Each treatment method described above should be carefully examined for its effect upon endotoxin through the use of a challenge sample. Challenge recovery is required by the USP and EP for samples that are treated. This approach involves analyzing the sample and comparing it to the sample to which a known amount of endotoxin is added. One should be careful not to add extremely high concentrations of endotoxin as a challenge when compared to the actual endotoxin contained in the sample. This approach would require large dilutions that would most likely not be used in the actual assay. A suggested range of endotoxin concentration to consider is 2–5 times the actual endotoxin contained in the sample. This will ensure that the signal of the challenge sample is readily distinguishable from the signal of the sample. The challenge recovery can then be calculated by the following:

$$\text{Recovered Endotoxin (EU/mL)} = \text{Challenge Sample (EU/mL)} - \text{Sample (EU/mL)}$$

$$\text{Recovery (\%)} = (\text{Recovered Endotoxin/Added Endotoxin}) \times 100$$

TABLE 10.6
Example Kinetic Turbidimetric Endotoxin Assay Data

Dilution Factor	PPC Recovery (%)	Final Result (EU/mL)
1:10	1 (inhibition)	0.182
1:100	94	Greater than highest calibration standard
1:1000	153	9.93
1:10,000	105	<10.0

Consideration should be given when choosing the assay method for endotoxin as the methods differ in their suitability for the detection of endotoxin. Some oligonucleotides can be tested by the gel-clot assay method after dilution of the sample. This generally applies to the simpler materials with large maximum valid dilution (MVD).

The kinetic turbidimetric and chromogenic methods are frequently chosen to test the more complex oligonucleotides owing to the increased assay sensitivity when compared to the gel-clot assay. The increased assay sensitivity allows for a greater dilution, which can be useful in overcoming inhibition from pH and/or alcohol used to dissolve/dilute the oligonucleotide.

Some general points to consider when starting an endotoxin assay for an oligonucleotide:

- A 1 or 10 mg/mL aqueous solution of the oligonucleotide is a good starting concentration for endotoxin testing. This concentration, along with the endotoxin limit and lysate sensitivity, is all used to calculate the MVD.
- For an initial screening of the oligonucleotide, a 10-fold dilution scheme with testing in duplicate at each dilution can be used.
- Following directions from the USP, the positive product control (PPC) is chosen from the middle of the calibration curve and assayed with the sample dilutions.
- Once the assay is completed, the data should be reviewed in the following fashion:
  - Many oligonucleotides cause inhibition of the LAL reaction at the lower dilutions. This will be seen through the PPC recovery of less than 50% of the added endotoxin concentration. For example, at a 1:10 dilution, a PPC recovery of 17% of the added endotoxin concentration indicates an inhibitory effect of the oligonucleotide at that dilution. Higher dilution should be examined until the PPC recovery is within the specification (for the USP the range is 50–200%).
  - Table 10.6 shows some example kinetic turbidimetric data for a 10 mg/mL oligonucleotide solution assay.

On the basis of the screening data above, smaller dilutions factors could be used to screen the sample solution as shown in Table 10.7.

TABLE 10.7
Example Kinetic Turbidimetric Endotoxin Assay Data Using a Smaller Dilution Factor

Dilution Factor	PPC Recovery (%)	Final Result (EU/mL)
1:2000	142	10.0
1:4000	130	8.26
1:8000	134	<8.0

10.10 SUMMARY

Oligonucleotides can be analyzed for bacterial endotoxin if sufficient method development is performed first. The features of oligonucleotides that make them unique to their purpose also make them a challenge to perform endotoxin testing. When attempting method development for bacterial endotoxin testing, sufficient consideration should be given to the ionic state of the test material and whether the material is polyionic, disrupting any micelles that may exist, and how tightly folded the conformer/complex may be. The more complicated oligonucleotides often require a more complex method development process. Many of the treatments previously listed can be combined to successfully test these oligonucleotides. With each treatment process, careful attention should be paid to the challenge recovery to ensure that the treatment is appropriate for use in the bacterial endotoxin assay and that any endotoxin that may be present in the oligonucleotide may still be detected.

REFERENCES

1. Howell, W. H. 1885. Observations upon the chemical composition and coagulation of the blood of *Limulus polyphemus*, *Callinectes hastatus*, and *Cucumaria sp*. Johns Hopkins University Circular. 43: 4–5.
2. Loeb, L. 1902. The blood cells and inflammatory processes of *Limulus*. *Journal of Medicinal Research* 2: 145–158.
3. Levin, J., and F. B. Bang. 1964. The role of endotoxin in the extracellular coagulation of Limulus blood. *Bulletin of Johns Hopkins Hospital* 115: 265–274.
4. Nakamura, S., T. Morita, S. Iwanaga, M. Niwa, and K. Takahashi. 1977. A sensitive substrate for the clotting enzyme in horseshoe crab hemocytes. *Journal of Biochemistry* 81: 1567–1569.
5. Barthel, K.W. 1974. *Limulus*: A Living Fossil. *Naturwisson-Schaften* 61: 428–433.
6. Greisman, S. E., and R. B. Hornick. 1969. Comparative pyrogenic reactivity of rabbit and man to bacterial endotoxin. *Proceedings for the Society of Experimental Biology and Medicine* 131: 1154–1158.
7. Levin, J. 1979. The reaction between bacterial endotoxin and amebocyte lysate. In *Biomedical Applications of the Horseshoe Crab (Limulidae)*, ed. E. Cohen, Progress in Clinical and Biological Research 29, 131–146. New York: Alan R. Liss.
8. Levin, J., and F. B. Bang. 1964. A description of cellular coagulation in the Limulus. *Bulletin of Johns Hopkins Hospital* 115: 337–345.
9. Levin, J., and F. B. Bang. 1968. Clottable protein in Limulus: its localization and kinetics of its coagulation by endotoxin. *Thrombosis et diathesis haemorrhagica* 19: 186–197.
10. Food and Drug Administration. 1987. Guideline on validation of the Limulus Amebocyte Lysate test as an end-product endotoxin test for human and animal parenteral drugs. Biological Products, and Medical Devices. http://www.fda.gov/downloads/BiologicsBloodVaccines/GuidanceComplianceRegulatoryInformation/Guidances/Blood/UCM080966.pdf.
11. Associates of Cape Cod, Inc. 2005. LAL Workshop for Bacterial Endotoxin Testing.

11 Analysis of Residual Solvents by Head Space Gas Chromatography

Sky Countryman
Phenomenex
Jose V. Bonilla
Girindus America, Inc.

CONTENTS

11.1 INTRODUCTION

Residual solvents can be defined as any solvent used or produced in the manufacturing or formulation of an oligonucleotide that is not completely removed during the process. If not removed, these solvents can be present in the final formulation of a drug product and administered to a patient when undergoing medical treatment. The solvents do not provide any therapeutic value and may even pose a potential health risk to the patient, depending on the solvents toxicity and the amount ingested, so they should be removed to the extent possible. In this chapter we will review the current testing requirements for residual solvents in oligonucleotide products as well as provide detailed instruction about how to perform this testing.

11.2 THEORY

The most common method for analyzing residual solvents in pharmaceutical products utilizes head space gas chromatography (HS-GC). In this section we will review some of the HS-GC theory as well as discuss other techniques that can be employed if HS-GC does not work. This section assumes some knowledge of gas chromatography theory and terminology; for more detailed information on the technique of gas chromatography, see Refs. 6, 7, 8, and 11.

There are three primary components to a GC system that must be optimized when developing a new method for residual solvent testing: (1) injection techniques, (2) detectors, and (3) the GC column.

11.2.1 Injection Techniques

11.2.1.1 Head Space

In a head space analysis, a sample (liquid or solid) is put into a sealed vial and heated to cause volatile components in the sample to partition into the gas phase. A defined volume of this gas is sampled and introduced into the GC system for quantitation. There are two main advantages of head space as an injection technique. The first is that only the volatile portion of the sample is introduced into the column. Drug products often contain nonvolatile components that can damage the GC column and cause problems with the analysis. The second main benefit is an increase in sensitivity. Reports have shown that head space injections can increase sensitivity up to 1 order of magnitude over traditional extraction techniques.[1]

The equilibrium distribution of the analyte between the liquid or solid phase and the gas phase is expressed by the partition coefficient K (Equation 11.1). Sensitivity of the testing procedure is largely affected by the partitioning of the analyte into the gas phase (Figure 11.1). The partitioning is influenced by a number of variables including (1) temperature, (2) sample matrix, and (3) sample volume.[2]

$$K = \frac{C_s}{C_g} \tag{11.1}$$

where C_s is the concentration of the analyte in the sample phase and C_g is the concentration of the analyte in the gas phase.

11.2.1.1.1 Temperature

Increasing the temperature helps decrease the *K* value and drives the analyte out of solution and into the head space. Higher temperatures can help improve *K* for polar analytes or those solvents

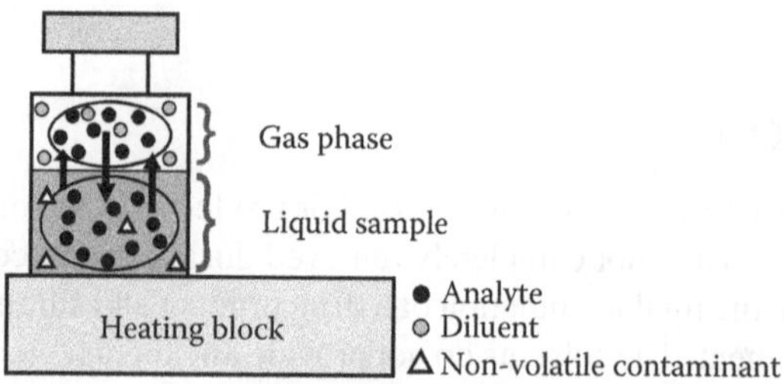

FIGURE 11.1 Partition coefficient (*K*) of a head space sample.

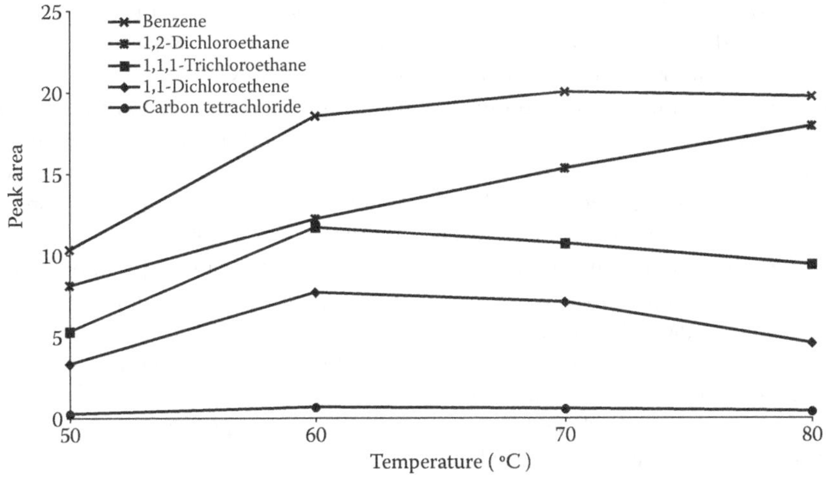

FIGURE 11.2 Effect of temperature on analyte sensitivity in HS-GC.

with higher boiling points. At higher temperatures, there is a risk of saturation of the head space with the diluent.

When performing head space testing, it is important to use a temperature below the boiling point of the diluent to prevent saturation from occurring. Saturation by the diluent can reduce sensitivity for the target analytes and cause interferences in the chromatography. The ideal temperature for a mixture of solvents will be a compromise that provides the required sensitivity for all components but may not provide maximum sensitivity for all. When starting new method development, try starting at 20°–30°C below the boiling point of the diluent and optimize up or down from there based on experimental results.

The example in Figure 11.2 shows the effect of temperature on five solvents diluted in water. Increasing the temperature from 50° to 60°C resulted in almost double the peak area, which greatly improved sensitivity for the assay. When temperature was increased to 70°C and then to 80°C, the response for 1,2-dichloroethane increased proportionally, but the response for 1,1,1,-dichloroethane

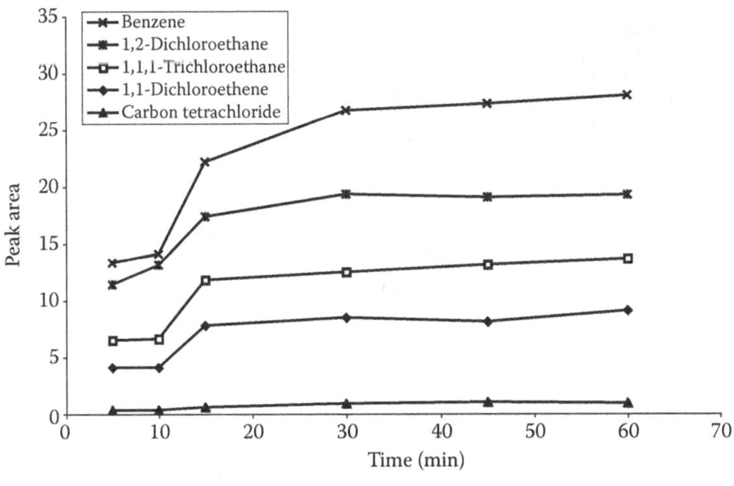

FIGURE 11.3 Effect of equilibration time on response.

and 1,1,-dichloroethene decreased. The response for benzene and carbon tetrachloride remained relatively unchanged, indicating that temperature had no more effect on *K* for those analytes. In this example, the best compromise for sensitivity for the assay was 60°C.

Accurate quantitation in HS-GC requires that the sample has had time to reach complete equilibrium in the gas phase. The required time for equilibration will be unique for each solvent and diluent. The time required to achieve that equilibrium must be determined experimentally by running samples heated at different time points. When no increase in response is observed with increased time, the equilibrium has been reached (Figure 11.3). When working with a mixture of compounds, the equilibration time is based on the analyte with the slowest partitioning time.

Returning to the example of the five solvents in water, the equilibrium for 1,2-dichloroethane, 1,1,1-trichloroethane, and 1,1-dichloroethene was achieved after about 15–20 min. However, it required about 30 min for benzene and carbon tetrachloride to reach equilibrium. Because it is important that all compounds reach equilibrium, the samples had to be heated for a minimum of 30 min before analysis could be done. Heating for longer than 30 min provided no further benefit in peak response and would have been wasted time.

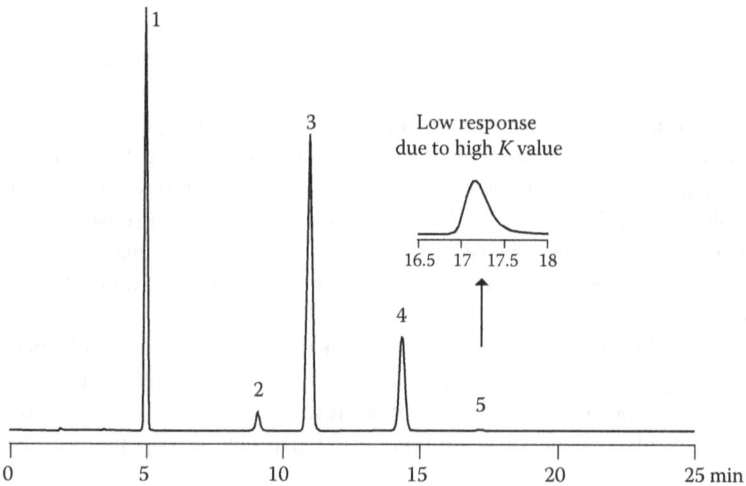

Column: Zebron ZB-624
Dimension: 30 m × 0.53 mm × 3.00 μm
Injection: Split 2:1 @ 140°C, 1 μL
Carrier Gas: Constant flow helium, 4.63 mL/min
Oven Program: 35°C for 20 min to 240°C at 35°C/min.
Detector: FID @ 260°C
Sample: Samples in water with 1g of anhydrous sodium sulfate; equilibrated at 80°C for 60 min

Peak Number	Analyte	Concentration (μg/mL)
1	Methylene chloride	10
2	Chloroform	60
3	Benzene	1
4	Trichloroethylene	20
5	1,4-Dioxane	2

FIGURE 11.4 Response of various solvents in water by HS-GC at 80°C.

11.2.1.1.2 Sample Matrix

The sample matrix can have a large impact on *K* for each solvent, which is why it is important to minimize sample to sample variation. Slight variation between batches of an oligo product or slight changes in the formulation can have an effect on *K*. One of the more common reasons this problem occurs is related to changes in the ionic strength of the matrix solution due to buffers or salts. To improve reproducibility, it is recommended that all samples be supersaturated so as to normalize the ionic strength. The choice of salt can influence *K* for a given analyte, so if poor results are seen with one salt, then others should be evaluated.[3]

Some salts may not be compatible with organic diluents. In these cases, other options should be explored in order to reduce matrix related sample variations. One option may be sample agitation during the heating process. Many of the commercial head space units include a shake option that can be programmed.

Another option is to change the sample diluent to one in which the analyte of interest is less soluble. The antifoaming agent 1,4-dioxane is very water soluble and does not like to partition into the gas phase (Figure 11.4). Increasing temperature from 40° to 80°C decreases the *K* value for 1,4-dioxane from 1618 to 288, which is still a relatively low volatility when compared to other solvents.[4] At this temperature, there is also a high concentration of water that is partitioned into the gas phase, which can then interfere with the analysis. When looking to analyze 1,4-dioxane by head space, water is not the ideal diluent. Using an organic diluent like DMSO or DMF may improve the partitioning of 1,4-dioxane.

11.2.1.1.3 Sample Volume

The final step in optimization of head space conditions should be to adjust the sample volume in order to achieve the required sensitivity. The volume of the sample and the size of the vial used for the head space analysis can affect the sensitivity of the assay.[5] Sample volume has the largest effect on sensitivity for more volatile compounds. When working with compounds with high *K* values, sample volume will have limited effect on the sensitivity. Most solvents are fairly volatile, with the exceptions of some alcohols or other more water soluble compounds like isopropanol, which has a *K* value of 825 at 40°C. The primary drawback of using larger sample sizes is the corresponding time required to achieve equilibrium.

The variation of sensitivity resulting from small changes in sample volume can also affect the precision of the assay. In general, care should be taken to keep sample volumes as consistent as possible. Volumetric glassware should be used for all dilutions, and each vessel should be properly capped to prevent evaporation. An internal standard can also be used to correct for small variations in sample volume.

Standard head space vials are typically 20 mL, but they are also available from various manufacturers in 6, 10, and 12 mL sizes as well. If the ratio of sample volume to vial size is kept constant, the sensitivity for the assay will be maintained regardless of the size of the vial. However, if the sample volume is kept constant and the vial size is reduced, there will be an increase in response for each analyte. Using a larger sample in a smaller vial can help overcome sensitivity issues, as long as the volume of head space in the vial is greater than the intended sampling volume.

11.2.1.2 Liquid Injections

Head space is a good technique for residual solvent analysis, but there are some formulations and/or analytes that can be problematic. In these cases, the next option to be explored is liquid injection. The advantage of this technique is that all solvents can be detected. There are several different types of liquid injection, but we will simply discuss split and splitless techniques. For additional techniques, see Refs. 6 and 7.

In a liquid injection, a relatively small sample amount is introduced (1–4 μL) into a heated injection port, where it is instantaneously vaporized. The resulting volume of gas is referred to as the expansion volume, and it is characteristic for each solvent. The gas expansion volume should not exceed the volume of the GC liner because this can cause reproducibility problems.

Contrary to popular belief, water is not bad for most GC columns (the exceptions are polyethylene glycol based phases such as WAX or FFAP, and nonbonded phases). The reproducibility issues that are often encountered have more to do with the gas expansion volume exceeding the liner volume. A typical 4 mm ID liner has a volume of about 900 μL. The expansion volume of a 1 μL injection DMSO at 250°C is only ~300 μL, which is well below the liner volume. The same 1 μL injection of water will expand to about 1100 μL of gas and exceeds the volume of the liner. When the expansion volume exceeds the liner volume, it can cause reproducibility problems and contamination throughout the system that will lead to sample carryover.

There are several strategies to overcoming the problem of water injections; the first is to simply inject less volume. Making a 0.5 μL injection will reduce the resulting gas volume to an acceptable volume. Another strategy is to use a pressure pulsed injection, where column flow is increased for a short time (30–60 s) during the injection to help transfer the sample onto the column and then reduce the flow to perform the analysis. According to the ideal gas law, the gas expansion volume for a solvent is reduced at higher column head pressures. If the pressure can be increased enough to control the volume to an acceptable level, then reproducible results should be possible. However, when working with water, the best way to control the expansion volume is to use a smaller injection volume.

11.2.1.3 Split versus Splitless Injections

In a split injection, the split valve is open during the injection and only a portion of the sample is injected onto the column, while a majority of the sample is vented to waste. Split injection techniques significantly reduce the amount of contamination that enters the column and decreases the residence time of the sample in the injection port. The split ratio should be adjusted to reduce the amount of contamination that is injected onto the column while still achieving the required sensitivity limits. Using higher split ratios can significantly prolong column life. Typical split ratios are from 10:1 up to 100:1.

A split injection technique should always be investigated first as it does not require some of the careful set up that a splitless injection does and it will reduce the system maintenance that is required. If sensitivity is an issue, then splitless injection techniques can be tested. Both techniques will require some routine inlet maintenance program to periodically replace the liner, septum, and clip 3–4 inches from the head of the column. A unique feature to the Agilent GC systems, the gold seal, can also become dirty and must be replaced periodically.

In a splitless injection, the split vent is closed during the injection and about 95% of the sample is transferred onto the column. Usually 30–60 seconds after the injection, the split vent is opened to flush out any remaining sample in the inlet. The advantage of this technique is that it provides the highest sensitivity because a vast majority of the sample is transferred onto the column. The main drawback is system contamination when working with dirty samples because all of the contaminants will be transferred onto the column. The sample also spends a longer amount of time in the heated inlet so this technique may exaggerate poor performance of thermally labile or active compounds, such as pyridine.

If choosing to use a splitless injection, it is important to optimize the splitless hold time and the starting oven temperature to achieve the best chromatographic performance.[7] If the starting oven temperature is set too high, the sample will not focus on the head of the column, resulting in peak distortion of the early eluting peaks. This occurs because the more volatile compounds in the sample do not recondense onto the head of the column, they immediately start traveling through the column toward the detector. The time required to transfer the sample onto the column in a splitless injection is much longer and can result in much broader peaks. When working with volatile diluents, such as methanol or acetonitrile, if the diluent is not properly focused, it can start to carry higher boiling solvents down the column with it and distort peaks through the entire chromatogram. A very clear symptom of this problem is a broad tailing solvent front and broad peaks before or after the main solvent front that gradually sharpen as retention time increases.

To overcome this problem, the oven temperature should be set at 20°–40°C below the boiling point of the lowest boiling analyte. If this temperature requires starting below about 40°C, it might require cryogenic cooling to adequately focus the sample. Cryogenics require special modifications to the GC system and are not recommended unless there is no other option. Alternatively, using a GC column with a thicker film can help trap highly volatile samples and might eliminate the need for cryogenic cooling. Increasing the column flow rate is another potential solution but can have detrimental effects on resolution of early eluting peaks. The use of a pressure pulsed injection can provide the increased flow during the sample injection, without the additional flow rate throughout the analysis. Again, split injections do not require this optimization.

Once the sample has been properly focused, the splitless hold time can then be optimized. The objective is to maintain splitless injection conditions long enough for a majority (~95%) of the sample to be transferred onto the column. Depending on the column flow rate and injection volume, most of the sample should be onto the column in ~30–60 s. After the sample is introduced onto the column, the split vent is opened and the injection port is flushed out to prevent carry over to the next sample. Holding splitless for too long can also cause peaks to tail or distort. To optimize this time, an experiment should be set up where repeated injections are made and the splitless hold time is varied in increments of 10–20 s. When the data are plotted, there will be a plateau where increased hold time resulted in very little increased signal. The point at which this plateau begins should be the time point when the split vent is turned on and the inlet is flushed out.

11.2.2 Detectors

The choice of detection method can influence the sensitivity and data obtained for a target analyte peak. There are two general types of detectors: nonspecific and specific. Nonspecific detectors give reasonable response for a wide range of compounds. Nonspecific detectors include flame ionization detector (FID), thermal conductivity detector (TCD), and mass selective detector (MSD). Specific detectors provide improved sensitivity for certain classes of compounds. Examples of specific detectors are electron capture detector (ECD) for halogenated compounds, nitrogen phosphorous detector (NPD) for nitrogen and phosphorous compounds, and MSD (tunable based on mass fragments). For the purposes of residual solvents, we will discuss only the FID and MSD. For more information on detectors, see Ref. 8.

11.2.2.1 Flame Ionization Detector

The FID works by pyrolyzing compounds in a hydrogen flame and detecting the resulting ions that are formed. The FID responds to any compound containing carbon-hydrogen bonds, which makes it ideal for most solvent analysis. The FID provides good sensitivity (low parts per million) for a wide range of compounds and is inexpensive and easy to operate.

If the compound does not contain carbon-hydrogen bonds, it will not be detected by the FID. This can be beneficial for compounds like water, which is a very commonly used solvent in head space analysis that could co-elute with low boiling solvents. However, this can cause problems with other solvents like carbon tetrachloride (Figure 11.2), which has a low response in the FID. Changes to the head space parameters had little effect on the response because the sensitivity problem is detector related. Changing to an electron capture detector or a mass spectrometer (MS) would improve detection limits for this compound.

11.2.2.2 Mass Spectrometer

There are several different types of commercially available MS systems and ionization techniques.[9] For the purposes of this discussion, we will limit ourselves to the electron impact quadrupole MS (EI-MS) system. The EI-MS uses a high-energy beam of electrons (70 eV) to ionize and fragment a

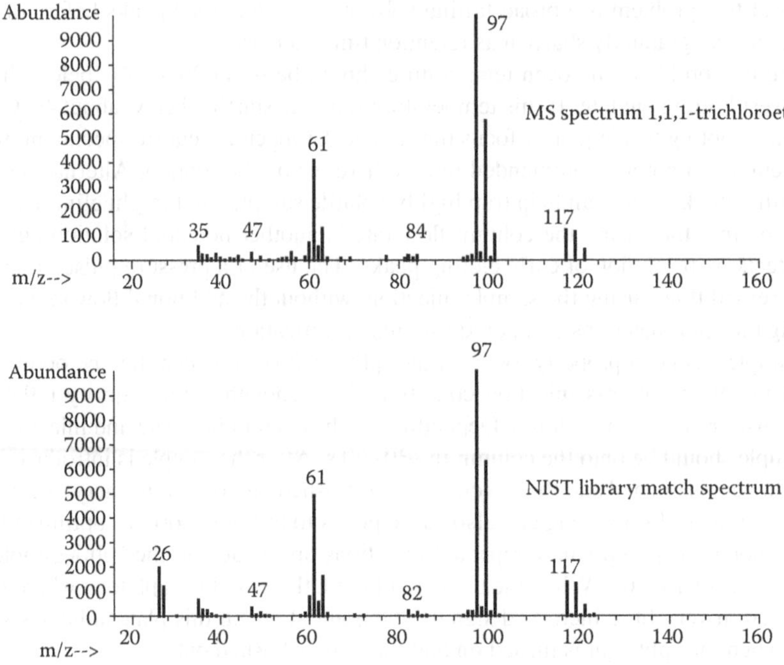

FIGURE 11.5 Mass spectrum of 1,1,1-trichloroethane by EI-MS.

compound. The resulting ions are accelerated to a detector plate. The quadrupole filters the masses based on a user defined range using a radio frequency. Only the ions of the appropriate mass are allowed to reach the detector; ions that do not have the appropriate mass are ejected. The current generated by the ions striking the detector plate is amplified by an electron multiplier and the signal can be displayed.

When working with complex samples that have co-eluting peaks, it can be challenging to achieve good quantitation and there is the potential for misidentification of unknown co-eluting solvents. A mass spectrometer provides specific information about the structure of the compound being analyzed. The mass and ratio of these ions produced during fragmentation to each other are unique and highly reproducible for each compound, allowing the compound to be identified based on a match to a reference library spectrum (Figure 11.5). The only exception is for structural isomers, such as xylenes, which have the same mass spectrum. Combining the qualitative information of the MS detector with the separation power of GC provides a very powerful combination for residual solvent analysis.[10]

11.2.3 The GC Column

GC is the process by which volatile compounds (gasses, liquids, and solids) are separated according to differences in boiling point and chemical properties. The separation is achieved by using a GC column that selectively retains analytes through various retention mechanisms as they partition into and out of the stationary phase. There are several different types of GC columns, but we will limit our discussion here to capillary columns. A capillary column is made from fused silica tubing that has been drawn into a long (5–105 m), thin (0.1–0.53 mm ID) tube and coated externally with polyimide resin to give it strength and flexibility. A liquid stationary phase is then coated on the inside walls of the capillary tubing and provides the separation characteristics of that column.

When doing new method development, choosing the right GC column phase and dimension will provide the best separation in the least amount of time. The separation characteristics of a column are based on four main variables: (1) length (*L*), (2) internal diameter (ID), (3) film thickness (*df*), and (4) stationary phase composition.

Once these have been selected, the temperature program can be adjusted to separate the target analytes, while keeping run time within the optimal range. The choice of carrier gas can also affect column efficiency but does not play a role in selectivity, so it will not be covered in this discussion.

Chromatographic resolution is described by the Master Resolution Equation:

$$R_s = \frac{\sqrt{N}}{4} \cdot \left(\frac{\alpha - 1}{\alpha}\right) \cdot \left(\frac{k}{1+k}\right) \tag{11.2}$$

where *N* is efficiency, α is selectivity, and *k* is the capacity factor.

Another way to think about Equation 11.2 is to consider there to be an efficiency, selectivity, and retention time component to every separation. With this in mind, the four column parameters listed above influence different aspects of Equation 11.2. By optimizing the right parameters, maximum resolution can be achieved without the need for a long run time. For more information on GC method development, see Ref. 11.

11.2.3.1 Column Length

The length of the GC column (*L*) affects column efficiency (*N*) and the analysis time (*k*). Longer columns give more separation power (higher *N*), but result in longer run times (higher *k*). When doing method development, attempts should be made to use the shortest column possible that provides the required resolution. The most commonly used length is 30 m and should provide adequate resolution for most analysis. If analyzing less than five solvents, a shorter 15 m column might be a good alternative that can further reduce analysis time.

The longer 60 m columns should only be considered when analyzing samples containing a large number of compounds. Trying to increase resolution of a critical pair by using a longer column can sometimes be counterproductive. If resolution between a critical pair is poor, it is often better to change the stationary phase (selectivity) to give better resolution than increase analysis time.

11.2.3.2 Internal Diameter

Column ID affects both loading capacity and column efficiency (*N*). The larger the column ID the greater the amount of sample that can be injected, but separation power is reduced. Because the residual solvents being analyzed are expected in the low parts per million (ppm) levels, there is not a great need for high sample capacity. Even if high concentration samples were expected, a split ratio could be used to reduce the on-column concentration allowing the smaller ID columns to be used. For most methods, a 0.25–0.32 mm ID column is the best balance between capacity and resolution.

However, when using head space as an injection technique, there can be an additional benefit to using larger 0.53 mm ID columns: the rate at which the sample is transferred to the column. The larger 0.53 ID have a volumetric flow rate through the column that is much higher than on a 0.25 or 0.32 mm ID column at the same linear velocity (cm/s). Head space injections of greater than 1 mL of gas onto the column are typical. A common flow rate on a 30 m × 0.25 mm ID column is about 1 mL/min, which means the sample will take 1 min to transfer onto the column. This long transfer time leads to band broadening, which affects resolution and sensitivity. A typical flow rate on a 30 m × 0.53 mm ID column is ~5–10 mL/min, which results in a much narrower sample band. Adding a small split ratio or using a pressure pulsed injection are other options to reduce the time required to transfer the sample onto the column.

To summarize, if using liquid injections, 0.25 mm ID columns give the best overall performance for most samples. If making head space injection, consider using a 0.32 or 0.53 mm ID column to more quickly transfer the sample onto the column. The 0.32 mm ID columns will give the highest column efficiency and still provide the relatively high volumetric flow rates that are ideal for head space injections.

11.2.3.3 Film Thickness

The film thickness (*dF*) on a GC column affects the retention (*k*), efficiency (*N*), column bleed, and activity level. With so many influences on the separation, it can be difficult to choose the appropriate film thickness. Thicker film columns will provide longer retention of volatile compounds and reduce activity for acidic or basic analytes. However, thick film columns also have the least separation power and will give more bleed at higher temperatures.

Because many residual solvents have low boiling points, using a slightly thicker stationary phase usually provides better results, especially when working with nonpolar phases. If the compounds being analyzed have boiling points >100°C, it might be better to use a column with a slightly thinner

TABLE 11.1
Commercially Available Phase for Use in Method Development

USP Designation	Phase Chemistry	Trade Names	Use
G3	50% Phenyl-50% dimethylpolysiloxane	ZB-50, DB-17, Rtx-50, CP-Sil 24 CB	Mid-polar phase good for separation of samples with both nonpolar and polar compounds.
G16	Polyethylene glycol (average MW 15,000)	ZB-WAX*plus*, Stabilwax, DB-Wax, CP-Wax 57 CB	Polar phase good for separation of alcohols and other oxygenated solvents. Only phase capable of separating meta/para xylene.
G25/G35	Polyethylene glycol TPA (Carbowax 20M terephthalic acid)	ZB-FFAP, DB-FFAP, StabilWax-DA	Specially deactivated Wax phase for the analysis of volatile free acids.
G27	5% Phenyl-95% dimethylpolysiloxane	ZB-5, DB-5, SPB-5, Rtx-5	Good for general use of semivolatile containing samples. Low activity for good peak shape for acid and basic compounds.
G38	Phase G1 plus a tailing inhibitor	ZB-1, DB-1, CP-Sil 5 CB, Rtx-1	Low polarity phase that separates compounds based on boiling point. Good for general use.
G42	35% Phenyl-65% dimethylpolysiloxane	ZB-35, DB-35, Rtx-35	Midpolar phase good for chlorinated aromatic compounds. Good confirmation column for G27 and G38 phases
G43	6% Cyanopropylphenyl-94% dimethylpolysiloxane	ZB-624, DB-624, Rtx-624	Specially designed phase for the analysis of low boiling solvents. Good starting point for new method development when working with residual solvents.
G46	14% Cyanopropylphenyl-86% dimethylpolysiloxane	ZB-1701, DB-1701	Good resolution of chlorine containing solvents and/or pesticides.
G48	Highly polar, partially cross-linked cyanopolysiloxane	BPX70, CP-Sil 88 CB, HP-88	Unique phase designed for resolution of cis/trans fatty acids

stationary phase to reduce analysis time. The thinner phase will also provide higher resolution and reduce column bleed.

11.2.3.4 Stationary Phase

The selectivity α provided by the GC column stationary phase has the largest effect on the separation power of the column. When it comes to choosing the appropriate stationary phase, it is best to try to match the phase to the compounds being analyzed: "like dissolves like." If the compounds are polar, better retention and resolution are typically achieved on polar stationary phases like the Zebron ZB-WAX or Zebron ZB-FFAP. If the compounds are more hydrophobic, then they may separate better on nonpolar phases like Zebron ZB-1 and Zebron ZB-5. When analyzing a mix of polarities, using a unique phase like a Zebron ZB-624 shows more difference in elution patterns between compounds than standard nonpolar or polar phases. Table 11.1 contains a list of commercially available phases and suggestions as to their use.

Column stability and bleed are also related to phase polarity; the more polar GC columns usually have higher bleed. If looking to analyze higher boiling solvents like DMSO, xylenes, or pyridine, using lower polarity phases will give better sensitivity. The lower polarity phases are also better suited for liquid injections that may contain nonvolatiles. Polar stationary phases are more easily damaged when analyzing dirty samples because high temperature cannot be used to remove contaminants. Care should be taken not to allow oxygen to enter a GC column, especially WAX and cyano-containing phases, as it can rapidly destroy the stationary phase when heated. All columns should be sealed when not installed in the GC system.

11.3 HISTORY OF RESIDUAL SOLVENT TESTING

In 1997, the International Conference on Harmonization (ICH) published a document entitled "Impurities: Guideline for Residual Solvents Q3C (R3)," which was immediately adopted by the three ICH regulatory bodies FDA, European Medicines Agency (EMEA), and Ministry of Health, Labor and Welfare (MHLW).[12] The driving force behind the Q3C guidelines is patient safety. To help ensure this, every effort should be made to remove solvents from the final product within practical limits of manufacturing technology. To help understand the potential health risk to an individual, Q3C has classified solvents based on their toxicity into three main classes:

Class 1: Solvents that should be avoided because of known human carcinogenic properties or deleterious environmental effects.
Class 2: Solvents that should be limited because of inherent but reversible toxicity.
Class 3: Solvents that are regarded as less toxic and of lower risk to human health.

Q3C provides concentration limits and/or permitted daily exposure limits (PDE) for 59 solvents. It also identifies an additional 10 solvents that may be of concern but for which there is no toxicological data available to assign an appropriate daily exposure limit. The limit for each solvent is based on the no observable effect level (NOEL) or the lowest observable effect level (LOEL), with large safety factors built in. In some cases, the NOEL or LOEL was based on toxicity data from animal studies, so additional correlation factors are used to account for differences in body mass and metabolism.

The Q3C guidance was adopted by the European Pharmacopeia (*Ph. Eur.*) in November 2003 into the fifth addition of the general Chapters (2004–2007) and applied only to new drug products on the market. The general monograph 2034 made the testing of residual solvents mandatory and an analytical method was provided in general Chapter 2.4.24.[13] In that same year, the new Q3C limits for residual solvents were proposed by the U.S. Pharmacopeia (USP) in the 29(4) edition of the Pharmaceutical Forum (PF).[14] In July 2008, the USP officially adopted ICH Q3C guidelines for the control of residual solvents. The revised chapter contained a testing methodology that was similar to the *Ph. Eur.* methodology, with some important modifications.

The FDA has made exceptions with regard to reporting residual solvent information for coating materials, colorants, flavors, capsules, and imprinting inks.[15] Because these products are known to contain high levels of solvents (in some cases they are a majority component), but they are generally used in very small amounts, it does not make sense to report the residual solvent content and they are excluded. The possible exception is when class 1 solvents are used in the manufacture of these components. In such cases, additional testing would be required.

11.4 OVERVIEW OF USP GENERAL CHAPTER <467>

The drug product manufacturer is responsible for identifying and controlling any solvents that have been used or produced in any component of its drug product. This includes excipients and/or other nonactive components in the final drug product. To help minimize the amount of testing, it is only required to test for the solvents that are likely to be present based on the manufacturing process. Chapter <467> also states that the supplier needs to provide certain information about the solvents used in their process to help reduce the amount of testing a company must perform.[16]

11.4.1 Class 1 Solvents

Class 1 solvents should be avoided at all costs. If they are used (or produced) in the manufacturing process, testing of the final product will be required to ensure they have been removed. Standard testing procedures for class 1 solvents are provided by the USP and *Ph. Eur.* If the therapeutic value of the drug product is high and it is impossible or impractical to produce the drug without the class 1 solvent, the level should be kept below the concentration limit, however this in itself will not guarantee acceptance. The regulatory body may still deny the application unless it provides a significant therapeutic benefit.

11.4.2 Class 2 Solvents: The Option Method

The levels of class 2 solvents should be reduced as much as possible to avoid causing irreversible damage to patients taking a drug product. To help reduce the amount of testing required on the part of pharmaceutical manufactures, QC3 provides companies with the option to calculate the amount of solvent present in the final drug product. If the levels are below the concentration limit (option 1) or the PDE limit (option 2), then no testing of the final drug product is required.

Standard testing procedures for class 2 solvents are provided by the USP and *Ph. Eur.*, with the exception of the following compounds: ethylene glycol, formamide, 2-ethoxyethanol, 2-methoxymethanol, sulfalone, and N-methylpyrrolidone. In addition, N,N-dimethylacetamide and N,N-dimethylformamide can also be problematic using the HS-GC method provided. In such cases, an alternative testing procedure must be developed.

11.4.3 Class 3 Solvents: Loss on Drying

The relatively low toxicity of the class 3 solvents makes them an ideal choice for manufacturing. In many cases, it may be possible to substitute a class 2 solvent with a class 3 solvent without causing problems in production. For example, hexane is a class 2 solvent, but pentane and heptane are both class 3 solvents. Because these solvents are relatively similar in chemical structure, it is likely they may be substituted for one another.

Using a class 3 solvent in place of a class 2 solvent can also significantly reduce the amount of testing that needs to be done. When only class 3 solvents are present, loss on drying (LOD) can be used to determine the level of solvent present as long as a LOD protocol is specified in the monograph. If the drug product fails the LOD test, it is possible to correct for moisture content and subtract that amount from the calculated concentration, again depending on the Monograph. If the substance being analyzed still fails after correction for moisture content, then a step must be added in the process to reduce the solvent level.

None of the pharmacopeias (USP, *Ph. Eur.*, and MHLW) offer testing protocols for class 3 solvents. In the event that class 3 solvent must be identified and quantitated, a lab will need to develop an appropriate analytical procedure. Many of the solvents used in the manufacture of oligonucleotides are class 3 solvents, so testing may become necessary.

11.4.4 USP Chapter <467> Residual Solvents by Gas Chromatography

The advantage of using a methodology like the one described in USP Chapter <467> is that it has already been validated and only requires a simple verification/qualification process to demonstrate suitability for a particular sample. The regulatory agencies such as the FDA, EMEA, and MHLW are also familiar with this methodology, so less justification may be needed regarding the testing process. Please note that if any adjustments are made to the methodology beyond allowable limits, the method must be re-validated.

Drug components or products containing class 1 and 2 residual solvents are analyzed by head space–gas chromatography (HS-GC) using a FID detection. Procedure A is a qualitative limit test to determine possible solvents that are in the sample. Procedure B is a confirmatory test using an alternative GC column stationary phase to verify the identity of the solvent. Procedure C is a quantitative test to determine the amount of solvent contained in the drug product, substance, or excipient. If the identity of the solvent is known, a company may choose to go right to procedure C to quantitate the amount of residual solvent present. Because this significantly reduces the amount of testing required, most companies favor this option.

A two-column approach is used to help reduce any misidentifications of solvents that arise from co-elutions on the GC column. There are 59 Class 1 and 2 solvents, 48 of which are detectable by head space injection, and there are several known co-elutions on each phase. The FID is unable to differentiate co-eluting peaks based on chemical structure. Because there are other unclassified solvents that may potentially interfere, there is a relatively high probability for misidentification if using only one GC column. By using another column of complementary selectivity, co-elutions on the first phase will be separated on the second phase. If the identified solvent has the correct retention time on both phases, the compounds identity can be made with much higher certainty.

All samples are introduced via head space injection using one of three conditions specified in the chapter. A laboratory is free to choose any of the three procedures listed, based on the one that gives the best performance. For those labs using syringe-based systems, some of the parameters do not apply. There are several commercially available head space systems by various manufacturers, and each system uses a slightly different technique to transfer a defined volume of gas onto the GC column. The type of head space autosampler being used can affect performance of the method. It is important to know what type of system a lab uses to ensure good results. It is advisable to contact the head space manufacturer when setting up a system for the first time to understand the specific requirements of that unit. Reference 2 also gives additional information about the different head space systems.

Examples of class 1 and 2 solvents run using procedure A are shown in Figures 11.6 through 11.8, for examples of procedure B see Ref. 10. The USP reference standards break the class 2 solvents into mix A and mix B to eliminate co-elutions. All preparations of standard solutions referred to in the <467> monograph are based on the USP reference standards. Labs may choose to purchase these standards directly from the USP, or they can be purchased from other commercial standard suppliers. If a standard is made up in-house, special attention should be made to the purity of the DMSO

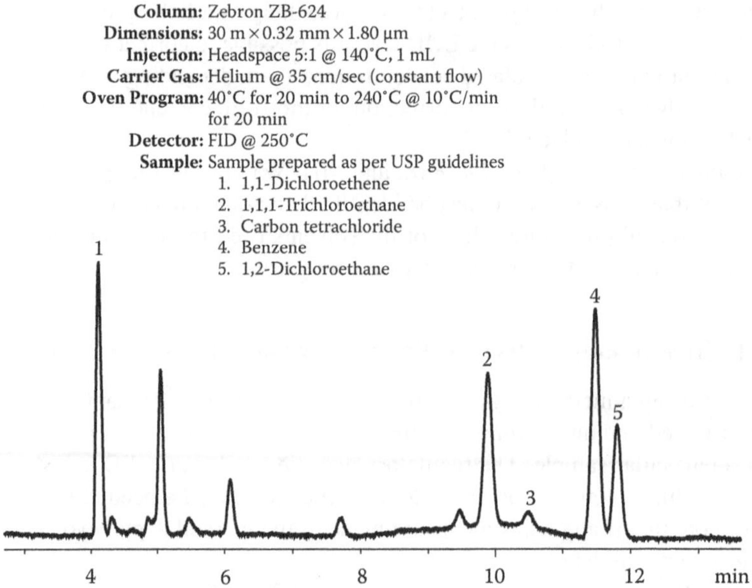

FIGURE 11.6 USP <467>: residual solvents procedure A—class 1.

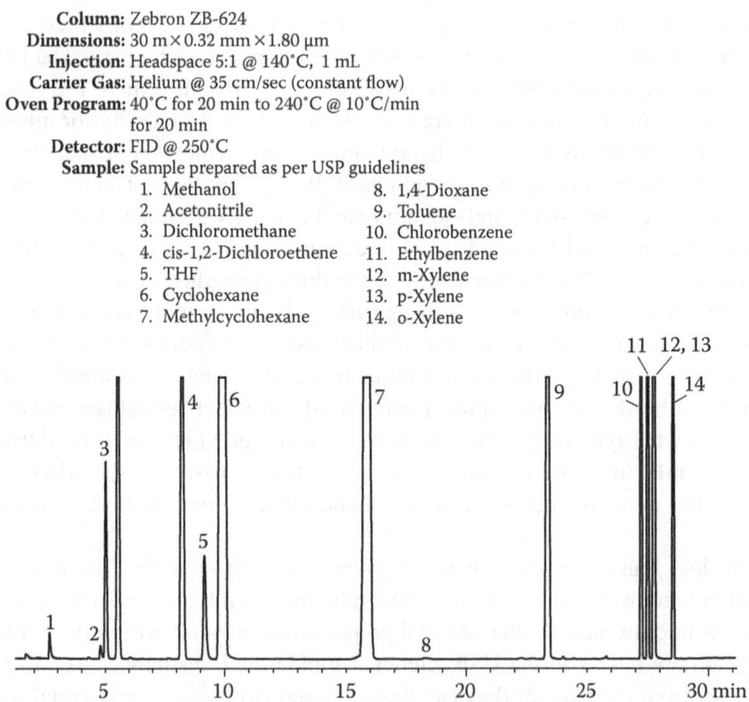

FIGURE 11.7 USP <467>: residual solvents procedure A—class 2 mix A.

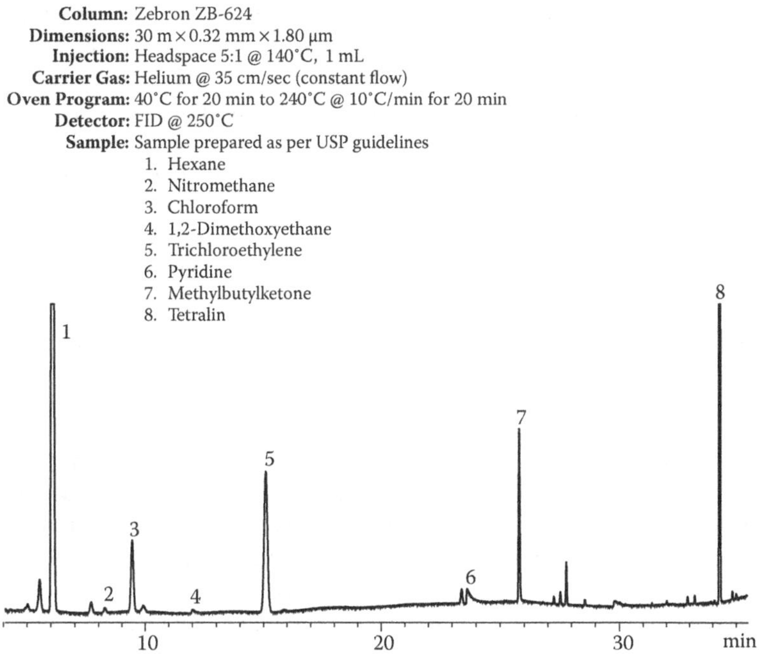

FIGURE 11.8 USP <467>: residual solvents procedure A—class 2 mix B.

used as the diluent to prevent contamination peaks in the chromatogram. The purity of DMSO varies greatly from manufacturer to manufacturer, and each new lot should be qualified before it is used for standard preparation.

The method identifies system suitability criteria that must be met in order to verify system performance (Table 11.2). Only after system suitability requirements have been met can samples be analyzed. To remain compliant with <467>, system suitability criteria must be met without changing any of the conditions listed in the monograph, with several minor exceptions. The USP lists chromatographic parameters that may be adjusted in order to meet system suitability requirements in Chapter <621>; the other Pharmacopeias have similar documents. If any adjustments are made outside of these parameters, then the method has been changed and the method validation must be performed.

If high-performance GC columns are used and there is a good system maintenance schedule in place, the resolution requirements are fairly easy to meet. However, GC column selectivity can be slightly different from manufacturer to manufacturer, so always confirm column performance when setting up a new method.

The signal-to-noise requirements for 1,1,1-trichloroethane and benzene in the class 1 Standard solution are easily met by a well-functioning GC system. The main problem encountered is meeting

TABLE 11.2
System Suitability Requirements for Chapter <467>

	Procedure A	Procedure B
Signal-to-noise: Class 1 standard solution	1,1,1 Trichloroethane ≥5	Benzene ≥5
Signal-to-noise: Class 1 system suitability solution	All peaks must be ≥3	All peaks must be ≥3
Resolution (R_s): Class 2 mixture A standard solution	Acetonitrile/methylene chloride ≥1.0	Acetonitrile/cis-dichloroethene ≥1.0

the signal-to-noise requirement for carbon tetrachloride in the class 1 system suitability solution under procedure A conditions. The class 1 system suitability solution combines 1.0 mL of the class 1 standard solution with 4.0 mL of a test solution containing the article being tested. This is done to evaluate how the article being tested affects the sensitivity of the head space analysis. In some cases, the article may interfere with the partition coefficient (*K*) of the solvents in solution. If this is the case, the article may not be suitable for analysis by the compendial methodology. This will be discussed in more detail later.

Carbon tetrachloride has very limited response in GC-FID owing to a lack of carbon-hydrogen bonds. Increasing the concentration injected onto the column results in only a small increase in peak response. Even though carbon tetrachloride is not typically used as a solvent in oligonucleotide manufacturing, the system suitability requirements of the method require the specified S/N ratio to be met. To meet these S/N requirements, most labs must decrease their split ratio from 5:1, down to 2:1 or even 1:1. Using a GC column with 0.32 mm internal diameter (ID) increases separation efficiency (N) and can help improve S/N ratios.

If there are still problems meeting the system suitability requirements and the identity of the analyte is known, it is possible to move directly to procedure C. In this case, a WAX phase can be used in procedure B for quantitation as long as there are not additional co-elutions. On the WAX column, carbon tetrachloride co-elutes with 1,1,1-trichloroethane so S/N cannot be determined. All other class 1 analytes show acceptable response by GC-FID.

After system suitability requirements have been met, samples can be analyzed. When working with unknown samples, start with procedure A to identify any potential solvents above the concentration limit. If a peak with a retention time corresponding to a class 1 or 2 residual solvent is observed and it is above the acceptable response, use procedure B to verify the identity of the solvent. If after running procedure B the peak area is above the limit then, procedure C must be used to determine the actual concentration of the compound.

Procedure C uses a standard addition method to calculate the amount of solvent present in the sample. Standard addition methods compare the peak area resulting from the compound to be tested to a solution containing the compound to be tested which has been spiked with a known amount of the same substance (Figure 11.9). The increase in area from the spiked test solution is used to calculate the concentration in the article being tested. For quantitation, it is assumed that the

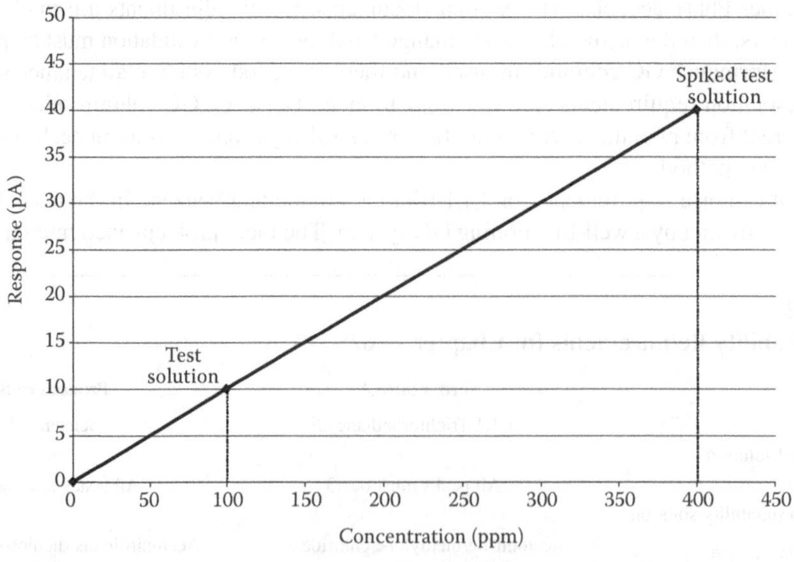

FIGURE 11.9 Example of a standard addition calibration curve.

response of the compound is linear and the calibration curve is forced through zero. Now, on the basis of the increased area observed in the spiked test sample a concentration value can be assigned. Applying this calculation to the article in Figure 11.9, the concentration of the solvent would be 300 ppm.

The monograph provides an equation for calculating the concentration in both water-soluble and water-insoluble articles. The equation factors in the weight of the article being tested, the concentration of the solvent of interest in the reference solution, and the differences in areas observed in the reference standard and the spiked test sample. The only difference between the two equations is the multiplication factor. The difference in the number is related to the mass of test article used to prepare the test stock solution. In the water-soluble method, 250 mg of substance is used, and in the water-insoluble method it is 500 mg. The ratio of the value to the mass is 0.20 for both, which normalizes the difference in mass of the test article.

Good sample preparation is critical to achieving accurate results. The monograph states that samples should be dissolved whenever possible and provides options for both water-soluble and water-insoluble products. The solubility applies to the drug product being analyzed, not the solvent being analyzed. In the water-soluble procedure, the substance to be analyzed is diluted with water to the appropriate concentration and analyzed. Water-insoluble compounds are first diluted with either DMF or DMSO and then diluted with water to achieve a test solution containing ~75% water. The use of water in the final test solution improves analyte partitioning for head space analysis.

If the sample cannot be dissolved, the chapter states it should be pulverized into a fine powder using a mortar and pestle. The finer the material can be pulverized the more accurate the analysis, but it is important to not let the samples sit in the open for too long. The volatile nature of these solvents means that they begin to evaporate immediately. When making up head space samples (liquid or solid), it is important to perform each step as fast as possible, while maintaining good accuracy. To help prevent loss due to evaporation, consider using gastight syringes for sample transfer steps in place of standard pipettes. Rubber sealing caps are available for most analytical glassware and the hypodermic needle of the gastight syringes can be used to pierce this seal for easy sample transfer.

If the sample will not dissolve and cannot be pulverized, it might be unreliable to use the procedure specified in <467>. In such cases, the chapter allows for very little flexibility in the sample

Column: Zebron ZB-WAX
Dimensions: 60 m × 0.53 mm × 1.00 μm
Injection: Split 20:1 @ 250°C, 1 μL
Carrier Gas: Helium @ 2 mL/min (constant flow)
Oven Program: 40°C for 10 min to 200°C @ 10°C/min
Detector: FID @ 250°C
Headspace sample equilibrated at 85°C for 30 min
Sample: 1. IPA
2. Ethanol
3. Acetonitrile
4. Lut
5. Pyridine

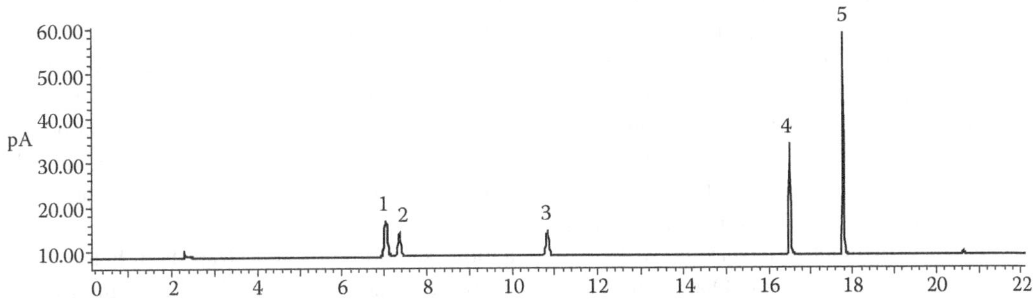

FIGURE 11.10 Analysis of residual solvents by HS-GC in an oligonucleotide product.

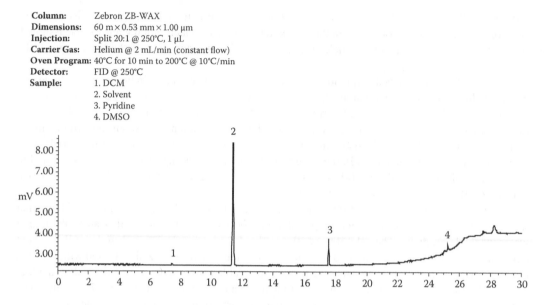

Column: Zebron ZB-WAX
Dimensions: 60 m × 0.53 mm × 1.00 µm
Injection: Split 20:1 @ 250°C, 1 µL
Carrier Gas: Helium @ 2 mL/min (constant flow)
Oven Program: 40°C for 10 min to 200°C @ 10°C/min
Detector: FID @ 250°C
Sample: 1. DCM
2. Solvent
3. Pyridine
4. DMSO

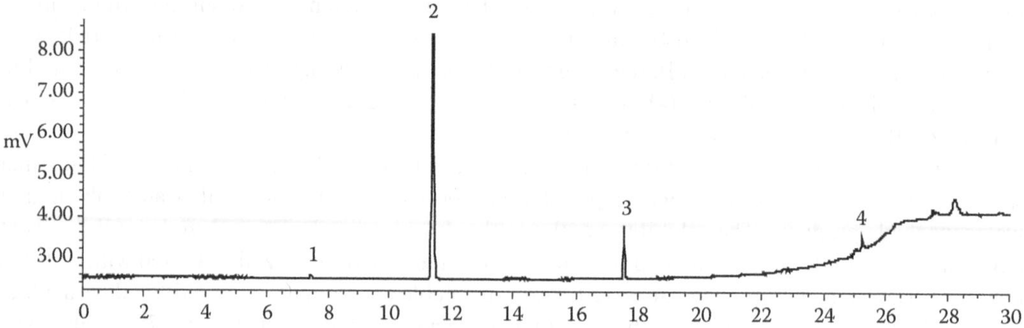

FIGURE 11.11 Analysis of residual solvents by direct injection in an oligonucleotide product.

preparation techniques. If an alternate sample preparation procedure must be used, then the process must be validated and <467> testing protocol is not applicable. Fortunately, oligonucleotides are freely water soluble so they should not pose any major challenges. However, excipients used in the formulation could still cause problems that require the use of an alternative procedure.

11.5 RESIDUAL SOLVENT TESTING ON OLIGONUCLEOTIDE PRODUCTS

In the case of oligonucleotide manufacturing, acetonitrile, ethanol, toluene, 2-propanol, and pyridine are the primary solvents used. In this case, the compendial method as written may be too long and laborious to make sense for routine production work. Generally speaking, there are two approaches to analyze oligonucleotides by GC for the presence of residual solvents: by head space–gas chromatography (HS-GC) and by direct sample injection into the GC.

When using HS-GC, the sample is dissolved in water at a concentration of 20 mg/mL of solvent, or greater, and placed in a head space vial and sealed carefully in order to avoid any possibility of head space leakage. Vials with the sample and corresponding calibration standards are loaded in the head space autosampler and equilibrated for a set time at a set temperature. A fixed gas volume of the head space from the sample and standards are sequentially injected into the GC for analysis. The amount of residual solvent in the sample is then calculated using the resulting calibration curves. A typical chromatogram for analysis by this approach is shown in Figure 11.10.

For analysis by direct injection GC, the sample is dissolved in acetonitrile or other appropriate solvent and placed in GC vials for analysis. The vial with the sample and the corresponding calibration standards are loaded in the GC autosampler to be analyzed by direct injection. Normally, the amount of (liquid) sample injected into the instrument is 1 microliter. The amount of residual solvent in the sample is then calculated using the resulting calibration curves. A typical chromatogram for analysis by direct injection is shown in Figure 11.11.

Because oligonucleotides are not volatile, precautions should be taken to trap them before they can damage the GC column. One strategy for this is to use a liner with a small piece of glass wool, which will trap any nonvolatiles. Another would be to use a guard column attached to the head of the GC columns using a small press-fit connector.

11.6 SUMMARY

The regulation of residual solvents in pharmaceutical products is based on a risk potential to patients. Solvents have been grouped according to their toxic potential and concentration limits and/or daily exposure limits have been given by ICH Q3C guidance. The major pharmacopeias have adopted these guidelines for identification and control of solvents in drug substances, excipients, and products.

In order for a company to understand its testing liability, it must first determine those solvents that are used in its internal processes. Survey and/or audit the suppliers to determine the solvents used in their process. Once a complete list of solvents has been compiled, a testing strategy can be developed. When possible, the option method should be used for class 2 solvents to reduce the testing that must be done.

A compendial testing methodology has been provided for class 1 and most class 2 solvents. When applicable, it is beneficial to use this procedure because it is already validated and can be implemented in a lab after a simple verification/qualification process. If the compendial method is not suitable for the product or there are other solvents that are not covered in the monograph, a new protocol must be developed and validated.

When doing new method development, there are several important GC parameters that can be optimized to improve method accuracy and performance. Head space is a good injection technique when available because it prevents nonvolatile components in a drug formulation from being injected on the column. The choice of detectors can impact analysis sensitivity and the separation required to achieve good quantitation. The use of MS gives the ability to identify unknown compounds.

Choosing the right GC columns can improve separation between closely eluting compounds and reduce analysis time. Shorter columns with smaller ID provide the highest separation power in the shortest amount of time. Film thickness can then be optimized to give adequate retention of low boiling compounds.

It is important to keep in mind that the driving force behind residual solvent analysis is patient safety. Understanding the solvents that are used or produced in the manufacture of a drug product is critical to ensuring this safety.

REFERENCES

1. Gan, J., S. Papiernik, and S. R. Yates. 1998. Static headspace and gas chromatographic analysis of fumigant residues in soil and water. *Journal of Agricultural and Food Chemistry* 46(3): 986–990.
2. Kolb, B., and L. S. Ettre. 2006. *Static Headspace-Gas Chromatography: Theory and Practice*. New York: John Wiley.
3. Naffaf, A., and J. Balla. 2000. Improved sensitivity of headspace gas chromatography for organic aromatic compounds. *Chromatographia Supplement* 51: S241–S248.
4. Kolb, B., C. Welter, and C. Bichler. 1992. Determination of partition coefficients by automatic equilibrium headspace gas chromatography by vapor phase calibration. *Chromatographia* 34(5–8): September/October.
5. Ettre, L. S., and B. Kolb. 1991. Headspace-gas chromatography: the influence of sample volume on analytical results. *Chromatographia* 32(½): July.
6. Grob, R., and E. Barry. 2004. *Modern Practice of Gas Chromatography*. New York: John Wiley.
7. Grob, R. L. 1995. *Modern Practice of Gas Chromatography*, 3rd ed. New York: John Wiley.
8. McNair, H., and J. Miller. 2009. *Basic Gas Chromatography*, 2nd ed. New York: John Wiley.
9. McMaster, M. 2008. *GC/MS A Practical Users Guide*, 2nd ed. New York: John Wiley.
10. Countryman, S. 2007. Understanding the revisions to USP monograph <467>: residual solvents. Phenomenex Inc. White Paper.
11. Jennings W., E. Mittlefehldt, and P. Stremple. 1997. *Analytical Gas Chromatography*, 2nd ed. San Diego, CA: Academic Press.
12. International Conference on Harmonization (ICH). 1997. Impurities: guideline for residual solvents on step 4. Paper presented at ICH on Technical Requirements for the Registration of Pharmaceuticals for Human Use, Q3C (R3), July 1997.

13. European Pharmacopoeia. 2004. Residual Solvents (5.4), Supplement 4.6, Directorate for the Quality of Medicines of the Council of Europe, Strasbourg, 4th ed., p. 3911.
14. Schniepp, S. 2009. Complying with the USP residual solvents requirements: industry perspectives. *American Pharmaceutical Review* September/October.
15. Food and Drug Administration. 2008. Residual solvents in ANDAs: questions and answers. http://www.fda.gov, accessed October 28, 2008.
16. Pharmacopoeia Convention Inc. 2009. USP <467>, general chapter on residual solvents, USP 32-NF 28, Rockville, MD: USP.

12 Determination of Extinction Coefficient

Veeravagu Murugaiah
Alnylam Pharmaceuticals

CONTENTS

12.1 INTRODUCTION

The accurate determination of oligonucleotides, both in drug substance and drug product, is of the utmost importance to ensure the quality of production and patient safety. With the current exponential increase in research activities in antisense and short interfering RNA (siRNA) therapeutics, a reliable method is necessary to quantitate dosing solutions accurately. The scale of manufacturing ranges from milligrams to kilograms. If the method of quantitation is not accurate, in addition to compromising patient safety because of incorrect dose, there may be financial implications as well due to inaccurate yield information.

Because of hygroscopicity and the electrostatic nature of oligonucleotides, gravimetric measurement is not suitable for minute quantities of samples. Routine methods for determination of the concentrations of oligonucleotide usually rely on measurements of UV absorbance, which is related by a fundamental property of the molecule called the molar extinction coefficient. The molar extinction coefficient, also known as molar absorptivity ε is a measurement of how strongly a chemical species absorbs light at a given wavelength. It is an intrinsic property of the species; the actual absorbance A of a sample is dependent on the path length l and the concentration C of the species via the Beer–Lambert law:

$$A = \varepsilon C l$$

Molar extinction coefficient of an oligonucleotide is a unique physical property determined by its sequence, temperature, pH, and nature of the buffer or solvent in which it is dissolved. In general, the extinction coefficient of an oligonucleotide is measured at a wavelength of 260 nm, ε_{260}, although it is more accurate to determine the molar extinction coefficient at the lambda maximum.

Optical density (OD) is an older terminology used by nucleic acid chemists. The optical density unit, or more commonly the OD_{260} unit, is an absorbance measurement of an oligonucleotide. Each of the bases in a nucleic acid strand has an absorbance at or near 260 nm, owing to their conjugated double bond systems. Because the exact base sequence and composition is known, the OD_{260} unit is a very accurate and convenient measure to quantitate an oligonucleotide. The OD_{260} unit is a normalized unit of measurement that is defined as the amount of oligonucleotide required to give an absorbance value of 1.0 at 260 nm in 1.0 mL of solution using a 1.0 cm path length cuvette. For an example, consider a siRNA duplex of molecular weight 14303.6 g/mole with a molar extinction coefficient of 296843 $M^{-1}cm^{-1}$. Using the relationship $A = \varepsilon Cl$, and a solution that gives an absorbance reading of 1.0 at 260 nm in 1.0 mL, OD_{260} is related as follows:

$$\text{nmole}/OD_{260} = (1/296843)*10^6 = 3.4$$

$$\mu g/OD_{260} = 14303.6*(1/296843) = 48.2$$

Rapid development of combinatorial chemistry methodology results in a vast number of sequences suitable for therapeutic use. Such an activity requires quick calculation of extinction coefficients and other thermodynamic parameters to design oligonucleotides. Several Web-based calculators, developed by oligonucleotide supply companies, universities, and health institutions are available for researchers. This chapter will review sources available for the theoretical calculation of extinction coefficients and provide experimental details for the direct determination of extinction coefficients of oligonucleotides.

12.2 CALCULATION OF ε_{260}

Chemical determination of nucleic acids in mammalian tissues and microorganisms was determined by the method of Schneider.[1] Nucleic acids were extracted with hot trichloracetic acid (TCA). Total nucleic acid (TNA) was determined both by absorbance of the extract at 260 nm and also from the phosphorus content of the extract. DNA was determined by reaction with p-nitrophenylhydrazine.[2] UV absorption was used as a rapid method of determining TNA. Acid hydrolysis with TCA denatures nucleic acids extensively and reproducibly. TNA in tissues and microorganisms were analyzed by TCA hydrolysis and UV absorption with correction for TCA absorption.[3,4]

Published molar extinction coefficients of mononucleotides and dinucleoside phosphates[5,6] form the basis of most modern calculators for extinction coefficients of synthetic oligonucleotides. The extinction coefficient of an oligonucleotide is normally calculated in one of four ways. With increasing complexity they are:

1. Regardless of base composition or sequence, the method assumes a nominal concentration of 33 μg/mL of single-strand oligonucleotide to have a unit absorbance.[7] This approach is conveniently used for the calculation of a crude yield of oligonucleotides.
2. The second method assumes that all nucleotides in an oligonucleotide structure have the same or similar molar extinction coefficients as the free nucleotide in solution. This approximation is only useful if the oligonucleotide does not assume any secondary structures.[8] The sum is computed over all types of bases present in single strands. Also, over all types of base pairs in double-stranded nucleic acids, extinction coefficient values of 4 mononucleotide phosphates and possible 16 dinucleoside monophosphates were used for the calculation.[9,10] Using the 260 nm mononucleotides and dinucleoside phosphates published values for ε_{260} (6) given in Table 12.1, the calculation of ε_{260} for a 9-mer oligonucleotide, GGC UCU UAG is shown below:

$$eG + eG + eC + eU + eC + eU + eU + eA + eG$$

$$11500 + 11500 + 7200 + 9900 + 7200 + 9900 + 9900 + 15400 + 11500 = 94000\ M^{-1}cm^{-1}$$

TABLE 12.1
Parameters for the Extinction Coefficients Oligonucleotides at 260 nm

Stack or monomer	ε_{260} $M^{-1}cm^{-1}$	Stack or monomer	ε_{260} $M^{-1}cm^{-1}$
pA	15400	GpA	25200
pC	7200	GpC	17400
pG	11500	GpG	21600
pU	9900	GpU	21200
pdT	8700	UpA	24600
ApA	27400	UpC	17200
AdC	21000	UpG	20000
ApG	25000	UpU	19600
ApU	24000	ApdT	23400
CpA	21000	CpdT	15700
CpC	14200	GpdT	20600
CpG	17800	UpdT	18200
CpU	16200	dTpdT	16800

3. In addition to the simple sum of nucleotides extinction coefficients, it is recommended to multiply the sum of the extinction coefficients of the individual bases by a factor of 0.9. This is due to base stacking interactions in the single strand, which suppresses the absorbance of DNA relative to the value calculated from the extinction coefficients of the individual nucleosides. This effect is greater for a duplex, and the multiplication factor for a self-complementary sequence is about 0.8. These figures are estimates for typical DNA sequences.[10,11]
4. The light absorptions of isolated nucleoside bases reported in literature are measured in solution in high dilution. They undergo marked changes when they are in close proximity to neighboring bases resulting in ordered secondary structures in oligonucleotides. In such ordered structures, the bases can stack face-to-face and thus share π-π electron interactions that profoundly affect the transition dipoles of the bases. Most importantly, base-base stacking results in a decrease in ε, known as hypochromicity. This fourth method considers the base composition and sequence order and makes a correction for the stacking factor. This is considered the most accurate method based on the nearest-neighbor (first-neighbor) model. Using the values in Table 12.1, calculation of ε_{260} for the 9-mer GGC UCU UAG with next-neighbor correction is shown below:

eGG + eGC + eCU +eUC + eCU + eUU + eUA + eAG – (eG + eC + eU + eC + eU + eU + eA)

$$21600 + 17400 + 2*16200 + 17200 + 19600 + 24600 + 25000 - (11500 + 7200 + 9900 + 7200 + 9900 + 9900 + 15400) = 86800\ M^{-1}cm^{-1}$$

Calculators based on the nearest-neighbor parameters are readily available to predict thermodynamic properties of oligonucleotides. The extinction coefficients for mononucleotides and dinucleotides of both DNA and RNA were determined at a wavelength of 260 nm, at 25°C, and at neutral pH in water.

Richard Owczarzy et al. have developed an integrated set of bioinformatics tools that predicts the properties of native and chemically modified nucleic acids and assist in their design.[12] The online tool, OligoAnalyzer 3.1, provides a feature for calculation of molar extinction coefficients for both single-stranded DNA and RNA.[13] Further, the nearest-neighbor model was applied to predict absorbance spectra illustrating the dependence of extinction coefficient as a function of wavelength.[12] Another online calculator has been developed for DNA and RNA, both single and double stranded, properties including molecular weight, solution concentration, melting temperature, and estimated absorbance coefficients.[14]

The calculated extinction coefficients differ as much as 10–20% of the values obtained experimentally. These differences may be a result of inaccuracies of the data utilized for the values of mononucleotide and dinucleotides phosphates and the assumption that the ε_{260} values are the same for the cognate DNA and RNA mononucleotides. The method for the determination of extinction coefficient reported by Kallansrud and Ward[15] relied on the hydrolysis of RNA by NaOH. Also, DNA was hydrolyzed with a mixture of snake venom phosphodiesterase (SVP) or SVP/DNase I. The authors found to have an agreement within 20% of calculated values. Hydrolysis based experimental determination of ε_{260} relies on complete hydrolysis of the oligonucleotides and the accuracy of extinction coefficients for the monomer nucleotides. Further, with the current advances in synthetic oligonucleotides with base modifications and conjugation chemistry to enhance nuclease resistance and drug delivery, these oligonucleotides may not be readily amenable to hydrolysis. Also, hydrolysis may result in degradation following an unknown pathway.

Cavaluzzi and Borer improved the accuracy of extinction coefficients by determining the concentration of the nucleoside-5′-monophosphates by ^{1}H-NMR[16] Cavaluzzi-Borer correction for extinction coefficients of DNA bases at 260 nm is incorporated into Integrated DNA Technologies (IDT) SciTools.[13]

12.3 EFFECT OF BUFFER

Results obtained for oligonucleotide extinction coefficient calculators are based on data from oligonulceotides dissolved in water at a pH of 7 of variable ionic strength, depending on which calculator one chooses to use. On the basis of practical experience, measurements made in water are not accurate. Figure 12.1 illustrates the variation of absorbance at 260 nm measured in a kinetic experiment of a siRNA in water, in 1xPBS, and in 0.9% saline, all with same concentration (2 μM). Absorbance obtained for siRNA in phosphate buffered saline (PBS) and 0.9% saline rapidly equilibrated and gave a constant absorbance reading. However, absorbance of the aqueous solution continued to increase with time over a period of 10 min and did not stabilize, indicating that the

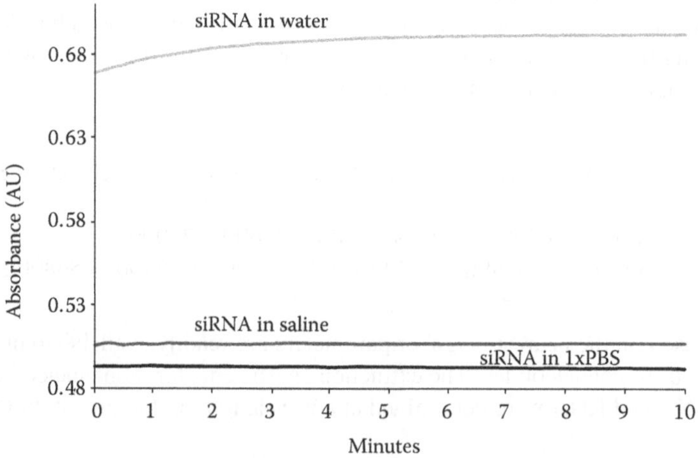

FIGURE 12.1 Plot of absorbance as a function of time: 2 μM siRNA in water, mean absorbance of 0.6872, %RSD of 0.90, 2 μM siRNA in 0.9% saline, mean absorbance of 0.5071, %RSD of 0.06, 2 μM siRNA in 1xPBS mean absorbance of 0.4930, %RSD of 0.06. All measurements were made at 260 nm and 25°C.

oligonucleotide had not attained equilibrium in aqueous medium. Precision of replicate measurements of absorbance of an aqueous oligonucleotide solution, prepared in a well-controlled quality control (QC) laboratory will vary over 1–2%, based on the interval between measurements. Measurements made in salt buffers are an order of magnitude more precise than in water, as shown in Figure 12.1.

A siRNA duplex in aqueous solution is partially denatured into sense and antisense strands. Because this is a dynamic equilibrium, total absorbance (sum of duplex and single strands) changes over time. The transition between an ordered, native structure and a disordered, denatured state in siRNA, resulting in hyperchromic effects, is clearly illustrated in the absorbance as a function of time in Figure 12.1. In the presence of salts, as found in saline or 1xPBS solutions, which are the most common solutions used for parenteral administration, the absorbance of siRNA is hypochromic relative to its constituent single strands. Replicate absorbance measurement made in salt solutions will provide more reproducible results. Because calculated values of extinction coefficients are good to 10–20% within experimental values, it is recommended to perform direct determination of ε_{260} in 0.9% saline. In general, use of 0.9% saline provides reliable data for most RNA duplexes.

12.4 CALCULATED VALUE VERSUS EXPERIMENTAL VALUE

Calculated values of ε_{260} obtained from Internet tools are satisfactory during the drug discovery phase, where it is most common to rank order of activity of various oligonucleotide sequences, and the absolute quantitation of the oligonucleotides is not necessary. Once an oligonucleotide is advanced to preclinical and clinical development, it is critical to determine the oligonucleotide concentration in the dosing solution accurately. Drug product "label claim" should be determined by reliable analytical methods to meet preestablished specification limits. An illustration on the impact of differences between calculated vs. experimental ε_{260} is given below.

ALN-RSV01 is being developed for the treatment of RSV infection.[17] This synthetic, double-stranded RNA oligonucleotide is formed by the hybridization of two partially complementary single-strand RNAs. Each of the single-strand RNAs is composed of 19 ribonucleotides and two thymidine units at the 3′ end. The 19 ribonucleotides of one of the strands hybridize with the complementary 19 ribonucleotides of the other strand, thus forming 19 ribonucleotide pairs and a bis-thymidine overhang at each side of the duplex. The composition of the sense and antisense strands is given below:

Sense: 5′– GGC UCU UAG CAA AGU CAA GTT –3′

Antisense: 5′– CUU GAC UUU GCU AAG AGC CTT –3′

The experimentally determined ε_{260} of the sense strand in water was found to be 183,100 ($M^{-1}cm^{-1}$). The calculated value for sense strand using OligoAnalyzer 3.1[18] was found to be 209,700 ($M^{-1}cm^{-1}$), which is 14.5% greater than the experimentally determined value. Using the method of Ribotask,[19] the calculated value for the sense strand form was found to be 220,600 ($M^{-1}cm^{-1}$), which is 20.5% greater than the experimentally determined value. For the antisense strand in water, the experimentally determined ε_{260} value was found to be 168,300 ($M^{-1}cm^{-1}$).

The calculated value for the antisense strand using OligoAnalyzer 3.1[18] was found to be 197,800 ($M^{-1}cm^{-1}$), which is 17.5% greater than the experimentally determined value.

The calculated value for the antisense strand using Ribotask[19] was found to be 209,100 ($M^{-1}cm^{-1}$), which is 24.2% greater than the experimentally determined value. Several other Web tools are available for similar comparisons of calculated values with experimental values.

The above comparisons clearly indicate the differences in calculated values resulting from various calculative methods, as well as directly determined values. In order to construct a duplex from single strands, molar concentrations of individual strands have to be determined, and mixtures of known molar ratios must be carefully made. If calculations are made based on the calculated values from different Web tools, the quality of the siRNA duplex formed may be compromised. Further, the final concentration of drug formulations based on erroneous calculation may not be accurate and fail to meet specifications for label claim.

12.5 DIRECT DETERMINATION OF EXTINCTION COEFFICIENT

The iterative process of improving the accuracy of extinction coefficient calculations, by various Web-based oligonucleotide calculators, depends on the accurate determination of extinction coefficients of the mononucleotide phosphates.[16] However, with the rapid increase in a variety of chimeric DNA-RNA sequences for siRNA therapeutics, not all extinction coefficients can be determined using online calculators. Even an approximate estimation of extinction coefficient, with 10–20% differences in true value, can have serious implications in the conclusion of toxicological safety data, as well as in errors in calculation of clinical dosing solutions. Such errors in calculations can also result in inaccurate yield results for the synthesis of oligonucleotides at a kilogram scale can also lead to significant increases in the cost of finished products.

Therefore, it is recommended to make direct determination of extinction coefficients on the lead sequences that are to be utilized in toxicological and clinical studies. Direct determination, as described below, does not rely on individual extinction coefficient values of mononucleotide dinucleoside phosphates. Lengthy hydrolysis of the sample is not required, however, for direct determination; reference standards of the highest purity should be synthesized and used for ε_{260} determination.

TABLE 12.2
Concentration and Absorbance Data of Direct Determination of ε_{260}

	Sample Weight, mg	% Moisture	Moisture Corrected Weight	Stock Concentration, M
Replicate 1	19.02	3.03	18.444	1.29E-05
Replicate 2	20.04	3.03	19.433	1.36E-05
Replicate 3	21.91	3.03	21.246	1.49E-05
	Aliquot, mL	**Final Volume, mL**	**Concentration, M**	**Absorbance, AU**
Dilution 1	5	100	6.45E-07	0.1945
Dilution 2	10	100	1.29E-06	0.386
Dilution 3	15	100	1.93E-06	0.5785
Dilution 4	10	50	2.58E-06	0.7765
Dilution 5	15	50	3.87E-06	1.151
	Aliquot, mL	**Final Volume, mL**	**Concentration, M**	**Absorbance, AU**
Dilution 1	5	100	6.79E-07	0.2035
Dilution 2	10	100	1.36E-06	0.4105
Dilution 3	15	100	2.04E-06	0.609
Dilution 4	10	50	2.72E-06	0.8125
Dilution 5	15	50	4.08E-06	1.2175
	Aliquot, mL	**Final Volume, mL**	**Concentration, M**	**Absorbance, AU**
Dilution 1	5	100	7.43E-07	0.222
Dilution 2	10	100	1.49E-06	0.444
Dilution 3	15	100	2.23E-06	0.667
Dilution 4	10	50	2.97E-06	0.8865
Dilution 5	15	50	4.46E-06	1.3175

12.5.1 Experimental Procedure

Weigh out 20.0 ± 2.0 mg of lyophilized oligonucleotide in a moisture-controlled glove box. Quantitatively transfer into a 100 mL class A volumetric flask and dissolve in 0.9% saline. It is important to determine moisture content (by Karl Fisher method) at the same time the sample is weighed for the ε_{260} determination. Excess salt in the sample may also contribute to the sample weight. Therefore, it is recommended to weigh out a separate sample for sodium analysis at the same time of weighing for ε_{260} determination. Perform moisture correction on the weighed out sample. If the sodium is not significantly higher than the theoretical value of sodium, there is no need to correct for excess sodium content. Calculate the concentration of this stock solution (approximately 0.2 mg/mL). Perform dilutions ($n = 5$) so that an absorbance reading of between 0.3 and 1.2 can be achieved for calibration standards (see the sample dilution scheme in Table 12.2). Set the temperature of the UV instrument to 25°C.

After at least 30 min, confirm the photometric accuracy of the UV instrument at 260 nm. Make triplicate measurements of each calibration solution, and plot the average reading against

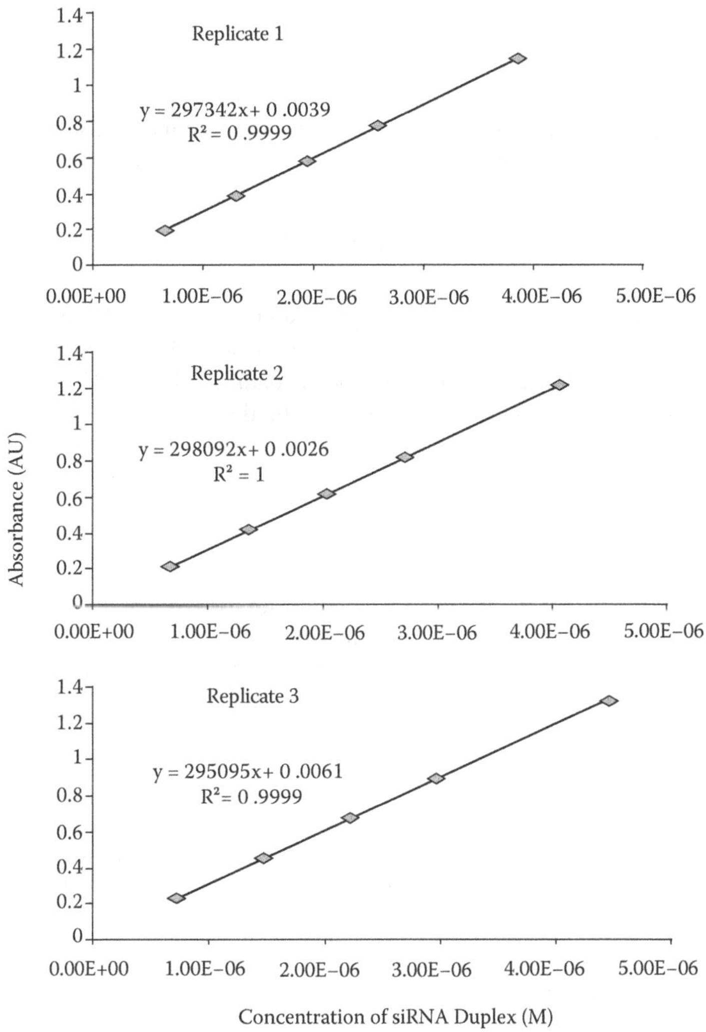

FIGURE 12.2 Calibration plots of triplicate molar extinction coefficient determination.

the concentration of the calibration standards. The slope of the calibration plot of UV absorbance against the concentration, expressed in mole/L, is the value of ε_{260}. Perform the above determination of ε_{260} in triplicate and average the ε_{260} to get a final value for ε_{260}.

12.5.2 Illustration of ε_{260} Calculation from Direct Measurement

A siRNA duplex of a molecular weight of 14303.6 g/mole and with a moisture content of 3.03% was weighed in triplicate and dissolved in 100 mL 0.9% saline.

For replicate 1 (see Table 12.2) the moisture-corrected sample weight

$$= 19.02*(100 - 3.03)/100 = 18.444 \text{ mg.}$$

The concentration of replicate 1 stock solution

$$= 18.444 \text{ mg}*(\text{g}/1000 \text{ mg})*(1/100 \text{ mL})*(1000 \text{ mL/L})*(\text{mole}/14303.6 \text{ g})$$

$$= 1.29\text{E}^{-05} \text{ M}$$

A serial dilution (n = 5) was performed, and triplicate measurements of absorbance at 260 nm, 25°C, and a path length of 1.0 cm for the colorimeter cuvette were made. The first dilution of stock solution of replicate 1 is 5 mL in 100 mL. The concentration of this diluted solution is

$$= (5/100)*\ 1.29\text{E}^{-05} = 6.45\text{E}^{-07}$$

and the mean absorbance measurement (n = 3) of this solution was 0.1945 AU. See Table 12.2 for a complete data set for five dilutions of three weighed replicates. The extinction coefficient of the siRNA at 260 nm was calculated from the slope of the data; see Figure 12.2 for calibration plots of five measurements for each replicate. The slopes of the three replicates were 297,342, 298,092, 295,095, and the average value of molar extinction coefficient was 296,843 M^{-1} cm^{-1}, with a relative standard deviation of 0.5%.

12.6 SUMMARY

This chapter enables the reader to evaluate the need for an experimentally determined extinction coefficient. Specific examples are given to illustrate differences between the various calculated values and experimentally determined values. A detailed experimental procedure is provided for the direct determination of the molar extinction coefficient for a siRNA oligonucleotide.

REFERENCES

1. Schneider, W. C. 1945. Phosphorus compound in animal tissues. I. Extraction and estimation of desoxypentose nucleic acid and of pentose nucleic acid. *Journal of Biological Chemistry* 161: 293–303.
2. Webb, J. M., and H. B. Levy. 1955. A sensitive method for the determination of deoxyribonucleic acid in tissues and microorganisms. *Journal of Biological Chemistry* 213: 107–117.
3. Webb, J. M. 1958. Studies on the determination of total nucleic acids by ultraviolet absorption methods. *Journal of Biological Chemistry* 230: 1023–1030.
4. Cota-Robles, E. H., A. G. Marr, and E. H. Nilson. 1958. Submicroscopic particles in extracts of azotobacter agilis. *Journal of Bacteriology* 75: 243–252.
5. Cantor, C. R., M. M. Warshaw, and H. Shapiro. 1970. Oligonucleotide interactions. III. Circular dichroism studies of the conformation of deoxyoligonucleotides. *Biopolymers* 9: 1059–1077.

6. Tataurov, A. V., Y. You, and R. Owczarzy. 2008. Predicting ultraviolet spectrum of single stranded and double stranded deoxyribonucleic acids. *Biophysical Chemistry* 133: 66–70.
7. Sambrook, J., E. F. Fritsch, and T. Maniatis. 1989. *Molecular Cloning: A Laboratory Manual*, 2nd ed., Cold Spring Harbor, N. H.: Cold Spring Harbor Laboratory Press.
8. Wallace, R. B., and C. G. Miyada. 1987. In *Methods of Enzymology*, eds. S. L. Berger and A. R. Kimmel, 152:438. San Diego: Academic Press.
9. Warshaw, M. M., and I. Tinoco, Jr. 1966. Optical properties of sixteen dinucleoside phosphates. *Journal of Molecular Biology* 20: 29–38.
10. Borer, P. N. 1975. Optical properties of nucleic acids, absorption, and circular dichroism. In *Handbook of Biochemistry and Molecular Biology, Nucleic Acids*, ed. G. D. Fasman, 3rd ed., 152:189. Boca Raton, FL: CRC Press.
11. Brown, T., and D. J. S. Brown. 1991. Modern machine-aided methods of synthesis. In *Oligonucleotides and Analogs, A Practical Approach*, ed. F. Eckstein, 20. Oxford: IRL Press, Oxford.
12. Owczarzy, R., A. V. Tataurov, Y. Wu, et al. 2008. IDT SciTools: a suite for analysis and design of nucleic acid oligomers. *Nucleic Acids Research* 36: 163–169.
13. Integrated DNA Technologies. IDT SciTools. http://www.idtdna.com/Scitools/Scitools.aspx.
14. Kibbe, W. A. 2007. OligoCalc: an online oligonucleotide properties calculator. *Nucleic Acids Research* 35: Web Server issue W43-W46 (http://basic.northwestern.edu/biotools/OligoCalc.html).
15. Kallansrud, G., and B. Ward. 1966. A comparison of measured and calculated single and double stranded oligodeoxynucleotide extinction coefficients. *Analytical Biochemistry* 236: 134–138.
16. Cavaluzzi, M. J., and P. N. Borer. 2004. Revised UV extinction coefficients for nucleoside-5′-monophosphates and unpaired DNA and RNA. *Nucleic Acids Research* 32: e12.
17. Alvarez, R., S. Elbashir, T. Borland, et al. 2009. RNA Interference-mediated silencing of the respiratory syncytial virus nucleocapsid defines a potent antiviral strategy. *Antimicrobial Agents and Chemotherapy* 53: 3952–3962.
18. Integrated DNA Technologies. IDT SciTools OligoAnalyzer 3.1. http://www.idtdna.com/analyzer/Applications/OligoAnalyzer/Default.aspx (accessed October 8, 2009).
19. RiboTask. Oligo Calculator (DNA, RNA & LNA). RiboTask APS. http://www.ribotask.dk/index.php?pid=94 (accessed October 8, 2009).

13 Structural Determination by NMR

Michele L. DeRider
Catalent Pharma Solutions

Doug Brooks
Regado Biosciences

Gary Burt
Girindus America, Inc.

CONTENTS

13.1 INTRODUCTION

Nuclear magnetic resonance spectroscopy (NMR) is an incredibly powerful technique owing to its quantitative nature, broad dynamic range, and nondestructive nature. Furthermore, the versatility that accompanies NMR's ability to provide atomic resolution is readily apparent for the study of oligonucleotides. NMR methods can be applied to starting materials, intermediates, drug substances, and drug products. Application of NMR to therapeutic oligonucleotides ranges from methods used for the structural characterization of reference standards and impurities to the use of validated methods used for the testing and release of starting materials and products.

This chapter provides several case studies relevant to the manufacturing, characterization, and quality control analysis of therapeutic oligonucleotides. Case studies presented include NMR applications to phosphoramidite starting materials, a phosphorothioate DNA single-strand oligonucleotide, a stem loop representing an aptamer, and a model DNA triplex. A theoretical exercise is presented for the annealing of two strands to form an siRNA duplex. The methods from the case studies can be extended to other classes of oligonucleotides and starting materials. The case studies were specifically selected to include the elements necessary for validation of the NMR methods for quality control testing in a regulated environment.

13.2 NOMENCLATURE

The nomenclature for atom numbering and specifying conformation will follow the IUPAC recommendations.[1] Figure 13.1 displays the recommended designators for the atoms of the common bases and the pentose rings. The three-dimensional structure of oligonucleotides can be presented several ways; the recommended way is to describe using torsional angles.[2] The torsional angles are displayed in Figure 13.2. The puckering of the sugar ring is typically described in terms of pseudorotation angle, the two common structures are C3′-endo (N-type) and C2′-endo (S-type).[3]

Oligonucleotides potentially possess three levels of structure: primary, secondary, and tertiary. Primary structure refers to the nucleotide sequence. Secondary structure is characterized by the local folding as a result of the hydrogen bonding patterns between the imino/amino resonances. Several hydrogen bonding partners are displayed in Figure 13.3. The Watson–Crick GC and AU base pairs and a model triplet base complex are demonstrated. More thorough descriptions of potential base pairs are discussed elsewhere.[4] DNA is most often double-stranded duplex. Helical regions can have different conformations, most commonly A, B, and Z. Furthermore, the two strands can hydrogen bond in either sense or antisense orientations. RNA commonly has more diverse secondary structural elements. In addition to duplex formation, features include single-stranded regions, hairpins, bulges, internal loops, and junctions. Less common secondary structures are triplexes and pseudoknots. Secondary structure is typically represented by two-dimensional drawings. Figure 13.4 displays several common secondary structural features. The tertiary structure is the three-dimensional shape, as defined by the atomic coordinates. Typically, this involves the interactions between distinct secondary structural elements.

13.3 THEORY

13.3.1 Chemical Shift

The basis of nuclear magnetic resonance (NMR) is that some nuclei can be oriented by a strong magnetic field and then absorb radiation at frequencies characteristic of that nuclei. The oriented nuclei are irradiated by a radio frequency field (RF) of a specific duration and strength to rotate the magnetization from the z axis into the transverse (xy) plane. The frequency of the radiation necessary for absorption of energy depends on both the type of nucleus and the chemical environment of the specific nucleus. The magnetization in the transverse plane decays with time, with the individual nuclei precessing at different rates that depend on their electronic and chemical environment. The precession causes an electrical current that is detected, producing a free induction decay (FID). The NMR spectrum is obtained by Fourier transformation of the FID. Each nuclei of the same element in different environments give rise to distinct spectral lines (referred to as chemical shift)—making it possible to observe signals from individual atoms. This ability to achieve atomic resolution makes NMR such a powerful technique. The primary factor impacting the chemical shift is the local magnetic field, which is influenced by the electron density of the neighboring atoms. In general, NMR

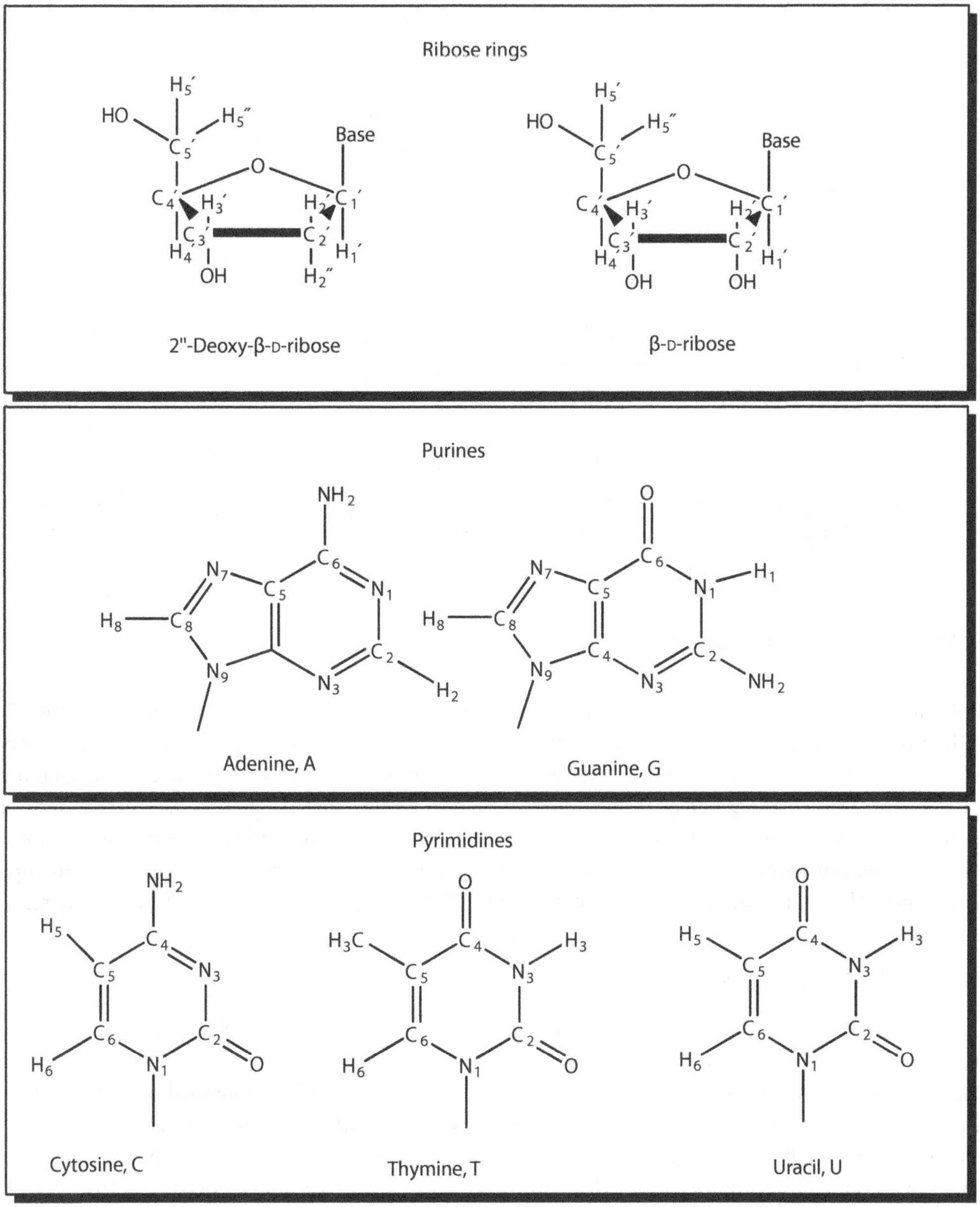

FIGURE 13.1 Nomenclature, structures, and atom numbering for the ribose rings and the common bases in DNA and RNA.

active nuclei adjacent to electronegative atoms are deshielded, and the resulting spectral line corresponding to that nucleus will be observed at a larger chemical shift (referred to as downfield). If the electron density about a proton nucleus is relatively high, the induced field due to electron motions will be stronger than if the electron density is relatively low. The shielding effect from a high electron density will therefore be larger, and a higher external field will be needed for the RF energy to excite the nuclear spin. The inductive effects of the substituents are only one factor in the chemical shift. For example, as expected based on the different neighboring atom, the H5 base protons of cytosine absorb at a different frequency than the H6 base protons. Furthermore, the individual H5 base protons of multiple cytosine nucleotides in the same RNA hairpin would also likely absorb

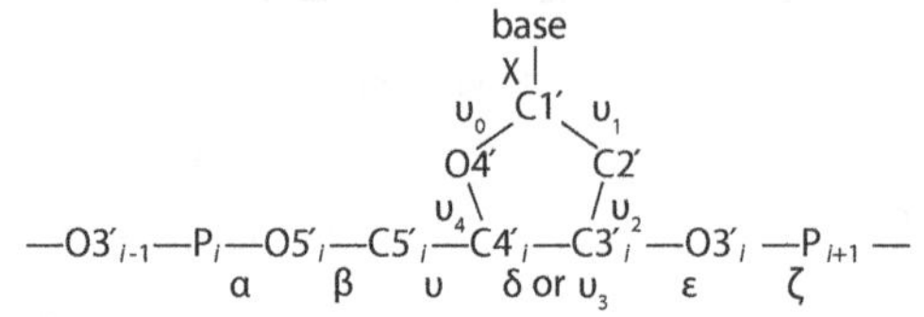

	Angle	Atoms
Backbone and glycosid torsion angles	α	O3′$_{i-1}$—P$_i$—O5′$_i$—C5′$_i$
	β	P$_i$—O5′$_i$—C5′$_i$—C4′$_i$
	γ	O5′$_i$—C5′$_i$—C4′$_i$—C3′$_i$
	δ	C5′$_i$—C4′$_i$—C3′$_i$—O3′$_i$
	ε	C4′$_i$—C3′$_i$—O3′$_i$—P$_{i+1}$
	ζ	C3′$_i$—O3′$_i$—P$_{i+1}$—O5′$_{i+1}$
	χ(Py)	O4′$_i$—C1′$_i$—N1$_i$—C2$_i$
	χ(Pu)	O4′$_i$—C1′$_i$—N9$_i$—C4$_i$

	Angle	Atoms
Sugar pucker	υ_0	C4′ — O4′ — C1′ — C2′
	υ_1	O4′ — C1′ — C2′ — C3′
	υ_2	C1′ — C2′ — C3′ — C4′
	υ_3	C2′ — C3′ — C4′ — O4′
	υ_4	C3′ — C4′ — O4′ — C1′

FIGURE 13.2 Torsional angles in oligonucleotides.

at slightly different frequencies also due to the slightly different chemical environments based on nucleotide sequence and/or structural differences. The approximate ranges of proton and carbon chemical shifts for oligonucleotides are shown in Figures 13.5 and 13.6, respectively. Chemical shift is expressed in parts per million (ppm) relative to a reference frequency. It is common to use an internal reference standard, such as 2,2-dimethylsilapentane-5-sulfonic acid (DSS) for protons in aqueous solutions. Indirect referencing of the other nuclei is also commonly performed if an appropriate internal or external standard is not available. This topic is discussed in detail elsewhere.[2]

13.3.2 NMR Active Nuclei

Atomic isotopes that are amenable to study by NMR spectroscopy are referred to as NMR-active nuclei. There are many elements that have at least one NMR-active isotope, the most commonly studied isotopes for oligonucleotides are ^{1}H, ^{13}C, ^{15}N, ^{19}F, and ^{31}P. The natural abundance of the NMR active nuclei must be taken into account. For example, the most studied because of both the sensitivity and abundance is ^{1}H (referred to as proton). Fluorine and phosphorous isotopes both have 100% natural abundances and thus are also readily studied using NMR spectroscopy. For nuclei with low natural abundance of the NMR active isotope, it is possible to isotopically enrich the sample. This is often referred to as spin labeling or simply labeling. Both ^{13}C and ^{15}N can be isotopically enriched using chemical methods for both DNA and RNA. RNA can be labeled using isotopically enriched ribonucleotide triphosphates for *in vitro* transcription using T7 RNA polymerase, reviewed elsewhere.[5] Although the ^{13}C isotope is only 1.1% abundant, natural abundance NMR spectroscopy is still routinely performed. The natural abundance of ^{15}N makes it generally not feasible to use for NMR analysis without isotope enrichment in oligonucleotides. NMR methods applicable to the study of therapeutic oligonucleotides represent a subset of the total NMR methods that have been developed to study the structure and function of oligonucleotides.[6,7] While many RNA and DNA structural studies are performed utilizing ^{13}C and ^{15}N labeled oligonucleotides, these labeled nucleosides are not generally available for therapeutic oligonucleotides as many of them are chemically modified and the labeled modified nucleotides are generally not available or prohibitively expensive. Because of these limitations, many of the multidimensional NMR methods

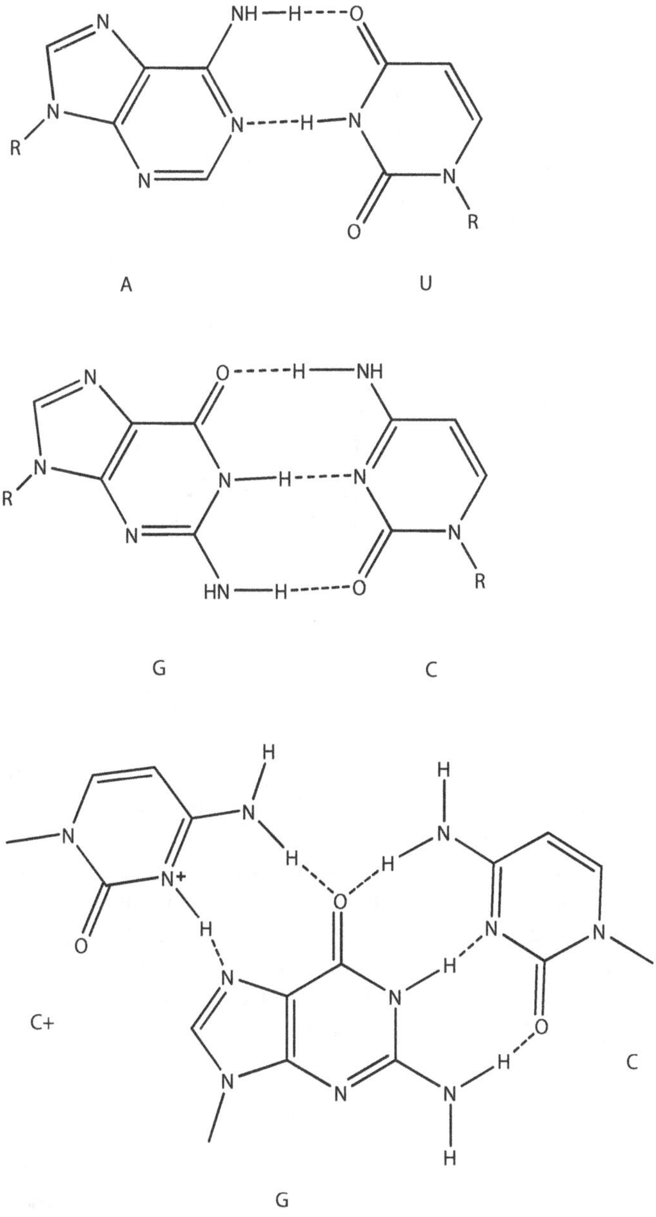

FIGURE 13.3 Schematic representation of example base pairs (AU and GC) and base triplet (CGC+). The dashed lines display hydrogen bonds.

are not directly applicable to therapeutic oligonucleotides. Many thorough reviews summarizing the process have been written.[4,8–10]

13.3.3 Scalar and Dipolar Interactions

NMR active nuclei can interact or perturb each other. There are two primary modes: through bonds (scalar) or through space (dipolar). One strength of NMR is the ability to utilize this phenomenon to work in multiple dimensions. Most of these experiments consist of the first dimension corresponding to the conventional spectrum, while the second dimension contains information about the

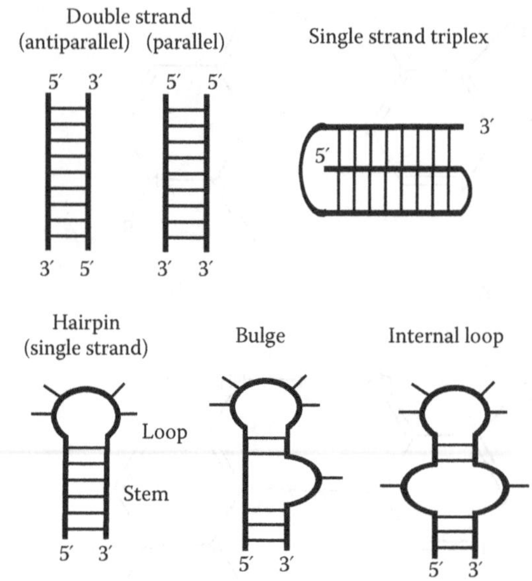

FIGURE 13.4 Common secondary structural elements of nucleic acids.

connections between that first dimension with either the resonances on the diagonal or other interacting NMR active nuclei. NMR experiments for biological systems routinely include three and four dimensions. Typically, those experiments require isotopic labeling. The multidimensional spectra not only spread the signals out over two or three dimensions but establish relationships between the nuclei giving rise to the different signals.

Scalar coupling arises from the interaction of different spin states through the chemical bonds of a molecule and results in the splitting of NMR signals. One important feature of spin-spin splitting is that it is independent of magnetic field strength. Therefore, increasing the magnetic field strength will increase the chemical shift difference between two peaks in hertz (not parts per million) but will not alter the coupling constant. To simplify a spectrum, especially with ^{13}C and ^{15}N NMR, it is common to employ decoupling, which consists of irradiation of the protons at their resonance

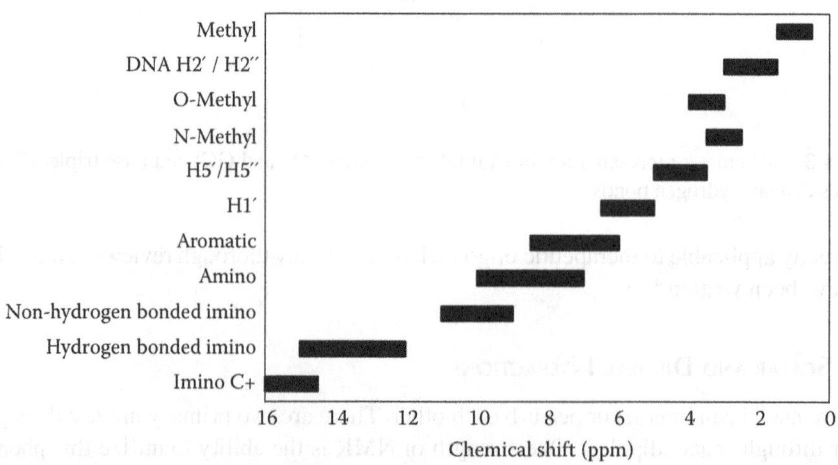

FIGURE 13.5 Approximate ^{1}H chemical shift ranges in DNA and RNA.

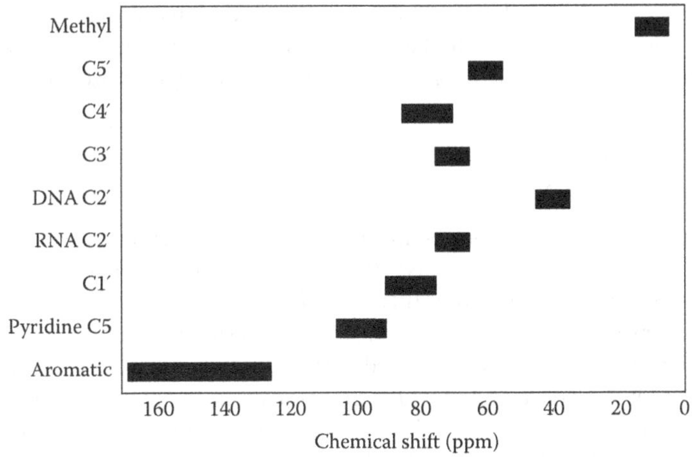

FIGURE 13.6 Approximately ^{13}C chemical shift ranges in DNA and RNA.

frequency to collapse the multiplet of the ^{13}C or ^{15}N resonance into a singlet. In oligonucleotide samples, the signal line width may be broader than the splitting, so the latter may not always be apparent. The measurement of couplings can provide valuable structural information.

The dipolar interaction of two NMR active nuclei is the cause of the nuclear Overhauser effect (NOE), which can be used to measure internuclear distances. Through-space connectivities between protons that are in close proximity to each other can be seen as correlations of the two-dimensional NOESY NMR spectrum. One-dimension NOE experiments can be performed by irradiating a resonance of interest and observing the decrease in signal of neighboring resonances.

13.4 PRACTICAL CONSIDERATIONS

13.4.1 Instrumentation

The core of the NMR spectrometer is a helium-cooled superconducting magnet. The field strength of the spectrometer is typically referred to as the resonant frequency of proton nucleus. For example, a 900 MHz spectrometer is a 21 Tesla magnet. A larger magnetic field results in a more sensitive signal and gives greater chemical shift dispersion. For large molecules, such as an oligonucleotide, it is generally regarded that magnets operating at least 500 MHz are required for the necessary sensitivity and signal dispersion. The probe is inserted in the hollow core of the magnet. The sample is placed in the probe for analysis. The probe is the most specialized part of the spectrometer, acting as the interface between a sample and the spectrometer. It is responsible for transmitting and receiving the radio frequency to and from the NMR sample. Most spectrometers have multiple probes designed for different uses. The type of nuclei to be observed and the required sensitivity dictates what type of probe should be utilized. Full discussion of the different aspects of probes is beyond this text; details have been summarized elsewhere.[11]

In general, probes utilized for liquid NMR samples are designed with layers of different coils, with inner coil (i.e., the coil closest to the NMR sample) being the most sensitive. For example, broadband or switchable probes are designed for the observation of lower-frequency nuclei such as ^{13}C and ^{31}P. The inner coil can be tuned over a frequency range to observe different nuclei. The outer coil is designed for proton detection. Although these probes can be utilized for proton detection sequences, the sensitivity is less than if an inverse probe is used. Inverse probes are designed for more sensitivity for proton detection with the inner coil optimized for protons and the outer coil(s).

Indirect probes are typically used in multidimensional spectra that utilize the coherence transfer from less sensitive nuclei (such as ^{13}C) to the proton dimension for increased sensitivity. An advance in probes is the use of cryogenically cooled probes that drastically increase the sensitivity of NMR spectrometers.

13.4.2 Sample Preparation

Many of the drug product formulations being evaluated for therapeutic-based oligonucleotides are highly amenable to NMR study. Common NMR solution conditions are approximately 1 mM oligonucleotide in 10–20 mM phosphate buffer (pH range 4.5–7), 50–200 mM NaCl. NMR oligonucleotide sample conditions in research also commonly include a small amount of $MgCl_2$ and/or EDTA. Other buffers can be used, preferably deuterated buffers for proton detection. It is possible to utilize buffers with the nuclei to be observed provided the resonances do not interfere with the resonances of interest. For example, phosphate buffer is often used for the collection of phosphorous spectra of oligonucleotides because the chemical shift of the buffer generally is not coincident with the resonances of oligonucleotides with phosphodiester or phosphorothioate backbones. It is typically not recommended to utilize buffers with protons for proton detection as complications are likely that impact the two-dimensional spectral quality based on the different dynamicproperties and relative intensities.

The buffer concentration should be kept low as high buffer concentrations may catalyze the hydrogen exchange of imino protons. The increased rate of exchange of imino resonances results in broadened resonances and thus decreased sensitivity and resolution. A small amount of lock solvent, typically deuterium oxide, can be added (less than 10%) directly to the formulation. NMR sample size depends on the instrumentation available. Typically most probes are designed for 5 mm width NMR tubes, where typical sample volumes are 500–700 μL, depending on the coil size of the probe. When the amount of sample is limited, restricted volume tubes reduce the sample volume by approximately one third. The restricted volume NMR tubes consist of a small volume inner microtube mounted in a standard size NMR tube. The small volume of sample solution is held directly in the region of the NMR tube length that the NMR coil "sees," without large portions of the sample solution being excluded from the measurement. In addition, the assemblies are magnetic susceptibility matched to the NMR solvent (most likely D_2O).

When it is necessary to do assignments other than the imino resonances, data collection of oligonucleotides is also typically performed on samples in 100% D_2O. The exchangeable proton resonances are exchanged with deuterium and thus would not be observed in the ^1H NMR spectrum. This decreases the complexity of the spectra and also confirms the differentiation between exchangeable and nonexchangeable. Furthermore, because of the slightly different dynamic properties of deuterium oxide relative to H_2O, the spectra can have better resolution. Deuterium exchange of oligonucleotides is often accomplished by several cycles of lyophilization between D_2O rinses.

After the preparation of the NMR sample, it is common to heat the sample to remove potential concatamers and other aggregates that may form. The sample may be either quickly or slowly annealed; typically, the fast annealing step favors the formation of hairpin structures and the slow anneal favors the formation of duplexes.

13.4.3 Sensitivity and Resolution

NMR is typically referred to as an insensitive method owing to the high concentrations and relatively large sample size required. The high concentration (typically 1–100 mM) of NMR samples is typically not limiting for therapeutics. In the formulations of many oligonucleotide drug products, the concentrations can be of the same order of magnitude required by NMR. NMR experiments achieve greater sensitivity by averaging many scans or transients. During the optimization of NMR experiments, the number of transients necessary to achieve the desired signal to noise is empirically determined. Signal-to-noise increases proportionally to the square root of the number of scans. Limits of detection and quantitation are determined by evaluation of peak relative to the average of the noise. As in other analytical methodologies, it is necessary to understand the limits of quantitation with respect to reporting.

The resolution of NMR experiments is dependent on many factors. There are sample considerations such as temperature, salt, concentration, viscosity, and molecular weight that impact line-width. Furthermore, there are sample-independent factors impacting resolution such as the strength of the magnetic field and homogeneity of the magnetic field (as achieved by shimming). Spectrometer resolution is therefore very different than for sample resolution. For spectrometer resolution, often a standard of chloroform in deuterated acetone is used and the line width is measured in terms of Hertz. As mentioned above, there are a number of factors that contribute to the resolution obtained for the sample of interest, and there are a number of ways to improve the effective resolution for resonances of interest although often at the cost of signal to noise. During acquisition, it is possible to increase the number of points, in either the direct or indirect dimension. Postacquisition processing of the spectra can be utilized to further increase resolution. Prior to Fourier transformation, it is common to multiply the FID by a mathematical function (termed apodization). Choice of the apodization function can increase resolution or sensitivity. The resolution can also be enhanced postacquisition using either zero-filling or linear prediction.

13.4.4 Water Suppression

In biological NMR it is often necessary to obtain proton spectra from samples dissolved in nondeuterated solvents. For oligonucleotides, observation of the imino protons is typically conducted in H_2O with a small amount of deuterium oxide for a lock. This allows for the detection of exchangeable protons such as imino and amino protons in RNA/DNA. The very high proton concentration of the nondeuterated solvent would generate an NMR signal over a thousand times more intense than any of the solute signals. To overcome this dynamic range issue, solvent suppression techniques should be used during data acquisition. Presaturation of the residual water resonance is the simplest, but the relative fast exchange of imino/amino resonances with water leads to saturation of their resonances by the transfer of saturation. Therefore, presaturation can only be used when the imino/amino resonances are not of interest. There are many selective excitation and solvent suppression techniques to reduce the large signal from the water such as jump return and watergate. It is possible to suppress multiple signals utilizing WET sequence. This is commonly utilized in LC-NMR systems to remove both the residual water resonance and the organic component of the mobile phase. A recent review of solvent suppression techniques utilized in NMR outlines nuances of solvent suppression.[12] After data collection, it is also possible to remove some of the residual solvent peak using different mathematical treatments of the FID.

13.4.5 Pulse Sequences

NMR methods applicable to the study of therapeutic oligonucleotides represent a subset of the total NMR methods that have been developed to study the structure and function of oligonucleotides. While many RNA and DNA structural studies are performed utilizing ^{13}C and ^{15}N labeled oligonucleotides, as previously mentioned, these labeled nucleosides are not generally available for therapeutic oligonucleotides. For therapeutic-based oligonucleotides, methods must rely on naturally abundant nuclei such as proton, carbon, phosphorous, and fluorine. A summary of common NMR experiments utilized in therapeutic oligonucleotides is displayed in Table 13.1.

COSY and TOCSY methods are utilized for the observation of scalar or though bond connectivity between neighboring protons. The COSY-type spectrum contains cross peaks between protons separated by three bonds, whereas the TOCSY-type spectrum contains cross peaks for all protons connected by carbon atoms, also known as the spin system. The ^{13}C-HSQC NMR experiment correlates proton lines with carbons connected by one bond, allowing the direct assignment of the protonated carbons. It is possible to utilize other scalar couplings between NMR active nuclei, such as long range ^{1}H-^{31}P correlations in the two-dimensional HETCOR. Scalar couplings can be utilized to aid in the assignment of resonances, but they also can yield quantitative information about bond angles as the magnitude of coupling between vicinal resonances is dependent on the dihedral angles between the two atoms.

The dipolar interaction of two NMR active nuclei is the cause of the nuclear Overhauser effect (NOE), which can be used to measure internuclear distances. Through-space connections between protons that are in close proximity to each other can be seen in the two-dimensional NOESY NMR spectrum. On the basis of the choice of instrument parameters, it is possible to observe correlation for protons within approximately a 5 Å range. This information is invaluable for the assignment of the chemical shifts. With care in the choice of NMR acquisition parameters, it is possible to measure the NOESY correlations to obtain distance restraints for NMR structure determination. A few parameters to consider are the choice of method for solvent suppression (Section 13.4.4), relaxation delay, and the time the magnetization transfer is allowed to proceed (i.e., the mixing time). The use of presaturation to suppress solvent signal should be avoided as it reduces the intensity of correlations involving protons that resonate near the water resonance and protons that are exchanging with

TABLE 13.1
Levels of Information Available from NMR Spectroscopy along with Typical NMR Experiments

Stage	NMR Experiments	Information
Identification (fingerprint)	^{1}H 1D ^{31}P or ^{19}F 1D ^{1}H 1D temperature studies	Determine the presence of secondary structure, including the number of base pairs Determine the presence of secondary structure Determine melting temperature
Impurity or conformational identification	^{1}H 1D ^{31}P 1D ^{19}F 1D LC-NMR or DOSY	Identify and/or quantitate the presence of multiple conformations and/or impurities
Imino assignment	2D NOESY 2D TOCSY/COSY	Identify base pairs and show connectivity (imino walk)
Base proton assignment of aromatic walk	^{13}C-HSQC D_2O NOESY D_2O TOCSY	Secondary structure. Aromatic to H1′ sugar sequential assignment (walk). H6 T/C and H8 G/A base assignments
Complete ^{1}H assignment (as possible); ^{13}C assignments as possible	Different temperatures and/or mixing times may be necessary (NOESY and TOCSY)	Sugar assignments as possible. Additional base protons (e.g., H2 of A bases). Carbon assignments as possible
Tertiary structure (rough)	Quantitative NOESY with different mixing times 2D ^{31}P,^{1}H HETCOR 2D COSY-variant	Distance restraints from full assignment of the correlations observed in the two-dimensional NOESY spectra Angle restraints from quantitative analysis of couplings
Tertiary structure (refined)	1D and/or 2D NMR spectra in aligning media	Residual dipolar couplings utilized to obtain long range restraints
Dynamics	1D or 2D T_1 or T_2 experiments such as inversion recovery and CPMG	Sequence dependence dynamics
Ligand interaction	Observation of chemical shift perturbation in NMR experiments or cross-correlation experiments	Oligonucleotide resonances in contact with a ligand.

the solvent (such as the imino/amino protons). The relaxation delay should be sufficiently long to avoid steady-state effects that would perturb the intensity of cross peaks. The length of the mixing time should chosen to be long enough to ensure adequate mixing, but not so long to allow the diffusion of magnetization between other nuclei. Spin diffusion artifacts would be seen as a correlation between two resonances that are much farther away than 5 Å, transferred via intervening spins. Interestingly, it is common to collect NOESY spectra with longer mixing times using known spin-diffusion effects to aid in the assignment process. The molecular weight of the system in place must also be considered. Small molecules have a NOESY cross peak opposite in sign from the diagonal peaks. Large molecules have a NOESY cross peak the same sign as the diagonal peaks. For medium-sized molecules, the NOESY cross peaks can be almost zero. A variant of the NOESY, called the ROESY or rotating frame Overhauser effect spectroscopy, can be utilized for systems such as that. The ROESY can also be helpful to avoid spin-diffusion effects.

13.4.6 Relaxation

Not only can NMR be used to determine structure, but it can also reveal the dynamic properties of the oligonucleotide of interest. There are two different physical processes responsible for the relaxation of the nuclear spin magnetization, termed T_1 and T_2. The T_1 is the longitudinal decay of the recovery of the *z* component (equilibrium). It is also referred to as spin-lattice relaxation time because a variety of relaxation mechanisms allow nuclear spins to exchange energy with their surrounding, the lattice, allowing the spin populations to equilibrate. The transverse relaxation time T_2 is the decay constant for the component perpendicular to the external magnetic field. T_2 relaxation corresponds to the dephasing of the *xy* components. Random fluctuations of the local magnetic field at each NMR active nuclei in turn lead to random variations in the instantaneous NMR precession frequency. As a result, the initial phase coherence of the nuclear spin disperses until eventually the phases are disordered and there is no net *xy* magnetization. The local fluctuations involve only the phases of other nuclear spins and it is often called spin-spin relaxation. Without isotopic labeling, typically, only one-dimensional relaxation data are utilized, but it is possible, albeit insensitive, to utilize two-dimensional natural ^{13}C abundance relaxation.[13] Sophisticated NMR methods utilizing the measurement of these relaxation parameters can be equated to molecular rotational and translational diffusion characteristics.

T_1 relaxation plays an important role and has a significant impact in selecting parameters for quantitative NMR experiments. In fact, the relaxation delays for most of the NMR experiments should be based on the T_1 of the compound. In quantitative studies, it is necessary for the nuclei being studied to fully relax between pulses. The length of the delay required depends on the pulse—use of 30° or 45° pulses instead of a 90° pulse require less of a delay. For example, using a 90° pulse and a relaxation delay of 5 times the T_1 corresponds to a quantization accuracy of 99.3%, whereas a 45° pulse with that delay corresponds to 99.8% accuracy.[14]

13.5 NMR METHODS USED FOR THE STRUCTURAL CHARACTERIZATION OF REFERENCE STANDARDS AND IMPURITIES

NMR analysis of oligonucleotides offers many different levels of analysis (Table 13.1). General characterization of therapeutic oligonucleotides can be carried out by one-dimensional NMR in an effort to develop a fingerprint for purposes of reference standard characterization and to develop an identity test. The next level of characterization is the identification of the secondary structure. The next level of structural analysis could be the determination of tertiary structure. This more thorough characterization involving the complete spectral assignment and structural determination as is performed without isotopically enriched ^{13}C and ^{15}N oligonucleotides is possible but time consuming. The degree to which assignments can be made is dependent upon the size and structural characteristics of the therapeutic oligonucleotide being developed. While not universal, it would be expected that several of the imino protons could be assigned to at least a nucleobase but not necessarily within a sequence-dependent context.

A further level of information available is dynamics of the compound. In addition, the interaction of the oligonucleotide with another compound can be monitored by observing changes in the chemical shifts by the addition of the ligand (referred to as chemical-shift mapping). The use of cross-saturation experiments that selectively irradiate the resonances of one component of the complex and then observe the saturation that was transferred to the other component by spin diffusion.[15]

The use of NMR in the study of therapeutic oligonucleotides with a RNA aptamer, a DNA triplex, and phosphorothioate DNA case studies. In addition, a potential use of NMR to aid in siRNA duplex formation is discussed. Aptamers are a promising class of oligonucleotide therapeutics. Aptamers bind with high affinity and specificity to targets derived from the SELEX process. A variety of stem and loops are often predicted with both Watson–Crick and noncanonical base pairs. These loops and stems can induce significant dispersion of the chemical shifts and provide for interesting NMR studies. Most aptamers are chemically modified, and the NMR methods suitable for the analysis of therapeutic based aptamers tend to rely on the availability of natural abundant nuclei. Aptamers present an additional challenge as they are typically PEGylated with high molecular polyethylene glycol. While the additional size of the PEGylated molecule may increase line widths due to reduced rotational diffusion properties, the larger issue is the abundance of the PEG methylene protons relative to the protons from the aptamer. A large PEG will contribute on a molar basis several thousand protons for each aptamer proton of interest. In this aptamer case study, the focus will be on a model stem loop structure without PEG. The NMR structure of the example triplex was well characterized by Feigon and coworkers.[16] It is 32-base DNA oligonucleotide that forms a stable intramolecular triple helical structure.

Analysis of starting materials used in the manufacture of therapeutic oligonucleotides is more in line with the analysis of small molecule NMR. However, of central importance to the quality of oligonucleotides is the quality of the phosphoramidites building blocks. In the case study dedicated to phosphoramidites, the importance of signal to noise is illustrated in developing NMR methods for the detection of impurities.

13.5.1 Identification Test

For siRNA and aptamers, differences in sequence can give rise to dispersion of the chemical shifts throughout the spectrum. The imino region of the spectrum is indicative of the hydrogen bonding and is influenced by sequence specific base stacking. The imino protons are best observed via NMR when the imino protons are protected from exchange with the bulk water (e.g., involved in hydrogen bonding). The approximate proton chemical shifts observed for oligonucleotides resonances are summarized in Figure 13.5. Likewise, the aromatic region will show similar dispersion in the chemical shifts. For unstructured regions, the chemical dispersion is reduced. Additional fingerprint information can be obtained with two-dimensional methods that rely on correlating the chemical shift dispersion through space (NOE) or on the basis of coherence transfer between proton and carbon (HSQC). Detection of protons involved in hydrogen bonding necessitates preparing samples in H_2O with the use of selective excitation methods to minimize the signal generated from the protons of water.

The sensitivity to secondary structure of the imino region is demonstrated by a one-dimensional ^1H NMR temperature study of a model aptamer, a 2′-O-methyl RNA 15-mer duplex. The fingerprint of chemical shifts generated for the imino protons involved in hydrogen bonding is governed by the neighboring base pairs and sugar conformation and the hydrogen bonded imino protons are detected as a series of peaks with chemical shifts from 10 to 15 ppm. The nonhydrogen bonded imino protons are observed as a broad resonance owing to the chemical exchange with water. The secondary structure present in the model compound is observed in the imino fingerprint region.

The temperature dependence of the spectra for the 15-mer 2′-O-methyl RNA sequence is shown in Figure 13.7. At 285 K, the imino region consists of three resonances from the three GC base pairs and two resonances from the GU wobble base pair. With increasing temperature, the resonances all broaden. Only the two internal GC imino resonances can still be observed at 312 K. At 316 K, the internal GC imino resonances are only faintly visible as two broad resonances. As the temperature is increased, two elements impact the spectra: first, the thermal denaturation of the hairpin and, second, the increased rate of exchange of the imino protons with water. Even without assignment of the base pairs, the one-dimensional ^{1}H NMR spectrum can be utilized as an identification test as the NMR spectra is reflective of the primary sequence and secondary structure. Identification tests can utilize other NMR active nuclei also.

13.5.2 Structural Characterization

The use of homonuclear NMR to assign resonances has been described in detail elsewhere.[4,8–10] The first step is to identify the base pairing imino–amino resonances to determine secondary structure. This is accomplished through the one-dimensional ^{1}H NMR spectra and the two-dimensional NOESY in H_2O. The imino and amino resonances in the two-dimensional NOESY have distinctive patterns easily identified. For example, GU base pairs can be identified by a distinctive imino-imino NOE cross peak between the two imino resonances in the wobble base pair. The imino-to-imino region of a two-dimensional NOESY of the model triplex is displayed in Figure 13.8. The next step in the assignment process is the assignment of the nonexchangeable proton resonances. Typically, valuable information can be obtained from the NOESY correlations observed between the imino–amino resonances. For traditional assignment procedures, the expected NOESY patterns for a specific helical form (A, B, Z) are utilized in this process. The next step is to identify the sequential arrangement. One way is to utilize the imino-imino NOESY cross peaks between sequential base pairs that can be observed sequentially and between strands. Sequential assignments can also be evaluated in the region from approximately 5 to 9 ppm. The spectral region contains the proton resonances for amino, aromatic base, and H1′. The sequential assignments can be made using the fact that each aromatic proton has NOESY correlations to two

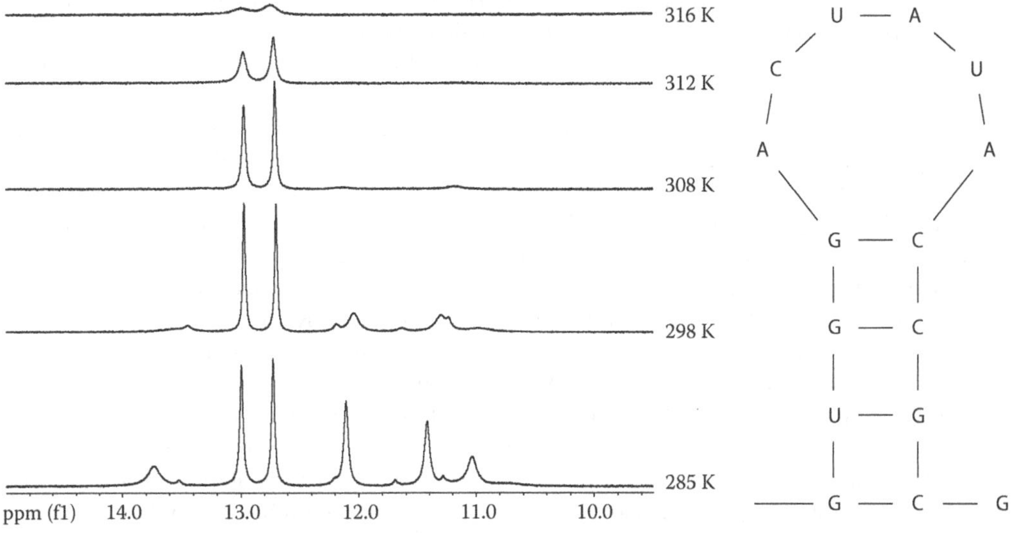

FIGURE 13.7 Imino region of the one-dimensional ^{1}H NMR spectra of the model aptamer collected with watergate water suppression at five temperatures.

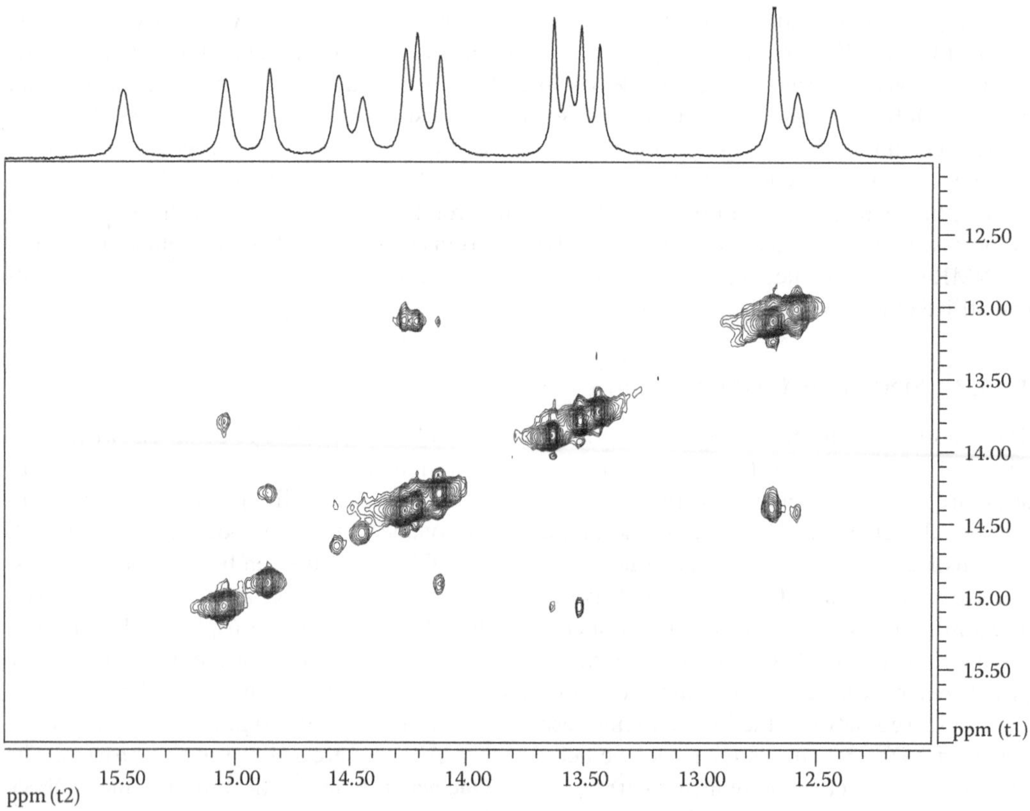

FIGURE 13.8 Imino–imino proton cross-peak region of the DNA triplex in a two-dimensional NOESY spectrum.

H1′ protons, its own and the preceding H1′ (i.e., 5′). Spectrum of this region is displayed in Figure 13.9. In addition, the use of natural abundance ^{13}C HSQC aids in the assignment process. Figure 13.10 displays the two regions of interest—the aromatic and ribose regions of the model aptamer. For example, the adenosine C2 resonances have distinct ^{13}C chemical shifts, as summarized in Figure 13.6.

The sugar spin systems are also assigned using both NOESY and two-dimensional COSY/TOCSY spectra. In DNA, assignment of the sugar resonances is more straightforward owing to the methyl group of thymidine. This region is displayed in the two-dimensional TOCSY spectrum of the DNA triplex in Figure 13.11. For unmodified RNA, the overlap of the sugar resonances makes assignment extremely difficult. For RNA containing a modified ribose, the resulting increased chemical shift dispersion simplifies the assignment of the sugar resonances. For example, there is a distinctive chemical shift from methyl protons of 2′-O-methyl RNA or the unique chemical shift of the 2′ proton in 2′F modified sugars (2′-deoxy-2′-fluoro). These assignments are typically made through the collection of TOCSY and COSY spectra. Care must be taken when setting up this experiment to ensure being able to observe the H1′ to H2′ correlation when the sugar pucker is in the *3′-endo*, as this coupling is extremely small and thus may not be observed. Aiding in the assignment of the sugar resonances is the use of two-dimensional natural abundance HSQC. The carbon resonances of the sugar are reasonably well dispersed, even for RNA. Use of NOESY correlations can also aid in the assignment of the sugar resonances. Care must also be taken when interpreting the NOESY spectrum as the correlations are through-space and subject to spin-diffusion concerns.

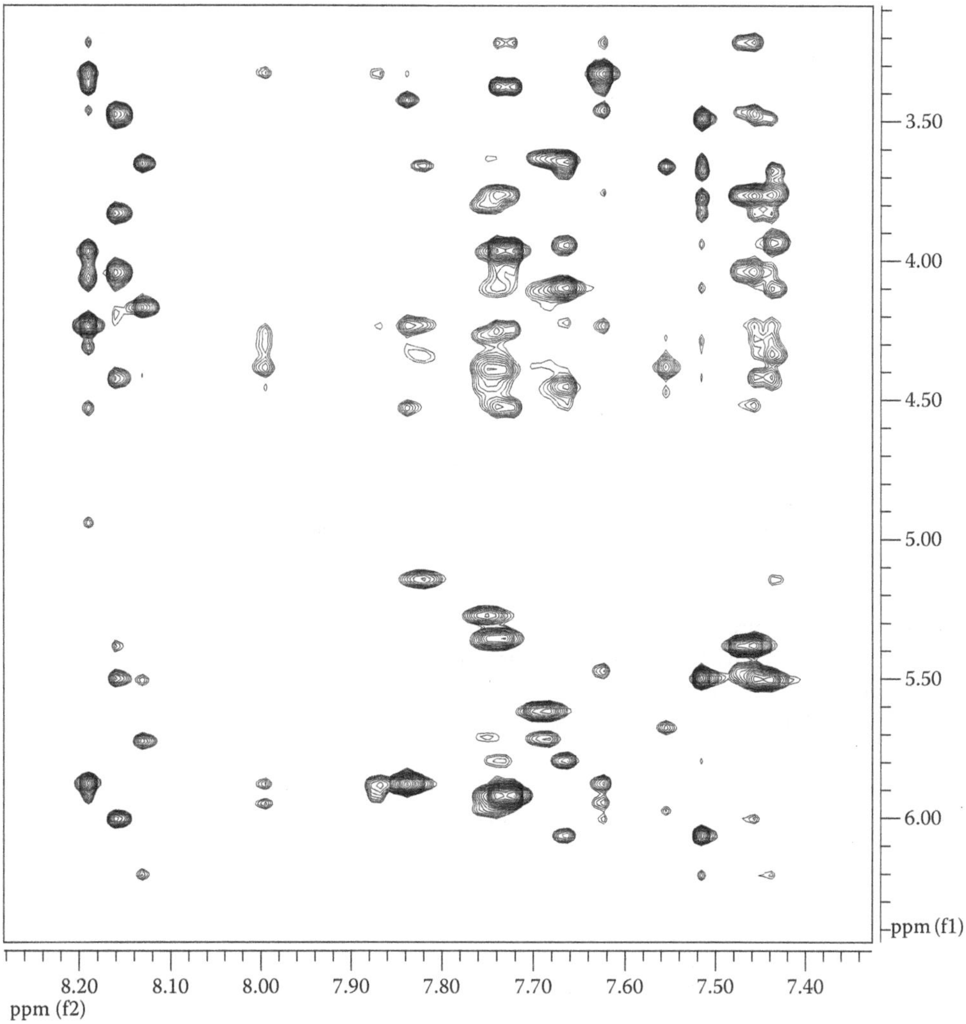

FIGURE 13.9 Aromatic-to-sugar NOESY cross-peak region of model aptamer is collected in 100% D_2O.

13.5.3 SiRNA Titration

The discovery of siRNA has led to its use as a powerful research tool for evaluating the regulation and expression of gene function. The discovery has also led to the development of a significant number of siRNA therapeutic oligonucleotides being evaluated in clinical trials. In some siRNA, phosphorothioate modifications are incorporated. These modifications would provide ^{31}P spectrum indicative of the phosphodiester and phosphorothioate linkages. Alternatively, siRNA incorporating ^{19}F as a modification against nucleases may be exploited—it has been proposed to utilize ^{19}F chemical shifts to observe the secondary structure of RNA.[17] A theoretical description of NMR applied to the annealing of siRNA duplexes is presented here. The advantages of ^{1}H NMR as a method for the analysis of siRNA duplex content and confirmation of duplex structure are discussed.

During the siRNA manufacturing process two equal amounts of the complementary single strands are annealed to form the siRNA duplex. The concentration of the single strands can be determined utilizing UV-visible spectroscopy combined with an HPLC determination of the purity of the single strands. Typically, one strand is titrated into the other strand until equal molar ratios or a slight excess of the antisense strand are reached. A number of nondenaturing methods are

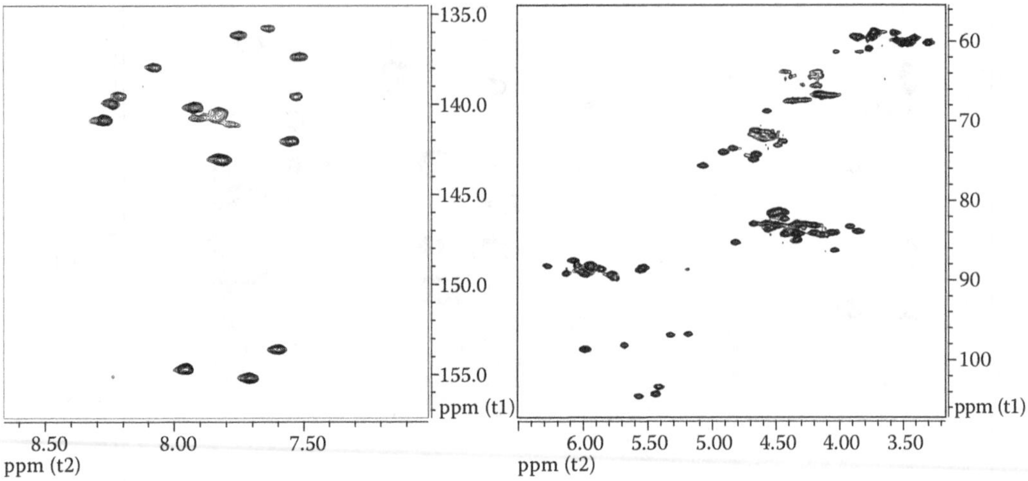

FIGURE 13.10 Two two-dimensional natural abundance ^{13}C HSQC spectra, displaying the (left) aromatic and (right) ribose regions of the RNA aptamer.

currently used to measure the duplex content. Methods include ion-exchange chromatography, size exclusion chromatography, HPLC, and capillary electrophoresis. Another approach is to perform nondenaturing size exclusion HPLC to identify excess single strand present. This is discussed in detail in Chapter 3. In this section, the principle for measuring the annealing by NMR spectroscopy is presented.

The method relies on detecting the hydrogen bonded imino protons involved in base pairing to form the duplex siRNA and the ability to detect imino protons from single stands that are not

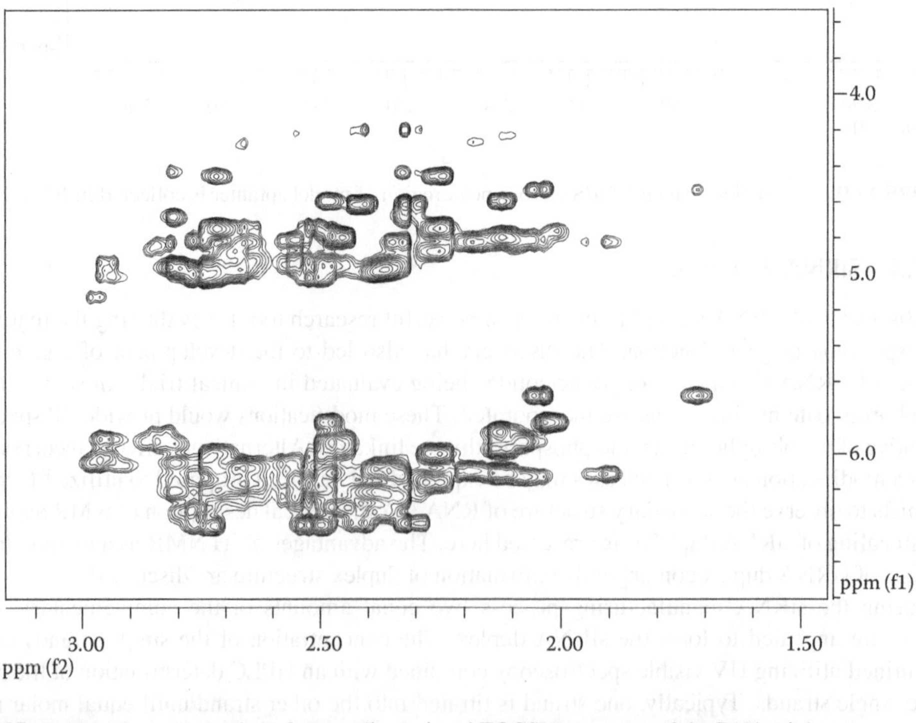

FIGURE 13.11 The sugar region of a two-dimensional TOCSY spectrum of the DNA triplex.

involved in hydrogen bonding. The method provides a unique option to directly measure the duplex content at a range of concentrations and in sample matrixes used for the delivery of drug products. Both single strands should be evaluated using one-dimensional ^{1}H NMR temperature studies. If one or both of the components displays the presence of secondary structure (as identified by the presence of resonances in the region from 12 to 16 ppm) this fact can be used as an aid in the titration process. The mixing process can be evaluated using one-dimensional ^{1}H NMR spectra. After the addition of a small amount of the less structured component into the first component, the sample should be heated and slow cooled. As mentioned previously, the slow cool process after annealing typically encourages the formation of duplexes over hairpins. The NMR spectrum of this sample should display small peaks in the imino region corresponding to the new complex forming. If the spectra appear to be a mixture of the two single strands, the secondary structure of the single strands may be too stable at those conditions. Varying the concentrations of the components and the salt may be necessary to favor duplex formation over hairpin. Small amounts of the titrating strand should be added and observed using one-dimensional ^{1}H NMR—the original imino resonances from the single-strand structure should be less intense and the new imino resonances from the complex should be more intense.

NMR is a powerful technique in measuring the duplex content at concentrations and in sample matrixes used for the delivery of drug products. The above chromatographic methods often require significant sample dilution and running buffers that can be different from the buffer used in the drug product solution. This is also the case for the measurement of thermal melting (T_m), which requires the use of diluted oligonucleotide solutions to keep the UV signals within range. While the concentration of sample necessary to achieve adequate signal to noise is usually a detriment to NMR methods, in the case of analyzing drug product solutions it becomes an advantage owing to the potential of concentration dependence on duplex stability and structure at the typical concentrations for drug product solutions. Thus, having a method capable of directly evaluating drug product solution in the native state is advantageous.

13.5.4 Phosphorothioate Oligonucleotides

Phosphorous NMR is an invaluable aid in analysis of oligonucleotides.[18,19] One-dimensional ^{31}P NMR can be used as an additional fingerprint for the purpose of identification. Unique structures such as loops and bulges will give the phosphorous backbone unique chemical shifts and are indicative of identity and structure. As mentioned previously, assignment strategies also exist for the use of two-dimensional correlated proton-phosphorus spectroscopy. The example utilized is the analysis of a phosphorothioate oligonucleotide by ^{31}P NMR, which allows the quantitation of both phosphodiester bonds and the (n-1) phosphorothioate related substance. It has been determined that impurities in the controlled pore glass (CPG) solid support can result in the generation of an oligonucleotide missing the 3′ terminal base but including a 3′ phosphorothioate. Recent improvement in the manufacture of supports has minimized this impurity. Because NMR signals are proportional to the number of nuclei, failure to fully sulfurize a single bond will yield a phosphodiester signal ~ 1/(n-1), where n is the number of residues present in the target molecule. Therefore, a single phosphodiester bond in a phosphorothioate oligonucleotide of 21-mer can be expected to yield an integrated signal of 0.05 (5%) of the total spectral signal.

Modern synthetic techniques, coupled with column purification of phosphorothioate oligonucleotides, provide the ability to achieve final product with less than 1% phosphodiester. Low-level impurities obligate the need to achieve a high signal to noise ratio. While ^{31}P is 100% abundant, the lower gyromagnetic ratio of phosphorous generates less signal than for protons and thus requires a greater number of scans to achieve the same signal intensity. This requires both adequate concentrations of oligonucleotide and long run times. A typical oligonucleotide ^{31}P NMR experiment may run for 12 hours or longer, with overnight runs generally providing phosphodiester sensitivity to 0.1% or better. High concentrations for oligonucleotide ^{31}P NMR are often utilized, commonly 60 to 120 mg/mL in D_2O or

10% D_2O in water. The chemical shift range for (n-1)PS is usually in the –45 to –43 ppm region. The phosphate phosphorous nuclei (PO) resonances for DNA are generally observed in the region from 0 to –1 ppm. Oligonucleotides containing LNA or other modified amidites may yield a slight chemical shift for PO, but these will always remain in the general region around 0 ± 10 ppm. A spectrum of a phosphorothioate DNA oligonucleotide containing small quantities of PO and (n-1)PS is presented in Figure 13.12.

The presence of peaks at 2.4 ppm and –7.3 ppm are likely indicative of the presence of phosphate salts and not oligonucleotide PO content. Therefore, one should be aware of any use of phosphate salt during late stage purification or ultrafiltration/diafiltration. As an example, a small amount of phosphate buffer, pH 7.6 (approximately 0.1% by weight) was added to a fully phosphorothioate oligonucleotide. The spectrum is provided in Figure 13.13. It can be seen that even a small amount of inorganic phosphate will yield significant signal. Careful evaluation of spectra will still allow quantitation of any phosphodiester impurity.

For quantitation of phosphodiester impurities present at low levels in phosphorothioate oligonucleotides it is necessary to obtain sufficient signal to noise, typically, a S/N of 10:1 for quantitation. For example, quantitation of a 0.1% impurity requires a minimum signal-to-noise ratio for the main component of 1000. This is based upon having a S/N for the impurity of at least 10 and the ratio between the impurity and the main component (1000). To perform the test quantitatively, one typically must wait 5*T_1 of the longest nuclei of interest so that all nuclei are at equilibrium prior to the initiation of the next acquisition.

13.5.5 Phosphoramidite Starting Materials

The quality of phosphoramidites can have an impact on purity and yield of oligonucleotide. Quality control of phosphoramidites is often performed by combining HPLC, HPLC-MS, and ^{31}P NMR,

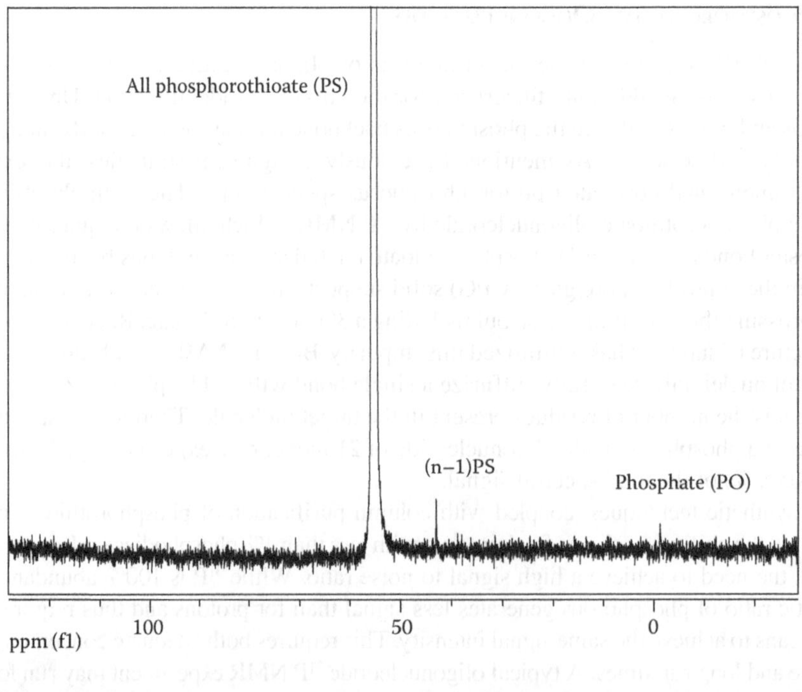

FIGURE 13.12 One-dimensional ^{31}P NMR spectrum of a phosphorothioate oligonucleotide containing small quantities of PO and (n-1)PS.

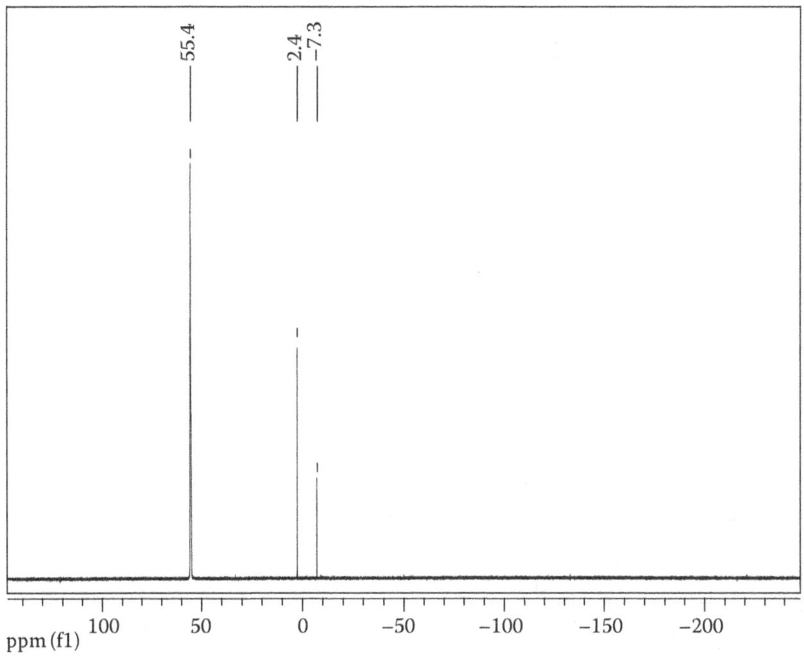

FIGURE 13.13 One-dimensional ^{31}P NMR spectrum of a phosphorothioate oligonucleotide spiked with 0.1% (by weight) of phosphate buffer.

and each method has certain advantages and disadvantages. While chromatography provides a good measure of amidite purity, current analytical procedures often rely on relative UV absorbance of parent amidite and related impurities. Impurities that lack a chromophore, such as lacking a purine or pyrimidine base, yield no UV signal and can be overlooked. Additionally, $P^{(III)}$ and $P^{(V)}$ impurities are not easily distinguished. The use of ^{31}P NMR enhances the ability to detect, distinguish, and quantitate these impurities potentially coming from the phosphitylation reagent. Table 13.2 summarizes structures of common phosphitylating agents and several products from their hydrolysis.

The analysis of phosphoramidites by ^{31}P NMR requires special precautions be taken during sample preparation to avoid hydrolysis and oxidation of the $P^{(III)}$ active phosphoramidites. Sample preparation techniques may vary, however; the use of deuterated chloroform, stored over silver, or modified by adding 0.5–1% triethylamine by volume, will yield amidite solutions suitable for overnight analysis. The use of open, nonstabilized deuterated chloroform is discouraged, as significant $P^{(V)}$ impurities are generated, both upon dissolution and over time. Chloroform will allow concentrations of greater than 250 mg/mL. Deuterated acetonitrile also will provide suitable spectra; however, the solubility of amidites may be greatly reduced, necessitating longer run times to obtain suitable signal to noise. Acquisition parameters shown in the following table have yielded spectra suitable for quantitation to ~0.05% impurity levels or better. Using proton decoupled spectra improve the signal to noise and assist with impurity quantitation. Use of proton coupled spectra is more useful for impurity identification, provided the impurity is present in large quantity or intentionally synthesized. Typical experimental conditions follow:

NMR: 400 MHz, Proton decoupled, 30° pulse
Amidite concentration: 250 to 300 mg/mL in $CDCl_3$
Number of scans: ≥1000
Acquisition time: 1.5 s
Relaxation delay: 2 s
Spectral width ≥ 250 ppm, offset ~75 to 100 ppm

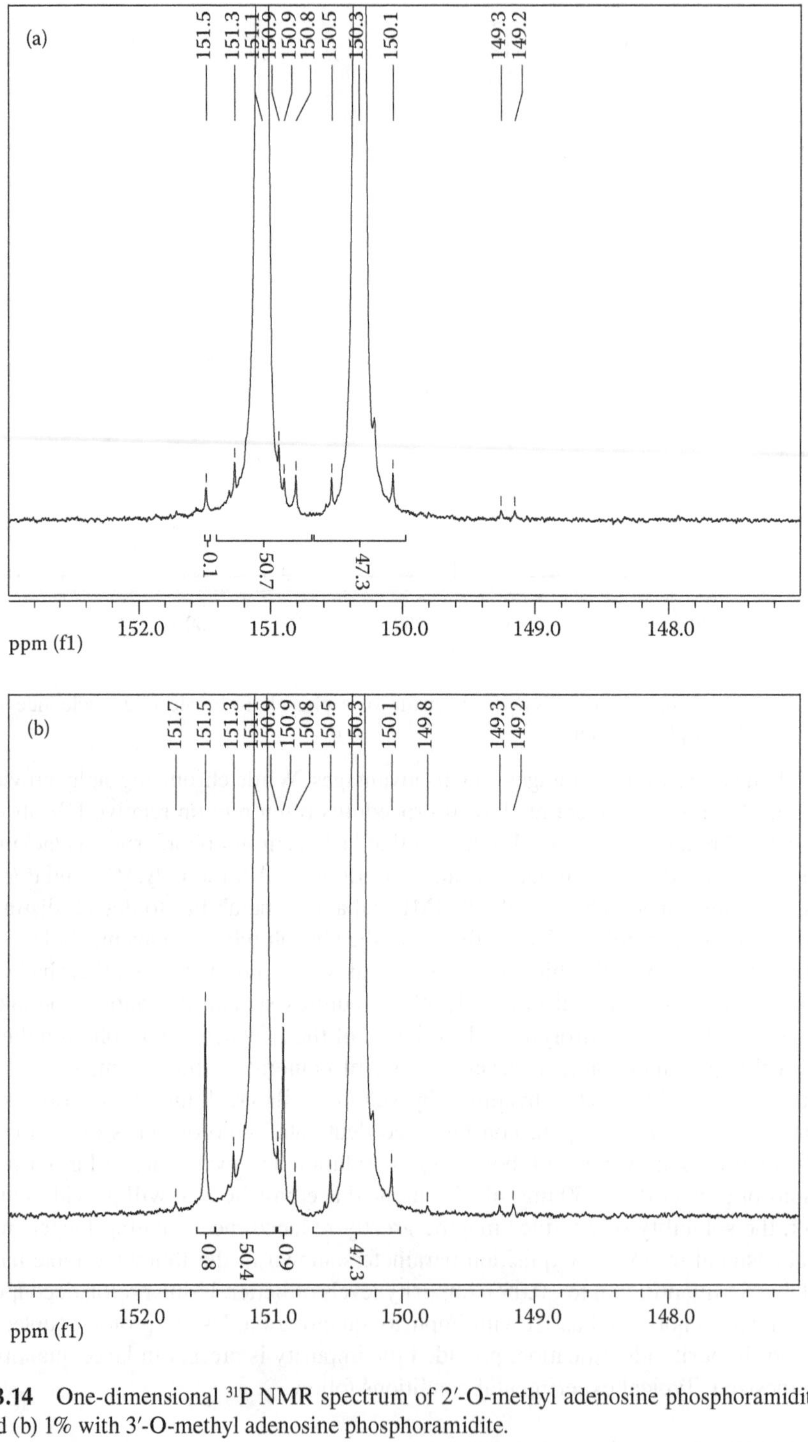

FIGURE 13.14 One-dimensional ^{31}P NMR spectrum of 2′-O-methyl adenosine phosphoramidite spiked at (a) 0.1% and (b) 1% with 3′-O-methyl adenosine phosphoramidite.

TABLE 13.2
Structures of Phosphitylation Agents and Products from Their Hydrolysis

Compound	~ δ_P (ppm)[a]
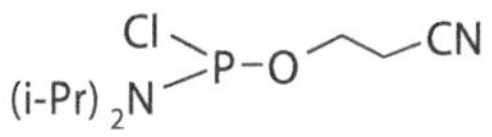	180.0
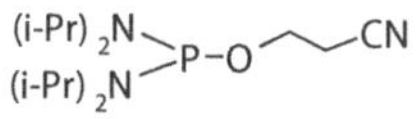	123.5
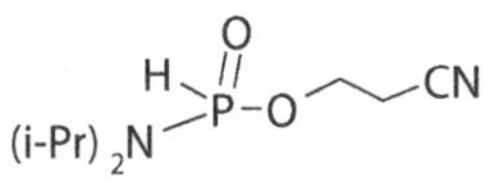	13.2
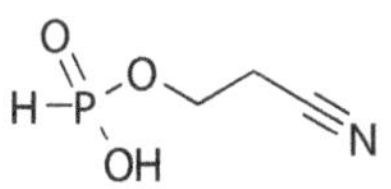	2.5

[a] All values determined in $CDCl_3$.

As an example, 3′-O-methyl adenosine was spiked at 0.1 and 1% into 2′-O-methyl adenosine phosphoramidite Figure 13.14.[4] The spectra show the two diastereomer $P^{(III)}$ resonances at 151.1 and 150.3 ppm. The spectra are expanded on the vertical axis by many orders of magnitude to allow display of smaller peaks. Satellite peaks from ^{13}C and ^{15}N are present at approximately ±0.125 ppm and +0.21/–0.26 ppm from the phosphorous main resonances. The satellite peaks at ±0.125 ppm may not be apparent when using NMR field strengths below 400 MHz. Other commonly observed impurities occur from 148 to 149 and 139 ppm. In the above example, the spiked impurity can be easily detected in the 1% spiked solution at 151.5 and 150.9 ppm (Figure 13.14b). Quantitation can be difficult owing to resonance overlap and challenges with integration. For example, in 0.1% spike (Figure 13.14a), the up field resonance for 3′-O-methyl adenosine at 150.9 ppm appears to be not much more than a shoulder of the main component. The reader is cautioned that the integrated values of impurity signals in close proximity to main peaks are likely to be overstated because of the main component peaks bleeding into the impurities. Often peak heights instead of intensities are utilized, as it is reasonably straightforward to correct for the height of the shoulder of the larger peak.

All amidites analyzed to date by the author have contained some level of $P^{(III)}$ impurities, ranging from below 0.05% to more than 1%. Freshly prepared high-quality amidites should be expected to contain no more than 0.3% $P^{(III)}$ or 2% $P^{(V)}$ species, although those containing slightly higher levels have yielded suitable syntheses of oligonucleotides. Amidites stored at –0°C will exhibit some increase in $P^{(V)}$ species over time. This varies based on packaging and moisture content. When dissolved and allowed to age, most amidite $P^{(V)}$ species will be observed in the 40 ppm to –5 ppm range. These are nonreactive species but represent the loss of active coupling agent and may decrease coupling efficiencies, impacting yields with the potential to increase n-1, n-1, etc. deletion sequences. Additionally, studies performed by Cramer and his colleagues (unpublished results) have demonstrated that an undesirable $P^{(III)}$ species may be created in amidite solutions, when amidite solutions prepared in anhydrous acetonitrile are stored with or without zeolyte drying agents at room temperature for extended periods. The greatest change was observed with 2′-O-methyl-(ibu) guanosine, demonstrating the signal at approximately 139 ppm linearly increasing to 2% increase over 20 days. The two charts in Figure 13.15 demonstrate the solution stability for 2′-O-methyl guanosine in acetonitrile and the impact on degradation rate with and without molecular sieve.

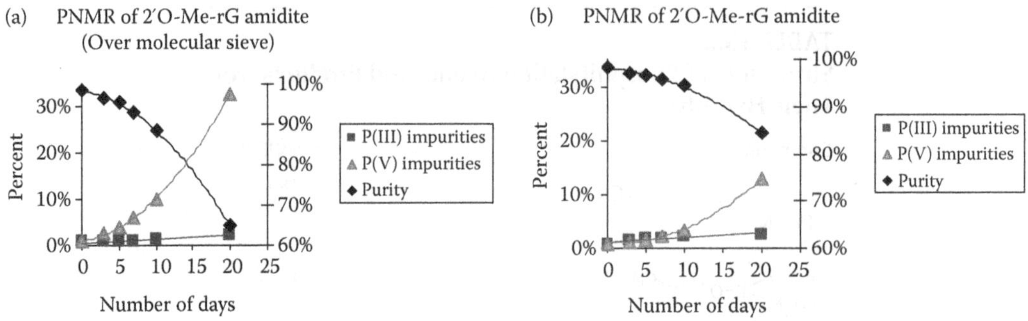

FIGURE 13.15 Solution stability for 2′-O-methyl-guanosine in acetonitrile (a) with and (b) without molecular sieves.

13.5.6 Tertiary Structure Determination

Therapeutic oligonucleotides provide a unique challenge in evaluating tertiary structure. On the one hand, without the use of labeled oligonucleotides many of the methods necessary for tertiary structure evaluation aren't amenable and the methods are limited to oligonucleotides of less than 5 KD. On the other hand, most therapeutic oligonucleotides are less than 5 KD. In the event that the evaluation of tertiary structure is feasible, the determination of tertiary structure of oligonucleotide proceeds by collection of as many structural restraints as possible. The restraints will include chemical shift, distance (from the NOESY spectra), and torsional angle (from the correlation spectra) restraints. These constraints are utilized by structure determination software to find energy minimized structures. For RNA, the number of constraints that can be developed from NOE and torsional restraints in the model is limited owing to the proton poor nature of RNA and the chemical shift overlap. In general, one obtains a general picture of the tertiary structure.

Further refinement of the NMR structure can be obtained through the measurement of residual dipolar coupling (RDC).[20] In solution, molecules tumble isotropically and the residual dipolar coupling is difficult to detect. The addition of polymeric matrices can enhance the anisotropy of the molecule in solution and hence give rise to measurable residual dipolar coupling. Alignment matrices include filamentous phages, phospholipid mixtures, polyacrylamide gels, and various lyotropic liquid crystalline systems leading to an incomplete averaging of spatially anisotropic dipolar coupling. As opposed to traditional NOE based structure determination; RDCs provide long-distance structural information. This is especially informative for elongated molecules, like oligonucleotides. Typically, local torsional information and short distances are not enough to constrain the structures; RDC measurements can provide information about the orientations of specific chemical bonds throughout a nucleic acid with respect to a single coordinate frame. It also provides information about the dynamics in molecules on time scales slower than nanoseconds.

13.5.7 Other Applications: LC-NMR, DOSY, and T_2-Edited Experiments

There are many other NMR applications suitable for the analysis of therapeutic oligonucleotides. Liquid chromatography systems can be coupled with NMR through the use of specially designed probes. LC-NMR can be used to study individual components of mixtures that are separated using chromatographic techniques. Performing LC-NMR gives the analyst the ability to associate NMR spectra from each chromatographic peak and thereby simplify the interpretation of the NMR spectra. Another method for the analysis of complex mixtures is the diffusion order spectroscopy or DOSY type experiments. The DOSY method allows measurement of the translational self-diffusion coefficient of molecules in solution. Spectra of mixtures can be separated depending on the value of their apparent diffusion coefficients. Diffusion measurement by NMR and especially the DOSY-type experiments are thus powerful analytical tools for the analysis of multicomponent mixtures, such as the identification of small molecule impurities in oligonucleotide formulations. DOSY is also routinely utilized to differentiate monomer from dimer structures of oligonucleotides through the use of reference oligonucleotide solutions of known molecular weights.

Another option for the identification of small molecule impurities utilizes the differences of the T_2 relaxation of the oligonucleotide and the small molecules. T_2 edited spectra can be collected by altering the CPMG delay (named after the famous T_2 method of Carr-Purcell-Meiboom-Gill). This method relies on filtering out the signal from large molecules and leaving the signal from small molecules. As mentioned above, NOESY peaks from small molecules are negative, and this too can be used for identifying small molecules from the oligonucleotide of interest.

13.6 SUMMARY

The applications of NMR to therapeutic oligonucleotides are as numerous as the chemical modifications, formulations, and indications that this class of therapeutics is poised to treat. A select number of examples were used in the chapter to illustrate the powerful nature of NMR, and while it was not possible to present data from all classes of therapeutic oligonucleotides, many of the illustrative points are directly applicable between classes. As the therapeutic utility of oligonucleotides develops, it is accordingly expected that new and novel NMR methodologies will be applied to meet the many challenges.

REFERENCES

1. IUPAC-IUB Joint Commission on Biochemical Nomenclature (JCBN). 1983. Abbreviations and symbols for the description of conformations of polynucleotide chains. Recommendations 1982. *European Journal of Biochemistry* 131: 9–15.
2. Markley, J. L., A. Bax, Y. Arata, C. W. Hilbers, R. Kaptein, B. D. Sykes, P. E. Wright, and K. Wüthrich. 1998. Recommendations for the presentation of NMR structures of proteins and nucleic acids. *Pure and Applied Chemistry* 70(1): 117–142.
3. Altona, C., and M. Sundaralingam. 1972. Conformation analysis of the sugar ring in nucleosides and nucleotides. New description using the concept of pseudorotation. *Journal of the American Chemical Society* 94: 8205.
4. Wijmenga, S. S., M. M. W. Mooren, and C. W. Hilbers. 1993. NMR of nucleic acids: from spectrum to structure. In *NMR of Macromolecules: A Practical Approach*, 217–288. New York: Oxford University Press.
5. Batey, R. T., J. L. Battiste, and J. R. Williamson. 1995. Preparation of isotopically enriched RNAs for heteronuclear NMR. *Methods in Enzymology* 261: 300–322.
6. Furtig, B., C. Richter, J. Wohnert, and H. Schwalbe. 2003. NMR spectroscopy of RNA. *ChemBioChem* 4: 936–962.
7. Latham, M. P., D. J. Brown, S. A. McCallum, and A. Pardi. 2005. NMR methods for studying the structure and dynamics of RNA. *ChemBioChem* 6: 1492–1505.
8. Wüthrich, K. 1986. *NMR of Proteins and Nucleic Acids*. John Wiley: New York.
9. Varani, G., F. Aboul-ela, and F. H.-T. Allain. 1996. NMR investigation of RNA structure. *Progress in Nuclear Magnetic Resonance Spectroscopy* 29: 51–127.
10. Feigon, J., V. Skelená, E. Wang, D. E. Gilbert, R. F. Macaya, and P. Schultze. 1992. ^{1}H NMR spectroscopy of DNA. *Methods in Enzymology* 221: 235–253.
11. Mukhopadhyay, R. 2007. Liquid NMR probes: oh so many choices. *Analytical Chemistry* 79(21): 7959–7963.
12. Zheng, G., and W. S. Price. 2010. Solvent signal suppression in NMR. *Progress in Nuclear Magnetic Resonance Spectroscopy* 56(3): 267–288.
13. Shajani, Z., and G. Varani. 2006. NMR studies of dynamics in RNA and DNA by ^{13}C relaxation. *Biopolymers* 86(5–6): 348–359.
14. Derome, A. 1987. *Modern NMR Techniques for Chemistry Research*. New York: Pergamon.
15. Takahashi, H., T. Nakanishi, K. Kami, Y. Arata, and I. Shimada. 2000. *Nature Structural Biology.* 7: 220–223.

16. Wang, E., S. Malek, and J. Feigon. 1992. Structure of a GTA triplet in an intramolecular DNA triplex. *Biochemistry* 31: 4838–4846.
17. Graber, D., H. Moroder, and R. Micura. 2008. ^{19}F NMR spectroscopy for the analysis of RNA secondary structure populations. *Journal of the American Chemical Society* 130: 17230–17231.
18. Gorenstein, D. G. ^{31}P NMR of DNA. 1992. *Methods in Enzymology* 211: 254–286.
19. Gorenstein, D. G. 1981. Nucleotide conformational analysis by ^{31}P nuclear magnetic resonance spectroscopy. *Annual Review of Biophysics and Bioengineering* 10: 355–386.
20. Bax, A., G. Kontaxis, and N. Tjandra. 2001. Dipolar couplings in macromolecular structure determination. *Methods in Enzymology* 339: 127–174.

14 Infrared Analysis of Oligonucleotides

Jose V. Bonilla
Girindus America, Inc.

CONTENTS

14.1 INTRODUCTION

Infrared spectroscopy (FTIR) is a well-established analytical technique for the characterization of a broad variety of products in the chemical and pharmaceutical industries. It is perhaps the most widely used method of applied spectroscopy for the positive identification of chemical compounds. In the specific case of oligonucleotides, this technique can be used for the identification of the raw materials used during synthesis as well as identification of the drug substance. In practice, however, it is observed that mass spectrometry, base composition, and sequencing tend to be preferred for characterization of the drug product. In any case, the FTIR technique is available as a highly reliable approach for laboratories not equipped with mass spectrometry capabilities.

Before going into the practical aspects of instrumentation and the analytical approaches used for sample handling and preparation, the basic principles of infrared spectroscopy will be reviewed.

14.2 THEORY AND BACKGROUND

Infrared spectroscopy is based on the principle that molecules have specific frequencies at which they rotate or vibrate corresponding to discrete energy levels. These resonant frequencies are determined by the shape of the molecule, the mass of the atoms, and the associated vibration energy. In order for vibrational energy in a molecule to be infrared active it must be associated with changes in the permanent dipole. The resonant frequencies are related to the strength of the bond and the mass of the atoms at either end of the bond. Thus the frequency of the vibrations can be associated with a particular bond type and chemical composition. The infrared spectrum is a unique physical property of molecular species, and the spectral information obtained can be directly correlated to the chemical structure of the material. For more detailed information and further understanding of the theory and principles of infrared spectroscopy, the reader is provided with additional references.[1–4]

Infrared spectroscopy is widely used in both research and routine analysis as a simple and reliable technique for qualitative and quantitative analysis. One of the most common uses of this technique is the positive identification of raw materials and drug substances by using the resulting spectra as a "molecular fingerprint," which is specific for each compound. This works quite well because most FTIR instruments may be set up to quickly and automatically report the identity of the product being analyzed by automatically matching the spectra to an existing database or library containing thousands of reference spectra.

Infrared spectroscopy has also been used in many instances to monitor chemical reactions during synthesis of oligonucleotides and related products. This technique can provide information about the product and the raw materials used for its synthesis.

The wavelength range covered in the fundamental infrared is from 2500 to 25,000 nm, although, in practice, it is more common to use the term frequency, expressed in wave number (4000–400 cm^{-1}) because this is directly related to the vibrational energy involved in the energy transitions of the molecules. The upper limit is more or less arbitrary and was originally based on the performance of early instruments. It adequately covers all fundamental absorptions, with the exception of the H-F stretching vibration, which is observed between 4100 and 4200 cm^{-1}. The lower limit in most cases is defined by a specific optical component on the instrumentation. The beam splitter that is used in modern FTIR instruments is made from a potassium bromide (KBr) substrate with a transmission cut-off just below 400 cm^{-1}. This range can be extended, but special optical materials are required for this purpose and are normally done as an option for high-end instruments.

At the molecular level, vibrational spectroscopy involves energy transitions measured from the ground state to an excited, higher energy state. In the fundamental region of the mid-infrared the measurement involves transitions from the ground state ($v = 0$) to the first excited stated ($v = 1$). Transitions to higher-energy levels do occur, and these give rise to what are known as overtones in the spectrum, occurring nominally at higher multiples of the fundamental frequency. These absorptions are weak compared to that of the fundamental region. They may be observed in the main spectral region of the mid-infrared, but they are also observed in higher-frequency ranges, in a region of the spectrum known as the near-infrared spectroscopy.

The infrared spectrum of a sample is produced by passing a beam of infrared light through the sample. Analysis of the transmitted light reveals how much energy was absorbed at each wavelength. From this information a transmittance or absorbance spectrum can be produced, showing the infrared wavelengths the sample absorbs. Analysis of the absorption characteristics reveals details about the molecular structure of the sample. Complex molecular structures lead to more absorption bands and more complex spectra.

The instruments used for this purpose vary in capability from relatively inexpensive models used for routine analysis to highly sophisticated models used in conjunction with microscopes for research and development purposes and for trace analyses of product contaminations or trace components in the finished product.

This chapter also describes how to identify the oligonucleotide product and the raw materials used for its synthesis according to their resulting fingerprint from their infrared spectra. This section demonstrates how to identify and characterize a product from its infrared spectrum and is intended for readers not particularly experienced in FTIR spectral interpretation. There are several well-established references that may be useful for general compound identification.[1–4]

In the case of oligonucleotide products, because each product may contain quite different base composition, the final identification must be performed by a comparison of the sample spectrum with the standard reference spectra of the product of interest if available. If standard reference spectra or a standard reference product is not available, one or two additional techniques must be used to confirm the product identification, for example, nuclear magnetic resonance (NMR), mass spectrometry (MS), or retention time by and appropriate high-performance liquid chromatography (HPLC) method.

14.3 INSTRUMENTATION

14.3.1 Fourier Transform Spectrometry

Infrared spectroscopy was developed as an analytical technique in the 1940s as a method for the quality control of petroleum products and a number of synthetic polymers. One of the first practical applications involved the determination of the different monomers used in the production of synthetic rubbers. The very first instruments used the principle of light dispersion from a prism, to generate the spectrum. That principle led to the name "dispersive instrumentation." The dispersive instrument evolved over time, and the use of the prism optics was eventually replaced by the diffraction grating, which provides a higher resolution spectrum. During the 1960s, another infrared measurement technology was developed on the basis of the optical measurement known as interferometry. This technique generates a unique signature where each frequency in the spectrum is uniquely modulated. This signature, known as the interferogram, is converted into the normal analytical infrared spectrum by a mathematical function called a Fourier transform. This technique is known as Fourier transform infrared or FTIR.[4]

FTIR is nowadays the foundation of modern analytical infrared instrumentation.[3] All sample spectra presented in this chapter were obtained with FTIR instrumentation. FTIR offers unique advantages over conventional dispersive infrared in terms of sample handling and data processing and as such is more versatile for a wide variety of sample analysis.

The instrument key components are the source, the beam splitter, and the detector (Figures 14.1 and 14.2). These key components may be modified to customize the instrument to meet the needs of specific applications. Different sources, beam splitters and detectors, may be used to extend the range of the instrument, either to shorter wavelengths (the near-infrared and visible) or to longer wavelengths (far-infrared). The most popular combination in the mid-infrared is a thermal source, such as an electrically heated metal filament, a ceramic or silicon carbide rod. Typical source

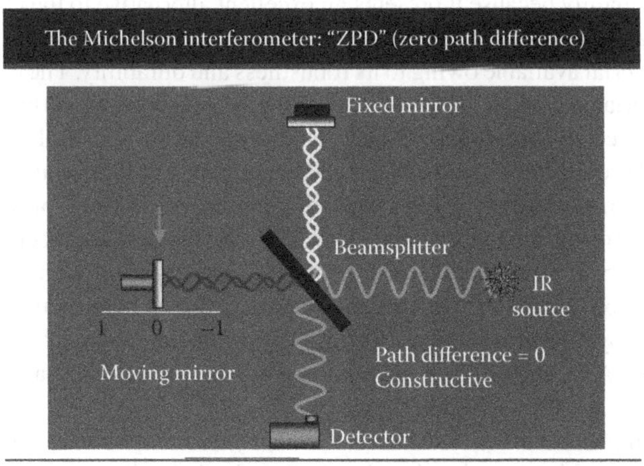

FIGURE 14.1 FTIR instrument diagram showing the instrument major components: IR source, interferometer, sample compartment, and detector. (Courtesy of Thermo Fisher Scientific.)

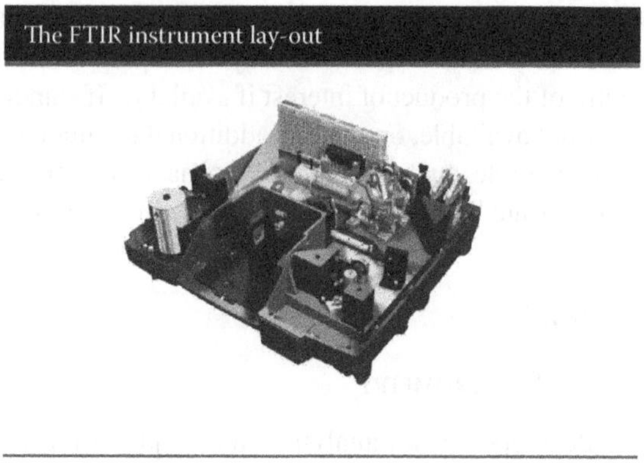

FIGURE 14.2 FTIR instrument with the cover removed showing all its major components. (Courtesy of Thermo Fisher Scientific.)

temperatures range from 1100 to 1500K. The standard beam splitter in the mid-infrared is germanium, deposited as a thin film on a potassium bromide (KBr) substrate. There are several detectors commercially available, the most popular being the deuterated triglycine sulfate (DTGS) pyroelectric detector. Other options include the lower cost and slightly lower sensitivity lithium tantallate detector and the higher performance (speed and sensitivity) photon detecting mercury cadmium telluride (MCT) detector. The latter requires cooling, normally cryogenically, with liquid nitrogen and is normally required for FTIR microscopy applications.

The instrument parameters used in the measurement of a sample are influenced by the hardware and software used in the instrument and the sampling method being used. For a given instrument, the sampling accessory often defines the spectrum quality, and, in turn, tends to define the instrument parameters that are used. For most analytical methods used, the normal parameters are 2 or 4 cm^{-1} spectral resolution. For some applications, such as microscopy, 8 cm^{-1} resolution is often sufficient with 32–128 averaged scans, the latter being sometimes defined by the desired time frame for the analysis.

14.3.2 Attenuated Total Reflectance

The technique of attenuated total reflectance (ATR) has in recent years revolutionized solid and liquid sample analyses mainly because it presents an excellent alternative to the challenging aspects of infrared analyses, namely the sample preparation and FTIR spectral reproducibility. Diamond is the best ATR crystal material available owing to its robustness and durability. The cost is higher than that of other crystal materials available in the market like zinc selenide and germanium, both of which can scratch and break easily with improper use. As normally done with all FTIR measurements, an infrared background is collected from the clean ATR crystal before sample analysis. The crystals are usually cleaned with an appropriate solvent soaked in a piece of tissue. Typically, the solvents used are water, methanol, isopropanol, or acetone. The ATR crystal must be checked for contamination and sample carryover before sample analysis. With time and experience any contamination on the ATR crystal becomes obvious in the background spectra obtained before sample analysis.

In ATR spectroscopy all that is required for analysis is that the sample of interest be brought into close contact with the ATR crystal. The infrared beam is passed into the sample through the ATR element.

Coarse or hard samples are not easy to analyze either by the compressed pellet or mull techniques, mainly because of difficulties associated with grinding. In these situations, the best approach is to use an accessory, such as diamond ATR. In general, these devices may be used for most forms of powder and liquid samples. This alternative procedure is very easy to use and requires essentially

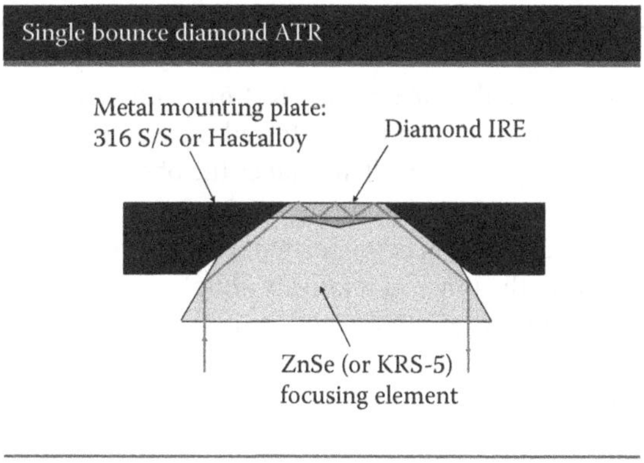

FIGURE 14.3 Diagram of a single bounce diamond ATR accessory. (Courtesy of Thermo Fisher Scientific.)

no sample preparation or very little sample preparation. Many types of ATR accessories are horizontally positioned and equipped with a pressure device. The use of pressure is recommended to ensure close contact between the sample and the ATR surface. The introduction of diamond-based ATR accessories makes it possible to handle most types of solid and liquid materials "as is" with no sample preparation.[5–14]

An additional advantage of the diamond-type ATR device specifically to the oligonucleotide field is sample size requirements. The sample size required for FTIR analysis using a diamond ATR is very small; a microgram of sample or a small grain of sample is all that is normally required. Because this is a nondestructive analysis, after the FTIR analysis is completed, the sample can be easily recovered and used for additional analyses or for other purposes. This can be a very attractive characteristic because samples in oligonucleotide research and development are normally very small and quite expensive. The fundamental principle of sample analysis by ATR is illustrated in Figure 14.3.

14.4 QUALITATIVE ANALYSIS

The most popular application of infrared spectroscopy is qualitative analysis in the identification of chemical compounds or chemical mixtures. It is perhaps the most powerful tool for identifying types of chemical bonds (functional groups). The wavelength of light absorbed is characteristic of the chemical bond. By analyzing the infrared absorption spectrum, the chemical bonds in a molecule can be determined. FTIR spectra of pure compounds are generally so unique that they can be used as a molecular "fingerprint." In general, organic compounds have very detailed spectra, while inorganic compounds are usually much simpler. For most materials, the spectrum of an unknown can be readily identified by comparison to a library of known compounds. There are several infrared spectral libraries commercially available through vendors of infrared instruments and also through online library search services for the purpose of spectra matching for positive identification of unknowns or verification of chemical identity. For less common products a library can be created for the purpose of positive identification of samples compared to the library of reference spectra. In some instances in order to identify less common materials, the infrared spectra can be combined with nuclear magnetic resonance, mass spectrometry, and other techniques described in detail in other chapters of this book to aid in the positive identification of a chemical product or raw material.

14.5 QUANTITATIVE ANALYSIS

Quantitative analysis by FTIR is an important approach to determine the quantity of a material present in a chemical mixture or to determine or monitor the degree or level of a chemical reaction.

By measuring a specific frequency (or selected frequencies) over time, changes in the character or quantity of a particular bond can be measured. This is especially useful in measuring the rate of a chemical reaction during manufacturing. This feature is quite valuable considering the speed of modern research instruments which can take infrared measurements across the whole range of interest as frequently as 32 times per second. This makes the observations of chemical reactions and chemical processes convenient, quick, and accurate.

Because the strength of the absorption is proportional to the concentration, infrared can be used for quantitative analyses. Normally, this is a relatively simple technique offered as a standard software option by most vendors and is useful for analyses in the concentration range of a few parts per million up to the percent levels. Quantitative analysis is possible as these measurements follow the Beer–Lambert law. This law provides a mathematical relationship between the infrared radiation absorbed by the sample and the sample concentration:

$$A = a \times b \times c$$

where A is absorbance, a is absorptivity, b is path length, and c is sample concentration.

The Beer–Lambert law states that absorbance is linearly proportional to sample concentration if sample path length and absorptivity remain constant. If the actual measurements are made in percent transmittance (which is not a linear function of concentration), then it should be converted to absorbance in order to perform the above calculations.

14.6 SAMPLE PREPARATION AND SAMPLE HANDLING

Depending on the type of sample being analyzed and the sample preparation being used, the time consuming step in infrared analysis can be sample preparation. Sample preparation is very important and should be done carefully to ensure the sample is not altered in any way that will negatively affect the quality of the resulting spectra. In the past, sample preparation techniques could be time consuming and destructive in nature. However, as pointed out in a previous section, the advent of ATR sampling devices allows for sample analysis to be done with essentially no sample preparation requirements and nondestructive sample analysis for both liquid and solid samples. This is a significant advantage for oligonucleotide analysis for reasons discussed in more detail in a later section.

The approach to sample preparation and sample handling depends on the physical state of the sample among other factors such as sample stability, chemical reactivity, hygroscopicity, sample toxicity, etc. The preferred sample preparation for FTIR analyses also depends on the type of accessories available for sample analysis, and it can be relatively simple and straightforward or highly complicated depending on the type of sample accessories available to the analyst.

14.6.1 Liquids

For liquid samples, the easiest way is to place one drop of sample on a slide or between two slides of sodium chloride. Sodium chloride salt is transparent to infrared light. The drop forms a thin film on the slide or a thin layer between the slides, which is then placed in the instrument sample compartment for analysis. Liquids are usually considered to be relatively easy for sample handling with the exception of low molecular weight, high volatility compounds. The traditional approach for liquids is to use a transmission cell, either in the form of a sealed container, as a thin sample sandwiched between two windows or as a film deposited on a FTIR slide. ATR-based methods[5–14] have become very popular and can be the preferred method for many applications because of ease of use, the ease of cleaning, and a resulting good spectral match. For volatile liquids, there is always the option to handle the sample in the vapor form within a heated gas cell. This approach is beyond the scope of this chapter.

An important issue to keep in mind is the presence of moisture in the sample. This is important because most of the common sample handling media used for TFIR measurements are those made from salt optics, which are usually made from KBr, NaCl, or CsI. These are very soluble in water and must not be used for aqueous media. There are, however, several materials and sampling devices that are not moisture sensitive and can be used to handle samples with high moisture content (aqueous solutions) with no problem at all. Samples containing water are best handled with barium or calcium fluoride, silver halide windows, or an ATR accessory.

Many samples exist as viscous liquids. A simple and convenient method for such nonvolatile materials is to produce film between a pair of transmission windows. If the material is very viscous, another approach is to form a thin film on a single window. For high viscosity samples it is possible to dissolve the sample in a suitable volatile solvent and evaporate the solvent with a stream of dry nitrogen resulting in a film of the material from solution on the surface of a single infrared window. This approach may also be used for solid (soluble) products that are easily dissolved in a volatile solvent such as hexane, toluene, or acetone. If this method is used, it is important to ensure that all traces of solvent are removed before analyzing, taking care that solvent residues are not trapped as microdroplets by viscous materials. Heating under an infrared lamp or in a vacuum oven can be used to assist in the efficient removal of the solvent.

14.6.2 Solids

Most solid materials must go through some sample preparation procedure before their infrared spectra can be collected. In many cases, sample preparation involves grinding of the sample and mixing it with an IR transparent material such as KBr or KCl and then pressing a pellet. This method of solids analysis can be time consuming, but it can result in good spectra free from interferences. The best method for preparation of solid samples involves mixing the sample (about 1–5% by weight) with an FTIR transparent material, normally KBr, and pressing a pellet. Generation of a pellet involves pressing the prepared KBr-sample mixture with a hydraulic or hand press into a hard disk. The pellet, which is ideally 0.5–1 mm in thickness, is then placed in a sample holder for analysis. Normally, the pellet technique provides good quality spectra with a wide spectral range and no interfering absorbance bands. Samples that do not grind well and/or are affected by solvents and mulling agents can be analyzed with high-pressure techniques. The accessory used for such applications utilizes two diamond anvils. Difficult samples are placed between the diamonds and compressed at high pressures to obtain the thickness necessary to acquire quality FTIR spectra. The high-pressure diamond cells require the use of a beam condenser or an infrared microscope. An alternate method for analysis of solid materials involves making a mull. Mulls are sample suspensions in Nujol (refined mineral oil) or Fluorolube (perfluorohydrocarbon). The process is based upon mixing 1–2 drops of the mulling agent with a ground sample until a uniform paste is formed. The paste is transferred onto a KBr or other IR transparent disk and placed in the sample compartment of the spectrometer for analysis. The advantage of this technique is that it is a relatively quick and simple procedure. The disadvantage is that often it contains interference from the mulling agent absorption. In both cases it is necessary to reduce the particles to a submicron particle size, and to disperse the finely ground material. This approach is intended to reduce the impact of light scattering from the sample particles.

14.7 SAMPLE ANALYSIS

As mentioned above, there are two predominant modes of sample analysis by FTIR: transmission mode and reflection mode.

14.7.1 Analysis of Samples in Transmission Mode

The sample holder is set in place for transmission analysis in the sample chamber, which is kept closed in order to be able to purge the sample chamber with dry nitrogen in order to minimize moisture interference.

For solid samples, a small amount is ground to a fine powder with a mortar and pestle and blended with dry KBr as described above in the sample preparation section. The blended material is placed in a pellet mini press, which consists of a barrel and two bolts. The bolts are tightened, and the resulting pellet is placed in the sample compartment and analyzed on the appropriate holder. Alternatively, a solid sample can be casted on a transparent slide, placed in the transmission compartment, and analyzed in a similar manner.

Liquid samples can be placed in a transmission cell, sandwiched between two slides or wet on a KBr slide, and placed in the transmission compartment for analysis in a similar manner.

14.7.2 Analysis of Samples Using an ATR Accessory

For ATR analysis, the ATR accessory is placed into the sample compartment. Solid or liquid samples are placed on the ATR window and require essentially no sample preparation. If the sample is solid, it must be pressed against the ATR window in order to assure a good sample contact with the crystal. Normally the pressure applied is torque limited in order to protect the crystal in the ATR viewing window. A small amount of powder sample (microgram) is placed on top of the diamond window and pressed with a pressure device with a preset torque to assure good contact with the diamond window. The instrument is normally set up to require background spectra before every sample. The sample spectra are automatically corrected for background. This approach is ideal for oligonucleotide products and amidite raw materials that tend to absorb water relatively fast. Because the measurement can be done in less than a minute, moisture absorption during analysis is not a problem. Additionally, because this is a nondestructive analysis, the sample can be recovered for further use if needed.

Sample spectra for some typical oligonucleotides in the powder state are shown in Figures 14.4 through 14.8. Sample spectra for amidite raw materials are shown in Figures 14.9 through 11.

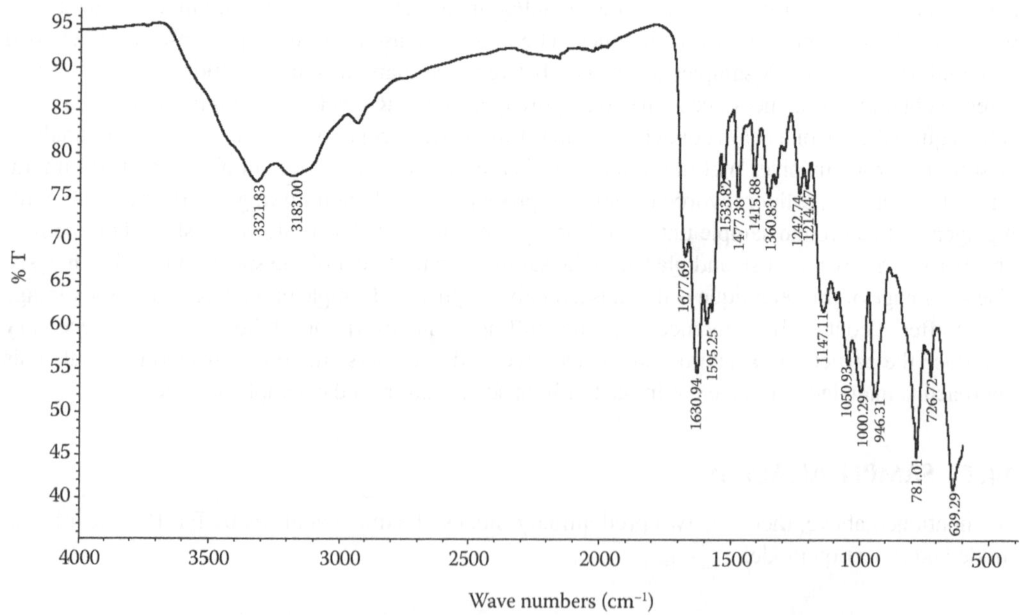

FIGURE 14.4 FTIR spectra of a 40-mer, oligonucleotide, bands around 1650 and 1550 cm^{-1}. These vibrational bands are consistent with the amide I carbonyl and the amide II N-H bend of the amide bond. Characteristic absorbance at 1100–1000 cm^{-1} can provide a simple and rapid diagnostic for phosphorous present.

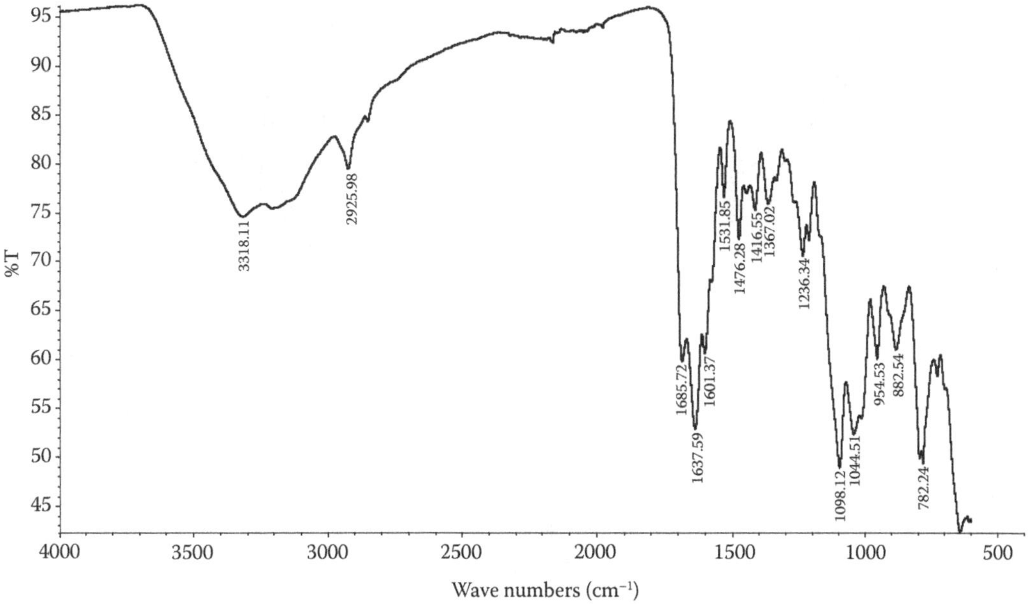

FIGURE 14.5 FTIR spectra of an aliphatic modified oligonucleotide showing the aliphatic content (C-H stretching) which occur around 2900 cm^{-1}. This is a short-chain oligonucleotide with a thiophosphoramidate chemical backbone, with a lipid molecule permanently attached to one end of the molecule.

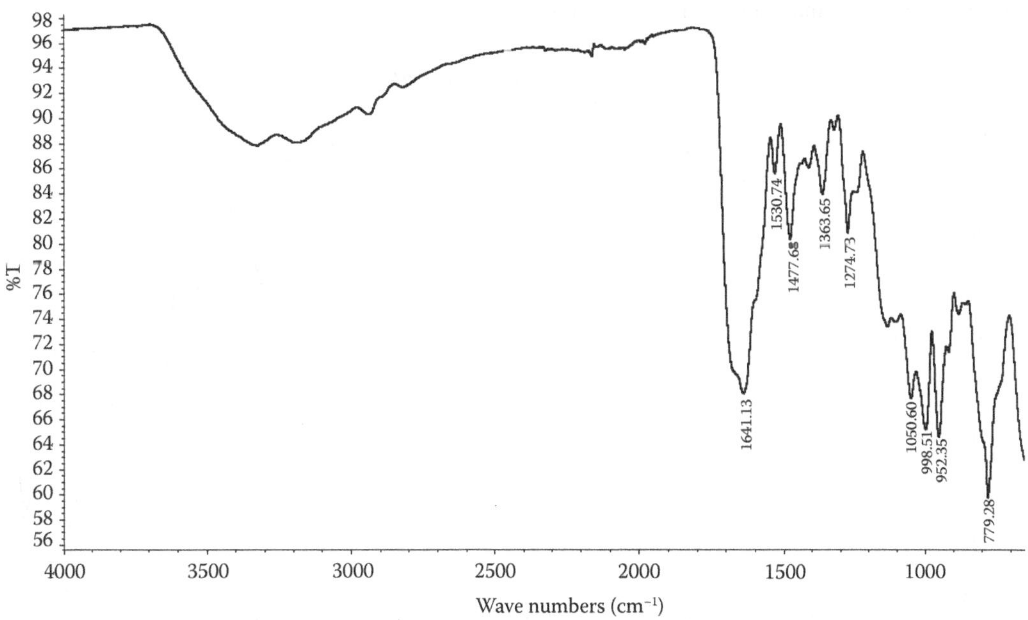

FIGURE 14.6 FTIR spectra of a 23-mer oligonucletide showing a strong absorption in the area of 1670–1630 cm^{-1}, typical of CO-amide groups and 1530 cm^{-1}, secondary amide.

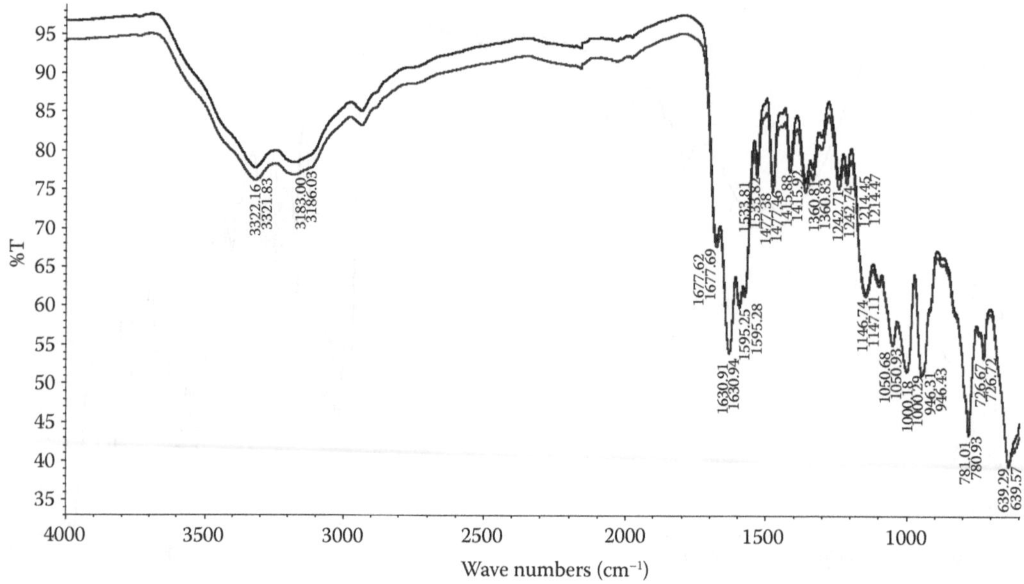

FIGURE 14.7 FTIR spectral comparison of an oligonucleotide product with its reference spectra. There is an excellent match of spectra which is normally observed for spectral comparison when performing product identification by FTIR. The top spectrum is the reference standard, the lower spectra is the sample spectra with a resulting spectral match quality of 100%.

For liquid samples a drop of sample is placed on the ATR crystal for analysis with no further sample preparation required. High volatility samples should be covered with a small cap to avoid or minimize evaporation.

In order to assure best instrument performance, the spectrometer is sealed and desiccated with an active desiccant and under constant purge with dry nitrogen.

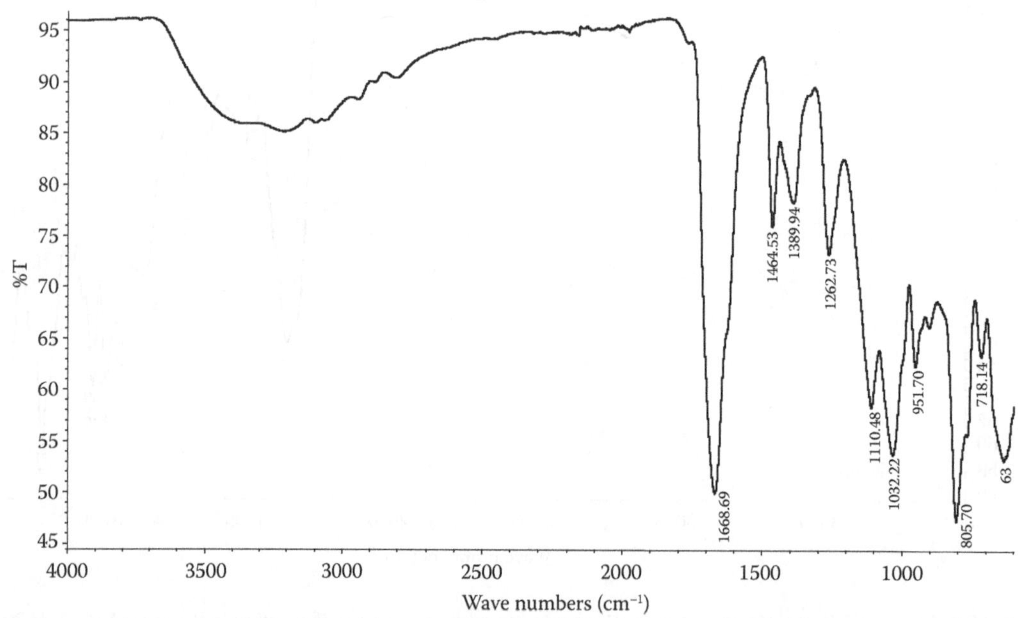

FIGURE 14.8 FTIR spectra of poly-U 20-mer oligonucleotide.

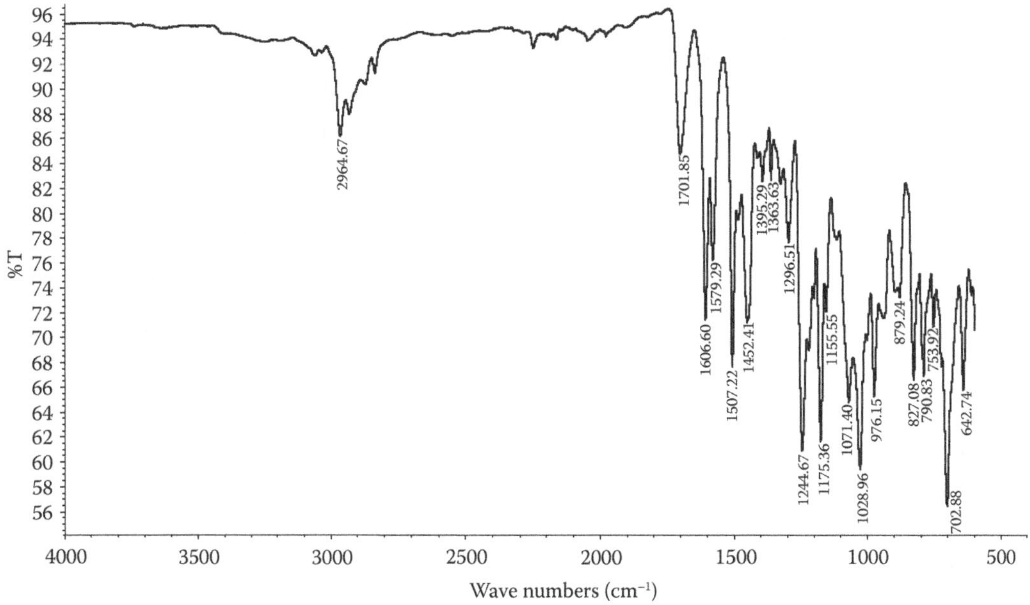

FIGURE 14.9 FTIR spectra of dA amidite.

14.8 DATA ANALYSIS

After a sample spectrum is obtained, the software is used to identify peaks (absorption or transmission bands). Comparison of the collected spectrum may also be made against library spectra, including quality control (QC) raw material or product libraries created in-house.

Most instrument suppliers provide software capabilities to do automatic library searches to compare unknowns against commercial libraries or in-house developed libraries. An additional function for unknown identification is "compare," which allows the user to directly compare and match

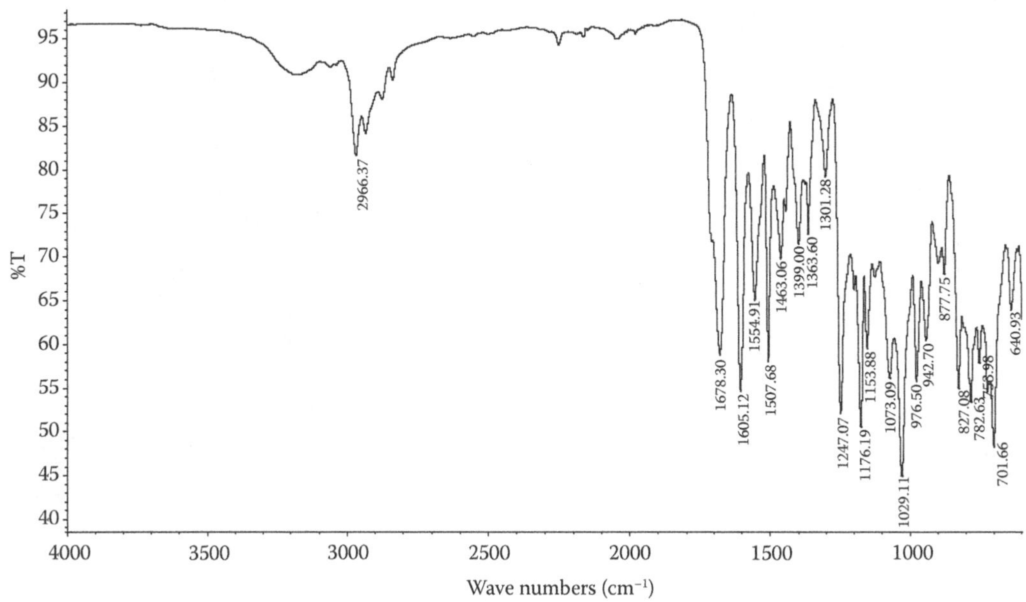

FIGURE 14.10 FTIR spectra of dG amidite.

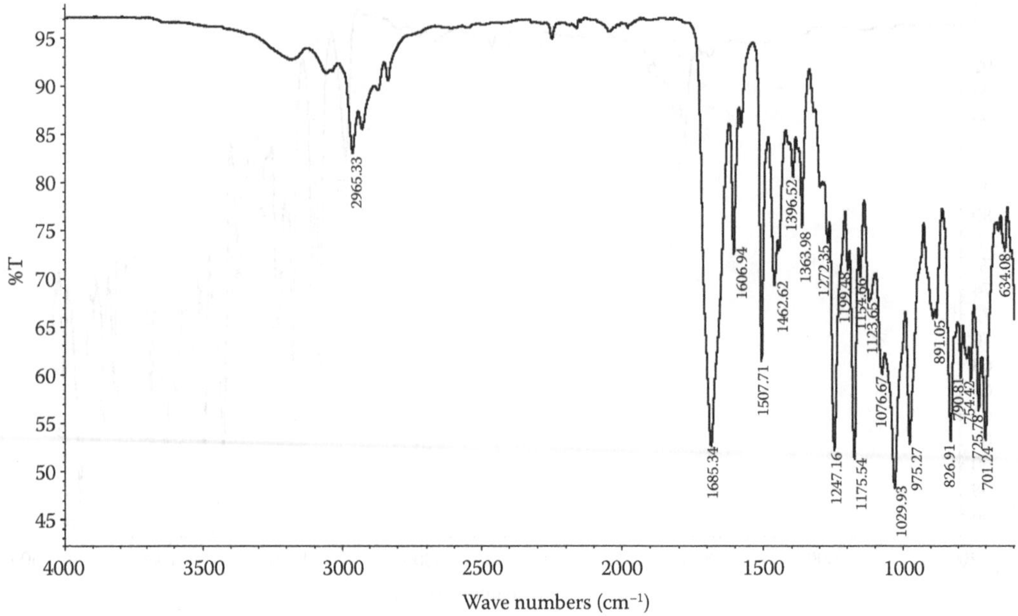

FIGURE 14.11 FTIR of dT amidite.

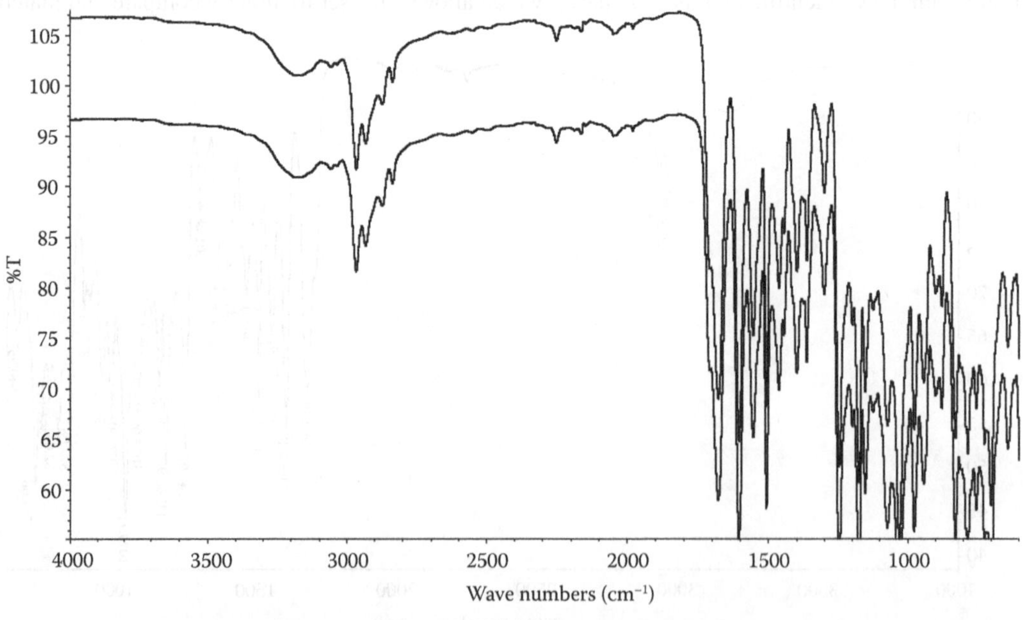

FIGURE 14.12 Spectral match quality of (top) dG amidite FTIR spectra versus (bottom) dG amidite reference spectra. Percent correlation is 100% which is an excellent spectral match.

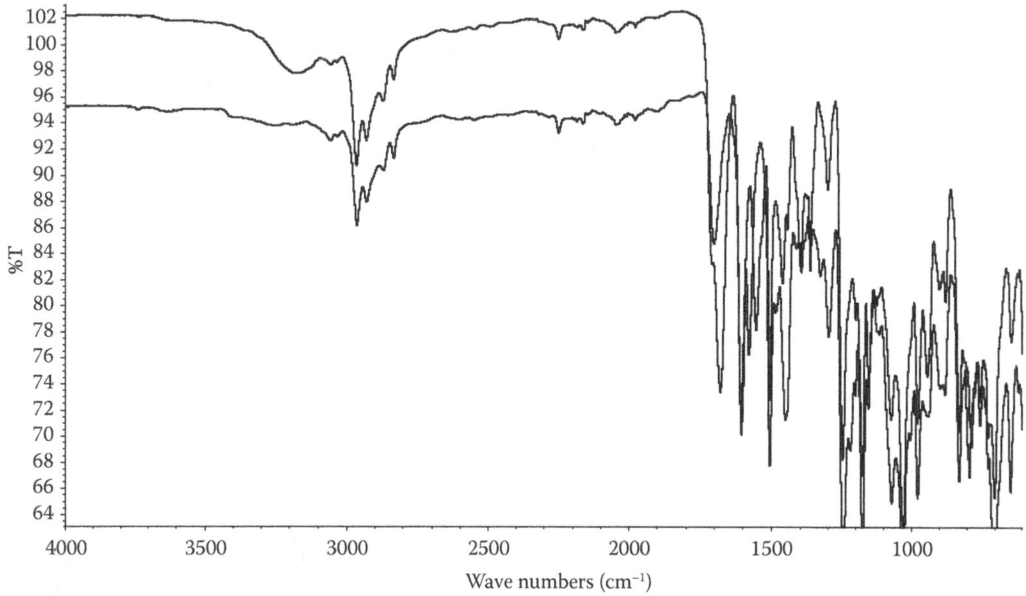

FIGURE 14.13 Spectral match quality of (top) dG amidite FTIR spectra versus (bottom) dA amidite spectra. Percent correlation is 94.3%.

unknown FTIR spectra against the acquired spectra of a certified standard. Spectral comparison results in a quality match normally given on percent basis, 100% being a perfect match with the reference spectra been compared against. In general in FTIR, a good spectral match will yield a result of 99% or better. Figures 14.7 and 14.12 show a good spectral match of a sample versus the corresponding reference spectra. Figure 14.13 shows a poor match of 94.3% for the two spectra being compared, dG amidite versus dA amidite. In this instance, the result clearly indicates that the sample being analyzed does not match the intended reference, while Figure 14.12 shows a 100% spectral match, a good positive identification. Although the normal approach for this analysis is spectral matching by comparison to a reference spectra, manual band assignment can be performed as well for structural elucidation. This aspect is beyond the scope of this chapter, and the reader is referred to a good reference on FTIR spectral band assignment for oligonucleotides.[15]

It is recommended that before adding a new spectrum to the in-house library, it should be verified against a certified reference library or cross-checked with another acceptable identity test such as mass spectrometry or NMR.

Most modern FTIR instruments are equipped with 21 CFR compliant software in order to satisfy current regulatory requirements and automatic or manual instrument qualification providing automat alignment for optimum performance. Instrument qualification is normally performed using certified polystyrene standards at two thicknesses (1.5 and 3.0 mil polystyrene).

14.9 ADDITIONAL EXAMPLES OF FTIR APPLICATIONS

The application of FTIR in the field of oligonucleotides is quite broad owing to instrument versatility, ease of use, and nondestructive sample analysis options. In addition to chemical identification of raw materials and finished products (specific examples provided above), FTIR is also used for the compositional analysis of chemical mixtures and to accurately determine the extent of chemical reactions. Some examples are given below to illustrate the power and versatility of this technique.

Results from work performed by Jeffery Mihaichuk et al.[16] show how FTIR spectroscopy can be used for the quantitation of nucleoside loading on a solid support. On the basis of a linear standard curve, the species absorbance at 1104 cm^{-1} was directly translated into a corresponding concentration. The concentration characterized the number of active phosphoramidite sites on the solid support to a level of 0.02 mmol/g. This is a useful example of the power of FTIR for quantitative analysis by monitoring a specific frequency or a set of frequency absorptions.

Miyamoto et al.[17] investigated the *in situ* hybridization and denaturation of DNA in aqueous solution using infrared absorption spectroscopy in the multiple internal reflection (MIR) geometry. It was demonstrated that conformational changes of DNA strands due to hybridization (binding of two complementary single-stranded DNAs) and denaturation (separation of double helix at elevated temperatures) are reflected in the infrared absorption spectra in the frequency region where vibrational modes of the bases of DNA appear. Comparison with results of *ab initio* cluster calculation showed that hybridization produces the specific CO carbonyl stretching vibration modes in the hydrogen-bonded base pairs. The ratio of absorbance of the CO stretching peak at 1690 cm^{-1} to the absorbance at 1660 cm^{-1} provided a definitive metric for determining DNA hybridization. The results suggest that FTIR is applicable to label-free, high-sensitive biosensors that provide insight about the gene expression and a variety of biological interactions such as DNA-protein interactions.

Work conducted by Carmona and Molina[18] indicate that Infrared spectroscopy can be used to study the assembly of a hairpin nucleotide sequence (nucleotides 3–30) of the 5′ non-coding region of the hepatitis C virus RNA (5′-GGCGGGGAUUAUCCCCGCUGUGAGGCGG-3′) with a RNA 20-mer ligand (5′-CCGCCUCACAAAGGUGGGGU-3′) in the presence of magnesium ion and spermidine. The resulting complex involved two helical structural domains: the first one was an intermolecular duplex stem at the bottom of the target hairpin and the second one is a parallel triplex generated by the intramolecular hairpin duplex and the ligand. Infrared spectroscopy indicated that N-type sugars were exclusively present in the complex.

Saprigin and coauthors[19] performed quantitative determination of surface coverage, film thickness, and molecular orientation of DNA oligomers covalently attached to aminosilane self-assembled monolayers using FTIR spectroscopy and x-ray photoelectron spectroscopy (XPS). Spectral variations between surface immobilized oligomers of the different nucleic acids were reported. Carbodiimide condensation was used for covalent attachment of phosphorylated oligonucleotides to silanized aluminum substrates. FTIR and XPS were used to characterize the surfaces after each modification step. FTIR of covalently bound DNA provided orientational information. Surface density and layer thickness were obtained from XPS data. The surface density of immobilized DNA was found to depend on base composition. Comparison of antisymmetric to symmetric phosphate stretching band intensities in reflection-absorption spectra of immobilized DNA and transmission FTIR spectra of DNA indicated that the sugar-phosphate backbone was predominantly oriented with the sugar-phosphate backbone lying parallel to the surface.

FTIR spectroscopy has been employed to segregated different oligonucleotide sequences.[20] In this work, the authors analyzed a number of simple oligonucleotides designed to contain various base compositions. These products were designed to contain 15 bases in length and various combinations of purines (adenine, guanine) or pyrimidines (cytosine, thymine). The spectra were analyzed using principal component analysis followed by linear discriminant analysis. The study shows that this approach clearly segregated different oligonucleotide sequences, even in the presence of a single base difference. The authors suggest that mid-IR spectroscopy might have future roles in interrogating polymorphic forms of DNA templates.

Work performed by Stefan Raddatz[21] shows how FTIR can be used to monitor the synthesis of new phosphoramidite building blocks and their use for the modification of oligonucleotides with hydrazides. The infrared spectra were obtained using an FTIR instrument equipped with an ATR unit.

FTIR has also been proven to be well suited for the characterization of nucleic acid conformations.[14] Results from these studies provide information on interactions involving specific groups and structural data concerning the nucleoside conformations.

Work performed by Lamar Dewald and coworkers at the Dow Chemical Company[22] has shown how an online attenuated total reflectance Fourier transform infrared (FTIR) analyzer can be used to study and optimize the operation of an oligo synthesizer. The FTIR was used to follow each reaction and wash step, monitor the reaction for completion, and collect timing information for reactor control and modeling. Results from this work were critical to the synthesizer's operation as the plant would experience yield loss and difficulties in product purification if any of the reactions or wash steps were incomplete. The spectra were used to develop the quantitative chemometric methods required to trend 15 components in the synthesizer. Amidites, solvents, coupling reagents, thiolation reagents, deblocking reagents, and intermediates were monitored in the process. Components that only appeared during the reaction transitions were identified and quantitated. The isolated, pure component spectra were used to develop models and a quantitative method for the transition components. FTIR results were used to optimize the operation of the synthesizer and to provide input into the synthesizer's feedback control system.

Considering the potential complexity of the resulting FTIR spectra for oligonucleotides and related products, it is actually surprising how in recent years, major research efforts have focused on the use of FTIR in the characterization of these biomolecules in different states including analysis of molecular structure, elucidation of conformation and orientation,[23–39] and monitoring of different synthesis reactions.[16,22,40]

14.10 SUMMARY

Infrared spectroscopy provides an excellent approach for sample identification of oligonucleotides and key raw materials involved in the synthesis of oligonucleotides. The application of FTIR in this field is quite broad owing to instrument versatility, ease of use, and nondestructive sample analysis options. In addition to chemical identification of raw materials and final drug substances, FTIR is also used for the compositional analysis of chemical mixtures and to accurately determine the extent of chemical reactions. Analysis is normally done by transmission or reflectance modes. Most samples measured by transmission techniques require lengthy sample preparation steps that could be a problem for the oligonucleotide samples and raw materials that are moisture and heat sensitive. However, the advent of the diamond ATR and other ATR options offer a great advantage in these analyses requiring little or no sample preparation for analysis. An additional advantage is that these ATR attachments require a very small amount of sample (microgram levels) for this analysis. Additionally, because this technique is nondestructive, the sample used for analysis by this technique can be used for further characterization by other complementary analytical techniques. Amenability to automation and microsampling offer significant advantages for this analytical approach.

ACKNOWLEDGMENTS

The author wishes to express gratitude to Girindus America, who supplied many of the product specimens to show the usefulness of this technique and for allowing the opportunity to engage in this publication.

REFERENCES

1. Smith, B. 1999. *Infrared Spectral Interpretation, A Systematic Approach.* Boca Raton, FL: CRC Press.
2. Bellamy, L. J. 1980. *Advances in Infrared Group Frequencies, Infrared Spectra of Complex Molecules.* New York: Chapman and Hall.
3. Steele, D. 1977. *Interpretation of Vibrational Spectra.* London: Chapman and Hall.
4. Lin-Vien, D., N. B. Colthup, W. G. Fateley, and J. G. Grasselli. 1991. *Infrared and Raman Characteristic Frequencies of Organic Molecules.* San Diego: Academic Press.

5. Coates, J. 2000. *Interpretation of Infrared Spectra, A Practical Approach, Encyclopedia of Analytical Chemistry*, 10815–37. New York: John Wiley.
6. Hind, A. R., S. K. Bhargava, and A. McKinnon. 2001. At the solid/liquid interface: FTIR/ATR—the tool of choice. *Advances in Colloid and Interface Science* 93: 91–114.
7. Oberg, A. K. A., and L. Fink. 1998. A new attenuated total reflectance Fourier transform infrared spectroscopy method for the study of proteins in solution. *Analytical Biochemistry* 25(1): 92–106.
8. Schwinte, P., J. C. Voegel, C. Picart, Y. Haikel, P. Schaaf, and B. Szalontai. 2001. Stabilizing effects of various polyelectrolyte multilayer films on the structure of adsorbed/embedded fibrinogen molecules: An ATR-FTIR study. *Journal of Physical Chemistry* 105(47): 11906–11916.
9. Stefan, I. C., D. Mandler, and D. A. Scherson. 2002. In situ FTIR-ATR studies of functionalized self-assembled bilayer interactions with metal ions in aqueous solutions. *Langmuir* 18(18): 6976–6980.
10. Bahng, M. K., N. J. Cho, J. S. Park, and K. Kim. 1998. Interaction of indolicidin with model lipid bilayers: FTIR-ATR spectroscopic study. *Langmuir* 14(2): 463–470.
11. Ding, F. X., H. B. Xie, B. Arshava, J. M. Becker, and F. Naider. 2001. ATR-FTIR study of the structure and orientation of transmembrane domains of the Saccharomyces cerevisiae alpha-mating factor receptor in phospholipids. *Biochemistry* 40(30): 8945–8954.
12. Gershevitz, O., and C. N. Sukenik. 2004. In situ FTIR-ATR analysis and titration of carboxylic acid-terminated SAMs. *Journal of the American Chemical Society* 126(2): 482–483.
13. Iwaki, M., A. Puustinen, M. Wikstrom, and P. R. Rich. 2003. ATR-FTIR spectroscopy of the P-M and F intermediates of bovine and Paracoccus denitrificans cytochrome c oxidase. *Biochemistry* 42(29): 8809–8817.
14. Voegelin, A., and S. J. Hug. 2003. Catalyzed oxidation of arsenic(III) by hydrogen peroxide on the surface of ferrihydrite: an in situ ATR-FTIR study. *Environmental Science Technology* 37(5): 972–978.
15. Tsuboi, M., and Y. Kyogoku. 1973. Infrared spectroscopy of nucleic acid components. In *Synthetic Procedures in Nucleic Acid Chemistry*, vol. 2, 215–265. New York: Wiley-Interscience.
16. Mihaichuk, J., C. Tompkins, and W. Pieken. 2002. A method to accurately determine the extent of solid-phase reactions by monitoring an intermediate in a nondestructive manner. *Analytical Chemistry* 74(6): 1355–1359.
17. Miyamoto, I., I. K.-I. Ishibashi, R.-T. Yamaguchi, Y. Kimura, H. Ishii, and M. Niwano. 2006. In situ observation of DNA hybridization and denaturation by surface infrared spectroscopy. *Journal of Applied Physics* 99: 094702.
18. Carmona, P., and M. Molina. 2002. Binding of oligonucleotides to a viral hairpin forming RNA triplexes with parallel GGC triplets. *Nucleic Acids Research* 30(6): 1333–1337.
19. Saprigin, A. V., C. W. Thomas, C. S. Dulcey, C. H. Patterson, Jr., and M. S. Spector. 2005. Spectroscopic quantification of covalently immobilized oligonucleotides. *Surface and Interface Analysis* 37(1): 24–32.
20. J. G. Kelly, P. L. Martin-Hirsch, and F. L. Martin. 2009. Discrimination of base differences in oligonucleotides using mid-infrared spectroscopy and multivariate analysis. *Analytical Chemistry* 81(13): 5314–5319.
21. Raddatz, S., J. Mueller-Ibeler, J. Kluge, L. Wäß, G. Burdinski, J. R. Havens, T. J. Onofrey, D. Wang, and M. Schweitzera. 2002. Hydrazide oligonucleotides: new chemical modification for chip array attachment and conjugation. *Nucleic Acids Research* 30(21): 4793–4802.
22. Dewald, L., M. B. Seasholtz, and R. Pell. 2008. On-line attenuated total reflectance Fourier transform infrared analysis of an oligonucleotides synthesizer effluent stream. Paper presented at 35th FACSS Meeting.
23. Birke, S. S., M. Moses, B. Kagalovsky, et al. 1993. Infrared CD of deoxy oligonucleotides. Conformational studies of 5′d(GCGC)3′, 5′d(CGCG)3′, 5′d(CCGG)3′, and 5′d(GGCC)3′ in low and high salt aqueous solution. *Biophysics Journal* 65(3): 1262–1271.
24. Olejniczak, A. B., A. Sut, A. E. Wróblewski, and Z. Leśnikowski. 2005. Infrared spectroscopy of nucleoside and DNA-oligonucleotide conjugates labeled with carborane or metallacarborane cage. *Vibrational Spectroscopy* 39(2): 177–185.
25. Sarkar, M., U. Dornberger, E. Rozners, H. Fritzsche, R. Strömberg, and A. Gräslund. 1997. FTIR spectroscopic studies of oligonucleotides that model a triple-helical domain in self-splicing group I introns. *Biochemistry* 36(49): 15463–15471.
26. Gue, M., V. Dupont, A. Dufour, and O. Sire. 2001. Bacterial swarming: A biochemical time-resolved FTIR-ATR study of Proteus mirabilis swarmcell differentiation. *Biochemistry* 40(39): 11938–11945.

27. Olejniczak, A. B., A. Sut, A. E. Wróblewski, and Z. J. Leśnikowski. 2005. Infrared spectroscopy of nucleoside and DNA-oligonucleotide conjugates labeled with carborane or metallacarborane cage. *Vibrational Spectroscopy* 39(2): 177–185.
28. Dagneaux, C., H. Gousset, A. K. Shchyolkina, M. Ouali, R. Letellier, J. Liquier, V. L. Florentiev, and E. Taillandier. 1996. Parallel and antiparallel A*A-T intramolecular triple helixes. *Nucleic Acids Research* 24: 4506–4512.
29. Dagneaux, C., J. Liquier, and E. Taillandier. 1995. Sugar conformations in DNA and RNA-DNA triple helices determined by FTIR spectroscopy. *Biochemistry* 34: 16618–16623.
30. Dagneaux, C., H. Porumb, R. Letellier, C. Malvy, and E. Taillandier. 1995. Triple helix forming oligonucleotides used as drugs: An FTIR, Raman and molecular modeling study. *Journal of Molecular Structure* 347: 343–350.
31. White, A. P., and J. W. Powell. 1995. Observation of the hydration-dependent conformation of the (dG)20•(dG)20•(dC)20 oligonucleotide triplex using FTIR spectroscopy. *Biochemistry* 34: 1137–1142.
32. Dagneaux, C., J. Liquier, and E. Taillandier. 1995. Sugar conformations in DNA and RNA-DNA triple helices determined by FTIR spectroscopy: Role of backbone composition. *Biochemistry* 34(51): 16618–16623.
33. Liquier, J., A. Akhebat, E. Taillandier, F. Ceolin, T. Huynh-Dinh, and J. Igolen. 1991. Characterization by FTIR spectroscopy of the oligoribonucleotide duplexes r(A-U)6 and r(A-U)8. *Spectrochimica Acta Part A: Molecular Spectroscopy* 47: 177–186.
34. Liquier, J., P. Coffinier, M. Firon, and E. Taillandier. 1991. Triple helical polynucleotidic structures: Sugar conformations determined by FTIR spectroscopy. *Journal of Biomolecular Structure Dynamics* 9: 437–445.
35. Tsuboi, M. 1969. Application of infrared spectroscopy to structure studies of nucleic acids. *Applied Spectroscopy Review* 3: 45–90.
36. Taillandier, E., and J. Liquier. 1992. Infrared spectroscopy of DNA. *Methods in Enzymology* 211: 307–335.
37. Akhebat, A., C. Dagneaux, J. Liquier, and E. Taillandier. 1992. Triple helical polynucleotide structures: An FTIR study of the C+•G•C triplet. *Journal of Biomolecular Structure Dynamics* 10: 577–588.
38. Ouali, M., R. Letellier, J. S. Sun, A. Akhebat, F. Adnet, J. Liquier, and E. Taillandier. 1993. Determination of G*G•C triple-helix structure by molecular modeling and vibrational spectroscopy. *Journal of the American Chemical Society* 115: 4264–4270.
39. White, A. P., and J. W. Powell. 1995. Observation of the hydration-dependent conformation of the (dG)20•(dG)20•(dC)20 oligonucleotide triplex using FTIR spectroscopy. *Biochemistry* 34: 1137–1142.
40. Li, S., J. X. Tang, M. J. Ji, P. Hou, P. F. Xiao, and N. Y. He. 2004. A novel substrate for in situ synthesis of oligonucleotides: Hydrolyzed microporous polyamide-6 membrane. *Chinese Chemical Letters* 15(7): 837–840.

27. Olejniczak, A. B., A. Sut, A. E. Wroblewski, and Z. J. Lesnikowski. 2005. Infrared spectroscopy of nucleoside and DNA-oligonucleotide conjugates labeled with carborane or metallacarborane cage. *Vibrational Spectroscopy* 39(2):177–185.
28. Dagneaux, C., H. Gousset, A. K. Shchyolkina, M. Ouali, R. Letellier, J. Liquier, V. L. Florentiev, and E. Taillandier. 1996. Parallel and antiparallel A*A-T intramolecular triple helices. *Nucleic Acids Research* 24:4506–4512.
29. Dagneaux, C., J. Liquier, and E. Taillandier. 1995. Sugar conformations in DNA and RNA-DNA triple helices determined by FTIR spectroscopy. *Biochemistry* 34:16618–16623.
30. Dagneaux, C., H. Porumb, R. Letellier, C. Malvy, and E. Taillandier. 1995. Triple helix forming oligonucleotides used as drugs: An FTIR, Raman and molecular modeling study. *Journal of Molecular Structure* 347:341–350.
31. White, A. P., and J. W. Powell. 1995. Observation of the hydration-dependent conformation of the (dG)20•(dG)20•(dC)20 oligonucleotide triplex using FTIR spectroscopy. *Biochemistry* 34:1137–1142.
32. Dagneaux, C., J. Liquier, and E. Taillandier. 1995. Sugar conformations in DNA and RNA-DNA triple helices determined by FTIR spectroscopy: Role of backbone composition. *Biochemistry* 34:16618–16623.
33. [illegible], A. Acheha, T. Theophanides, F. [illegible], Hayes, [illegible], and [illegible]. 199[illegible]. Characterization by FTIR spectroscopy of the oligoribonucleotide duplexes r(A-U)6 and r(A-U)8. *Spectrochimica Acta A Molecular Spectroscopy* 4[illegible]:171–180.
34. Liquier, J., P. Coffinier, M. Firon, and E. Taillandier. 1991. Triple helical polynucleotidic structures: Sugar conformations determined by FTIR spectroscopy. *Journal of Biomolecular Structure & Dynamics* 9:437–445.
35. Tsuboi, M. 1969. Application of infrared spectroscopy to structure studies of nucleic acids. *Applied Spectroscopy Reviews* 3:45–90.
36. Taillandier, E., and J. Liquier. 1992. Infrared spectroscopy of DNA. *Methods in Enzymology* 211:307–335.
37. Akhebat, A., C. Dagneaux, J. Liquier, and E. Taillandier. 1992. Triple helical polynucleotidic structures: An FTIR study of the C+•G•C triplet. *Journal of Biomolecular Structure & Dynamics* 10:577–588.
38. Ouali, M., R. Letellier, J. S. Sun, A. Akhebat, F. Adnet, J. Liquier, and E. Taillandier. 1993. Determination of G*G•C triple-helix structure by molecular modeling and vibrational spectroscopy. *Journal of the American Chemical Society* 115:4264–4270.
39. White, A. P., and J. W. Powell. 1995. Observation of the hydration-dependent conformation of the (dG)20•(dG)20•(dC)20 oligonucleotide triplex using FTIR spectroscopy. *Biochemistry* 34:1137–1142.
40. Li, S. J., X. Zhao, M. [illegible], B. P. Hou, P. F. Xiao, and N. Y. He. 2004. A novel substrate for in situ synthesis of oligonucleotides: Hydrolyzed microporous polyamide membrane. *Chinese Chemical Letters* 15(?):[illegible].

15 Stability Indicating Methods for Oligonucleotide Products

Veeravagu Murugaiah
Alnylam Pharmaceuticals

CONTENTS

15.1 PREFACE

In this chapter on stability indicating methods (SIMs) for product stability studies, the salient features of SIMs such as sample generation, method development, and method validation will be discussed. A SIM is a quantitative analytical procedure used to detect a decrease in the amount of the drug substance present in the drug product due to degradation. Hence, specificity and mass balance of the methods will be discussed at length. Because this chapter is intended to provide insight and instructions on the development of SIMs, detailed experimental design and method optimization are given for select examples of short interfering RNA (siRNA), as both the drug substance and the drug product formulated in a simple buffer. In addition, a novel example of the quantitative analysis of two siRNA duplexes in a drug product formulation is provided. The hygroscopicity data provide information on the physical state of the oligonucleotides and may be useful in designing the storage conditions for the oligonucleotide. Further, stability monitoring of oligonucleotide drug substance and drug product is illustrated with select stability protocols. Several statistical techniques are used for the evaluation of stability data for determination of shelf life, which are discussed in the references cited.

The applications described in this chapter may also be used to design stability studies for DNA duplexes. Further, in a simple form, the approaches described in this chapter may be used to structure stability studies of single stranded DNA or RNA oligonucleotides.

15.2 INTRODUCTION

Stability testing is an important part of the process of drug product development. The purpose of stability testing is to provide evidence on how the quality of a drug substance or drug product varies with time under the influence of a variety of environmental factors such as temperature, humidity, and light, and enables recommendation of storage conditions, retest periods, and shelf lives to be established. The two main aspects of drug product stability that play an important role in shelf life determination consist of active drug assay and degradation products, generated during the stability study. The assay of drug substance and drug product in the stability program needs to be determined using stability-indicating methods (SIMs), as recommended by the International Conference on Harmonization (ICH) guidelines[1,2] and FDA guidelines.[3] Hence, the stability-indicating nature of the chromatographic methods needs to be demonstrated prior to use in a stability study.

A SIM is a quantitative analytical procedure used to detect a decrease in the amount of the drug substance present due to degradation. Anion-exchange chromatography (AEX), size exclusion chromatography (SEC), and ion-pair reversed phase high-performance chromatography (IP-RP HPLC) methods comprise the most suitable SIMs for the analysis of RNA, DNA, and modified oligonucleotides. According to FDA guidelines, a SIM is defined as a validated analytical procedure that accurately and precisely measures drug substance or drug product free from potential interferences like process related impurities, degradation products, and excipients. Hence, there are three components necessary for implementing a SIM: (1) generation of samples to mimic potential degradation products and process related impurities, (2) development of a SIM that provides satisfactory resolution, and (3) validation of the method.

15.3 GENERATION OF SAMPLES

In the case of oligonucleotides, process impurities such as deletion or addition sequences could be readily synthesized based on knowledge of the sequence under study. These impurities may be used to verify the resolution with respect to full length product. Refer to Chapter 2 for more on this topic. Degradation products of oligonucleotides that one might anticipate to be formed during storage and or transport under different environmental conditions, however, are not readily predictable. Also, the nature of the degradation products that may be formed during a well-designed stability

TABLE 15.1
Common Conditions Used in Forced Degradation Studies

Study	Conditions	Comments
Acidic stress	0.1 N HCl	Modified oligonucleotides may need stronger acid solution
Basic stress	0.1 N NaOH	Modified oligonucleotides may need stronger base solution
Oxidative stress	0.3% H_2O_2	Some formulation may need 3% H_2O_2
Thermal stress	60°–80°C	As high as 100°C may be used
Photolysis (UV)	1000 Watt h/m^2	Refer to ICH Q1B
Photolysis (fluorescence)	6 × 10^6 lux h	Refer to ICH Q1B

program is not readily known. Therefore, SIMs are developed by stressing the drug substance under conditions exceeding those normally used for accelerated stability testing. The stress testing, also referred to as forced degradation, demonstrates specificity of the SIMs and provides information on the degradation pathways and products that could form during storage. Stressing drug substances in both lyophilized, solid form, and in formulated solution forms generate samples that contain the products most likely to form under the most realistic storage conditions. Such samples are used, in turn, to develop the SIM. The goal of the SIM is to obtain baseline resolution of *all* the resulting products, drug substance, and degradation products. However, for large molecules, such as oligonucleotides, this is not always possible. Table 15.1 lists some common conditions used in designing forced degradation studies for oligonucleotides.[4,5] The stress conditions may be modified to achieve a target degradation of 10–30% within a reasonable incubation time of one week.

15.4 DEVELOPING HPLC METHOD

The details of method development and optimization of AEX-HPLC, IP-RP-HPLC, and SEC-HPLC are explained in detail in Chapters 1, 2, and 3.[6] The samples, generated using a properly designed and executed forced degradation experiment, are used to verify and improve the resolution of the methods to meet the expectations of the SIMs. During method development, selectivity can be manipulated by any one or a combination of different factors that include solvent composition, type of column stationary phase, and mobile phase buffers and pH, as discussed in Chapters 1, 2, and 3.

The sample subjected to stress by exposure to acidic, basic, oxidative, photolytic, and thermal conditions are analyzed at regular time points during the study to check the status of the degradation. Samples should be stored in appropriate vessels that allow sampling at time intervals while protecting and preserving the integrity of the sample. It should be noted that acid and base solutions should be neutralized before analysis. Generally, the goal of these studies is to degrade the drug substance by 5–20%. Any more than this could lead to the formation of irrelevant degradation products. It should be noted that not all stress conditions could lead to degradation products. Modified oligonucleotides are designed for greater *in vivo* stability. With a novel approach to enhance drug delivery, oligonucleotides are made into liposome formulations. Such drug substances and drug products might require a longer incubation time and/or a stronger stress agent to produce target degradation.

15.4.1 Specificity of SIMs

An important key factor to evaluate during SIM development is specificity. The specificity of a method is its ability to assess unequivocally the analyte of interest in the presence of potential interferences.[7–9] Specificity is evaluated by resolution, peak shape, and tailing factor. Also, chromatograms of forced degradation samples are compared with control drug substance chromatograms by overlaying the two chromatograms and seeking new degradation peaks. The modern photodiode array detector (PDA) is a powerful technology for evaluating specificity.[10–14] USP and ICH guidelines recommend PDA detection or mass spectrometry (MS) to establish specificity. However, in the case of oligonucleotides, use of PDA detectors for the establishment of peak purity have not proven useful, because the process related impurities and degradation products tend to have very similar UV profiles as that of the drug being studied.

It is a difficult task in chromatography to ascertain whether the peaks within a sample chromatogram are pure or consist of more than one compound. This is a common situation in forced degradation samples of double-stranded oligonucleotides as deletion sequences that differ by unit nucleotide in a duplex may co-elute with one another. Therefore, procedures for detecting impure peaks should be used that are achieved by hyphenated technologies. For characterization of the degradation pathways of oligonucleotides, mass spectrometric detection has proven to be the method of choice, because of its ability to provide structural identification of impurities in the drug substance and drug products.[15,16]

The SIMs developed, as described in this chapter, could be directly or indirectly coupled to MS detection to achieve additional specificity for stability monitoring. In the case of IP-RP-HPLC, by using MS compatible ion-pairing reagents, these methods can be directly coupled with MS detection, thereby allowing structural characterization of degradation products. Forced degradation samples contain salts from treatment with acids and bases and need to be desalted. A new approach for automated desalting and analyzing samples by LC-MS is given in Chapter 2. Refer to Chapters 1 and 4 for more on RP-HPLC and MS detection.

15.4.2 Mass Balance

Mass balance correlates the measured loss of a parent drug or drug substance to the measured increase in the amount of degradation products. It is a good quality control check on analytical methods to show that all degradation products are adequately detected and do not interfere with quantitation of the parent drug or drug substance that is a characteristic of SIM. Regulatory agencies use mass balance to assess the appropriateness of the analytical method as a SIM and to determine whether all degradant products have been accounted for.[1–3,17]

Under certain circumstances, such as in the analysis of oligonucleotides, it is not feasible to estimate mass balance directly from the amount of degradation products formed but rather the percent parent drug substance lost. The percentage drug substance lost, which corresponds to the amount of degradation products formed, is the ultimate quantity that should be used to reconcile mass balance. The amount of degradation products is converted to the corresponding percent drug substance degraded by means of the ratio of area-percent of the drug substance relative to that of the degradation products. A fundamental assumption made in this calculation is that the response factors of the degradation products are the same as that of the drug substance at the wavelength of measurement of 260 nm. Thus, the operative entity used in reconciling mass balance is the percent of drug substance degraded. For example, in the analysis of a duplex, loss in area-percent of sense strand and antisense strand is monitored at suitable interval of the forced degradation study as a function of the total area. If the total area remains the same within experimental error, then one would consider that mass balance is preserved.

15.5 VALIDATION OF SIMS

Often the assay and impurity testing can be performed using a single HPLC method. However, the assay and purity determinations may also be separate methods. Full validation of the analytical method is not required to support first-in-man studies, but suitability of the analytical methods needs to be established through assessment of specificity and precision. Limit of quantitation (for impurities) is important at the earliest stages, because verification of stability hinges on a suitable method for separating impurities from the active ingredient and at least quantifying the impurities relative to the drug substance. Detailed discussions of method validation are beyond the scope of this chapter; refer to Refs. 18 and 19 for more on this topic. Refer to Chapter 19 for more of the regulatory expectations on method validation.

15.6 DEVELOPMENT OF SIMS—SELECT APPLICATION

15.6.1 Short Interfering RNA

Short interfering RNA (siRNAs) are double-stranded RNAs of 21–23 nucleotides. Chemically synthesized siRNAs usually contain modified backbone, sugars, and/or bases to provide resistance to nucleases,[20,21] reduce off-target effects, and limit undesired immune stimulation. A sample of siRNA duplex is subjected to acidic, basic, oxidative, and thermal stress, and analyzed by denaturing AEX-HPLC that was optimized for the resolution of the sense and antisense strands. The protocol for the stress conditions for this experiment is provided in Table 15.2 and the details of the individual procedures are given below.

15.6.1.1 Stressing Procedures

15.6.1.1.1 Acid Hydrolysis

Transfer 350 μL of nuclease free water and 50 μL of a 2 mg/mL solution of duplex siRNA into a 1.7 mL autosampler vial (make replicate preparations, n = 5). Add 300 μL of 0.1 M hydrochloric acid, cap the vial, and vortex to mix for approximately 1 min. Allow to sit for approximately 30 min. Uncap the first vial, and add 300 μL of 0.1 M sodium hydroxide (neutralizing reagent) to the vial. Hermetically seal the vial and vortex to mix. Analyze by AEX-HPLC and repeat the above procedure at suitable intervals to obtain sufficient degradation for the SIM.

15.6.1.1.2 Base Hydrolysis

Transfer 350 μL of nuclease free water and 50 μL of a 2 mg/mL solution of duplex siRNA into a 1.7 mL autosampler vial (make replicate preparations, n = 5). Add 300 μL of 0.1 M sodium hydroxide, cap the vial, and vortex mix for approximately 1 min. Allow to sit for approximately 30 min. Uncap the first vial and add 300 μL of 0.1 M hydrochloric acid (neutralizing reagent) to the vial. Hermetically seal the vial and vortex to mix. Analyze by AEX-HPLC and repeat the above procedure at suitable intervals to obtain sufficient degradation.

15.6.1.1.3 Oxidative Stress

Transfer 350 μL of nuclease free water and 50 μL of a 2 mg/mL solution of duplex siRNA into a 1.7 mL autosampler vial (make replicate preparations, n = 5). Add 300 μL of 0.35% hydrogen peroxide, cap the vial, and vortex to mix for approximately 1 min. Allow to sit for approximately 30 min. Uncap the first vial and add 300 μL of nuclease free water to the vial. Hermetically seal the vial and vortex to mix. Analyze by AEX-HPLC and repeat the above procedure at suitable intervals

TABLE 15.2
Force Degradation Study Protocol to Demonstrate Mass Balance

Stress Type	Stress Reagent	Stress Reagent (μL)	Neutralizing Reagent	Neutralizing (μL)	Final Volume (μL)
Acid	0.1 N HCl	300	0.1 N NaOH	300	1000
Base	0.1 N NaOH	300	0.1 N HCl	300	1000
Oxidation	0.35% H_2O_2	300	NF water	300	1000
Thermal	75°C	N/A	NF water	600	1000

to obtain sufficient degradation. If 0.35% hydrogen peroxide does not give sufficient degradation, increase the concentration to 3.5% hydrogen peroxide.

15.6.1.1.4 Thermal Stress

Transfer 350 μL of nuclease free water and 50 μL of a 2 mg/mL solution of duplex RNA into a 1.7 mL autosampler vial (make replicate preparations, n = 5). Hermetically seal the vial and heat in a hot plate at 75°C for 1 hour. Remove from hot plate, uncap the first vial, and add 600 μL of nuclease free water. Hermetically seal the vial and vortex to mix. Analyze by AEX-HPLC and repeat the above procedure at suitable intervals to obtain sufficient degradation.

15.6.1.1.5 Control

Transfer 950 μL of nuclease free water and 50 μL of a 2 mg/mL solution of duplex RNA into a 1.7 mL autosampler vial. Mix well and analyze by AEX-HPLC. Use these data as the positive control for drug substance, duplex RNA.

15.6.1.2 Results and Discussion

Under denaturing conditions, siRNA duplexes are denatured into sense and antisense strands as shown in Figure 15.1. Both the control chromatogram and the stressed sample chromatograms were integrated. Area counts and percent degradation of sense strand and antisense strand and total area counts were tabulated for all chromatograms generated at different intervals during this degradation study. The total area count is a means of evaluating mass balance in the forced degradation study. Assuming that the response factors for drug substance main peaks and degradation peaks to be the same, the total area counts should remain the same for different stages of incubation under the stress conditions used. A significant drop in the total area count means that the sample has been extensively stressed, and such a sample is not suitable to evaluate the SIM. The loss of peak area could arise from the loss of chromospheres due to depurination of the base by stressing agents.

15.6.1.2.1 Acid Hydrolysis

Figure 15.1 is an overlay of the control sample and the acid stressed sample for 75 min incubation, and the extent of degradation is shown in Table 15.3. Degradation of this sample for a longer period will produce exhaustive degradation, resulting in minor impurities eluting in the void: refer to data for 4 hours in Table 15.3.

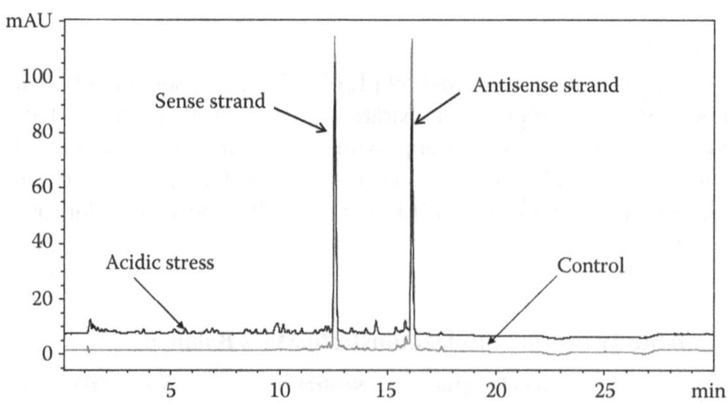

FIGURE 15.1 Chromatogram of a siRNA duplex: System: Agilent 1100 HPLC; column: Dionex DNAPac PA-200; column temperature: 30°C; buffer A: 20 mM sodium phosphate, 10% CH_3CN pH at 11; buffer B: 20 mM sodium phosphate, 1 M NaBr, 10% CH_3CN pH at 11; gradient: 30% B to 50% B in 20 min, ramp and hold at 100% B for 3 min and equilibrate 30% B for 5 min; flow rate 1.0 mL/min; detection at 260 nm; overlay of control sample and acid stressed sample incubated for 75 min at ambient condition.

TABLE 15.3
Mass Balance of Acidic Degradation as a Function of Incubation Time

Incubation Time	Sense Strand (Area-Percent)	Sense % Degradation	Antisense Strand (Area-Percent)	Antisense % Degradation	Total Area Count
Control	43.3	N/A	42.9	N/A	1985
5 min	43.7	–0.9	42.6	0.7	1888
45 min	40.6	6.2	34.3	20.0	1843
75 min	37.8	12.7	28.7	33.0	1867
4 hours	28.0	35.3	12.4	71.0	1296

15.6.1.2.2 Base Hydrolysis

Figure 15.2 is an overlay of the control sample and the base stressed sample for 30 min. As shown in Table 15.4 within experimental error, forced degradation samples are conserved up to about 20 min. Degradation for a longer period would have produced exhaustive degradation resulting in minor impurities eluting in the void. It should be noted that the extent of degradation is sequence dependent.

15.6.1.2.3 Oxidative Stress

Figure 15.3 is an overlay of the control sample and the oxidative stressed sample for 21 hours. As shown in Table 15.5, the duplex is resistant to oxidation by hydrogen peroxide. It took about 72 hours to achieve the targeted amount of degradation.

15.6.1.2.4 Thermal Stress

Figure 15.4 is an overlay of the control sample and the thermally stressed sample at 75°C for 23 hours. As shown in Table 15.6, the sample used is resistant to thermal stress.

15.7 SEQUENCE SPECIFIC ASSAYS

Weight-by-weight (wt/wt) quantitative assay of oligonucleotide is required for (1) estimation of the yield of synthesis, (2) design of drug formulation, and (3) assessment of drug substance and/or drug product stability. Routine methods for the determination of the concentrations of oligonucleotide, C, usually rely on measurements of UV absorbance (at 260 nm), A_{260}, which is related by a fundamental property of the molecule called the molar extinction coefficient, ε_{260}. The actual absorbance

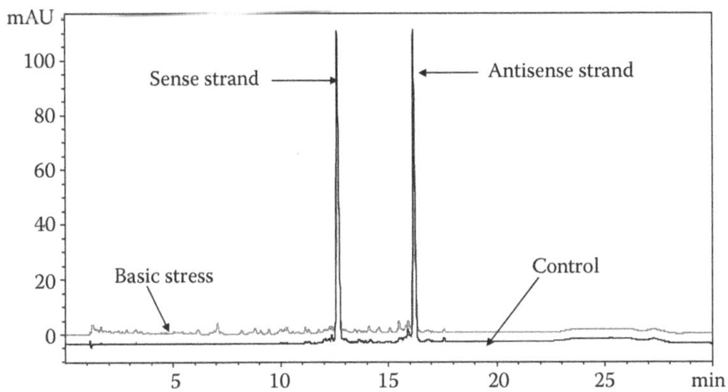

FIGURE 15.2 Overlay of control sample and base stressed sample incubated for 30 min at ambient condition. Refer to Figure 15.1 for chromatographic conditions.

TABLE 15.4
Mass Balance of Basic Degradation as a Function of Incubation Time

Incubation Time	Sense Strand (Area-Percent)	Sense % Degradation	Antisense Strand (Area-Percent)	Antisense % Degradation	Total Area Count
Control-A	44.3	N/A	43.9	N/A	1972
5 min	42.1	5.0	40.3	8.2	1916
15 min	41.1	7.2	40.3	8.2	1906
30 min	39.9	9.9	30.9	29.6	1835
60 min	37.9	14.4	23.2	47.2	1702

A of a sample is dependent on the path length *l* and the concentration *C* of the species via the Beer–Lambert law:

$$A_{260} = \varepsilon_{260} Cl$$

The molar extinction coefficient may be directly used for the quantitative determination of oligonucleotide or in combination with a chromatographic technique.

15.7.1 Direct Determination

Direct determination of an assay of a lyophilized oligonucleotide (ODN) requires knowledge of a molar extinction coefficient of the highest purity reference standard. A detailed account of the determination of a molar extinction coefficient is provided in Chapter 12. For direct assay of ODN, it is necessary to correct for moisture and purity of the compound. If the ODN is a duplex, such as a siRNA, the duplex in native form will invariably have a slight excess of single strand impurity. The excess single-strand impurity is determined by SEC. Refer to Chapter 3 for more on SEC. An accurate determination of assay of siRNA will warrant a correction for moisture and SEC purity.

15.7.2 Experimental Procedure

Weigh 20 ± 2 mg of duplex and quantitatively transfer into a 100 mL volumetric flask with 0.9% saline. Add 0.9% saline to 100 mL mark and mix well. From this stock solution, transfer 1.0 mL into 10 mL volumetric flask and make up to 10 mL mark with 0.9% saline. Perform this dilution

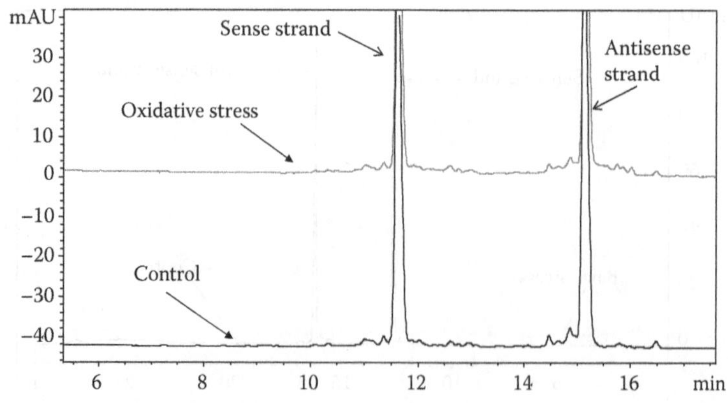

FIGURE 15.3 Overlay of control sample and oxidative stressed sample incubated for 21 hours at ambient condition. See Figure 15.1 for chromatographic conditions.

TABLE 15.5
Mass Balance of Oxidative Degradation as a Function of Incubation Time

Incubation Time	Sense Strand (Area-Percent)	Sense % Degradation	Antisense Strand (Area-Percent)	Antisense % Degradation	Total Area Count
Control	44.0	N/A	43.8	N/A	1980
45 min	43.4	1.4	43.4	0.9	1957
4.0 hours	42.7	3.0	42.7	2.5	1626
21.0 hours	41.7	5.2	38.3	12.6	1804
72.0 hours	40.4	8.2	36.0	15.8	1723

in triplicate. Mix well and determine the UV absorbance of the diluted solutions at 260 nm using a UV spectrophotometer set at 25°C with a cuvette of 1 cm path length. Sample weighing for moisture determination of the duplex by Karl Fischer titration should be performed at the same time. This will reduce the error associated with moisture absorption of the duplex under different humidity conditions.

15.7.3 Assay Calculations

Consider a duplex of molecular weight 14420.6 with a molar extinction coefficient ε_{260} of 303,603 $M^{-1}cm^{-1}$ measured in 0.9% saline. A 21.2 mg aliquot of this duplex was transferred into a 100 mL volumetric flask and dissolved in 0.9% saline. From this stock solution, 1 mL aliquots, in triplicate, were transferred to separate 10 mL flasks and diluted to volume with saline. UV measurements of triplicate dilutions were 0.339, 0.337, and 0.338 AU, giving a mean absorbance of 0.338 AU. The moisture content of the duplex was 6.3% by KF titration.

$$\text{Moisture corrected duplex weight} = \text{sample weight} \times (100 - \%\text{-moisture content})/100$$

$$\text{Moisture corrected duplex weight} = 21.2 \times (100 - 6.3)/100$$

$$= 19.9 \text{ mg}$$

$$\text{Concentration of stock solution} = \text{DF} \times A/\varepsilon_{260}$$

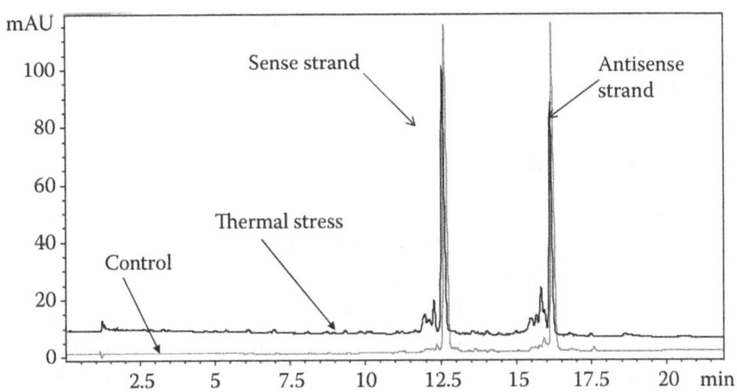

FIGURE 15.4 Overlay of control sample and thermally stressed sample incubated for 23 hours at 75°C. Refer to Figure 15.1 for chromatographic conditions.

TABLE 15.6
Mass Balance of Thermal Degradation as a Function of Incubation Time

Incubation Time	Sense Strand (Area-Percent)	Sense % Degradation	Antisense Strand (Area-Percent)	Antisense % Degradation	Total Area Count
Control	43.3	N/A	42.9	N/A	1985
2 hours	42.1	2.8	41.2	4.0	1966
3.5 hours	42.0	3.0	40.5	5.6	1950
4 hours	41.4	4.4	39.7	7.5	2014
23 hours	35.4	18.2	28.6	33.3	1905
64 hours	21.8	49.7	11.8	72.5	1918

where DF is the dilution factor, A is the absorbance of the diluted duplex solution, and ε_{260} is the molar extinction coefficient at 260 nm. In this example, 1 mL was diluted to 10 mL, DF = 10.

$$\text{Concentration of stock solution} = 10 \times 0.338/303603$$
$$= 1.11\ \text{E}^{-05}\ \text{M}$$
$$= 1.11\ \text{E}^{-05}\ \text{M} \times 14420.6\ \text{mg/mL}$$
$$= 0.16\ \text{mg/mL}$$

$$\text{Sample stock solution} = 19.9/100 = 0.199\ \text{mg/mL}$$

$$\text{Assay of the duplex} = 100 * 0.16/0.199 = 80.4\%\ (\text{wt/wt})$$

The net assay of the drug substance is obtained by making a correction for SEC purity. For example, if the duplex has an SEC purity of 97.0%, then the final assay of the duplex is 80.4 × (97.0/100) = 78.0%.

15.7.4 Merits of Direct Determination

The advantage of the above direct determination of assay is that this method is simple and less time consuming. However, it is not very sensitive for degradation determinations of a drug substance. In the case of using UV alone to assess total oligonucleotide content, the degradation products also tend to absorb at the detection wavelength of 260 nm. If one multiplies the total oligonucleotide content, as determined by UV, by the SEC area-percent, the resultant assay value, although closer to a true value, is still not a responsive indicator of stability, as the SEC method in itself is not very sensitive to chemical changes in the duplex.

15.7.5 Chromatographic Determination

Chromatographic methods developed should achieve a suitably retained peak with a retention factor or capacity factor, k', of 4–10. This range allows a suitable time space in the chromatogram for degradation products to elute before or after the active (major) peak. In the analysis of oligonucleotides, in addition to the retention factor, one should consider selectivity of the method relative to major single strands of interest in the duplex. Size exclusion chromatography is the only method available that maintains the integrity of the duplex. However, its utility as a stability indicating method is limited to only single duplexes.

The complexity of quality control analysis is increased by the use of multiple oligonucleotides in a single pharmaceutical product for the simultaneous silencing of multiple genes. Examples include

TABLE 15.7
Sequences of siRNA Used in This Study[a]

Target	Strand Type	Sequence	Molecular Weight
KSP	Sense (A-19562)	5′ucGAGAAucuAAAcuAAcuT*T3′	6764.4
KSP	Antisense (A-19563)	5′AGUuAGUUuAGAUUCUCGAT*T3′	6677.1
VEGF	Sense (A-3981)	5′GcAcAuAGGAGAGAuGAGCU*U3′	6871.3
VEGF	Antisense (A-3982)	5′AAGCUcAUCUCUCCuAuGuGCu*G3′	7307.5

[a] The lowercase letters in the sequence represent 2′-*O*-methyl-modified nucleotides; asterisks near the 3′-end represent a phosphorothioate linkage.

the use of two, single-strand, antisense oligonucleotides in a sterile solution for pulmonary applications and the use of two siRNA duplex oligonucleotides in a liposome formulation.[22,23] The analysis of a drug substance, consisting of two or more duplexes, by quantitative SEC is not always possible. In these cases, denaturing methods such as AEX-HPLC or IP-RP-HPLC is a reasonable option for analysis of multiple duplexes in a formulation. A case study of a drug substance consisting of two siRNA duplexes in a 1:1 molar ratio is used to illustrate a stability indicating AEX-HPLC method.

15.7.6 Mixed siRNA Duplexes

Two duplexes selected for this illustration are ALN-12115 and ALN-3133. The ALN-12115 consists of sense (A-19562) and antisense (A-19563) strands, and ALN-3133 consists of sense (A-3981) and antisense (A-3982) strands as given in Table 15.7. When four single strands are mixed, it is possible *a priori* to have six nonidentical pairs of association. Integrity of the duplexes in the mixture can be established by determining pair wise T_m determination. Refer to Table 15.8 for the results. It was found that only two definite duplexes exist in the mixture as indicated by well-defined inflections for T_m as shown in Figure 15.5. The duplex, ALN-12115, has a T_m value of 67.8°C, and ALN-3133 has a T_m value of 81.3°C. T_m values indicate the stability of the duplexes that gives insight for sequence specific method development. Refer to Chapter 6 for more on T_m.

15.7.7 Denaturing AEX-HPLC of siRNA

Denaturation of duplexes of ODNs into two single strands is achieved by an informed selection of temperature, pH, and columns and by optimization of the selected method. In general, combinations

TABLE 15.8
Summary of Cross Hybridization T_m Data

Sequence Combination	T_m	Comments
A-3981 sense strand	45.1°C	Hairpin structure
A-3982 antisense strand	48.1°C	Hairpin structure
A-19562 sense strand	No inflection	N/A (not applicable)
A-19563 antisense strand	No inflection	N/A
A-3981+A-19562	No inflection	N/A
A-3981+A-19563	No inflection	N/A
A-3982+A-19562	No inflection	N/A
A-3982+A-19563	No inflection	N/A
A-3981+A-3982	81.3°C	Formation of duplex ALN-3133
A-19562+A-19563	67.8°C	Formation of duplex ALN-12115

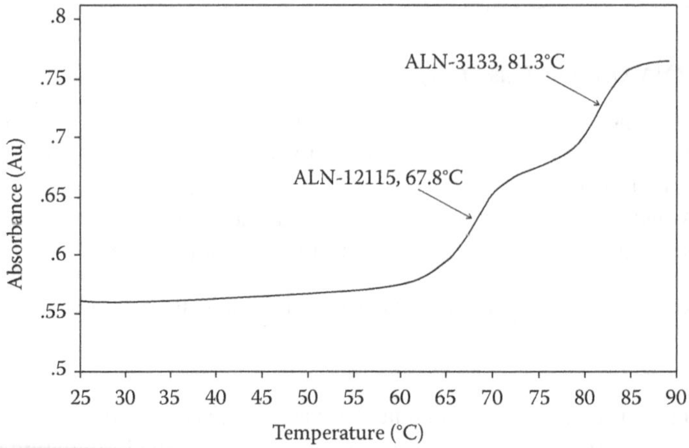

FIGURE 15.5 T_m trace of a mixture of all four single strands resulting in two well-defined inflexions matching the T_m values of duplexes ALN-12115 and ALN-3133.

of low pH with high temperature or a high pH with low temperature are used. Two columns suitable for testing these methods are the Dionex DNAPac PA-100 and the Dionex DNAPac PA-200. An evaluation of these columns for the dual duplex assay is provided below to illustrate method and column selection.

15.7.8 AEX-HPLC with PA-200 Column at 75°C and pH 8

As shown in Figure 15.6a, the duplex ALN-12115, which has a T_m of 67.8°C, is readily denatured into the corresponding single strands and gives rise to baseline resolved peaks. The ALN-3133 has a T_m of 81.3°C and, because of its inherent thermodynamic stability, it is not denatured at 75°C under the pH 8 chromatographic conditions; refer to Figure 15.6b. Two duplexes are combined in a 1:1 molar ratio to form the drug substance, ALN-VSPDS01. As shown in Figure 15.6c, the chromatographic conditions are not suitable for a stability indicating method. Further optimization of the gradient conditions did not improve the resolution. All four single strands, belonging to the drug substance, should be well resolved to monitor degradation peaks during a stability program.

15.7.9 AEX-HPLC with PA-200 Column at 35°C and pH 11

As shown in Figure 15.7a, the duplex ALN-12115 is denatured into its corresponding main single strands. The peaks are baseline resolved. As shown in Figure 15.7b, the duplex ALN-3133 is denatured into its corresponding main single strands. The peaks are baseline resolved as well. However, as shown in Figure 15.7c, when both duplexes are present in ALN-VSPDS01, two of the peaks co-elute and result in three peaks instead of the expected four peaks. This chromatographic condition is also not suitable as a stability indicating method.

15.7.10 AEX-HPLC with PA-100 Column at 35°C and pH 11

As shown in Figure 15.8a, the duplex ALN-12115 is denatured into its corresponding main single strands. The peaks are baseline resolved. As shown in Figure 15.8b, the duplex ALN-3133 is denatured into its corresponding main single strands. The peaks are baseline resolved as well. As shown in Figure 15.8c, when both duplexes are present in ALN-VSPDS01, all four peaks are baseline resolved and sufficiently separated from one another to provide a stability indicating method.

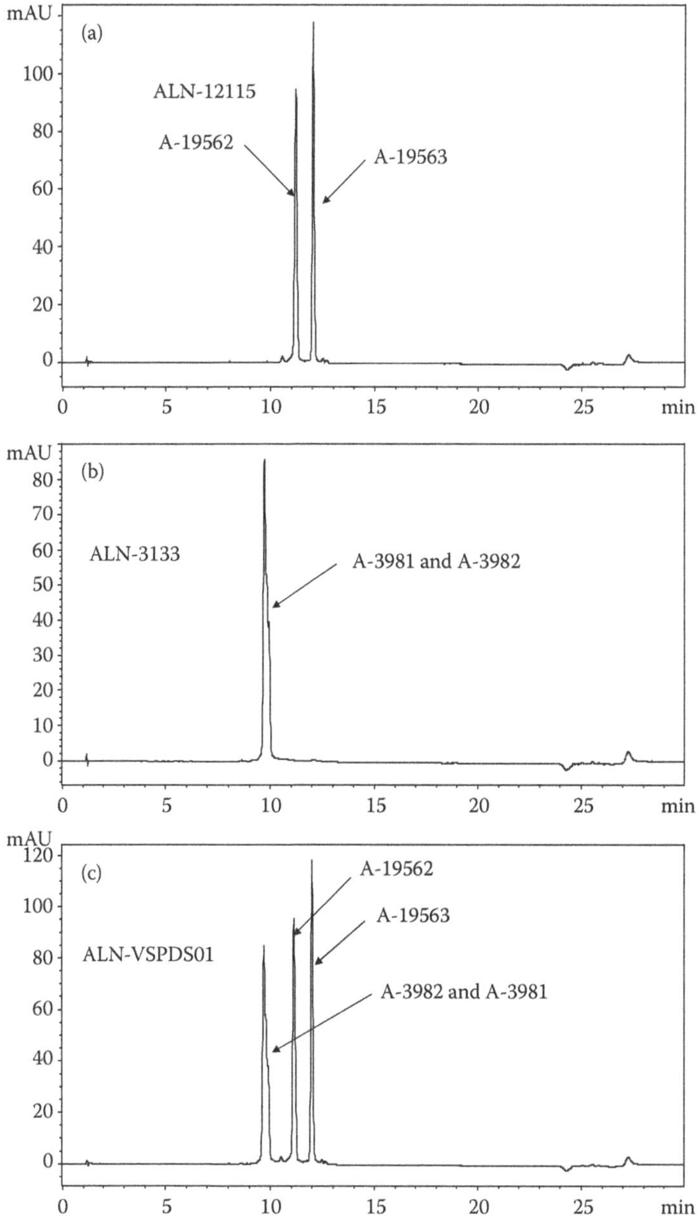

FIGURE 15.6 (a) Chromatogram of ALN-12115, (b) chromatogram of ALN-3133, and (c) chromatogram of ALN-VSPDS01. System: Agilent 1100 HPLC; column: Dionex DNAPac PA-200; column temperature: 75°C; buffer A: 10% CH_3CN, 25 mM Tris-HCl buffer pH at 8; buffer B: 10% CH_3CN, 25 mM Tris-HCl buffer pH at 8 and 1 M NaBr; gradient: 30–60% B in 22 min; flow rate 1.0 mL/min and detection at 260 nm. Sequence information given in Table 15.7.

15.7.11 Chromatographic Assay of a Duplex

In order to determine the concentration of a duplex in a formulation or in a lyophilized drug substance, it is necessary to have a high-purity reference standard of the same sequence. These reference standards are utilized to construct calibration curves for quantitation. Because the oligonucleotides are moisture sensitive and disperse under electrostatic conditions, it is difficult to weigh small

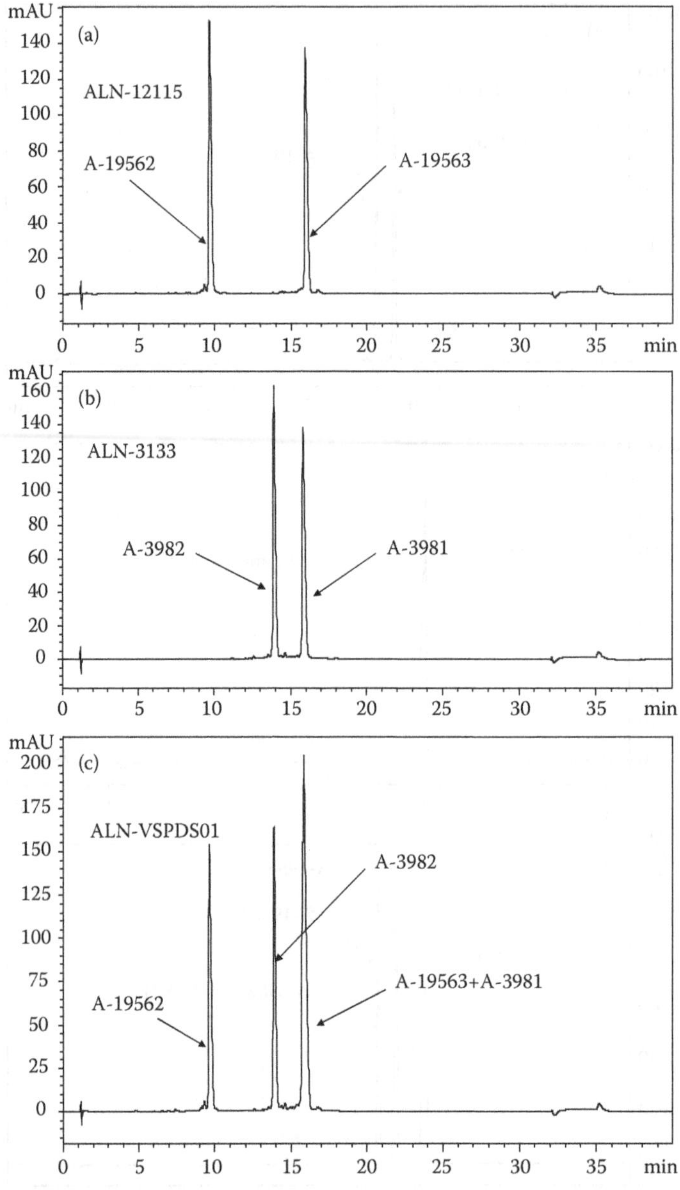

FIGURE 15.7 (a) Chromatogram of ALN-12115, (b) chromatogram of ALN-3133, and (c) chromatogram of ALN-VSPDS01. System: Agilent 1100 HPLC; column: Dionex DNAPac PA-200; column temperature: 35°C; buffer A: 20 mM sodium phosphate, 10% CH_3CN pH at 11; buffer B: 20 mM sodium phosphate, 1 M NaBr, 10% CH_3CN pH at 11; gradient: 35% B to 47% B in 13 min then to 52% B in next 17 min; flow rate 1.0 mL/min and detection at 260 nm.

quantities for accurate determination. One approach for routine analysis is to determine the molar extinction coefficient ε_{260} of the high-purity reference standard in 0.9% saline. Refer to Chapter 12 for details of ε_{260} determination. With a known value of ε_{260}, it is recommended to make a stock solution of reference standard. From a stock standard concentration, using the UV absorbance measurement at 260 nm, make aliquots equivalent to approximately 2 mg single-use vials and lyophilize them. Stability of these reference standard solutions should be monitored to ensure suitability for use over the desired period.

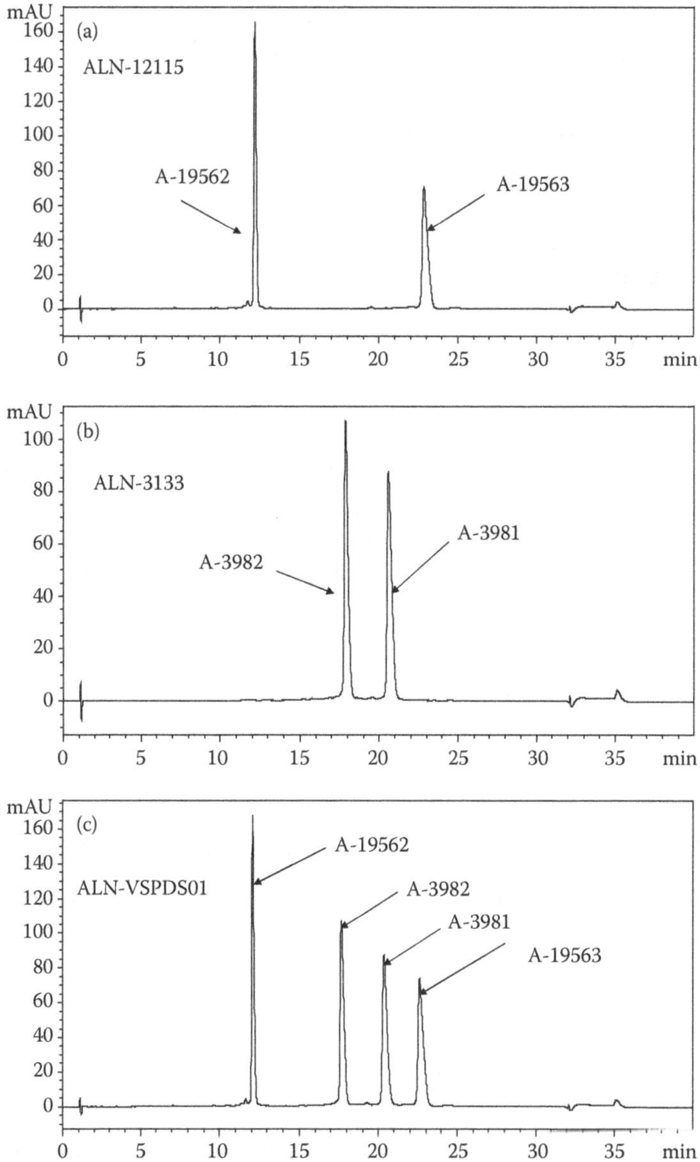

FIGURE 15.8 (a) Chromatogram of ALN-12115, (b) chromatogram of ALN-3133, and (c) chromatogram of ALN-VSPDS01. System: Agilent 1100 HPLC; column: Dionex DNAPac PA-100; column temperature: 35°C; buffer A: 20 mM sodium phosphate, 10% CH_3CN pH at 11; buffer B: 20 mM sodium phosphate, 1 M NaBr, 10% CH_3CN pH at 11; gradient: 35% B to 47% B in 13 min then to 52% B in next 17 min; flow rate 1.0 mL/min and detection at 260 nm.

15.7.12 Concentration Determination

High-purity reference standards for calibration are made as single-use vials of 2 mg of the duplex of interest. ALN-12115 is used here for the purpose of illustration. Prepare a stock standard solution by dissolving the contents of one vial (2 mg) in 1 mL of 0.9% saline and transfer to a 10 mL volumetric flask. Add sufficient saline to make up to 10 mL mark and mix well. Take one mL of this stock solution and dilute to 10 mL with 0.9% saline and measure the UV absorbance at 260 nm. From this value determine the concentration of stock standard solution of ALN-12115.

15.7.13 Calibration Curves for ALN-12115

Prepare a series of dilute standards using 750, 625, 500, 375, and 250 μL of stock standard solution and adding water to each to make up to 1000 μL final volume. Use the experimental conditions given in Figure 15.8 to analyze these calibration standards using a 25 μL injection. Construct a calibration curve of peak area of A-19562, as well as peak area of A-19563 versus concentration of ALN-12115. Response of both sense and antisense peaks is used to get an estimate of the concentration of the duplex ALN-12115.

15.7.14 Sample Analysis

To determine the ALN-12115 content in ALN-VSPDS01, weigh 30 ± 1 mg of ALN-VSPDS01 test sample and transfer to 100 mL flask and dissolve in water. Perform a secondary dilution of 3 mL in a 10 mL flask and fill to volume with water and mix well. Make triplicate injections of 25 μL of the sample solution. Find the area of the A-19562 and A-19563 peaks from the chromatographic run described in Figure 15.8.

15.7.15 Data for Calculation

Molecular weight of ALN-12115, MW = 14320.7 (as the sodium form of the duplex)

Molar extinction coefficient in 0.9% saline at 260 nm, $\varepsilon_{260} = 293998\ M^{-1}cm^{-1}$

UV absorbance at 260 nm of the diluted stock solution of ALN-12115, $A = 0.3777$ AU

$$\text{Concentration of the diluted standard solution} = (A) \times (\text{MW})/\varepsilon_{260}$$

$$= 0.3777 \times 14320.7/293998$$

$$= 0.0184\ \text{mg/mL}$$

Concentration of the stock standard solution = 10 × 0.0184 mg/mL, where 10 is the dilution factor, is equal to 0.184 mg/mL.

Data of the calibration set and peak area counts of A-19562 and A-19563 are given in Table 15.9. The results of linear regression analysis of the two plots of area of A-19562 and A-19563 versus concentration of ALN-12115 are given Table 15.10.

TABLE 15.9
Calibration Data

Volume of Standard (μL)	Volume of Water (μL)	Concentration of Standard (mg/mL)	A-19562 Area	A-19563 Area
750	250	0.138	1297	1271
625	375	0.115	1066	1051
500	500	0.092	861	850
375	625	0.069	630	633
250	750	0.146	425	416

TABLE 15.10
Regression Analysis Data

Parameter	Plot of A-19562 Area versus Concentration of ALN-12115	Plot of A-19563 Area versus Concentration of ALN-12115
Slope	9478	9252
Intercept	−16.2	−7.0
R^2	0.9996	0.9998

The moisture-corrected sample weight of ALN-VSPDS01 was 28.2 mg, which was dissolved in 100 mL water. This stock solution was diluted 3 in 10, and triplicate analyses were performed. The mean peak area of A-19562 from three injections was 430 area units. Using the slope and intercept, from the plot of A-19562 peak area versus concentration of ALN-12115, the concentration of ALN-12115 in the sample is calculated according to a linear equation as follows (with reference to data in Table 15.10):

$$\begin{aligned}[\text{ALN-12115}] &= [(\text{peak area} - \text{intercept})/\text{slope}] \times \text{DF} \\ &= [(430 - (-16.2))/9478.261] \times 10/3 \\ &= 0.157\ \text{mg/mL}\end{aligned}$$

The test sample concentration is 28.2 mg/100 mL = 0.282 mg/mL

Weight/weight assay of ALN-12115 in the test sample using A-19562 peak area response = 100 × (0.157/0.282) = 55.7%.

The above calculation is repeated with mean area response of A-19563 peak and an average of wt/wt assay is calculated.

15.8 STABILITY MONITORING OF OLIGONUCLEOTIDES

15.8.1 Additional Characterization—Hygroscopicity

Understanding the characteristics of the solid-state of oligonucleotides is necessary to determine the storage conditions to successfully execute a stability program. Most oligonucleotides in development are amorphous in nature and also have the tendency to absorb moisture during storage. Hygroscopicity is a measure of the potential for moisture uptake by oligonucleotide under thermal and relative humidity conditions. Dynamic vapor sorption (DVS) is a gravimetric vapor sorption technique using an ultra-balance.[24] An automated vapor sorption balance (DVS-1000, Surface Measurements Systems, London, UK) was used to study the water sorption behavior of the sample. The temperature of the sample chamber was maintained at 25°C. About 1.5 mg of the ALN-RSV01 sample was dried at 0% relative humidity (RH) for 10.5 hours, under dry nitrogen flow (flow rate 200 mL/min) until a constant weight was attained. The sample was then exposed to the desired RH (30% followed by 90%). The sample weight was recorded at 1 min intervals over a period of approximately 5 hours at 30% RH.

The RH was then increased to 90%, and the sample weight was recorded for approximately 12 hours. ALN-RSV01 had an initial water content of 4.5% w/w. When the sample was next exposed to 30% RH, the water content after 30 min was 12.0%. Storage at this RH for a total of 5 hours resulted in a total water content of 12.3%. When the RH was increased to 90%, the ALN-RSV01 sample rapidly sorbed water and the weight gain after 30 min was 56.5%. Storage at this RH for a total of 12 hours resulted in a weight gain of 64.3%.

15.8.2 Drug Substance Storage

The water sorption isotherm demonstrates that the drug substance is hygroscopic. In order to mitigate the potential adverse effects of hygroscopicity, it is recommended that the drug substance is packaged in wide mouth high-density polyethylene (HDPE) bottles with polypropylene screw closures. An additional precaution may be taken by placing the HDPE bottles into foil bags, which serve as a secondary moisture barrier. The experience with siRNA without extensive modification in the sequence suggests that the indicated container closure system and a storage condition of –20°C effectively limits moisture uptake to levels below 15%. Under these conditions, no significant changes in quality are expected to occur for 2 years or longer.

15.8.3 Drug Product Storage

Various approaches are taken to formulate oligonucleotides to achieve systemic delivery. Long-term studies should be undertaken in packaging that is intended for storage and distribution. Containers and closure systems used in primary stability studies of finished products should be those intended for marketing.

One simple formulation of siRNA is to use phosphate buffered saline at physiological pH conditions. Such formulations may be packaged in a vial (2 mL, 13 mm, type I clear borosilicate, treated glass) with a Teflon stopper and a 13 mm, flip-off aluminum seal.[25]

Different strategies have been used to deliver siRNA *in vivo*,[26–28] including liposomal delivery systems. Liposomes are used for intravenous delivery of several clinically approved drug products[29,30] and dramatically extend the circulation lifetime of the drug. Cationic liposomes are formed from saturated lipids with hydrophilic properties such as cholesterol, cationic lipids, and polyethylene glycol.[31–34] The container closure system for such liposomal formulations should be chosen to protect the sterile product from microbiological contamination. All components are standard items for parenteral products.

15.8.4 Stability Protocols

Stability testing is the primary tool used to assess expiration dating and storage conditions for pharmaceutical products. Many protocols have been used for stability testing, but most in the industry are now standardizing using the recommendations of the International Conference on Harmonization (ICH). These guidelines were developed as a cooperative effort between regulatory agencies and industry officials from Europe, Japan, and the United States.

Stability testing includes long-term studies, where the product is stored at room temperature and humidity conditions, as well as accelerated studies where the product is stored under conditions of high heat and humidity. Proper design, implementation, monitoring, and evaluation of the studies are crucial for obtaining useful and accurate stability data. Stability studies are linked to the establishment and assurance of safety, quality, and efficacy of the drug product from early phase development through the lifecycle of the drug product. Stability data for the drug substance are used to determine optimal storage and packaging conditions for bulk lots of the material. The stability studies for the drug product are designed to determine the expiration date (or shelf life). In order to assess stability, the appropriate physical, chemical, biological, and microbiological testing must be performed. Usually, this testing is a subset of the release testing. The SIMs discussed at the beginning of this chapter are used for chromatographic purity and "label claim" assay of the drug substance and drug products.

Stability testing must be continued throughout clinical trials to support the safety, quality, and efficacy of materials released for clinical trials. Stability data must be submitted as part of the IND filing prior to initiating the phase 1 clinical trial. Prior to the first phase 1 stability study, the preclinical studies should provide information on the appropriate long-term condition and the appropriate container/closure system. ICH Q1A provides the guidance for the design of clinical stability studies.[1] Selection of batches, the container closure system, specifications, testing frequency, and storage conditions are the most important issues to consider when designing a stability study.

The container closure system must be evaluated for compatibility with the drug substance and drug product to ensure that the container does not contribute to degradation or contamination.

The testing frequency represents the minimum data required for filing. It may be advisable to pull and test a 1-month sample for each storage condition to ensure that the study is proceeding as expected.

During phase 1, it may be necessary to evaluate multiple formulations, dosage strengths, and container closure systems. Use of bracketing and/or matrixing can frequently reduce the resource allocation for these studies. These two design approaches are discussed in ICH Q1D.[35] Bracketing uses the extremes to provide data for the entire study. For example, if dosage strengths of 10, 25, 50, and 100 mg are to be evaluated, the study may include testing of all strengths at the initial and final time points with only the 10 and 100 mg strengths being tested at the intermediate time points. Matrixing might be used to evaluate the same strength in multiple container/closure systems by selecting only certain container closure systems for testing at each time point. This selection is usually done in a random fashion.

Table 15.11 summarizes the minimum requirements for stability data at the time of an NDA submission. The studies must continue until the long-term stability study is completed for the shelf life and retest period proposed in the NDA submission. Temperature cycling studies and in-use stability studies may be needed for certain types of formulations (particularly formulations in salt buffers and liposomes). In early phase 3 studies, one should expect to be placing the batches on stability (at least three drug substance and drug product lots) that will be used for filing the NDA. These may be the validation batches if process validation is performed early enough. Process validation may be performed near the end of phase 3, but adequate stability data for these batches may not be available at the time of filing.

15.8.5 Stability Protocol for Oligonucleotide Drug Substance

For lyophilized oligonucleotide siRNA drug substances, T_m measurements are often made in early batches to ascertain whether they provide meaningful information. Samples are stored under both recommended and accelerated storage conditions and tested for appearance, assay, and impurity profile by stability-indicating methods such as SEC and AEX-HPLC or IP-RP-HPLC.

TABLE 15.11
ICH Q1A Summary of Stability Parameters

Study	Storage Condition	Minimum Time Period[a] (Months)
General case: long term	25°C ± 2°C/60% RH ± 5% RH or 30°C ± 2°C/65% RH ± 5% RH	12
General case: intermediate	30°C ± 2°C/65% RH ± 5% RH	6
General case: accelerated	40°C ± 2°C/75% RH ± 5% RH	6
Refrigeration: long term	5°C ± 3°C	12
Refrigeration: accelerated	25°C ± 2°C/60% RH ± 5% RH	6
Freezer: long term	−20°C ± 5°C	12

[a] Must cover retest or shelf life period at a minimum and includes storage, shipment, and subsequent use.

15.8.6 Stability Protocol for a Drug Product

Stability protocols for a salt-buffered oligonucleotide or a liposome formulation may include testing of vials in a normal position, as well as in an "inverted" position. In general, liposome formulations are not to be frozen. The stability storage conditions cover refrigeration (2°–8°C) for general storage, ambient temperature for handling excursions, and 40°C for accelerated degradation studies. In addition to the SIMs discussed in this chapter, analytical methods to characterize lipid components in the liposome should be performed according to the ICH guidance on liposome products.[36,37]

15.9 EVALUATION OF STABILITY DATA

Shelf life and retest periods may be determined statistically with adequate quantitative data. The guideline presented in ICH Q1E address the evaluation of stability data that should be submitted in registration applications for new molecular entities and associated drug products.[38] The guidelines provide recommendations on establishing retest periods and shelf lives for drug substances and drug products intended for storage at or below "room temperature."

A stepwise approach to stability data evaluation and when and how much extrapolation can be considered for a proposed retest period or shelf life is given in the guidance. Further, information on how to analyze long-term data for appropriate quantitative test attributes are given in the guidance. Detailed statistical analysis of stability data is beyond the scope of this chapter.[39–47]

15.10 CONCLUSION

Proper implementation of a SIM relies on three critical aspects: generation of the sample, method development, and method validation. The use of a properly designed and executed forced-degradation study will generate a representative sample that will, in turn, help to ensure that the resulting method adequately reflects long-term stability. New technology and hyphenated advanced detector techniques are extremely valuable in providing increased levels of resolution and specificity and faster analysis times resulting in higher sample throughput. Although specificity does play a central role, all of the remaining pertinent validation parameters must also be evaluated in order to properly validate a SIM in a regulated environment and ultimately ensure the method accomplishes its intended purpose. Stability testing is interwoven through the entire fabric of the drug product lifecycle. On the basis of the examples provided in this chapter, it is clear that customized method development and stability programs need to be carefully designed for a complete product development program.

REFERENCES

1. ICH Q1A: Stability testing of new drug substances and products. Paper presented at International Conference on Harmonization of Technical Requirements for the Registration of Drugs for Human Use, February 2003, in Geneva, Switzerland.
2. ICH Q1B: Stability testing: photostability testing of new drug substances and products. Paper presented at International Conference on Harmonization of Technical Requirements for the Registration of Drugs for Human Use, November 1996, in Geneva, Switzerland.
3. Guidance for Industry, Drug Stability Guidelines, U.S. Department of Health and Human Services, Food and Drug Administration, Center for Veterinary Medicine, December 9, 2008.
4. Rhodes, C., and J. T. Carstensen, eds. 2000. *Drug Stability: Principles and Practices*, vol. 107, 3rd ed. New York: Marcel Dekker.
5. Alsante, K. M., A. Ando, R. Brown, J. Ensing, T. D. Hatajik, W. Kong, and Y. Tsuda. 2007. The role of degradant profiling in active pharmaceutical ingredients and drug products. *Advanced Drug Delivery Reviews* 59: 29–37.
6. Snyder, L., J. J. Kirkland, and J. Glajch. 1997. *Practical HPLC Method Development*, 2nd ed. New York: Wiley-Interscience.

7. U.S. Pharmacopeia. 2002. 25/National Formulary 20, Chapter 1225, 2256–9. Rockville, MD: U.S. Pharmacopeial Convention.
8. ICH Q2B: Validation of analytical procedures: methodology. Paper presented at International Conference on Harmonization of Technical Requirements for the Registration of Drugs for Human Use, May 1997, Geneva, Switzerland.
9. Orr, J., M. E. Swartz, and I. S. Krull. 2003. Validation of impurity methods, Part I. *LCGC* 21(12): 1146.
10. Huber, L., and S. George. 1993. *Diode-Array Detection in High-Performance Liquid Chromatography*. New York: Marcel Dekker.
11. Warren, W. J., W. A. Stanick, M. V. Gorenstein, and P. M. Young. 1995. HPLC analysis of synthetic oligonucleotide. *BioTechniques* 18(2): 282–297.
12. Minkiewicz, P., J. Dziuba, M. Darewicz, and D. Nalecz. 2006. Application of high-performance liquid chromatography on-line with ultraviolet/visible spectroscopy in food science. *Polish Journal of Food and Nutritional Sciences* 15(56): 145–153.
13. Gorenstein, M. V., J. B. Li, and J. Van Antwerp. 1994. Detecting coeluting impurities by Spectral comparison. *LCGC* 12: 768–772.
14. Szabo, I., and M. H. Maguire. 1991. On-line recognition and quantitation of coeluting hypoxanthine and guanine in reversed-phase high-performance liquid chromatography of placental tissue extracts: photodiode-array detection and spectral analysis of coeluting peaks. *Analytical Biochemistry* 215: 253–260.
15. Wilson, I. D., J. K. Nicholson, J. Castro-Perez, et al., High resolution "ultra performance" liquid chromatography coupled to a-TOF mass spectrometry as a tool for differential metabolic pathway profiling in functional genomic studies. *Journal of Proteome Research* 4(2): 591–598.
16. Gilar, M. 2001. Analysis and purification of synthetic oligonucleotides by reversed-phase high-performance liquid chromatography with photodiode array and mass spectrometry detection. *Analytical Biochemistry* 298: 196–206.
17. Nussbaum, M. A., P. J. Jansen, and S. W. Baertschi. 2005. Role of "mass balance" in pharmaceutical stress testing. In *Pharmaceutical Stress Testing: Understanding Drug Degradation*, ed. S.W. Baertschi, 181–204. New York: Taylor and Francis.
18. Food and Drug Administration. 2000. Guidance for industry, analytical procedures and method validation, U.S. Department of Health and Human Services.
19. ICH Q2B: Validation of analytical procedures: methodology. Paper presented at International Conference on Harmonization of Technical Requirements for the Registration of Drugs for Human Use, May 1997, Geneva, Switzerland.
20. Beaucage, S. L., and R. P. Iyer. 1993. The functionalization of oligonucleotides via phosphoramidite derivatives. *Tetrahedron* 49: 1925–1963.
21. Sproat, B. S. RNA synthesis using 2′-O-(tert-butyldimethylsilyl) protection. In *Methods in Molecular Biology, Oligonucleotide Synthesis: Methods and Applications*, ed. P. Herdewijn, 17–31. Totowa, NJ: Humana Press.
22. Wang, S., Z. Shi, W. Liu, J. Jules, and X. Feng. 2006. Development and validation of vectors containing multiple siRNA expression cassettes for maximizing the efficiency of gene silencing. *BMC Biotechnology* 6: 50.
23. Pfister, E. L., L. Kennington, J. Straubhaar, et al. 2009. Five siRNAs targeting three SNPs may provide therapy for three-quarters of Huntington's disease patients, *Current Biology* 19: 774–778.
24. Hassel, R., and N. D. Hesse. Investigation of pharmaceutical stability using dynamic vapor sorption analysis, TA Instruments, 109 Lukens Drive, New Castle DE 19720, USA.
25. West Pharmaceutical Services. http://www.westpharma.com.
26. Manoharan, M. 2004. RNA interference and chemically modified small interfering RNAs. *Current Opinion in Chemical Biology* 8: 570–579.
27. Corey, D. R. 2007. Chemical modification: the key to the clinical application of RNA interference? *Journal of Clinical Investigations* 117: 3615–3622.
28. Akhtar, S., and I. F. Benter. 2007. Nonviral delivery of synthetic siRNAs in vivo, *Journal of Clinical Investigations* 117: 3623–3632.
29. Lappalainen, K., R. Miettinen, J. Kellokoski, I. Jaaskelainen, and S. Syrjanen. 1997. Intracellular distribution of oligonucleotides delivered by cationic liposomes: light electron microscopic study. *Journal of Histochemistry and Cytochemistry* 45: 265–274.
30. Fenske, D. B., and P. R. Cullis. 2005. Entrapment of small molecules and nucleic acid-based drugs in liposomes. *Methods in Enzymology* 391: 7–40.
31. Torchilin, V. P. 2005. Recent advances with liposomes as pharmaceutical carriers. *Nature Reviews of Drug Discovery* 4: 145–160.

32. Heyes, J., L. Palmer, K. Bremner, and I. MacLachlan. 2005. Cationic lipid saturation influences intracellular delivery of encapsulated nucleic acids. *Journal of Controlled Release* 107: 276–287.
33. Torcilin, V. P. 2006. Recent approaches to intracellular delivery of drugs and DNA and organelle targeting. *Annual Reviews of Biomedical Engineering* 8: 343–375.
34. Maurer, N., K. F. Wong, H. Stark, L. Louie, D. McIntosh, T. Wong, P. Scherrer, S. Semple, and P. R. Cullis. 2001. Spontaneous entrapment of polynucleotides upon electrostatic interaction with ethanol-destabilized cationic liposomes. *Biophysics Journal* 80: 2310–2326.
35. ICH Q1D. 2003. Bracketing and matrixing designs for stability testing of new drug substances and products.
36. Akinc, A., M. Goldberg, J. Qin, et al. 2009. Development of lipidoid–siRNA formulations for systemic delivery to the liver. *Molecular Therapy* 17: 872–879.
37. Food and Drug Administration. 2002. Guidance for industry on liposome drug products, 1–15.
38. ICH Q1E. 2003. ICH evaluation of stability data. Paper presented at International Conference on Harmonization of Technical Requirements for the Registration of Drugs for Human Use, Geneva, Switzerland, February 2003.
39. Carstensen, J. T. 1977. *Stability and Dating of Solid Dosage Forms: Pharmaceutics of Solids and Solid Dosage Forms*, 182–185. New York: Wiley-Interscience.
40. Ruberg, S. J., and J. W. Stegeman. 1991. Pooling data for stability studies: testing the equality of batch degradation slopes. *Biometrics* 47: 1059–1069.
41. Ruberg, S. J., and J. C. Hsu. 1992. Multiple comparison procedures for pooling batches in stability studies. *Technometrics* 34: 465–472.
42. Shao, J., and S. C. Chow. 1994. Statistical inference in stability analysis. *Biometrics* 50: 753–763.
43. Murphy, J. R., and D. Weisman. 1990. Using random slopes for estimating shelf-life. *Proceedings of American Statistical Association of the Biopharmaceutical Section* 1: 196–200.
44. Yoshioka, S., Y. Aso, and S. Kojima. 1997. Assessment of shelf-life equivalence of pharmaceutical products. *Chemical and Pharmaceutical Bulletin* 45: 1482–1484.
45. Chen, J. J., H. Ahn, and Y. Tsong. 1997. Shelf-life estimation for multifactor stability studies. *Drug Information Journal* 31: 573–587.
46. Fairweather, W., T. D. Lin, and R. Kelly. 1995. Regulatory, design, and analysis aspects of complex stability studies. *Journal of Pharmaceutical Sciences* 84: 1322–1326.
47. Shein, C.-C. 2007. *Statistical Design and Analysis of Stability Studies*. Boca Raton, FL: CRC Press.

16 Analysis by Hydrophilic Interaction Chromatography

Renee N. Easter and Patrick A. Limbach
University of Cincinnati

CONTENTS

16.1 INTRODUCTION

The separation and characterization of oligonucleotides and their synthesis side- and by-products pose several challenges for analytical method development. Traditionally, such separations have been done using reversed phase high-performance liquid chromatography (RP-HPLC) owing to the versatility of that analytical technique. Because oligonucleotides are typically highly charged anions in solution, ion-pairing reagents can be added to the RP-HPLC mobile phase to improve resolution. Alternatively, separations based on ion-exchange mechanisms, such as anion-exchange chromatography (AEX-HPLC), have also been popular and useful approaches for the separation of oligonucleotides. Both RP-HPLC and AEX-HPLC are the subject of other chapters in this handbook, and the interested reader is referred there for additional details.

In this chapter, a relatively newer HPLC approach for oligonucleotide separation, hydrophilic interaction chromatography (HILIC), is discussed. HILIC was introduced as early as 1975 as a fast and simple way to separate carbohydrates.[1] HILIC is proposed to be an improved HPLC technique for the separation of polar compounds. Traditionally, polar compounds exhibit relatively poor retention in RP-HPLC, when retained at all. As is noted in the recent reviews on HILIC, applications have primarily focused on the separation of proteins, peptides, sugars, and modified proteins such as glycosolated or phophorylated proteins. However, the high polarity of oligonucleotides suggests this class of compounds should also benefit from the HILIC technique.

The roadmap for this chapter first starts with a review of HILIC theory and operating principles. In particular, when possible we shall compare HILIC operating principles with those from the more traditional RP-HPLC. That discussion will then be followed by the few examples of HILIC separations of oligonucleotides that are available in the literature, as well as some work done in our own lab. These data will serve to illustrate the state of the art for HILIC separations of oligonucleotides at the present time and provide the basis for discussing future areas that require development. Next, we shall briefly summarize the variety of HILIC stationary phases commercially available, which might provide chromatographer's interested in HILIC of oligonucleotides a reasonable starting point for their own method development work. We then conclude this chapter by discussing the characteristics of oligonucleotides and how those do and do not bode well for HILIC-based separations coupled with mass spectrometry. In short, we hope to provide you, the reader, with some key information that would motivate your own application of this technique.

16.2 HILIC THEORY AND OPERATING PRINCIPLES

Although sometimes referred to as "reverse reversed phase chromatography" and often compared with traditional normal phase chromatography, HILIC is a unique chromatographic technique that cannot be characterized by one simple model of separation. In addition to the multiple separation mechanisms believed to be involved, a number of experimental variables including temperature, mobile phase pH, mobile phase modifiers, and stationary phase functional groups all play important roles in determining the success (or not) of HILIC-based separations.

HILIC is a chromatographic technique that is based on the interaction between the analyte of interest and a hydrophilic stationary phase. One of the defining characteristics of HILIC is that water serves as the strong eluent.[2] Thus it is quite common to compare HILIC to RP-HPLC, where the organic solvent is utilized as the strong eluent—necessary to disrupt the hydrophobic interactions between the analyte and the hydrophobic stationary phase. There are two mechanisms believed to be important in HILIC: a liquid-liquid extraction mechanism occurring near the solvated stationary phase, and Columbic interactions and/or hydrogen bonding that can occur between the analyte and the stationary phase. In contrast, hydrophobic interactions related to the partitioning and adsorption of the analyte are all thought to be involved, at some level, in RP-HPLC separations.

HILIC stationary phases are typically hydrophilic materials that enable the formation of a water-rich layer at the surface of the stationary phase. This water-rich layer on the stationary phase of the column was first described by Gregor and coworkers.[3] When this layer is formed, polar analytes can partition into the water-rich layer. As the percent water of the mobile phase increases, the analytes will partition out of the water-rich layer near the stationary phase and elute from the column. This partitioning of the analyte between two liquid phases leads to a liquid-liquid extraction mechanism for analyte separation in HILIC.

To clarify what constitutes a HILIC-based separation, Alpert proposed that the following conditions must both be present in the separation:[2] water is the stronger eluent and retention must be by partitioning. To differentiate between a separation involving partitioning and one that also involves surface adsorption, like in RP-HPLC, Equations 16.1 and 16.2 can be used.

$$\log k' = \log k'_{w} - S\varphi \tag{16.1}$$

where k'_w is the capacity factor for the weaker eluent, φ is the volume fraction (concentration) of the stronger eluent in the binary mobile phase mixture, and S is the slope obtained when $\log k'$ is plotted against φ and then fitted to a linear regression model.[1]

$$\log k' = \log k'_{B} - A_s/\eta_B \log N_B \tag{16.2}$$

where $\log k'_B$ is the solute retention factor with pure B, A_s and η_B are the cross-sectional areas occupied by the solute molecules on the surface and the B molecules, respectively, and N_B is the mole fraction of the stronger eluent.[1]

Equation 16.1 is traditionally used to describe the partitioning mechanism as found in RP-HPLC. Equation 16.2 relates retention to mole fraction, X_b, of the stronger eluent in normal phase chromatography. By using both equations and plotting $\log k'$ versus the linear and logarithmical function of

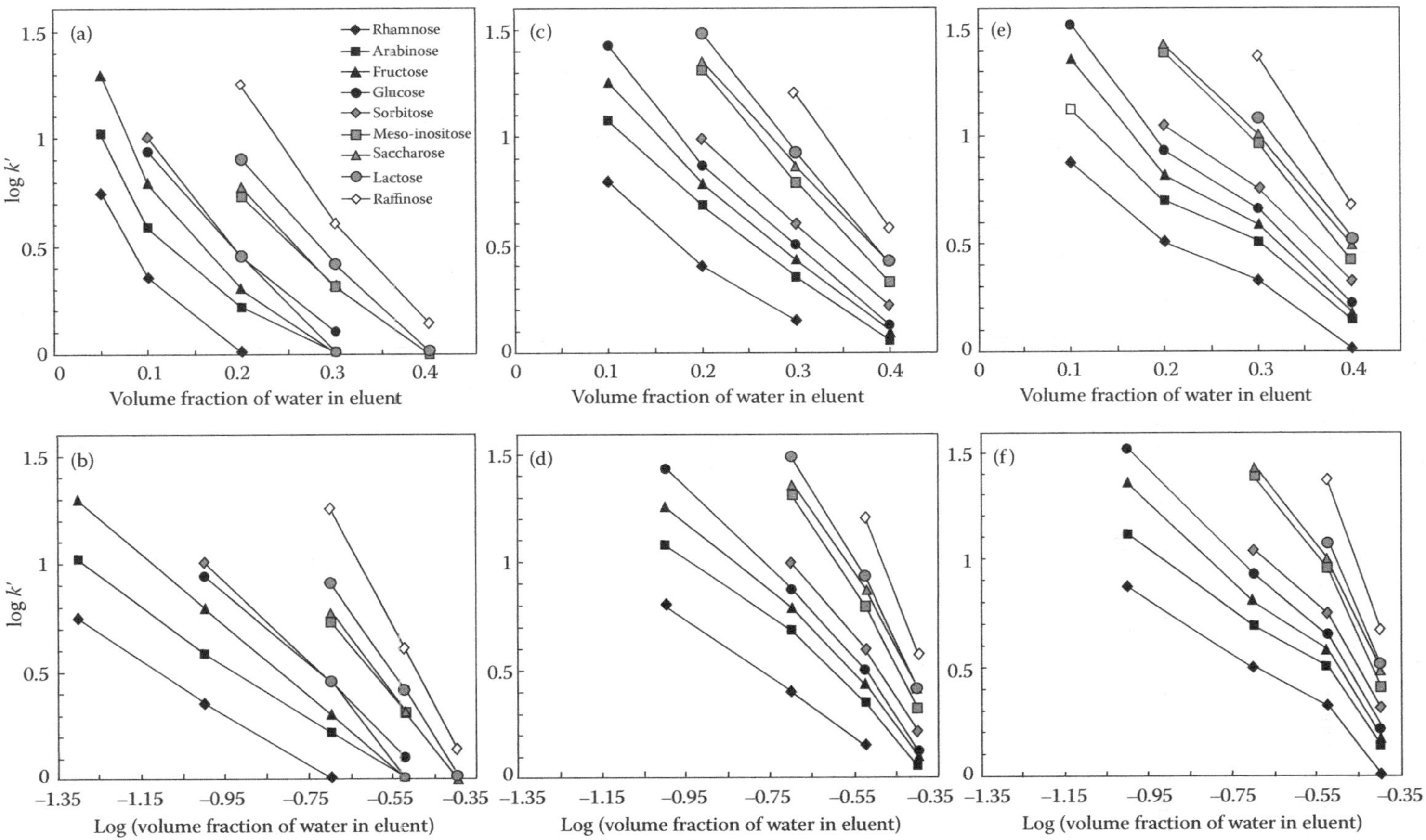

FIGURE 16.1 Dependence of log k' on the water contents in ACN/water eluents on LiChrosorb Si 100 reacted with (a + b) amino-, (c + d) diamino-, and (e + f) triaminosilane, all plotted on the same scale. The upper plots (a, c, and e) were made from data extracted from Fig. 1 of Ref. 16, whereas the lower curves (b, d, and f) contain the same data plotted against the logarithm of the water fraction in the eluent. The linear fit of the log-log model suggest an adsorptive mechanism up to 30% water content. Above 30% the column appears to become saturated with water and a partitioning mechanism begins to take hold. (From Hemstrom, P. and K. Irgum, *Journal of Separation Science*, 29, 1784–1821, 2006. Copyright Wiley-VCH Verlag GmbH & Co. KGaA. Reproduced with permission.)

the water content of the binary mobile phase used in a HILIC separation, a partitioning or adsorptive mechanism can be determined for that separation (Figure 16.1). If it is a partition mechanism, then the term HILIC can properly be used. However, in practice, the term HILIC mixed mode is usually used to show that both mechanisms have a hand in the separation.

HILIC separations can be performed by isocratic elution or by a gradient elution between the organic and aqueous mobile phases. Typically, the organic mobile phase is acetonitrile owing to its good miscibility with water and low viscosity, which leads to lower column pressure. However,

TABLE 16.1
Commercial HILIC Columns Listing Column Characteristics and Specifications

Column	Pore Diameter	Pore Size (μm)	Length (cm)	Stationary Phase	Functional Group	Advantages
Epic HILIC HC	0.5–15 mm	3, 5	1–60	Silica	Polyhydroxylated polymer	High capacity
Epic HILIC RP	120 Å	3, 5	1–60	Silica	Polyhydroxylated polymer and ODS groups	Combination of HILIC and RP
Epic HILIC Pl	120 Å	3, 5	1–60	Silica	Aromatic amine	Polar amine compounds
Epic HILIC Fl	120 Å	3, 5	1–60	Silica	Fluorinated based groups	Separates halogenated, polar amines, and aromatic compounds
Luna HILIC	200 Å	3, 5	5–25	Silica	Cross-linked diol	Provides a hydrophilic and neutral stationary phase
Venusil	100 Å	5	3–25	Silica	Urea group	Good for small molecules
Obelisc N	100 Å	2, 5, 10	1–25	Silica	Charged silica pores	Combination of ion exchange, HILIC, and NP
Polyhydroxyethyl A	60–300 Å	5	3.5–20	Silica	Polyhydroxyethyl A	Uses less organic solvent
XBridge HILIC	130 Å	2.5, 3.5, 5	1–60	Silica	BEH (bis(triethoxysilyl) ethane monomer)	Improved peak shape, retention, and column life
Atlantis	100 Å	3	10	Silica	none	Longer column lifetime
Acquity HILIC UPLC	130 Å	1.7	3–15	Silica	BEH AMIDE	Wide pH range
zic-HILIC	100, 200 Å	3.5, 5	3–15	Silica	Polymeric resin	Zwitterionic surface
zic-pHILIC	100 Å	5	5–15	Porous polymer particles	Polymeric resin	Higher pH range
Halo HILIC	90 Å	2.7	5	Silica	Fused core technology	Rapid separation time
TSK-gel Amide 80	80 Å	3, 5, 10	5–30	Silica	Amide functional group	Novel H-bonding mechanism

because of the scarcity and expense of acetonitrile, other organic solvents such as acetone, isopropanol, propanol, ethanol, dioxane, and dimethylformamide have also been used as the organic component. Ammonium formate and ammonium acetate are the two most common salts used in the aqueous phase owing to their excellent solubility in both the organic and aqueous components and their suitability with mass spectrometry. Buffer concentrations typically range from 5 to 20 mM in order to maintain a good solubility of salt in the mobile phase. Phosphate buffers have low solubility and can cause precipitation on the column bed. Ion-pairing agents such as trifluoroacetic acid are typically avoided during HILIC separation of peptides because they decrease the partitioning mechanism, thus creating a reverse phase-based separation of the analyte. Moreover, such ion-pairing agents have also been shown to decrease the sensitivity when mass spectrometry is used as the detection method.

Most manufacturers recommend that at least 60% of the organic component be maintained during separation and at least 5% of the aqueous component be maintained as well. Retention is significantly affected by the organic content of the mobile phase. Most gradient separations have a starting point of 90% organic/10% aqueous and go down to the recommended 60% organic. If there is no analyte retention under gradient conditions, then isocratic elution can be investigated. For isocratic separations, the typical mobile phase is composed of 75% acetonitrile and 25% aqueous buffer. The pH of the buffer is chosen based on the analyte class being separated, and it is important to note that analyte separation in HILIC is sensitive to even slight changes in pH.

16.3 HILIC TECHNIQUES

Published methods using HILIC for the separation of oligonucleotides are rare. This lack of publications arises, in large part, owing to the well-established success of methods based on RP-HPLC (with or without ion-pairing agents) or AEX-HPLC. However, as detailed in Section 16.4, there arises a more recent motivation for exploring HILIC separation of oligonucleotides especially in the context of developing a HILIC-based LC-MS approach for separation and characterization of oligonucleotide samples. Thus, here we shall review those applications of HILIC for the separation of oligonucleotides.

A significant number of HPLC column vendors now offer columns that can be used in HILIC (Table 16.1). To date, the only stationary phases that have been used for the separation of oligonucleotides are the cross-linked diol, polyhydroxyethyl A, and amide. Results obtained from those stationary phases are discussed below.

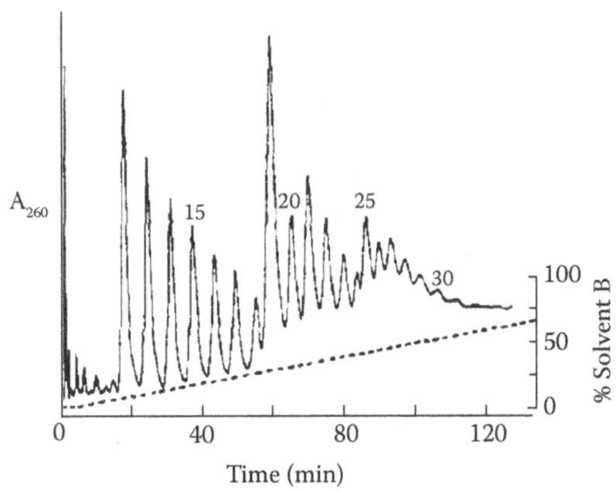

FIGURE 16.2 Separation of oligonucleotides ranging from 12- to 30-mers. Separation was performed under gradient conditions using a polyhydroxyethyl A column. Flow rate was 1 mL min^{-1}. (Reprinted from Alpert, A. J., *Journal of Chromatography,* 499, 177–196. Copyright 1990, with permission from Elsevier.)

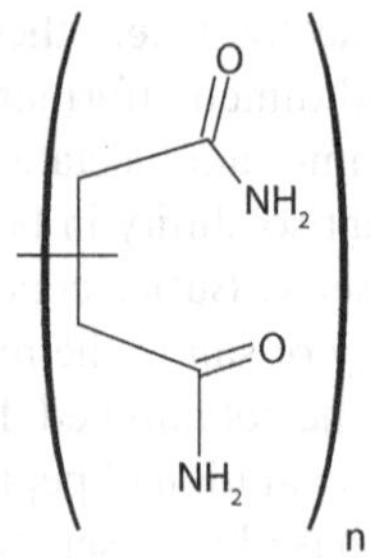

FIGURE 16.3 Structure of polyhydroxyethyl A.

16.3.1 Separation of Oligonucleotides on Polyhydroxyethyl A Column

Alpert[2] was able to separate oligonucleotides (polythymidylic acid) ranging from 12-mers to 30-mers as seen in Figure 16.2. The separation was done on a polyhydroxyethyl A column (3 µm particle size, 100 × 4.6 mm). The structure of polyhydroxyethyl A (Figure 16.3) provides a neutral material for the separation of a wide range of analytes. It also promotes sharper peaks with less organic solvent. Eluent A was 40 mM triethylammonium phosphate (TEAP, pH 5.0) with 70% acetonitrile, and eluent B was 75 mM TEAP with 85% acetonitrile. The oligonucleotides were separated under gradient conditions of 0–100% B in 200 min.

From the separation, Alpert was able to conclude that the retention increased as the size of the oligonucleotide increased and retention was also very sensitive to ionic strength of the mobile phase. Different bases also caused differences in retention. The retention of adenine was 3 times greater than the retention for thymidine. This separation of oligonucleotides and other analytes from his work allowed him to hypothesize about the mechanism that controlled HILIC separations. While this publication provides the historical precedent for using HILIC to separate oligonucleotides, until recently no further studies resulted that followed up on this approach or that explored in more detail whether a HILIC mechanism for oligonucleotide separation would have analytical advantages.

16.3.2 Separation of Oligonucleotides Using Capillary Monolithic Columns

The first contemporary report of oligonucleotide separations using HILIC arrived in 2006. Holdšvendová and coworkers tested hydroxymethyl methacrylate-based monolithic columns that they designed to separate oligonucleotides by capillary HILIC.[4] The authors were able to prepare the columns by using a monomer mixture of *N*-(hydroxymethyl) methycrylamide (HMMAA) and ethylene dimethacrylate (EDMA) (Figure 16.4). The monomer mixture was optimized by testing three different porogenic systems consisting of different ratios of propane-1-ol and butane 1,4-diole (Table 16.2).

They demonstrated the successful separation of three oligonucleotides (15-, 19-, and 20-mer) using monolithic columns prepared in this manner. The separation was performed under gradient conditions using 100 mM triethylammonium acetate (TEAA) in water (eluent A) and 100 mM

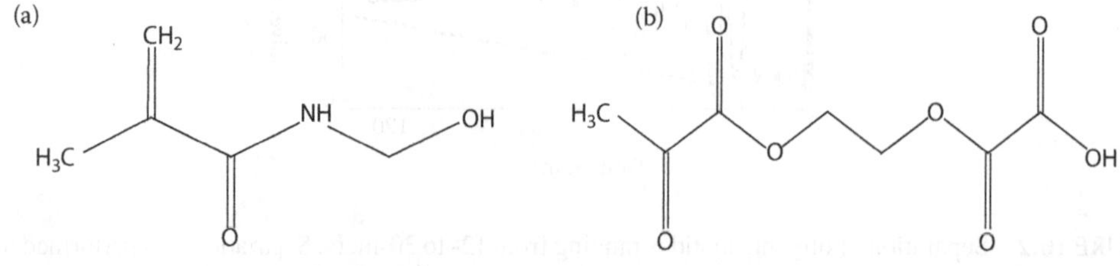

FIGURE 16.4 Structures of (a) HMMAA and (b) EDMA.

TABLE 16.2
Column Ratios for the Three Prepared Monolithic Columns[a]

	Column 1	Column 2	Column 3
Butane-1,4-diole	24.0	35.8	7.2
Propane-1-ol	47.6	35.8	64.4
EDMA	12.4	12.4	12.4
HMMAA	16.0	16.0	16.0

Source: Reprinted from Holdsvendova, P. et al., *Journal of Biochemical and Biophysical Methods*, 70, 23–29. Copyright 2007, with permission from Elsevier.

[a] The HMMAA and EDMA were kept constant while the ratio of butane-1,4-diole and propane-1-ol were varied.

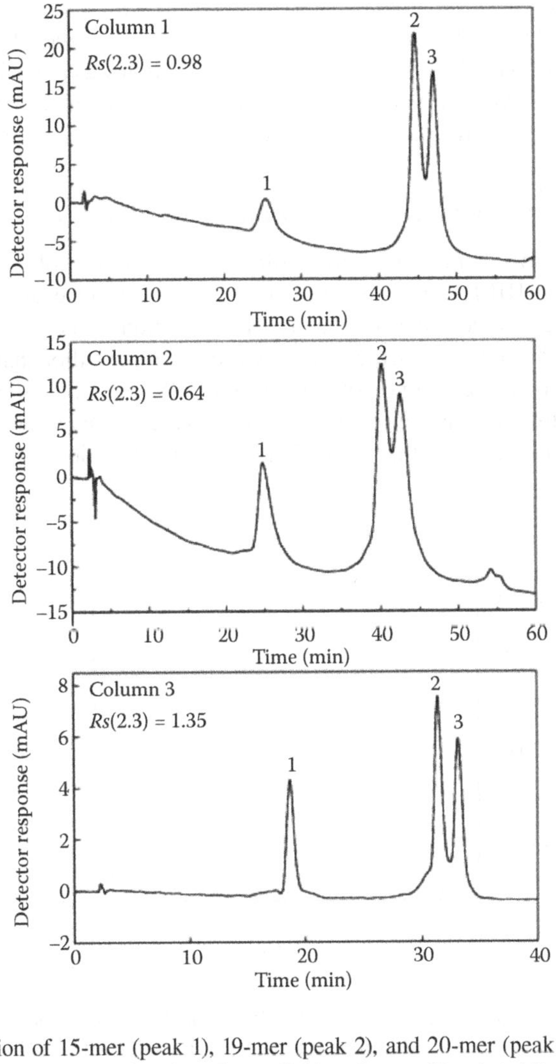

FIGURE 16.5 Separation of 15-mer (peak 1), 19-mer (peak 2), and 20-mer (peak 3) oligonucleotides using three different hydroxymethylmethacrylate-based monolithic columns. (Reprinted from Holdsvendova, P. et al., *Journal of Biochemical and Biophysical Methods*, 70, 23–29. Copyright 2007, with permission from Elsevier.)

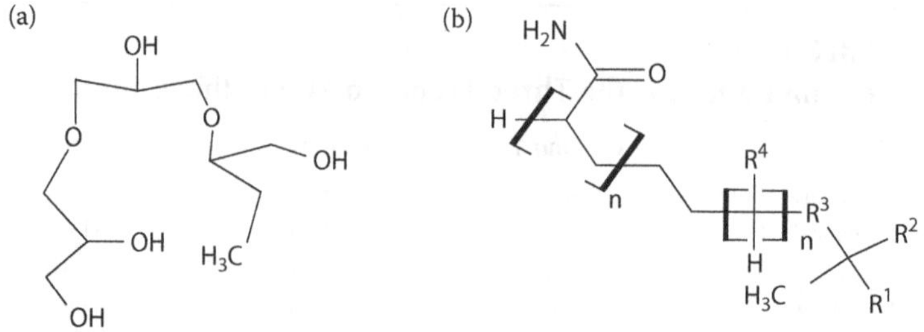

FIGURE 16.6 Structures of (a) cross-linked diol and (b) TSK-gel amide 80.

TEAA in acetonitrile (eluent B). The gradient started at 85% B and linearly decreased at 0.25% min^{-1} for column 1, which was prepared using 24% butane-1,4-diole and 47.6% propane-1-ol. For column 2, prepared with 35.8% butane-1,4-diole and 35.8% propane-1-ol, the gradient started at 85% B with a linear decrease at 0.33% min^{-1}. With column 3, prepared with only 7.2% butane-1,4-diole and 64.4% propane-1-ol, the gradient started at 94% B with a linear decrease at 0.12% min^{-1} (Figure 16.5). Column 3 offered nearly base line separation between the 19- and 20-mer oligonucleotides with a run time of only 35 min. This column composition was made in triplicate and the column tested with the oligonucleotides for reproducibility of the column. Relative standard deviations (%RSDs) for oligonucleotides retention times were determined to be less than 5%.

16.3.3 Separation of Oligonucleotides Using Amide Column

In our own lab, we have separated polythymidilic oligonucleotides of various lengths (10-, 15-, 20-, and 30-mer) using HILIC.[5] In this study, two columns were evaluated for their ability to separate oligonucleotides under conventional HILIC conditions. The first column was the Luna HILIC column from Phenomenex, which utilizes a cross-linked diol that can be solvated by a water layer (Figure 16.6a). The second column was the TSK-gel amide 80 column, which offers a unique hydrogen bonding mechanism (Figure 16.6b). Both columns allowed for the separation of oligonucleotide mixtures; however, the Luna HILIC column separation suffered from peak splitting

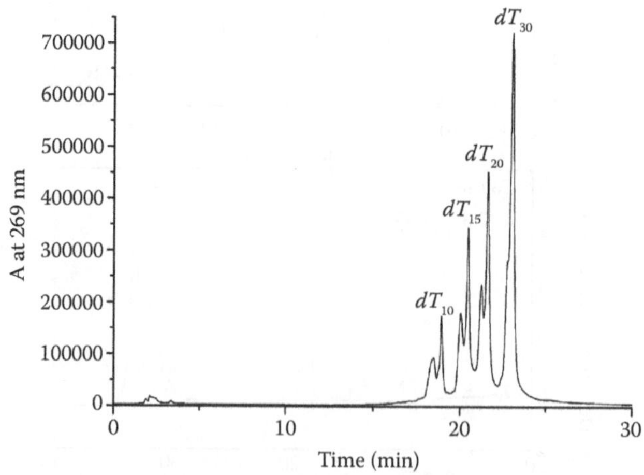

FIGURE 16.7 Separation of dT_{10}, dT_{15}, dT_{20}, and dT_{30} oligonucleotides on a Luna HILIC column under gradient conditions. Mobile phase A was 5 mM ammonium acetate in water with a pH of 5.8 and mobile phase B was acetonitrile. The gradient started at 15% A and ramped up to 30% at 6% min^{-1} followed by another ramp from 30 to 60% A at 2% min^{-1}. Column regeneration period was 40 min.

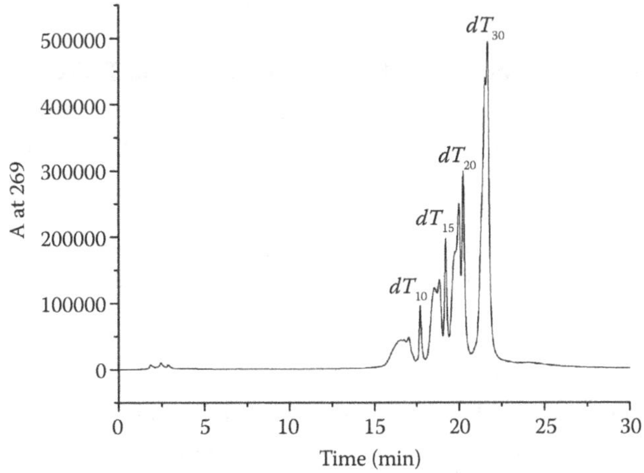

FIGURE 16.8 Separation of dT_{10}, dT_{15}, dT_{20}, and dT_{30} oligonucleotides on a Luna HILIC column under gradient conditions illustrating peak splitting problems. Mobile phase and gradient conditions are identical to those in Figure 16.7. Severe peak splitting is present for these analytes on this stationary phase during a 10 min column regeneration period.

(Figure 16.7) and long column reconditioning times in between chromatographic runs (Figure 16.8). The long reconditioning time was found to be necessary to regenerate the water layer on the surface of the cross-linked diol stationary phase. At insufficient reconditioning periods, the oligonucleotides can differentially partition into and out of the (partial) water layer at the stationary phase, leading to the problems seen during separation.

In contrast, results obtained from the TSK-gel amide column did not suffer from peak splitting, and the reconditioning period necessary to prepare the stationary phase for a subsequent injection could be as small as 10 min (Figure 16.9). Thus, we further explored the utility of this stationary phase for the reproducible separation of oligonucleotides. Figures of merit for a representative separation of polythymidilic acids using this column are provided in Table 16.3. While the resolution

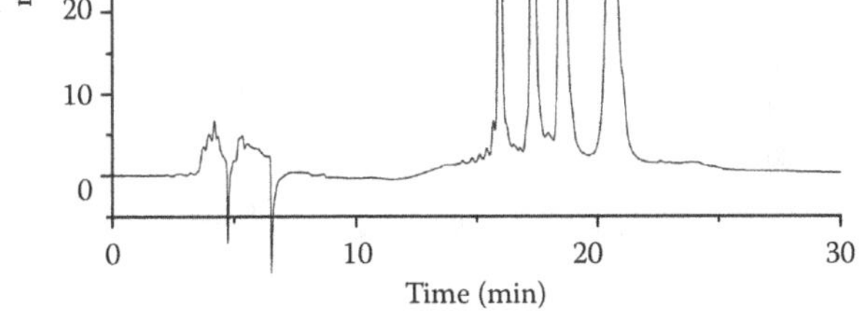

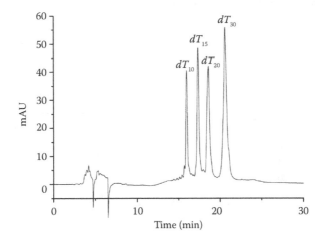

FIGURE 16.9 Separation of dT_{10}, dT_{15}, dT_{20}, and dT_{30} oligonucleotides on TSK-gel amide 80 column. Mobile phase A was 5 mM ammonium acetate in water with a pH of 5.8 and mobile phase B was acetonitrile. The gradient started at 15% A and ramped up to 30% at 6% min^{-1} followed by another ramp from 30 to 60% A at 2% min^{-1}.

TABLE 16.3
Figures of Merit for the Separation of Oligonucleotides on the TSK-Gel Amide 80 Column

Analyte	Capacity Factor	Resolution	Average Retention Time	%RSD	Average Area	%RSD
dT_{10}	3.22	58.63	16.2	1.03	89	4.55
dT_{15}	3.58	5.73	17.6	1.02	124	6.16
dT_{20}	3.91	4.06	18.8	0.94	135	1.49
dT_{30}	4.43	4.94	20.9	0.87	233	3.56

was adequate to separate this simple mixture of polythymidilic acids, further improvements in the technique will be required to handle more complex samples of oligonucleotides. However, in contrast to the crossed-diol stationary phase, this column exhibited reasonable reproducibility in the separation of this mixture of oligonucleotides.

16.3.4 Separation of Oligonucleotides Using Electrostatic Repulsion Hydrophilic Interaction Chromatography

Oligonucleotides have also been separated using a modified form of HILIC. Electrostatic repulsion hydrophilic interaction chromatography (ERLIC) is a combination of ion exchange and HILIC. This mode allows for the isocratic separation of oligonucleotides that, under typical HILIC conditions, require the use of buffers such as sodium phosphate that are incompatible with mass spectrometry. Alpert has used a cation-exchange column, polysulfoethyl A, to separate nucleotides and oligonucleotides using ERLIC.[6] The mobile phase consisted of 30 mM TEAP with 80% acetonitrile at pH 3.0.

The pH optimization studies in this work serve to illustrate the mechanism of ERLIC. At pH 6.0, the phosphate groups on the nucleotides and oligonucleotides obtain a second negative charge, which causes an electrostatic repulsion sufficient enough to limit retention between the analyte and stationary phase. As the pH is lowered, retention increases because the phosphate groups become neutralized, allowing the oligonucleotides sample to partition between the stationary phase and the

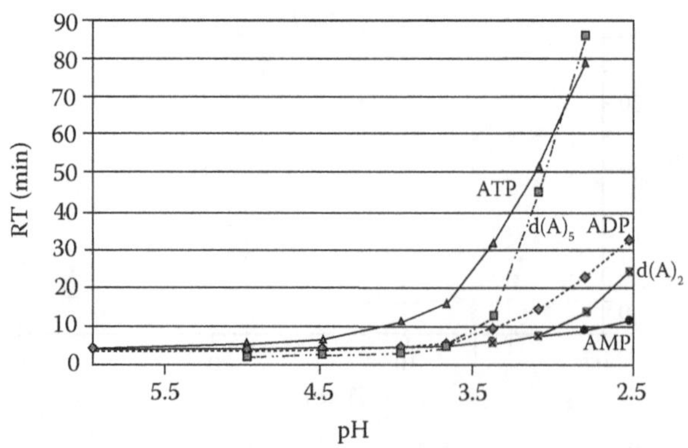

FIGURE 16.10 Effects of pH on the separation of nucleotides and oligonucleotides using ERLIC. Isocratic separation with 30 mM TEAP, pH as noted, with 80% acetonitrile. (Reprinted with permission from Alpert, A. J., *Analytical Chemistry*, 80, 62–76. Copyright 2008, American Chemical Society.)

mobile phase (Figure 16.10). ERLIC so far appears to be a promising development in the field of HILIC, with particular applications for separation of oligonucleotide-related compounds.

16.4 MOTIVATION FOR USING HILIC COUPLED WITH MASS SPECTROMETRY

The analysis of oligonucleotides and nucleic acids by mass spectrometry is a rather difficult process owing to the high molecular weight and thermal lability of these molecules. Two of the major problems in such analyses are the need for a "soft" ionization method capable of generating intact molecular ions and the limited upper mass range of most mass analyzers. In the mid 1980s, Fenn and coworkers demonstrated that electrospray ionization (ESI) could be used to analyze molecules with molecular weights (M_r) larger than the mass-to-charge ratio (*m/z*) limit of the mass analyzer.[7] Later work on oligonucleotides opened the door for accurate and high resolution analysis of these compounds by ESI-MS.[8]

A necessary requirement for ESI is that the analyte molecules be charged in solution. The negatively charged phosphate backbone of oligonucleotides allows for negative-ion mode analysis of ESI-generated ions. Transfer of ions from solution phase to the gas phase is accomplished by generating an electric field between a spraying needle, which is held at a high negative potential, and a counter electrode held at ground or a positive potential some distance from the needle. The solution being sprayed exits the needle as a conical distribution of droplets ("Taylor cone"), each containing excess negative charge. A heated drying gas, such as nitrogen, is typically used to assist evaporation of the solvent sheath from the ion. The desolvated, multiply charged ion is then introduced into the mass spectrometer for analysis.[9]

A number of applications of ESI to oligonucleotide and nucleic acid analysis have been reported since the introduction of this technique. In addition, the ESI source can be readily coupled with a number of separation techniques such as liquid chromatography,[10,11] capillary electrophoresis,[12] and capillary electrochromatography[13] for the analysis of mixtures of oligonucleotides. The most popular LC-MS method for the analysis of oligonucleotides involves the use of ion-pairing agents in the HPLC mobile phase and nonpolar stationary phases, such as C18.[14,15]

Traditionally, LC separation of oligonucleotides, especially those analyzed in trityl-off condition, has been performed using reverse phase ion-pairing chromatography. The ion-pairing agents are typically volatile salts of ammonium, with triethylammonium acetate being among the most popular due to its compatibility with mass spectrometry detection. As discussed in Chapter 1, ion-pairing agents interact with the negatively charged phosphodiester backbone, allowing for an increase in hydrophobic-based separations on a reverse phase column.

While low amounts of ion-pairing agents can be tolerated by mass spectrometry, when oligonucleotides are to be separated and analyzed by LC-MS, there is a trade-off between LC separation and MS detection. An increase in ion-pairing agent concentration will increase the resolution obtained during oligonucleotide separations, but the cost is a reduction in sensitivity for ESI-MS detection. Apffel and coworkers proposed adding HFIP to the LC mobile phase to overcome the limitation of reduced MS sensitivity. The details of HFIP effects on LC-ESI-MS analysis of oligonucleotides are described in Chapter 4. Of particular interest here, even using optimized conditions of HFIP, ion-pairing reagent, and organic modifier, the LC-ESI-MS characterization of complex mixtures of oligonucleotides and/or mixtures of longer ($n > 20$-mer) oligonucleotides is challenging due to the trade-offs between LC resolution and MS sensitivity mentioned above.

Optimal LC-MS methods provide baseline separation of components via chromatography with minimal increase in mass spectral background, ion suppression, cation adducts to analytes, and damage to the instrument. Online LC-MS analysis with ESI provides an inherent advantage by concentrating analytes as they elute from the column. However, the continuous introduction of the column eluent lends an inherent disadvantage of the coupling of these systems as buffer salts and ion-pairing agents sprayed into the instrument, especially at high amounts, increase the need for frequent cleaning, degrade instrument components, and decrease instrument performance.

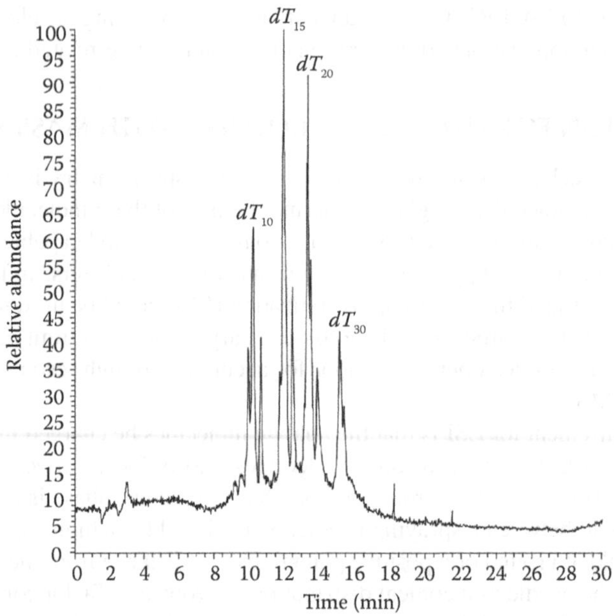

FIGURE 16.11 Online HILIC LC-ESI-MS of dT_{10}, dT_{15}, dT_{20}, and dT_{30} oligonucleotides on a Luna HILIC column. Mobile phase A was 5 mM ammonium acetate in water with a pH of 5.8 and mobile phase B was acetonitrile. The gradient started at 15% A and ramped up to 30% at 6% min^{-1} followed by another ramp from 30 to 60% A at 2% min^{-1}. The representative total ion chromatogram is shown and peak splitting, common for this stationary phase, is observed in this example.

It was noted fairly early in the development of HILIC that the use of higher relative amounts of organic modifier should provide for a separation method that would be more compatible with ESI-MS, because ESI-MS sensitivity is determined, in larger part, by the aqueous content of the sample. A larger aqueous content, because of the relatively high surface tension of water, requires smaller droplets and/or higher electrospray voltages for efficient droplet desolvation. As will be discussed below, the relatively low aqueous content in HILIC should promote more efficient droplet desolvation during ESI-MS, thus leading to improved MS sensitivity in coupled LC-MS approaches. For this reason, there is some interest in developing HILIC for oligonucleotide separations, which could possibly lead to an online LC-MS method for oligonucleotide separation and detection that offers at least comparable LC resolution to RP methods requiring ion-pairing agents, with improved MS sensitivity.

Because oligonucleotides are negatively charged biopolymers at pH values above 3, owing to the pKa of the phosphodiester backbone, for further discussion we will assume a negatively charged analyte is the starting material for all separation. As HILIC operates based on a partitioning mechanism, between the aqueous layer solvating the stationary phase and the liquid mobile phase, a true HILIC separation of oligonucleotides would require the partitioning of the sample between these two phases. Unfortunately, oligonucleotides are strongly solvated in solution, primarily to reduce electrostatic repulsive interactions between the deprotonated oligonucleotide backbone. This means that oligonucleotides naturally prefer the aqueous phase, and any partitioning mechanism would have to provide a liquid phase environment more conducive to the oligonucleotide than the liquid layer of the stationary phase. While liquid phases containing sufficiently high concentrations of salt might be suitable, as noted above, such liquid phases reduce compatibility with downstream analysis by ESI-MS. Thus, even for HILIC separation of oligonucleotides, ion-pairing agents are required to interact with the phosphodiester backbone thereby promoting a separation that can be governed by the nucleobase properties (i.e., hydrophilicity) rather than an ion-exchange mechanism.

Initial work performed in our lab coupling HILIC with electrospray ionization mass spectrometry (ESI-MS) suggests that this chromatographic approach has potential applications in this field. Figure 16.11 is a representative total ion chromatogram of a mixture of polythymidilic acids (dT_n values) that were separated and analyzed by HILIC-MS. The analytes are eluted over a narrow elution window (5 min, corresponding to a 5% change in aqueous content), and this elution window is significantly smaller than what is typically observed for the same mixture of oligonucleotides when separated using ion-pairing reversed phase conditions with HFIP. Although these results do not represent optimized conditions for separation and ESI-MS analysis, they illustrate that HILIC, when appropriately developed, will potentially be applicable to online LC-MS separation and analysis of oligonucleotides.

16.5 CONCLUSIONS

Hydrophilic interaction chromatography for the separation of oligonucleotides offers the promise of a new separation approach, which potentially could lead to a more sensitive LC-MS method. Although oligonucleotide separation by HILIC is still a new area of development, there is much that remains to be explored. Only a few of the commercially available stationary phases and columns (Table 16.1) have been investigated for their applicability for oligonucleotide separation. Further, although the few published results examining oligonucleotide separation using HILIC incorporate some type of ammonium-based salt, mobile phase conditions also deserve additional investigation and optimization. Because HILIC, when coupled with MS, does not require the use of an organic modifier such as HFIP, this approach remains an attractive option to develop for possible high sensitivity LC-MS characterization of oligonucleotides.

ACKNOWLEDGMENTS

Financial support for this work was provided by the National Science Foundation (CHE 0602413 cofunded by the MPS/CHE and BIO/IDBR Divisions, and by the MPS Office of Multidisciplinary Activities) and the University of Cincinnati.

REFERENCES

1. Hemström, P., and K. Irgum. 2006. Hydrophilic interaction chromatography. *Journal of Separation Science* 29: 1784–1821.
2. Alpert, A. J. 1990. Hyrophilic-interaction chromatography for the separation of peptides, nucleic acids and other polar compounds. *Journal of Chromatography* 499: 177–196.
3. Gregor, H. P., F. C. Collins, and M. J. Pope. 1951. Studies on ion-exchange resins. III. Diffusion of neutral molecules in a sulfonic acid cation-exchange resin. *Journal of Colloid Science* 6: 304–322.
4. Holdšvendová, P., J. Suchankova, M. Buncek, V. Backovska, and P. Coufal. 2006. Hydroxymethyl methacrylate-based monolithic columns designed for the separation of oligonucleotides in hydrophilic-interaction capillary liquid chromatography. *Journal of Biochemistry and Biophysical Methods* 70: 23–29.
5. Easter, R. N., K. K. Kröning, J. A. Caruso, and P. A. Limbach. Separation and identification of oligonucleotides by novel hydrophilic interaction liquid chromatography (HILIC) inductively coupled plasma mass spectrometry (ICPMS). Unpublished results.
6. Alpert, A. J. 2008. Electrostatic repulsion hydrophilic interaction chromatography for isocratic separation of charged solutes and selective isolation of phosphopeptides. *Analytical Chem*istry 80: 62–76.
7. Fenn, J. B., M. Mann, C. K. Meng, S. F. Wong, and C. M. Whitehouse. 1989. Electrospray ionization for mass spectrometry of large biomolecules. *Science* 246: 64–71.
8. Covey, T. R., R. F. Bonner, B. I. Shushan, and J. Henion. 1988. The determination of protein, oligonucleotide, and peptide molecular weights by ion-spray mass spectrometry. *Rapid Communications in Mass Spectrometry* 2: 249–256.
9. Gaskell, S. J.1997. Electrospray: principles and practice. *Journal of Mass Spectrometry* 32: 677–688.

10. Apffel, A., J. A. Chakel, S. Fischer, K. Lichtenwalter, and W. S. Hancock. 1997. New procedure for the use of high-performance liquid chromatography-electrospray ionization mass spectrometry for the analysis of nucleotides and oligonucleotides. *Journal of Chromatography A* 777: 3–21.
11. Apffel, A., J. A. Chakel, S. Fischer, K. Lichtenwalter, and W. S. Hancock. 1997. Analysis of oligonucleotides by HPLC-electrospray ionization mass spectrometry. *Analytical Chemistry* 69: 1320–1325.
12. Barry, J. P., J. Muth, S.-J. Law, B. L. Karger, and P. Vouros. 1996. Analysis of modified oligonucleotides by capillary electrophoresis in a polyvinylpyrrolidone matrix coupled with electrospray mass spectrometry. *Journal of Chromatography A* 732: 159–166.
13. Ding, J., and P. Vouros. 1997. Capillary electrochromatography and capillary electrochromatography—mass spectrometry for the analysis of DNA adduct mixtures. *Analytical Chemistry* 69: 379–384.
14. Huber, C. G., and H. Oberacher. 2001. Analysis of nucleic acids by on-line liquid chromatography-mass spectrometry. *Mass Spectrometry Review* 20: 310–343.
15. Oberacher, H., H. Niederstätter, B. Casetta, and W. Parson. 2006. Some guidelines for the analysis of genomic DNA by PCR-LC-ESI-MS. *Journal of the American Society of Mass Spectrometry* 17: 124–129.
16. Orth, P., and H. Engelhardt. 1982. Separation of sugars on chemically modified silica gel. *Chromatographia* 15: 91–96.

17 Determination of Base Composition

Hüseyin Aygün
BioSpring GmbH

CONTENTS

17.1 INTRODUCTION

Base composition analysis is a widely used analytical method in the fields of molecular biology and biochemistry. Erwin Chargaff first recognized that DNA base composition is characteristic for the species it is derived from. Nowadays, DNA base composition, expressed as the G/C content, is one of the most important criteria in microbial taxonomy. Also, chemically synthesized oligonucleotides are examined via base composition analysis. It not only enables determination of oligonucleotide identity, but it also gives important information about side products and process related impurities. A number of procedures are available for determining the base composition. Among these procedures, the most widely used technique includes the enzymolysis of the nucleic acid to its constituents followed by HPLC quantitation of the individual nucleosides.

17.2 APPLICATION

17.2.1 Base Composition of Genomic DNA

Base distribution within genomic DNA is a pivotal attribute for the taxonomical classification of organisms and viruses. Particularly, the G/C content plays an important role as an indicator in this context.[1–4] Almost all eukaryotic genomes contain "rare" bases (<1%) in addition to the four

TABLE 17.1
Some Examples of "Rare" Bases Found in Genomic DNA

Name	Reference
N^6-Methyl-adenine	14, 15, 7
N^4-Methylcytosine	14, 15, 7
3-Methyl-thymidine	16, 17
O^6-Methyl-deoxyguanosine	18, 11, 19, 12
7-Methyl-deoxyguanosine	11, 12
8-Oxo-7,8-dihydro-2′-deoxyguanosine	20, 21
8-Hydroxy-2′-deoxyguanosine	11, 13
8-Oxo-7,8-dihydro-2′-deoxyadenosine	20, 21

standard bases. The most common of the latter is the well-known 5-methyl-cytosine, an epigenetic modification primarily found in the so-called CpG motifs.[5,6] Apart from these bases, a number of additionally modified bases can be found by base composition analysis within genomic DNA.[7] Some of these modifications could have arisen owing to oxidative stress and thus might serve as important analytical biomarkers to determine the degree of cell damage. Current techniques for the analysis of base composition detect even femtomolar amounts of such biomarkers.[8–13] A few examples of such "rare" bases are listed in Table 17.1.

17.2.2 Analysis of Oligonucleotides

Analysis of base composition is also an important analytical tool for the ascertainment of product quality and of identity of chemically synthesized oligonucleotides.[22,33] Hydrolysis of a sequence not only enables precise identification of single base composition, but synthesis process failures and chemical modifications are also analytically more accessible after complete digestion of the oligonucleotide.[24,25] Moreover, base composition analysis can also serve as a fingerprint when comparing oligonucleotides from different production lots, the so called "interlot variation" with one another.[23]

Even though several chemical improvements have made oligonucleotide synthesis a highly efficient method these days, the oligonucleotide still comes into contact with several very reactive chemicals. Some of these chemicals are able to modify the oligonucleotide sequence in different ways. Within this context, the ability to assess accurately the degree and nature of chemical modification is especially useful for evaluation of new and existing manufacturing strategies. Examples for these types of modifications are depurinations,[26] and backbone modifications, such as the incorporation of *chloral* during detritylation[27] or of the DMT cation to incompletely oxidized P(III) species.[25] Acid-catalyzed depurination of the oligonucleotide can arise, for example, during each detritylation step (i.e., by use of trichloroacetic acid). Such apurinic sites, although not expected to survive basic treatment without strand brake, provide short, 3′-modified nucleosides after enzymolysis. In addition to a phosphate group, these nucleosides contain a sugar residue. These types of by-products are easily distinguishable from other residual nucleosides by RP-HPLC.[24] However, depurinated sequences can also fragment due to thermal stress. A side-product of this fragmentation process (4-oxo-pentenal) causes chemical modification of bases such as cytosine within the oligonucleotide sequence.[28] In addition to the impurities caused by the production process itself, reactive components within the phosphoramidites can be easily detected by analysis of base composition. A further advantage in the process of analysis method design is that modification of applied procedures is simple, thereby enabling sequencing of oligonucleotides.[23,29,30] Just recently, base composition analysis of highly modified oligonucleotides was successfully accomplished.[30] The oligonucleotides used in this process were terminally protected by inverted-abasic caps (5′ to 5′ or 3′ to 3′), contained

2′-fluoro-cytidine and 2′-fluoro-uridine in addition to several internal 2′-deoxynucleotides. A two-step incubation procedure with nuclease P1, snake venom phosphodiesterase, and calf intestinal alkaline phosphatase was used for complete digestion of the highly modified oligonucleotide. Analysis of composition was performed following chromatographic separation of individual components using a HSS C18 column (Waters) and a formiate-buffering system at pH 4.7 and 269 nm. In a further example, base composition analysis of oligonucleotides with 2′-O-methyl-modifications was also successful.[29] Oligonucleotides containing a phosphorothioate backbone pose an especially challenging problem during analysis of base composition when using enzymatic methods. An easy procedure to make these oligonucleotides accessible for analysis is to chemically convert the phosphorothioate into a phosphate. This conversion of phosphorothioates can be achieved by a chemical pre-oxidation step.[31,32] A mixture of THF/water/1-methylimidazole/iodine has been shown to be very efficient in that respect.[23,27,33,34]

The presence of residual protective groups after incomplete deprotection can also be readily detected during analysis of base composition. These groups are well tolerated by the enzymes and therefore provide nucleosides with additional exocylic amino protecting groups, such as benzoyl and isobutyryl.[17,35] Owing to their late retention time, such process related cleavage products are easily distinguishable from the other nucleosides.

Synthetic nucleic acid libraries are used nowadays in Systematic Evolution of Ligands by Exponential enrichment (SELEX) experiments in which highly variable initial pools of up to 10^{16} sequences are screened by *in vitro* selection against a target.[36–39] These libraries are randomized at several positions inside the sequence aiming to cover a wide range of variations. Special phosphoramidite-mixtures containing the individual bases in their reactive form are mostly used in the process of building such libraries. These mixtures are usually comprised in such way that during the coupling process all bases are available for an extension reaction with similar probability.[40–42] Different concentrations of individual phosphoramidites compensate for differences in reactivity, thus making them chemically “equiactive.” A simple analysis of the complete library is very desirable at this point, because a faulty distribution of the four bases could lead to a shift of the entire library. This issue can be elegantly resolved using the method of analysis as presented in the following section.

17.3 METHOD OVERVIEW

Erwin Chargaff was the first to develop micromethods for the accurate analysis of nucleobases, and hence the base composition of nucleic acids. The first studies investigating base composition used strong acids to enable complete hydrolysis.[43–46] In contrast to enzymatic hydrolysis, bases are individually released using this method. Unfortunately, incubation conditions often cause chemical alterations of the individual bases, thus hampering respective chromatographic correlation and evaluation. Considerably milder acting acids, such as formic acid, were subsequently used on the basis of these initial investigations,[9,47] thereby significantly reducing the amount of side reactions. However, this technology was not refined even after the discovery of enzymatic, more gentle methods of hydrolysis. Nonetheless, it is still applied for analysis of oxidatively modified bases.[11,12,20]

Enzymatic methods for the release of nucleosides or nucleotides dominate current procedures of base composition analysis. Hydrolysis using a nuclease (Figure 17.1) such as nuclease P1 or snake venom phosphodiesterase (SVP) yields hydrophilic nucleotides. The easiest way to analyze nucleotides is to use an anion-exchange resin as the stationary phase for HPLC.[48] Nucleosides are obtained if a phosphatase, such as bacterial alkaline phosphatase (BAP) or calf intestinal phosphatase (CIP), is employed in addition to the nuclease. The specific buffer conditions can be adjusted in a manner that both reactions occur simultaneously (“*one-pot*”). Nucleosides are most easily separated using reverse phase chromatography, such as by using a suitable C18 stationary phase.[9,35,49,50] This technique is also transferable to LC-MS, if the appropriate buffering system such as *ammonium formiate* or *triethylammonium formiate* is used.[27–30] UV detectors are routinely utilized for the

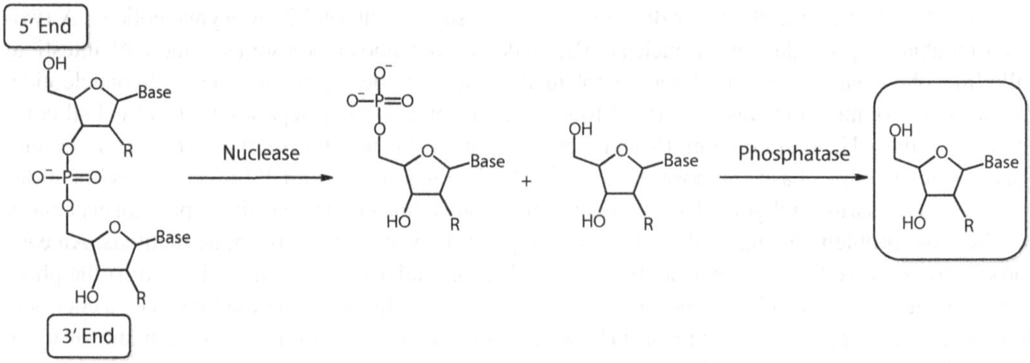

FIGURE 17.1 Release of nucleosides by enzymolysis of a dimer.

detection of nucleotides or nucleosides in HPLC. The UV detection limit for nucleotides can be further improved by use of more sensitive fluorescence detectors. Nucleotides obtained after respective digestion are specifically labeled with reactive fluorescent dyes for that purpose. Sonoki et al.[51,52] developed a respective method for taxonomical analysis of microorganisms based on the G/C content of the latter. Nucleotides produced by nuclease P1 digestion of DNA are labeled in order to enable subsequent separation and fluorescence detection via RP-HPLC. The 5-dimethylaminonaphthalene-1-[*N*-(2-aminoethyl)]sulfonamide (dansylEDA) is used for this labeling, which acts specifically at the 5′-phosphate of the individual nucleotides. This procedure offers further advantages in addition to considerably increased sensitivity. First, differences in running characteristics and peak profiles of the individual nucleotides improve after dansylEDA labeling. Second, uniform labeling balances differences of extinction coefficients, thus simplifying calculation of base composition.

Base composition of oligonucleotides can also be determined using capillary gel electrophoresis. This technique is also based on enzymatic digestion of DNA. Nucleotides produced in such a manner can then be separated via high-performance capillary gel electrophoresis (HPCGE) or via capillary zone electophoresis (CZE), and subsequently quantified.[53,54]

Irrespective of chromatographic techniques, there are so-called direct procedures for determination of base composition. The spectral properties of double stranded DNA are most frequently put to use for direct analysis. The ratio of UV absorption (260 versus 280 nm) was demonstrated to be a possible way of A/T content determination of genomic material.[55] Another spectrophotometric method deducing values such as the G/C content is based on the interaction of DNA with circular polarized light, the so-called circular dichroism.[56] In addition, determination of melting temperature (T_m) of a double-stranded DNA also enables to calculate base composition of a sample. The thermal denaturation causes the chromaticity of double stranded DNA to change around 260 nm. A simple linear equation reflects the relationship between T_m and the G/C content:[57]

$$\text{G and C base composition [\%]} = 2.44 \times (T_m - 66.0)$$

17.4 PRACTICAL CONSIDERATIONS

The following section contains a simple protocol for the determination of base composition. The described procedure is generally applicable and is also suitable for oligonucleotides containing additional modifications. This protocol is based upon a nuclease digestion of the oligonucleotide (DNA or RNA), followed by an enzymatic dephosphorylation step. After their hydrolysis, the individual components of the oligonucleotide are subsequently analyzed via C18 reverse phase chromatography.[17,35,50,58,59] Triethylammonium acetate (TEAA) pH 6.8–7.5 and acetonitrile (ACN) are used as standard running buffers (Figure 17.2) but different buffer systems can also be used,

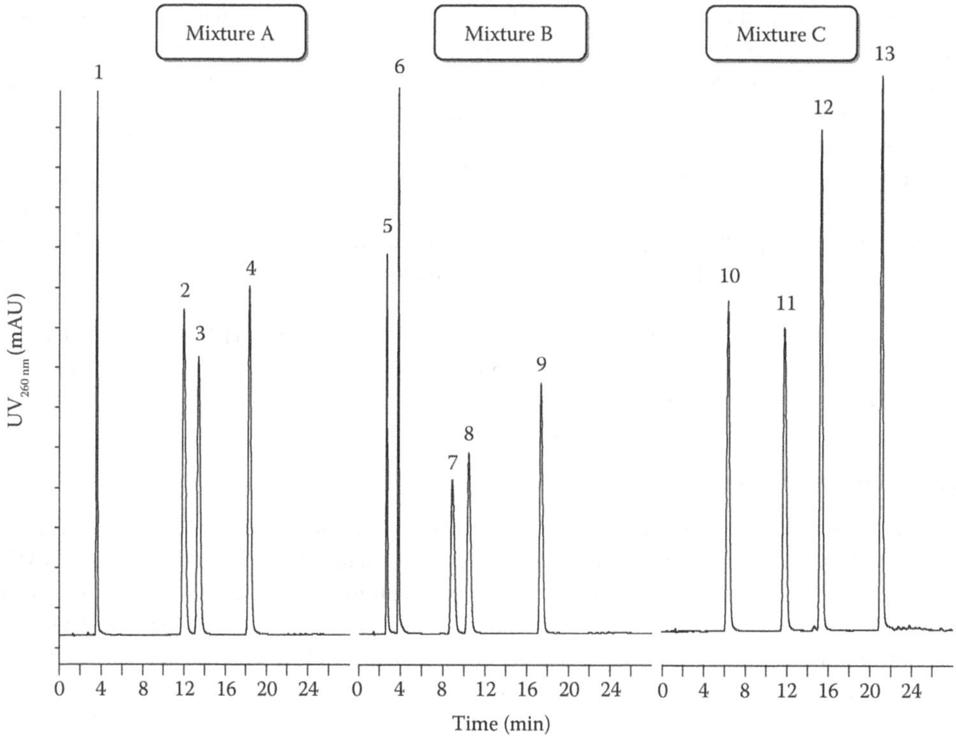

FIGURE 17.2 Highly pure nucleosides analyzed by RP-HPLC. A 4.6 mm × 50 mm XBridge 2.5 μm column (Waters) was used applying a two-step gradient of 0.1M TEAA (buffer A) and 10% ACN (buffer B) at 260 nm. In detail the gradient was 3 min 0% buffer B, 10 min 0% → 30% buffer B, 10 min 30% → 100% buffer B, and 5 min 100 % buffer B. The flow rate was set to 0.5 mL/min at 22°C; 5 μl of a 1 mg/ml solution of three different nucleoside mixtures (A– C) were injected. Mixture A was 1 = 2′-deoxycytidine, 2 = 2′-deoxyguanosine, 3 = thymidine, and 4 = 2′-deoxyadenosine; mixture B was 5 = cytidine, 6 = uridine, 7 = inosine, 8 = guanosine, and 9 = adenosine; mixture C was 10 = 2′-O-methyl-cytidine, 11 = 2′-O-methyl-uridine, 12 = 2′-O-methyl-guanosine, and 13 = 2′-O-methyl-adenosine.

depending on what is to be analyzed. Further simplifying the evaluation procedure is the fact that several of the individual components are commercially available in very good quality. If the extinction coefficients of the monomers are also known, the areas of chromatographically separated single components can be put in relation to one another, thereby ultimately revealing base composition of the oligonucleotide. It is crucial for a successful analysis of base composition of a given oligonucleotide that it can be disassembled completely into its monomeric components. In contrast to enzymatic sequencing of oligonucleotides, where a single incomplete oligonucleotide digest is sufficient for adequate analysis, complete hydrolysis of the oligonucleotide is necessary for accurate determination of base composition. Modifications at single nucleotides (sugar and base modifications), or sequence-related super structures[60,61] can strongly influence the enzymatic degradation reaction under the given incubation conditions. This can result in an incomplete degradation and thereby cause a faulty base composition analysis. Therefore, a second analysis monitoring completeness of degradation should always be performed, verifying the results of the first incubation. In case the oligonucleotide additionally contains phosphorothioates as a form of backbone modification, it can be helpful to chemically convert these modifications into phosphates prior to enzymolysis.[23,27,33] Furthermore, oligonucleotide purity also critically influences the quality of the composition results. In addition to the very common truncated sequences, oligonucleotides can also contain several further by-products from the synthesis and further processing (see above). Such by-products are also captured by enzymolysis and can therefore interfere considerably with the analysis of base composition. Moreover, it is important to run a parallel negative control without oligonucleotide in order to enable exact determination of peak areas of possible contaminants. Enzyme preparations often contain UV-active components at 260 nm, which can consequently influence peak area calculation of individual nucleosides in different ways. Oligonucleotides purified via HPLC and containing sodium as a counterion are best suited for enzymatic degradation. Triethylammonium should be avoided at this point. A simple salt exchange using sodium acetate / ethanol precipitation is a quick and easy option to improve oligonucleotide quality for an enzymatic degradation. This method is especially suitable if rapid sample throughput is desired. Salt exchange using gel filtration also works very well in this context. For this purpose, one-way cartridges, i.e., PD MiniTrap G-10 (GE-Healthcare), or chromatographic desalting columns, such as HiPrep or HiTrap columns (GE-Healthcare), can be used. Compared to precipitation, salt exchange via gel filtration is the method of choice for sample preparation.

17.4.1 Use of Enzymes

Several different enzymes can be used for the enzymatic degradation of an oligonucleotide.[17,62–64] A mixture of snake venom phosphodiesterase (SVP) and bacterial alkaline phosphatase (BAP) has proven to be most successful in this regard. Snake venom phosphodiesterase is a 3′ → 5′-exonuclease which preferably attacks and hydrolizes phosphates of the oligonucleotide backbone at the 3′-end. Single nucleotides are the predominant degradation products of these phosphodiesterases. Bacterial

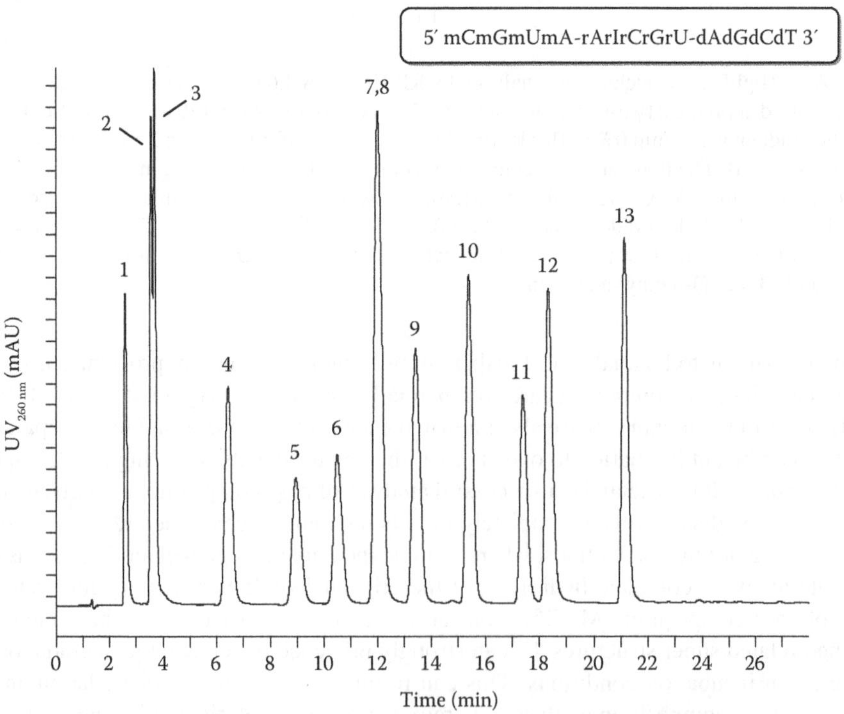

FIGURE 17.3 The 13-mer (5′ mCmGmUmA-rArIrCrGrU-dAdGdCdT 3′) digested with fresh SVP (USB) and BAP (Sigma-Aldrich), 24 hours at 37°C (pH 9.2), separation of nucleosides via ultrafree-0.5 spin columns (Millipore). Analyzed by HPLC with a 4.6 mm × 50 mm XBridge 2.5 μm column (Waters) using 0.1 M TEAA / 10% ACN at 260 nm (see also Figure 17.4): 1 = cytidine, 2 = 2′-deoxycytidine, 3 = uridine, 4 = 2′-O-methylcytidine, 5 = inosine, 6 = guanosine, 7 = 2′-deoxyguanosine, 8 = 2′-O-methyl-uridine, 9 = thymidine, 10 = 2′-O-methyl-guanosine, 11 = adenosine, 12 = 2′-deoxyadenosine, and 13 = 2′-O-methyl-adenosine.

TABLE 17.2
Different Nucleases That Can Be for Base Composition Analysis

Enzyme	Specificity	Phosphates	Reference
RNAse V1	Double-strand-specific for RNA	5′	66
Nuclease P1	Single-strand-specific for DNA or RNA	5′	67
Nuclease S1	Single-strand-specific for DNA or RNA	5′	67
Mung bean nuclease	Single-strand-specific for DNA or RNA	5′	68
DNAse I	Nonspecific DNA endonuclease	5′	67
DNA exonuclease III	Double-strand-specific for DNA	5′	67
Benzonase nuclease	Single- and double-strand-specific nuclease	5′	64
Nuclease Bal 31	Single- and double-strand-specific nuclease	5′	67

alkaline phosphatase, a phosphomonoesterase made from *Escherichia coli*, subsequently removes the 5′-monophosphate overhangs, thereby releasing individual nucleosides. Snake venom phosphodiesterase is commercially available predominantly as a lyophilized powder (i.e., by USB). It is a very robust enzyme that can be stored also as a solution (e.g., at 1 mg/mL) at –20°C for a very long time. Bacterial alkaline phosphatase is commercially available as a stabilized suspension either in ammonium sulfate (Sigma-Aldrich) or in glycerol (Invitrogen). Excess ammonium sulfate is separated from the enzyme-containing supernatant by centrifugation prior to use. The supernatant can be diluted and stored at –20°C for a long time. Solutions in glycerol can be employed directly.

Most buffer systems can be used for enzymatic digestion if they contain adequate amounts of magnesium (≥5 mM). The pH can be varied from pH 7 to 11; the incubation temperature should optimally be set at 25°–37°C for the combined incubation of SVP and BAP. An incubation period between 2 and 24 hours is sufficient for a complete digestion, depending on the activity of the enzyme used (Figure 17.3). As mentioned previously, it is very important to control for the completeness of the digestion (this is unlike oligonucleotide sequencing). Best suited for this purpose is a standard method for each particular oligonucleotide. All enzymes should be deactivated prior to analysis of the hydroylsate by HPLC. This step is especially important if oligonucleotides are to be analyzed with the same equipment. Alternatively, enzymes can be separated from nucleosides via ethanol precipitation or ultrafiltration, using, for example, the so-called microcon or ultrafree-0.5 spin-columns (Millipore).

Snake venom phosphodiesterase (phosphodiesterase I) is a very robust enzyme with a wide spectrum of enzymatic activity. Nonetheless, several types of modifications can cause difficulties leading to an incomplete digestion of the oligonucleotide. It can therefore be helpful to perform incubation with an alternative or additional nuclease, such as nuclease S1 or nuclease P1[65] in some instances. Table 17.2 offers a short overview. Simultaneous use of several enzymes can be advantageous if the oligonucleotide to be analyzed is a highly modified sequence or if it is an aptamer. In these two cases, incubation conditions (incubation duration, temperature, pH, and enzymatic activity used) must be individually adjusted.

17.4.2 Experimental Procedure

17.4.2.1 Reagents

Nuclease P_1: from *Penicillium citrinum* ≥ 200 U/mg (Sigma/Aldrich, catalog number N8630)
Phosphodiesterase I: from *Crotalus adamanteus* ≥ 20 U/mg (USB, catalog number 20240Y)
Bacterial Alkaline Phosphatase: from E. coli C90 buffered solution 150 U/μL (Invitrogen, catalog number 18011-015)

Buffer I: 10 mM of $MgCl_2$, 75 mM Tris-HCl pH 8.9
Buffer II: 20 mM NaOAc, 100 mM NaCl, 5 mM $ZnCl_2$ (adjusted to pH 5.6 with diluted HOAc)

17.4.2.2 Enzyme Mixes

NP1-Mix: Dissolve nuclease P_1 with buffer II to get 0.5 U/μL, vortex for 10 s. Reconstituted product is stable for 2–3 weeks at 4°C.
PDE-Mix: Dissolve phosphodiesterase I with buffer I to get 0.1 U/μL, vortex for 10 s. Store reconstituted product at 4°C.
BAP-Mix: dilute bacterial alkaline phosphatase solution with water to 15 U/μL, vortex for 10 s. Store diluted solution at 4°C.

17.4.2.3 Procedure for Standard Oligonucleotides

Prior to digestion the oligonucleotide solution should be desalted. For this purpose, one-way cartridges, i.e., PD MiniTrap G-10 (GE-Healthcare), or chromatographic desalting columns, such as HiPrep or HiTrap columns (GE-Healthcare), can be used.

1. Evaporate 0.4–1.0 OD of oligonucleotide to dryness under vacuum in a microcentrifuge tube.
2. Dissolve the pellet in 85 μL buffer I.
3. Add first 5 μL of PDE-Mix solution (0.5 U), vortex 10 s.
4. Add second 10 μL BAP-Mix solution (150 U), vortex 10 s.
5. Incubate at 37°C for 18 hours (use shaker).
6. Separate enzyme from free nucleosides using ultrafree-0.5 spin columns.
7. Add 50 μL of water to spin column, spin down.
8. Add 50 μL of water to spin column, spin down.
9. Collect 200 μL of nucleoside solution and evaporate to dryness.
10. Dissolve pellet in 50 μL water, vortex for 30s, spin down.
11. Inject 25 μL (~0.2–0.5 OD per injection based on oligonucleotide) on HPLC.
12. A second analysis should be performed monitoring completeness of degradation, verifying the results of the first incubation.

17.4.2.4 Comments

- Run a digest blank to establish an absorbance baseline profile. Use same procedure for blank preparation.
- Use a reference oligonucleotide to check for enzymatic activity. Use this reference oligonucleotide always.
- To check for deaminase activity of your enzymes, perform a digestion using commercially available nucleosides instead of an oligonucleotide (Sigma-Aldrich).

17.4.2.5 Procedure for Modified "Protected" Oligonucleotides

Prior to digestion the oligonucleotide solution should be desalted. For this purpose, one-way cartridges, i.e., PD MiniTrap G-10 (GE-Healthcare), or chromatographic desalting columns, such as HiPrep or HiTrap columns (GE-Healthcare), can be used.

1. Evaporate 0.4–1.0 OD of desalted oligonucleotide to dryness under vacuum.
2. Dissolve pellet in 45 μL buffer II.
3. Add first 5 μL of NP1-Mix solution (2.5 U), vortex 10 s.
4. Incubate 2 hours at 37°C (use shaker).

TABLE 17.3
Gradients for HPLC Analysis of Digested Oligonucleotides

Step	Time[a] (min)	Gradient[a]
I	3	0% buffer B
II	10	0% → 30% buffer B
III	10	30% → 100% buffer B
IV	5	100 % buffer B
V	10	100% buffer B → 100% buffer C
VI	8	0% buffer B

[a] 4.6 mm × 50 mm XBridge 2.5 µm column (Waters); buffer A = 0.1M TEAA; buffer B = 10% ACN; buffer C = 80% ACN; flow rate = 0.5 mL/min; UV = 260 nm; temperature = 22°C.

5. Incubate 2 hours at 60°C (use shaker).
6. Evaporate to dryness under vacuum.
7. Proceed with "Procedure for Standard Oligonucleotides."

17.4.2.6 HPLC Method

After enzymolysis, the individual components of the oligonucleotide are subsequently analyzed via C18 reverse phase HPLC chromatography (Table 17.3). Triethylammonium acetate (TEAA) pH 6.8–7.5 and acetonitrile (ACN) are used as standard running buffers. For optimum resolution around 0.2, OD should be injected per run.

The easiest way to convert the peak areas obtained by HPLC chromatography into base composition values is to use extinction coefficients from published literature. These coefficients are used to perform a normalization of the chromatographic peak areas. For practical purposes, this can be implemented by introducing cofactors. These cofactors are relative extinction coefficients related to a reference value. If, for example, the extinction coefficient of dC is used as a reference coefficient, the cofactor of dA is 2.11, relative to this reference coefficient (Table 17.4).

The calculation of base composition is initiated by first performing a normalization of the integrated peak areas employing the cofactors.

$$(\text{integrated peak area [mAU} \times \text{s]})/(\text{cofactor}) = \text{normalized peak area [mAU} \times \text{s]}$$

TABLE 17.4
Use of Cofactors for Base Composition Analysis

Nucleoside	Extinction Coefficient (ε)[a] (M^{-1} cm^{-1})	Cofactor (Cf)
dA	15,200	2.07
dG	11,750	1.60
dT	8,750	1.19
dC	7,350	1.00

[a] Measured at pH 7.0, 25°C, 260 nm.[9]
[b] $\varepsilon_{dA,dG,dT,dC} / \varepsilon_{dC} = Cf_{dA,dG,dT,dC}$.

The base composition in percent for each nucleoside (dN) can easily be determined using these normalized peak areas and the respective total peak area.

$$\text{Sum of all normalized peak areas [mAU} \times \text{s] = total peak area [mAU} \times \text{s]}$$

$$\text{dN [\%]} = 100\% \times \text{(normalized peak area dN [mAU} \times \text{s])/(total peak area [mAU} \times \text{s])}$$

Finally, this value is multiplied with the length of the oligonucleotide. This ensures a better comparison between base composition obtained and number of nucleosides within a given sequence.

$$\text{(dN [\%]} \times \text{length of the oligonucleotide)/(100\%) = Base composition}$$

Table 17.5 depicts an analysis of the described type for an exemplary 30-mer DNA oligonucleotide. The calculation of cofactors in the above example was based on published extinction coefficients.[9,60,69–71] However, extinction coefficients can differ substantially from standard values, depending on the HPLC system used and on the measurement conditions applied. These coefficients can change, for example, depending on the employed mobile and/or stationary phase. Moreover, variables such as pH, temperature, and the applied HPLC gradient can significantly influence the extinction coefficient of an individual nucleoside. More robust results can be obtained if a calibration of the system with commercially available nucleoside standards (Sigma-Aldrich) is performed. The use of the above mentioned cofactors for calculation is omitted in that case. A so-called response factor is obtained instead of the cofactor, thus creating the basis for subsequent reliable quantification of nucleosides.

Additional nucleosidic by-products not correlated to the tested oligonucleotide often emerge in the HPLC after digestion. These by-products are usually caused by residual enzymatic activity that can be present in the enzyme preparations, depending on the respective manufacturer or manufacturing conditions. Such unwanted residual activity can lead to chemical modifications of the nucleosides released during the incubation step. As a result, calculation of base composition can deviate substantially. An example for such unwanted residual enzymatic activity is adenosine deaminase, which converts adenosine (adenine) into inosine (hypoxanthine). Adenosine deaminase activity (Figure 17.4) is frequently found in bacterial alkaline phosphatase (BAP).[23,50,72,73] A simple

TABLE 17.5
Base Composition Analysis of a 30-mer Oligonucleotide

Nucleoside[a]	Integrated Peak Area[b] (mAU × s)	Normalized Peak Area[b] (mAU × s)	Base Composition	Found[c]	Actual[c]
dA	170,540	82,386	16.8%	5.4	5
dG	210,101	131,313	26,8%	8.0	8
dT	120,555	101,307	20.6%	6.2	6
dC	175,400	175,400	35.8%	10.7	11
Total		490,406	100%	30.3	30

[a] Sequence digested with fresh SVP (USB) and BAP (Invitrogen), 18 hours at 37°C, separation of enzymes via ultrafree-0.5 spin columns (Millipore).

[b] Analyzed by HPLC with a 4.6 mm × 50 mm XBridge 2.5 μm column (Waters) using 0.1 M TEAA/10% ACN at 260 nm.

[c] Sequence: 5′ CCC ACG TGG TTT GGA TAA CCC CGG ATC CGC 3′; sequence length, 30-mer; actual composition, $A_5G_8T_6C_{11}$.

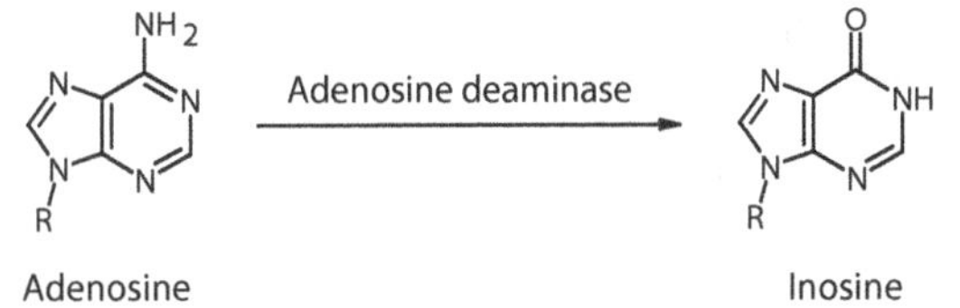

FIGURE 17.4 Unwanted residual enzymatic adenosine deaminase activity.

test that should be performed to rule out such unwanted enzymatic activity is the direct incubation of the enzyme preparations to be used with a nucleoside standard or a reference oligonucleotide. If residual activity cannot be suppressed, the degradation products can also be implemented into calculation by including respective calibration standards or extinction coefficients.[23]

17.5 SUMMARY

Determination of base composition is widely utilized for the analysis of nucleic acids. Aside from enabling taxonomic characterization of organisms, this procedure offers a wide variety of applications within the context of oligonucleotide analysis. Analysis of base composition not only answers simple identity related questions, but also those regarding potential by-products and interlot variations. The protocols described in this chapter enable the reader to apply this analytical method in a simple and straightforward manner and to modify it as needed to meet specific requirements.

REFERENCES

1. Chargaff, E., and H. S. Shapiro. 1955. Remarks on deoxyribonuclease. *Experimental Cell Research* Suppl. 3: 64–71.
2. Tamaoka, J., and K. Komagata. 1984. Determination of DNA base composition by reversed-phase high-performance liquid chromatography. *FEMS Microbiology Letters* 25: 125–128.
3. Katayama-Fujimura, Y., Y. Komatsu, H. Kurashi, and T. Kaneko. 1984. Estimation of DNA base composition by high performance liquid chromatography of its nuclease PI hydrolysate. *Agricultural and Biological Chemistry* 48: 3169–3172.
4. van Regenmortel, M. H. V., C. M. Fauquet, D. H. L. Bishop, et al. 2000. Virus taxonomy: The classification and nomenclature of viruses. In *The Seventh Report of the International Committee on Taxonomy of Viruses*. San Diego, CA: Academic Press.
5. Costello, J. F., and C. Plass. 2001. Methylation matters. *Journal of Medical Genetics* 38: 285–303.
6. Johnston, J. W., K. Harding, D. H. Bremner, G. Souch, J. Green, P. T. Lynch, B. Grout, and E. E. Benson. 2005. HPLC analysis of plant DNA methylation: a study of critical methodological factors. *Plant Physiology and Biochemistry* 43: 844–853.
7. Ratel, D., J. L. Ravanat, F. Berger, and D. Wion. 2006. N6-methyladenine: the other methylated base of DNA. *Bioessays* 28: 309–315.
8. Gehrke, C. W., R. A. McCune, M. A. Gama-Sosa, M. Ehrlich, and K. C. Kuo. 1984. Quantitative reversed-phase high-performance liquid chromatography of major and modified nucleosides in DNA. *Journal of Chromatography* 28(301): 199–219.
9. Johnson, J. L. 1985. Determination of DNA base composition. In *Methods in Microbiology*, vol. 18, 1–31. London: Academic Press.
10. McCabe, D. R., K. Hensley, and I. N. Acworth. 1999. Method for the detection of nucleosides, bases, and hydroxylated adducts using gradient HPLC with coulometric array and ultraviolet detection. *Journal of Medicinal Food* 2: 209–214.
11. Kaur, H., and B. Halliwell. 1996. Measurement of oxidized and methylated DNA bases by HPLC with electrochemical detection. *The Biochemical Journal* 15: 21–23.
12. Zhang, F., M. J. Bartels, L. H. Pottenger, B. B. Gollapudi, and M. R. Schisler. 2006. Simultaneous quantitation of 7-methyl- and O6-methylguanine adducts in DNA by liquid chromatography-positive electrospray tandem mass spectrometry. *Journal of Chromatography B* 833: 141–148.

13. Kelly, M. C., B. White, and M. R. Smyth. 2008. Separation of oxidatively damaged DNA nucleobases and nucleosides on packed and monolith C18 columns by HPLC-UV-EC. *Journal of Chromatography B* 863: 181–186.
14. Ehrlich, M., G. G. Wilson, K. C. Kuo, and C. W. Gehrke. 1987. N4-methylcytosine as a minor base in bacterial DNA. *Journal of Bacteriology* 169: 939–943.
15. Ehrlich, M., M. A. Gama-Sosa, L. H. Carreira, L. G. Ljungdahl, K. C. Kuo, and C. W. Gehrke. 1985. DNA methylation in thermophilic bacteria: N4-methylcytosine, 5-methylcytosine, and N6-methyladenine. *Nucleic Acids Research* 13: 1399–1412.
16. McBride, L. J., J. S. Eadie, J. W. Efcavitch, and W. A. Andrus. 1987. Base modification and the phosphoramidite approach. Nucleosides, *Nucleotides and Nucleic Acids* 6: 297–300.
17. Eadie, J. S., L. J. McBride, J. W. Efcavitch, L. B. Hoff, and R. Cathcart. 1987. High-performance liquid chromatographic analysis of oligodeoxyribonucleotide base composition. *Analytical Biochemistry* 165(2): 442–447.
18. Fowler, K. W., G. Buechi, and J. M. Essigmann. 1982. Synthesis and characterization of an oligonucleotide containing a carcinogen-modified base: O6-methylguanine. *Journal of the American Chemical Society* 104: 1050–1054.
19. Delaney, J. C., and J. M. Essigmann. 2001. Effect of sequence context on O(6)-methylguanine repair and replication in vivo. *Biochemistry* 40: 14968–14975.
20. Acworth, I. N., D. R. McCabe, and T. J. Maher. 1997. The analysis of free radicals, their reaction products, and antioxidants. In *Oxidants, Antioxidants, and Free Radicals,* eds. S. I. Baskin and H. Salem, 23–77. Washington, DC: Taylor and Francis.
21. Bourdat, A. G., D. Gasparutto, and J. Cadet. 1999. Synthesis and enzymatic processing of oligodeoxynucleotides containing tandem base damage. *Nucleic Acids Research* 27: 1015–1024.
22. Kambhampati, R. V., Y. Y. Chiu, C. W. Chen, and J. J. Blumenstein. 1993. Regulatory concerns for the chemistry, manufacturing, and controls of oligonucleotide therapeutics for use in clinical studies. *Antisense Research and Development* 3: 405–410.
23. Schuette, J. M., D. L. Cole, and G. S. Srivatsa. 1994. Development and validation of a method for routine base composition analysis of phosphorothioate oligonucleotides. *Journal of Pharmaceutical and Biomedical Analysis* 12: 1345–1353.
24. Morgan, R. L., J. E. Celebuski, and J. R. Fino. 1993. Base composition analysis of oligonucleotides containing apurinic sites. *Nucleic Acids Research* 21: 4574–4576.
25. Capaldi, D. C., H. J. Gaus, R. L. Carty, et al. 2004. Formation of 4,4′-dimethoxytrityl-C-phosphonate oligonucleotides. *Bioorganic and Medicinal Chemistry Letters* 14: 4683–4690.
26. Septak, M. 1996. Kinetic studies on depurination and detritylation of CPG-bound intermediates during oligonucleotide synthesis. *Nucleic Acids Research* 24: 3053–3058.
27. Gaus, H., P. Olsen, K. V. Sooy, C. Rentel, B. Turney, K. L. Walker, J. V. McArdle, and D. C. Capaldi. 2005. Trichloroacetaldehyde modified oligonucleotides. *Bioorganic and Medicinal Chemistry Letters* 15: 4118–4124.
28. Rentel, C., X. Wang, M. Batt, C. Kurata, J. Oliver, H. Gaus, A. H. Krotz, J. V. McArdle, and D. C. Capaldi. 2005. Formation of modified cytosine residues in the presence of depurinated DNA. *Journal of Organic Chemistry* 70: 7841–7845.
29. Farand, J., and M. Beverly. 2008. Sequence confirmation of modified oligonucleotides using chemical degradation, electrospray ionization, time-of-flight, and tandem mass spectrometry. *Analytical Chemistry* 80: 7414–7421.
30. Farand, J., and F. Gosselin. 2009. De novo sequence determination of modified oligonucleotides. *Analytical Chemistry* 81: 3723–3730.
31. Connolly B. A., B. V. Potter, F. Eckstein, A. Pingoud, and L. Grotjahn. 1984. Synthesis and characterization of an octanucleotide containing the EcoRI recognition sequence with a phosphorothioate group at the cleavage site. *Biochemistry* 23: 3443–3453.
32. Agrawal, S., J. Goodchild, M. P. Civeira, A. H. Thornton, P. S. Sarin, and P. C. Zamecnik. 1988. Oligodeoxynucleoside phosphoramidates and phosphorothioates as inhibitors of human immunodeficiency virus. *Proceedings of the National Academy of Sciences of the United States of America* 85: 7079–7083.
33. Porritt, G. M., and C. B. Reese. 1990. Use of the 2,4-dinitrobenzyl protecting group in the synthesis of phosphorodithioate analogues of oligodeoxyribonucleotides. *Tetrahedron Letters* 31: 1319–1322.
34. Wyrzykiewicz, T. K., and D. L. Cole. 1994. Sequencing of oligonucleotide phosphorothioates based on solid-supported desulfurization. *Nucleic Acids Research* 22: 2667–2669.

35. Andrus, A., and R. G. Kuimelis. 2001. Base composition analysis of nucleosides using HPLC. *Current Protocols in Nucleic Acid Chemistry* chapter 10: unit 10.6.
36. Dai X., A. De Mesmaeker, and G. F. Joyce. 1995. Cleavage of an amide bond by a ribozyme. *Science* 267: 237–240.
37. Gold, L., B. Polisky, O. Uhlenbeck, and M. Yarus. 1995. Diversity of oligonucleotide functions. *Annual Review of Biochemistry* 64: 763–797.
38. Huizenga, D. E., and J. W. Szostak. 1995. A DNA aptamer that binds adenosine and ATP. *Biochemistry* 34: 656–665.
39. Klussmann, S. 2006. *The Aptamer Handbook. Functional Oligonucleotides and Their Applications*, 1st ed. Weinheim: Wiley-VCH.
40. Bartel, D. P., and J. W. Szostak. 1993. Isolation of new ribozymes from a large pool of random sequences. *Science* 261: 1411–1418.
41. Unrau, P. J., and D. P. Bartel. 1998. RNA-catalysed nucleotide synthesis. *Nature* 395: 260–263.
42. Herdewijn, P. 2005. *Oligonucleotide Synthesis—Methods and Applications*. Totawa, NJ: Humana Press.
43. Chargaff, E., B. Magasanik, E. Vischer, C. Green, R. Doniger, and D. Elson. 1950. Nucleotide composition of pentose nucleic acids from yeast and mammalian tissues. *Journal of Biological Chemistry* 186: 51–67.
44. Wyatt, G. R. 1951. The purine and pyrimidine composition of deoxypentose nucleic acids. *Biochemical Journal* 48: 584–590.
45. Wyatt, G. R. 1955. In *The Nucleic Acids*, vol. 1, ed. E. Chargaff and J. N. Davidson, 243–65. New York: Academic Press.
46. Huang, P. C., and E. Rosenberg. 1966. Determination of DNA base composition via depurination. *Analytical Biochemistry* 16: 107–113.
47. Pollack, Y., A. L. Katzen, D. T. Spira, and J. Golenser. 1982. The genome of Plasmodium falciparum. I: DNA base composition. *Nucleic Acids Research* 10: 539–546.
48. Krstulovic, A. M., G. Zweig, and J. Sherma. 1987. *Handbook of Chromatography*, vol. 1. Boca Raton, FL: CRC Press.
49. Brown, P. R. 1983. Current high performance liquid chromatographic methodology in analysis of nucleotides, nucleosides, and their bases. II. *Cancer Investigation* 1: 527–536.
50. Manderville, R. A., and A. M. Kropinski. 2009. Approaches to the compositional analysis of DNA. *Methods in Molecular Biology* 502: 11–17.
51. Sonoki, S., S. Hisamatsu, and A. Kiuchi. 1993. High-performance liquid chromatographic determination of DNA base composition with fluorescence detection. *Nucleic Acids Research* 21: 2776.
52. Sonoki, S., J. Lin, and S. Hisamatsu. 1998. Liquid chromatographic determination of 5-methylcytosine in DNA with fluorescence detection. *Analytica Chimica Acta* 365: 213–217.
53. Dinh, T. Q., C. N. Sridhar, and N. Dattagupta. 1996. Base composition analysis of phosphorothioate oligomers by capillary gel electrophoresis. *Journal of Chromatography A* 744: 341–346.
54. Hua, N. P., and T. Naganuma. 2007. Application of CE for determination of DNA base composition. *Electrophoresis* 28: 366–372.
55. Fredericq, E., A. Oth, and F. Fontaine. 1961. The ultraviolet spectrum of deoxyribonucleic acids and their constituents. *Journal of Molecular Biology* 3: 11–17.
56. Maiti, M., and R. Nandi. 1987. Spectropolarimetric determination of the guanine-cytosine content of DNA. *Analytical Biochemistry* 164: 68–71.
57. De Ley, J. 1970. Reexamination of the association between melting point, buoyant density, and chemical base composition of deoxyribonucleic acid. *Journal of Bacteriology* 101: 738–754.
58. Li, B. F. L., and P. F. Swann. 1989. Synthesis and characterization of oligodeoxynucleotides containing O6-methyl-, O6-ethyl-, and O6-isopropylguanine. *Biochemistry* 28: 5779–5786.
59. Singhal, R. P., P. Landes, N. P. Singhal, L. W. Brown, P. J. Anevski, and J. A. Toce. 1989. High-performance liquid chromatography for trace analysis of DNA and kinetics of DNA modification. *Biochromatography* 4: 78–88.
60. Crothers, D. M., A. V. Bloomfield, and T. Ignacio. 2000. *Nucleic Acids: Structures, Properties, and Functions*. Sausalito, CA: University Science Books.
61. Blackburn, G. M., M. J. Gait, and D. Loakes. 2005. *Nucleic Acids in Chemistry and Biology*, 3rd ed. Cambridge: Royal Society of Chemistry.
62. Crain, P. F. 1990. Preparation and enzymatic hydrolysis of DNA and RNA for mass spectrometry. *Methods in Enzymology* 193: 782–790.

63. Donald, C. E., P. Stokes, G. O'Connor, and A. J. Woolford. 2005. A comparison of enzymatic digestion for the quantitation of an oligonucleotide by liquid chromatography-isotope dilution mass spectrometry. *Journal of Chromatography B* 817: 173–182.
64. Quinlivan, E. P., and J. F. Gregory III. 2008. DNA digestion to deoxyribonucleoside: A simplified one-step procedure. *Analytical Biochemistry* 373: 383–385.
65. Fasman, G. D. 1989. *Practical Handbook of Biochemistry and Molecular Biology*. Boca Raton, FL: CRC Press.
66. Auron, P. E., L. D. Weber, and A. Rich. 1982. Comparison of transfer ribonucleic acid structures using cobra venom and S1 endonucleases. *Biochemistry* 21: 4700–4706.
67. Sambrook, J., E. F. Fritsch, and T. Maniatis. 1989. *Molecular Cloning: A Laboratory Manual*, 2nd ed. Cold Spring Harbor, NY: Cold Spring Harbor Laboratory Press.
68. Laskowski, M., Sr. 1980. Purification and properties of the mung bean nuclease. *Methods in Enzymology* 65: 263–276.
69. Puglisi, J. D., and I. Tinoco, Jr. 1989. Absorbance melting curves of RNA. *Methods in Enzymology* 180: 304–325.
70. Gray, D. M., S. H. Hung, and K. H. Johnson. 1995. Absorption and circular dichroism spectroscopy of nucleic acid duplexes and triplexes. *Methods in Enzymology* 246: 19–34.
71. Cavaluzzi, M. J., and P. N. Borer. 2004. Revised UV extinction coefficients for nucleoside-5′-monophosphates and unpaired DNA and RNA. *Nucleic Acids Research* 32: e13.
72. Kuo, K. C., R. A. McCune, C. W. Gehrke, R. Midgett, and M. Ehrlich. 1980. Quantitative reversed-phase high performance liquid chromatographic determination of major and modified deoxyribonucleosides in DNA. *Nucleic Acids Research* 8: 4763–4776.
73. Singhal, R. P., and J. P. Landes. 1988. High-performance liquid chromatographic analysis of DNA composition and DNA modification by chloroacetaldehyde. *Journal of Chromatography* 458: 117–128.

18 Analysis of Metals in Oligonucleotides

Michael P. Murphy
Intertek Analytical Services

CONTENTS

18.1 INTRODUCTION

Two interconnected forces drive the need for analysis of oligonucleotides for metals content. The first is the need for process monitoring during the isolation or generation of samples. The second is the regulatory requirements associated with oligonucleotides. Accurate quantitation of metal content in oligonucleotides serves several different purposes depending on the end use of the material being tested.

Trace metals may be objectionable if the sample is intended to be administered to patients as part of course of treatment. Metals have been shown to catalyze degradation of oligonucleotides, thus affecting sample stability and causing unforeseen expense regarding the need to study and identify the degradation products if the oligonucleotide is to be used for pharmaceutical purposes. The presence of metals, even if neither toxic nor degrading to the sample, may be objectionable merely because of a visual effect on the sample, causing a sample batch to be rejected for flakes, spots, incorrect color, etc.[1]

18.2 SOURCES OF METALS

If one deliberately adds a metallic impurity to the sample then its origin is obvious. The presence of cesium is almost always because of purification from density layer ultracentrifugation using cesium chloride. Silicon may be introduced from contact with glass vessels. Titanium can be introduced from contact with production equipment. Sodium is the typical counterion for most oligonucleotides. The analysis of sodium will be discussed later in this chapter.

However, most metal impurities are either from residual manufacturing process impurities or environmental in nature (contaminated water supply, storage vessels, air supply, etc.). Trace metals analysis, especially if one is in pharmaceutical or clinical environment, can quickly become a daunting exercise.

There are two fundamental issues to be considered when planning the experiments. First, sample contamination is often difficult to prevent. The common metal elements (silicon, sodium, aluminum, boron, iron, zinc, etc.) are present in robust quantities in the laboratory and on the analysts themselves. Thus, extra care is required to prevent systematic contamination from clothing, hair, cosmetics, jewelry, airborne dust, tracked soil, flasks, and containers in the laboratory. Second, the chemist will eventually be asked to identify and quantify any or all metals in the sample. The challenge will be to find a suitable method for all elements using a minimal quantity of sample. Good laboratory technique can minimize contamination issues and metals analyses have a great advantage in that all possible impurities (the elements themselves) are known to us and can be purchased for use as quantitative standards. The effort expended to avoid contamination is directly related to the affect of the contaminant upon the sample or its intended end use.

It should be noted that process impurities could occur from many unanticipated sources. Any contact of the sample with stainless steel will allow for trace contamination with iron, copper, chromium, nickel, and molybdenum in various ratios depending on the grade of steel. Contact with glass will expose the sample to aluminum, strontium, cesium, boron, silicon, and other elements depending on the grade of glass. Samples stored in plastic vessels have potential exposure to multiple elements depending on the care taken to clean and leach the storage vessels. Gaskets of reaction vessels, liners of bottles or centrifuge tubes, and pipette tips used to transfer solutions can all contribute to contamination.[2,3]

18.3 BACKGROUND OF THE ICP-MS TECHNIQUE

The inductively coupled plasma-mass spectrometry (ICP-MS) technique is an extremely sensitive analytical tool for trace metal analyses. This can be seen in the instrument detection limits presented in Table 18.1. Generally, it will be the technique of choice for metal analysis with certain exceptions (most notably sodium) because the actual sample, the oligonucleotide, is often available only in small quantities. Although this chapter is not a primer on ICP-MS technology, a brief description of the technique is in order because it will help to clarify the rationale for decisions which need to be made during sample preparation and analysis. There are many good resources for ICP-MS information.[4–6] ICP-MS is a mature technology with commercial instruments being available since the 1980s. The technology offers the benefit of sensitivity for most elements from sub part per trillion to parts per billion in solution. This is on par or superior to graphite furnace atomic absorption for most elements and several orders of magnitude superior over standard inductively coupled plasma-optical emission spectrometry (ICP-OES). Because the analysis is typically performed with a quadrupole for mass discrimination and mass detector, heavier elements tend to be more sensitive than lighter elements. The situation is more complex, of course, depending on ionization potentials, isotope abundances, interfering species, etc., but the basic trend here is a useful rule of thumb. Modern instrumentation offers a wide dynamic range from part per trillion to part per million in solution. Data handling systems typically allow for convenient use of both external and internal standards for the analysis. Depending on the desired analytes, complex solution matrices can be analyzed.

The ICP-MS experiment has several basic parts. The sample must be delivered to the instrument for analysis. This can be accomplished by solubilizing the sample and analyzing the solution, laser ablation of a solid sample followed by analysis of the ablated material, thermal vaporization of the sample, or even flow injection/hydride generation of the analytes. A typical modern ICP-MS instrument is shown in Figure 18.1. Typically, the oligonucleotide will be dissolved in a suitable solvent or acid digested and the resulting solution analyzed.

The prepared sample solution is aerosolized using a nebulizer. A nebulizer may be obtained in many different forms in order to best work with certain sample solution types (high efficiency, high

TABLE 18.1
ICP-MS Instrument Detection Limits

Element	Symbol	Detection limit (μg/L)	Element	Symbol	Detection limit (μg/L)
Aluminum	Al	0.005[a]	Molybdenum	Mo	0.001
Antimony	Sb	0.0009	Neodymium	Nd	0.0004
Arsenic	As	0.0006[b]	Nickel	Ni	0.0004[c]
Barium	Ba	0.00002[d]	Niobium	Nb	0.0006
Beryllium	Be	0.003	Palladium	Pd	0.0005
Bismuth	Bi	0.0006	Phosphorous	P	0.1[a]
Boron	B	0.003[c]	Platinum	Pt	0.002
Bromine	Br	0.2	Potassium	K	0.0002[d]
Cadmium	Cd	0.00009[d]	Protactinium	Pr	0.00009
Calcium	Ca	0.0002[d]	Rhenium	Re	0.0003
Cerium	Ce	0.0002	Rhodium	Rh	0.0002
Cesium	Cs	0.0003	Rubidium	Rb	0.0004
Chlorine	Cl	12	Ruthenium	Ru	0.0002
Chromium	Cr	0.000 [d]	Samarium	Sm	0.0002
Cobalt	Co	0.0009	Scandium	Sc	0.004
Copper	Cu	0.0002[c]	Selenium	Se	0.0007[b]
Dysprosium	Dy	0.0001[f]	Silicon	Si	0.03[a]
Erbium	Er	0.0001	Silver	Ag	0.002
Europium	Eu	0.00009	Sodium	Na	0.0003[c]
Fluorine	F	372	Strontium	Sr	0.00002[d]
Gadolinium	Gd	0.0008[g]	Sulfur	S	28[j]
Gallium	Ga	0.0002	Tantalum	Ta	0.0005
Germanium	Ge	0.001[h]	Tellurium	Te	0.0008[k]
Gold	Au	0.0009	Terbium	Tb	0.00004
Hafnium	Hf	0.0008	Thallium	Tl	0.0002
Holmium	Ho	0.00006	Thorium	Th	0.0004
Indium	In	0.0007	Thulium	Tm	0.00006
Iodine	I	0.002	Tin	Sn	0.0005[a]
Iridium	Ir	0.001	Titanium	Ti	0.003[l]
Iron	Fe	0.0003[d]	Tungsten	W	0.005
Lanthanum	La	0.0009	Uranium	U	0.0001
Lead	Pb	0.00004[d]	Vanadium	V	0.0005
Lithium	Li	0.001[c]	Ytterbium	Yb	0.0002[m]
Lutetium	Lu	0.00005	Yttrium	Y	0.0002
Magnesium	Mg	0.0003[c]	Zinc	Zn	0.0003[d]
Manganese	Mn	0.00007[d]	Zirconium	Zr	0.0003
Mercury	Hg	0.16[i]			

Source: From PerkinElmer, Inc. Atomic spectroscopy: A guide to selecting the appropriate technique and system. 2008. With permission.

Note: Unless otherwise noted, ICP-MS detection limits were determined using an ELAN 9000 equipped with Ryton™ spray chamber, Type II Cross-Flow nebulizer and nickel cones. All detection limits were determined using 3-s integration times and a minimum of eight measurements. Letters following an ICP-MS detection limit value refer to the use of specialized conditions or a different model instrument as follows: a, Run on ELAN DRC in standard mode using Pt cones and quartz sample introduction system; b, Run on ELAN DRC in DRC mode using Pt cones and quartz sample introduction system; c, run on ELAN DRC in standard mode in Class-100 Clean Room using Pt cones and quartz sample introduction system; d, Run on ELAN DRC in DRC mode in Class-100 Clean Room using Pt cones and quartz sample introduction system; e, Using C-13; f, Using Dy-163; g, Using Gd-157; h, Using Ge-74; i, Using Hg-202; j, Using S-34; k, Using Te-125; l, Using Ti-49; m, Using Yb-173.

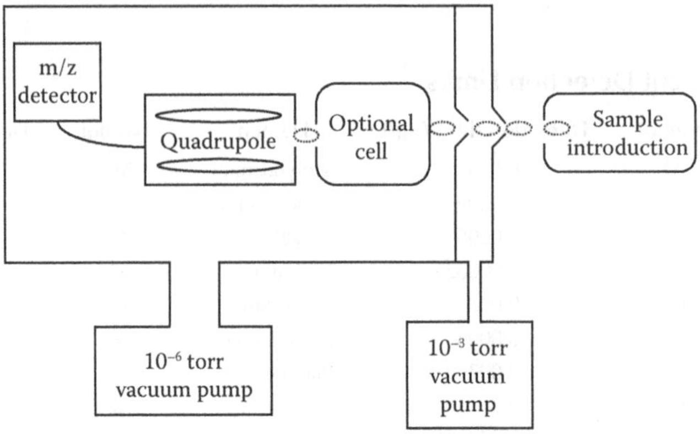

FIGURE 18.1 Schematic of a typical modern ICP-MS instrument

dissolved solids, etc.). Regardless of the exact mechanism employed, all nebulizers are designed to produce a fine aerosol of the sample solution. The aerosol is injected into a spray chamber whose purpose is to allow the most uniform dispersion of aerosol possible to enter the instrument and rejecting both the larger and smaller droplets. The aerosol is swept into the plasma torch by a stream of argon gas. The ICP typically uses an argon gas plasma (other gases can be used but argon is the most common), whose temperature varies from 8000 to 10,000 K in most instrument configurations.[7] The plasma has the important role of generating elemental ions for the ICP-MS detector and also serves to greatly reduce polyatomic clusters in the sample solution. Therefore, unlike chromatography, the ICP-MS experiment is totally destructive of the sample. Any organic compounds not completely destroyed in the sample preparation process are further consumed in the plasma. The plasma dries the aerosol, breaks apart molecular compounds, and converts some of the atoms to ions. The conversion to ions is critical because either a quadrupole or magnetic field sector is used as a mass filter separating the elements on the basis of their mass to charge ratio (*m/z*). This is easier than driving an atom or ion into an excited state and then measuring the intensity of light given off when the atom or ion returns to the ground state, as in ICP-OES spectrometry. Thus, elements such as rubidium and cesium are easily analyzed by ICP-MS even though they are very insensitive by ICP-OES. However, elements whose ionization energies are greater than argon cannot be detected by ICP-MS when using argon plasma.

The detector is physically impacted by the ions; each impact causes a cascade of electrons to form along the chain of dynodes on the detector with the result of greatly amplifying the signal caused by each ion impact into an electrical signal, which is recorded by the instrument data system. The instrument data system converts the signal into concentration based on the parameters the analyst has chosen for the experiment.

Because all of the possible metallic impurities are known, external calibration standards can easily be prepared containing the elements of interest. These standard solutions are analyzed to establish a calibration curve (intensity versus concentration) and the measured intensity of the sample solution is compared to this calibration curve to determine the concentration of each element in the original sample.

Unfortunately, it is not possible to provide a single set of analytical instructions for the user. One reason is there are multiple data handling packages in existence, distinct for each vendor. Even when provided with an analytical method from the literature, unless the user has the same equipment as the author, there will always be some trial and error attempting to find equivalent instrument systems on the user's equipment. Another reason is the pace of improvement in instrument design. The major manufacturers Agilent Technologies, GBC Scientific, PerkinElmer, Thermo-Elemental, Varian Inc., Spectro, and others routinely improve their products. Instruments featuring advances in time of flight (TOF) are challenging quadrupole instruments for biological applications; magnetic sector instruments (i.e., "high-resolution" instruments) are powerful advances over less expensive quadrupole systems but are still out of reach for many laboratories. Most ICP-MS instruments sold today are either reaction cell or collision cell technology, PerkinElmer providing an instrument that combines standard mode, reaction cell, and collision cell in a single instrument. Spectro has introduced a simultaneous instrument with a multichannel detector system. With this pace of both instrument improvement and variety the analytical instructions need to be a broad brush-stroke unless a specific instrument is being utilized.

18.4 SAMPLE PREPARATION

Sample preparation and analytical conditions are intimately connected. Often the sample preparation procedure is chosen to be compatible with the sample introduction system available to the analyst. The choice of technique for sample preparation is then dependent on the analytes sought for analysis, the history of the oligonucleotide sample, and the structure of the oligonucleotide. Generally, there are only four means of sample preparation: dissolution in an appropriate solvent, chemical degradation (acid digestion, often using a closed microwave system), thermal degradation (ashing), and fusion with high-purity inorganic compounds (sodium carbonate, lithium metaborate, etc.). There may be other steps refining the process such as filtration, solid phase extraction, chelation, multiphase extraction, etc. but most other steps are performed to either concentrate the analyte or remove potentially interfering species from the solution. Unless a laser ablation system is being employed, ultimately, the analyte of interest must be brought into solution so it can be delivered to the ICP-MS plasma for analysis. With the possible exception of inorganic silicon analyses the oligonucleotides will be either dissolved in solvent or acid digested to prepare the samples for analysis. When performing sample preparation at the trace and ultra-trace level for inorganic elements the potential for contamination is ever present. In their text, Howard and Statham illustrate the many sources of contamination possible from seemingly innocuous labware.[4]

The most common choices for sample containers are polyethylene and fluorinated plastics such as Teflon® or PFA for sample preparation and storage. All plastic-ware needs to be soaked in dilute nitric acid before initial use. This is performed to leach residual metals from the plastic-ware. Surprisingly, large quantities of contaminants can be leached from plastics. Moody and Lindstrom had shown aluminum as high as 30 μg/g in high-density polyethylene and antimony at 0.6 μg/g in polypropylene. They had tested nine polymers for assorted elements for total composition as well as extractable/leachable studies. Several authors have studied this phenomenon. Glassware is to be avoided if at all possible. The ambient level of impurities in glassware can be orders of magnitude greater than that present in the samples they are used to prepare.

High-purity plastic-ware can be obtained from vendors such as Sarstedt, Savillex, SCP Scientific, and others. The SCP Digiprep® 50 mL sample tubes can be used for low-temperature preparations (approximately 110°C or less) and can be used as sample holders for most autosampler instruments. This has the advantage of minimizing sample transfer and therefore exposure to possible contaminants. The Savillex vessels are convenient if higher temperatures are required.

The most common approaches to sample preparation are dissolution of the oligonucleotide in dilute acid and digestion with nitric acid or nitric acid/peroxide. Most elements can be captured for analysis using either technique. However, in sample preparation, there are no panaceas. There is no single preparation scheme that is optimum for all elements and all samples. For example, the nitric acid or nitric/peroxide scheme could present problems if the analyte of interest were Os. The OsO_4 compound is far more volatile than OsO_2 or Os and therefore will yield erroneously high analytical results when analyzed against standards prepared from OsO_2.

The most common preparation used is simple dissolution in dilute nitric acid (~0.1–0.2 M nitric acid) at room temperature. If necessary the sample may be warmed to sub-boiling temperatures (<100°C) to help degrade the organic compounds present while retaining volatile elements. If the sample does not completely dissolve, the solution can be filtered with ion-chromatography grade syringe filters (polyethersulphone) after testing for potential retain or release of analytes or preferably centrifuged to clarify the solution. If the analyte of interest is silicon and the sample requires filtration, it will be necessary to use equipment that does not include silicon-lubricated plungers on the syringe barrel.

Often, dilute surfactant, such as Triton X-100, is added to the sample solutions.[8,9] This will improve the precision of the sample analysis, especially if using a nonwetting sample introduction system such as PFA. The standards will need to be matrix matched for acid and surfactant content for best results. Of course, as with all reagents, addition of extra reagents increases the risk of inadvertent contamination of the sample.

Because oligonucleotides normally contain functional groups such as phosphates, ether linkages and amine, or amide bonds, these compounds normally respond well to acid digestion. The functional group opens the molecule up to attack by the acid. The sample is transferred to the plastic vessel and then the trace metals grade or ultra-trace metal grade acid added to the vessel. The sample can then be heated in a heating block at <100°C if volatile elements such as mercury or ruthenium are being analyzed. It is possible to be too gentle with the sample. Simon et al. has shown that oligonucleotides with certain substitutions can be stable in solutions of pH 1 at 37°C for hours; therefore, a little heating to advance the degradation is desirable.[10] It may be tempting to cap the vessel while heating in order to retain the vapors of the more volatile elements. This is not without risk because most caps are made of a different material than the vessels themselves. These caps can then leach unintended metals into the sample preparation solution. Often plastics are color coded, and this color is often imparted using an inorganic compound. If the sample requires more vigorous digestion to destroy the sample and release the metals for analysis, then microwave digestion is warranted. The choice between a sealed system and an open system once again depends on which elements are being sought.

The ICP-MS instrument typically supports analysis of solutions with total dissolved solids not greater than 0.2% of the solution. The instrument will also display a mass bias affect. Essentially, higher mass analytes are more intense than lower mass analytes at the same concentration. This can be important if the sample is high in a heavy analyte such as lead. Matrix matching analytical standards to the same solution matrix will alleviate most of these issues. When full matrix matching cannot be achieved, internal standards can often alleviate the problems associated with the mismatch. The choice of internal standard is usually a combination of similar sample mass and first ionization potential. In this manner, multiple elements may be required to act as internal standards across the entire mass range. Most vendors can supply premixed standard solutions with elements covering the entire mass range to simplify this process (e.g., Inorganic Ventures supplies a mix of analytes containing isotopically enriched ^{6}Li, as well as Bi, In, Sc, Tb, and Y; other vendors supply similar internal standard element sets).

Mercury analysis performed using cold vapor atomic absorption methodology normally requires the addition of $KMnO_4$ in order to assure the conversion of organic mercury to the oxide and thus make it available for analysis when a reducing agent is added to the sample solution. This extra reagent is not generally necessary if the mercury is to be determined using the ICP-MS instrument in normal aspiration mode. If a hydride generation system is being used, then the permanganate should be added prior to reduction to assure that all of the mercury is being captured.

Although no single sample preparation scheme is perfect for all elements, some compromises can be made. For example, if Os is sought and a hydrochloric acid extraction is performed, most other elements will also readily form soluble chloride salts. Silver would be a notable exception, because it is unlikely the hydrochloric acid concentration would be in great enough concentration to resolubilize the silver chloride. Arsenic would also be an exception if the analysis were being performed using normal aspiration because the ArCl interference would swamp the As signal. If it were necessary to use the hydrochloric acid solution for arsenic analysis (e.g., the sample is a scarce research sample and there is no other sample available), then the arsenic can be determined using an ICP-MS instrument capable of collision cell sample introduction system.[11] Alternatively, arsine generation from the hydrochloric acid solution and a flow injection-ICP-MS experiment can be performed with very sensitive results.

A topic not addressed here is speciation experiments. Because sample preparation is key in speciation work, in order to avoid accidental conversion of an analyte from one species to another prior to analysis, the topic is too broad to be examined here. The European Virtual Institute for Speciation Analysis (EVISA) Web site (http://www.speciation.net) is an excellent resource for the required techniques.

Chromatographic cleanup of samples is perhaps the most interesting technique used to prepare samples for analysis. This allows for the analysis of the oligonucleotide without having to first purify the sample from any other organic impurities present. Obviously, this is a powerful technique for biological fluids or synthetic mixtures.

If the objective of the analysis is metal content that is covalently linked to the oligonucleotides, then it is advantageous to make use of well-known chromatographic methods to separate the oligonucleotide and its bound metals from other organic constituents in the mixture. In this manner, binding constants and purity through doped binding sites on the molecule can be determined. The chromatographic separation followed by direct analysis of covalently bound metals can be used to identify a given compound structure through the mole ratio of metal analytes found.

In summary, there are multiple ways to prepare the sample for analysis, but advance planning with regard to desired analytes and quantitation limit is required to avoid wasting precious sample with unnecessary or nonproductive preparations. There are two main purposes of sample preparation. The first is to look for metal content in the sample. Once the sample has been digested, this becomes a routine analytical problem with analytical conditions determined by the available instrumentation. The other purpose is to use the metals bound to the molecule as a means of studying the structure/properties of the compound itself. In these cases, sample preparation is not concerned with destruction of the sample but isolation of the sample of interest from other potential sources of metals, which may interfere with the analysis or cause false positive readings.

18.5 ICP-MS ANALYSIS

The ICP-MS platform is extremely flexible when it comes to sample analysis. Sample introduction system options include choices in nebulizer design, spray chamber design, composition of injectors, torches, addition of auxiliary gases such as oxygen to aid in analyses featuring organic solvent systems, ultrasonic nebulizers, and collision/reaction cell technology to reduce polyatomic interferences as well as plasma conditions (cool plasma versus hot plasma conditions that can help with space charge affects and other interferences). Because the sample preparation scheme will tend to be less complex for the oligonucleotides exotic systems such as Peltier cooled spray chambers or direct injection, high-efficiency nebulizers are not normally required for good analytical results. However, this is not to say that interesting analyses cannot be performed with oligonucleotides.

Once the sample preparation scheme has been chosen on the basis of the analytical requirements, the samples can be analyzed. If the samples are digested, the prepared sample solutions are now no different than any other prepared sample solution and existing methodology in the literature can be applied to these solutions (within the limitations introduced by the preparation scheme chosen, e.g., hydrochloric acid matrix is a poor choice if Ag analysis is desired, etc.). Because the prepared sample solutions routine analytical conditions can be applied for the determination of metal content in the sample. Because the digestion results in prepared sample solutions that are similar to those analysts most often encounter a laboratory's routine analytical conditions can be applied for the metals analysis.

When biomolecules are examined for covalently bonded metals, the metals are generally referred to as either "elemental tags" or "marker tags." In these cases the interest is not in impurities present in the biomolecules but in the elucidation of the biomolecule's structure. Wrobel et al. made use of covalently bonded osmium and phosphorous analysis by ICP-MS to evaluate the methylation of DNA in terms of methylated cytosine versus nonmethylated cytosine.[12] The phosphorous concentration was used to calculate the total number of mer units in the DNA sample. Good correlation was found with results using more typical HPLC–DAD analyses.

Fluorescent flow cytometry results compared well to ICP-MS data for *in situ* hybridized mRNA samples by Ornatsky et al.[13] The flow cytometry and ICPMS analyses were carried out simultaneously from a single sample flow source. The ICP-MS results were shown to be specific and accurate. This form of analysis circumvented the problem of the parent compound and the possible degradation products both exhibiting fluorescence in the same sample solution. The phosphorous and sulfur concentrations of a 16- to 24-mer oligonucleotide were studied by Lokits et al. using ICP-MS.[14] Using this sample, they studied various nebulizer/spray chamber combinations finding that absolute detection limits of 0.17 pg of phosphorous and 0.16 pg of sulfur could be obtained.

Other work has shown good correlation between immunoassay results and ICP-MS results for oligonucleotides using the elemental tag method of analysis.[15,16] It was demonstrated that lanthanum labeled antibodies retained the same binding activity as nonlabeled antibodies. These labeled samples were then used for ICP-MS analyses to study binding activity in cell lines. The net result was a marked increase in sample throughput along with good analytical accuracy equivalent to the ELISA immunoassay results.

Similar work was done by Zhang et al. using Au nanoparticles for the labeling of antibodies.[15] Au nanoparticles, approximately 15 nm in size, were used to monitor the α-fetoprotein in solution over a range of 0.016–6.8 μg/L in serum. The results showed that the concept of using tagged element ICP-MS analysis could be used as a sensitive immunoassay technique.

Kerr and Sharp made use of streptavidin nanogold conjugate to label biotin containing oligonucleotides.[17] Purification was performed by HPLC analysis followed by ICP-MS analysis. The authors worked out conditions such that a single gold nanoparticle was associated with a single biotinylated DNA molecule. The results showed that ICP-MS analyses could be used to establish mass balance information for the labeled compound ensuring that all compounds in the reaction are identified. The organic compound is not important once the analytical sample has been aliquoted, because the instrument is only measuring the metallic element and not the total molecule in solution, and therefore sample stability is not an issue.

Not to be outdone by HPLC, separation studies by Bruchert and Bettmer et al. used gel electrophoresis separations coupled to an ICP-MS spectrometer to study phosphorous in DNA samples after acid digestion of the samples.[18,19] They studied various DNA samples with good precision (<3% RSD), but not every DNA sample responded to the test. They also showed the technique worked equally well on 100 base pair samples as well as ~3 Mbase pair samples. The authors then turned to studying 8-mer sample interaction with cisplatin. Again, using gel electrophoresis to isolate the complexes. The ICP-MS data were used to generate kinetic rate constants for two different oligonucleotides.

18.6 REGULATORY CONSIDERATIONS

The international regulatory situation is in flux with regard to metallic impurities. The ICH Guideline Q6a states in Section 3.3.1 that ". . . other inorganic impurities may be determined by other appropriate procedures, e.g., atomic absorption spectroscopy" and guideline Q6b in Section 6.2.1 states "Downstream-derived impurities include, but are not limited to, enzymes, chemical and biochemical processing reagents (e.g., cyanogen bromide, guanidine, oxidizing and reducing agents), inorganic salts (e.g., heavy metals . . .". However, little guidance is provided for actual analytes to be tested and analytical methodology. The United States Pharmacopoeia (USP) is currently seeking to replace the limit test, Chapter <231> "Heavy Metals" with an instrumental version of the analysis. The proposed chapters are <232> "Elemental Impurities—Limits" and <233> "Elemental Impurities—Procedures." The advantage of these new chapters would be that accurate quantitation is sought for the elements of interest. Chapter <232> would base the limits of each element on the intended use of the drug product through the permissible daily exposure as dictated by the means of delivery to the patient. Chromium would have a daily permissible limit of 250 μg per day for orally administered drugs, but only 25 micrograms per day in the form of a parenteral drug, for

TABLE 18.2
Comparison of Current and Proposed USP Element Lists with EMEA Element List

USP <231> Heavy Metals Elements[a]	Proposed USP Chapter <232> Elements[b]	EMEA Elements[c]
Ag	—	—
As	As	—
Bi	—	—
Cd	Cd	—
—	Cr	Cr
Cu	Cu	Cu
		Fe
Hg	Hg	—
—	Ir	Ir
—	Mn	Mn
Mo	Mo	Mo
—	Ni	Ni
—	Os	Os
Pb	Pb	—
—	Pd	Pd
—	Pt	Pt
—	Rh	Rh
—	Ru	Ru
Sb	—	—
Sn	—	—
—	V	V
—	—	Zn

[a] U.S. Pharmacopeial Convention, USP 32–NF 27, chapter <232>.
[b] Pharmacopeial Forum, vol. 36(1): Jan–Feb 2010.
[c] Guideline on the Specification Limits for Residues of Metal Catalysts or Metal Reagents, Doc. Ref. EMEA/CHMP/SWP/4446/2000, London, 21 February 2008.

example. Thus the analyst will need to know the intended final dosage form and the patient exposure to calculate the required limit for each element in the original sample. Chapter <233> attempts to provide a framework for analytical testing and validation but is as yet incomplete. These two chapters increase the number of tested elements from 10 in Chapter <231> to 16 in Chapter <232> (originally, 31 elements were proposed in the Stimulus for Revision). The European Medicines Agency (EMEA) currently specifies 14 elements. The EMEA limits are also exposure based and the document gives a detailed justification for those values. A simple comparative list of analytes is presented in Table 18.2. Because the required element list may yet again be altered, prudence would dictate the analyst be prepared to quantitate all elements listed and perhaps a few more as well.

18.7 SODIUM ANALYSIS

Sodium (Na) is the usual counterion for oligonucleotides. Thus, it may be used as a means to monitor the size of the molecule and the availability of unreacted sites. Trace level Na is difficult with standard quadrupole ICP-MS instruments. Normally, reaction or collision cell technology is required along with exacting laboratory procedures to prevent environmental contamination. However, the

Na tends to be in the percentage by weight class of concentration in oligonucleotides. A 1 mg sample with a concentration of 5% Na wt/wt prepared into 50 mL of dilute acid would yield an analytical solution of 1 µg/mL, well within the bounds of ICP-OES, ICP-MS, and flame AA analyses.

Because such a small quantity of sample is required, the normal matrix issues caused by samples will be minor. The assay can be performed by any of the three techniques mentioned although all still have minor analytical challenges. Sodium by flame AA is often best performed in the presence of an ionization suppressant such as cesium chloride. Analysis of Na by ICP-OES is often a challenge for instruments that are strictly axial viewing mode. The radial viewing mode tends to yield more accurate results. The ICP-MS instrument may actually be better suited depending on the configuration at your facility. Sodium is relatively insensitive, compared to heavier ions, and therefore a final concentration of 1 ppm Na should not exceed the linear range of the analyte.

The sample, 1–2 mg of sample, if homogenous, larger sizes if not homogenous, should be taken up in dilute nitric acid and heated, if needed, to either dissolve or digest the organic sample. Dilute the sample to at least 50 mL or larger if possible. The internal standard may or may not be necessary because the solution will be very dilute. Whether you add an internal standard or not the calibration standards should be matrix matched to the acid concentration in the analytical sample solution as closely as possible. The best analytical results are normally achieved when the sample solution concentration is less than the highest calibration standard but greater than the lowest calibration standard. Recovery of added Na to the sample will help to guide the analyst in selection of the most appropriate analytical conditions. It is common to add the spike after samples have been digested to assess the effect of the sample matrix on the analyte values. However, a final more rigorous approach is to spike the sample prior to digestion. In this manner the effect of the sample digestion scheme can be tested as well as the residual sample matrix.

18.8 PHOSPHOROUS ANALYSIS

Many oligonucleotides contain phosphorous in the polymer as available phosphate sites.[20–22] These sites can be used to study the number of mer units in the oligonucleotide. The ICP experiment will not be able to discern inorganic phosphorous from organic phosphorous. With that in mind the sample can be treated the same as for sodium and analyzed. The standard quadrupole instrument, in standard mode, is not very sensitive toward phosphorous compared to other atoms. A reaction or collision cell may increase the sensitivity of phosphorous mainly by virtue of the lower background noise one finds in these instruments. This type of instrument would be required to accurately quantitate low levels of phosphorous. A sample preparation scheme similar to Na can be employed but with larger sample sizes (~25 mg).

18.9 SUMMARY

The use of ICP-MS analysis on oligonucleotides has been demonstrated to yield absolute number of mer values for samples of unknown size. Good correlation with standard ELISA and flow cytometry results show the tagged element approach is a robust and rapid method for analysis of oligonucleotides that can be tagged. The technique has been shown to be useful for the rapid generation of kinetic binding constants for samples. And, of course, the technique is useful as a straightforward assay of metallic impurities in the oligonucleotide samples.

Analytes of specific interest to oligonucleotides such as sodium and phosphorous can be analyzed by ICP-MS, assuming some care is used with regard to choice of analytical conditions.

REFERENCES

1. Becht, S., X. Gu, and X. Ding. 2007. Vaccine characterization using advanced technology. *The Biopharm International Guide*, August 2007: 1–7.

2. Inorganic Ventures. "Container-material-properties. http://www.inorganicventures.com/tech/trace-analysis/container-material-properties (accessed March 2010).
3. Howard, A. G., and P. J. Statham. 1993. *Inorganic Trace Analysis: Philosophy and Practice*, 43–44. New York: John Wiley.
4. Beauchemin, D. 2006. Inductively coupled plasma mass spectrometry. *Analytical Chemistry* 78: 4111–4136.
5. PerkinElmer, Inc. "The 30-minute guide to ICP-MS." http://www.perkinelmer.com (accessed March 2010).
6. Thomas, R. 2008. *Practical Guide to ICP-MS a Tutorial for Beginners*, 2nd ed. Boca Raton, FL: CRC Press.
7. Sneddon, J., and M. D. Vincent. 2008. ICP-OES and ICP-MS for the determination of metals: application to oysters. *Analytical Letters* 41: 1291–1303.
8. Chen, W.-Y., and Y.-C. Chen. 2003. Reducing the alkali cation adductions of oligonucleotides using sol-gel-assisted laser desorption/ionization mass spectrometry. *Analytical Chemistry* 75(16): 4223–4228.
9. St'astnia', M., I. Nemcova, and J. Zyka. 1999. ICP-MS for the determination of trace elements in clinical samples. *Analytical Letters* 32(13): 2531–2543.
10. Simon, L., P. Miller, and P. Ts'O. 1999. Oligonucleotide analogues having improved stability at acid PH. US Patent 6,005,094, filed Dec. 27, 1994, and issued Dec. 21, 1999.
11. Pick, D., M. Leiterer, and J. Einax. 2010. Reduction of polyatomic interferences in biological material using dynamic reaction cell ICP-MS. *Microchemical Journal* doi:10.1016/jmicroc.2010.01.008.
12. Wrobel, K., J. A. L. Figueroa, S. Zaina, G. Lund, and K. Wrobel. 2010. Phosphorus and osmium as elemental tags for the determination of global DNA-methylation—A novel application of high performance liquid chromatography inductively coupled plasma mass spectrometry in epigenetic studies. *Journal of Chromatography B* 878: 609–614.
13. Ornatsky, O. I., V. I. Baranov, D. R. Bandura, S. D. Tanner, and J. Dick. 2006. Messenger RNA detection in leukemia cell lines by novel metal-tagged in situ hybridization using inductively coupled plasma mass spectrometry. *Translational Oncogenomics* 1: 1–9.
14. Lokits, K. E., P. A. Limbach, and J. A. Caruso. 2009. Interfaces for capillary LC with ICPMS detection: A comparison of nebulizers/spray chamber configurations. *Journal of Analytical Atomic Spectrometry* 24: 528–534.
15. Hu, S., R. Liu, S. Zhang, Z. Huang, Z. Xing, and X. Zhang. 2009. A new strategy for highly sensitive immunoassay based on single-particle mode detection by inductively coupled plasma mass spectrometry. *Journal of the American Society of Mass Spectrometry* 20: 1096–1103.
16. Razumienko, E., O. Ornatsky, R. Kinach, M. Milyavsky, E. Lechman, V. Baranov, M. A. Winnik, and S. D. Tanner. 2008. Element-tagged immunoassay with ICP-MS detection: Evaluation and comparison to conventional immunoassays. *Journal of Immunological Methods* 336: 56–63.
17. Kerr, S. L., and B. Sharp. 2007. Nano-particle labeling of nucleic acids for enhanced detection by inductively-coupled plasma mass spectrometry (ICP-MS). *Chemical Communications* no. 43: 4537–4539.
18. Brüchert, W., R. Krüger, A. Tholey, M. Montes-Bayón, and J. Bettmer. 2008. A novel approach for analysis of oligonucleotide-cisplatin interactions by continuous elution gel electrophoresis coupled to isotope dilution inductively coupled plasma mass spectrometry and matrix-assisted laser desorption/ionization mass spectrometry. *Electrophoresis* 29: 1451–1459.
19. Brüchert, W., and J. Bettmer. 2006. DNA quatification approach by GE-ICP-SFMS and complementary total phosphorus determination by ICP-SFMS. *Journal of Analytical Atomic Spectrometry* 21: 1271–1276.
20. Donald, C. E., P. Stokes, G. O'Connor, and A. J. Woolford. 2005. A comparison of enzymatic digestion for the quantitation of an oligonucleotide by liquid chromatography-isotope dilution mass spectrometry. *Journal of Chromatography B* 817: 173–182.
21. Yang, I., M.-S. Han, Y.-H. Yim, E. Hwang, and S.-R. Park. 2004. A strategy for establishing accurate quantitation standards of oligonucleotides: quantitation of phosphorus of DNA phosphodiester bonds using inductively coupled plasma-optical emission spectroscopy. *Analytical Biochemistry* 335: 150–161.
22. Wrobel, K., J. A. L. Figueroa, S. Zaina, G. Lund, and K. Wrobel. 2010. Phosphorus and osmium as elemental tags for the determination of global DNA-methylation—A novel application of high performance liquid chromatography inductively coupled plasma mass spectrometry in epigenetic studies. *Journal of Chromatography B* 878: 609–614.

19 Regulatory Considerations for the Development of Oligonucleotide Therapeutics

G. Susan Srivatsa
ElixinPharma

CONTENTS

19.1 BACKGROUND

Oligonucleotides represent a novel class of therapeutics with unique mechanisms of action. A typical oligonucleotide is a short chain DNA or RNA molecule, usually manufactured by chemical synthesis utilizing automated synthesizers. Although most oligonucleotides in clinical development are chemically similar, DNA, RNA, or LNA, often with chemical modifications to afford nuclease stability in biological matrices, the mechanisms of action vary widely. As a consequence, a broad range of oligonucleotides are currently in clinical development for a variety of clinical indications. The types of oligonucleotides in preclinical or clinical development include antisense oligonucleotides, immunostimulatory oligonucleotides, DNA duplex decoys, small interfering RNAs (siRNAs), ribozymes, microRNAs, aptamers, and spiegelmers.

Antisense oligonucleotides are short chain DNA strands that are complementary to the messenger RNA target. They strongly and selectively bind to their messenger RNA to inhibit or downregulate production of harmful or overexpressed proteins. Antisense molecules are also capable of binding to endogenous enzymes such as RNase H resulting in degradation of the molecular target. Immunostimulatory oligonucleotides are short chain DNA strands that activate innate immune responses in a highly selective fashion by activating toll-like receptors (TLRs) found in certain immune cells. This enhances the response of T cells involved in specific disease instead of the entire immune system thereby averting an autoimmune response.

DNA duplex decoys are short chain, double-stranded DNA molecules that act as decoys to effectively attenuate the activity of transcription factors that are implicated in many human inflammatory diseases.

siRNAs (short interfering RNA), which are RNA duplexes of approximately 20–25 base units in length work by the RNA interference mechanism in inhibiting the expression of mRNA or viral RNA. Ribozymes are RNA strands that behave similarly to enzymes in promoting catalytic reactions. They act as "molecular scissors" by selectively binding to and cleaving messenger RNAs thereby preventing the production of harmful proteins. Micro RNAs are single-stranded RNA molecules of approximately 20–25 base units in length that act by regulating gene expression and preventing translation of messenger RNAs to form proteins.

Aptamers are single-stranded RNA or DNA molecules that adopt well-defined three-dimensional conformations and function similarly to monoclonal antibodies. They selectively and strongly bind to target proteins, thereby inhibiting their action. Spiegelmers, also known as mirror image aptamers (from the German word Spiegel for mirror), utilize L-ribonucleic acids in place of the natural D-ribonucleic acid affording the resultant spiegelmer enzymatic resistance to nucleases in biological matrices such as plasma.

The first investigational new drug application (IND) for an oligonucleotide drug candidate was filed in 1992 by Isis Pharmaceuticals for the treatment of human papilloma virus (HPV). Since then, hundreds of oligonucleotide-based candidates have advanced to clinical development. To date, there are only two oligonucleotide-based drugs that have received approval by the U.S. Food and Drug Administration (FDA) for commercial marketing. The first oligonucleotide drug to receive market approval was Vitravene™, a 21-based single-stranded DNA phosphorothioate oligonucleotide for the treatment of cytomegalo viral (CMV) retinitis. Vitravene is an antisense molecule that blocks translation of viral mRNA by binding to a coding segment of a key CMV gene. Subsequently, another oligonucleotide drug, Macugen™, a PEGylated aptamer, was approved for the treatment of age-related macular degeneration (AMD). The aptamer in Macugen™ is an inhibitor of the vascular endothelial growth factor (VEGF) protein, which upon binding to VEGF effectively blocks angiogenesis and prevents leakage of fluid and blood in the retina. Both of these drugs are approved for local administration by intravitreal administration. Although there are many oligonucleotide drugs currently in development for parenteral administration, some in late stage clinical development, there are no approved oligonucleotide drugs to date for systemic indications.

In spite of two approved drugs and the large numbers of other oligonucleotide drug candidates in various stages of clinical development in the United States, Canada, Europe, and other parts of the world, there are no formal regulatory guidelines available from any regulatory agency for the development of this class of therapeutics. As a result of advances in basic biology, there has been an explosive growth in the number of oligonucleotides that have been promoted to clinical development. In an effort to bridge the gap between the need for general guidance for the chemical and pharmaceutical development of this class of therapeutics and the lack of definitive guidance documents, this chapter attempts to outline the type of chemistry, manufacturing, and controls information required in the quality section of an IND for the initiation and support of clinical trials in the United States.[1] Also included are detailed descriptions of the type of specific information expected for the various classes of oligonucleotides currently in development as well as a general discussion of bioanalysis to support pharmacokinetics and metabolism studies.

The information provided in this chapter is based on three sources: regulatory guidelines from the U.S. FDA and International Conference on Harmonization (ICH) for assuring quality of new chemical entities; presentations by FDA and other regulatory authorities at national and international conferences regarding their general expectations for the quality section of oligonucleotide applications; and the author's personal experience regarding generally accepted practices by the industry and regulatory agencies for the manufacturing and quality control of synthetic oligonucleotides. It is cautioned that the regulatory practices by the various authorities are evolving rapidly to keep pace with developments in clinical development, and therefore this document represents a snapshot in time for current regulatory policies. As additional experience is gathered, both on the part of industry and the regulatory agencies, review practices and subsequent requirements for synthetic oligonucleotides are expected to evolve in line with additional knowledge.

19.2 REGULATORY GUIDANCE DOCUMENTS

To date, there are no formal regulatory guidelines specifically addressing this class of therapeutics. However, regulatory agencies have been responsive to the request of the oligonucleotide industry in offering some unofficial guidance documents for the basic expectations for characterization and quality control of oligonucleotides.[2,3]

Oligonucleotides are manufactured by chemical synthesis using commercially available automated synthesizers. In spite of their large size (molecular weights of *ca.* 7000 or larger), oligonucleotides are more similar to small molecule drugs known as new chemical entities (NCEs) than biologics because they are manufactured by organic chemical synthesis. Therefore, it has been general practice for regulatory agencies to apply regulatory guidelines intended for small molecules to oligonucleotide therapeutics. These include all quality-related regulatory guidelines issued by the U.S. FDA and the ICH.

Some ICH guidelines intended for NCEs, ICH Q3A,[4] and ICH Q6A[5] specifically exclude oligonucleotides. At times, this is misinterpreted to mean that these guidelines do not apply to oligonucleotides. On the contrary, it is often the case that the guideline is not considered comprehensive enough to apply to oligonucleotides. The spirit of these guidelines still applies to oligonucleotides, but there is some flexibility in cases where certain elements may not specifically apply. For example, in ICH Q6A, which specifically excludes oligonucleotides, drug substances are expected to be characterized for polymorphism. Most oligonucleotides are amorphous in nature, and therefore polymorphism does not apply. Similarly, there may be additional specification requirements for oligonucleotides that are not expected for a small molecule NCEs as described in ICH Q6A but is expected for oligonucleotides. An example of this is the possible need for molecular sequencing as an assurance of identity or bioactivity assays, in the case of aptamers.

Because of the similarities in the manufacturing processes between peptides and oligonucleotides—both are manufactured by solid phase chemical synthesis followed by preparative chromatographic purification and downstream processing—it has been helpful to apply best industry practices utilized by the peptide arena to oligonucleotides.

19.3 CHARACTERIZATION OF OLIGONUCLEOTIDE DRUG SUBSTANCES

19.3.1 Nomenclature and Structure

Description of an oligonucleotide drug substance begins with a listing of all the names used for that molecule during development. These include laboratory code names, chemical name, IUPAC name, International Non-Proprietary Name (INN), other nonproprietary names such as United States Adopted Name (USAN) or British Adopted Name (BAN), etc., and Chemical Abstracts Service Registry Number (CAS).

In addition to commonly used naming systems for small molecules, oligonucleotides have other naming conventions that describe their unique sequence in a simplified way.[6] The example shown below is for a random RNA sequence with 2-*O*-methyl and 2-fluoro modifications in addition to dT groups at the 3′ end and a phosphorothioate between the two deoxythimidines. The sequence may be described using the following variation of standard oligonucleotide nomenclature, where A = adenosine, C = cytidine, C_m = 2′-*O*-methylcytidine, C_f = 2′- fluorocytidine, G = guanosine, U = uridine, and dT = deoxythymidine.

5′-C_mAC_f-AUC-CAC-GAG-ACC-AGC-CdT-ps-dT- 3′

Another commonly accepted oligonucleotide nomenclature is as follows:

(2′-O-Methyl)cytidylyl-adenosylyl-(2′-fluoro)cytidylyl-adenosylyl-uracilyl-cytidylyl-cytidylyl-adenosylyl-cytidylyl-guanosylyl-adenosylyl-guanosylyl-adenosylyl-cytidylyl-cytidylyl-adenosylyl-guanosylyl-cytidylyl-cytidylyl-thymidylyl-(3′→5′*O,O*-phosphorothioyl)-thymidine, 20-sodium salt.

Oligonucleotides may also be named using a convention in which the heterocyclic base names are given three letter abbreviations:

Cyd_m-Ado-Cyd_f-Ado-Uro-Cyd-Cyd-Ado-Cyd-Guo-Ado-Guo-Ado-Cyd-Cyd-Ado-Guo-Cyd-Cyd-Thd-ps-Thd

Molecular structure, along with molecular weight of the sodium salt and the free acid, are also described.

19.3.2 Proof of Structure

Elucidation of the structure of oligonucleotides is often performed utilizing techniques that are commonly used for small molecules as well as those that are unique for oligonucleotides. Proof of structure is usually established by a combination of techniques, which when taken together provide definitive structure information. As such, it is important to note that the sponsor is not expected to perform *all* of the characterization tests discussed, but those that contribute meaningfully to confirming of the purported structure.

One of the most important aspects of proof of structure is the confirmation of molecular weight and molecular sequence. Molecular weight (Chapter 4) of an oligonucleotide is usually confirmed by high-resolution mass spectrometry such as electrospray ionization (ESI-MS), although matrix assisted laser desorption ionization–time of flight–mass spectrometry (MALDI-TOF-MS) has also been deemed acceptable provided it has adequate mass resolution for the intended application.

Definitive establishment of molecular sequence was problematic for antisense molecules, which typically included phosphorothioate modification to afford nuclease resistance in biological matrices. The introduction of the phosphorothioate modification rendered conventional enzyme-based sequencing techniques ineffective. This was successfully overcome with oxidation of the phosphorothioate to the native phosphodiester using a pre-oxidation step followed by enzymatic digestion. The digestion product was then monitored by MALDI-TOF-MS to confirm the entire sequence.[7] Other approaches for sequence analysis of chemically modified oligonucleotides include chemical cleavage methods such as Maxam-Gilbert[8] and direct sequencing by ESI-MS-MS.[9] These techniques are discussed in more detail in Chapter 5. Direct sequencing of a siRNA duplex by MALDI-TOF mass spectrometry after acid hydrolysis was recently reported.[10]

Base composition analysis (Chapter 17), conceptually similar to amino acid analysis for peptides, has been used in demonstrating the correct stoichiometric equivalents of the various bases in the target sequence. This analysis provides additional confirmatory evidence to mass spectrometry for establishing the correct sequence. In base composition analysis, the oligonucleotide is subjected to either chemical or enzymatic digestion followed by simple chromatography to quantitate the various bases. For sequences that are enzymatically resistant such as phosphorothioates, a pre-oxidation step can be used prior to enzymatic digestion and chromatographic analysis.[11]

Spectroscopic analysis is also used to provide supportive information for demonstrating proof of structure. It is recognized that for large molecules such as oligonucleotides, the spectra are quite complex and definitive proof of structure is not possible by using spectroscopic methods alone. Commonly used spectroscopic methods include Fourier transform infrared spectroscopy (FT-IR) (Chapter 14) and nuclear magnetic resonance spectroscopy (NMR) (Chapter 13).

FT-IR spectral frequencies provide evidence for the presence of P=O bonds, amino groups, carbon-carbon double bonds, and carbonyl groups of nucleobases.[12,13] Proton NMR[14] and carbon NMR[15] may be used to confirm the presence of resonances consistent with the DNA or RNA strands. Phosphorous NMR may used to demonstrate the backbone composition and nature of the internucleotide linkage.[16] In the case of phosphorothioates, a characteristic resonance at 57 ppm may be observed, providing evidence of P=S bonds. Ultraviolet spectroscopy (UV) is often used to confirm an absorbance maximum in the range of 255–260 nm consistent with the presence of a DNA or RNA oligomer. It is common to record the UV spectrum in water and a phosphate buffer at neutral pH.

A characterization test that is unique to oligonucleotides is the measurement of the hybridization temperature, also known as the thermal melt temperature (T_m) (Chapter 6). T_m measures the affinity of an oligonucleotide to its complementary strand by UV spectroscopy. The higher the T_m, the more stable the resultant duplex oligonucleotide. T_m can be very sensitive to subtle changes in the sequence such as a base deletion or a base substitution, resulting in a measurable difference in the melting temperature.[17] Therefore, this test is a simple, yet powerful means to provide supportive evidence of the correct sequence. For antisense oligonucleotides that function by binding to their messenger RNA target, this has served as a useful surrogate activity assay. For siRNAs, which are RNA duplexes, the T_m test has been used to demonstrate duplex identity as well as stability. In the case of aptamers, which must adopt a well-defined conformation for pharmacological activity, T_m has been a useful test to confirm correct conformation in solution.

Oligonucleotide chain length is demonstrated by a variety of techniques including polyacrylamide gel electrophoresis (PAGE), capillary gel electrophoresis (CGE) (Chapter 7), or other HPLC methods (IP-RP-HPLC and AX-HPLC) (Chapters 1 and 2). For chemically modified oligonucleotides such as DNA phosphorothioates, AX-HPLC has proven equivalent or superior to ^{31}P-NMR for providing definitive evidence of the presence of predominantly phosphorothioate backbone. For most antisense DNA phosphorothioates of 20 or 21 base units in length, AX-HPLC has demonstrated exquisite resolution to separate the fully phosphorothioated sequence from the full length monophosphodiester analog, which arises from sulfurization failure at a single location.[18]

In the case of double-stranded oligonucleotides such as DNA duplex decoys or siRNA, structure elucidation of each strand is performed prior to confirming features related to the duplex. As a first step, mass spectrometry of the duplex to demonstrate the presence of the correct single strands is useful. T_m may also be used to confirm the presence of the correct duplex structure as subtle differences in the sequence of either strand will have a significant impact on the T_m. Chromatographic methods such as size exclusion HPLC (SEC) (Chapter 3) or nondenaturing electrophoresis methods (PAGE, CGE) may be used to confirm the presence of a duplex structure. Circular dichroism (CD) has also been used to confirm the presence of a classical A form signature of DNA and RNA duplexes.[19]

For siRNA, H-NMR spectroscopy to record the resonances of imino protons of U and G has been helpful for establishing duplex structure. For oligonucleotides in the single-strand form, the imino protons rapidly exchange with the deuterium atoms in the solvent, and thus no signals are observed in the ^{1}H-NMR spectrum. When imino protons are involved in base pair forming hydrogen bonds, as in the case of duplex structure, exchange with deuterium is slow and distinct resonances are observed. As a result, presence of the signals for the imino protons in the ^{1}H-NMR spectrum confirms the duplex structure.[20] In the case of larger oligonucleotides such as aptamers, a bioactivity assay is also used as additional evidence of correct structure and conformation.

19.3.3 Physical–Chemical Characteristics

The physical-chemical characterization of oligonucleotides is similar to that of small molecules. Oligonucleotides are generally lyophilized powders and are amorphous in nature. They are highly electrostatic owing to the high degree of charge and are also highly hygroscopic. They are readily soluble in water at solubilities higher than 300 mg/mL and often result in a deliquescent gel upon saturation.

As part of the characterization of new oligonucleotide sequences, it is essential to determine the solubility in aqueous solutions and other organic media that the oligonucleotide may come in contact with during manufacturing and downstream processing. Also important is the solubility of the oligonucleotide candidate in formulation buffers used for toxicology and clinical supplies. PEGylated oligonucleotide solutions of high concentrations are often viscous solutions that may be difficult to handle and use for clinical dosing. Therefore, it is recommended to record both the solubility and viscosity of PEGylated oligonucleotides in order to confirm that the concentrations needed to support preclinical and clinical dosing are not too viscous for use.

It is common to record the pH of a 1% (w/w) solution of an oligonucleotide. The pKa measurements are also useful; however, it may be difficult to determine a unique pKa value for the molecule owing to the presence of a large number of ionizable protons.[21]

Oligonucleotides are highly hygroscopic, partly attributable to their amorphous nature. In order to understand the extent of sensitivity to ambient moisture, it is recommended that formal studies be conducted to characterize the hygroscopicity of the oligonucleotide candidate under simulated laboratory handling conditions. The most commonly used technique is dynamic vapor sorption (DVA) analysis in which the response of a sample subjected to controlled temperature and humidity conditions is measured gravimetrically. A plot of the change in sample weight as a function of temperature and humidity is recorded. As quality control testing includes a number of tests that are dependent on accurate sample weight, a good understanding of the hygroscopicity of the molecule can prove very useful in assessing the operating limitations of quantitative analysis. Similarly, recording the moisture content of all batches of oligonucleotides is important for assessing the freeze dry step in manufacturing.

Oligonucleotides have a characteristic absorbance in the vicinity of 260 nm. It is helpful to determine the molar extinction coefficient of the oligonucleotide in water, relevant buffers and the formulation matrix. Experimentally determined molar extinction coefficient in relevant buffers allows quick assessment of oligonucleotide content by UV spectroscopy in a given sample. There are various approaches to determination of molar extinction coefficient. These are covered in detail in Chapter 12.

19.3.4 Impurities in Oligonucleotides

Impurities that may be present in oligonucleotide drug substances are described as oligonucleotide impurities, organic impurities, inorganic impurities including heavy metals, and residual solvents.

19.3.4.1 Oligonucleotide Impurities

Four common classes of oligonucleotide impurities are likely to be present: oligonucleotide sequences of shorter chain length (deletion sequences), longer chain length (addition sequences), phosphodiesters resulting from sulfurization failures in the case of oligonucleotides with phosphorothioate modifications, and modified sequences. Thorough characterization of the impurities present in oligonucleotide drug substances are most efficiently performed by the use of LC-MS analysis. This is discussed in more detail in Chapter 4.

Deletion sequences (n-x) arise mainly from incomplete coupling, although they can also be produced by incomplete oxidation and cleavage of the phosphate triester during subsequent detritylation under acidic conditions. This leads to the loss of the newly coupled nucleoside. In most cases the sequence resulting from either the coupling or oxidation steps is capped prior to the start of the next coupling cycle. This produces a shortmer oligonucleotide or truncated sequence that will not be available for coupling. As a consequence, these nonreactive oligonucleotides will be removed during the downstream processing steps, i.e., preparative chromatographic purification. In the situation where the failure is not capped, an internal failure or deletion sequence will result. These impurities contribute only a very small amount to the overall impurity profile. Most purity methods are capable of detecting (n-x) impurities in sequences in the range of 20 bases. By far, the most common deletion sequence impurities found in the final drug substance are the (n–1) sequences as the higher-order deletion sequences are more effectively removed during preparative chromatographic purification. Complete site-specific analysis of the deletion sequences is not easily achieved;[22–24] therefore, control of "total n–1 deletion sequences" has thus far been considered adequate. For longer sequences such as aptamers, it may be difficult to achieve single base resolution to detect (n–1) sequences in the sample.

Addition sequences (n+x) are produced during the coupling step by acid-catalyzed deprotection of the 5′-OH group of the incoming phosphoramidite. This results in a double coupling whereby the particular phosphoramidite is added twice during a single coupling step.[25,26] Similar to n-x sequences, most purity methods have the selectivity needed to resolve n+x impurities from full length oligonucleotide products of 20 base units.

In the case of phosphorothioate oligonucleotides, sulfurization failures result in the presence of one or more phosphodiesters in place of the phosphorothioate. For sequences that function by the antisense mechanism of action, these mono- or diphosphodiester impurities are expected to bind to the mRNA complement equally well and therefore are considered pharmacologically active. Nevertheless, anion-exchange HPLC can be used to separate and control the monophosphodiesters from the fully sulfurized full length sequence for most antisense oligonucleotides of approximately 20 base units in length.

Another class of impurities consists of full length oligonucleotides that contain some of the protecting groups such as isobutyryl, benzoyl, or cyanoethyl adducts. These are formed as a result of incomplete deprotection of the oligonucleotide. In general, these impurities may be detected by the chromatographic techniques commonly employed for purity analysis.

Another class of impurities, depurinated oligonucleotides, have been detected in some sequences. Because of structural similarity to the oligonucleotide of interest, it has been difficult to achieve chromatographic separation of these impurities. However, these impurities are readily determined by mass spectrometric analysis (Chapter 4).

Some oligonucleotides have shown a tendency to aggregate or form multimers. Fortunately, because of the larger size, it has been relatively easy to detect aggregated forms, if present, by conventional purity methods and in the case of larger oligonucleotides such as aptamers, by SEC (Chapter 3).

19.3.4.2 Organic Impurities

Nonoligonucleotide-related organic impurities include organic molecules such as starting materials, by-products, intermediates, and reagents. Because of the extensive wash steps involved in each step of the synthesis process as well as chromatographic purification and downstream processing, it is not common to find these impurities in purified oligonucleotide drug substances.

19.3.4.3 Inorganic Impurities

There are no heavy metal catalysts or other metal-based reagents used in the manufacture of oligonucleotides. Thus, purified oligonucleotides are generally free of heavy metal impurities. Any metals present are introduced by raw materials and the equipment used during manufacture. Therefore, oligonucleotides are usually tested for specific metals such as chromium and copper by highly sensitive techniques such as ICP-MS (inductively coupled plasma with mass spectrometric detection) (Chapter 18).

During manufacturing, oligonucleotides are usually purified by preparative anion-exchange chromatography. The resultant purified oligonucleotide is subjected to ultrafiltration to remove salts such as sodium bromide, sodium chloride, and sodium phosphate. Although this desalting step is extremely effective and well controlled by the use of conductivity meters during processing, the resultant purified oligonucleotides are often tested for residual salts in the final drug substance. The most common technique for the analysis of residual salts is by ion chromatography, but this analysis can also be performed using capillary zone electrophoresis.[27]

19.3.4.4 Residual Solvents

Residual solvents used in the synthesis process are effectively removed during the purification step. Therefore, the only solvents that are generally detected in the purified drug substance are those used during purification or downstream processing. Still, oligonucleotide drug substances are tested for residual solvents by capillary gas chromatography with direct, or more commonly, headspace injection (Chapter 11).

19.4 ASSURING OLIGONUCLEOTIDE QUALITY THROUGH CONTROL OF THE MANUFACTURING PROCESS

The quality of the oligonucleotide drug is assured through systematic control of factors that affect overall quality including control of raw materials, adequate in process controls, and final product testing. In addition, special attention is paid to other factors that may affect final product quality, as defined in ICH Q7A, such as facilities, equipment, personnel, and systems.[28]

The manufacturing process for a typical DNA or RNA oligonucleotide consists of chemical synthesis of the oligonucleotide by conventional nonstereoselective solid phase synthesis using 3′-*O*-(2-cyanoethyl) phosphoramidite chemistry with the 5′-hydroxyls protected with 4,4′-dimethoxytriphenylmethyl (dimethoxytrityl, DMT) groups and tert-butyldimethylsilyl (TBDMS) protection on the 2′-hydroxyls of the ribose nucleosides. Assembly of an oligonucleotide chain by the phosphoramidite method on a solid support such as controlled pore glass (CPG), polystyrene (PS) or other polymeric support is shown in Figure 19.1. More recently, universal supports such as Unylinker™ have been successfully used.[29] Each elongation cycle consists of 5′- hydroxyl deprotection, coupling, oxidation and capping. Each coupling reaction is carried out by activation of the appropriate amidites and reaction with the free 5′-hydroxyl group of a support-immobilized protected nucleotide or oligonucleotide. After the appropriate number of cycles, the crude oligonucleotide is cleaved from the solid support followed by deprotection of the relevant protecting groups. The resulting crude oligonucleotide is purified using preparative chromatography followed by desalting and freeze drying to yield the oligonucleotide drug substance. A process flow diagram for a generic single-stranded DNA/RNA oligonucleotide is depicted schematically in the Figure 19.2.

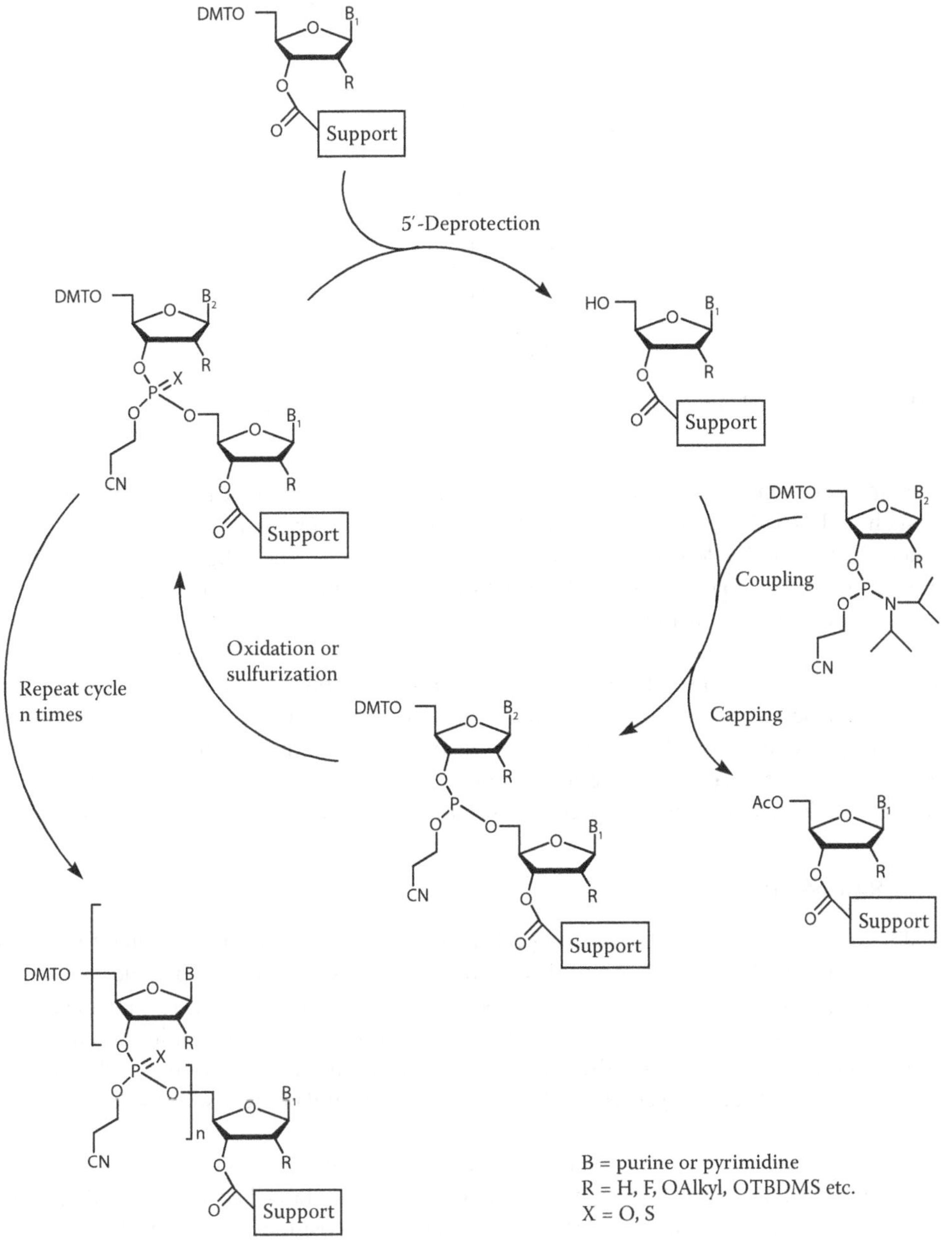

FIGURE 19.1 Cycle of steps in the solid phase synthesis of an oligonucleotide.

In the case of phosphorothioate oligonucleotides, replacement of the nonbridging oxygen at each phosphorus in the backbone by a sulfur results in R_p and S_p diastereoisomers. For a 20-base phosphorothioate sequence, this represents 524,288 (2^{19}) possible diastereoisomers, well beyond the resolution capability of any analytical method to date. Although there have been literature reports of controlling the stereochemistry around the phosphorous during synthesis, for the purposes of practical drug development, it has been sufficient to demonstrate process control of nonstereoselective or stereorandom synthesis.[2,3,30,31]

The process description in a regulatory application generally includes a process flow diagram, chemical description of various steps in the process, purification details, and descriptions of

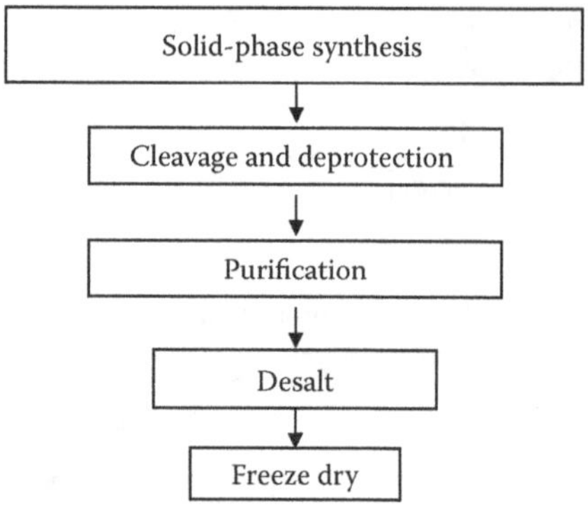

FIGURE 19.2 A general process flow diagram for the manufacture of a single-strand oligonucleotide.

downstream processing steps such as desalting and freeze-drying. In the case of duplex oligonucleotides such as DNA duplex decoys or siRNA, the annealing step to form the final duplex drug substance should also be described. For PEGylated aptamers or other oligonucleotide conjugates, manufacturing of the oligonucleotide along with the description of the conjugation reaction, and details of further purification and downstream processing of the resultant oligonucleotide-conjugate should be presented.

The manufacturing site used for manufacture of clinical supplies is specified along with a statement of compliance with current good manufacturing practices (cGMP). All sites involved in the manufacturing, quality control testing, stability testing, labeling, packaging, and distribution are also listed along with their respective responsibilities. Usually, the responsibilities of the various groups are well defined in a quality agreement prior to initiation of manufacturing operations.

19.4.1 Raw Materials

The quality of the starting materials and reagents play a critical role in the overall quality of the final purified oligonucleotide drug substance. Therefore, special attention should be paid to ensure and document the quality of the starting materials used in oligonucleotide synthesis. Starting materials for oligonucleotide synthesis generally include the appropriate phosphoramidite building blocks, the solid support, and if applicable, the sulfurization reagent. In the case of oligonucleotide conjugates, the conjugate such as polyethylene glycol (PEG) and the linker are also considered as starting materials.

Because of their potential impact on the overall quality of the resultant oligonucleotide, the quality of the starting materials should be well controlled and described in the regulatory application. Description of the solid support should include composition (e.g., controlled pore glass, polystyrene, etc.), particle size, pore size, chemical description of the linker, spacer, and the nucleoside that is loaded on the support and DMT loading. Critical reactive impurities such as trivalent phosphorus species in phosphoramidite starting materials should be tested for and controlled by ^{31}P-NMR (Chapter 13) and/or HPLC. Nonreactive impurities such as hydrolysis/oxidation products or phosphate triesters may also be monitored and controlled. Structural characterization of the impurities in phosphoramidite starting materials by LC-MS is becoming more common and has been helpful in delineating the correlation between the quality of the amidite monomer and final product quality.

Critical attributes of other raw materials, reagents, and solvents should be addressed via quality specifications and controlled prior to use. The source of all raw materials (plant, animal, or synthetic) should be described and TSE/BSE (transmissible spongiform encephalopathies including bovine spongiform encephalopathy) certification obtained from respective vendors.

19.4.2 In-Process Controls

An essential part of assuring quality of the final drug substance is appropriate controls during various stages of manufacturing. Before initiation of synthesis, synthesis parameters are verified and the synthesis program documented. Identity of all starting materials such as phosphoramidites is confirmed prior to use. Quality of the crude oligonucleotide is tested after cleavage from the solid support and subsequent deprotection. Effectiveness of the deprotection reaction is also monitored and recorded. During development, analysis of fractions from preparative chromatographic purification runs is used to make pooling decisions. If appropriate, mock pools are made and analyzed to provide additional assurance of desired quality. Completion of desalting is monitored using conductivity readings. Typically, vacuum and temperature is monitored and controlled throughout the lyophilization cycle. The entire manufacturing operation may be streamlined as experience is gained about the manufacturing process and the process parameters are established to yield product of consistently high quality.

For duplex oligonucleotides, the purified single strands are annealed to form the duplex drug substance. The amount of duplex in the final drug substance, and therefore the purity of the final duplex, is influenced by the ability to bring together precisely one-to-one molar equivalents of the sense and the antisense strands during the annealing step of drug substance manufacturing. In order to afford control of the annealing process, it is necessary to determine the molar extinction coefficient of the two strands in the annealing medium. The annealing reaction may be monitored and controlled using a variety of techniques including RP-HPLC (Chapter 1), AX-HPLC (Chapter 2), and SEC (Chapter 3). Post annealing, the duplex drug product is moved forward to lyophilization.

Similarly, in the case of conjugated oligonucleotides such as PEGylated aptamers, the oligonucleotide is moved forward to the PEGylation step and the resultant oligonucleotide conjugate is purified prior to desalting and lyophilization. In-process controls should be in place to monitor the PEGylation reaction as well as the preparative chromatographic purification step. Commonly performed in-process control testing during oligonucleotide manufacturing is described in Table 19.1.

19.4.3 Validation of Oligonucleotide Manufacturing Process

During late stage development in preparation for commercialization, it is necessary to validate the oligonucleotide manufacturing process to ensure consistent performance during commercial marketing.[32,33]

Validation of an oligonucleotide manufacturing process follows the essential elements of those for small molecules but with additional considerations relating to solid phase synthesis.[34] These include appropriate qualification studies performed on all equipment used in manufacturing including the solid phase synthesizer, purification systems, and downstream processing. In the case of

TABLE 19.1
General Description of in Process Controls for Single-Stranded Synthetic Oligonucleotides

Process Step	Test	Purpose
After solid phase synthesis, cleavage and deprotection	UV	Determine yield
	HPLC, MS	Determine amount of full length oligonucleotide and overall quality of crude
During preparative chromatographic purification	UV	Determine yield
	HPLC	Analyze fraction pools by HPLC toward pooling decisions
Post ultrafiltration	Conductivity	Confirm desalting
Freeze-drying	Various methods	Quality control release testing

automated solid phase synthesizers, appropriate qualification of both the hardware and software should be performed. Qualification of the automated synthesizer includes the computerized phosphoramidite monomer and reagent delivery system through appropriate IQ, OQ, PQ (installation, operation, and performance qualification) studies. Performance qualification of the phosphoramidite delivery sequence may be performed by synthesis of test sequences containing all possible amidites and analysis of resultant product using validated methods. In addition to the starting materials, key raw materials and reagents are identified and the impact of their quality on the final product considered and controlled. For oligonucleotides, critical raw materials include the solid supports, phosphoramidite monomers, deblock reagent, and capping reagents. Other critical reagents include linkers and reagents for sulfurization, conjugation, and other chemistries.

Critical process parameters are identified and optimum operating ranges are established for all critical steps of the process. For oligonucleotides, these include various chemical reactions of the synthesis cycle to consistently yield oligonucleotide of appropriate quality; efficiency of the cleavage reaction from the solid support; detritylation and other deprotection steps; chromatographic purification; pooling strategies; reprocessing of side fractions; desalting and ultimately, lyophilization. Validation is often carried out as unit operations, thereby allowing flexibility for additional validation if a specific unit operation is scaled up or further optimized post validation.

In the case of oligonucleotide duplexes such as DNA duplex decoys or siRNAs, the annealing step is critical and should be appropriately validated. For oligonucleotide conjugates such as PEGylated oligonucleotides, the conjugation step and subsequent purification should be well characterized and validated. Noncritical process parameters should be justified with appropriate rationale or experimental data. Stability of intermediates at logical stop points in the process should be assessed using validated methods to guide limits on storage times. If a multiproduct facility is utilized for the production of intended commercial supply, then appropriate cleaning validations should be performed to document adequacy of cleaning procedures. Although highly sensitive general techniques such as total organic carbon analysis (TOC) are commonly used, if appropriate, sequence specific tests should be considered.

Although most oligonucleotide manufacturing facilities are not considered sterile, appropriate controls should be in place to ensure that the resultant drug product, usually a sterile solution intended for parenteral administration, will meet the strict requirements for endotoxins (Chapter 10), bioburden, and sterility.

19.5 QUALITY CONTROL TESTING OF OLIGONUCLEOTIDE DRUG SUBSTANCES

Assuring the quality of the final oligonucleotide drug substance is a culmination of appropriate controls placed on the manufacturing equipment, starting materials, reagents, and solvents; adequate in process controls and finally, testing the purified drug substance using methods that have been demonstrated to be suitable for use.

Quality control specifications for oligonucleotides generally follow the requirements for NCEs as described in ICH Q6A[5] and relevant compendia.[35] However, because of the large size of most oligonucleotides, there are other requirements that are specific to this class of molecules. Generic specifications along with commonly used techniques for the three most common classes of oligonucleotides in development are summarized in Tables 19.2 through 19.4. These are discussed in more detail in subsequent sections. The general information provided in these tables may be adapted to other classes of oligonucleotides.

Sponsors are expected to provide a rationale and justification for the tests and limits proposed in quality control specifications. For tests where limits are not proposed, it is advisable to explain the reasons why limits are not needed or not being proposed. In some cases, it may be appropriate to include a discussion of specification tests not proposed and the rationale for omitting such a test (e.g., bioassays).

TABLE 19.2
Recommended Quality Control Strategy for Single-Stranded Oligonucleotides

Attribute	Commonly Used Methods
Appearance	Visual
Identity, molecular weight	Mass spectrometry
Identity, sequence confirmation	ESI-MS-MS
	Enzymatic or chemical digestion and MALDI-TOF
Identity, counterion	IC, AA, ICP-OES, or ICP-MS
Purity	HPLC (anion exchange, reversed phase)
	LC-MS
	CGE
Assay	UV multiplied by purity
	HPLC
	CGE
pH of aqueous solution	pH meter
Moisture content	Karl Fischer titration
Residual solvents	Capillary gas chromatography
Heavy metals	ICP-MS
Bacterial endotoxins	USP
Bioburden	USP

TABLE 19.3
Recommended Quality Control Strategy for Double-Stranded Oligonucleotides

Attribute	Commonly Used Methods
In-process control of single strands	
Identity, molecular eight	Mass spectrometry
Identity, molecular sequence	ESI-MS-MS
	Enzymatic or chemical digestion and MALDI-TOF
Purity, for process-related impurities	HPLC (anion exchange, reversed phase)
oligonucleotide impurities	CGE
Quality control of the duplex drug substance	
Appearance	Visual
Identity, for presence of correct single strands	Mass spectrometry
	HLPC
Identity, confirmation of duplex	Nondenaturing HPLC
	Nondenaturing CGE
	T_m
Identity, counterion	IC, AA, ICP-OES, or ICP-MS
Purity, for amount of duplex, excess single-strand	Nondenaturing HPLC
and unhybridized impurities	CGE
Assay	UV multiplied by duplex purity
	Nondenaturing HPLC
	Nondenaturing CGE
Moisture content	Karl Fischer titration
Residual solvents	Capillary gas chromatography
Heavy metals	ICP-MS
Bacterial endotoxins	USP
Bioburden	USP

TABLE 19.4
Recommended Quality Control Strategy for PEGylated Aptamers

Attribute	Commonly Used Methods
In-process control of purified aptamer oligonucleotide	
Identity, molecular weight	Mass spectrometry
Identity, molecular sequence (if possible)	ESI-MS-MS
	Enzymatic or chemical digestion and MALDI-TOF
Purity, for process-related oligonucleotide impurities	HPLC (anion exchange, reversed phase)
Quality Control of the PEGylated aptamer	
Appearance	Visual
Identity, presence of PEGylated aptamer	HPLC
Identity, molecular weight	SEC
Identity, counterion	IC, AA, ICP-OES, or ICP-MS
Purity, for presence of unPEGylated oligonucleotide and presence of aggregates	SEC including polydispersity HPLC (anion exchange, reversed phase, size exclusion)
Assay	UV multiplied by purity HPLC
Moisture content	Karl Fischer titration
Residual solvents	Capillary gas chromatography
Heavy metals	ICP-MS
Bacterial endotoxins	USP
Bioburden	USP

19.5.1 Identity

Appearance of the oligonucleotide is generally recorded prior to initiation of other testing. It is fundamentally necessary to establish the identity of each batch of oligonucleotide. The most commonly used technique is high-resolution mass spectrometry such as ESI-MS; however, MALDI-TOF-MS is also acceptable, provided it has sufficient mass resolution for the intended application. Molecular sequencing is an important test of identity. This can be performed by either chemical or enzymatic degradation followed by MS; in source fragmentation approaches by MALDI-TOF-MS and ESI-MS-MS. Complete sequencing of chemically modified oligonucleotides may be difficult to achieve because of their inherent resistance to commonly used enzymes. In these cases, it is acceptable to utilize a multitude of techniques to establish definitive sequence.

For duplex drug substances, mass spectrometric analysis provides evidence for the presence of the correct single strands in the molecule. Evidence of duplex structure is often confirmed by chromatographic (SEC) or nondenaturing electrophoretic analysis such as polyacrylamide gel electrophoresis (PAGE) of capillary gel electrophoresis (CGE). T_m can also serve as a reliable method for assuring identity of a given duplex. The identity of the counter ion, usually sodium, is readily established by ion chromatography (IC), atomic absorption spectroscopy (AA), inductively coupled plasma with optical emission spectroscopy (ICP-OES), or with MS detection (ICP-MS).

19.5.2 Purity

Process related impurities in oligonucleotide drug substances are usually determined by HPLC or CGE. Various modes of separation are used including anion exchange (Chapter 2), ion pair reversed phase (Chapter 1), hydrophilic interaction (Chapter 16), size exclusion (Chapter 3), and electrophoretic methods (Chapter 7). More recently, LC-MS has emerged as the technique of choice for structural characterization of the process related impurities in synthetic oligonucleotides (Chapter 4). The major objective of a purity method for synthetic oligonucleotides is to separate the process related impurities from the full length oligonucleotide drug. Length-selective methods are particularly important to ensure adequate resolution of the addition and deletion sequences from the full length oligonucleotide. Methods should also have the specificity to resolve organic adducts or partially deprotected sequences from the product. For phosphorothioate sequences, presence of phosphodiesters should be determined, usually by anion exchange HPLC or MS. Although ^{31}P-NMR is commonly used, it often lacks the sensitivity required to detect small amounts of phosphodiester species present in the drug (Chapter 13).

For oligonucleotide duplexes, the purity of the two individual strands is established as part of in-process control testing prior to the annealing step. Once the impurity profile is established at this stage, it is generally not necessary to retest for the same process related impurities in the annealed duplex. However, it is important to confirm that the annealing step in itself does not increase the amount of these impurities controlled at the single-strand stage. Thus the impurity analysis at the duplex stage can be limited to confirming completion of the annealing step and controlling the amount of free single strands in the sample. This is generally performed using chromatographic (SEC) or nondenaturing electrophoretic (CGE) methods. The amount of duplex in the final drug substance, and therefore the purity of the final duplex, are influenced by the ability to bring together precisely one-to-one molar equivalents of the sense and the antisense strands during the annealing step of drug substance manufacturing. Excess of either strand should be detectable by the impurity profile method in addition to those single-strand impurities (e.g., internal deletion sequences in either strand) that do not hybridize with the complementary strands.

For oligonucleotide conjugates, such as PEGylated aptamers, purity of the oligonucleotide is established as part of in-process controls prior to the PEGylation reaction. This is particularly important, as the resolution of the chromatographic methods are often insufficient to adequately determine the oligonucleotide process related impurities in the final PEGylated molecule whose molecular weights can approach *ca.* 50 kD. Purity analysis of PEGylated oligonucleotides should have the specificity to detect aggregation or other high-order structures. Additional control of final quality of the PEGylated aptamers is assured through rigorous quality control of the incoming PEG entity for purity and polydispersity. PEG is inherently unstable at high temperatures, resulting in formation of peroxides which can, in turn, impact stability of the final PEGylated oligonucleotide. Therefore, adequate care should be taken during handling and storage of the PEG raw material to ensure long-term chemical stability. Testing for residual peroxide in the starting PEG raw material may be useful in controlling quality.[36–38]

Limits for impurities are based on the levels of impurities qualified by toxicology studies, batch analysis, as well as process and analytical capabilities.[4,39] Even if there is insufficient information to set limits for impurities at the early development stage, it is important to record the levels of process related impurities in the format described in the ICH guideline Q3A.[5] These include specified and unspecified impurities as well as total impurities. Because of the complexity of oligonucleotides, the commonly used reporting, identification, and qualification thresholds for small molecules do not apply to oligonucleotides. Instead, these limits are generally proposed by the sponsor and considered by the regulatory authorities on a case-by-case basis. Their assessment will be based on the clinical indication, route of administration, total dose, frequency of dosing, process and analytical limitations, and any other information that may relate to the overall safety of the intended drug substance.

19.5.3 Strength or Assay

A specific, stability-indicating method is ideal for the analysis of content of an oligonucleotide drug substance. This may include analysis of the sample by the most discerning and specific analytical method such as chromatography or electrophoresis against a well-defined reference standard. In practice, however, it has been problematic to achieve reproducible assay values for oligonucleotide drug substances.

Because of the highly hygroscopic and electrostatic nature of most oligonucleotides, quantitative assay measurements are analytically challenging. The drug substance assay measurement by the conventional approach requires accurate weighing of highly hygroscopic/electrostatic solids, determination of the moisture content of the standard and the sample, followed by HPLC/CGE analysis.[40] While careful attention is paid to simultaneous weighing of both the standard and sample for moisture and assay, any change in moisture value of any of the samples will impact the final assay value. Therefore, the error in the final assay value is a cumulative error associated with the weighing of the reference standard and samples; moisture analysis of the reference standard and sample; analytical dilutions and the analysis. In an effort to curb the error associated with the handling and analysis of the reference standard, some sponsors have established a solution-based reference standard and have frozen aliquots that are used as needed. Fortunately, most oligonucleotides have shown to be stable in a buffered solution under frozen conditions, making this approach practical in a development setting. Still, if this approach is used, it is important to monitor the stability of the frozen reference standard solution over time to ensure its integrity for use.

In early stage development, it has been common practice to utilize a simpler UV procedure that does not require analysis against a reference standard at all. Weight/weight percent assay of the oligonucleotide drug substance is determined by multiplication of the total oligonucleotide content as determined by UV (using an experimentally determined molar extinction coefficient) by the amount of full length oligonucleotide in the sample (area-percent), as measured by the most selective separation method such as HPLC or CGE. This approach has generally been accepted by regulatory authorities until additional experience is gained in performing accurate and reproducible assays. Approaches for the assay of oligonucleotides are discussed in detail in Chapter 9.

In order to minimize or eliminate the negative impact of inconsistent drug substance assay values on the label claim of the resultant drug product, most sponsors have designed their drug product manufacturing processes without the need for the drug substance assay values at all. In effect, the bulk drug product formulation is made in approximately 50–70% of the total batch volume; an in-process control sample is taken and tested for assay using a simple UV analysis that may (or may not) be multiplied by the full length oligonucleotide content (area-percent) from the drug substance quality control certificate of analysis (C of A). The bulk formulation is then adjusted with the amount of formulation buffer needed to achieve target label claim. In practice, this approach has worked extremely well and completely obviates the need for the assay values reported in the drug substance C of A in achieving label claim of the drug product.

For oligonucleotide duplexes such as siRNA, assay values are determined on the duplex drug substance as part of quality control analysis. During the annealing step of the siRNA manufacturing process, there is a need for accurate assessment of the single-strand content for the sense and antisense strands so that equimolar amounts of the two single strands can be combined to form the duplex drug substance without significant excess of either strand. At this step, the oligonucleotide content of each strand is determined accurately, usually by UV analysis as part of in-process control. In early stage development, this mixing step is titrated using in-process analysis of the resultant duplex to ensure the highest duplex content achievable. As additional experience is gained with the process as well as analytics, it may be possible to assure reproducibly high-quality duplex during routine manufacturing without the need for titration.

Assays of conjugated systems pose additional challenges. In the case of PEGylated oligonucleotides, the contribution of the PEG entity to the total molecular weight of the sample (as much as 80% of the total weight of the drug) and the polydispersity and variability around the PEG molecular weight (as much as 10% by specification), results in large errors in the assay values if reported on a PEGylated oligonucleotide sodium salt basis. Thus, for most PEGylated oligonucleotides, it has been generally acceptable to report assay values for the oligonucleotide portion of the molecule only.

19.5.4 Quality

Other tests that contribute to the overall quality of oligonucleotide drug substances may include, as appropriate, appearance, pH of an aqueous solution, moisture content by Karl Fischer titration, residual solvents by capillary gas chromatography,[41,42] and heavy metals by ICP-MS.[43,44] Microbiological attributes such as bioburden and bacterial endotoxins[45] are tested if the drug is intended to be formulated as a sterile solution. Some oligonucleotides have shown to be inhibitory in the USP gel clot test. In these situations, various strategies may be used to overcome inhibition, thereby allowing for testing of endotoxins to the levels required by the specific clinical dosing regimen. For a more detailed discussion, refer to Chapter 10. For drug substances that may be used to support other dosage forms such as oral or topical administration, other tests may be required at the drug substance stage to ensure the quality of the resultant dosage form.

19.5.5 Bioassays

There has been considerable controversy around the need for a bioassay as part of quality control testing of oligonucleotides, in part due to the large size of oligonucleotides, *ca.* 7000 Da, as compared to small molecules. The issue has been clouded by the sheer diversity of the types of oligonucleotides in development, from short single strands such as antisense or immunostimulatory oligonucleotides, double strands such as siRNA, or larger, more complex oligonucleotides such as aptamers, which, in some cases are bound to large biopolymers such as PEG. There has been some question of whether a specific conformation is necessary for activity so that chemical testing alone to assure correct sequence is not sufficient to ensure biological activity. This is certainly the case for aptamers, which are required to adopt a specific conformation for biological activity.

In the simplest analysis, regardless of the type of oligonucleotide or the intended mechanism of action, all oligonucleotides in development today are manufactured by organic chemical synthesis. They are synthesized by well-controlled sequential coupling of the appropriate phosphoramidite monomers to an elongating oligonucleotide that is covalently bound to a solid support. The sequence assembly is performed using commercially available, automated, computer-controlled oligonucleotide synthesizers. The absolute identity of the resulting oligonucleotides is assured by a combination of molecular weight and sequence confirmation by high-resolution mass spectrometry. In view of the totally chemical synthesis of the oligonucleotides along with the comprehensive chemical analysis of the molecule at every stage of manufacturing to ensure sequence identity and authenticity, it is reasonable to argue that oligonucleotides are more similar to small molecules than a biologic, and therefore a bioassay should not be necessary as a quality control release test of the oligonucleotides.

However, this argument in itself is insufficient. It is still important to demonstrate that the oligonucleotide as formulated will act as intended to confer biological activity. In the case of short-strand oligonucleotides such as those used for antisense and immunostimulatory applications, the regulatory authorities are generally in agreement that a bioassay is not required.[2,3] Similarly, for duplexes such as siRNA, the duplex nature of the drug is confirmed through a variety of tests as part of quality control analysis, and therefore a bioassay has not been required to date.

In the case of aptamers, the activity of the oligonucleotide is very much dependent upon the conformation of the aptamer. Therefore, at this time, nearly all regulatory authorities require a bioassay as part of release testing of an aptamer at the drug product stage. One concern faced by the sponsors is the insensitivity and poor reproducibility of bioactivity assays as compared with chemical tests. As we gather more experience in the development of aptamers, it may be possible to validate the manufacturing and formulation processes such that, within well-controlled ranges of manufacturing and formulation components, the aptamer is assured of the correct conformation and thereby eliminate the need for a bioassay. Also, it may be possible to develop a surrogate, highly sensitive biochemical test that is responsive to the conformational aspect of aptamers. As with any novel approach to assuring quality, such ideas should be proactively discussed with the regulatory agencies prior to implementation.

19.6 ANALYTICAL METHODS USED FOR OLIGONUCLEOTIDE ANALYSIS

19.6.1 Suitability of Methods and Validation

As for small molecule drugs, the analytical methods used for testing clinical supplies should be demonstrated to be suitable for their intended use. Most of the analytical methods used for oligonucleotide analysis tend to be compendial or generic for this class of molecules. The generic methods include tests for appearance, pH, moisture content, residual solvents, heavy metals, endotoxins and bioburden. Molecular weight determination and sequencing, if performed by mass spectrometry, may be acceptable provided appropriate system suitability requirements are met prior to sample analysis. Sequence/compound specific analytical methods such as purity and assay should be well characterized prior to use for quality control release testing and stability monitoring. Other methods that are unique to the oligonucleotide of interest should also be well understood.

As development continues, all methods should be validated to regulatory standards as described in relevant USP and ICH guidelines.[46–48]

19.6.2 Reference Standards

It is recommended that a well-characterized reference standard be established early in the development program. It is often the case that the purest batch of oligonucleotide available is set aside as the reference standard. If a sponsor chooses to make a separate reference standard batch, it is usually prepared using a similar manufacturing process (although not strictly necessary), but tighter fractions may be pooled during preparative chromatography to yield a standard of high purity. It is also acceptable to implement a second purification step to achieve a high-purity reference standard. Some sponsors perform proof of structure and structure elucidation analysis on the reference standard in addition to the standard quality control testing. If it is not possible to establish a high-purity reference standard early in the program, it is recommended that a secondary or working reference standard be designated such that all testing is related back to this working standard.

As development continues, it is appropriate to consider setting up impurity standards, which are mixtures of authentic process related impurities, which may be run as a system suitability standard to confirm adequate resolution of the chromatography or electrophoresis system prior to sample analysis. Other types of standards include sensitivity standards to confirm system response factors, assay standard to determine concentration, identity standards to confirm proof of identity, and standards for biological activity.

All reference standards should have associated documentation, be stored under controlled conditions, and have their stability monitored on a periodic basis. As information is gathered on the stability of the reference standards, it is common practice to establish expiration dates.

19.7 STABILITY

It is important to understand the chemical reactivity of the oligonucleotide drug substances intended for clinical development. Prior to initiation of any stability studies, it is necessary to demonstrate the stability-indicating nature of the methods to be used for stability monitoring (Chapter 15). This is most often assessed by analysis of samples that have been subjected to stress conditions such as acidic, basic, oxidative, thermal, and photochemical stress.

For most oligonucleotides, particularly single-stranded molecules, the purity and assay methods used for quality control release testing are often suitable for stability monitoring. In the case of duplexes such as siRNA, stability monitoring should include both a nondenaturing method to assess duplex integrity as well as a denaturing method to assess changes to the individual strands, if any, over time.

Once the stability-indicating nature of the methods is well established and documented, other basic studies of chemical reactivity may be performed. These include stability of the oligonucleotide as a function of pH, often as a prelude to dosage form development.

For the purposes of inclusion in regulatory submissions it is common to perform stability studies under accelerated storage conditions, usually consistent with those recommended by ICH guidelines. In addition, photostability testing is recommended to assure stability to ambient lighting and other harsh lighting conditions.[49–53] Although the ICH stability guidelines are not intended for development stage products, adherence to these conditions assures gathering supportive data that may be suitable for inclusion in regulatory applications at a later date. On the strength of stability at accelerated conditions as well as short-term stability studies under storage conditions of as little as 1–3 months, it is possible get a good assessment of the stability of the drug product over the course of the length of early clinical trials.

19.8 OLIGONUCLEOTIDE DRUG PRODUCTS

Most oligonucleotide drug products under development to date are simple buffered solutions intended for local or systemic delivery. The methods used for quality control testing and stability monitoring of such products are similar to those for the drug substance. Additional dosage form related tests may include compendial tests such as pH, osmolality, particulate matter, volume in container, endotoxins, bioburden, and sterility.[5,54]

19.9 OLIGONUCLEOTIDES IN COMPLEX DELIVERY SYSTEMS

In addition to simple sterile solutions, oligonucleotides have been successfully formulated as topical gels or creams, oral tablets for local delivery, microspheres dendrimers, liposomes, and nanoparticles. In recent years, there has been a move toward development of novel delivery systems for oligonucleotide drugs for improving nuclease resistance, targeting to sites of interest thereby lowering toxicity, improving pharmacokinetics and lowering cost of goods. Many of these systems are rapidly emerging from research laboratories and entering clinical development.

The explosive growth in the area of delivery systems for oligonucleotide drugs has prompted additional interest in the regulatory requirements for entering the clinic with such systems. Characterization and testing requirements for various common dosage forms for oral,[55] topical,[56] and pulmonary[57,58] applications have been described in detail by regulatory authorities. Unlike the case of simple sterile solutions, analytical methods for the testing of oligonucleotides in these complex formulations often require sophisticated sample preparation procedures to overcome interferences from the formulation excipients and appropriate validation to ensure accurate and reproducible analysis of purity, label claim, and stability.[59,60] It is anticipated that these novel oligonucleotide drug products will need to comply, at least in spirit, with the expectations already described in current guidance documents. For nanoparticles, liposomes, dendrimers, and other similar novel delivery systems, the FDA guideline for liposomal products can serve as an excellent resource.[61] Finally, if novelty of a dosage form is such that none of the existing guidance documents may be reasonably applied, it is recommended that the sponsor company meet with the appropriate regulatory authority to reach agreement on their plans for assuring quality and control of the product.

19.10 BIOANALYSIS OF OLIGONUCLEOTIDES

One of the most challenging aspects of analysis of oligonucleotides is quantitation of these molecules and their possible metabolism products in biological matrices. Although the type of techniques used for such analyses are similar to those used for drug product quality control, there are significant additional challenges related to overcoming matrix effects and achieving adequate detection limits. Substantial progress has been made in addressing the need for robust analytical methods to support pharmacokinetics and drug metabolism studies for the various classes of oligonucleotides (see Chapter 8). Prior to use in supporting preclinical and clinical development, these methods should be validated in conformance with FDA requirements.[62]

19.11 SUMMARY

Despite the nearly 20 years since the first oligonucleotide drug candidate reached the clinic and after more than a decade since the market approval of the first oligonucleotide drug, the regulatory expectations for such products are still evolving. This is, in part, because of the various unique mechanisms of action of this class of molecule, resulting in quite a diverse group of molecular entities for which a unified set of regulatory expectations is difficult to achieve. At the same time, considerable progress has been made in the development of novel analytical technologies that have advanced our understanding of the chemistry of this unique class of drugs. It is hoped that the concepts presented in this chapter will be useful in forming a basis for the regulatory strategy for an oligonucleotide product while assuring patient safety and drug efficacy.

REFERENCES

1. Food and Drug Administration (FDA). 1995. Guidance for industry: Content and format of investigational new drug applications (INDs) for phase I studies of drugs, including well-characterized, therapeutic, biotechnology-derived products.
2. Kambhampati, R., Y-Y. Chiu, C.W. Chen, et al.1993. Regulatory concerns for the chemistry, manufacturing and controls of oligonucleotide therapeutics for use in clinical studies. *Antisense Research and Development* 3: 405–410.
3. Kambhampati, R. 2007. Points to consider for the submission of chemistry, manufacturing and controls (CMC) information in oligonucleotide-based therapeutic drug applications. Center for Drug Evaluation Research, Office of New Drug Quality Assessment, US Food and Drug Administration, DIA Industry and Health Authority Conference on Oligonucleotide Therapeutics.
4. International Conference on Harmonization (ICH) Q3A. 2008. Impurities in new drug substances.
5. ICH Q6A 1999. Specifications: Test procedures and acceptance criteria for new drug substances and new drug products: chemical substances.
6. Sober, H. A., ed. 1969. *Handbook of Biochemistry Selected Data for Molecular Biology*. Cleveland, OH: Chemical Rubber Company.
7. Schuette, J., U. Pieles, S. D. Maleknia, et al. 1995. Sequencing analysis of phosphorothioate oligonucleotides via matrix assisted laser desorption ionization time-of-flight mass spectrometry. *Journal of Pharmaceutical and Biomedical Analysis* 13: 1195–1203.
8. Maxam, A. M., and W. Gilbert. 1977. A new method for sequencing DNA. *Proceedings of the National Academy of Sciences of the United States of America* 74(2): 560–564.
9. Little, D. P., et al. 1994. Rapid sequencing of oligonucleotides by high-resolution mass spectrometry. *Journal of the American Chemical Society*, 116(11): 4893–4897.
10. Bahr, U., H. Aygun, and M. Karas. 2009. Sequencing of single and double stranded RNA oligonucleotides by acid hydrolysis and MALDI mass spectrometry. *Analytical Chemistry* 81(8): 3173–3179.
11. Schuette, J., G. S. Srivatsa, and D. L. Cole. 1994. Development and validation of a method for routine base composition analysis of phosphorothioate oligonucleotides. *Journal of Pharmaceutical and Biomedical Analysis* 12: 1345–1353.
12. Emsley, J., and D. Hall. 1976. 31P n.m.r. and vibrational spectra of phosphorous compounds. In *The Chemistry of Phosphorus*, 78–109. New York: Harper & Row.
13. Tsuboi, M., and Y. Kyogoku. 1973. Chapter 6. In *Synthetic Procedures in Nucleic Acid Chemistry: Physical and Physicochemical Aids in Characterization and Determination of Structure*, ed. W. W. Zorbach, Vol. 2, 215–265. New York: John Wiley.
14. Wuthrich, K. 1986. *NMR of Proteins and Nucleic Acids*, 1–224. New York: John Wiley.
15. Stone, M. P., S. A. Winkle, G. D. McFarland, et al. 1985. 13C-NMR of ribosyl A-A-A, A-A-G, and A-U-G: synthesis and assignment. *Biophysical Chemistry* 23: 129–138.

16. Pretsch, E., P. Buhlmann, and C. Affolter. 2000. *Structure Determination of Organic Compounds: Tables of Spectral Data*. New York: Springer-Verlag.
17. Frier, S. 1993. Hybridization and antisense drugs. In *Antisense Research and Applications*, eds. S. Crooke and B. Lebleu, 67–82. Boca Raton, FL: CRC Press.
18. Srivatsa, G. S., P. Klopchin, M. Batt, et al. 1997. Selectivity of anion exchange HPLC and capillary gel electrophoresis for the analysis of phosphorothioate oligonucleotides. *Journal of Pharmaceutical and Biomedical Analysis* 16: 619–630.
19. Blackburn, M., and M. Gait. 1996. Chapter 10. In *Nucleic Acids in Chemistry and Biology*. New York: Oxford University Press.
20. Furtig, B., C. Richter, J. Wohnert, et al. 2003. NMR spectroscopy of RNA. *Chembiochemistry* 4(10): 936–962.
21. Sanger, W. 1985. Chapter 5. In *Principles of Nucleic Acid Structure*. New York: Springer-Verlag.
22. Chen, D., and G. S. Srivatsa. 1999. n–1 deletion sequence analysis of oligonucleotides by hybridization to immobilized specific probes. *Nucleic Acid Research* 27(2): 389–395.
23. Temsamani, J., M. Kubert, and S. Agrawal. 1995. Sequence identity of the n–1 product of a synthetic oligonucleotide. *Nucleic Acids Research* 23(11): 1841–1844.
24. Fearon, K. L., J. T. Stults, B. J. Bergot, L. M. Christensen, and A. M. Raible. 1995. Investigation of the 'n–1' impurity in phosphorothioate oligodeoxynucleotides synthesized by the solid-phase ß-cyanoethyl phosphoramidite method using stepwise sulfurization. *Nucleic Acids Research* 23: 2754–2761.
25. Krotz, A. H., P. G. Klopchin, K. L. Walker, G. S. Srivasta, D. L. Cole, and V. T. Ravikumar. 1997. On the formation of longmers in phosphorothioate oligodeoxyribonucleotide synthesis. *Tetrahedron Letters* 38: 3875–3878.
26. Kurata, C., K. Bradley, H. Gaus, N. Luu, I. Cedillo, V. T. Ravikumar, K. Van Sooy, J. V. McArdle, and D. C. Capaldi. 2006. Characterization of high molecular weight impurities in synthetic phosphorothioate oligonucleotides. *Bioorganic and Medicinal Chemistry Letters* 16: 607–614.
27. Chen, D., P. Klopchin, J. Parsons, and G. S. Srivatsa. 1997. Determination of sodium acetate in antisense oligonucleotides by capillary zone electrophoresis. *Journal of Liquid Chromatography and Related Technologies* 20(8): 1185–1195.
28. ICH Q7A. 2001. Good manufacturing practice guidance for active pharmaceutical ingredients.
29. Ravikumar, V. T., K. R. Kumar, P. Olsen, et al. 2008. UnyLinker: An efficient and scaleable synthesis of oligonucleotides utilizing a universal linker molecule: A novel approach to enhance the purity of drugs. *Organic Process Research and Development* 12(3): 399–410.
30. Wyrzykiewicz, T. A., and D. L. Cole. 1995. Stereo-reproducibility of the phosphoramidite method in the synthesis of phosphorothioate oligonucleotides. *Bioorganic Chemistry* 23: 33–41.
31. Ravikumar, V., and D. L. Cole. 2002. Development of 2′-Methoxyethyl phosphorothioate oligonucleotide as antisense drugs under stereochemical control. *Organic Process Research and Development* 6: 798–806.
32. FDA. 1998. Guidance for industry: Process validation: General principles and practices.
33. ICH Q9. 2001. Quality risk management.
34. Srivatsa, G. S., A. Scozzari, and D. L. Cole. 2000. Special issues for synthetic antisense oligonucleotides. In *Biopharmaceutical Process Validation*, eds. G. Sofer and D. Zabriskie, 309–327, New York: Marcel Dekker, Inc.
35. ICH Q4B. 2008. Evaluation and recommendation of pharmacopoeial texts for use in the ICH regions. Annexes 1–10.
36. Jiang, Z.-Y., J. V. Hunt, and S. P. Wolff. 1992. A simple Fe++-oxidation method for detection of lipid peroxide in low density lipoprotein and liposomes. *Analytical Biochemistry* 202: 384–389.
37. Kumar, V., and V. Kalonia. 2003. Removal of peroxides in polyethylene glycols by vacuum drying: Implications in the stability of biotech and pharmaceutical formulations. *AAPS PharmSciTech* 7(3): Article 62.
38. Krotz, A., R. Mehta, and G. Hardee. 2005. Peroxide mediated desulfurization of phosphorothioates and its prevention. *Journal of Pharmaceutical Sciences* 94(2): 341–352.
39. ICH Q3B (R2). 2006. Impurities in new drug products.
40. Srivatsa, G. S., M. Batt, J. S. Schuette, et al. 1994. Assay of phosphorothioate oligonucleotides in pharmaceutical formulations by capillary gel electrophoresis (QCGE). *Journal of Chromatography A* 680: 469–477.
41. FDA. 2009. Guidance for industry: Residual solvents in drug products marketed in the United States.
42. ICH Q3C. 1997. Impurities: Residual solvents.
43. USP. 2006. General chapter on inorganic impurities: Heavy metals stimuli article.
44. European Agency for the Evaluation of Medicinal Products (EMEA). 2002. Note for guidance on specification limits for residues of metal catalysts.

45. FDA. 1987. Guideline on the validation of the LAL test as an end-product endotoxin test for human and animal parenteral drugs, biological products, and medical devices.
46. USP 34. 2011. General chapter on methods validation.
47. ICH Q2A. 1995. Text on validation of analytical procedures.
48. ICH Q2B. 1996. Validation of analytical procedures: Methodology.
49. ICH Q1A (R2). 2003. Stability testing of new drug substances and products.
50. ICH Q1B. 1996. Photostability testing of new drug substances and products.
51. ICH Q1C. 1996. Stability testing of new dosage forms.
52. ICH Q1D. 2003. Bracketing and matrixing designs for stability testing of new drug substances and products.
53. ICH Q1E. 2004. Evaluation of stability data.
54. ICH Q8 (R2). 1997. Pharmaceutical development.
55. FDA. 1997. Guidance for industry: Dissolution testing of immediate release solid oral dosage forms.
56. FDA. 1998. Guidance for industry: Topical and dermatological drug product NDAs and ANDAs—in vivo bioavailability, bioequivalence, in vitro release, and associate studies.
57. FDA. 2002. Guidance for industry: Nasal spray and inhalation solution, suspension, and spray drug products—chemistry, manufacturing and controls documentation.
58. FDA. 2002. Guidance for industry: Metered dose inhaler (MDI) and dry powder inhaler (DPI) drug products.
59. D. Chen, D. Cole, and G. S. Srivatsa. 2000. Determination of free and encapsulated oligonucleotides in liposome formulated drug product. *Journal of Pharmaceutical and Biomedical Analysis* 22(5): 791–801.
60. Murugaiah, V., W. Zedalis, G. Lavine, et al. 2010. Reversed-phase high-performance liquid chromatography method for simultaneous analysis of two liposome-formulated short interfering RNA duplexes. *Analytical Biochemistry* 401(1): 61–67.
61. FDA. 2002. Guidance for industry: Liposome drug products.
62. FDA. 2001. Guidance for industry: Bioanalytical method validation.

Index

F

G

H

I

N

O

P

Q

R

S